P9-CNE-976

HOLT

MIDDLE SCHOOL Math

Course 1

Jennie M. Bennett
David J. Chard
Audrey Jackson
Jim Milgram
Janet K. Scheer
Bert K. Waits

HOLT, RINEHART AND WINSTON
A Harcourt Education Company
Austin • Orlando • Chicago • New York • Toronto • London • San Diego

Staff Credits

Editorial

Lila Nissen, *Editorial Vice President*
Robin Blakely, *Associate Director*
Joseph Achacoso, *Assistant Managing Editor*
Threasa Boyar, *Editor*

Student Edition
April Warn, *Senior Editor*
Katie Seawell, *Editor*

Teacher's Edition
Kelli Flanagan, *Senior Editor*
Ronald Fowler, *Associate Editor*
Monica Robinson, *Associate Editor*

Ancillaries
Mary Fraser, *Executive Editor*
Higinio Dominguez, *Associate Editor*

Technology Resources
John Kerwin, *Executive Editor*
Robyn Setzen, *Senior Editor*
Patricia Platt, *Senior Technology Editor*
Manda Reid, *Technology Editor*

Copyediting
Denise Nowotny, *Copyediting Supervisor*
Patrick Ricci, *Copyeditor*

Support
Jill Lawson, *Senior Administrative Assistant*
Benny Carmona, III, *Editorial Coordinator*

Design

Book Design
Marc Cooper, *Design Director*
Tim Hovde, *Senior Designer*
Lisa Woods, *Designer*
Teresa Carrera-Paprota, *Designer*
Bruce Albrecht, *Design Associate*
Ruth Limon, *Design Associate*
Holly Whittaker, *Senior Traffic Coordinator*

Teacher's Edition
José Garza, *Designer*
Charlie Taliaferro, *Design Associate*

Cover Design
Pronk & Associates

Image Acquisition
Curtis Riker, *Director*
Tim Taylor, *Photo Research Supervisor*
Stephanie Friedman, *Photo Researcher*
Elaine Tate, *Art Buyer Supervisor*
Sam Dudgeon, *Senior Staff Photographer*
Victoria Smith, *Staff Photographer*
Lauren Eischen, *Photo Specialist*

New Media Design
Ed Blake, *Design Director*

Media Design
Dick Metzger, *Design Director*
Chris Smith, *Senior Designer*

Graphic Services
Kristen Darby, *Director*
Eric Rupprath, *Ancillary Designer*
Linda Wilbourn, *Image Designer*

Prepress and Manufacturing

Mimi Stockdell, *Senior Production Manager*
Susan Mussey, *Production Supervisor*
Rose Degollado, *Senior Production Coordinator*
Sara Downs, *Production Coordinator*
Jevara Jackson, *Senior Manufacturing Coordinator*
Ivania Lee, *Inventory Analyst*
Wilonda Ieans, *Manufacturing Coordinator*

2005 Printing

Printed in the United States of America

ISBN 0-03-070982-2

4 5 6 7 8 9 048 10 09 08 07 06 05

Authors

Jennie M. Bennett, Ed.D., is the Instructional Mathematics Supervisor for the Houston Independent School District and president of the Benjamin Banneker Association.

David J. Chard, Ph.D., is an Assistant Professor and Director of Graduate Studies in Special Education at the University of Oregon. He is the President of the Division for Research at the Council for Exceptional Children, is a member of the International Academy for Research on Learning Disabilities, and is the Principal Investigator on two major research projects for the U.S. Department of Education.

Audrey Jackson is a Principal in St. Louis, Missouri, and has been a curriculum leader and staff developer for many years.

Jim Milgram, Ph.D., is a Professor of Mathematics at Stanford University. He is a member of the Achieve Mathematics Advisory Panel and leads the Accountability Works Analysis of State Assessments funded by The Fordham and Smith-Richardson Foundations. Most recently, he has been named lead advisor to the Department of Education on the implementation of the Math-Science Initiative, a key component of the No Child Left Behind legislation.

Janet K. Scheer, Ph.D., Executive Director of Create A Vision™, is a motivational speaker and provides customized K-12 math staff development. She has taught internationally and domestically at all grade levels.

Bert K. Waits, Ph.D., is a Professor Emeritus of Mathematics at The Ohio State University and co-founder of T³ (Teachers Teaching with Technology), a national professional development program.

Consulting Authors

Paul A. Kennedy is a Professor in the Mathematics Department at Colorado State University and has recently directed two National Science Foundation projects focusing on inquiry-based learning.

Mary Lynn Raith is the Mathematics Curriculum Specialist for Pittsburgh Public Schools and co-directs the National Science Foundation project PRIME, Pittsburgh Reform in Mathematics Education.

Reviewers

Thomas J. Altonjy
Assistant Principal
Robert R. Lazar Middle School
Montville, NJ

Jane Bash, M.A.
Math Education
Eisenhower Middle School
San Antonio, TX

Charlie Bialowas
District Math Coordinator
Anaheim Union High School District
Anaheim, CA

Lynn Bodet
Math Teacher
Eisenhower Middle School
San Antonio, TX

Louis D'Angelo, Jr.
Math Teacher
Archmere Academy
Claymont, DE

Troy Deckebach
Math Teacher
Tredyffrin-Easttown Middle School
Berwyn, PA

Mary Gorman
Math Teacher
Sarasota, FL

Brian Griffith
Supervisor of Mathematics, K–12
Mechanicsburg Area School District
Mechanicsburg, PA

Ruth Harbin-Miles
District Math Coordinator
Instructional Resource Center
Olathe, KS

Kim Hayden
Math Teacher
Milford Jr. High School
Milford, OH

Susan Howe
Math Teacher
Lime Kiln Middle School
Fulton, MD

Paula Jenniges
Austin, TX

Ronald J. Labrocca
District Mathematics Coordinator
Manhasset Public Schools
Manhasset, NY

Victor R. Lopez
Math Teacher
Washington School
Union City, NJ

George Maguschak
Math Teacher/Building Chairperson
Wilkes-Barre Area
Wilkes-Barre, PA

Dianne McIntire
Math Teacher
Garfield School
Kearny, NJ

Kenneth McIntire
Math Teacher
Lincoln School
Kearny, NJ

Francisco Pacheco
Math Teacher
IS 125
Bronx, NY

Vivian Perry
Edwards, IL

Vicki Perryman Petty
Math Teacher
Central Middle School
Murfreesboro, TN

Jennifer Sawyer
Math Teacher
Shawboro, NC

Russell Sayler
Math Teacher
Longfellow Middle School
Wauwatosa, WI

Raymond Scacalossi
Math Chairperson
Hauppauge Schools
Hauppauge, NY

Richard Seavey
Math Teacher–Retired
Metcalf Jr. High
Eagan, MN

Sherry Shaffer
Math Teacher
Honeoye Central School
Honeoye Falls, NY

Gail M. Sigmund
Math Teacher
Charles A. Mooney Preparatory School
Cleveland, OH

Jonathan Simmons
Math Teacher
Manor Middle School
Killeen, TX

Jeffrey L. Slagel
Math Department Chair
South Eastern Middle School
Fawn Grove, PA

Karen Smith, Ph.D.
Math Teacher
East Middle School
Braintree, MA

Bonnie Thompson
Math Teacher
Tower Heights Middle School
Dayton, OH

Mary Thoreen
Mathematics Subject Area Leader
Wilson Middle School
Tampa, FL

Paul Turney
Math Teacher
Ladue School District
St. Louis, MO

Welcome to Holt Middle School Math North Carolina Edition!

Have you ever wondered if anyone really uses the math that you learn in school? As you work through this book, you will find out that they certainly do. Not only that, but people are using math almost right in your backyard.

We have found the places where math is being used in your home state. In each chapter, you will "visit" one or two North Carolina locations, or you will learn about a specific aspect of North Carolina's culture, history, or industry to see how math is being used near you. It will be like taking a math tour of North Carolina without leaving your classroom!

Chapter 1 Learn about the elevations of the scenic Great Smoky Mountains.

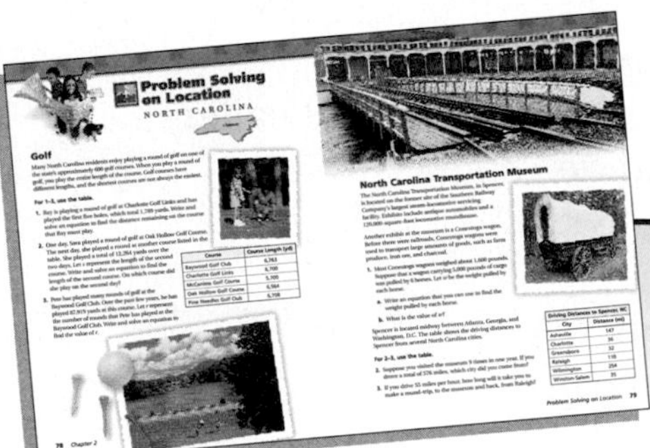

Chapter 2 Calculate some travel distances at the North Carolina Transportation Museum.

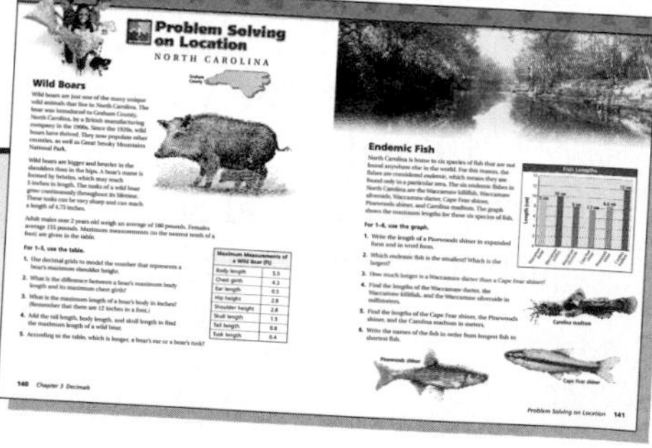

Chapter 3 Six species of fish are endemic to North Carolina. Learn to describe them using decimals.

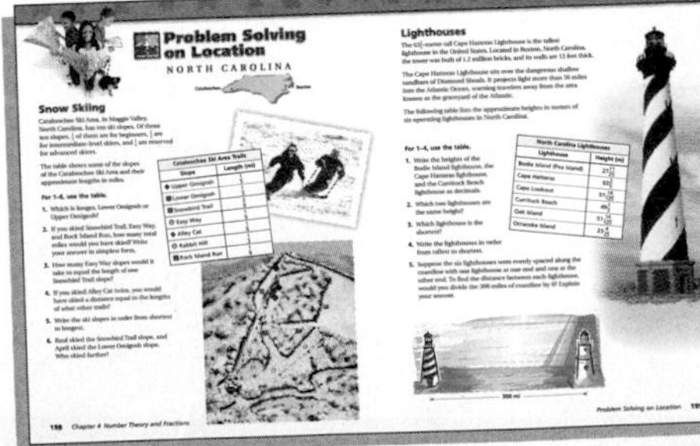

Chapter 4 Use fractions to measure the dimensions of North Carolina's lighthouses.

Chapter 5 Check out some of North Carolina's hiking and canoe trails. Use fractions to record the distances that can be traveled.

Chapter 6 Seagrove, North Carolina, is known for its pottery. Analyze data about pottery.

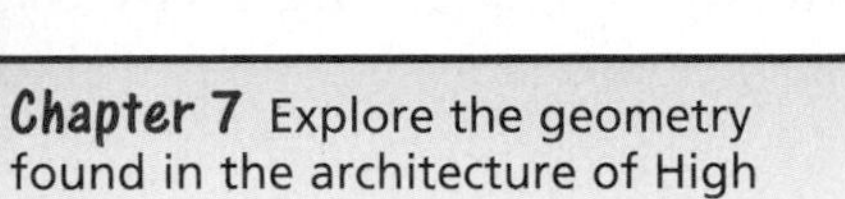

Chapter 7 Explore the geometry found in the architecture of High Point's Showplace Convention Center.

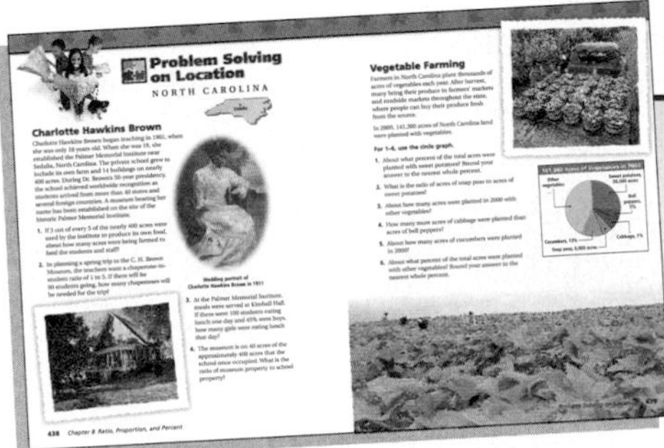

Chapter 8 North Carolina farmers plant thousands of acres of vegetables each year. Learn more about the size of their harvests.

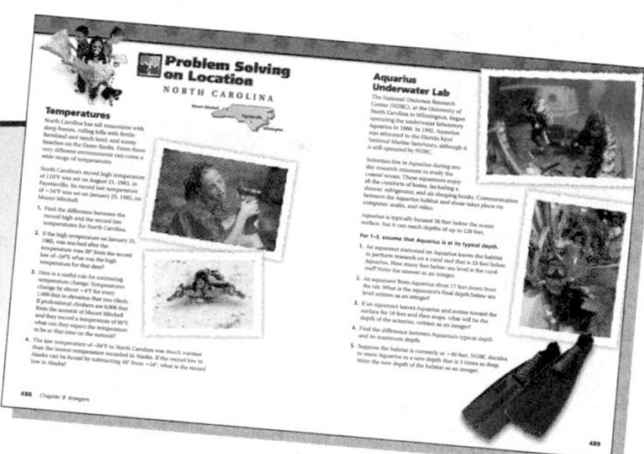

Chapter 9 Use integers to describe the depth of *Aquarius,* the first underwater lab. *Aquarius* is operated from the University of North Carolina at Wilmington.

Chapter 10 Asheville's Historical Significant Events Imagery contains satellite images of weather events. Investigate the area of regions affected by certain kinds of weather.

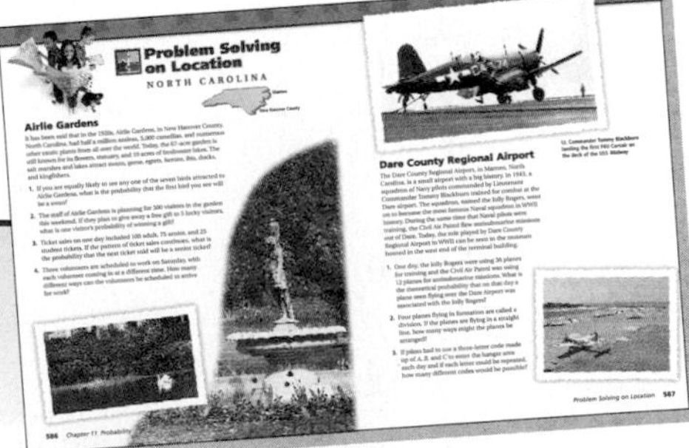

Chapter 11 Use probability to determine which sights you are likely to see at Airlie Gardens, in New Hanover County.

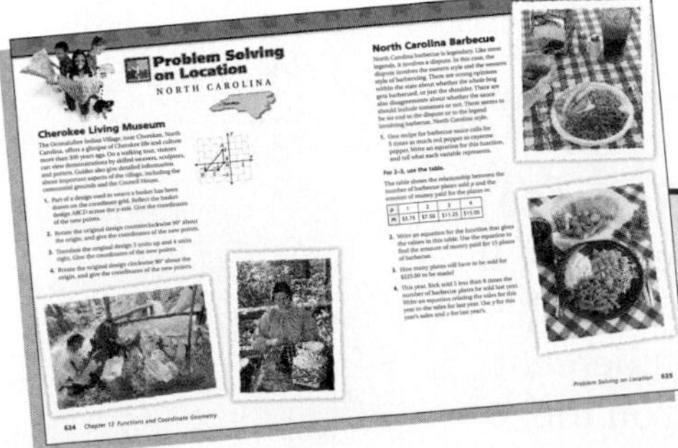

Chapter 12 Examine some of the geometric patterns in weaving at the Cherokee Living Museum.

North Carolina End-of-Grade (EOG) Test for the Middle Grades

The North Carolina End-of-Grade (EOG) Test is a multiple-choice test. To answer questions on the EOG test, you will fill in an answer sheet. It is very important to fill in your answer sheet correctly. When shading in circles, make your marks heavy and dark. Fill in the circles completely, but do not shade outside the circles. Do not make any stray marks on your answer sheet.

Read each question carefully and work the problem. Choose your answer from among the answer choices given, and fill in the corresponding circle on your answer sheet. If your answer is not one of the choices, read the question again. Be sure that you understand the problem. Check your work for possible errors.

Some questions on the EOG test are calculator inactive, and others are calculator active. You may not use a calculator on the calculator inactive questions, but you may on the calculator active questions.

Sample Question

Questions on the EOG test may require an understanding of number and operations, algebra, geometry, measurement, and data analysis and probability. Drawings, grids, or charts may be included for certain types of questions. Try the following practice question to prepare for taking a multiple choice test. Choose the best answer from the choices given.

In a group of 30 students, 27 are middle school students and the others are high school students. If one person is selected at random from this group, what is the probability that the person selected will be a high school student?

A $\frac{1}{30}$

B $\frac{1}{10}$

C $\frac{3}{10}$

D $\frac{9}{10}$

Think About the Solution

First consider the total number of people in the group (30). If 27 out of 30 are middle school students, how many of them are high school students? (3) If one person is selected at random, there is a probability of 3 out of 30. This can be written as a ratio (3:30), a fraction $\left(\frac{3}{30}\right)$, a decimal (0.1), or a percent (10%). None of these solutions is listed as one of the choices, so you must look for a solution that is equivalent. The fraction $\frac{3}{30}$ can be simplified to $\frac{1}{10}$. Since $\frac{1}{10}$ is given as one of your answer choices, B is the correct response.

Indicate your response by filling in the circle that contains B.

Test-Taking Tip

Sometimes you can find the best solution to a test question by understanding what is wrong with some of the choices. Read the sample question again. Why are A, C, and D incorrect?

Response A is $\frac{1}{30}$.

You might think A is correct because there are 30 people and you are selecting one. However, this answer indicates that only 1 of the 30 people is a high school student. Since that is not what the problem states, A cannot be correct.

Response C is $\frac{3}{10}$.

This answer indicates that 3 out of 10 people are high school students. The numerator is correct, since there are three high school students in the group. However, the denominator must show the relationship 3 out of 30. C is not correct.

Response D is $\frac{9}{10}$.

If you chose D, read the problem again. The problem asks you to find the probability that a high school student will be selected. Answer D would be the best choice if you wanted to find the probability that a *middle school* student will be selected, but D does not match the question that was asked.

Practice, Practice, Practice

On the following pages, you will find a practice standardized test. This test has been designed to resemble the EOG test and the types of questions it contains. Use this test to practice answering these questions, as well as to review some of the math that you will be tested on. The more comfortable and familiar you can become with the EOG test, the better your chances of success!

EOG Test Practice

Directions: Read each question and choose the correct response. You may *not* use a calculator on this part of the test.

1. Sam gathered 12 buckets of clams, Kristi gathered 9 buckets, and Mario gathered 6 buckets. They sold the clams for $3 a bucket. How much money did they earn?

A $9
B $78
C $81
D $87

2. Which is the prime factorization of 36?

A 3×6
B $2 \times 2 \times 9$
C $2 \times 2 \times 3 \times 3$
D $2 \times 2 \times 2 \times 3$

3. What is the value of 7^3?

A 21
B 73
C 343
D 2,187

4. The base of a triangle is 14 inches and its height is 6 inches. What is the triangle's area?

A 84 square inches
B 42 square inches
C 35 square inches
D 21 square inches

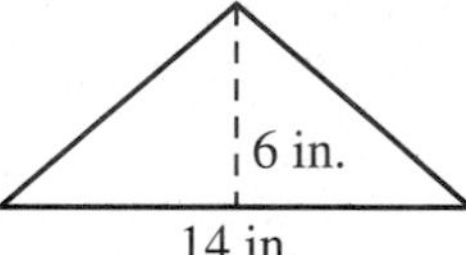

5. The sixth-grade class held a car wash. The students charged $3.25 to wash a car ($c$) and $5.75 to wash a van ($v$). The amount of money they raised can be represented by the expression $3.25c + 5.75v$. If they washed 42 cars and 13 vans, how much money did they raise?

A $55.00
B $74.75
C $136.50
D $211.25

6. Keith has 3 bookcases that are each 1.6 meters wide. He wants to place them side by side along his bedroom wall. What is the length that the bookcases will measure when they are side by side?

A 3.6 meters

B 4.096 meters

C 4.6 meters

D 4.8 meters

7. Which number is less than $8\frac{3}{8}$?

A 8.3

C 8.4

B $8\frac{5}{7}$

D $8\frac{2}{3}$

8. Nancy, Bob, Elizabeth, Chris, Trevor, and Jen all want to run in a three-legged race. If a team consists of two people, how many different teams can there be?

A 3

C 360

B 15

D 720

9. What is the value of $(3 + 5) \times 2^2 - 4$?

A 28

C 18

B 19

D 0

10. At the town park, there is a circular jogging path all along the outer edge of the park. The path is 9.4 kilometers.

How many kilometers is the path from the center of the park? Use 3.14 as an approximation for π and round your answer to the nearest tenth of a kilometer.

A 3.0 kilometers

B 2.9 kilometers

C 1.5 kilometers

D 1.4 kilometers

Go On

EOG Test Practice

11. Howard made a fishing net in the shape of a parallelogram. A drawing of his fishing net is shown below. What is the area of Howard's fishing net?

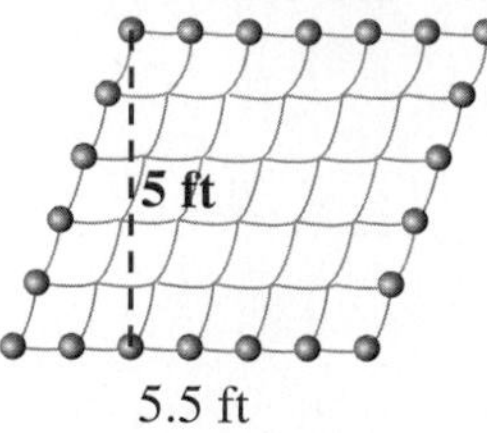

A 10.5 square feet

B 13.75 square feet

C 27.5 square feet

D 55 square feet

12. How many outcomes are possible if you toss a penny, a nickel, and a dime?

A 3

B 6

C 8

D 9

13. At the sandwich shop, customers have a choice of wheat, white, or rye bread. They can choose chicken, tuna, turkey, or ham for a filling. How many ways are there to select one bread and one filling?

A 3

B 4

C 7

D 12

14. The community theater group sold 300 tickets to the play for $9.75 each and 650 tickets for $5.75 each. How much money did the group make in ticket sales?

A $2,925.00

B $3,737.50

C $6,662.50

D $9,262.50

15. The planet Venus is about 1.0823×10^8 kilometers from the Sun. How many kilometers does 1.0823×10^8 represent?

A 10,823 kilometers

B 108,230 kilometers

C 10,823,000 kilometers

D 108,230,000 kilometers

16. Laura's route to the doctor's office is shown below. She travels from home to the store, to the doctor's office, and then straight home. She drives a total of 42 kilometers. What is the distance from the doctor's office to her home?

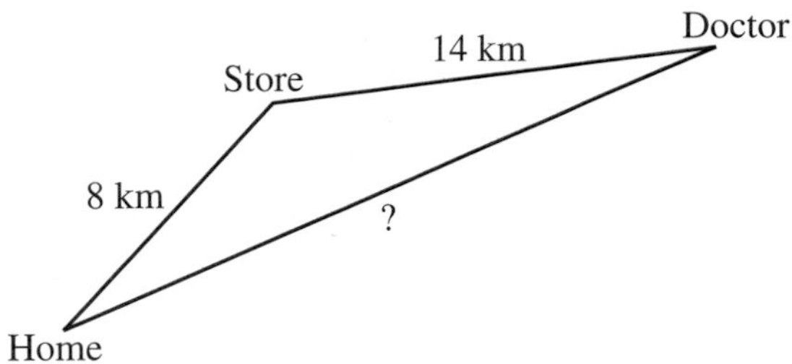

A 14 kilometers

B 20 kilometers

C 34 kilometers

D 48 kilometers

17. Charlie has a piece of wood that is $8\frac{1}{2}$ feet long. For a project, he needs $\frac{3}{4}$ foot segments. How many segments of this length can he cut?

A 6

B $6\frac{3}{8}$

C 11

D $11\frac{1}{3}$

18. Zeb wants to move into an apartment. He has narrowed his choices down to the two apartments whose floor plans are shown below. Zeb wants to choose the apartment with the *greater* amount of floor space.

What is the area of the floor space of the apartment that Zeb should choose?

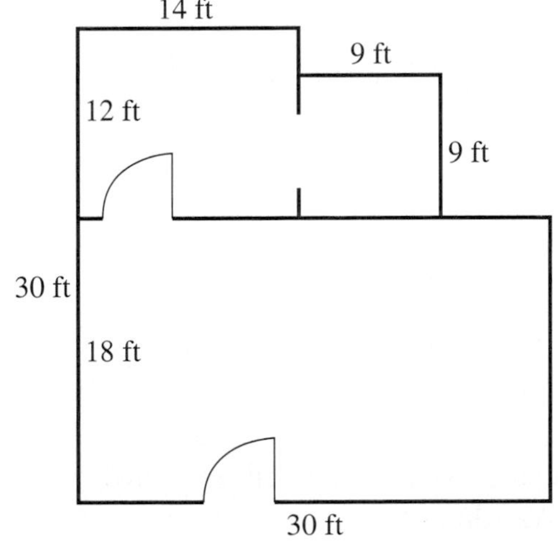

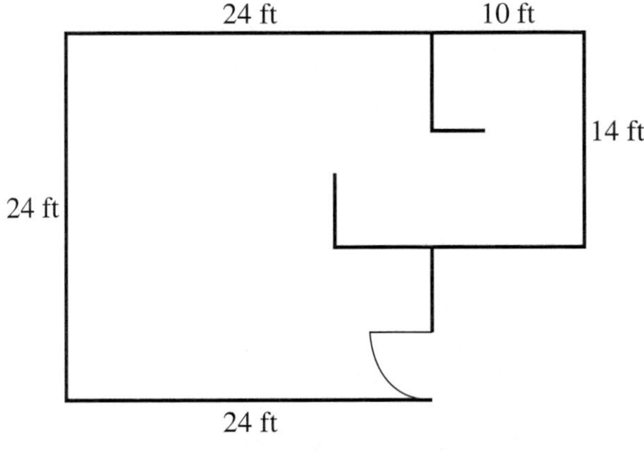

A 789 square feet

B 836 square feet

C 981 square feet

D 986 square feet

Go On

EOG Test Practice

19. Which of the following shows the numbers in order from *least* to *greatest*?

A $\frac{8}{3}$, 2.4, $\frac{27}{10}$, 100%

B 2.4, $\frac{27}{10}$, 100%, $\frac{8}{3}$

C 100%, 2.4, $\frac{8}{3}$, $\frac{27}{10}$

D $\frac{27}{10}$, 100%, 2.4, $\frac{8}{3}$

20. Which of the following numbers is divisible by 4?

A 100,231

B 215,412

C 303,899

D 400,002

21. On average, Earth is 92,900,000 miles from the Sun. Pluto is about 4,551,400,000 miles from the Sun. How much farther is Pluto from the Sun than Earth is?

A 4.4585×10^7 miles

B 4.4585×10^9 miles

C 4.5514×10^9 miles

D 4.6443×10^9 miles

22. Andrew cooked $10\frac{1}{2}$ quarts of soup. Each serving is $\frac{3}{8}$ quart. How many servings did Andrew make?

A 28

B 38

C 42

D 84

23. Wanda is cutting a 5-yard ribbon into pieces. If each piece will be $\frac{5}{6}$ yard long, how many pieces of this length can Wanda cut?

A 4

B 5

C 6

D 8

Directions: Read each question and choose the correct response. You may use a calculator on this part of the test.

24. Line 1 and line 2 are parallel. Which two angles are congruent?

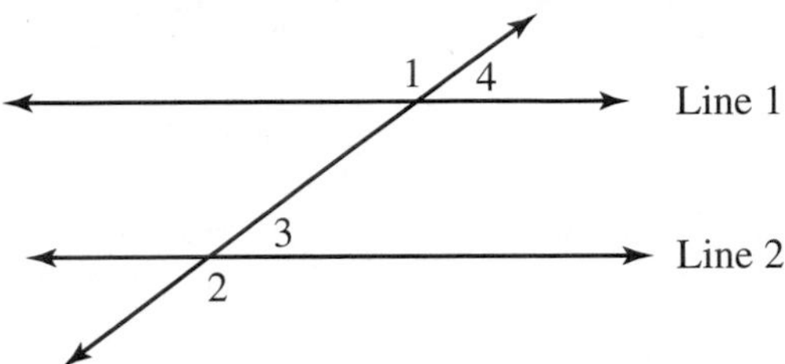

A 1 and 3

B 2 and 3

C 2 and 4

D 3 and 4

25. Becky has $100 in her bank account. She withdraws *d* dollars, leaving a balance of $63. Which equation describes this situation?

A $d - 100 = 63$

B $100 - d = 63$

C $d - 63 = 100$

D $63 - d = 100$

26. The bar graph shows the number of crayons of each color in a bucket.

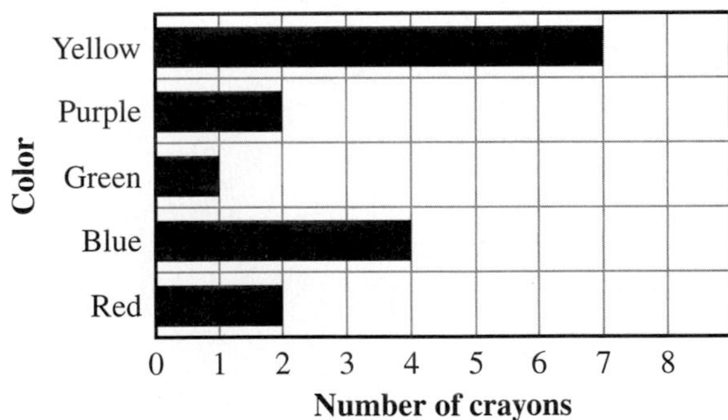

If you select a crayon at random, what is the probability that you will choose a purple crayon?

A $\frac{1}{4}$

B $\frac{1}{8}$

C $\frac{1}{16}$

D $\frac{7}{8}$

Go On

EOG Test Practice

27. Which number *best* represents the location of point *C* on the number line below?

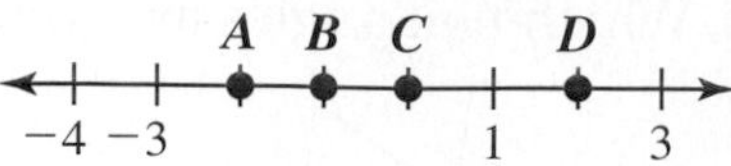

A 2   C $^{-}1$

B 0   D $^{-}2$

28. Your height (h) must be at least 42 inches to ride the roller coaster at the amusement park. Which mathematical statement represents this situation?

A $h \leq 42$

B $h > 42$

C $h = 42$

D $h \geq 42$

29. Which is the *best* estimate of the measure of angle *A*?

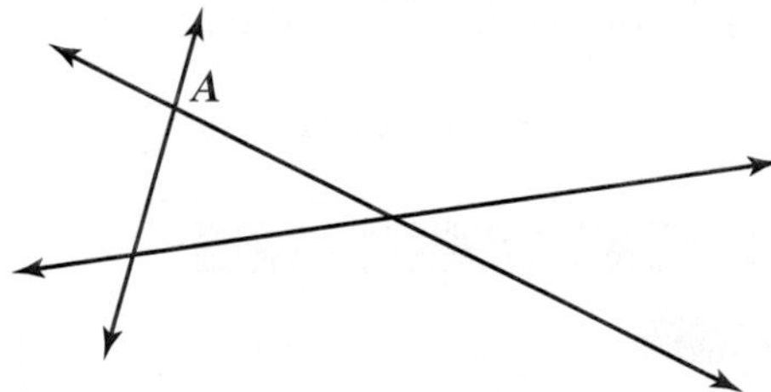

A 95°

B 85°

C 45°

D 135°

30. Which line segment represents a radius?

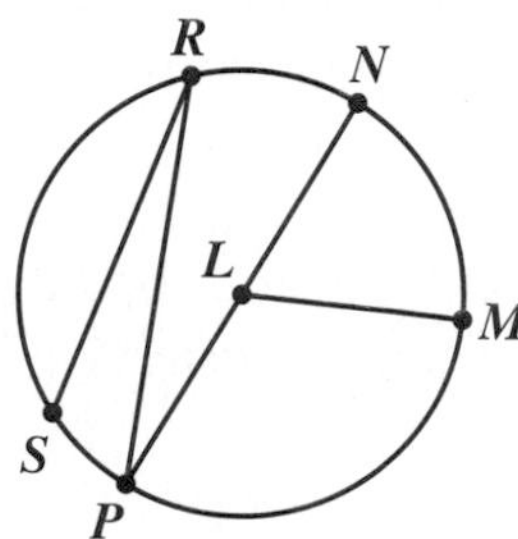

A $\overline{RS}$

B $\overline{NP}$

C $\overline{LM}$

D $\overline{PR}$

31. Which angle is acute?

A

C 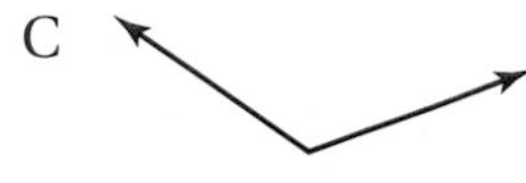

B 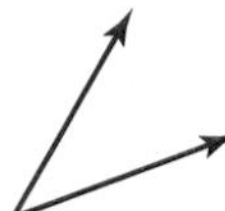

D

32. A drawer contains 8 red socks and 2 blue socks. What is the probability of removing a red sock and then removing a second red sock without replacing the first one?

A $\frac{28}{45}$

C $\frac{14}{25}$

B $\frac{3}{25}$

D $\frac{1}{5}$

33. The figure below will be translated two units to the right and then reflected across the *y*-axis.

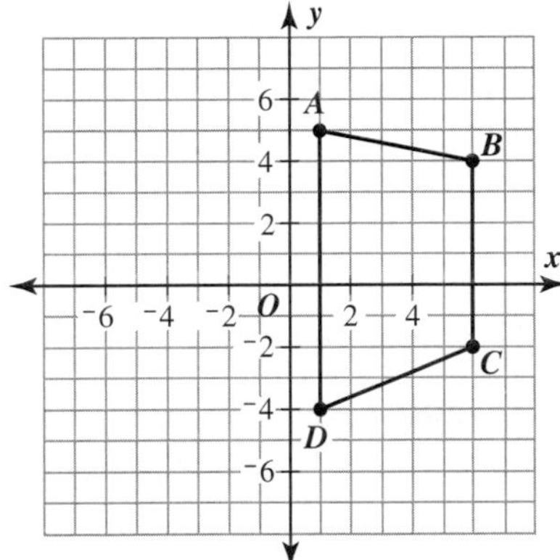

Which coordinates *best* represent the new location of point *A*?

A ($^-3$, 5)

C (1, 5)

B (3, 5)

D (1, $^-5$)

34. Danielle has a spinner on which the probability of spinning a 2 is $\frac{1}{5}$. What is the probability of *not* spinning a 2?

A $\frac{1}{5}$

C $\frac{4}{5}$

B $\frac{2}{5}$

D 1

Go On

EOG Test Practice

35. What is the measure of angle *ABD*?

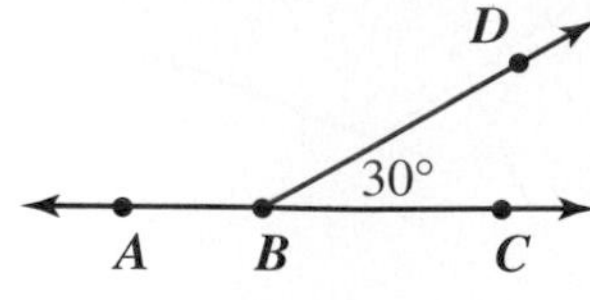

A 60°
B 70°
C 100°
D 150°

36. What percent of this grid is shaded?

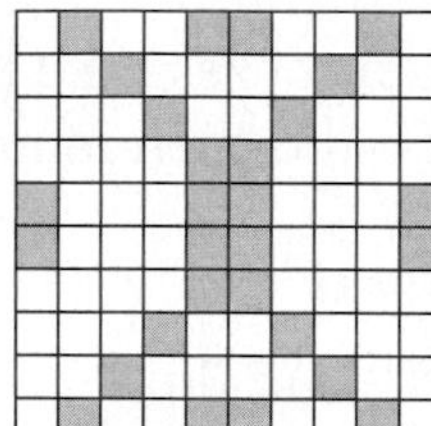

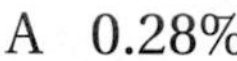
A 0.28%
B 2.8%
C 28%
D $\frac{28}{100}\%$

37. Which angle is complementary to angle *ADE*?

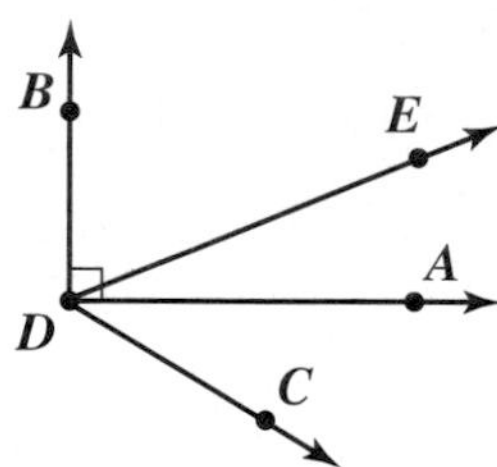

A angle *CDE*
B angle *BDE*
C angle *BDC*
D angle *ADB*

38. Suppose you roll a number cube whose sides are labeled 1 to 6. What is the probability that you will roll a 1 or a 4?

A $\frac{1}{3}$
B $\frac{1}{9}$
C $\frac{1}{12}$
D $\frac{1}{36}$

39. Given the expression $36 \div (9 - 3) + \frac{1}{3} \times 6^2$, which operation should you perform first?

A $(9 - 3)$
B 6^2
C $\frac{1}{3} \times 6^2$
D $36 \div 9$

40. Which angle could be supplementary to an angle that measures 142°?

A
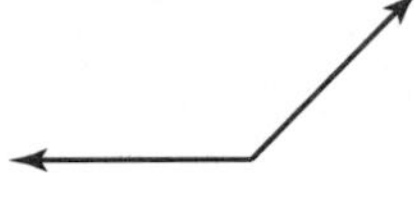

C

B
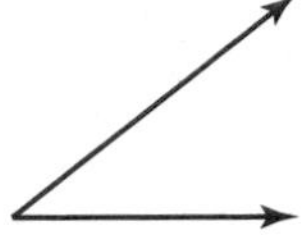

D

41. $\left(\frac{2}{5} + \frac{3}{5}\right) + \frac{1}{5} = \frac{2}{5} + \left(\frac{3}{5} + \frac{1}{5}\right)$ is an example of which property?

A Commutative Property of Addition

B Associative Property of Addition

C Distributive Property

D Addition Property of Equality

42. What is the measure of angle *A*?

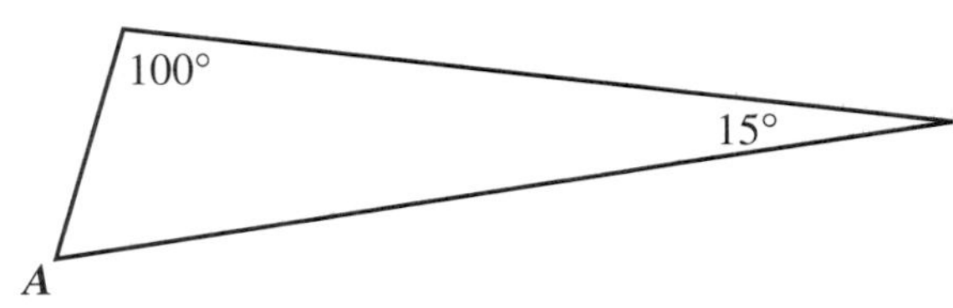

A 180°

B 115°

C 65°

D 15°

43. Which of the shapes below is a hexagon?

A
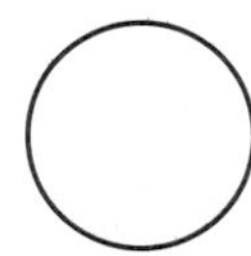

C
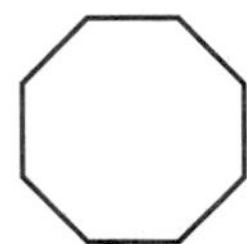

B
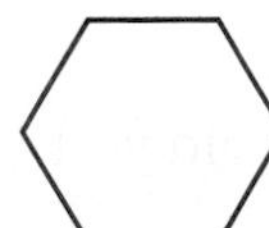

D
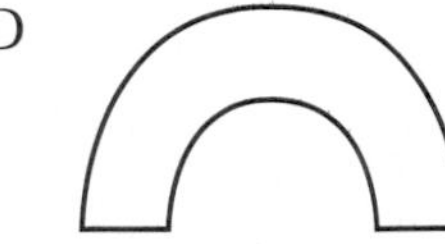

Go On

EOG Test Practice

44. Which figure below has only obtuse angles?

A 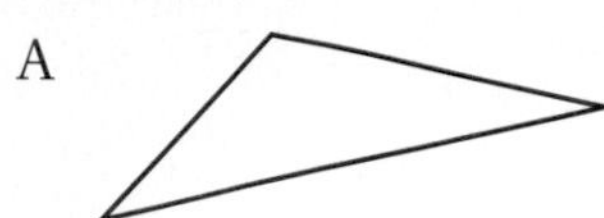

B 

C 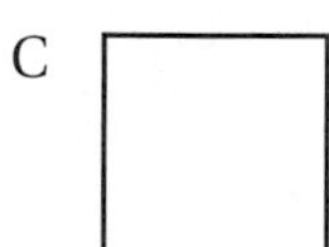

D 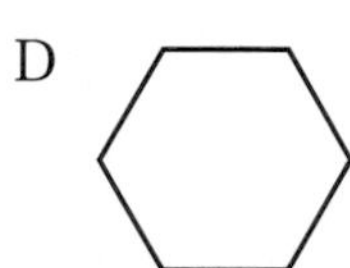

45. Which of the following shows the steps to evaluate $\left(6\frac{1}{2} + \frac{1}{4}\right) \times 5 - 3^4$ in the correct order?

A Add inside the parentheses, find the value of 3^4, multiply, subtract.

B Find the value of 3^4, add inside the parentheses, subtract, multiply.

C Add inside the parentheses, multiply, subtract, find the value of 3^4.

D Add inside the parentheses, find the value of 3^4, subtract, multiply.

46. What is the measure of angle Y?

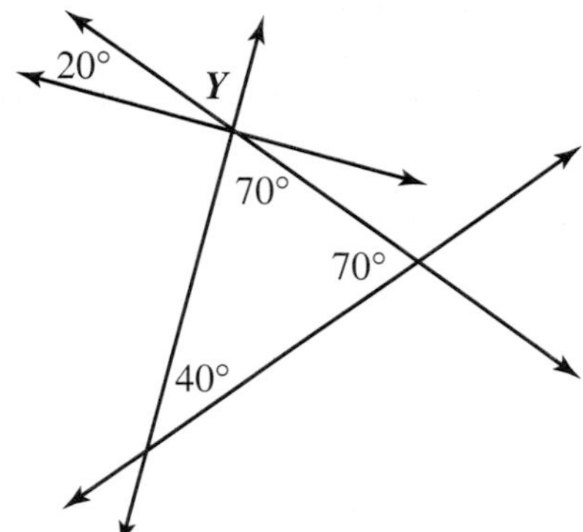

A 20°

B 40°

C 70°

D 90°

47. The drawing shows triangle *ABC*.

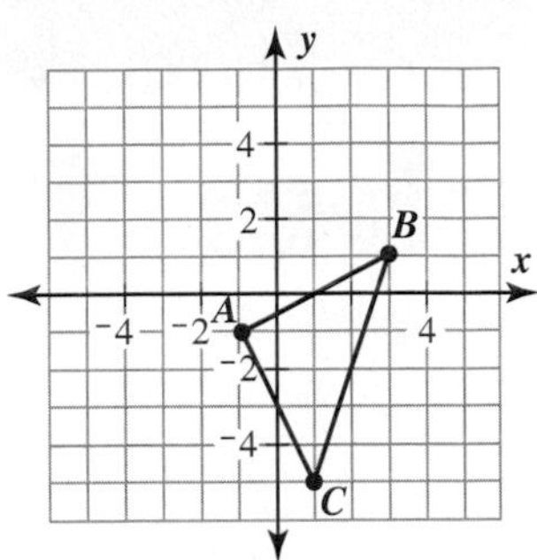

If triangle *ABC* were translated 3 units to the right, what would be the coordinates of the new triangle's vertices?

A (6, 1), (4, 5), (2, 1)

B (6, 1), (4, $^-5$), (2, $^-1$)

C ($^-6$, 1), ($^-4$, 5), ($^-2$, 1)

D ($^-6$, $^-1$), ($^-4$, $^-5$), ($^-2$, $^-1$)

48. Sari tossed a coin ten times and recorded her results in the table below.

COIN-TOSSING EXPERIMENT

Heads	Tails
~~IIII~~ II	III

Which of the following statements is true?

A Sari's experimental probability of tossing heads is $\frac{1}{2}$.

B Sari's experimental probability of tossing heads is less than her experimental probability of tossing tails.

C The theoretical probability of tossing heads is less than Sari's experimental probability of tossing heads.

D Sari's experimental probability of tossing tails is $\frac{7}{10}$.

Stop

Problem Solving Handbook

Number Toolbox

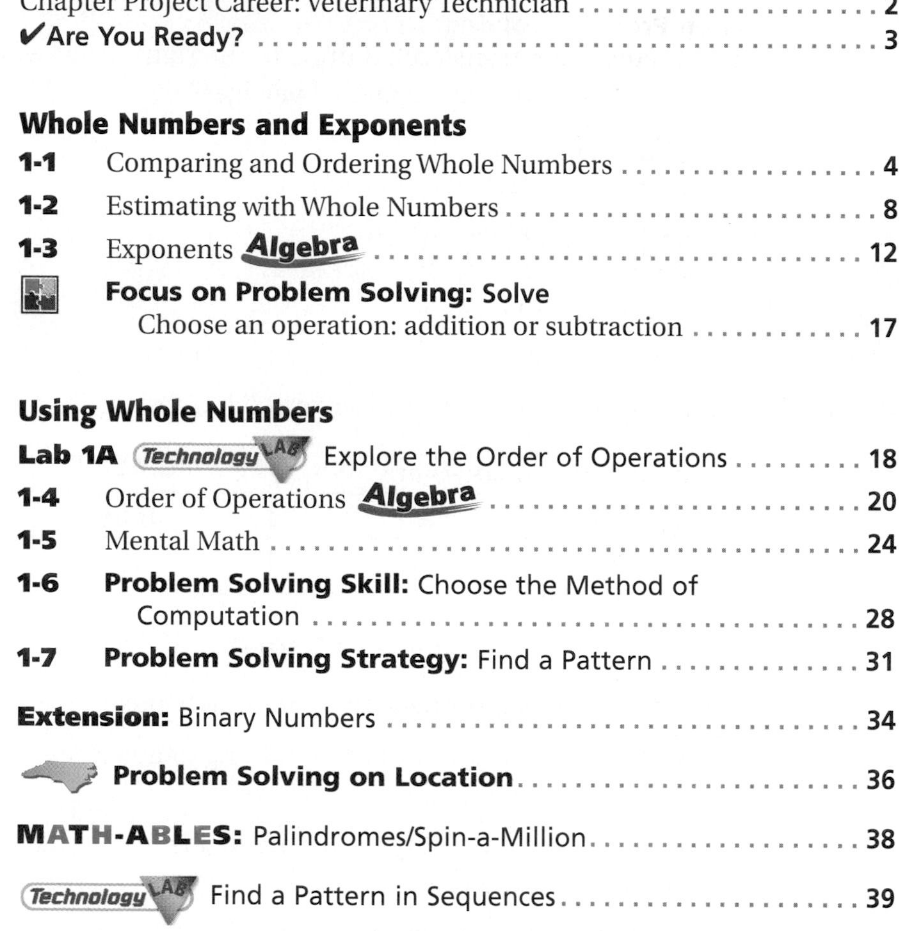

Student Help

internet connect
Homework Help Online
6, 10, 14, 22, 26, 29, 32
KEYWORD: MR4 HWHelp

Assessment

EOG Test Preparation Online . . . . KEYWORD: MR4 TestPrep

Algebra *Indicates algebra included in lesson development*

Introduction to Algebra

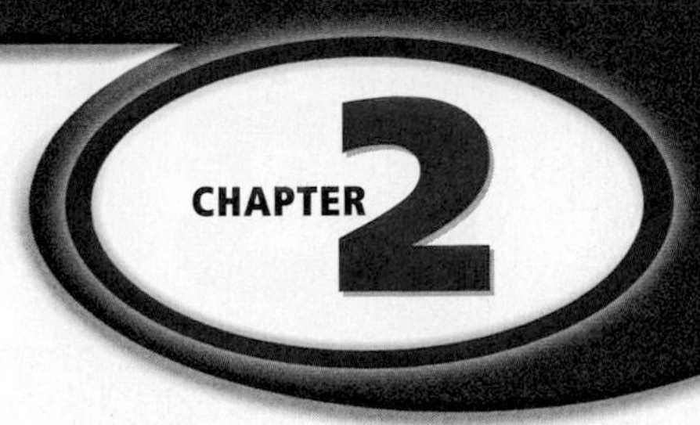

Decimals

Algebra *Indicates algebra included in lesson development*

Number Theory and Fractions

Assessment

EOG Test Preparation Online .... KEYWORD: MR4 TestPrep

Student Help

internet connect

Homework Help Online

154, 158, 162, 169, 174, 180, 184, 190, 194

KEYWORD: MR4 HWHelp

Fraction Operations

Interdisciplinary LINKS

Student Help

internet connect go.hrw.com

Homework Help Online

214, 218, 224, 228, 234, 238, 244, 248, 254, 258

KEYWORD: MR4 HWHelp

Assessment

go.hrw.com EOG Test Preparation Online .... KEYWORD: MR4 TestPrep

Algebra *Indicates algebra included in lesson development*

Collect and Display Data

Assessment

EOG Test Preparation Online . . . . . KEYWORD: MR4 TestPrep

Interdisciplinary LINKS

Student Help

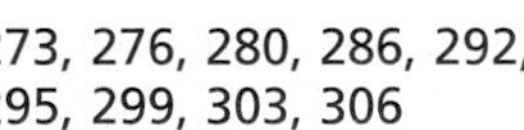

Homework Help Online

KEYWORD: MR4 HWHelp

Plane Geometry

Assessment

Interdisciplinary LINKS

Student Help

internet connect go.hrw.com

Homework Help Online

324, 328, 334, 338, 346, 350, 354, 358, 363, 367, 371, 374

KEYWORD: MR4 HWHelp

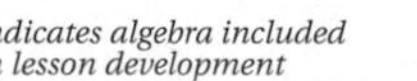 *Indicates algebra included in lesson development*

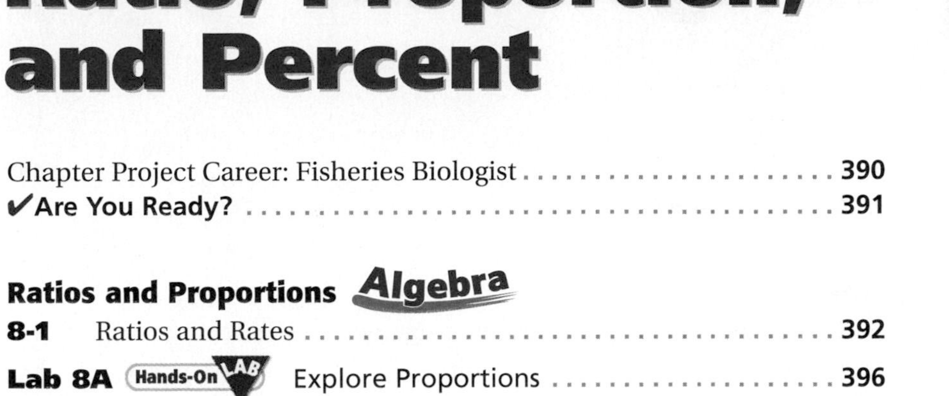

Ratio, Proportion, and Percent

Assessment

EOG Test Preparation Online . . . . . . KEYWORD: MR4 TestPrep

Interdisciplinary LINKS

Student Help

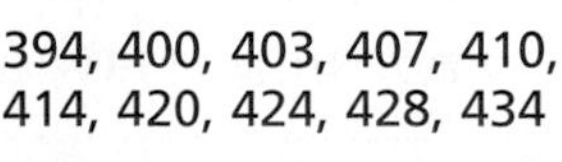

Homework Help Online

394, 400, 403, 407, 410, 414, 420, 424, 428, 434

KEYWORD: MR4 HWHelp

Integers

Interdisciplinary LINKS

Student Help

internet connect go.hrw.com

Homework Help Online

452, 456, 460, 467, 471, 474, 478, 484

KEYWORD: MR4 HWHelp

Assessment

go.hrw.com **EOG Test Preparation Online** .... KEYWORD: MR4 TestPrep

Algebra *Indicates algebra included in lesson development*

Perimeter, Area, and Volume

Assessment

go.hrw.com EOG Test Preparation Online . . . . KEYWORD: MR4 TestPrep

Interdisciplinary LINKS

Student Help

internet connect go.hrw.com

Homework Help Online

502, 506, 512, 518, 526, 532, 536, 540

KEYWORD: MR4 HWHelp

Probability

Interdisciplinary LINKS

Student Help

internet connect go.hrw.com

Homework Help Online

556, 560, 566, 572, 576, 582

KEYWORD: MR4 HWHelp

Assessment

go.hrw.com EOG Test Preparation Online .... KEYWORD: MR4 TestPrep

Algebra *Indicates algebra included in lesson development*

Functions and Coordinate Geometry

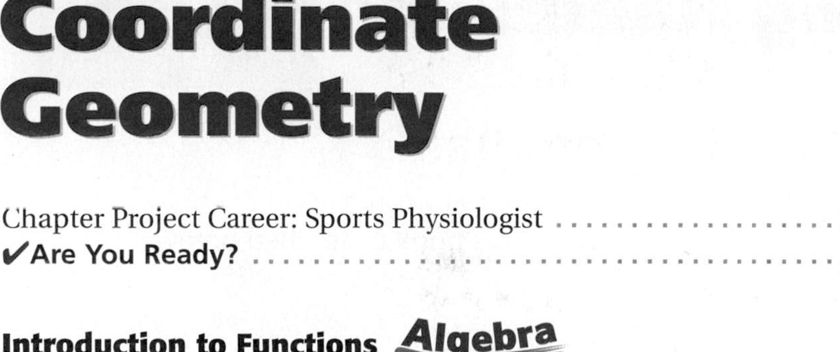
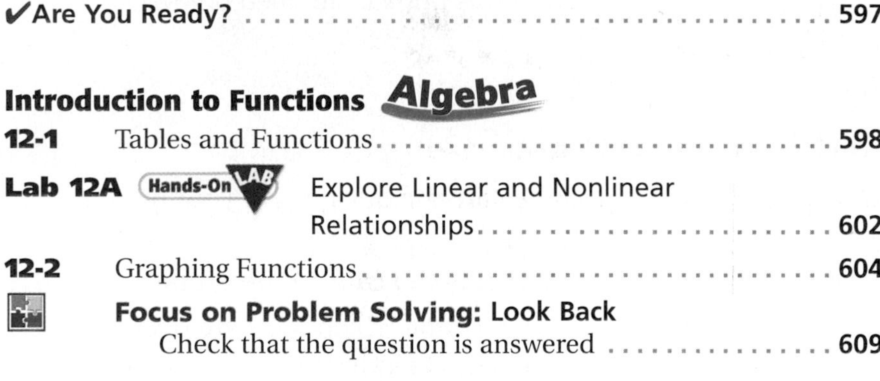
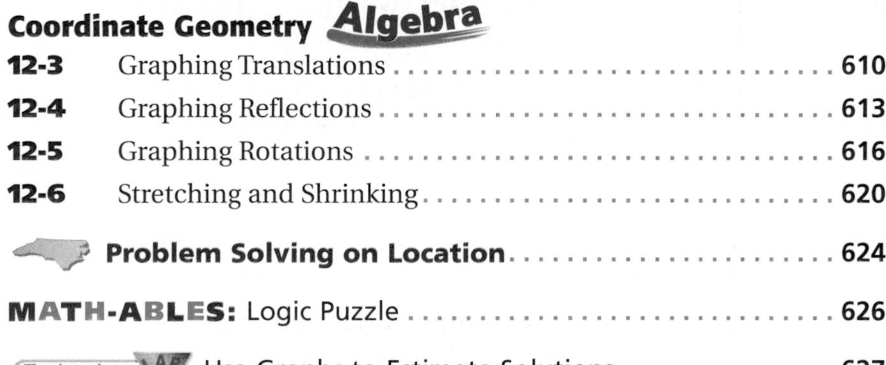
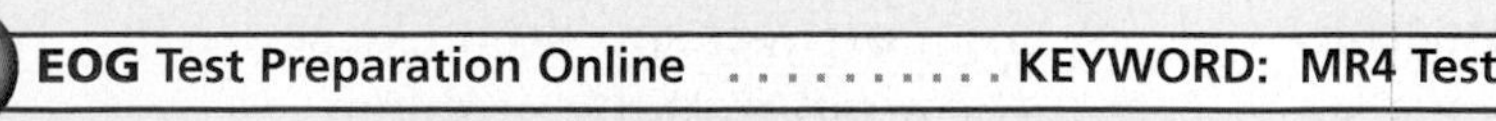
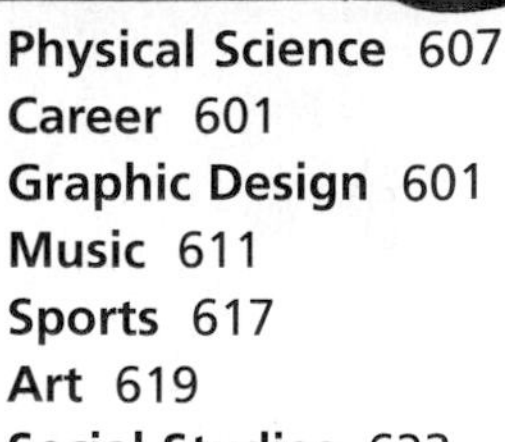

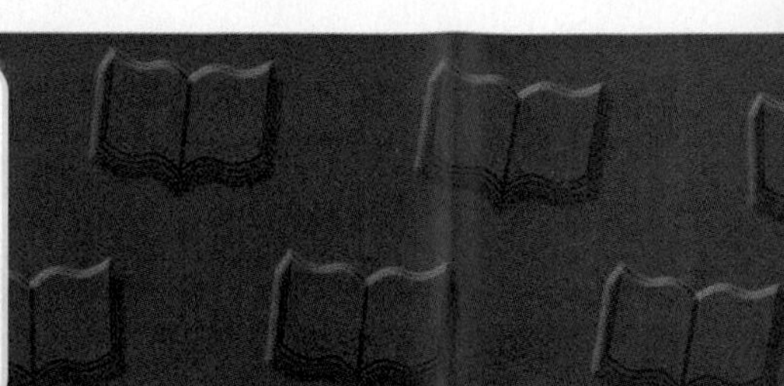

Problem Solving Handbook

The Problem Solving Plan

In order to be a good problem solver, you first need a good problem-solving plan. The plan used in this book is detailed below.

UNDERSTAND the Problem

■ **What are you asked to find?**	Restate the question in your own words.
■ **What information is given?**	Identify the facts in the problem.
■ **What information do you need?**	Determine which facts are needed to answer the question.
■ **Is all the information given?**	Determine whether all the facts are given.
■ **Is there any information given that you will not use?**	Determine which facts, if any, are unnecessary to solve the problem.

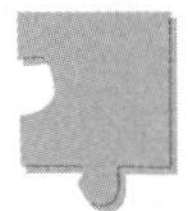

Make a PLAN

■ **Have you ever solved a similar problem?**	Think about other problems like this that you successfully solved.
■ **What strategy or strategies can you use?**	Determine a strategy that you can use and how you will use it.

SOLVE

■ **Follow your plan.**	Show the steps in your solution. Write your answer as a complete sentence.

LOOK BACK

■ **Have you answered the question?**	Be sure that you answered the question that is being asked.
■ **Is your answer reasonable?**	Your answer should make sense in the context of the problem.
■ **Is there another strategy you could use?**	Solving the problem using another strategy is a good way to check your work.
■ **Did you learn anything while solving this problem that could help you solve similar problems in the future?**	Try to remember the problems you have solved and the strategies you used to solve them.

Using the Problem Solving Plan

During summer vacation, Nicholas will visit first his cousin and then his grandmother. He will be gone for 5 weeks and 2 days, and he will spend 9 more days with his cousin than with his grandmother. How long will he stay with each family member?

UNDERSTAND the Problem

Identify the important information.

- Nicholas's visits will total 5 weeks and 2 days.
- He will spend 9 more days with his cousin than with his grandmother.

The answer will be how long he will stay with each family member.

Make a PLAN

You can draw a diagram to show how long Nicholas will stay. Use boxes for the length of each stay. The length of each box will represent the length of each stay.

SOLVE

Think: There are 7 days in a week, so 5 weeks and 2 days is 37 days in all. Your diagram might look like this:

Cousin	? days | 9 days	= 37 days	
Grandmother	? days		

Cousin	14 days | 9 days	$37 - 9 = 28$	*Subtract 9 days from the total days.*
Grandmother	14 days	$28 \div 2 = 14$	*Divide this number by 2 for the 2 places he will visit.*

So Nicholas will stay with his cousin for 23 days and with his grandmother for 14 days.

LOOK BACK

Twenty-three days is 9 days longer than 14 days. The total of the two stays is 23 + 14, or 37 days, which is the same as 5 weeks and 2 days. This solution fits the description of Nicholas's trip given in the problem.

Draw a Diagram

When problems involve objects, distances, or places, you can **draw a diagram** to make the problem easier to understand. You will often be able to use your diagram to solve the problem.

Problem Solving Strategies

Draw a Diagram	Make a Table
Make a Model	Solve a Simpler Problem
Guess and Test	Use Logical Reasoning
Work Backward	Use a Venn Diagram
Find a Pattern	Make an Organized List

All city blocks in Sunnydale are the same size. Tina starts her paper route at the corner of two streets. She travels 8 blocks south, 13 blocks west, 8 blocks north, and 6 blocks east. How far is she from her starting point when she finishes her route?

Understand the Problem

Identify the important information.

- Each block is the same size.
- You are given Tina's route.

The answer will be the distance from her starting point.

Make a Plan

Use the information in the problem to **draw a diagram** showing Tina's route. Label her starting and ending points.

End Start
6 blocks east
8 blocks north
8 blocks south
13 blocks west

Solve

The diagram shows that at the end of Tina's route she is 13 − 6 blocks from her starting point.

$13 - 6 = 7$

When Tina finishes, she is 7 blocks from her starting point.

Look Back

Be sure that you have drawn your diagram correctly. Does it match the information given in the problem?

PRACTICE

1. Laurence drives a carpool to school every Monday. He starts at his house and travels 4 miles south to pick up two children. Then he drives 9 miles west to pick up two more children, and then he drives 4 miles north to pick up one more child. Finally, he drives 5 miles east to get to the school. How far does he have to travel to get back home?

2. The roots of a tree reach 12 feet into the ground. A kitten is stuck 5 feet from the top of the tree. From the treetop to the root bottom, the tree measures 32 feet. How far above the ground is the kitten?

Make a Model

If a problem involves objects, you can sometimes **make a model** using those objects or similar objects to act out the problem. This can help you understand the problem and find the solution.

Problem Solving Strategies

Draw a Diagram	Make a Table
Make a Model	Solve a Simpler Problem
Guess and Test	Use Logical Reasoning
Work Backward	Use a Venn Diagram
Find a Pattern	Make an Organized List

Alice has three pieces of ribbon. Their lengths are 7 inches, 10 inches, and 12 inches. Alice does not have a ruler or scissors. How can she use these ribbons to measure a length of 15 inches?

Understand the Problem

Identify the important information.

- The ribbons are 7 inches, 10 inches, and 12 inches long.

The answer will show how to use the ribbons to measure 15 inches.

Make a Plan

Measure and cut three ribbons or strips of paper to **make a model.** One ribbon should be 7 inches long, one should be 10 inches long, and one should be 12 inches long. Try different combinations of the ribbons to form new lengths.

Solve

When you put any two ribbons together end to end, you can form lengths of 17, 19, and 22 inches. All of these are too long.

Try placing the 10-inch ribbon and the 12-inch ribbon end to end to make 22 inches. Now place the 7-inch ribbon above them. The remaining length that is **not** underneath the 7-inch ribbon will measure 15 inches.

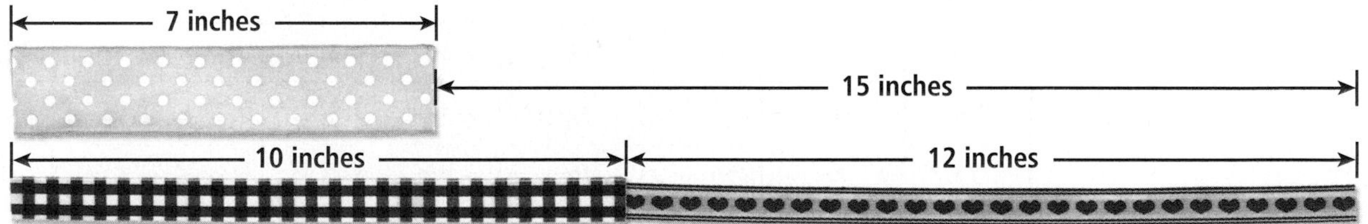

Look Back

Use another strategy. Without using ribbon, you could have **guessed** different ways to add or subtract 7, 10, and 12. Then you could have **tested** to see if any of these gave an answer of 15:

$10 + 12 - 7 = 15$

PRACTICE

1. Find other lengths that you can measure with the three pieces of ribbon.
2. Andy stacks four cubes, one on top of the other, and paints the outside of the stack (not the bottom). How many faces of the cubes are painted?

Guess and Test

If you do not know how to solve a problem, you can always make a **guess**. Then **test** your guess using the information in the problem. Use the result to make a better guess. Repeat until you find the correct answer.

Problem Solving Strategies

Draw a Diagram	Make a Table
Make a Model	Solve a Simpler Problem
Guess and Test	Use Logical Reasoning
Work Backward	Use a Venn Diagram
Find a Pattern	Make an Organized List

There were 25 problems on a test. For each correct answer, 4 points were given. For each incorrect answer, 1 point was subtracted. Tania answered all 25 problems. Her score was 85. How many correct and incorrect answers did she have?

Understand the Problem

Identify the important information.

- There were 25 problems on the test.
- A correct answer received 4 points, and an incorrect answer lost 1 point.
- Tania answered all of the problems and her score was 85.

The answer will be the number of problems that Tania got correct and incorrect.

Make a Plan

Start with a **guess** for the number of correct answers. Then **test** to see whether the total score is 85.

Solve

Make a first guess of 20 correct answers.

Correct	Incorrect	Score	Result
20	5	$(20 \times 4) - (5 \times 1) = 80 - 5 = 75$	Too low—guess higher
23	2	$(23 \times 4) - (2 \times 1) = 92 - 2 = 90$	Too high—guess lower
22	3	$(22 \times 4) - (3 \times 1) = 88 - 3 = 85$	Correct ✓

Tania had 22 correct answers and 3 incorrect answers.

Look Back

Notice that the guesses made while solving this problem were not just "wild" guesses. Guessing and testing in an organized way will often lead you to the correct answer.

PRACTICE

1. The sum of Joe's age and his younger brother's age is 38. The difference between their ages is 8. How old are Joe and his brother?

2. Amy bought some used books for $4.95. She paid $0.50 each for some books and $0.35 each for the others. She bought fewer than 8 books at each price. How many books did Amy buy? How many cost $0.50?

Work Backward

Some problems give you a sequence of information and ask you to find something that happened at the beginning. To solve a problem like this, you may want to start at the end of the problem and **work backward.**

Problem Solving Strategies

Draw a Diagram	Make a Table
Make a Model	Solve a Simpler Problem
Guess and Test	Use Logical Reasoning
Work Backward	Use a Venn Diagram
Find a Pattern	Make an Organized List

Jaclyn and her twin sister, Bailey, received money for their birthday. They used half of their money to buy a video game. Then they spent half of the money they had left on a pizza. Finally, they spent half of the remaining money to rent a movie. At the end of the day, they had $4.50. How much money did they have to start out with?

Understand the Problem

Identify the important information.

- The girls ended with $4.50.
- They spent half of their money at each of three stops.

The answer will be the amount of money they started with.

Make a Plan

Start with the amount you know the girls have left, $4.50, and **work backward** through the information given in the problem.

Solve

Jaclyn and Bailey had $4.50 at the end of the day.

They had twice that amount before renting a movie. $2 \times \$4.50 = \$9$

They had twice that amount before buying a pizza. $2 \times \$9 = \$18$

They had twice that amount before buying a video game. $2 \times \$18 = \$36$

The girls started with $36.

Look Back

Using the starting amount of $36, work from the beginning of the problem. Find the amount they spent at each location and see whether they are left with $4.50.

Start:	$36
Video game:	$\$36 \div 2 = \$18$
Pizza:	$\$18 \div 2 = \$9$
Movie rental:	$\$9 \div 2 = \$4.50$ ✓

PRACTICE

1. The Lauber family has 4 children. Chris is 5 years younger than his brother Mark. Justin is half as old as his brother Chris. Mary, who is 10, is 3 years younger than Justin. How old is Mark?
2. If you divide a mystery number by 4, add 8, and multiply by 3, you get 42. What is the mystery number?

Find a Pattern

In some problems, there is a relationship between different pieces of information. Examine this relationship and try to **find a pattern.** You can then use this pattern to find more information and the solution to the problem.

Problem Solving Strategies

Draw a Diagram	Make a Table
Make a Model	Solve a Simpler Problem
Guess and Test	Use Logical Reasoning
Work Backward	Use a Venn Diagram
Find a Pattern	Make an Organized List

Students are using the pattern at right to build stairways for a model house. How many blocks are needed to build a stairway with seven steps?

Understand the Problem The answer will be the total number of blocks in a stairway with seven steps.

Make a Plan Try to **find a pattern** between the number of steps and the number of blocks needed.

Notice that the first step is made of one block. The second step is made of two blocks, the third step is made of three blocks, and the fourth step is made of four blocks.

Step	Number of Blocks in Step	Total Number of Blocks in Stairway
2	2	$1 + 2 = 3$
3	3	$1 + 2 + 3 = 6$
4	4	$1 + 2 + 3 + 4 = 10$

To find the total number of blocks, add the number of blocks in the first step, the second step, the third step, and so on.

Solve The seventh step will be made of seven blocks. The total number of blocks will be $1 + 2 + 3 + 4 + 5 + 6 + 7 = 28$.

Look Back Use another strategy. You can **draw a diagram** of a stairway with 7 steps. Count the number of blocks in your diagram. There are 28 blocks.

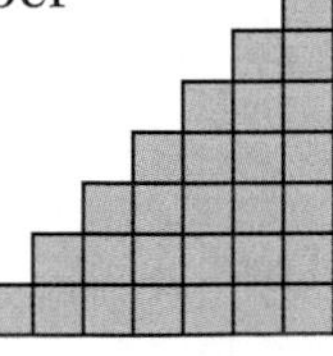

PRACTICE

1. A cereal company adds baseball cards to the 3rd box, the 6th box, the 11th box, the 18th box, and so on of each case of cereal. In a case of 40 boxes, how many boxes will have baseball cards?
2. Describe the pattern and find the missing numbers.

   1; 4; 16; 64; 256; ▒; ▒; 16,384

Make a Table

When you are given a lot of information in a problem, it may be helpful to organize that information. One way to organize information is to **make a table.**

Problem Solving Strategies

Draw a Diagram	**Make a Table**
Make a Model	Solve a Simpler Problem
Guess and Test	Use Logical Reasoning
Work Backward	Use a Venn Diagram
Find a Pattern	Make an Organized List

Mrs. Melo's students scored the following on their math test: 90, 80, 77, 78, 91, 92, 73, 62, 83, 79, 72, 85, 93, 84, 75, 68, 82, 94, 98, and 82. An A is given for 90 to 100 points, a B for 80 to 89 points, a C for 70 to 79 points, a D for 60 to 69 points, and an F for less than 60 points. Find the number of students who scored each letter grade.

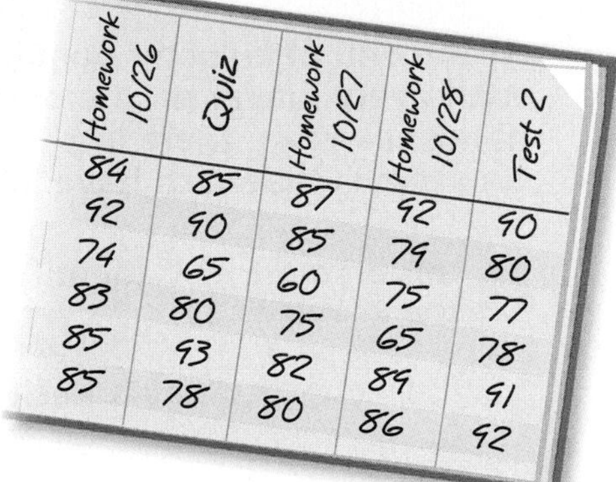

Homework 10/26	Quiz	Homework 10/27	Homework 10/28	Test 2
84	85	87	92	90
92	90	85	79	80
74	65	60	75	77
83	80	75	65	78
85	93	82	89	91
85	78	80	86	92

Understand the Problem

Identify the important information.

- You have been given the list of scores and the letter grades that go with each score.

The answer will be the number of each letter grade.

Make a Plan

Make a table to organize the scores. Use the information in the problem to set up your table. Make one row for each letter grade.

Solve

Read through the list of scores. As you read each score, make a tally in the appropriate place in your table. There are 20 test scores, so be sure you have 20 tallies in all.

Letter Grade	Number
A (90–100)	𝍸 I
B (80–89)	𝍸 I
C (70–79)	𝍸 I
D (60–69)	II
F (below 60)	

Mrs. Melo gave out six A's, six B's, six C's, two D's, and no F's.

Look Back

Use another strategy. Another way you could solve this problem is to **make an organized list.** Order the scores from least to greatest, and count how many scores are in each range.

62, 68, 72, 73, 75, 77, 78, 79, **80, 82, 82, 83, 84, 85,** 90, 91, 92, 93, 94, 98

D C B A

PRACTICE

1. The debate club has 6 members. Each member will debate each of the other members exactly once. How many total debates will there be?

2. At the library, there are three story-telling sessions. Each one lasts 45 minutes, with 30 minutes between sessions. If the first session begins at 10:00 A.M., what time does the last session end?

Solve a Simpler Problem

Sometimes a problem contains large numbers or requires many steps. Try to **solve a simpler problem** that is similar. Solve the simpler problem first, and then try the same steps to solve the original problem.

Problem Solving Strategies

Draw a Diagram	Make a Table
Make a Model	**Solve a Simpler Problem**
Guess and Test	Use Logical Reasoning
Work Backward	Use a Venn Diagram
Find a Pattern	Make an Organized List

At the end of a soccer game, each player shakes hands with every player on the opposing team. How many handshakes are there at the end of a game between two teams that each have 20 players?

Understand the Problem

Identify the important information.

- There are 20 players on each team.
- Each player will shake hands with every player on the opposing team.

The answer will be the total number of handshakes exchanged.

Make a Plan

Solve a simpler problem. For example, suppose each team had just one player. Then there would only be one handshake between the two players. Expand the number of players to two and then three.

Solve

When there is 1 player, there is $1 \times 1 = 1$ handshake. For 2 players, there are $2 \times 2 = 4$ handshakes. And for 3 players, there are $3 \times 3 = 9$ handshakes.

If each team has 20 players, there will be $20 \times 20 = 400$ handshakes.

Players Per Team	Diagram	Handshakes
1		1
2		4
3		9

Look Back

If the pattern is correct, for 4 players there will be 16 handshakes and for 5 players there will be 25 handshakes. Complete the next two rows of the table to check these answers.

PRACTICE

1. Martha has 5 pairs of pants and 4 blouses that she can wear to school. How many different outfits can she make?

2. What is the smallest 5-digit number that can be divided by 50 with a remainder of 17?

Use Logical Reasoning

Sometimes a problem may provide clues and facts that you must use to answer a question. You can **use logical reasoning** to solve this kind of problem.

Problem Solving Strategies

Draw a Diagram	Make a Table
Make a Model	Solve a Simpler Problem
Guess and Test	**Use Logical Reasoning**
Work Backward	Use a Venn Diagram
Find a Pattern	Make an Organized List

Kevin, Ellie, and Jillian play three different sports. One person plays soccer, one likes to run track, and the other swims. Ellie is the sister of the swimmer. Kevin once went shopping with the swimmer and the track runner. Match each student with his or her sport.

Understand the Problem

Identify the important information.

- There are three people, and each person plays a different sport.
- Ellie is the sister of the swimmer.
- Kevin once went shopping with the swimmer and the track runner.

The answer will tell which student plays each sport.

Make a Plan

Start with clues given in the problem, and **use logical reasoning** to find the answer.

Solve

Make a table with a column for each sport and a row for each person. Work with the clues one at a time. Write "yes" in a box if the clue applies to that person. Write "no" if the clue does not apply.

	Soccer	Track	Swim
Kevin		no	no
Ellie			no
Jillian			

- Ellie is the sister of the swimmer, so she is not the swimmer.
- Kevin went shopping with the swimmer and the track runner. He is not the swimmer or the track runner.

So Kevin must be the soccer player, and Jillian must be the swimmer. This leaves Ellie as the track runner.

Look Back

Compare your answer to the clues in the problem. Make sure none of your conclusions conflict with the clues.

PRACTICE

1. Karin, Brent, and Lola each ordered a different slice of pizza: pepperoni, plain cheese, and ham-pineapple. Karin is allergic to pepperoni. Lola likes more than one topping. Which kind of pizza did each person order?

2. Leo, Jamal, and Kara are in fourth, fifth, and sixth grades. Kara is not in fourth grade. The sixth-grader is in chorus with Kara and has the same lunch time as Leo. Match the students with their grades.

Use a Venn Diagram

You can **use a Venn diagram** to display relationships among sets in a problem. Use ovals, circles, or other shapes to represent individual sets.

Problem Solving Strategies

Draw a Diagram	Make a Table
Make a Model	Solve a Simpler Problem
Guess and Test	Use Logical Reasoning
Work Backward	**Use a Venn Diagram**
Find a Pattern	Make an Organized List

Robert is taking a survey to see what kinds of pets students have. He found that 70 students have dogs, 45 have goldfish, and 60 have birds. Some students have two kinds of pets: 17 students have dogs and fish, 22 students have dogs and birds, and 15 students have birds and goldfish. Five students have all three kinds of pets. How many students in the survey have only birds?

Understand the Problem List the important information.

- You know that 70 students have dogs, 45 have goldfish, and 60 have birds.

The answer will be the number of students who have only birds.

Make a Plan **Use a Venn diagram** to show the sets of students who have dogs, students who have goldfish, and students who have birds.

Solve Draw and label three overlapping circles. Work from the inside out. Write "5" in the area where all three circles overlap. This represents the number of students who have a dog, a goldfish, and a bird.

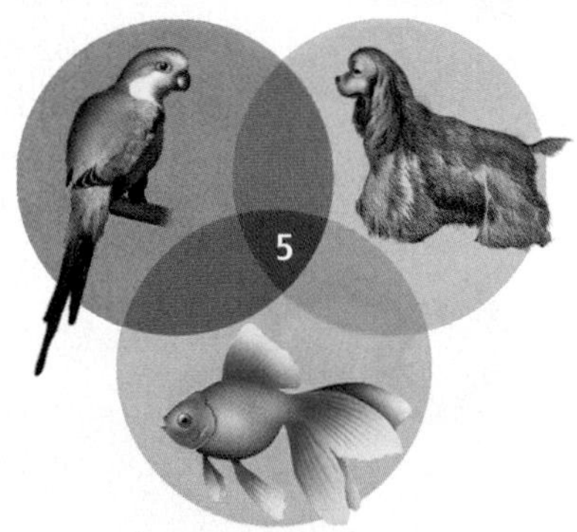

Use the information in the problem to fill in other sections of the diagram. You know that 60 students have birds, so the numbers within the bird circle will add to 60.

So 18 students have only pet birds.

Look Back When your Venn diagram is complete, check it carefully against the information in the problem. Make sure your diagram agrees with the facts given.

PRACTICE

1. How many students have only dogs?

2. How many students have only goldfish?

Make an Organized List

In some problems, you will need to find how many different ways something can happen. It is often helpful to **make an organized list.** This will help you count the outcomes and be sure that you have included all of them.

Problem Solving Strategies

Draw a Diagram	Make a Table
Make a Model	Solve a Simpler Problem
Guess and Test	Use Logical Reasoning
Work Backward	Use a Venn Diagram
Find a Pattern	**Make an Organized List**

In a game at an amusement park, players throw 3 darts at a target to score points and win prizes. If each dart lands within the target area, how many different total scores are possible?

Understand the Problem

Identify the important information.

- A player throws three darts at the target.

The answer will be the number of different scores a player could earn.

Make a Plan

Make an organized list to determine all possible outcomes and score totals. List the value of each dart and the point total for all three darts.

Solve

You can organize your list by the number of darts that land in the center. All three darts could hit the center circle. Or, two darts could hit the center circle and the third could hit a different circle. One dart could hit the center circle, or no darts could hit the center circle.

3 Darts Hit Center	2 Darts Hit Center	1 Dart Hits Center	0 Darts Hit Center
$10 + 10 + 10 = 30$	$10 + 10 + 5 = 25$	$10 + 5 + 5 = 20$	$5 + 5 + 5 = 15$
	$10 + 10 + 2 = 22$	$10 + 5 + 2 = 17$	$5 + 5 + 2 = 12$
		$10 + 2 + 2 = 14$	$5 + 2 + 2 = 9$
			$2 + 2 + 2 = 6$

Count the different outcomes. There are 10 possible scores.

Look Back

You could have listed outcomes in random order, but because your list is organized, you can be sure that you have not missed any possibilities. Check to be sure that every score is different.

PRACTICE

1. A restaurant has three different kinds of pancakes: cinnamon, blueberry, and apple. If you order one of each kind, how many different ways can the three pancakes be stacked?

2. How many ways can you make change for a quarter using dimes, nickels, and pennies?

Chapter 1

Number Toolbox

TABLE OF CONTENTS

African Plant-Eating Animals		
Animal	Weight (lb)	Daily Food Intake (lb)
Buffalo	1,500	45
Elephant	11,000	660
Giraffe	2,500	75
Hippopotamus	5,500	90
Zebra	950	30

Career *Veterinary Technician*

Do you like caring for animals? Veterinary technicians perform many of the same tasks for veterinarians as nurses do for doctors. Veterinary technicians also do research that can help animals. To care for animals, technicians must know what the animals need to eat and how they behave with other types of animals. Large plant-eating animals, many of which live in Africa, need to eat specific kinds of grasses and trees. The table above shows the approximate weight of some animals and the approximate amount of food the animals eat each day.

internet connect

Chapter Opener Online
go.hrw.com
KEYWORD: MR4 Ch1

ARE YOU READY?

Choose the best term from the list to complete each sentence.

1. The answer in a multiplication problem is called the ___?___.
2. 5,000 + 400 + 70 + 5 is a number written in ___?___ form.
3. A(n) ___?___ tells about how many.
4. The number 70,562 is written in ___?___ form.
5. Ten thousands is the ___?___ of the 4 in 42,801.

place value
estimate
product
expanded
standard
period

Complete these exercises to review skills you will need for this chapter.

✔ Compare Whole Numbers

Compare. Write <, >, or =.

6. 245 ▒ 219 **7.** 5,320 ▒ 5,128

8. 64 ▒ 67 **9.** 784 ▒ 792

✔ Round Whole Numbers

Round each number to the nearest hundred.

10. 567 **11.** 827 **12.** 1,642 **13.** 12,852

14. 1,237 **15.** 135 **16.** 15,561 **17.** 452,801

Round each number to the nearest thousand.

18. 4,709 **19.** 3,399 **20.** 9,825 **21.** 26,419

22. 12,434 **23.** 4,561 **24.** 11,784 **25.** 468,201

✔ Whole Number Operations

Add, subtract, multiply, or divide.

26. 18×22 **27.** $135 \div 3$ **28.** $247 + 96$ **29.** $358 - 29$

✔ Evaluate Whole Number Expressions

Evaluate each expression.

30. $3 \times 4 \times 2$ **31.** $20 + 100 - 40$

32. $5 \times 20 \div 4$ **33.** $6 \times 12 \times 5$

Comparing and Ordering Whole Numbers

Learn to compare and order whole numbers using place value or a number line.

The midyear world population in 1990 was 5,283,755,345 people. The world population by midyear 2010 is projected to be 6,823,634,553 people.

You can use place value to read and understand large numbers. In the place value chart below, 3 has a value of 3 millions, 3 ten thousands, or 3 ones, depending on its position in the number.

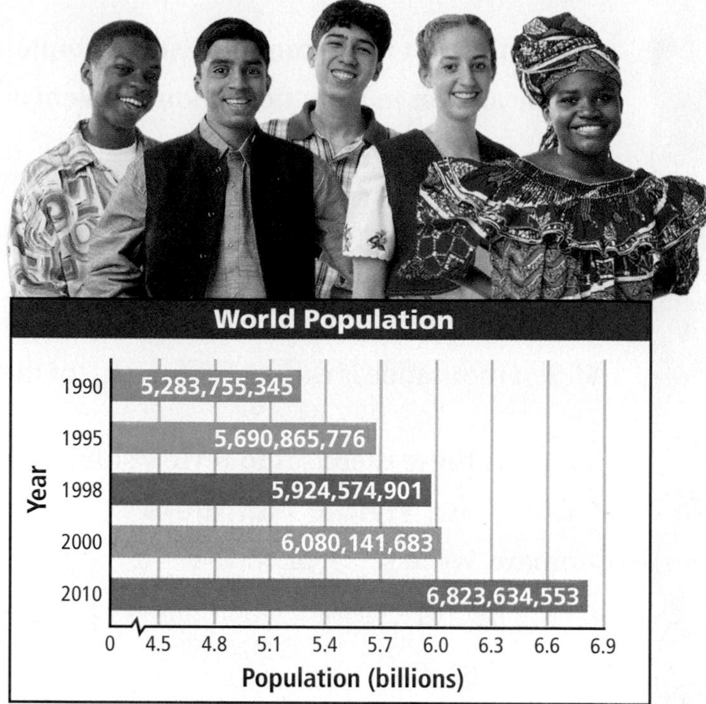

Source: U.S. Bureau of the Census, International Data Base, 2000

Place Value

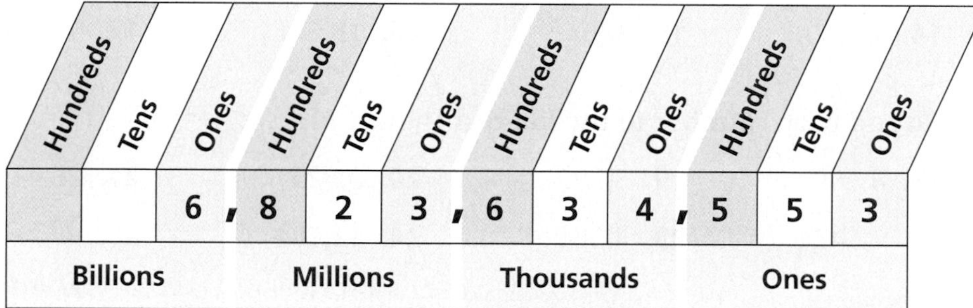

Standard form: 6,823,634,553

Expanded form: 6,000,000,000 + 800,000,000 + 20,000,000 + 3,000,000 + 600,000 + 30,000 + 4,000 + 500 + 50 + 3

Word form: six billion, eight hundred twenty-three million, six hundred thirty-four thousand, five hundred fifty-three

EXAMPLE 1 Using Place Value to Compare Whole Numbers

Belgium's 2001 population was 10,258,762 people. The Czech Republic's 2001 population was 10,264,212 people. Which country had more people?

Belgium:	1 0, 2 5 8, 7 6 2
Czech Republic:	1 0, 2 6 4, 2 1 2

Start at the left and compare digits in the same place value position. Look for the first place where the values are different.

50 thousand is less than 60 thousand.
10,258,762 is less than 10,264,212.

The Czech Republic had more people.

To order numbers, you can compare them using place value and then write them in order from least to greatest. You can also graph the numbers on a number line. As you read the numbers from left to right, they will be ordered from least to greatest.

EXAMPLE 2 Using a Number Line to Order Whole Numbers

Remember!

< means "is less than."
3 < 5   120 < 504
> means "is greater than."
17 > 9   212 > 83

Order the numbers from least to greatest.
923; 835; 1,266

Graph the following numbers on a number line:
The number 923 is between 900 and 1,000.
The number 835 is between 800 and 900.
The number 1,266 is between 1,200 and 1,300.

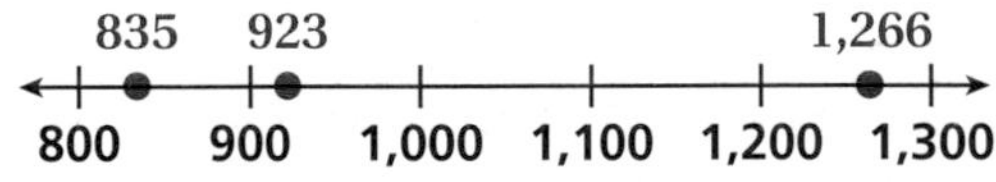

The numbers are ordered when you read the number line from left to right.

The numbers in order from least to greatest are 835, 923, and 1,266.

Think and Discuss

1. **Give** the place value of the digit 3 in each of the following numbers: 2,307,912; 2,370,912; 2,703,912.
2. **Read** each of the following numbers: 937,052; 3,012,480; 8,135,712,004.
3. **Look** at the bar graph at the beginning of the lesson. In which years was the population between 5,500,000,000 and 6,500,000,000?

1-1 Exercises

1.03

FOR EOG PRACTICE
see page 636

internet connect
Homework Help Online
go.hrw.com Keyword: MR4 1-1

GUIDED PRACTICE

See Example 1

1. Mount McKinley, in Alaska, is 20,320 feet tall. Mount Aconcagua, in Argentina, is 22,834 feet tall. Which mountain is taller?
2. The area of the Caribbean Sea is 971,400 square miles. The area of the Mediterranean Sea is 969,100 square miles. Which sea is smaller in area?

See Example 2

Order the numbers from least to greatest.

3. 726; 349; 642
4. 513; 915; 103
5. 497; 1,264; 809

INDEPENDENT PRACTICE

See Example 1

6. The attendance in 1999 at a theme park was 17,459,000 people. The attendance in 1999 at a water park was 15,200,000 people. Which park had the higher attendance?
7. According to the table, which river is longer, the Missouri or the Mississippi?
8. A New York City driving range reported 413,497 golf balls were hit by customers last year. A Philadelphia range reported customers hit 408,959 golf balls. Which range had more golf balls hit?

River	Length (mi)
Mississippi	2,340
Missouri	2,315
Ohio	618
Red	1,290
Rio Grande	1,900

See Example 2

Order the numbers from least to greatest.

9. 367; 597; 279
10. 619; 126; 480
11. 946; 705; 810
12. 423; 1,046; 805
13. 1,523; 2,913; 111
14. 1,764; 1,359; 666

PRACTICE AND PROBLEM SOLVING

Compare. Write < or >.

15. 46,495 ▒ 46,594
16. 162,648 ▒ 126,498
17. 3,654 ▒ 3,645
18. 512,105 ▒ 512,099
19. 29,448 ▒ 29,488
20. 913,203 ▒ 913,600

Order the numbers from greatest to least.

21. 591; 924; 341
22. 601; 533; 823; 149
23. 291; 911; 439; 747
24. 2,649; 3,461; 1,947
25. 5,349; 5,389; 5,480
26. 7,467; 7,239; 7,498

Before calculators and computers were invented, the abacus was used to solve math problems. On an abacus, rods represent place values, and beads that slide on the rods are used as counters.

go.hrw.com
KEYWORD: MR4 Abacus

CNN Student News

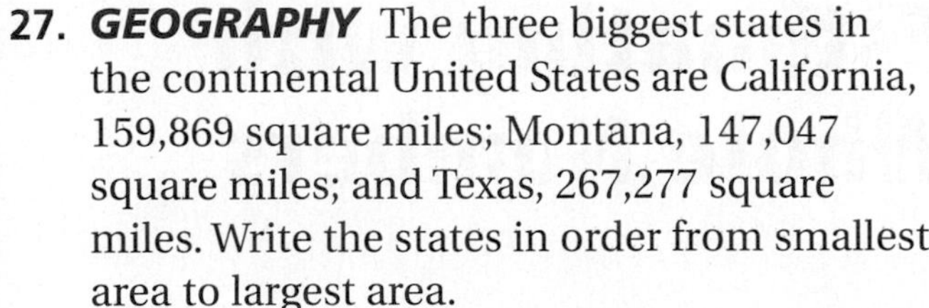

27. ***GEOGRAPHY*** The three biggest states in the continental United States are California, 159,869 square miles; Montana, 147,047 square miles; and Texas, 267,277 square miles. Write the states in order from smallest area to largest area.

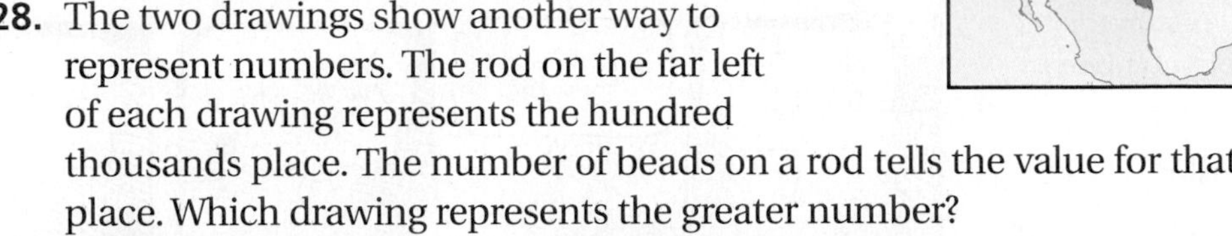

28. The two drawings show another way to represent numbers. The rod on the far left of each drawing represents the hundred thousands place. The number of beads on a rod tells the value for that place. Which drawing represents the greater number?

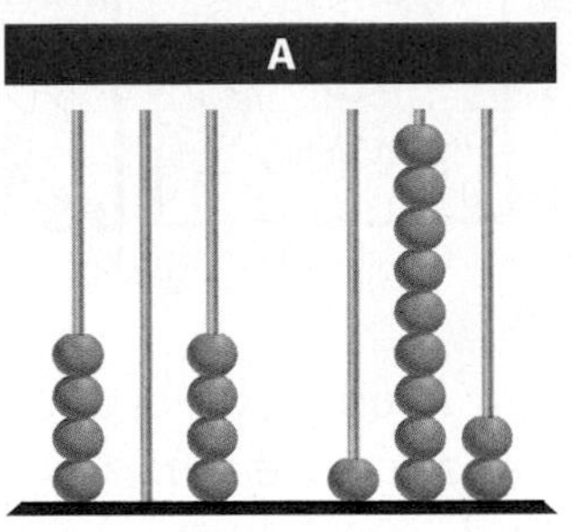

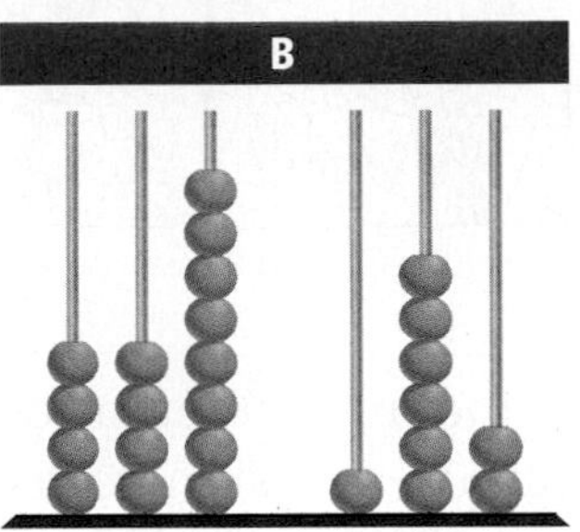

29. ***WHAT'S THE ERROR?*** A student said 19,465,405 is greater than 19,465,425. Explain the error. Write the statement correctly.

30. ***WRITE ABOUT IT*** Explain how you would compare 19,465,146 and 19,460,146.

31. ***CHALLENGE*** In Roman numerals, letters represent numbers. For example, I = 1, V = 5, X = 10, L = 50, and C = 100. Letters in Roman numerals are written next to each other; this is how the value of the number is shown. To read the numbers below, add the values of all of the letters. What numbers do the following represent?

a. CLX **b.** LVI **c.** CIII

Spiral Review

Write the value of the red digit in each number. (Previous course)

32. 649,809 **33.** 349,239 **34.** 27,463 **35.** 16,239

Write each number in word form. (Previous course)

36. 1,645 **37.** 24,498 **38.** 306,927 **39.** 4,605,926

Write each number in standard form. (Previous course)

40. two hundred thirty-four thousand, six hundred seventy-nine

41. fifteen million, nine hundred three thousand, one hundred eight

42. **EOG PREP** Which number has a 4 in the thousands place? (Previous course)

A 10,400 B 14,307 C 3,742,619 D 8,405,361

1-2 Estimating with Whole Numbers

Learn to estimate with whole numbers.

Vocabulary

compatible number

underestimate

overestimate

Sometimes in math you do not need an exact answer. Instead, you can use an estimate. Estimates are close to the exact answer but are usually easier and faster to find.

When estimating, you can round the numbers in the problem to *compatible numbers.* **Compatible numbers** are close to the numbers in the problem, and they can help you do math mentally.

EXAMPLE 1 Estimating a Sum or Difference by Rounding

Remember!

When rounding, look at the digit to the right of the place to which you are rounding.

- If that digit is 5 or greater, round up.
- If that digit is less than 5, round down.

Estimate each sum or difference by rounding to the place value indicated.

A **5,439 + 7,516; thousands**

$$\begin{array}{r} 5{,}000 \\ +\ 8{,}000 \\ \hline 13{,}000 \end{array}$$

Round 5,439 down.
Round 7,516 up.

The sum is about 13,000.

B **62,167 − 47,511; ten thousands**

$$\begin{array}{r} 60{,}000 \\ -\ 50{,}000 \\ \hline 10{,}000 \end{array}$$

Round 62,167 down.
Round 47,511 up.

The difference is about 10,000.

An estimate that is less than the exact answer is an **underestimate**.

An estimate that is greater than the exact answer is an **overestimate**.

EXAMPLE  2 Estimating a Product by Rounding

Remember!

The area of a rectangle is found by multiplying the length by the width.

$A = \ell \times w$

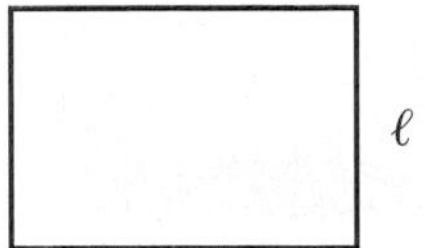

The sixth-grade class is preparing to paint a mural on one wall of the school. First the students need to paint the entire area of the wall white. The wall is a rectangle 9 feet tall and 27 feet wide.

One quart of paint will cover an area of 100 square feet. How many quarts of white paint should the students buy?

First find the area of the wall in square feet.

$9 \times 27 \rightarrow 9 \times 30$ *__Overestimate__ the area of the wall.*

$9 \times 30 = 270$ *The actual area is __less than__ 270 square feet.*

If one quart of paint will cover 100 square feet, then 2 quarts will cover 200 square feet. Three quarts of paint will cover 300 square feet.

The students should buy three quarts of paint.

EXAMPLE  3 Estimating a Quotient Using Compatible Numbers

Mrs. Byrd will drive 120 miles to take Becca to the state fair. She can drive 65 mi/h. About how long will the trip take?

To find how long the trip will be, divide the miles Mrs. Byrd has to travel by how many miles per hour she can drive.

miles ÷ miles per hour

$120 \div 65 \rightarrow 120 \div 60$ *120 and 60 are compatible numbers.* *__Underestimate__ the speed.*

$120 \div 60 = 2$ *Because she __underestimated__ the speed, the actual time will be __less than__ 2 hours.*

It will take Mrs. Byrd about two hours to reach the state fair.

Think and Discuss

1. **Suppose** you are buying items for a party and you have $50. Would it be better to overestimate or underestimate the cost of the items?
2. **Describe** situations in which you might want to estimate.

1-2 Exercises

1.04c, 1.07

FOR EOG PRACTICE
see page 636

internet connect
Homework Help Online
go.hrw.com Keyword: MR4 1-2

GUIDED PRACTICE

See Example 1
Estimate each sum or difference by rounding to the place value indicated.

1. 4,689 + 2,469; thousands

2. 50,498 − 35,798; ten thousands

See Example 2
3. The graph shows the number of bottles of water used in three bicycle races last year. If the same number of riders enter the races each year, estimate the number of bottles that will be needed for races held in May over the next five years.

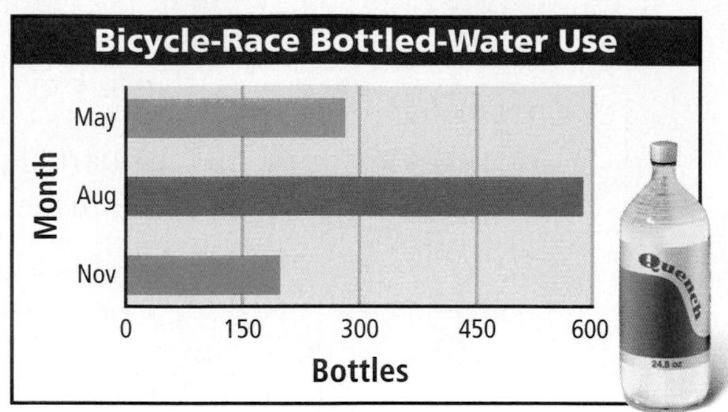

See Example 3
4. If a local business provided half the bottled water needed for the August bicycle race, about how many bottles did the company provide?

INDEPENDENT PRACTICE

See Example 1
Estimate each sum or difference by rounding to the place value indicated.

5. 6,570 + 3,609; thousands

6. 49,821 − 11,567; ten thousands

7. 3,912 + 1,269; thousands

8. 37,097 − 20,364; ten thousands

See Example 2
9. The recreation center has provided softballs every year to the city league. Use the table to estimate the number of softballs the league will use in 5 years.

See Example 3
10. The recreation center has a girls' golf team with 8 members. About how many golf balls will be available to each girl on the team?

Recreation Center Balls Supplied

Sport	Number of Balls
Basketball	21
Golf	324
Softball	28
Table tennis	95

PRACTICE AND PROBLEM SOLVING

Estimate each sum or difference by rounding to the greatest place value.

11. 152 + 269

12. 797 − 234

13. 6,152 − 3,195

14. 9,179 + 2,206

15. 82,465 − 38,421

16. 38,347 + 17,039

17. 639,069 + 283,136

18. 777,060 − 410,364

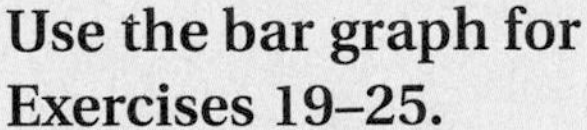

Social Studies LINK

Use the bar graph for Exercises 19–25.

19. On one summer day there were 2,824 sailboats on Lake Erie. Estimate the number of square miles available to each boat.

20. If the areas of all the Great Lakes are rounded to the nearest thousand, which two of the lakes would be the closest in area?

21. About how much larger is Lake Huron than Lake Ontario?

22. The Great Lakes are called "great" because of the huge amount of fresh water they contain. Estimate the total area of all the Great Lakes combined.

23. ***WHAT'S THE QUESTION?*** Lake Erie is about 50,000 square miles smaller. What is the question?

24. ***WRITE ABOUT IT*** Explain how you would estimate the areas of Lake Huron and Lake Michigan to compare their sizes.

25. ***CHALLENGE*** Estimate the average area of the Great Lakes.

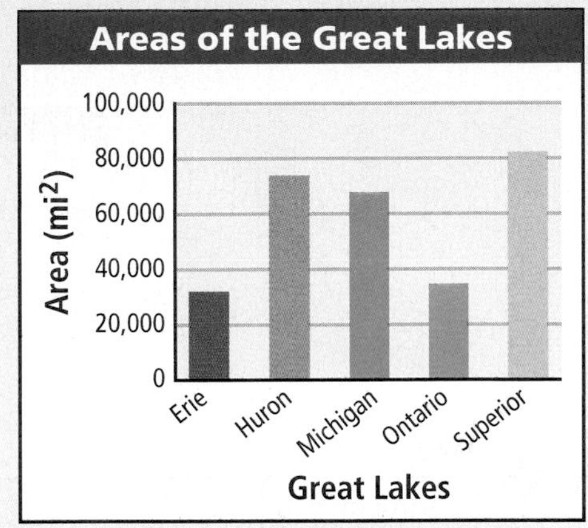

Area includes the water surface and drainage basin within the United States and Canada.

Spiral Review

Find each product or quotient. (Previous course)

26. 148 ÷ 4 **27.** 523 × 46 **28.** 1,054 ÷ 31

29. 223 × 16 **30.** 522 ÷ 18 **31.** 1,107 ÷ 27

Write each number in standard form. (Lesson 1-1)

32. 3,000 + 200 + 70 + 3 **33.** 10,000 + 500 + 20 + 1

34. 500,000 + 60,000 + 300 + 3 **35.** 70,000 + 7

36. **EOG PREP** Which of the following numbers is the standard form of eight hundred twenty-three thousand seven? (Lesson 1-1)

A 800,237 B 823,700 C 823,007 D 8,237,000

1-3 Exponents

Learn to represent numbers by using exponents.

Vocabulary

exponent

base

exponential form

The most recent eruption of Mount Vesuvius took place in 1944.

Since 1906, the height of Mount Vesuvius in Italy has increased by 7^3 feet. How many feet is this?

The number 7^3 is written with an exponent. An **exponent** tells how many times a number called the **base** is used as a factor.

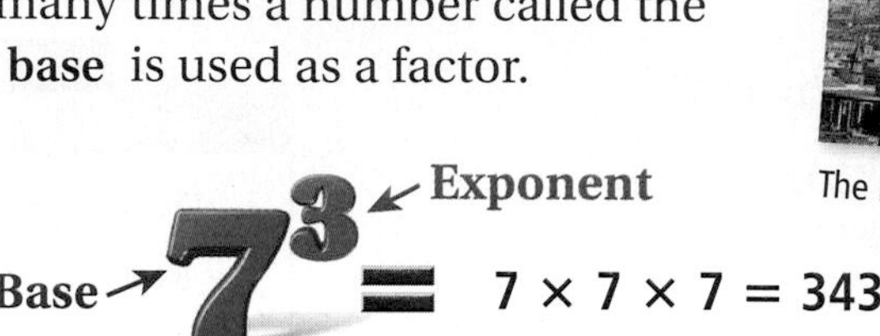

$7 \times 7 \times 7 = 343$

So the height of Mount Vesuvius has increased by 343 ft.

A number is in **exponential form** when it is written with a base and an exponent.

Exponential Form	Read	Multiply	Value
10^1	"10 to the 1st power"	10	10
10^2	"10 squared," or "10 to the 2nd power"	10×10	100
10^3	"10 cubed," or "10 to the 3rd power"	$10 \times 10 \times 10$	1,000
10^4	"10 to the 4th power"	$10 \times 10 \times 10 \times 10$	10,000

EXAMPLE 1 Writing Numbers in Exponential Form

Write each expression in exponential form.

A $4 \times 4 \times 4$

4^3 *4 is a factor 3 times.*

B $9 \times 9 \times 9 \times 9 \times 9$

9^5 *9 is a factor 5 times.*

EXAMPLE 2 Finding the Value of Numbers in Exponential Form

Find each value.

A 3^7

$3^7 = 3 \times 3 \times 3 \times 3 \times 3 \times 3 \times 3$

$= 2{,}187$

B 6^4

$6^4 = 6 \times 6 \times 6 \times 6$

$= 1{,}296$

EXAMPLE 3 PROBLEM SOLVING APPLICATION

In case Dana's school closes, a phone tree is used to contact each student's family. The secretary calls 3 families. Then each family calls 3 other families, and so on. How many families will be notified during the 6th round of calls?

1 Understand the Problem

The **answer** will be the number of families called in the 6th round.

List the **important information:**

- The secretary calls 3 families.
- Each family calls 3 families.

2 Make a Plan

You can draw a diagram to see how many calls are in each round.

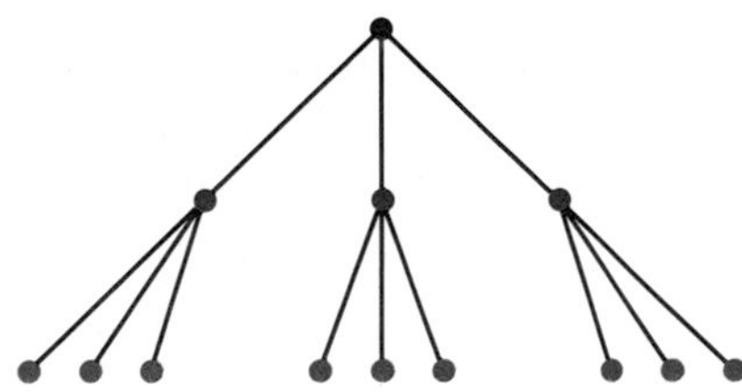

3 Solve

Notice that in each round, the number of calls is a power of 3.

1st round: 3 calls $= 3 = 3^1$

2nd round: 9 calls $= 3 \times 3 = 3^2$

So during the 6th round there will be 3^6 calls.

$3^6 = 3 \times 3 \times 3 \times 3 \times 3 \times 3 = 729$

During the 6th round of calls, 729 families will be notified.

4 Look Back

Drawing a diagram helps you visualize the pattern, but the numbers become too large for a diagram after the third round of calls. Solving this problem by using exponents can be easier and faster.

Think and Discuss

1. **Read** each number: 4^8, 12^3, 3^2.
2. **Give** the value of each number: 7^1, 13^2, 3^3.

1-3 Exercises

1.05, 1.06

FOR EOG PRACTICE

see page 637

internet connect

Homework Help Online
go.hrw.com Keyword: MR4 1-3

GUIDED PRACTICE

See Example 1 **Write each expression in exponential form.**

1. $8 \times 8 \times 8$
2. 7×7
3. $4 \times 4 \times 4 \times 4$
4. $5 \times 5 \times 5 \times 5 \times 5$

See Example 2 **Find each value.**

5. 4^2
6. 3^3
7. 5^4
8. 8^2

See Example 3

9. At Russell's school, one person will contact 4 people and each of those people will contact 4 other people, and so on. How many people will be contacted in the fifth round?

INDEPENDENT PRACTICE

See Example 1 **Write each expression in exponential form.**

10. $2 \times 2 \times 2 \times 2 \times 2 \times 2$
11. $9 \times 9 \times 9 \times 9$
12. $1 \times 1 \times 1$
13. $6 \times 6 \times 6 \times 6 \times 6$
14. $7 \times 7 \times 7 \times 7 \times 7 \times 7 \times 7$
15. 3×3

See Example 2 **Find each value.**

16. 2^4
17. 3^5
18. 6^2
19. 9^2
20. 8^3
21. 1^4
22. 16^2
23. 10^8

See Example 3

24. To save money for a video game, you put one dollar in an envelope. Each day for 5 days you double the number of dollars in the envelope from the day before. How much have you saved after 5 days?

PRACTICE AND PROBLEM SOLVING

Write each expression as repeated multiplication.

25. 16^3
26. 22^2
27. 31^6
28. 46^5
29. 4^1
30. 1^9
31. 17^6
32. 8^5

Find each value.

33. 10^6
34. 73^1
35. 9^4
36. 80^2
37. 19^2
38. 2^9
39. 57^1
40. 5^3

Compare. Write $<$, $>$, or $=$.

41. 6^1 ▒ 5^1
42. 9^2 ▒ 20^1
43. 10^1 ▒ $1{,}000{,}000^1$
44. 2^2 ▒ 3^2
45. 5^3 ▒ 11^2
46. 10^7 ▒ 10^8

Life Science 

You are able to grow because your body produces new cells. New cells are made when old cells divide. Single-celled bodies, like bacteria, divide by *binary fission*, which means "splitting into two parts."

47. In science lab, Carol has a dish containing 4^5 cells. How many cells are represented by this number?

48. A certain colony of bacteria triples in length every 15 minutes. Its length is now 1 mm. How long will it be in 1 hour? (*Hint:* There are four cycles of 15 minutes in 1 hour.)

Use the bar graph for Exercises 49–53.

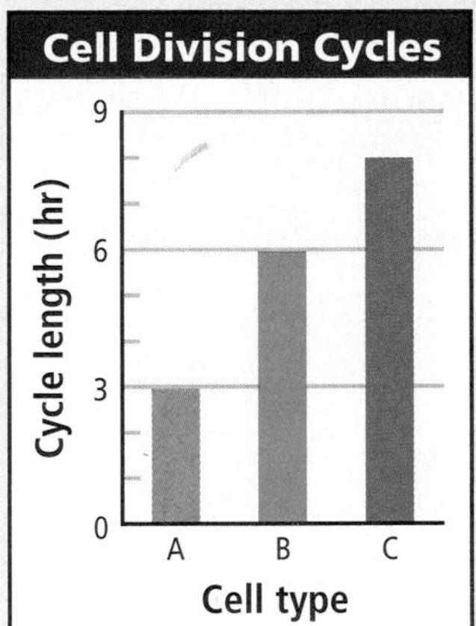

49. Determine how many times cell type A will divide in a 24-hour period. If you begin with one type A cell, how many cells will be produced in 24 hours?

50. Determine how many times cell type B will divide in a 24-hour period. If you begin with one type B cell, how many cells will be produced in 24 hours?

51. Determine how many times cell type C will divide in a 24-hour period. If you begin with one type C cell, how many cells will be produced in 24 hours?

52. ***WRITE ABOUT IT*** Explain how to find the number of type A cells produced in 48 hours.

53. ***CHALLENGE*** How many hours will it take one C cell to divide into at least 100 C cells?

This plant cell shows the anaphase stage of mitosis. Mitosis is the process of nuclear division in complex cells called eukaryotes.

Spiral Review

Write each number in expanded form. (Lesson 1-1)

54. 269

55. 1,354

56. 32,498

57. 416,798

Round each number to the given place value. (Lesson 1-2)

58. 131; hundreds

59. 796; tens

60. 2,369; thousands

61. 16,497; ten thousands

62. 319,020; ten thousands

63. 1,649,045; millions

64. **EOG PREP** Which number is the closest estimate for 817 + 259? (Lesson 1-2)

A 10,000 B 2,000 C 800 D 1,100

Chapter 1

Mid-Chapter Quiz

LESSON 1-1 (pp. 4–7)

Solve.

1. Which number is greater, 12,563,284 or 12,587,802?
2. Which number is greater, 783,100,570 or 780,223,104?
3. In 1998, there were 67,011,180 U.S. households with cable TV. In 1999, there were 67,592,000 U.S. households with cable TV. In which year did more U.S. households have cable TV?
4. In 2001, a university sold 1,981,299 tickets to its football games. In 2000, the same university sold 1,881,702 tickets. During which year were more tickets sold?

Order the numbers from least to greatest.

5. 1,052; 1,803; 1,231
6. 683; 542; 631
7. 2,305; 2,524; 3,012
8. 4,302; 5,019; 3,825
9. 4,344; 3,344; 3,444
10. 10,463; 14,063; 10,643

LESSON 1-2 (pp. 8–11)

Estimate each sum or difference by rounding to the place value indicated.

11. 61,582 + 13,281; ten thousands
12. 86,125 − 55,713; ten thousands
13. 7,903 + 2,654; thousands
14. 34,633 − 32,087; thousands
15. 1,896,345 + 3,567,194; hundred thousands
16. 56,129,482 − 37,103,758; ten millions
17. Marcus wants to make a stone walkway in his garden. The rectangular walkway will be 3 feet wide and 18 feet long. Each 2-foot by 3-foot stone covers an area of 6 square feet. How many stones will Marcus need?
18. Jenna's sixth-grade class is taking a bus to the zoo. The zoo is 156 miles from the school. If the bus travels an average of 55 mi/h, about how long will it take the class to get to the zoo?

LESSON 1-3 (pp. 12–15)

Write each expression in exponential form.

19. 7 × 7 × 7
20. 5 × 5 × 5 × 5
21. 3 × 3 × 3 × 3 × 3 × 3
22. 10 × 10 × 10 × 10
23. 1 × 1 × 1 × 1 × 1
24. 4 × 4 × 4 × 4

Find each value.

25. 3^3
26. 2^4
27. 6^2
28. 8^3
29. 1^8
30. 4^2
31. 5^4
32. 9^1

Focus on Problem Solving

Solve

• Choose the operation: addition or subtraction

Read the whole problem before you try to solve it. Determine what action is taking place in the problem. Then decide whether you need to add or subtract in order to solve the problem.

If you need to combine or put numbers together, you need to add. If you need to take away or compare numbers, you need to subtract.

Action	Operation	Picture
Combining Putting together	Add	
Removing Taking away	Subtract	
Comparing Finding the difference	Subtract	

Read each problem. Determine the action in each problem. Choose an operation in order to solve the problem. Then solve.

Most hurricanes that occur over the Atlantic Ocean, the Caribbean Sea, or the Gulf of Mexico occur between June and November. Since 1886, a hurricane has occurred in every month except April.

Number of Out-of-Season Hurricanes Since 1886	
Month	**Number**
Jan	1
Feb	1
Mar	1
May	14
Dec	10

Use the table for problems 1 and 2.

1. How many out-of-season hurricanes have occurred in all?
2. How many more hurricanes have occurred in May than in December?
3. There were 14 named storms during the 2000 hurricane season. Eight of these became hurricanes, and three others became major hurricanes. How many of the named storms were not hurricanes or major hurricanes?

Explore the Order of Operations

Use with Lesson 1-4

internet connect
Lab Resources Online
go.hrw.com
KEYWORD: MR4 Lab1A

Look at the expression $3 + 2 \cdot 8$. To evaluate this expression, decide whether to add first or multiply first. Knowing the correct *order of operations* is important. Without this knowledge, you could get an incorrect result.

Activity 1

Evaluate $3 + 2 \cdot 8$ two different ways.

Add first, and then multiply by 8.	$3 + 2 = \mathbf{5}$ $\mathbf{5} \cdot 8 = 40$
Multiply first, and then add 3.	$2 \cdot 8 = \mathbf{16}$ $\mathbf{16} + 3 = 19$

Now evaluate $3 + 2 \cdot 8$ using a graphing or scientific calculator.

The result, 19, shows that this calculator multiplied first, even though addition came first in the expression.

If there are no parentheses, then multiplication and division are done before addition or subtraction. If the addition is to be done first, parentheses *must* be used.

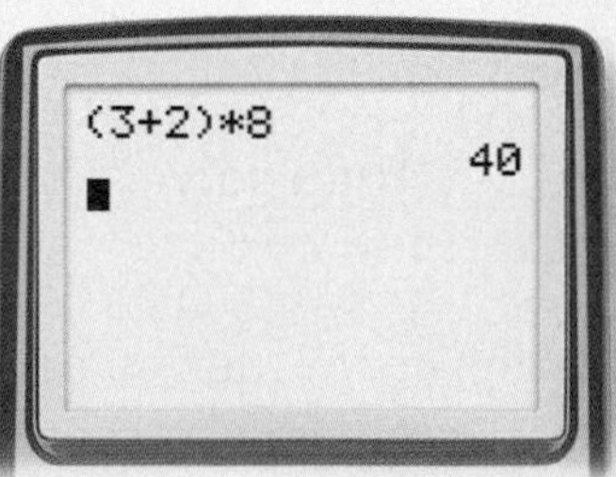

When you evaluate $(3 + 2) \cdot 8$ on a calculator, the result is 40. Because of the parentheses, the calculator adds before multiplying.

Graphing and scientific calculators follow a logical system called the algebraic order of operations. The order of operations tells you to multiply and divide before you add or subtract.

Think and Discuss

1. In $4 + 15 \div 5$, which operation do you perform first? How do you know?

2. Tell the order in which you would perform the operations in the expression $8 \div 2 + 6 \cdot 3 - 4$.

Try This

Evaluate each expression with pencil and paper. Check your answer with a calculator.

1. $4 \cdot 12 - 7$ **2.** $15 \div 3 + 10$ **3.** $4 + 2 \cdot 6$ **4.** $10 - 4 \div 2$

Activity 2

What should you do if the same operation appears twice in an expression? Use a calculator to decide which subtraction is done first in the expression $7 - 3 - 2$.

If $7 - 3$ is done first, the value of the expression is $4 - 2 = 2$.

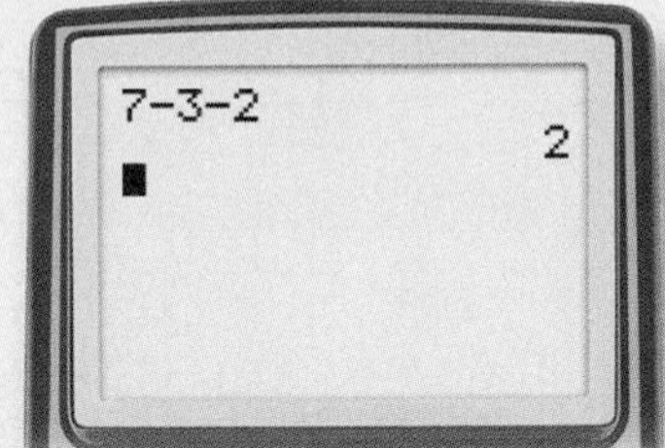

If $3 - 2$ is done first, the value of the expression is $7 - 1 = 6$.

On the calculator, the value of $7 - 3 - 2$ is 2. The subtraction on the left, $7 - 3$, is done first.

Addition and subtraction (or multiplication and division) are done from left to right.

Think and Discuss

1. In $15 + 5 + 4$, does it matter which operation you perform first? Explain.

2. Does it matter which operation you perform first in $15 - 5 + 4$? Explain.

Try This

Evaluate each expression. Check your answer with a calculator.

1. $8 - 6 - 1$ **2.** $20 \div 5 \div 2$ **3.** $3 \cdot 6 \cdot 2$ **4.** $19 + 6 + 5$

Activity 3

Without parentheses, the expression $8 + 2 \cdot 10 - 3$ equals 25. Insert parentheses to make the value of the expression 22.

What happens if you add first?	What happens if you subtract first?
$(8 + 2) \cdot 10 - 3$	$8 + 2 \cdot (10 - 3)$
$10 \cdot 10 - 3$	$8 + 2 \cdot 7$
$100 - 3$	$8 + 14$
97	22

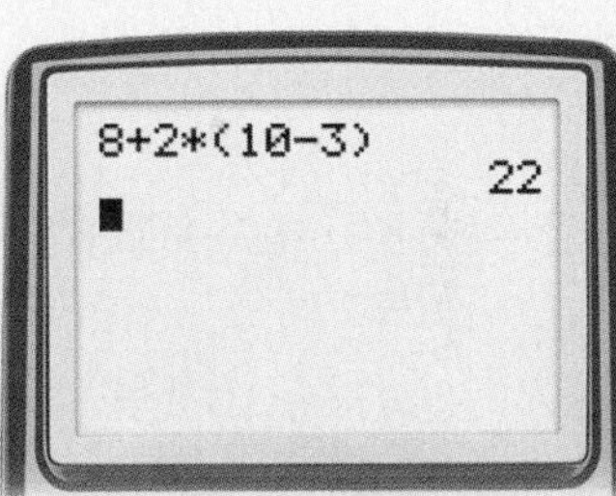

For the expression to equal 22, the subtraction must be done first.

Think and Discuss

1. To evaluate $13 + 5 \cdot 255$ on a calculator, you type $13 + 5$ and then press the $\cdot$ key. But before you can type in the 255, the display changes to 18!

a. Does this calculator follow the correct order of operations? Why?

b. How could you use this calculator to evaluate $13 + 5 \cdot 255$?

Try This

Insert parentheses to make the value of each expression 12.

1. $56 - 40 + 4$ **2.** $3 - 1 \cdot 10 - 4$ **3.** $18 \div 2 + 1 + 6$

1-4 Order of Operations

Learn to use the order of operations.

Vocabulary

numerical expression

evaluate

order of operations

A **numerical expression** is a mathematical phrase that includes only numbers and operation symbols.

Numerical Expressions	$4 + 8 \div 2 \times 6$	$371 - 203 + 2$	$5{,}006 \times 19$

When you **evaluate** a numerical expression, you find its value.

Erika and Jamie each evaluated $3 + 4 \times 6$. Their work is shown below. Whose answer is correct?

Helpful Hint

The first letters of these words can help you remember the order of operations.

Please	*Parentheses*
Excuse	*Exponents*
My	*Multiply/*
Dear	*Divide*
Aunt	*Add/*
Sally	*Subtract*

When an expression has more than one operation, you must know which operation to do first. To make sure that everyone gets the same answer, we use the **order of operations**.

ORDER OF OPERATIONS

1. Perform operations in **parentheses.**
2. Find the values of numbers with **exponents.**
3. **Multiply** or **divide** from left to right as ordered in the problem.
4. **Add** or **subtract** from left to right as ordered in the problem.

$3 + 4 \times 6$	*There are no parentheses or exponents. Perform the multiplication first.*
$3 + 24$	*Add.*
27	*Erika has the correct answer.*

EXAMPLE 1 Using the Order of Operations

Evaluate each expression.

A $9 + 12 \times 2$

$9 + 12 \times 2$	*There are no parentheses or exponents.*
$9 + 24$	*Multiply.*
33	*Add.*

B $4 \times 3^2 + 8 - 16$

$4 \times 3^2 + 8 - 16$	*There are no parentheses.*
$4 \times 9 + 8 - 16$	*Find the values of numbers with exponents.*
$36 + 8 - 16$	*Multiply.*
$44 - 16$	*Add.*
28	*Subtract.*

C $8 \div (1 + 3) \times 5^2 - 2$

$8 \div (1 + 3) \times 5^2 - 2$	
$8 \div 4 \times 5^2 - 2$	*Perform operations within parentheses.*
$8 \div 4 \times 25 - 2$	*Find the values of numbers with exponents.*
$2 \times 25 - 2$	*Divide.*
$50 - 2$	*Multiply.*
48	*Subtract.*

EXAMPLE 2 *Consumer Application*

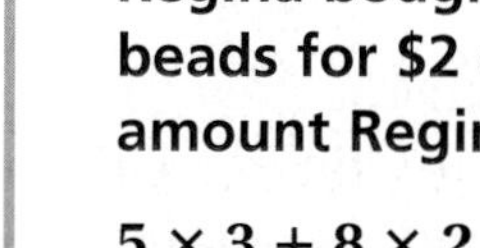
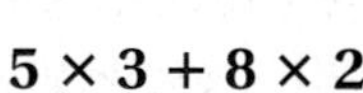

Regina bought 5 carved wooden beads for $3 each and 8 glass beads for $2 each. Evaluate the following expression to find the amount Regina spent for beads.

$\mathbf{5 \times 3 + 8 \times 2}$

$5 \times 3 + 8 \times 2$

$15 + 16$

31

Regina spent $31 for beads.

Think and Discuss

1. **Explain** why $6 + 7 \times 10 = 76$ but $(6 + 7) \times 10 = 130$.
2. **Tell** how you can add parentheses to the numerical expression $2^2 + 5 \times 3$ so that 27 is the correct answer.

1-4 Exercises

FOR EOG PRACTICE

see page 638

internet connect

Homework Help Online

go.hrw.com Keyword: MR4 1-4

1.04a, 1.07, 5.01e

GUIDED PRACTICE

See Example 1

Evaluate each expression.

1. $36 - 18 \div 6$

2. $7 + 24 \div 6 \times 2$

3. $11 + 2^3 \times 5$

4. $62 - 4 \times (15 \div 5)$

5. $5 \times (28 \div 7) - 4^2$

6. $5 + 3^2 \times 6 - (10 - 9)$

See Example 2

7. Coach Milner fed the team after the game by buying 24 Big Burger Deals for \$4 each and 7 Super Big Burger Deals for \$6 each. Evaluate the expression for the cost of the food: $24 \times 4 + 7 \times 6$.

INDEPENDENT PRACTICE

See Example 1

Evaluate each expression.

8. $9 + 27 \div 3$

9. $2 \times 7 - 32 \div 8$

10. $45 \div (3 + 6) \times 3$

11. $100 \div 5^2 + 7 \times 3$

12. $4^2 + 48 \div (10 - 4)$

13. $6 \times 2^2 + 28 - 5$

14. $6^2 - 12 \div 3 + (15 - 7)$

15. $21 \div (3 + 4) \times 9 - 2^3$

16. $5 + 3 \times 2 + 12 \div 4$

17. $(3^2 + 6 \div 2) \times (36 \div 6 - 4)$

See Example 

18. The nature park has a pride of 5 adult lions and 3 cubs. The adults eat 8 lb of meat each day and the cubs eat 4 lb. Use the expression to find the amount of meat consumed each day by the lions: $5 \times 8 + 3 \times 4$.

19. Angie read 4 books that were each 150 pages long and 2 books that were each 325 pages long. Evaluate the expression $4 \times 150 + 2 \times 325$ to find the total number of pages Angie read.

PRACTICE AND PROBLEM SOLVING

Evaluate each expression.

20. $12 + 3 \times 4$

21. $25 - 21 \div 3$

22. $60 \div (10 + 2) \times 4^2 - 23$

23. $10 \times (28 - 23) + 7^2 - 37$

24. $72 \div 9 - 2 \times 4$

25. $12 + (1 + 7^2) \div 5$

26. $(15 - 6)^2 - 34 \div 2$

27. $(2 \times 4)^2 - 3 \times (5 + 3)$

Add parentheses so that each equation is correct.

28. $2^3 + 6 - 5 \times 4 = 12$

29. $7 + 2 \times 6 - 4 - 3 = 53$

30. $3^2 + 6 + 3 \times 3 = 36$

31. $5^2 - 10 + 5 + 4^2 = 36$

32. $2 \times 8 + 5 - 3 = 23$

33. $9^2 - 2 \times 15 + 16 - 8 = 11$

Social Studies LINK

Archaeologists study cultures of the past by uncovering items from ancient cities. An archaeologist has chosen a site in Mexico for her team's next dig. She divides the location into rectangular plots and labels each plot so that uncovered items can be identified by the plot in which they were found.

Archaeologists uncovered pieces of pottery at the La Ventilla site in Mexico.

34. To prepare for the dig, the archaeologist must order a cover for the plot where the team is currently digging. Evaluate the expression $3 \times (2^2 + 6)$ to find the area of each plot in square meters.

35. In the first week, the archaeology team digs down 2 meters and removes a certain amount of dirt. Evaluate the expression $3 \times (2^2 + 6) \times 2$ to find the volume of the dirt removed from the plot in the first week.

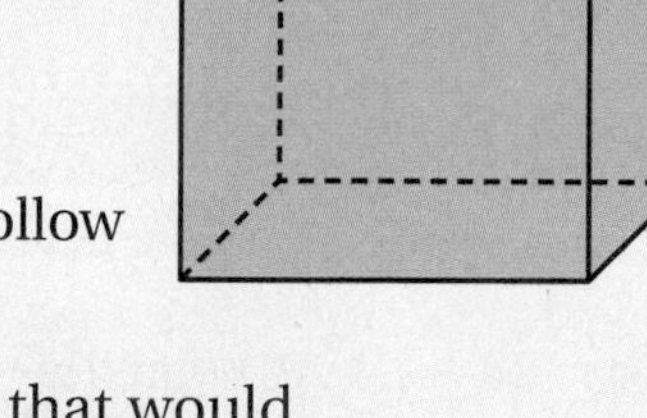

36. Over the next two weeks, the archaeology team digs down an additional 2^3 meters. Evaluate the expression $3 \times (2^2 + 6) \times (2 + 2^3)$ to find the total volume of dirt removed from the plot after 3 weeks.

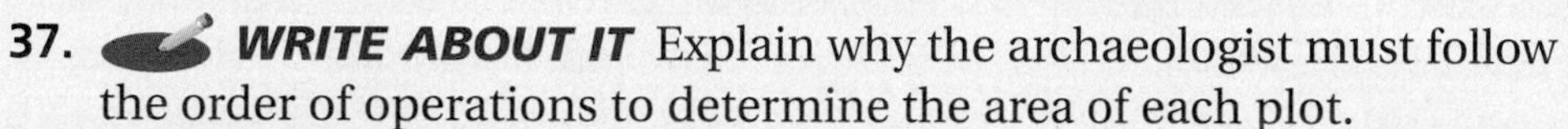

37. ***WRITE ABOUT IT*** Explain why the archaeologist must follow the order of operations to determine the area of each plot.

38. ***CHALLENGE*** Write an expression for the volume of dirt that would be removed if the archaeologist's team were to dig down an additional 3^2 meters after the first three weeks.

Spiral Review

Order the numbers from least to greatest. (Lesson 1-1)

39. 8,452; 8,732; 8,245 **40.** 984; 1,010; 991 **41.** 12,681; 12,751; 11,901

Estimate each sum or difference by rounding to the place value indicated. (Lesson 1-2)

42. 2,488 + 1,934; thousands **43.** 83,057 − 29,475; ten thousands **44.** 9,346 + 12,745; thousands

Find each value. (Lesson 1-3)

45. 11^2 **46.** 5^3 **47.** 9^1 **48.** 2^5

49. **EOG PREP** Which of the following is the value of the expression $3^4 + 9^1$? (Lesson 1-3)

A 21 B 81 C 90 D 82

1-5 Mental Math

Learn to use number properties to compute mentally.

Vocabulary

Commutative Property

Associative Property

Distributive Property

Mental math means "doing math in your head." Shakuntala Devi is extremely good at mental math. When she was asked to multiply 7,686,369,774,870 by 2,465,099,745,779, she took only 28 seconds to multiply the numbers mentally and gave the correct answer of 18,947,668,177,995,426,462,773,730!

Most people cannot do calculations like that mentally. But you can learn to solve some problems very quickly in your head.

Many mental math strategies use number properties that you already know.

COMMUTATIVE PROPERTY (Ordering)

Words	Numbers
You can add or multiply numbers in any order.	$18 + 9 = 9 + 18$ $15 \times 2 = 2 \times 15$

ASSOCIATIVE PROPERTY (Grouping)

Words	Numbers
When you are only adding or only multiplying, you can group any of the numbers together.	$(17 + 2) + 9 = 17 + (2 + 9)$ $(12 \times 2) \times 4 = 12 \times (2 \times 4)$

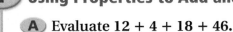

EXAMPLE 1 Using Properties to Add and Multiply Whole Numbers

A Evaluate $12 + 4 + 18 + 46$.

$12 + 4 + 18 + 46$	*Look for sums that are multiples of 10.*
$12 + 18 + 4 + 46$	*Use the Commutative Property.*
$(12 + 18) + (4 + 46)$	*Use the Associative Property to make*
$30 + 50$	*groups of compatible numbers.*
80	*Use mental math to add.*

B Evaluate $5 \times 12 \times 2$.

$5 \times 12 \times 2$	*Look for products that are multiples of 10.*
$12 \times 5 \times 2$	*Use the Commutative Property.*
$12 \times (5 \times 2)$	*Use the Associative Property to group*
12×10	*compatible numbers.*
120	*Use mental math to multiply.*

DISTRIBUTIVE PROPERTY

Words	Numbers
When you multiply a number times a sum, you can	
• find the sum first and then multiply, or	$6 \times (10 + 4) = 6 \times 14$ $= 84$
• multiply by each number in the sum and then add.	$6 \times (10 + 4) = (6 \times 10) + (6 \times 4)$ $= 60 + 24$ $= 84$

When you multiply two numbers, you can "break apart" one of the numbers into a sum and then use the Distributive Property.

EXAMPLE 2 Using the Distributive Property to Multiply

Use the Distributive Property to find each product.

Helpful Hint

Break the greater factor into a sum that contains a multiple of 10 and a one-digit number. You can add and multiply these numbers mentally.

A 4×23

$4 \times 23 = 4 \times (20 + 3)$	*"Break apart" 23 into 20 + 3.*
$= (4 \times 20) + (4 \times 3)$	*Use the Distributive Property.*
$= 80 + 12$	*Use mental math to multiply.*
$= 92$	*Use mental math to add.*

B 8×74

$8 \times 74 = 8 \times (70 + 4)$	*"Break apart" 74 into 70 + 4.*
$= (8 \times 70) + (8 \times 4)$	*Use the Distributive Property.*
$= 560 + 32$	*Use mental math to multiply.*
$= 592$	*Use mental math to add.*

Think and Discuss

1. **Give examples** of the Commutative Property and the Associative Property.
2. **Name** some situations in which you might use mental math.

1-5 Exercises

FOR EOG PRACTICE

see page 638

internet connect
Homework Help Online
go.hrw.com Keyword: MR4 1-5

1.04a, 1.07, 5.01

GUIDED PRACTICE

See Example 1

Evaluate.

1. 13 + 9 + 7 + 11
2. 19 + 18 + 11 + 32
3. 5 × 14 × 4
4. 4 × 16 × 5

See Example 2

Use the Distributive Property to find each product.

5. 5 × 24
6. 8 × 52
7. 4 × 39
8. 6 × 14
9. 3 × 33
10. 2 × 78
11. 9 × 12
12. 2 × 87

INDEPENDENT PRACTICE

See Example 1

Evaluate.

13. 15 + 17 + 3 + 5
14. 14 + 7 + 16 + 13
15. 5 × 25 × 2
16. 2 × 32 × 10

See Example 2

Use the Distributive Property to find each product.

17. 3 × 36
18. 4 × 42
19. 6 × 71
20. 2 × 94
21. 5 × 25
22. 6 × 62
23. 7 × 21
24. 8 × 41

PRACTICE AND PROBLEM SOLVING

Use mental math to find each sum or product.

25. 8 + 13 + 7 + 12
26. 2 × 25 × 4
27. 5 × 8 × 12
28. 5 + 98 + 95
29. 11 + 75 + 25
30. 8 × 11 × 5

Multiply using the Distributive Property.

31. 9 × 17
32. 4 × 27
33. 11 × 18
34. 7 × 51
35. 2 × 28
36. 9 × 42
37. 5 × 55
38. 3 × 78
39. 4 × 85
40. 6 × 36
41. 8 × 24
42. 11 × 51

43. ***BUSINESS*** Janice wants to order disks for her computer. She needs to find the total cost, including shipping and handling. If Janice orders 7 disks, what will her total cost be?

Description	Number	Unit Cost with Tax	Price
Computer Disk	7	$24.00	
		Shipping & Handling	$7.00
		Total	

Life Science LINK

Poison-dart frogs are members of the family Dendrobatidae, which includes about 170 species. Many are brightly colored, with yellow, red, green, orange, blue, or black markings.

44. ***LIFE SCIENCE*** Poison-dart frogs can breed underwater, and the females lay from 4 to 30 eggs. What would be the total number of eggs if four female poison-dart frogs each laid 27 eggs?

45. Paul is writing a story for the school newspaper about the landscaping done by his class. The students planted 15 vines, 12 hedges, 8 trees, and 35 flowering plants. How many plants were used in the project?

46. Rickie wants to buy 3 garden hoses at the home center clearance sale. How much will they cost?

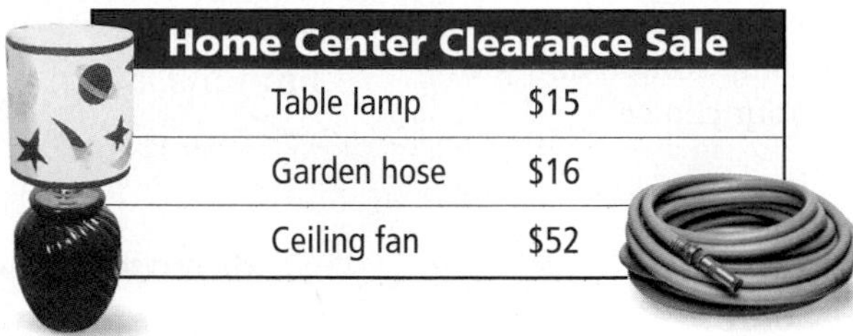

Home Center Clearance Sale	
Table lamp	\$15
Garden hose	\$16
Ceiling fan	\$52

47. The boys in Josh's family are saving money to buy 4 ceiling fans at the home center sale. How much will they need to save?

48. ***EARTH SCIENCE*** The temperature on Sunday was 58°F. The temperature is predicted to rise 4°F on Monday, then rise 2°F more on Tuesday, and then rise another 6°F by Saturday. What is the predicted temperature on Saturday?

49. ***WHAT'S THE ERROR?*** A student wrote $5 + 24 + 25 + 6 = 5 + 25 + 24 + 6$ by the Associative Property. What error did the student make?

50. ***WRITE A PROBLEM*** Give a problem that you could simplify using the Commutative and Associative Properties. Then, show the steps to solve the problem and label the Commutative and Associative Properties.

51. ***WRITE ABOUT IT*** Why can you simplify $5(50 + 3)$ using the Distributive Property? Why can't you simplify $5(50) + 3$ using the Distributive Property?

52. ***CHALLENGE*** Explain how you could find the product of $5^2 \times 112$ using the Distributive Property. Evaluate the expression.

Spiral Review

Estimate each sum or difference to the place value indicated. (Lesson 1-2)

53. 12,876 + 17,986; thousands

54. 72,876 + 15,987; ten thousands

Evaluate each expression. (Lesson 1-3)

55. 3^4

56. 2^5

57. 4^3

58. 11^0

59. **EOG PREP** What is the correct value of $3^2 + (9 \div 3 - 2)$? (Lesson 1-4)

A 7 B 10 C 18 D 4

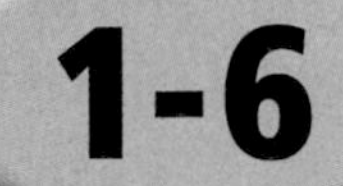

1-6 Choose the Method of Computation

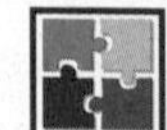

Problem Solving Skill

Learn to choose an appropriate method of computation and justify your choice.

Earth has one moon. Scientists have determined that other planets in our solar system have as many as 39 moons. Mercury and Venus have no moons at all.

EXAMPLE 1 *Astronomy Application*

How many known moons are in our solar system?

It might be hard to keep track of all of these numbers if you tried to add mentally. But the numbers themselves are small. You can use paper and pencil.

Planet	Moons
Mercury	0
Venus	0
Earth	1
Mars	2
Jupiter	39
Saturn	30
Uranus	21
Neptune	8
Pluto	1

Source: NASA, 2002

$$\begin{array}{r} 1 \\ 2 \\ 39 \\ 30 \\ 21 \\ 8 \\ +\ 1 \\ \hline 102 \end{array}$$

There are 102 known moons in our solar system.

EXAMPLE 2 *Astronomy Application*

The average temperature on Earth is 59°F. The average temperature on Venus is 867°F. How much hotter is Venus's average temperature?

Venus temperature − Earth temperature

867 − 59

These numbers are small, and 59 is close to a multiple of 10. You can use mental math.

$(867 + 1) - (59 + 1)$ *Think: Add 1 to 59 to make 60. Add 1 to 867 to compensate.*

$868 - 60$

808

The average temperature on Venus is 808°F hotter than the average temperature on Earth.

EXAMPLE 3 *Astronomy Application*

Every day, about 120 tons of cosmic dust—debris from outer space—enter Earth's atmosphere. How many tons of cosmic dust enter Earth's atmosphere each year?

tons per day × **days per year** *Think: There are 365 days in a year.*

120 × 365

These numbers are not compatible, so mental math is not a good choice.

You could use paper and pencil. But finding a product of 3-digit numbers requires several steps. Using a calculator will probably be faster.

Carefully enter the numbers on a calculator. Record the product.

120 × 365 = 43,800

Each year, about 43,800 tons of cosmic dust enter Earth's atmosphere.

Think and Discuss

1. **Give an example** of a situation in which you would use mental math to solve a problem. When would you use paper and pencil?
2. **Tell** how you could use mental math in Example 2 if the problem were 867 + 59.

1-6 Exercises

FOR EOG PRACTICE
see page 639

internet connect
Homework Help Online
go.hrw.com Keyword: MR4 1-6

1.04a, 1.07

GUIDED PRACTICE

See Example 

1. What is the total number of astronauts who have space flight experience?

U.S.	Germany	France	Canada	Japan	Italy	Russia
244	9	8	7	5	3	88

See Example 

2. In the 2000 Summer Olympic Games, 929 medals were given. The U.S. team brought home the most medals, 97. How many medals were not won by the U.S. team?

See Example 

3. A factory produces 126 golf balls per minute. How many golf balls can be produced in 515 minutes?

INDEPENDENT PRACTICE

See Example 1

4. A carnival has a coin-toss game. The highest score is a total of all the squares on the board. What is that score?

6	9	5
10	20	8
3	7	4

See Example 2

5. It takes Mars 687 days to complete one revolution around the Sun. It takes Venus only 225 days to revolve around the Sun. How many more days does it take Mars to revolve around the Sun than Venus?

See Example 3

6. If each store in a chain of 108 furniture stores sells 135 sofas a year, what is the total number of sofas sold?

PRACTICE AND PROBLEM SOLVING

Evaluate the expression, and state the method of computation you used.

7. 5 + 24 + 7 + 1 + 64 + 2 + 8
8. 16 + 2 + 4 + 13 + 5 + 1 + 14
9. 828 × 623
10. 742 − 167
11. 41 + 169
12. 499 − 201
13. 57 × 198
14. 338 + 12
15. 3,813 × 117
16. 337 − 124

17. A satellite travels 985,200 miles per year. How many miles will it travel if it stays in space for 12 years?

18. ***WHAT'S THE QUESTION?*** An astronaut has spent the following minutes training in a tank that simulates weightlessness: 2, 15, 5, 40, 10, and 55. The answer is 127. What is the question?

19. ***WRITE ABOUT IT*** Explain how you can decide whether to use pencil and paper, mental math, or a calculator to solve a subtraction problem.

20. ***CHALLENGE*** A list of possible astronauts was narrowed down by two committees. The first committee selected 93 people to complete a written form. The second selected 31 of those people to come to an interview. If 837 were not asked to complete a form, how many were on the original list?

Spiral Review

Write each expression in exponential form. (Lesson 1-3)

21. 4 × 4 × 4 × 4
22. 2 × 2 × 2 × 2 × 2
23. 10 × 10 × 10

Evaluate each expression. (Lesson 1-4)

24. 4 × 14 + 12 ÷ 2
25. $16 \div 4^2 + 15 - 2$
26. $5 + 2^2 (12 \div 3)$

27. **EOG PREP** Which expression does *not* have the same value as 7 × (34 + 23)? (Lesson 1-5)

A 7 × 57 B (7 × 34) + (7 × 23) C 7 × 34 + 23 D (7 × 50) + (7 × 7)

Find a Pattern

Problem Solving Strategy

Learn to find patterns and to recognize, describe, and extend patterns in sequences.

Vocabulary

perfect square

sequence

term

Whole numbers raised to the second power are called **perfect squares**. This is because they can be represented by objects arranged in the shape of a square.

1 4 9 16

The perfect squares can be written as a sequence. A **sequence** is an ordered set of numbers. Each number in the sequence is called a **term**. In a sequence, there is often a pattern between one term and the next.

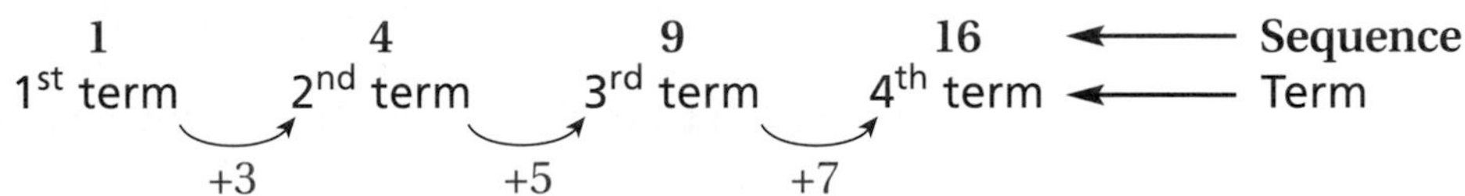

Helpful Hint

Look for a relationship between the 1st term and the 2nd term. Check if this relationship works between the 2nd term and the 3rd term, and so on.

You can use this pattern to find the fifth and sixth terms in the sequence. To get the fifth term, add 9. To get the sixth term, add 11.

1, 4, 9, 16, ▢, ▢, . . .

$16 + 9 = \mathbf{25}$ $\quad$ $25 + 11 = \mathbf{36}$

So the next two perfect squares are 25 and 36.

EXAMPLE 1 Extending Sequences with Addition and Subtraction

Identify a pattern in each sequence and name the next three terms.

A **3, 15, 27, 39, ▢, ▢, ▢, . . .**

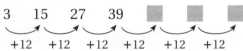

A pattern is to add 12 to each term to get the next term.

$39 + \mathbf{12} = 51$ $\quad$ $51 + \mathbf{12} = 63$ $\quad$ $63 + \mathbf{12} = 75$

So 51, 63, and 75 will be the next three terms.

B **4, 15, 8, 19, 12, 23, 16, ▢, ▢, ▢, . . .**

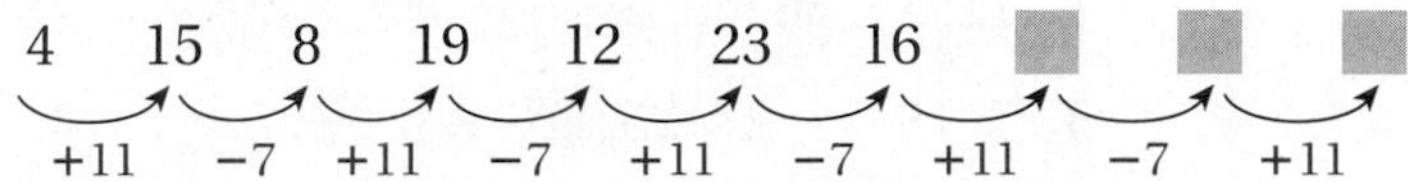

A pattern is to add 11 to one term and subtract 7 from the next.

$16 + \mathbf{11} = 27$ $\quad$ $27 - \mathbf{7} = 20$ $\quad$ $20 + \mathbf{11} = 31$

So 27, 20, and 31 will be the next three terms.

EXAMPLE 2 Completing Sequences with Multiplication and Division

Identify a pattern in each sequence and name the missing terms.

A **256, 128, 64, ■, 16, ■, . . .**

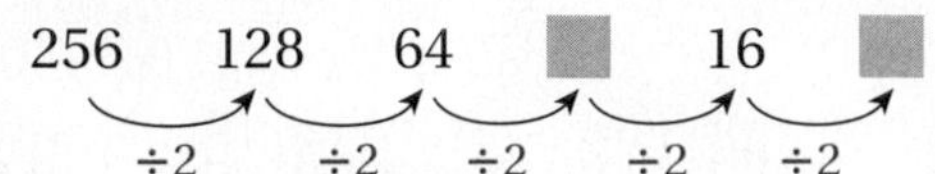

A pattern is to divide each term by 2 to get the next term.

$64 \div 2 = 32 \qquad 16 \div 2 = 8$

So 32 and 8 are the missing terms.

B **1, 6, 2, 12, 4, ■, 8, 48, ■, 96, . . .**

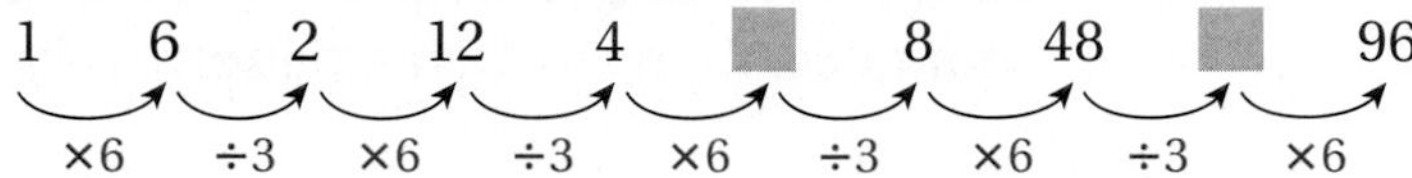

A pattern is to multiply one term by 6 and divide the next by 3.

$4 \times 6 = 24 \qquad 48 \div 3 = 16$

So 24 and 16 are the missing terms.

Think and Discuss

1. **Tell** how you could check whether the next two perfect squares in the sequence 1, 4, 9, 16, . . . are 25 and 36.
2. **Explain** how to find the next term in the sequence 8, 4, 2, ■, ...

1-7 Exercises

1.07

FOR EOG PRACTICE
see page 639

internet connect
Homework Help Online
go.hrw.com Keyword: MR4 1-7

GUIDED PRACTICE

See Example 1 **Identify a pattern in each sequence and name the next three terms.**

1. 12, 24, 36, 48, ■, ■, ■, . . .
2. 105, 90, 75, 60, 45, ■, ■, ■, . . .
3. 7, 18, 16, 27, 25, ■, ■, ■, . . .
4. 44, 38, 42, 36, 40, ■, ■, ■, . . .

See Example 2 **Identify a pattern in each sequence and name the missing terms.**

5. 2, 6, ■, 54, 162, ■, . . .
6. 80, 8, 40, ■, 20, 2, ■, 1, . . .
7. 1, 6, 3, ■, 9, 54, ■, 162, . . .
8. 1,024, 256, ■, 16, ■, 1, . . .

INDEPENDENT PRACTICE

See Example 1

Identify a pattern in each sequence and name the next three terms.

9. 9, 19, 30, 42, 55, ■, ■, ■, . . .

10. 95, 94, 92, 89, ■, ■, ■, . . .

11. 50, 55, 47, 52, 44, ■, ■, ■, . . .

12. 5, 3, 6, 4, 7, ■, ■, ■, . . .

See Example 2

Identify a pattern in each sequence and name the missing terms.

13. 1, 2, 6, ■, 120, ■, 5,040, . . .

14. 600, 300, ■, 30, 6, ■, . . .

15. 400, 100, ■, 50, 100, ■, 50, . . .

16. 120, 60, 180, ■, 270, ■, . . .

PRACTICE AND PROBLEM SOLVING

Use the pattern to write the first five terms of the sequence.

17. Start with 1; multiply by 3.

18. Start with 12; add 12.

19. Start with 100; subtract 7.

20. Start with 2; square each term.

21. The temperature was 45°F on Monday, 48°F on Tuesday, and 51°F on Wednesday. If the pattern continues, what temperature will it be on Friday?

22. ***CHOOSE A STRATEGY*** The * shows where a piece is missing from the pattern. What piece is missing?

A y B B C y D 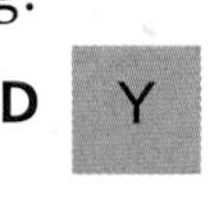

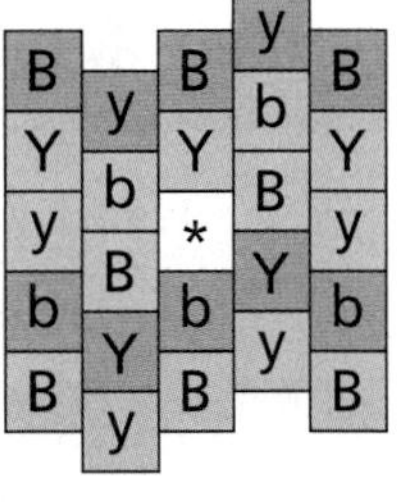

23. ***WRITE ABOUT IT*** How can you know whether a number is a perfect square?

24. ***CHALLENGE*** Find the missing terms in the following sequence: ■, 2^3, 27, 4^3, 125, ■, 343, . . .

Spiral Review

Estimate each sum or difference by rounding to the nearest thousands place. (Lesson 1-2)

25. 5,237 − 1,586

26. 915,178 + 451,836

27. 39,187 − 24,999

Find each value. (Lesson 1-3)

28. 8^5

29. 5^3

30. 3^8

31. 4^4

32. **EOG PREP** Which is the correct value of $7^2 - (4 \times 7 - (3 - 2)) \div 3$? (Lesson 1-5)

A 7 B 105 C 40 D 41

EXTENSION

Binary Numbers

Learn to investigate the binary number system.

Vocabulary

base-10 system

binary number system

Our number system is called the **base-10 system** because each place value is 10 times greater than the place value to the right. Base-10 numbers contain the digits 0 through 9.

10^4	10^3	10^2	10^1	10^0
Ten thousands	Thousands	Hundreds	Tens	Ones
2	5	6	0	1

$$25{,}601 = 20{,}000 + 5{,}000 + 600 + 0 + 1$$
$$= (2 \times 10{,}000) + (5 \times 1{,}000) + (6 \times 100) + (0 \times 10) + (1 \times 1)$$

In the **binary number system**, each place value is 2 times greater than the place value to the right. Binary numbers contain only the digits 0 and 1.

2^4	2^3	2^2	2^1	2^0
Sixteens	Eights	Fours	Twos	Ones
1	0	1	1	1

$$10111 = (1 \times 16) + (0 \times 8) + (1 \times 4) + (1 \times 2) + (1 \times 1)$$
$$= 16 + 0 + 4 + 2 + 1$$
$$= 23$$

10111 in the binary system is equal to 23 in the base-10 system.

EXAMPLE 1 Converting Binary Numbers to Base 10

Find the base-10 value for each binary number.

A 1111

$$1111 = (1 \times 8) + (1 \times 4) + (1 \times 2) + (1 \times 1)$$
$$= 8 + 4 + 2 + 1$$
$$= 15$$

B 11001

$$11001 = (1 \times 16) + (1 \times 8) + (0 \times 4) + (0 \times 2) + (1 \times 1)$$
$$= 16 + 8 + 0 + 0 + 1$$
$$= 25$$

C 10101

$$10101 = (1 \times 16) + (0 \times 8) + (1 \times 4) + (0 \times 2) + (1 \times 1)$$
$$= 16 + 0 + 4 + 0 + 1$$
$$= 21$$

To write a base-10 number as a binary number, "break apart" the base-10 number as a sum of powers of 2. Start with the highest power of 2 that is not more than the base-10 number.

EXAMPLE 2 Converting Base-10 Numbers to Binary

Find the binary number for each base-10 number.

A 9

$$9 = 8 + 0 + 0 + 1$$
$$= (1 \times 8) + (0 \times 4) + (0 \times 2) + (1 \times 1)$$
$$= 1001$$

B 13

$$13 = 8 + 4 + 0 + 1$$
$$= (1 \times 8) + (1 \times 4) + (0 \times 2) + (1 \times 1)$$
$$= 1101$$

C 27

$$27 = 16 + 8 + 0 + 2 + 1$$
$$= (1 \times 16) + (1 \times 8) + (0 \times 4) + (1 \times 2) + (1 \times 1)$$
$$= 11011$$

EXTENSION Exercises

Find the base-10 value for each binary number.

1. 101 **2.** 100 **3.** 111 **4.** 1000

5. 1011 **6.** 11111 **7.** 10 **8.** 10001

Write the expanded form of each binary number.

9. 11010 **10.** 10111 **11.** 11110 **12.** 11101

13. 10100 **14.** 10000 **15.** 10010 **16.** 11100

Find the binary number for each base-10 number.

17. 12 **18.** 6 **19.** 18 **20.** 10

21. 1 **22.** 14 **23.** 22 **24.** 19

Compare. Write <, >, or =. The number on the left is a base-10 number, and the number on the right is a binary number.

25. 20 ▢ 111 **26.** 24 ▢ 11000

27. 3 ▢ 101 **28.** 15 ▢ 1001

Problem Solving on Location

NORTH CAROLINA

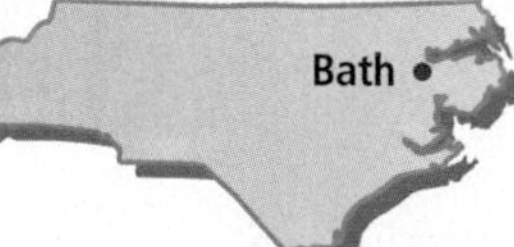

Population

In 1705, North Carolina's first town, Bath, was created. By 1708, Bath had a population of about 50 people. Since then, the number and size of North Carolina cities have certainly grown. The table gives recent populations for eight cities in North Carolina.

For 1–6, use the table.

1. List the cities in order from least population to greatest population.
2. How many more people live in Raleigh than live in Wilmington?
3. Estimate the total number of people in Greensboro and Cary combined.
4. Suppose that $\frac{1}{2}$ of the people living in High Point attended a city council meeting. How many people would have been at the meeting?
5. If the city of Charlotte grew by an additional 22,500 people per year, what would the population be in 5 years?
6. Suppose the population of Raleigh doubled every 20 years. What would its population be in 40 years?

City	Population
Fayetteville	127,558
Raleigh	273,203
Wilmington	75,629
Winston-Salem	173,568
Greensboro	208,210
Cary	92,972
High Point	77,586
Charlotte	526,245

The Great Smoky Mountains

The Great Smoky Mountains were named by early explorers who believed that the mist on the mountain ridges looked like smoke. Today millions of people visit Great Smoky Mountains National Park each year to enjoy the beautiful scenery. Many come during autumn to see the trees in the park turn their bright autumn colors.

The table shows the elevations of the three highest peaks in Great Smoky Mountains National Park.

Peak	Elevation (ft)
Mount Guyot	6,621
Clingmans Dome	6,643
Mount LeConte	6,593

For 1–3, use the table.

1. How much taller is Mount Guyot than Mount LeConte?
2. How many inches tall is Clingmans Dome? (Remember that there are 12 inches in a foot.)
3. Suppose that you could climb 400 feet in 1 hour. About how long would it take you to climb to the top of Mount Guyot?

The Laurel Falls Trail in the Great Smoky Mountains is 4 miles long and goes by the 75-foot Laurel Falls. At the beginning of the trail, the elevation is 2,250 feet. By the end of the trail, the elevation has reached 4,000 feet.

4. What is the difference in elevation between the beginning of the trail and the end of the trail?
5. If you walked the Laurel Falls Trail in 2 hours, about how many minutes did it take you to walk each mile?

MATH-ABLES

Palindromes

A *palindrome* is a word, phrase, or number that reads the same forward and backward.

Examples:

race car   Madam, I'm Adam.   3710173

You can turn almost any number into a palindrome with this trick.

Think of any number.
Now add that number in reverse.

$$\begin{array}{r} 283 \\ +\ 382 \\ \hline 665 \end{array}$$

Use the sum to repeat the previous step and keep repeating until the final sum is a palindrome.

$$\begin{array}{r} 665 \\ +\ 566 \\ \hline 1{,}231 \end{array}$$

$$\begin{array}{r} 1{,}231 \\ +\ 1{,}321 \\ \hline 2{,}552 \end{array}$$

It took only three steps to create a palindrome by starting with the number 283. What happens if you start with the number 196? Do you think you will ever create a palindrome if you start with 196? One man who started with 196 did these steps until he had a number with 70,928 digits and he still had not created a palindrome!

Spin-a-Million

The object of this game is to create the number closest to 1,000,000.

Taking turns, spin the pointer and write the number on your place-value chart. The number cannot be moved once it has been placed.

After six turns, the player whose number is closest to one million wins the round and scores a point. The first player to get five points wins the game.

internet connect

Go to ***go.hrw.com*** for a spinner and place value chart.
KEYWORD: MR4 Game1

Find a Pattern in Sequences

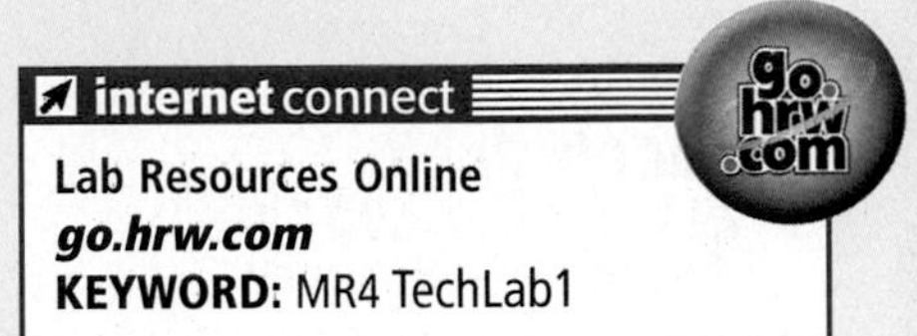

The numbers 4, 7, 10, 13, 16, 19, … form a sequence. To continue the sequence, identify a pattern. Here is a possible pattern:

$$4, \quad 4 + 3 = 7, \quad 7 + 3 = 10, \quad 10 + 3 = 13, \ldots$$

Activity

Use a spreadsheet to generate the first seven terms of the sequence above.

To start with 4, type **4** in cell A1.

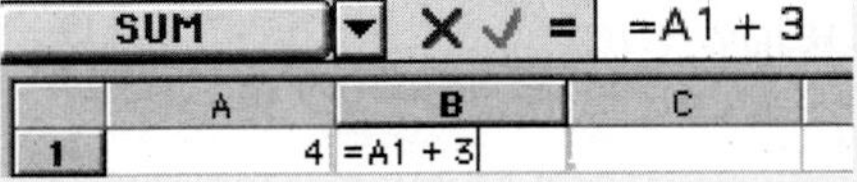

To add 3 to the value in cell A1, type **=A1 + 3** in cell B1.

Press ENTER.

To continue the sequence, click the square in the lower right corner of cell B1, hold down the mouse button, and drag the cursor across through cell G1.

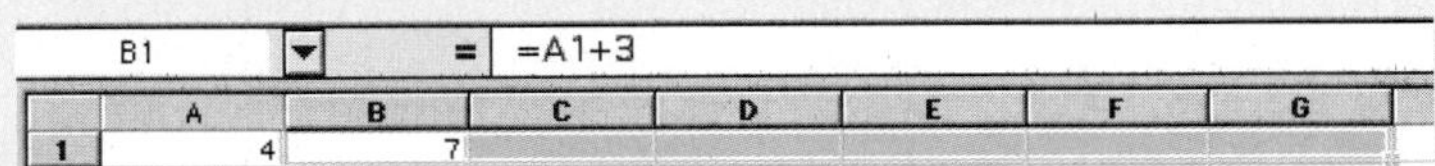

When you release the mouse button, A1 through G1 will list the first seven terms of the sequence.

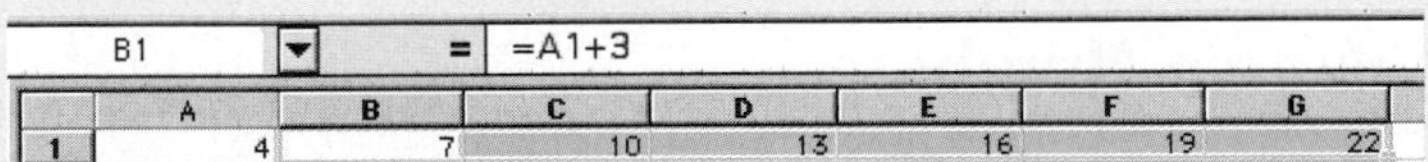

Think and Discuss

1. How do you use a sequence's pattern when you use your spreadsheet to generate the terms?

Try This

Identify a pattern in each sequence. Then use a spreadsheet to generate the first 12 terms.

1. 9, 14, 19, 24, 29, 34, …
2. 7, 13, 19, 25, 31, 37, …

Chapter 1

Study Guide and Review

Vocabulary

Associative Property 24
base 12
Commutative Property 24
compatible number 8
Distributive Property 25
evaluate 20
exponent 12
exponential form 12
numerical expression 20
order of operations 20
overestimate 8
perfect square 31
sequence 31
term 31
underestimate 8

Complete the sentences below with vocabulary words from the list above. Words may be used more than once.

1. An ordered set of numbers is called a(n) __?__. Each number in a sequence is called a(n) __?__.

2. In the expression 8^5, 8 is the __?__, and 5 is the __?__.

3. The __?__ is a set of rules used to evaluate an expression that contains more than one operation.

4. When you __?__ a numerical expression, you find its value.

1-1 Comparing and Ordering Whole Numbers (pp. 4–7)

EXAMPLE

■ **Order the numbers from least to greatest.**
4,913; 4,931; 4,391

4,913 $4,913 < 4,931$ 4,931 $4,391 < 4,931$
4,931   4,391

4,913 $4,391 < 4,913$
4,391

$4,391 < 4,913 < 4,931$

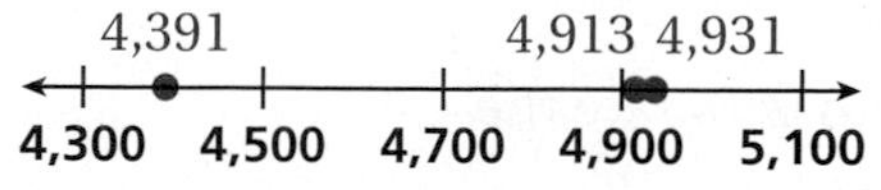

EXERCISES

Order the numbers from least to greatest.

5. 8,731; 8,737; 8,735; 8,740

6. 53,341; 53,337; 53,456; 53,452

7. 87,091; 8,791; 87,901; 81,790

8. 26,551; 25,615; 2,651; 22,561

9. 96,361; 96,631; 93,613; 91,363

10. 10,101; 11,010; 10,110; 11,110

1-2 Estimating with Whole Numbers (pp. 8–11)

EXAMPLES

- Estimate the sum 837 + 710 by rounding to the hundreds place.
  800 + 700 = 1,500
  The sum is about 1,500.
- Estimate the quotient of 148 and 31.
  150 ÷ 30 = 5
  The quotient is about 5.

EXERCISES

Estimate by rounding to the place value indicated.

11. 4,671 − 3,954; thousands

12. 3,123 + 2,987; thousands

13. 53,465 − 27,465; ten thousands

14. Ralph has 38 photo album sheets with 22 baseball cards in each sheet. About how many baseball cards does he have?

1-3 Exponents (pp. 12–15)

EXAMPLES

- **Write 6 × 6 in exponential form.**
  6^2 *6 is a factor 2 times.*

Find each value.

- 5^2
  $5^2 = 5 \times 5$
  $= 25$
- 6^3
  $6^3 = 6 \times 6 \times 6$
  $= 216$

EXERCISES

Write each expression in exponential form.

15. 5 × 5 × 5

16. 3 × 3 × 3 × 3

17. 7 × 7 × 7 × 7 × 7

18. 8 × 8

19. 4 × 4 × 4 × 4

20. 1 × 1 × 1

Find each value.

21. 4^4

22. 2^4

23. 3^3

24. 1^5

25. 5^3

26. 10^2

1-4 Order of Operations (pp. 20–23)

EXAMPLE

- **Evaluate $8 \div (7 - 5) \times 2^2 - 2 + 9$.**

$8 \div (7 - 5) \times 2^2 - 2 + 9$	
$8 \div 2 \times 2^2 - 2 + 9$	*Subtract in parentheses.*
$8 \div 2 \times 4 - 2 + 9$	*Simplify the exponent.*
$4 \times 4 - 2 + 9$	*Divide.*
$16 - 2 + 9$	*Multiply.*
$14 + 9$	*Subtract.*
23	*Add.*

EXERCISES

Evaluate each expression.

27. 9 × 8 − 13

28. 21 ÷ 3 + 4

29. 6 + 4 × 5

30. 19 − 12 ÷ 6

31. 30 ÷ 2 − 5 × 2

32. $(7 + 3) \div 2 \times 3^2$

33. $8 \times (7 + 5) \div 4^2 + 9 \div 3$

34. $3^2 \times 5 \div (10 \times 3 \div 2)$

1-5 Mental Math (pp. 24–27)

EXAMPLES

Find each sum or product.

- **4 + 13 + 6 + 7**
  4 + 6 + 13 + 7
  (4 + 6) + (13 + 7)
  10 + 20
  30

- **5 × 9 × 6**
  5 × 6 × 9
  (5 × 6) × 9
  30 × 9
  270

- **Use the Distributive Property to find the product.**
  3 × 16
  3 × 16 = 3 × (10 + 6)
  = (3 × 10) + (3 × 6)
  = 30 + 18
  = 48

EXERCISES

Find each sum or product.

35. 9 + 5 + 1 + 15
36. 8 × 13 × 5
37. 31 + 16 + 19 + 14
38. 6 × 12 × 15
39. 17 + 12 + 8 + 3
40. 16 × 5 × 4
41. 11 + 23 + 27 + 39
42. 13 × 5 × 2

Use the Distributive Property to find each product.

43. 7 × 24
44. 9 × 15
45. 6 × 34
46. 8 × 19
47. 8 × 27
48. 5 × 33
49. 4 × 13
50. 9 × 47

1-6 Choose the Method of Computation (pp. 28–30)

EXAMPLE

- The average annual rainfall in Washington, D.C., is 39 inches. How much rain does Washington, D.C., average in 8 years?

  You may not be able to quickly multiply the numbers in your head, but the numbers are not so big that you must use a calculator. Use pencil and paper to find the answer.
  39 × 8 = 312 inches

EXERCISES

51. The average high temperature for Washington, D.C., in January is 42°F. The record high temperature for Washington, D.C., is 104°F. How much hotter is the record temperature than the average high temperature in January?

52. There are 6 members on Lynn's chess team. If Lynn wants to give each member 31 mini chocolate bars—one for each day of the month—how many chocolate bars will she need?

1-7 Find a Pattern (pp. 31–33)

EXAMPLES

- **Find the next two terms in the sequence.**
  1, 3, 4, 7, ▇, ▇, . . .
  1 + 3 = 4   4 + 7 = 11
  3 + 4 = 7   7 + 11 = 18
  The next two numbers are 11 and 18.

EXERCISES

Find the next two terms in each sequence.

53. 1, 5, 6, 11, ▇, ▇, . . .
54. 1, 4, 7, 10, ▇, ▇, . . .
55. 1, 3, 4, 12, 13, 39, ▇, ▇, . . .
56. 2, 4, 8, 16, ▇, ▇, . . .

Chapter 1

Chapter Test

Solve.

1. Which number is greater, 16,880,953 or 16,221,773?
2. Which number is greater, 22,481,093 or 23,662,840?

Order the numbers from least to greatest.

3. 801; 798; 921
4. 4,835; 7,505; 4,310

Estimate each sum or difference by rounding to the place value indicated.

5. 8,743 + 3,198; thousands
6. 62,524 − 17,831; ten thousands

Estimate.

7. Kaitlin's family is planning a trip from Washington, D.C., to New York City. New York City is 227 miles from Washington, D.C., and the family can drive an average of 55 mi/h. About how long will the trip take?

Write each expression in exponential form.

8. $4 \times 4 \times 4 \times 4 \times 4$
9. $10 \times 10 \times 10$
10. $6 \times 6 \times 6 \times 6$

Find each value.

11. 2^3
12. 5^2
13. 4^4
14. 11^2

Evaluate each expression.

15. $12 + 8 \div 2$
16. $3^2 \times 5 + 10 - 7$
17. $12 + (28 - 15) + 4 \times 2$

Find each sum or product.

18. $15 + 23 + 47 + 5$
19. $5 \times 48 \times 2$
20. $44 + 18 + 12 + 6$

Use the Distributive Property to find the product.

21. 3×32
22. 52×6
23. 24×5
24. 81×6
25. At 5:00 A.M., the temperature was 41°F. By noon, the temperature was 69°F. By how many degrees did the temperature increase?

Identify a pattern in each sequence and name the missing terms.

26. 8, 22, 36, 50, ■, ■, ■, . . .
27. 2, 3, 5, 8, 12, ■, ■, ■, . . .

Chapter Test

Chapter 1

Performance Assessment

Show What You Know

Create a portfolio of your work from this chapter. Complete this page and include it with your four best pieces of work from Chapter 1. Choose from your homework or lab assignments, mid-chapter quiz, or any journal entries you have done. Put them together using any design you want. Make your portfolio represent what you consider your best work.

Short Response

1. Four people go to see a movie, and each person recommends it to 4 other friends, who then go to see it. Then each of those 4 recommends the movie to 4 other friends. If the pattern continues, how many rounds are needed for at least 1 million people to have seen the movie? Show the steps necessary to find the answer.
2. Create your own sequence using exponents. Explain how you created the sequence, and give the rule for extending the pattern.
3. Create an order-of-operations problem of at least five steps that results in an answer of 7. Be sure to include parentheses and an exponent, and use each operation only once. Show how to get the answer, and explain the steps you used to simplify the expression.

Extended Problem Solving

4. A movie's *gross* is the amount of money the movie makes at the box office, not counting the cost of advertising and distributing. The first four movies of the *Star Wars* series are among the top 20 grossing films worldwide.
   a. Use the information in the table to calculate the total gross of the four films. Show your work.
   b. What is the average gross of the four movies? Show your work.
   c. Explain why you agree or disagree with the following statement. "The first four *Star Wars* movies were more successful in the United States than outside the United States."

Box Office Sales (1977–1999)

Movie	U.S. Gross (millions)	Foreign Gross (millions)
Star Wars: Episode 1: The Phantom Menace	$431	$492
Star Wars	$461	$323
Return of the Jedi	$309	$264
The Empire Strikes Back	$290	$223

internet connect
EOG Test Practice Online
go.hrw.com Keyword: MR4 TestPrep

Getting Ready for EOG

Chapter 1

Cumulative Assessment, Chapter 1

1. Which number is the *greatest*?

A 6,568,217 C 6,701,953
B 6,739,549 D 6,589,211

2. What is five billion, two hundred fifty-two million, six hundred thousand, three hundred eleven in standard form?

A 5,252,603,011 C 5,252,600,311
B 52,526,311 D 5,252,060,311

TEST TAKING TIP!
An exponent tells how many times a number called the base is used as a factor.

3. What is the value of 4^3?

A 12 C 64
B 16 D 81

4. What is $5 \times 5 \times 5$ written in exponential form?

A 125 C 3^5
B $100 + 20 + 5$ D 5^3

5. What is the value of $7 \times 3 + 2$?

A 23 C 35
B 12 D 42

6. What is the value of $8^2 - (12 + 3) \times 2$?

A 98 C 110
B 34 D 58

7. The equation $6 \times 3 \times 4 = 3 \times 6 \times 4$ is an example of which property?

A Associative C Distributive
B Commutative D Exponential

8. The bar graph shows the number of miles Jan biked each day last week. How many total miles did she bike?

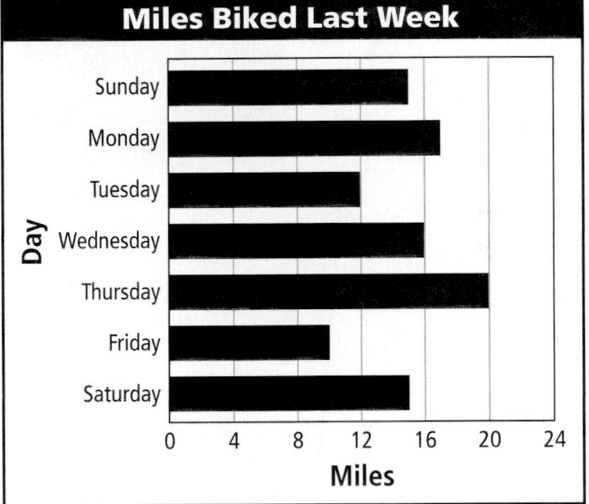

A 15 miles
B 17 miles
C 105 miles
D 110 miles

9. ***SHORT RESPONSE*** Find a pattern in the sequence 3, 6, 12, 24, 48, ... Use your pattern to find the next two terms.

10. ***SHORT RESPONSE*** Explain what 5^3 means. What is the value of 5^3?

Getting Ready for EOG

Chapter

2

Introduction to Algebra

TABLE OF CONTENTS

internet connect go.hrw.com

Chapter Opener Online
go.hrw.com
KEYWORD: MR4 Ch2

Number of Cars Traveling in Each Direction

	North	South	East	West
6–8 A.M.	114	36	48	57
8–10 A.M.	97	52	57	52
10 A.M.–noon	35	24	65	56
noon–2 P.M.	23	109	61	56
2–4 P.M.	18	138	70	72
4–6 P.M.	11	54	47	40

Traffic Engineer

Have you ever wondered why traffic moves quickly through one intersection but slowly through another? Traffic engineers program stoplights so that vehicles can move smoothly through intersections. There are many variables at a traffic intersection—the number of vehicles that pass, the time of day, and the direction in which each vehicle travels are examples. Traffic engineers use this information to control the timing of stoplights. The table lists traffic movement through a given intersection during a given weekday.

ARE YOU READY?

Choose the best term from the list to complete each sentence.

1. Multiplication is the ___?___ of division.
2. The ___?___ of 12 and 3 is 36.
3. The ___?___ of 12 and 3 is 15.
4. Addition, subtraction, multiplication, and division are called ___?___.
5. The answer to a division problem is called the ___?___.

dividend
factor
inverse
operations
product
quotient
sum

Complete these exercises to review skills you will need for this chapter.

✔ Whole Number Operations

Add.

6. $28 + 15$ **7.** $71 + 38$ **8.** $1{,}218 + 430$ **9.** $2{,}218 + 1{,}135$

Subtract.

10. $72 - 35$ **11.** $98 - 45$ **12.** $1{,}642 - 249$ **13.** $3{,}408 - 1{,}649$

Multiply.

14. 6×13 **15.** 8×15 **16.** 16×22 **17.** 20×35

Divide.

18. $9\overline{)72}$ **19.** $7\overline{)84}$ **20.** $16\overline{)112}$ **21.** $23\overline{)1{,}472}$

✔ Inverse Operations

Use the inverse operation to check each equation.

22. $14 + 5 = 19$ **23.** $24 + 19 = 43$ **24.** $125 + 219 = 344$

25. $19 - 3 = 16$ **26.** $37 - 14 = 23$ **27.** $242 - 196 = 46$

28. $8 \times 6 = 48$ **29.** $25 \times 5 = 125$ **30.** $14 \times 34 = 476$

31. $12 \div 6 = 2$ **32.** $72 \div 8 = 9$ **33.** $252 \div 12 = 21$

34. $52 \times 4 = 208$ **35.** $400 \div 16 = 25$ **36.** $19 \times 17 = 323$

2-1 Variables and Expressions

Learn to identify and evaluate expressions.

Vocabulary
variable
constant
algebraic expression

Inflation is the rise in prices that occurs over time. For example, you would have paid about $7 in the year 2000 for something that cost only $1 in 1950.

With this information, you can convert prices in 1950 to their equivalent prices in 2000.

1950	2000
$1	$7
$2	$14
$3	$21
$$p$	$$p \times 7$

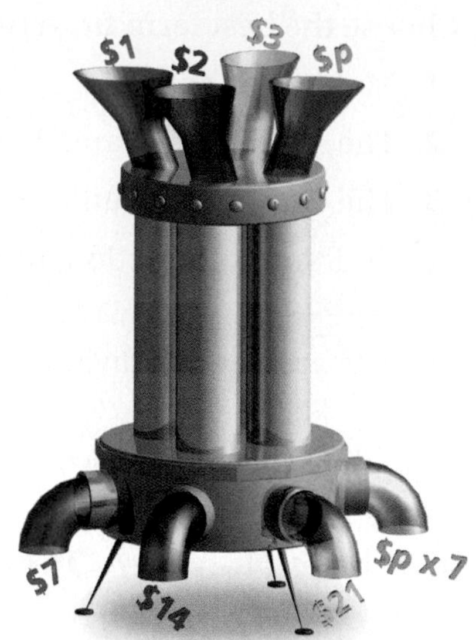

A **variable** is a letter or symbol that represents a quantity that can change. In the table above, p is a variable that stands for any price in 1950. A **constant** is a quantity that does not change. For example, the price of something in 2000 is always 7 times the price in 1950.

An **algebraic expression** contains one or more variables and may contain operation symbols. So $p \times 7$ is an algebraic expression.

Algebraic Expressions	NOT Algebraic Expressions
$150 + y$	$85 \div 5$
$35 \times w + z$	$10 + 3 \times 5$

To evaluate an algebraic expression, substitute a number for the variable and then find the value.

EXAMPLE 1 Evaluating Algebraic Expressions

Evaluate each expression to find the missing values in the tables.

A

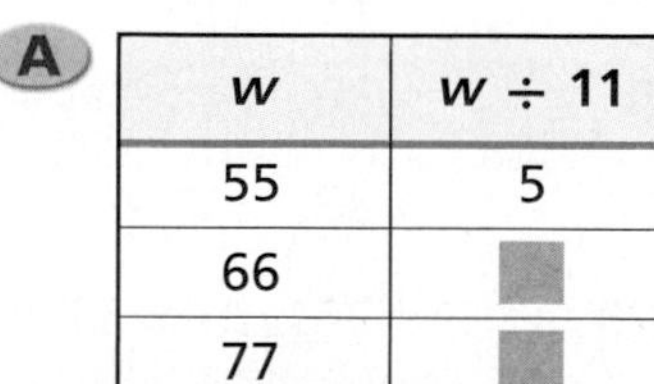

w	$w \div 11$
55	5
66	
77	

Substitute for w in $w \div 11$.

$w = 55;\ 55 \div 11 = 5$

$w = 66;\ 66 \div 11 = 6$

$w = 77;\ 77 \div 11 = 7$

The missing values are 6 and 7.

Evaluate each expression to find the missing values in the tables.

B

n	$4 \times n + 6$
1	10
2	■
3	■

Substitute for n in $4 \times n + 6$.
Use the order of operations.
$n = 1; 4 \times 1 + 6 = 10$
$n = 2; 4 \times 2 + 6 = 14$
$n = 3; 4 \times 3 + 6 = 18$

The missing values are 14 and 18.

Writing Math

When you are multiplying a number times a variable, the number is written first. Write "$3x$" and not "$x3$." Read $3x$ as "three x."

Multiplication and division expressions can be written without using the symbols × and ÷.

Instead of . . .	You can write . . .
$x \times 3$	$x \cdot 3$ $x(3)$ $3x$
$35 \div y$	$\frac{35}{y}$

EXAMPLE 2 Finding an Expression

Find an expression for each table.

A

x	■
6	48
7	56
8	64

$6 \cdot 8 = 48$
$7 \cdot 8 = 56$
$8 \cdot 8 = 64$

An expression is $x \cdot 8$, or $8x$.

B

x	■
12	1
24	2
36	3

$12 \div 12 = 1$
$24 \div 12 = 2$
$36 \div 12 = 3$

An expression is $\frac{x}{12}$.

Think and Discuss

1. **Name** a quantity that is a variable and a quantity that is a constant.
2. **Tell** whether each expression is an algebraic expression.

a. $54 \div 2$ **b.** $45 + x$ **c.** $16y$ **d.** $24 \div 12$

2-1 Exercises

5.01, 5.02

FOR EOG PRACTICE

see page 640

internet connect

Homework Help Online

go.hrw.com Keyword: MR4 2-1

GUIDED PRACTICE

See Example 1 Evaluate each expression to find the missing values in the tables.

1.

n	$n + 7$
38	45
49	
58	

2.

x	$12x$
8	96
9	
10	

See Example 2 Find an expression for each table.

3.

x	
50	45
45	40
40	35

4.

w	
23	32
33	42
43	52

INDEPENDENT PRACTICE

See Example 1 Evaluate each expression to find the missing values in the tables.

5.

x	$4x$
50	200
100	
150	

6.

n	$2n - 2$
1	
6	
7	

See Example 2 Find an expression for each table.

7.

x	
0	0
72	9
88	11

8.

n	
15	40
25	50
35	60

9.

x	
8	6
4	2
2	0

10.

n	
50	500
75	750
100	1,000

PRACTICE AND PROBLEM SOLVING

Why does a Polish coin show Australia and a kangaroo? This coin honors Pawel Edmund Strzelecki, a Polish explorer who explored and mapped much of Australia.

Evaluate each expression for the given value of the variable.

11. $3h + 2$ for $h = 10$

12. $2x$ for $x = 15$

13. $4p - 3$ for $p = 20$

14. $\frac{c}{7}$ for $c = 56$

15. $3x + 17$ for $x = 13$

16. $5p$ for $p = 12$

17. The zloty is the currency in Poland. In 2002, 1 U.S. dollar was worth 4 zlotys. How many zlotys were equivalent to 8 U.S. dollars?

18. Use the graph to complete the table.

Cups of Water	Number of Lemons
8	■
12	■
w	■

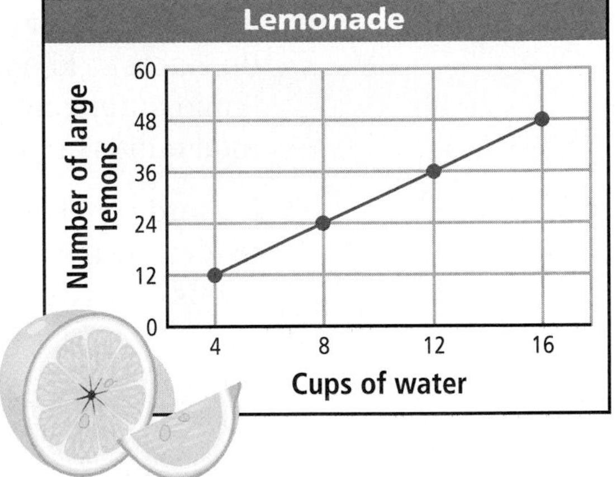

19. ***WHAT'S THE ERROR?*** A student evaluated the expression $x \div 2$ for $x = 14$ and gave an answer of 28. What did the student do wrong?

20. ***WRITE ABOUT IT*** A friend asks you to think of a number, double it, and then add 5. Write an algebraic expression to describe your friend's directions, and make a table of possible values.

21. ***CHALLENGE*** Using the algebraic expression $3n - 5$, what is the smallest whole-number value for n that will give you a result greater than 100?

Spiral Review

Order the numbers from least to greatest. (Lesson 1-1)

22. 798; 648; 923

23. 1,298; 876; 972

24. 1,498; 2,163; 1,036

Estimate each sum or difference by rounding to the place value indicated. (Lesson 1-2)

25. 17,281 + 23,008; thousands

26. 412,243 − 124,539; hundred thousands

Write each expression in exponential form. (Lesson 1-3)

27. $3 \times 3 \times 3$

28. $5 \times 5 \times 5 \times 5 \times 5 \times 5$

29. $10 \times 10 \times 10 \times 10$

30. **EOG PREP** Which is the missing term in the sequence 3, 7, 15, 31, 63, ■, 255, . . . ? (Lesson 1-7)

A 67 B 95 C 103 D 127

2-2 Translate Between Words and Math

Problem Solving Skill

Learn to translate between words and math.

The earth's core is divided into two parts. The inner core is solid and dense, with a radius of 1,228 km. Let c stand for the thickness in kilometers of the liquid outer core. What is the total radius of the earth's core?

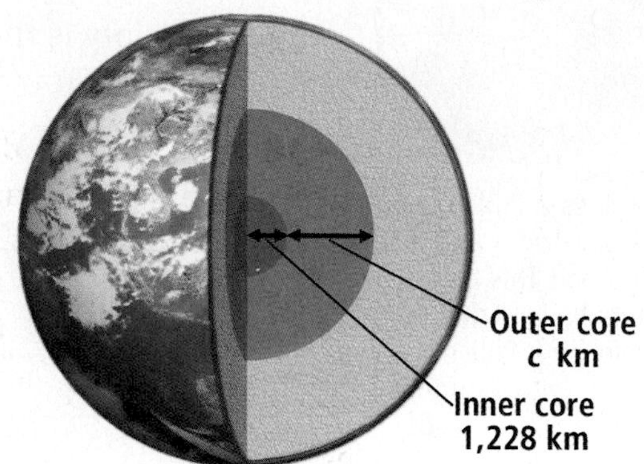

In word problems, you may need to identify the action to translate words to math.

Action	Put together or combine	Find how much more or less	Put together groups of equal parts	Separate into equal groups
Operation	Add	Subtract	Multiply	Divide

To solve this problem, you need to *put together* the measurements of the inner core and the outer core. To put things together, add.

$$1{,}228 + c$$

The total radius of the earth's core is $1{,}228 + c$ km.

EXAMPLE 1 *Social Studies Applications*

A The Nile River is the world's longest river. Let n stand for the length in miles of the Nile. The Amazon River is 4,000 miles long. Write an expression to show how much longer the Nile is than the Amazon.

To *find how much longer,* subtract the length of the Amazon from the length of the Nile.

$$n \quad - \quad 4{,}000$$

The Nile is $n - 4{,}000$ miles longer than the Amazon.

B Let s represent the number of senators that each of the 50 states has in the U.S. Senate. Write an expression for the total number of senators.

To *put together 50 equal groups of s,* multiply 50 times s.

$$50s$$

There are $50s$ senators in the U.S. Senate.

There are several different ways to write math expressions with words.

Operation	+	−	×	÷
Numerical Expression	$37 + 28$	$90 - 12$	8×48 or $8 \cdot 48$ or $(8)(48)$ or $8(48)$ or $(8)48$	$327 \div 3$ or $\frac{327}{3}$
Words	• 28 added to 37 • 37 plus 28 • the sum of 37 and 28 • 28 more than 37	• 12 subtracted from 90 • 90 minus 12 • the difference of 90 and 12 • 12 less than 90 • take away 12 from 90	• 8 times 48 • 48 multiplied by 8 • the product of 8 and 48 • 8 groups of 48	• 327 divided by 3 • the quotient of 327 and 3
Algebraic Expression	$x + 28$	$k - 12$	$8 \cdot w$ or $(8)(w)$ or $8w$	$n \div 3$ or $\frac{n}{3}$
Words	• 28 added to x • x plus 28 • the sum of x and 28 • 28 more than x	• 12 subtracted from k • k minus 12 • the difference of k and 12 • 12 less than k • take away 12 from k	• 8 times w • w multiplied by 8 • the product of 8 and w • 8 groups of w	• n divided by 3 • the quotient of n and 3

EXAMPLE 2 Translating Words into Math

Write each phrase as a numerical or algebraic expression.

A **287 plus 932**

$287 + 932$

B ***b* divided by 14**

$b \div 14$ or $\frac{b}{14}$

EXAMPLE 3 Translating Math into Words

Write two phrases for each expression.

A $a - 45$

- a minus 45
- take away 45 from a

B $(34)(7)$

- the product of 34 and 7
- 34 multiplied by 7

Think and Discuss

1. **Tell** how to write each of the following phrases as a numerical or algebraic expression: 75 less than 1,023; the product of 125 and z.
2. **Give** two examples of "$a \div 17$" expressed with words.

2-2 Exercises

5.02

FOR EOG PRACTICE
see page 641

internet connect
Homework Help Online
go.hrw.com Keyword: MR4 2-2

GUIDED PRACTICE

See Example 1

1. The Big Island of Hawaii is the largest Hawaiian island, with an area of 4,028 mi^2. The next biggest island is Maui. Let m represent the area of Maui. Write an expression for the difference between the two areas.

See Example 2

Write each phrase as a numerical or algebraic expression.

2. 279 minus 125
3. the product of 15 and x

See Example 3

Write two phrases for each expression.

4. $r + 87$
5. 345×196
6. $476 \div 28$
7. $d - 5$

INDEPENDENT PRACTICE

See Example 1

8. California has 21 more seats in the U.S. Congress than Texas has. If t represents the number of seats Texas has, write an expression for the number of seats California has.
9. Let x represent the number of television show episodes that are taped in a season. Write an expression for the number of episodes that are taped in 5 seasons.

See Example 2

Write each phrase as a numerical or algebraic expression.

10. 25 less than k
11. the quotient of 325 and 25
12. 34 times w
13. 675 added to 137
14. the sum of 135 and p
15. take away 14 from j

See Example 3

Write two phrases for each expression.

16. $h + 65$
17. $243 - 19$
18. $125 \div n$
19. $342(75)$
20. $\frac{d}{27}$
21. $45 \cdot 23$
22. $629 + c$
23. $228 - b$

PRACTICE AND PROBLEM SOLVING

Translate each phrase into a numerical or algebraic expression.

24. 13 less than z
25. 15 divided by d
26. 874 times 23
27. m multiplied by 67
28. the sum of 35, 74, and 21
29. 319 less than 678

Science LINK

The graph shows the number of U.S. space exploration missions from 1956 to 2000.

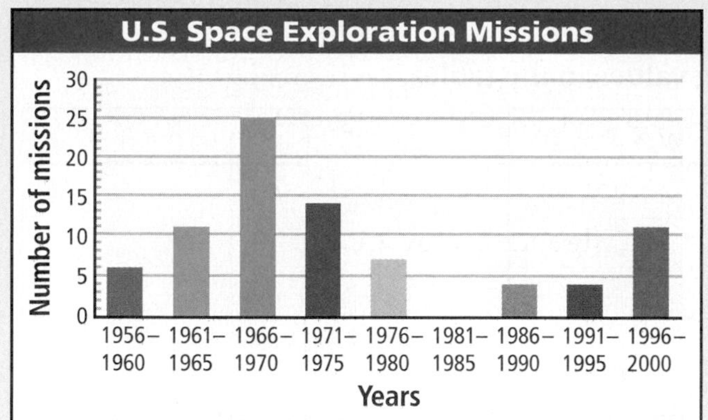

30. Between 1966 and 1970, the Soviet Union had m fewer space missions than the United States. Write an algebraic expression for this situation.

31. Let d represent the number of dollars that the United States spent on space missions from 1986 to 1990. Write an expression for the cost per mission.

32. ***WRITE A PROBLEM*** Use the data in the graph to write a word problem that can be answered with a numerical or algebraic expression.

33. ***WHAT'S THE QUESTION?*** The answer from the graph is $6 + 11 + 25 + 14 + 7 + 4 + 4 + 11$. What is the question?

34. ***WRITE ABOUT IT*** Let p stand for the number of missions between 1996 and 2000 that had people aboard. What operation would you use to write an expression for the number of missions without people? Why? Use the action in the problem to explain your answer.

35. ***CHALLENGE*** Write an expression for the following: two more than the number of missions from 1971 to 1975, minus the number of missions from 1986 to 1990. Then evaluate the expression.

Spiral Review

Compare. Write <, >, or =. (Lesson 1-1)

36. 1,256,589 ▒ 1,265,598

37. 2,568,987,254 ▒ 2,568,987,254

Find each value. (Lesson 1-3)

38. 5^4 **39.** 2^5 **40.** 7^2 **41.** 11^3

42. **EOG PREP** $5^2 + (7 + 2) - 8 \cdot 3 = ?$ (Lesson 1-4)

A 10 **B** 18 **C** 33 **D** 78

Chapter 2

Mid-Chapter Quiz

LESSON 2-1 (pp. 48–51)

Evaluate each expression to find the missing values in the tables.

1.

y	$23 + y$
17	40
27	■
37	■

2.

w	$w \times 3 + 10$
4	22
5	■
6	■

3.

x	$x \div 8$
40	5
48	■
56	■

Find an expression for each table.

4.

t	■
3	9
4	12
5	15

5.

y	■
36	4
45	5
54	6

6.

n	■
10	2
15	7
20	12

LESSON 2-2 (pp. 52–55)

7. The small and large intestines are part of the digestive system. The small intestine is longer than the large intestine. Let n represent the length in feet of the small intestine. The large intestine is 5 feet long. Write an expression to show how much longer the small intestine is than the large intestine.

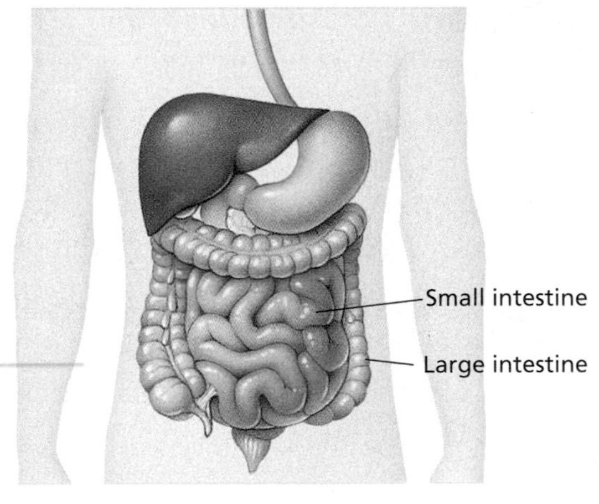

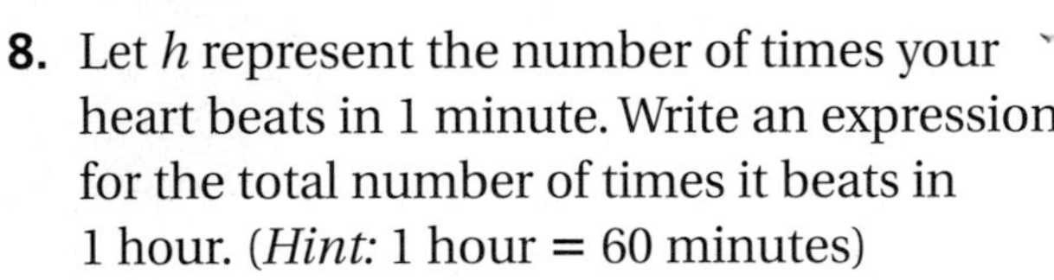

8. Let h represent the number of times your heart beats in 1 minute. Write an expression for the total number of times it beats in 1 hour. (*Hint:* 1 hour = 60 minutes)

Write each phrase as a numerical or algebraic expression.

9. 719 plus 210

10. t multiplied by 7

11. the sum of n and 51

12. 14 subtracted from p

13. the quotient of 510 and n

14. the product of 52 and z

15. q divided by 3

16. the difference of 25 and g

Write two phrases for each expression.

17. $n + 19$

18. $12 \cdot 13$

19. $72 - x$

20. $\frac{t}{12}$

21. $15s$

22. $27 \div 9$

23. $43 - z$

24. $93 \div k$

25. $20 + d$

Focus on Problem Solving

Understand the Problem

- **Identify too much or too little information**

Problems often give too much or too little information. You must decide whether you have enough information to work the problem.

Read the problem and identify the facts that are given. Can you use any of these facts to arrive at an answer? Are there facts in the problem that are not necessary to find the answer? These questions can help you determine whether you have too much or too little information.

If you cannot solve the problem with the information given, decide what information you need. Then read the problem again to be sure you haven't missed the information in the problem.

Copy each problem. Circle the important facts. Underline any facts that you do not need to answer the question. If there is not enough information, list the additional information you need.

1. The reticulated python is one of the longest snakes in the world. One was found in Indonesia in 1912 that was 33 feet long. At birth, a reticulated python is 2 feet long. Suppose an adult python is 29 feet long. Let f represent the number of feet the python grew since birth. What is the value of f?

2. The largest flying flag in the world is 7,410 square feet and weighs 180 pounds. There are a total of 13 horizontal stripes on it. Let h represent the height of each stripe. What is the value of h?

3. The elevation of Mt. McKinley is 20,320 ft. People who climb Mt. McKinley are flown to a base camp located at 7,200 ft. From there, they begin a climb that may last 20 days or longer. Let d represent the distance from the base camp to the summit of Mt. McKinley. What is the value of d?

4. Let c represent the cost of a particular computer in 1981. Six years later, in 1987, the price of the computer had increased to $3,600. What is the value of c?

2-3 Equations and Their Solutions

Learn to determine whether a number is a solution of an equation.

Vocabulary
equation
solution

An **equation** is a mathematical statement that two quantities are equal. You can think of a correct equation as a balanced scale.

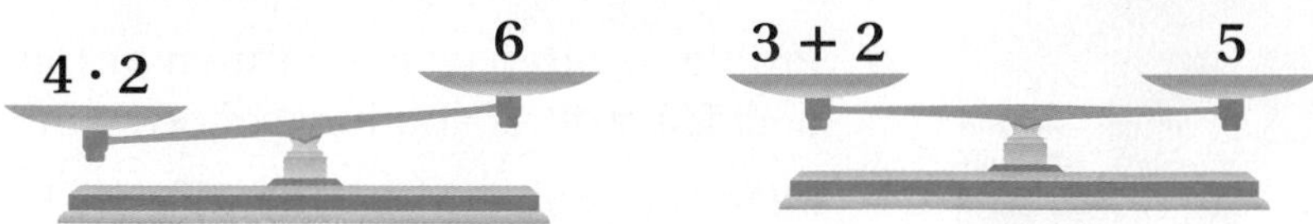

Equations may contain variables. If a value for a variable makes an equation true, that value is a **solution** of the equation.

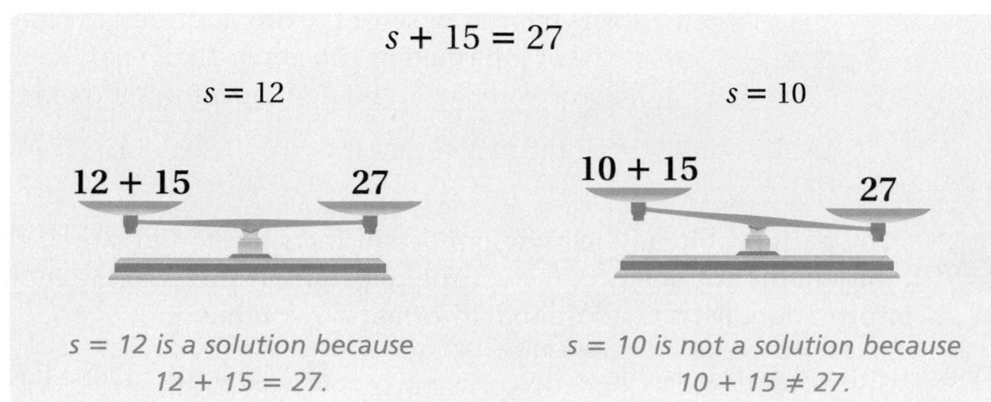

Reading Math

The symbol $\neq$ means "is not equal to."

EXAMPLE 1 **Determining Solutions of Equations**

Determine whether the given value of each variable is a solution.

A $a + 23 = 82$ **for** $a = 61$

$a + 23 = 82$

$61 + 23 \stackrel{?}{=} 82$ *Substitute 61 for a.*

$84 \stackrel{?}{=} 82$ *Add.*

84 82

Since $84 \neq 82$, 61 is not a solution to $a + 23 = 82$.

Determine whether the given value of each variable is a solution.

B $60 \div c = 6$ for $c = 10$

$60 \div c = 6$

$60 \div 10 \stackrel{?}{=} 6$ *Substitute 10 for c.*

$6 \stackrel{?}{=} 6$ *Divide.*

Because $6 = 6$, 10 is a solution to $60 \div c = 6$.

You can use equations to check whether measurements given in different units are equal.

For example, there are 12 inches in one foot. If you have a measurement in feet, multiply by 12 to find the measurement in inches: $12 \cdot \text{feet} = \text{inches}$, or $12f = i$.

If you have one measurement in feet and another in inches, check whether the two numbers make the equation $12f = i$ true.

EXAMPLE 2 *Life Science Application*

One science book states that a manatee can grow to be 13 feet long. According to another book, a manatee may grow to 156 inches. Determine if these two measurements are equal.

$12f = i$

$12 \cdot 13 \stackrel{?}{=} 156$ *Substitute.*

$156 \stackrel{?}{=} 156$ *Multiply.*

Because $156 = 156$, 13 feet is equal to 156 inches.

Think and Discuss

1. **Tell** which of the following is the solution to $y \div 2 = 9$: $y = 14$, $y = 16$, or $y = 18$. How do you know?
2. **Give an example** of an equation with a solution of 15.

2-3 Exercises

5.03

FOR EOG PRACTICE

see page 642

internet connect

Homework Help Online

go.hrw.com Keyword: MR4 2-3

GUIDED PRACTICE

See Example 1

Determine whether the given value of each variable is a solution.

1. $c + 23 = 48$ for $c = 35$
2. $z + 31 = 73$ for $z = 42$
3. $96 = 130 - d$ for $d = 34$
4. $85 = 194 - a$ for $a = 105$
5. $75 \div y = 5$ for $y = 15$
6. $78 \div n = 13$ for $n = 5$

See Example 2

7. An almanac states that the Minnehaha Waterfall in Minnesota is 53 feet tall. A tour guide said the Minnehaha Waterfall is 636 inches tall. Determine if these two measurements are equal.

INDEPENDENT PRACTICE

See Example 1

Determine whether the given value of the variable is a solution.

8. $w + 19 = 49$ for $w = 30$
9. $d + 27 = 81$ for $d = 44$
10. $g + 34 = 91$ for $g = 67$
11. $k + 16 = 55$ for $k = 39$
12. $101 = 150 - h$ for $h = 49$
13. $89 = 111 - m$ for $m = 32$
14. $116 = 144 - q$ for $q = 38$
15. $92 = 120 - t$ for $t = 28$
16. $80 \div b = 20$ for $b = 4$
17. $91 \div x = 7$ for $x = 12$
18. $55 \div j = 5$ for $j = 10$
19. $49 \div r = 7$ for $r = 7$

See Example 2

20. Kent earns \$6 per hour at his after-school job. One week, he worked 12 hours and received a paycheck for \$66. Determine if Kent was paid the correct amount of money. (*Hint:* \$6 • hours = total pay)

PRACTICE AND PROBLEM SOLVING

Determine whether the given value of the variable is a solution.

21. $93 = 48 + u$ for $u = 35$
22. $112 = 14 \times f$ for $f = 8$
23. $13 = m \div 8$ for $m = 104$
24. $79 = z - 23$ for $z = 112$
25. $64 = l - 34$ for $l = 98$
26. $105 = p \times 7$ for $p = 14$
27. $94 \div s = 26$ for $s = 3$
28. $v + 79 = 167$ for $v = 88$
29. $m + 36 = 54$ for $m = 18$
30. $x - 35 = 96$ for $x = 112$
31. $12y = 84$ for $y = 7$
32. $7x = 56$ for $x = 8$
33. $3x = 150$ for $x = 65$
34. $20k = 115$ for $k = 9$

Replace each ▇ with a number that makes the equation correct.

35. 4 + 1 = ▇ + 2

36. 2 + ▇ = 6 + 2

37. ▇ − 5 = 9 − 2

38. 5(4) = 10(▇)

39. 3 + 6 = ▇ − 4

40. 12 ÷ 4 = 9 ÷ ▇

41. Rebecca has 17 one-dollar bills. Courtney has 350 nickels. Do the two girls have the same amount of money? (*Hint*: First find how many nickels are in a dollar.)

42. ***CHOOSE A STRATEGY*** What should replace the question mark to keep the scale balanced?

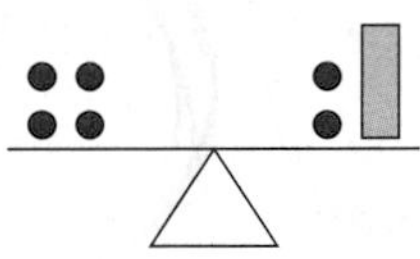

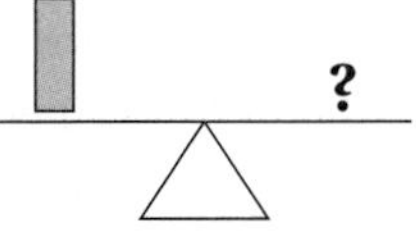

A

B 

C 

D

43. ***WRITE A PROBLEM*** Write an equation using a variable and information from the graph.

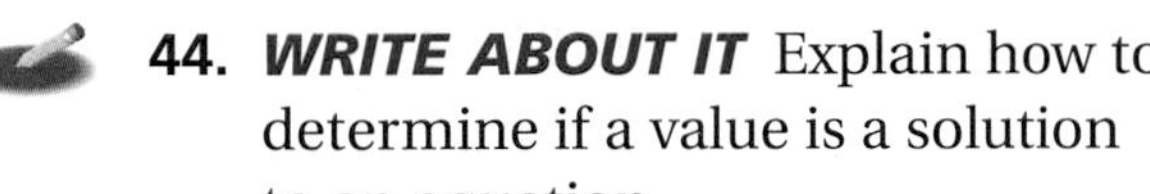

44. ***WRITE ABOUT IT*** Explain how to determine if a value is a solution to an equation.

45. ***CHALLENGE*** Is $n = 4$ a solution for $n^2 + 79 = 88$? Explain.

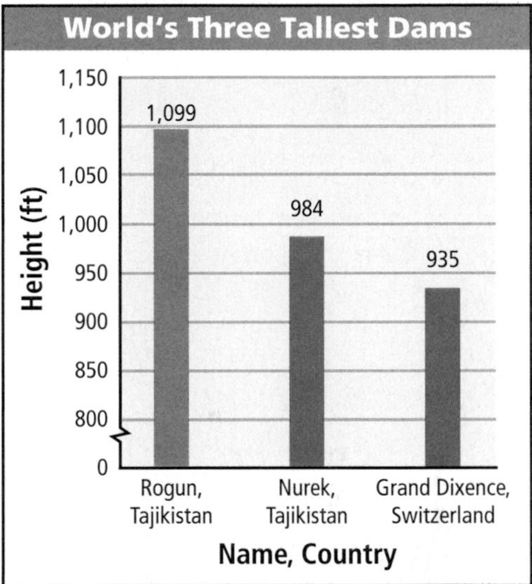

Spiral Review

Find each value. (Lesson 1-3)

46. 5^3

47. 3^4

48. 2^6

49. 6^3

Evaluate each expression. (Lesson 1-4)

50. $50 - 2 \times 10$

51. $16 \div (6 + 2) + 2^2$

52. $20 \div (19 - 9) \times 3^2 - 5$

53. $(8 + 7) \div 3 \times (19 - 4)$

54. **EOG PREP** Which of the following phrases can be represented by $79x$? (Lesson 2-2)

A 79 divided by x

B The difference of 79 and x

C The product of 79 and x

D The sum of 79 and x

2-4 Solving Addition Equations

Learn to solve whole-number addition equations.

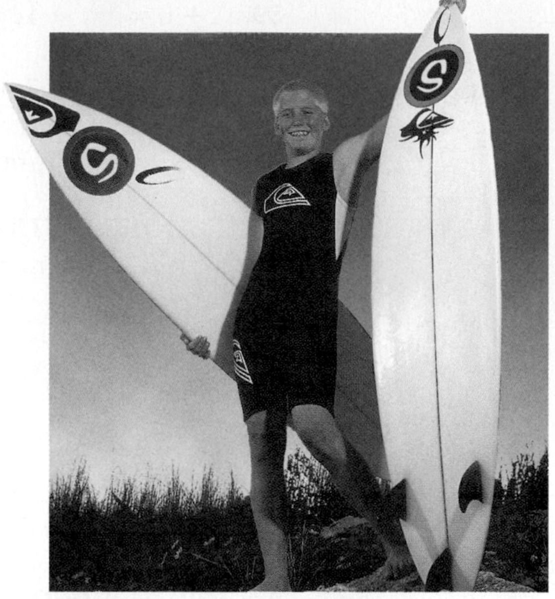

Some surfers recommend that the length of a beginner's surfboard be 14 inches more than the surfer's height. If a surfboard is 82 inches, how tall should the surfer be to ride it?

The height of the surfer *combined* with 14 inches equals 82 inches. To combine amounts, you need to add.

Let h stand for the surfer's height. You can use the equation $h + 14 = 82$.

$h + 14$ 82

The equation h + 14 = 82 can be represented as a balanced scale.

To find the value of h, you need h by itself on one side of a balanced scale.

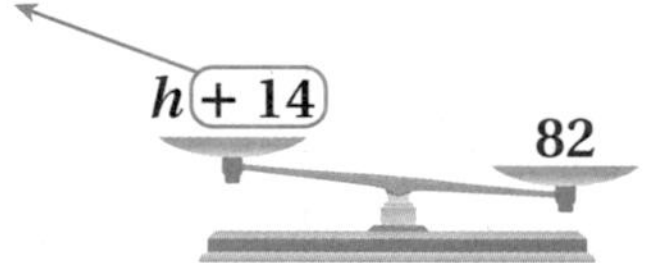

To get h by itself, first take away 14 from the left side of the scale. Now the scale is unbalanced.

To rebalance the scale, take away 14 from the other side.

A surfer using an 82-inch surfboard should be 68 inches tall.

Taking away 14 from both sides of the scale is the same as subtracting 14 from both sides of the equation.

$$\begin{aligned} h + 14 &= 82 \\ \underline{-\ 14} &\quad \underline{-\ 14} \\ h \quad &= 68 \end{aligned}$$

Subtraction is the inverse, or opposite, of addition. If an equation contains addition, solve it by subtracting from both sides to "undo" the addition.

EXAMPLE 1 Solving Addition Equations

Solve each equation. Check your answers.

A $x + 62 = 93$

$x + 62 = 93$	*62 is added to x.*
$\underline{-62} \quad \underline{-62}$	*Subtract 62 from both sides to undo*
$x = 31$	*the addition.*

Check $x + 62 = 93$

$31 + 62 \stackrel{?}{=} 93$	*Substitute 31 for x in the equation.*
$93 \stackrel{?}{=} 93$ ✔	*31 is the solution.*

B $81 = 17 + y$

$81 = 17 + y$	*17 is added to y.*
$\underline{-17} \quad \underline{-17}$	*Subtract 17 from both sides to undo*
$64 = y$	*the addition.*

Check $81 = 17 + y$

$81 \stackrel{?}{=} 17 + 64$	*Substitute 64 for y in the equation.*
$81 \stackrel{?}{=} 81$ ✔	*64 is the solution.*

EXAMPLE 2 *Social Studies Application*

Dyersberg, Newton, and St. Thomas are located along Ventura Highway, as shown on the map. Find the distance d between Newton and Dyersberg.

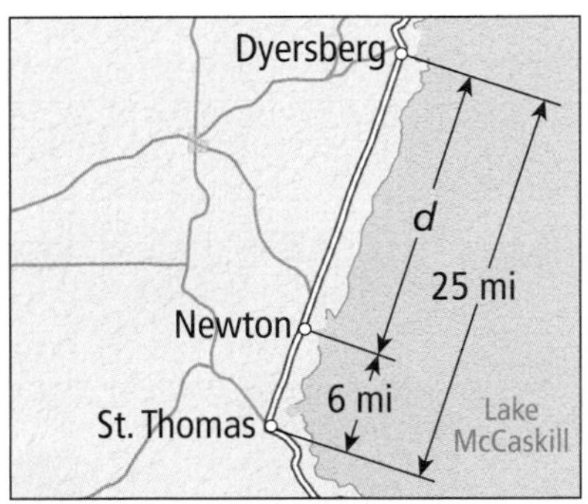

distance between Dyersberg and St. Thomas	=	distance between Newton and St. Thomas	+	distance between Newton and Dyersberg
25	=	6	+	d

$25 = 6 + d$	*6 is added to d.*
$\underline{-6} \quad \underline{-6}$	*Subtract 6 from both sides to undo*
$19 = d$	*the addition.*

The distance between Newton and Dyersberg is 19 miles.

Think and Discuss

1. **Tell** whether the solution of $c + 4 = 21$ will be less than 21 or greater than 21. Explain.
2. **Describe** how you could check your answer in Example 2.

2-4 Exercises

FOR EOG PRACTICE

see page 642

internet connect
Homework Help Online
go.hrw.com Keyword: MR4 2-4

5.03

GUIDED PRACTICE

See Example 1

Solve each equation. Check your answers.

1. $x + 54 = 90$
2. $49 = 12 + y$
3. $n + 27 = 46$
4. $22 + t = 91$
5. $31 = p + 13$
6. $c + 38 = 54$

See Example 2

7. Lou, Michael, and Georgette live on Mulberry Street, as shown on the map. Lou lives 10 blocks from Georgette. Georgette lives 4 blocks from Michael. How many blocks does Michael live from Lou?

INDEPENDENT PRACTICE

See Example 1

Solve each equation. Check your answers.

8. $x + 19 = 24$
9. $10 = r + 3$
10. $s + 11 = 50$
11. $b + 17 = 42$
12. $12 + m = 28$
13. $z + 68 = 77$
14. $72 = n + 51$
15. $g + 28 = 44$
16. $27 = 15 + y$

See Example 2

17. What is the length of a killer whale?

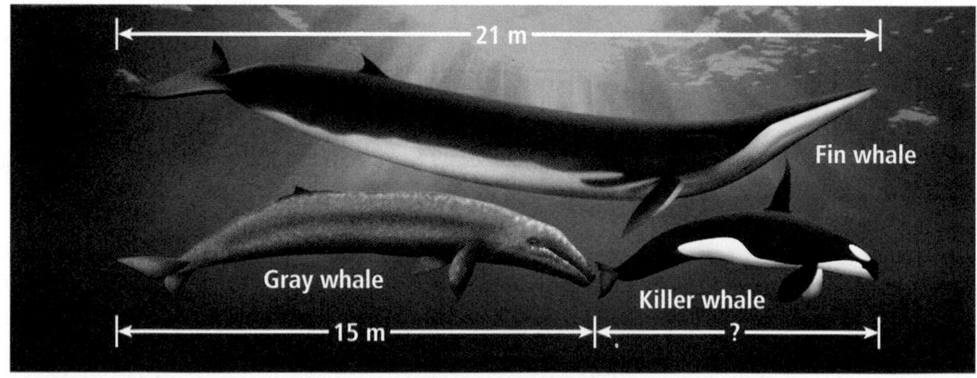

PRACTICE AND PROBLEM SOLVING

Solve each equation.

18. $x + 12 = 16$
19. $n + 32 = 39$
20. $23 + q = 34$
21. $52 + y = 71$
22. $73 = c + 35$
23. $93 = h + 15$
24. $125 = n + 85$
25. $87 = b + 18$
26. $12 + y = 50$
27. $t + 17 = 43$
28. $k + 9 = 56$
29. $25 + m = 47$

Popular items like the ball and the mood rings above are made of heat-sensitive materials. Changes in temperature, such as the warmth of your body heat, cause these materials to change color.

30. ***PHYSICAL SCIENCE*** Temperature can be measured in degrees Fahrenheit, degrees Celsius, or kelvins. To convert from degrees Celsius to kelvins, add 273 to the Celsius temperature. Complete the table.

	Kelvins (K)	°C + 273 = K	Celsius (°C)
Water Freezes	273	°C + 273 = 273	
Room Temperature	296	°C + 273 = 296	
Body Temperature	310		
Water Boils	373		

31. Alyssa has a temperature that is 2°C higher than normal body temperature. If Alyssa's temperature is 39°C, what is normal body temperature?

32. ***HISTORY*** In 1520, the explorer Ferdinand Magellan tried to measure the depth of the ocean. He weighted a 370 m rope and lowered it into the ocean. This rope was not long enough to reach the ocean floor. Suppose the depth at this location was 1,250 m. How much longer would Magellan's rope have to have been to reach the ocean floor?

33. ***WRITE A PROBLEM*** Use data from your science book to write a problem that can be solved using an addition equation. Solve your problem.

34. ***WRITE ABOUT IT*** Why are addition and subtraction called inverse operations?

35. ***CHALLENGE*** In the magic square at right, each row, column, and diagonal has the same sum. Find the values of x, y, and z.

7	61	x
y	37	1
31	z	67

Spiral Review

Estimate each sum or difference by rounding to the place value indicated. (Lesson 1-2)

36. 6,832 + 2,078; thousands

37. 52,854 − 25,318; ten thousands

38. 49,135 − 12,798; thousands

39. 78,497 + 19,980; ten thousands

Evaluate each expression. (Lesson 1-4)

40. $3^3 - (15 - 8) + 4 \times 5$

41. $17 \times (5 - 3) + 2^4 \div 8$

42. $81 - 4 \times 3 + 18 \div (6 + 3)$

43. **EOG PREP** Which equation below is an example of the Associative Property of Addition? (Lesson 1-5)

A $(7 \times 2) + 8 = 16 + 6$

B $2 \times 3 \times 5 = 2 \times 5 \times 3$

C $(5 + 7) + 9 = 5 + (7 + 9)$

D $4 \times (2 + 3) = (4 \times 2) + (4 \times 3)$

2-5 Solving Subtraction Equations

Learn to solve whole-number subtraction equations.

When John F. Kennedy became president of the United States, he was 43 years old, which was 8 years younger than Abraham Lincoln was when Lincoln became president. How old was Lincoln when he became president?

Let a represent Abraham Lincoln's age.

Kennedy was President from 1961 to 1963.

Lincoln was President from 1861 to 1865.

Abraham Lincoln's age	$-$	8	$=$	John F. Kennedy's age
a	$-$	8	$=$	43

Remember that addition and subtraction are inverse operations. When an equation contains subtraction, use addition to "undo" the subtraction. Remember to add to both sides of the equation.

$$\begin{aligned} a - 8 &= 43 \\ +8 \quad &\quad +8 \\ \hline a \quad &= 51 \end{aligned}$$

Abraham Lincoln was 51 years old when he became president.

EXAMPLE 1 Solving Subtraction Equations

A **Solve $p - 2 = 5$. Check your answer.**

$$\begin{aligned} p - 2 &= 5 \\ +2 \quad &\quad +2 \\ \hline p \quad &= 7 \end{aligned}$$

2 is subtracted from p.
Add 2 to both sides to undo the subtraction.

Check $p - 2 = 5$

$7 - 2 \stackrel{?}{=} 5$ *Substitute 7 for p in the equation.*

$5 \stackrel{?}{=} 5$ ✔ *7 is the solution.*

B **Solve $40 = x - 11$. Check your answer.**

$$\begin{aligned} 40 &= x - 11 \\ +11 & \quad\;\; +11 \\ \hline 51 &= x \end{aligned}$$

11 is subtracted from x.
Add 11 to both sides to undo the subtraction.

Check $40 = x - 11$

$40 \stackrel{?}{=} 51 - 11$ — *Substitute 51 for x in the equation.*

$40 \stackrel{?}{=} 40$ ✔ — *51 is the solution.*

C **Solve $x - 56 = 19$. Check your answer.**

$$\begin{aligned} x - 56 &= 19 \\ +56 & \quad +56 \\ \hline x &= 75 \end{aligned}$$

56 is subtracted from x.
Add 56 to both sides to undo the subtraction.

Check $x - 56 = 19$

$75 - 56 \stackrel{?}{=} 19$ — *Substitute 75 for x in the equation.*

$19 \stackrel{?}{=} 19$ ✔ — *75 is the solution.*

Think and Discuss

1. **Tell** whether the solution of $b - 14 = 9$ will be less than 9 or greater than 9. Explain.
2. **Explain** how you know what number to add to both sides of an equation containing subtraction.

2-5 Exercises

FOR EOG PRACTICE
see page 643

internet connect
Homework Help Online
go.hrw.com Keyword: MR4 2-5

5.03

GUIDED PRACTICE

See Example 1

Solve each equation. Check your answers.

1. $p - 8 = 9$
2. $3 = x - 16$
3. $a - 13 = 18$
4. $15 = y - 7$
5. $n - 24 = 9$
6. $39 = d - 2$

INDEPENDENT PRACTICE

See Example 1

Solve each equation. Check your answers.

7. $y - 18 = 7$
8. $8 = n - 5$
9. $a - 34 = 4$
10. $c - 21 = 45$
11. $a - 40 = 57$
12. $31 = x - 14$
13. $28 = p - 5$
14. $z - 42 = 7$
15. $s - 19 = 12$

PRACTICE AND PROBLEM SOLVING

Solve each equation.

16. $r - 57 = 7$	**17.** $11 = x - 25$	**18.** $8 = y - 96$
19. $a - 6 = 15$	**20.** $q - 14 = 22$	**21.** $f - 12 = 2$
22. $18 = j - 19$	**23.** $109 = r - 45$	**24.** $d - 8 = 29$
25. $g - 71 = 72$	**26.** $p - 13 = 111$	**27.** $13 = m - 5$
28. $20 = n - 4$	**29.** $45 = k - 19$	**30.** $t - 60 = 121$

31. ***GEOGRAPHY*** Mt. Rainier, in Washington, has a higher elevation than Mt. Shasta. The difference between their elevations is 248 feet. What is the elevation of Mt. Rainier? Write an equation and solve.

U.S. Mountains

Mountain	Height (ft)
Yale (CO)	14,196
Shasta (CA)	14,162
Sill (CA)	14,153
Augusta (AK)	14,070

32. ***SOCIAL STUDIES*** In 2000, the population of Seoul, South Korea, was 16 million less than the population of Tokyo, Japan. The population of Seoul was 10 million. Solve the equation $10 = t - 16$ to find the population of Tokyo.

33. ***WRITE ABOUT IT*** Suppose $n - 15$ is a whole number. What do you know about the value of n? Explain.

34. ***WHAT'S THE ERROR?*** Look at the student paper at right. What did the student do wrong? What is the correct answer?

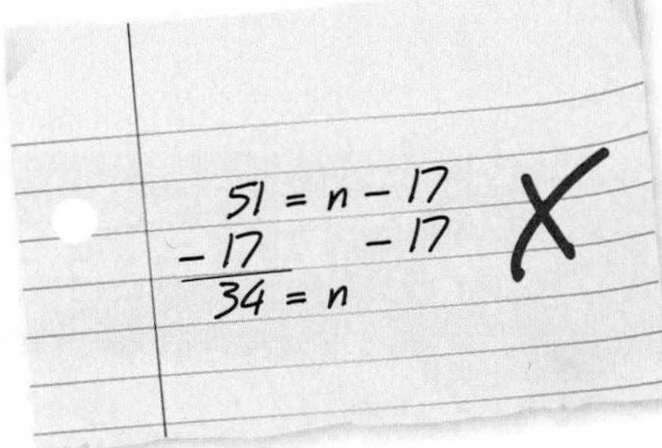

35. ***CHALLENGE*** Write "the difference between n and 16 is 5" as an algebraic equation. Then find the solution.

Spiral Review

Find each value. (Lesson 1-3)

36. 5^3 **37.** 9^2 **38.** 3^4 **39.** 12^2

Label each as a numerical expression, an algebraic expression, or an equation. (Lesson 1-4, 2-1)

40. $75 \div 3$ **41.** $x + 18 = 54$ **42.** $16 \times 12 + (8 - y)$

43. **EOG PREP** For which equation is $b = 8$ a solution? (Lesson 2-3)

A $13 - b = 8$ B $8 + b = 21$ C $b - 13 = 21$ D $b + 13 = 21$

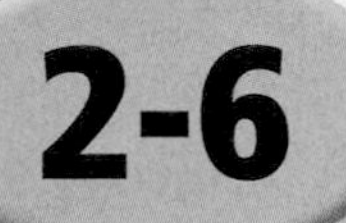

2-6 Solving Multiplication Equations

Learn to solve whole-number multiplication equations.

Armadillos are always born in groups of 4. If you count 32 babies, what is the number of mother armadillos?

To put together equal groups of 4, multiply. Let m represent the number of mother armadillos. There will be m equal groups of 4.

You can use the equation $4m = 32$.

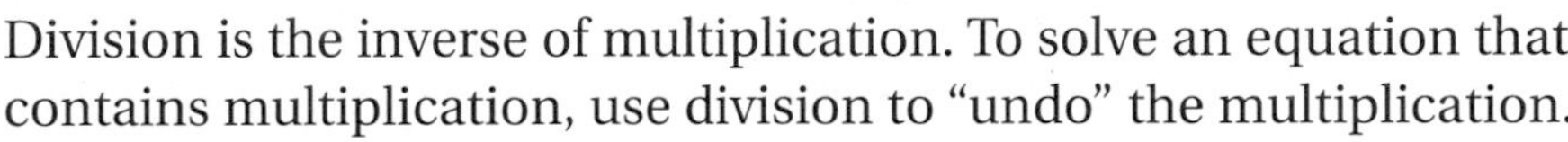

Division is the inverse of multiplication. To solve an equation that contains multiplication, use division to "undo" the multiplication.

Remember!

$4m$ means "$4 \times m$."

$$4m = 32$$
$$\frac{4m}{4} = \frac{32}{4}$$
$$m = 8$$

There are 8 mother armadillos.

EXAMPLE 1 Solving Multiplication Equations

Solve each equation. Check your answers.

A $3x = 12$

$3x = 12$ — *x is multiplied by 3.*

$\frac{3x}{3} = \frac{12}{3}$ — *Divide both sides by 3 to undo the multiplication.*

$x = 4$

Check $3x = 12$

$3(4) \stackrel{?}{=} 12$ — *Substitute 4 for x in the equation.*

$12 \stackrel{?}{=} 12$ ✔ — *4 is the solution.*

B $8 = 4w$

$8 = 4w$ — *w is multiplied by 4.*

$\frac{8}{4} = \frac{4w}{4}$ — *Divide both sides by 4 to undo the multiplication.*

$2 = w$

Check $8 = 4w$

$8 \stackrel{?}{=} 4(2)$ — *Substitute 2 for w in the equation.*

$8 \stackrel{?}{=} 8$ ✔ — *2 is the solution.*

EXAMPLE 2 PROBLEM SOLVING APPLICATION

The area of a rectangle is 36 square inches. Its length is 9 inches. What is its width?

1 Understand the Problem

The **answer** will be the width of the rectangle in inches.

List the **important information:**

- The area of the rectangle is 36 square inches.
- The length of the rectangle is 9 inches.

Draw a diagram to represent this information.

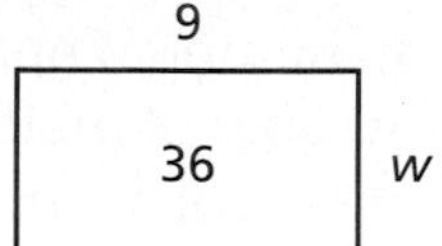

2 Make a Plan

You can write and solve an equation using the formula for area. To find the area of a rectangle, multiply its length by its width.

$$A = \ell w$$
$$36 = 9w$$

w

ℓ

3 Solve

$36 = 9w$ *w is multiplied by 9.*

$\frac{36}{9} = \frac{9w}{9}$ *Divide both sides by 9 to undo the multiplication.*

$4 = w$

So the width of the rectangle is 4 inches.

4 Look Back

Arrange 36 identical squares in a rectangle. The length is 9, so line up the squares in rows of 9. You can make 4 rows of 9, so the width of the rectangle is 4.

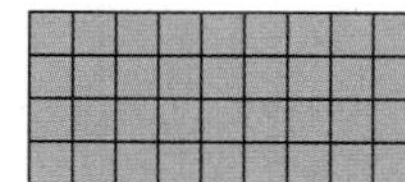

Think and Discuss

1. **Tell** what number you would use to divide both sides of the equation $15x = 60$.
2. **Tell** whether the solution of $10c = 90$ will be less than 90 or greater than 90. Explain.

2-6 Exercises

5.03

FOR EOG PRACTICE

see page 643

internet connect

Homework Help Online
go.hrw.com Keyword: MR4 2-6

GUIDED PRACTICE

See Example 1

Solve each equation. Check your answers.

1. $7x = 21$
2. $27 = 3w$
3. $90 = 10a$
4. $56 = 7b$
5. $3c = 33$
6. $12 = 2n$

See Example 2

7. The area of a rectangular deck is 675 square feet. The deck's width is 15 feet. What is its length?

INDEPENDENT PRACTICE

See Example 1

Solve each equation. Check your answers.

8. $12p = 36$
9. $52 = 13a$
10. $64 = 8n$
11. $20 = 5x$
12. $6r = 30$
13. $77 = 11t$
14. $14s = 98$
15. $12m = 132$
16. $9z = 135$

See Example 2

17. Colorado is almost a perfect rectangle on a map. Its border from east to west is 387 mi, and its area is 104,247 mi^2. Estimate the length of Colorado's border from north to south.

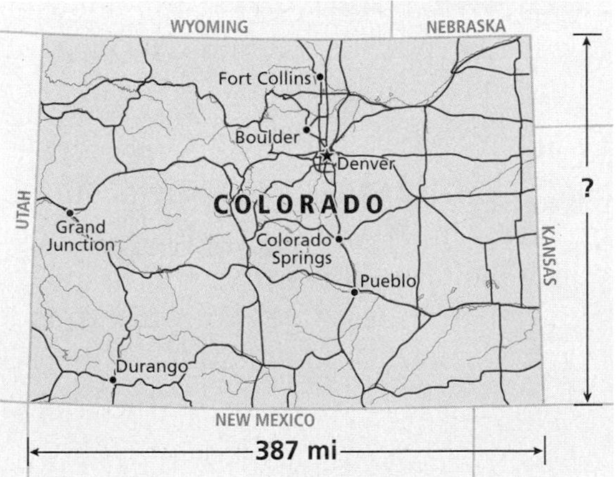

PRACTICE AND PROBLEM SOLVING

Solve each equation.

18. $5y = 35$
19. $18 = 2y$
20. $54 = 9y$
21. $15y = 120$
22. $4y = 0$
23. $22y = 440$
24. $3y = 63$
25. $z - 6 = 34$
26. $6y = 114$
27. $161 = 7y$
28. $135 = 3y$
29. $y - 15 = 3$
30. $81 = 9y$
31. $4 + y = 12$
32. $7y = 21$
33. $a + 12 = 26$
34. $10x = 120$
35. $36 = 12x$

Life Science LINK

This photo shows a horsefly magnified to twelve times its actual size.

Arthropods make up the largest group of animals on Earth. They include insects, spiders, crabs, and centipedes.

Arthropods have segmented bodies. In centipedes and millipedes, all of the segments are identical.

36. Centipedes have 2 legs per segment. They can have from 30 to 354 legs. Find a range for the number of segments a centipede can have.

37. Millipedes have 4 legs per segment. The record number of legs on a millipede is 752. How many segments did this millipede have?

Many arthropods have compound eyes. Compound eyes are made up of tiny bundles of identical light-sensitive cells.

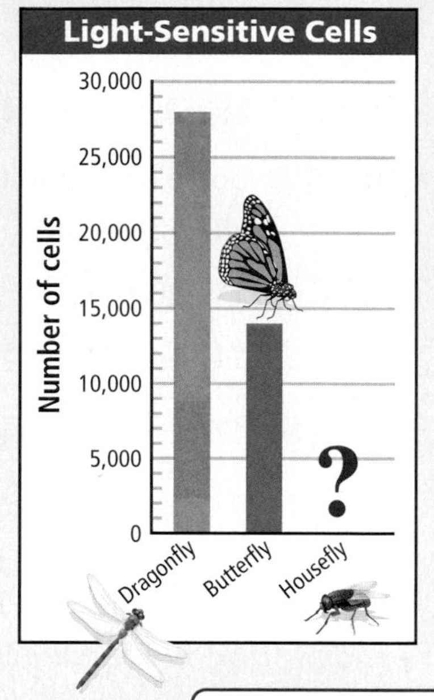

38. A dragonfly has 7 times as many light-sensitive cells as a housefly. How many of these cells does a housefly have?

39. Find how many times more light-sensitive cells a dragonfly has than a butterfly.

40. ***WRITE ABOUT IT*** A trapdoor spider can pull with a force that is 140 times its own weight. What other information would you need to find the spider's weight? Explain.

41. ***CHALLENGE*** There are about 6 billion humans in the world. Scientists estimate that there are a billion billion arthropods in the world. About how many times larger is the arthropod population than the human population?

go.hrw.com
KEYWORD: MR4 Arthropod
CNN Student News

Spiral Review

Compare. Write <, >, or =. (Lesson 1-1)

42. 56,902 ▒ 56,817 **43.** 14,562 ▒ 14,581 **44.** 1,240,518 ▒ 1,208,959

Evaluate each expression. (Lesson 1-4)

45. $7 + 3 \times 4 - 2$ **46.** $81 \div 3 - (5 + 6) \times 2$ **47.** $3^1 \times 7 + (18 - 5) \times 4$

48. **EOG PREP** Which value is a solution to the equation $n + 53 = 82$? (Lesson 2-3)

A 135 B 125 C 29 D 27

2-7 Solving Division Equations

Learn to solve whole-number division equations.

Japanese pearl divers go as deep as 165 feet underwater in search of pearls. At this depth, the pressure on a diver is much greater than at the water's surface. Water pressure can be described using equations containing division.

Multiplication is the inverse of division. When an equation contains division, use multiplication to "undo" the division.

EXAMPLE 1 Solving Division Equations

Solve each equation. Check your answers.

A $\frac{y}{5} = 4$

$\frac{y}{5} = 4$ — *y is divided by 5.*

$5 \cdot \frac{y}{5} = 5 \cdot 4$ — *Multiply both sides by 5 to undo the division.*

$y = 20$

Check

$\frac{y}{5} = 4$

$\frac{20}{5} \stackrel{?}{=} 4$ — *Substitute 20 for y in the equation.*

$4 \stackrel{?}{=} 4$ ✔ — *20 is the solution.*

B $12 = \frac{z}{4}$

$12 = \frac{z}{4}$ — *z is divided by 4.*

$4 \cdot 12 = 4 \cdot \frac{z}{4}$ — *Multiply both sides by 4 to undo the division.*

$48 = z$

Check

$12 = \frac{z}{4}$

$12 \stackrel{?}{=} \frac{48}{4}$ — *Substitute 48 for z in the equation.*

$12 \stackrel{?}{=} 12$ ✔ — *48 is the solution.*

EXAMPLE 2 *Physical Science Application*

Pressure is the amount of force exerted on an area. Pressure can be measured in pounds per square inch, or psi.

The pressure at the surface of the water is half the pressure at 30 ft underwater.

$$\text{pressure at surface} = \frac{\text{pressure at 30 ft underwater}}{2}$$

The pressure at the surface is 15 psi. What is the water pressure at 30 ft underwater?

Let p represent the pressure at 30 ft underwater.

$15 = \frac{p}{2}$ *Substitute 15 for pressure at the surface. p is divided by 2.*

$2 \cdot 15 = 2 \cdot \frac{p}{2}$ *Multiply both sides by 2 to undo the division.*

$30 = p$

The water pressure at 30 ft underwater is 30 psi.

Think and Discuss

1. **Tell** whether the solution of $\frac{c}{10} = 70$ will be less than 70 or greater than 70. Explain.
2. **Describe** how you would check your answer to Example 2.
3. **Explain** why $13 \cdot \frac{x}{13} = x$.

2-7 Exercises

FOR EOG PRACTICE
see page 643

internet connect
Homework Help Online
go.hrw.com Keyword: MR4 2-7

5.03

GUIDED PRACTICE

See Example 1

Solve each equation. Check your answers.

1. $\frac{y}{4} = 3$
2. $14 = \frac{z}{2}$
3. $\frac{r}{9} = 7$

See Example 2

4. Irene mowed the lawn and planted flowers. The amount of time she spent mowing the lawn was one-third the amount of time it took her to plant flowers. It took her 30 minutes to mow the lawn. Let p represent the amount of time she spent planting flowers. Find the amount of time Irene spent planting flowers.

INDEPENDENT PRACTICE

See Example 1

Solve each equation. Check your answers.

5. $\frac{d}{3} = 12$
6. $\frac{c}{2} = 13$
7. $7 = \frac{m}{7}$

See Example 2

8. The area of Danielle's garden is one-twelfth the area of her entire yard. The area of the garden is 10 square feet. Let y represent the area of the yard. Find the area of the yard.

PRACTICE AND PROBLEM SOLVING

Find the value of c in each equation.

9. $\frac{c}{12} = 8$
10. $4 = \frac{c}{9}$
11. $\frac{c}{15} = 11$
12. $14 = \frac{c}{5}$
13. $\frac{c}{4} = 12$
14. $\frac{c}{4} = 15$
15. $30 = \frac{c}{6}$
16. $49 = \frac{c}{3}$
17. $\frac{c}{24} = 18$

18. The Empire State Building is 381 m tall. At the Grand Canyon's widest point, 76 Empire State Buildings would fit end to end. Write and solve an equation to find the width of the Grand Canyon at this point.

19. ***EARTH SCIENCE*** You can find the distance of a thunderstorm in kilometers by counting the number of seconds between the lightning flash and the thunder and then dividing this number by 3. If a storm is 5 km away, how many seconds will you count between the lightning flash and the thunder?

20. ***WRITE A PROBLEM*** Write a problem about money that can be solved with a division equation.

21. ***WRITE ABOUT IT*** Use a numerical example to explain how multiplication and division undo each other.

22. ***CHALLENGE*** Let m represent the amount of money Janine had on Monday. She spent half of it on Tuesday and half of what was left on Wednesday. On Thursday she had \$2. How much money did she have on Monday?

Spiral Review

Identify a pattern in each sequence and name the next two terms. (Lesson 1-7)

23. 2, 6, 18, 54, ▒, ▒, …
24. 2, 5, 9, 14, 20, 27, ▒, ▒, …

Evaluate each expression for the given value of the variable. (Lesson 2-1)

25. $2y + 6$ for $y = 4$
26. $\frac{z}{5}$ for $z = 40$
27. $7r - 3$ for $r = 18$

28. **EOG PREP** Which is an algebraic expression for the product of y and 4? (Lesson 2-2)

A $4y$
B $4 + y$
C $\frac{y}{4}$
D $y - 4$

EXTENSION

Inequalities

Learn to solve and graph whole-number inequalities.

Vocabulary

inequality

Reading Math

$<$ means "is less than."
$>$ means "is greater than."
$\leq$ means "is less than or equal to."
$\geq$ means "is greater than or equal to."

An **inequality** is a statement that two quantities are not equal.

$15 > 3$ $\quad 12 \leq 29 \quad$ $41 \geq 18$ $\quad 17 < 90$

An inequality may contain a variable, as in the inequality $x > 3$. Values of the variable that make the inequality true are solutions of the inequality.

x	$x > 3$	Solution?
0	$0 \overset{?}{>} 3$	No; 0 is **not** greater than 3, so 0 is not a solution.
3	$3 \overset{?}{>} 3$	No; 3 is **not** greater than 3, so 3 is not a solution.
4	$4 \overset{?}{>} 3$	Yes; 4 is greater than 3, so 4 is a solution.
12	$12 \overset{?}{>} 3$	Yes; 12 is greater than 3, so 12 is a solution.

This table shows that an inequality may have more than one solution. You can use a number line to show all of the solutions.

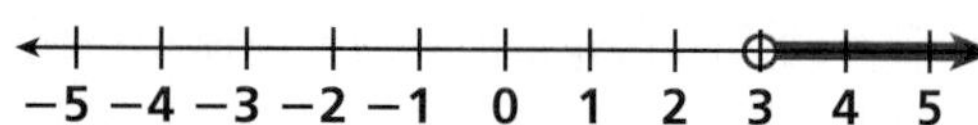

EXAMPLE 1 Graphing Inequalities

Graph the solutions to $w \leq 4$ on a number line.

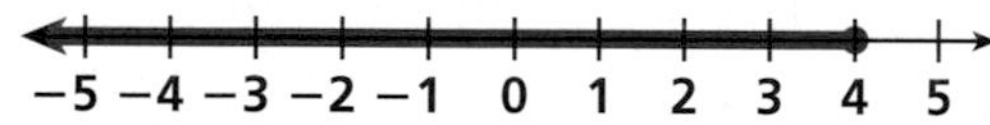

The closed circle on the point 4 shows that 4 is a solution.

You can solve inequalities in the same way that you solved equations.

EXAMPLE 2 Solving and Graphing Inequalities

Solve each inequality. Graph the solutions on a number line.

A $y + 7 < 9$

$$\begin{aligned} y + 7 &< 9 && \textit{7 is added to y.} \\ \underline{-7} &\quad \underline{-7} && \textit{Subtract 7 from both sides to undo the addition.} \\ y &< 2 \end{aligned}$$

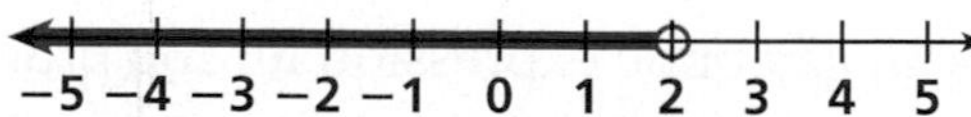

The open circle on the point 2 shows that 2 is not a solution.

Solve each inequality. Graph the solutions on a number line.

B $2m \geq 12$

$2m \geq 12$ — *m is multiplied by 2.*

$\frac{2m}{2} \geq \frac{12}{2}$ — *Divide both sides by 2 to undo the multiplication.*

$m \geq 6$

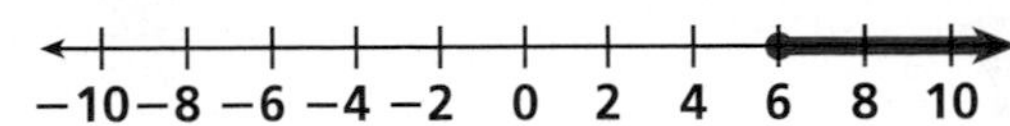

The closed circle on the point 6 shows that 6 is a solution.

EXTENSION

Exercises

Graph the solutions to each inequality on a number line.

1. $w \leq 0$

2. $x > 5$

3. $z \geq 9$

4. $7 < t$

5. $m > 2$

6. $4 \geq q$

7. $a \leq 8$

8. $6 > x$

9. $y < 3$

Solve each inequality.

10. $3t \leq 27$

11. $y - 5 \geq 0$

12. $x + 4 < 10$

13. $2c > 2$

14. $\frac{d}{6} \geq 1$

15. $r + 9 \leq 23$

16. $15n < 75$

17. $4 + r \leq 7$

18. $f - 11 > 16$

19. $2k < 8$

Write an inequality for each sentence. Then graph your inequality.

20. c is less than or equal to two.

21. p is greater than 11.

22. At some lakes, people who fish must throw back any trout that is less than 10 inches long. Write an inequality that represents the lengths of trout that may be kept.

23. ***GEOGRAPHY*** Mt. McKinley is the highest point in the United States, with an altitude of 20,320 ft. Let a be the altitude of any other U.S. location. Write an inequality relating a to Mt. McKinley's altitude.

24. ***WHAT'S THE ERROR?*** A student graphed $x > 1$ as shown. What did the student do wrong? Draw the correct graph.

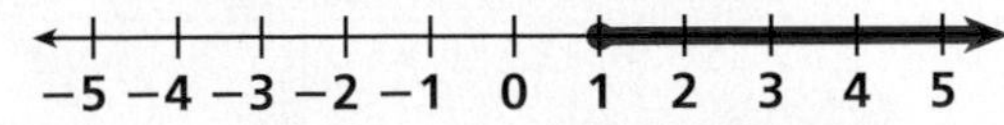

Problem Solving on Location

NORTH CAROLINA

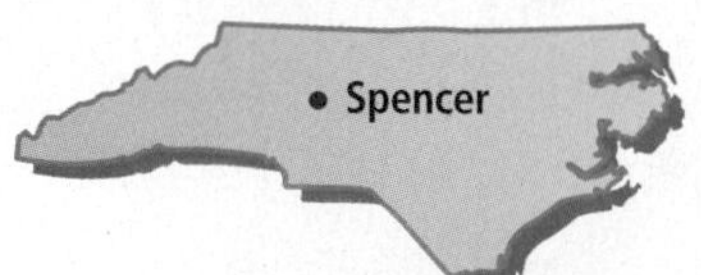

Golf

Many North Carolina residents enjoy playing a round of golf on one of the state's approximately 600 golf courses. When you play a round of golf, you play the entire length of the course. Golf courses have different lengths, and the shortest courses are not always the easiest.

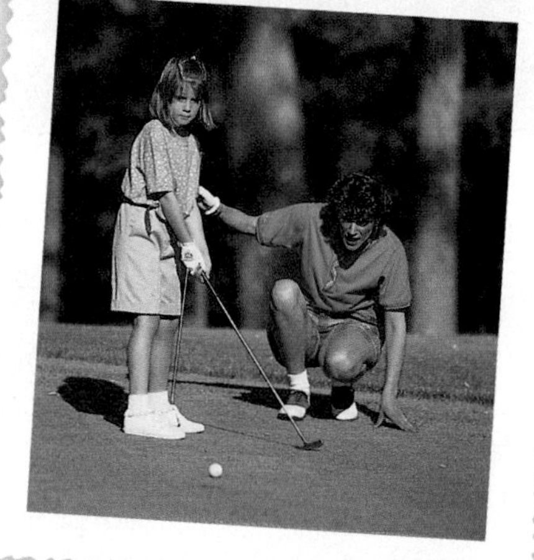

For 1–3, use the table.

1. Ray is playing a round of golf at Charlotte Golf Links and has played the first five holes, which total 1,789 yards. Write and solve an equation to find the distance remaining on the course that Ray must play.
2. One day, Sara played a round of golf at Oak Hollow Golf Course. The next day, she played a round at another course listed in the table. She played a total of 12,264 yards over the two days. Let *s* represent the length of the second course. Write and solve an equation to find the length of the second course. On which course did she play on the second day?
3. Pete has played many rounds of golf at the Baywood Golf Club. Over the past few years, he has played 87,919 yards at this course. Let *r* represent the number of rounds that Pete has played at the Baywood Golf Club. Write and solve an equation to find the value of *r*.

Course	Course Length (yd)
Baywood Golf Club	6,763
Charlotte Golf Links	6,700
McCanless Golf Course	5,700
Oak Hollow Golf Course	6,564
Pine Needles Golf Club	6,708

North Carolina Transportation Museum

The North Carolina Transportation Museum, in Spencer, is located on the former site of the Southern Railway Company's largest steam-locomotive servicing facility. Exhibits include antique automobiles and a 120,000-square-foot locomotive roundhouse.

Another exhibit at the museum is a Conestoga wagon. Before there were railroads, Conestoga wagons were used to transport large amounts of goods, such as farm produce, iron ore, and charcoal.

1. Most Conestoga wagons weighed about 1,600 pounds. Suppose that a wagon carrying 5,000 pounds of cargo was pulled by 6 horses. Let w be the weight pulled by each horse.
   a. Write an equation that you can use to find the weight pulled by each horse.
   b. What is the value of w?

Spencer is located midway between Atlanta, Georgia, and Washington, D.C. The table shows the driving distances to Spencer from several North Carolina cities.

Driving Distances to Spencer, NC	
City	Distance (mi)
Asheville	147
Charlotte	36
Greensboro	32
Raleigh	110
Wilmington	254
Winston-Salem	35

For 2–3, use the table.

2. Suppose you visited the museum 9 times in one year. If you drove a total of 576 miles, which city did you come from?
3. If you drive 55 miles per hour, how long will it take you to make a round-trip, to the museum and back, from Raleigh?

MATH-ABLES

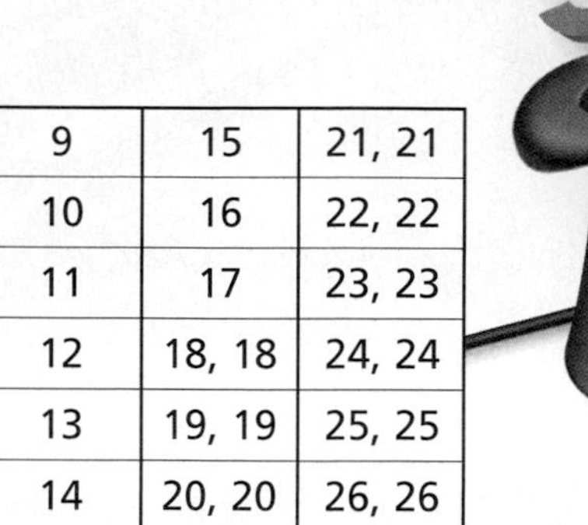

Math Magic

Guess what your friends are thinking with this math magic trick.

Copy the following number charts.

1	10	19
2, 2	11, 11	20, 20
4	13	22
5, 5	14, 14	23, 23
7	16	25
8, 8	17, 17	26, 26

3	12	21
4	13	22
5	14	23
6, 6	15, 15	24, 24
7, 7	16, 16	25, 25
8, 8	17, 17	26, 26

9	15	21, 21
10	16	22, 22
11	17	23, 23
12	18, 18	24, 24
13	19, 19	25, 25
14	20, 20	26, 26

Step 1: Ask a friend to think of a number from 1 to 26.
Example: Your friend thinks of 26.

Step 2: Show your friend the first chart and ask how many times the chosen number appears. Remember the answer.
Your friend says the chosen number appears twice on the first chart. 2

Step 3: Show the second chart and ask the same question. Multiply the answer by 3. Add your result to the answer from step 2. Remember this answer.
Your friend says the chosen number appears twice. $3 \cdot 2 = 6$
The answer from step 2 is 2. $6 + 2 = 8$

Step 4: Show the third chart and ask the same question. Multiply the answer by 9. Add your result to the answer from step 3. The answer is your friend's number.
Your friend says the chosen number appears twice. $9 \cdot 2 = 18$
The answer from step 3 is 8. $18 + 8 = 26$ ← *Your friend's number*

How does it work?

Your friend's number will be the following:

(answer from step 2) + (3 · answer from step 3) + (9 · answer from step 4)

This is an expression with three variables: $a + 3b + 9c$. A number will be on a particular chart 0, 1, or 2 times, so a, b, and c will always be 0, 1, or 2. With these values, you can write expressions for each number from 1 to 26.

a	b	c	$a + 3b + 9c$
1	0	0	$1 + 3(0) + 9(0) = \mathbf{1}$
2	0	0	$2 + 3(0) + 9(0) = \mathbf{2}$
0	1	0	$0 + 3(1) + 9(0) = \mathbf{3}$

Can you complete the table for 4–26?

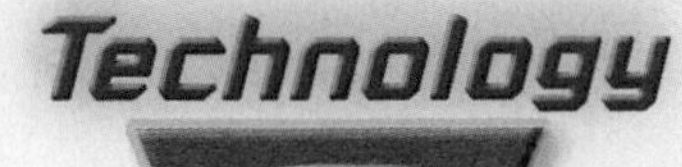

Evaluate Expressions

internet connect
Lab Resources Online
go.hrw.com
KEYWORD: MR4 TechLab2

You can use a graphing calculator to evaluate algebraic expressions. A graphing calculator is especially helpful when you are evaluating an expression for many values of the variable.

Activity

Evaluate $4x + 5$ for $x = 0, 1, 2, 3, 4, 5, 6, 7, 8, 9,$ and 10.

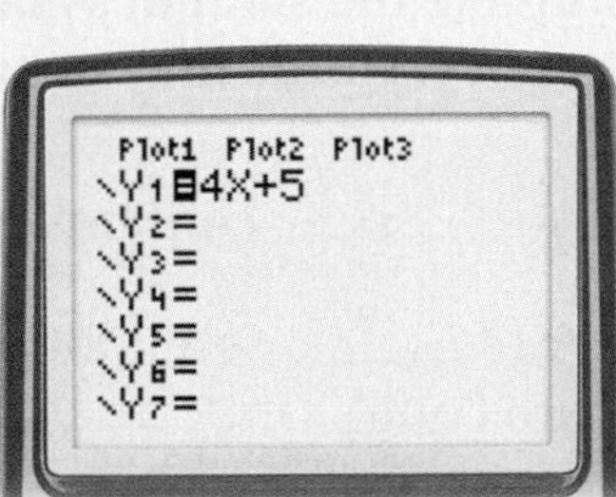

1 Enter $4x + 5$ into the **Y=** menu. Y= 4 X,T,θ,n + 5.

2 Press 2nd WINDOW (TBLSET) to access the **TABLE SETUP** menu. **TblStart** tells which value of x the table should begin with. Make sure this is set to 0, because 0 is the smallest value of x for which you must evaluate the expression. **ΔTbl** gives the difference between successive x-values. **ΔTbl = 1** means that the x-values in the table will increase by 1.

3 Press 2nd GRAPH (TABLE) to see the table. The **Y1** column shows the value of $4x + 5$ for several x-values. For example, when $x = 0$, $4x + 5 = 5$. Use the arrow keys to find the answers for other x-values.

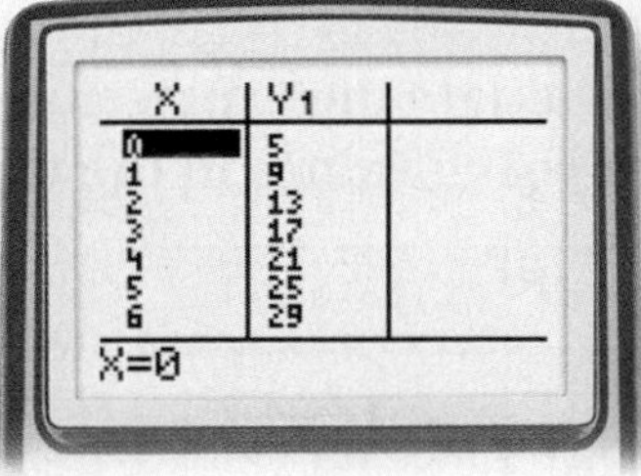

Think and Discuss

1. In the table, one row has 5 in the **X** column and 25 in the **Y1** column. What does this mean?

2. When does $y = 41$?

3. Would you use a calculator to evaluate $x + 2$ for $x = 5$? Explain why or why not.

Try This

Evaluate each expression for the given x-values.

1. $3x + 8$; $x = 4, 5,$ and 9

2. $45x + 67$; $x = 8, 10,$ and 12

3. $4x + 7$; $x = 0, 1, 2, 3, 4,$ and 5

4. $30 + 25x$; $x = 5, 10, 15,$ and 20

Chapter 2

Study Guide and Review

Vocabulary

Complete the sentences below with vocabulary words from the list above. Words may be used more than once.

1. A(n) __?__ contains one or more variables.
2. A(n) __?__ is a mathematical statement that says two quantities are equal.
3. In the equation $12 + t = 22$, t is a __?__.
4. A(n) __?__ is a quantity that does not change.

2-1 Variables and Expressions (pp. 48–51)

EXAMPLE

■ **Evaluate the expression to find the missing values in the table.**

n	$3n + 4$
1	7
2	■
3	■

$n = 1$ $3 \times 1 + 4 = 7$
$n = 2$ $3 \times 2 + 4 = 10$
$n = 3$ $3 \times 3 + 4 = 13$

The missing values are 10 and 13.

■ **Find an expression for the table.**

x	■
4	16
5	20
6	24

$4 \cdot 4 = 16$
$5 \cdot 4 = 20$
$6 \cdot 4 = 24$

An expression is $x \cdot 4$, or $4x$.

EXERCISES

Evaluate each expression to find the missing values in the tables.

5.

y	$y \div 7$
56	8
49	■
42	■

6.

k	$k \times 4 - 6$
2	2
3	■
4	■

Find an expression for each table.

7.

p	■
9	54
10	60
11	66

8.

s	■
18	9
36	18
48	24

2-2 Translate Between Words and Math (pp. 52–55)

EXAMPLES

Write each phrase as a numerical or algebraic expression.

- 617 minus 191
  $617 - 191$
- d multiplied by 5
  $5d$ or $5 \cdot d$ or $(5)(d)$

Write two phrases for each expression.

- $a \div 5$
  - a divided by 5
  - the quotient of a and 5
- $67 + 19$
  - the sum of 67 and 19
  - 19 more than 67

EXERCISES

Write each phrase as a numerical or algebraic expression.

9. 15 plus b

10. the product of 6 and 5

11. 9 times t

12. the quotient of g and 9

Write two phrases for each expression.

13. $4z$

14. $54 \div 6$

15. $3 - y$

16. $y - 3$

17. $15 + x$

18. $\frac{m}{20}$

19. $5{,}100 + 64$

2-3 Equations and Their Solutions (pp. 58–61)

EXAMPLE

- **Determine whether the given value of the variable is a solution.**

  $f + 14 = 50$ for $f = 34$

  $34 + 14 \stackrel{?}{=} 50$ *Substitute 34 for f.*

  $48 \neq 50$ *Add.*

  34 is not a solution.

EXERCISES

Determine whether the given value of each variable is a solution.

20. $28 + n = 39$ for $n = 11$

21. $12t = 74$ for $t = 6$

22. $y - 53 = 27$ for $y = 80$

23. $96 \div w = 32$ for $w = 3$

24. $15x = 90$ for $x = 8$

25. $x - 61 = 17$ for $x = 75$

2-4 Solving Addition Equations (pp. 62–65)

EXAMPLE

- **Solve the equation $x + 18 = 31$.**

$$\begin{aligned} x + 18 &= 31 \quad \text{18 is added to x.} \\ -18 &\quad -18 \quad \text{Subtract 18 from both sides to undo the addition.} \\ x &= 13 \end{aligned}$$

EXERCISES

Solve each equation.

26. $4 + x = 10$

27. $n + 10 = 24$

28. $c + 71 = 100$

29. $y + 16 = 22$

30. $44 = p + 17$

31. $94 + w = 103$

32. $23 + b = 34$

33. $56 = n + 12$

34. $39 = 23 + p$

35. $d + 28 = 85$

2-5 Solving Subtraction Equations (pp. 66–68)

EXAMPLE

■ **Solve the equation.**

$$\begin{aligned} c - 7 &= 16 \\ +7 \quad & \quad +7 \\ c \quad &= 23 \end{aligned}$$

7 is subtracted from c.
Add 7 to each side to undo the subtraction.

EXERCISES

Solve each equation.

36. $28 = k - 17$
37. $d - 8 = 1$
38. $p - 55 = 8$
39. $n - 31 = 36$
40. $3 = r - 11$
41. $97 = w - 47$
42. $12 = h - 48$
43. $9 = p - 158$

2-6 Solving Multiplication Equations (pp. 69–72)

EXAMPLE

■ **Solve the equation.**

$6x = 36$ *x is multiplied by 6.*

$\frac{6x}{6} = \frac{36}{6}$ *Divide both sides by 6 to undo the multiplication.*

$x = 6$

EXERCISES

Solve each equation.

44. $5v = 40$
45. $27 = 3y$
46. $12c = 84$
47. $18n = 36$
48. $72 = 9s$
49. $11t = 110$
50. $7a = 56$
51. $8y = 64$

2-7 Solving Division Equations (pp. 73–75)

EXAMPLE

■ **Solve the equation.**

$\frac{k}{4} = 8$ *k is divided by 4.*

$4 \cdot \frac{k}{4} = 4 \cdot 8$ *Multiply both sides by 4 to undo the division.*

$k = 32$

EXERCISES

Solve each equation.

52. $\frac{r}{7} = 6$
53. $\frac{t}{5} = 3$
54. $6 = \frac{y}{3}$
55. $12 = \frac{n}{6}$
56. $\frac{z}{13} = 4$
57. $20 = \frac{b}{5}$
58. $\frac{n}{11} = 7$
59. $10 = \frac{p}{9}$

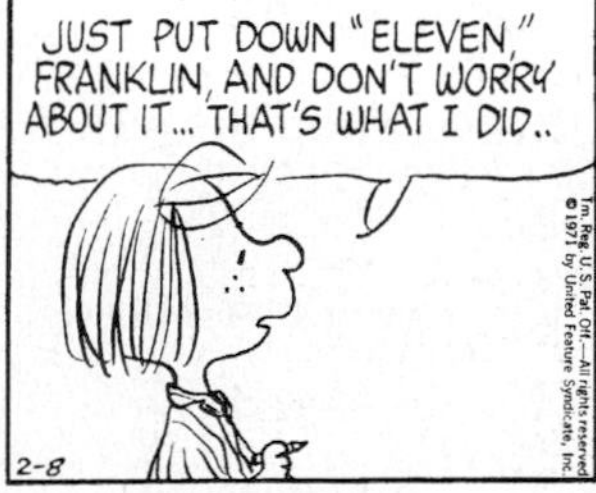

Chapter 2

Chapter Test

Evaluate each expression to find the missing values in the tables.

1.

a	$a + 18$
10	28
12	■
14	■

2.

y	$y \div 6$
18	3
30	■
42	■

3.

n	$n \div 5 + 7$
10	9
20	■
30	■

Find an expression for each table.

4.

s	■
36	3
48	4
60	5

5.

t	■
6	30
7	35
8	40

6.

b	■
100	79
75	54
50	29

7. There are more reptile species than amphibian species. Let n represent the number of living reptile species. There are 3,100 living species of amphibians. Write an expression to show how many more reptile species there are than amphibian species.

Write each phrase as a numerical or algebraic expression.

8. 26 more than n
9. g multiplied by 4
10. the quotient of 180 and 15
11. the difference of 100 and 17

Write two phrases for each expression.

12. (14)(16)
13. $n \div 8$
14. $p + 11$

Determine whether the given value of the variable is a solution.

15. $5d = 70$ for $d = 12$
16. $29 = 76 - n$ for $n = 46$
17. $108 \div a = 12$ for $a = 9$
18. $15 + m = 27$ for $m = 12$

Solve each equation. Check your answers.

19. $a + 7 = 25$
20. $121 = 11d$
21. $3 = t - 8$
22. $6 = \frac{k}{9}$
23. Air typically has about 4,000 bacteria per cubic meter. If your bedroom is 30 cubic meters, about how many bacteria would you expect there to be in the air in your bedroom?

Chapter 2

Performance Assessment

Show What You Know

Create a portfolio of your work from this chapter. Complete this page and include it with your four best pieces of work from Chapter 2. Choose from your homework or lab assignments, mid-chapter quiz, or any journal entries you have done. Put them together using any design you want. Make your portfolio represent what you consider your best work.

Short Response

1. Write the phrase "the quotient of n and 6" as an algebraic expression. Then evaluate the expression for $n = 72$.

2. The Morgans went on a car trip. They averaged 60 miles per hour. If h represents the number of hours they drove, write an expression for the distance the Morgans traveled in h hours.

3. Determine whether $x = 13$ is a solution for the equation $3x = 36$. If it is not a solution, solve the equation to find the value of x.

Extended Problem Solving

4. Three cities are located along the same highway. The distance between Artsville and Charlestown is 35 miles. Artsville is 18 miles from Burgston.

Pine Lake
9 mi
Deville
Charlestown
35 mi
d
Burgston
18 mi
Artsville

a. Let d represent the distance between Burgston and Charlestown. Write an equation that could be used to find the value of d.

b. To find the distance between Burgston and Charlestown, solve the equation you wrote in part **a.** Show your work.

c. An almanac states that Charlestown is 29,920 yards from Burgston. Is this the same as the distance you found in part **b**? Explain your answer in words. (*Hint:* There are 1,760 yards in a mile.)

internet connect
EOG Test Practice Online
go.hrw.com Keyword: MR4 TestPrep

Chapter 2

Getting Ready for EOG

Cumulative Assessment, Chapters 1–2

1. Which is a solution to the equation $8a = 48$?

A 8
B 7
C 6
D 5

2. What is the value of 9^1?

A 0
B 1
C 9
D 19

3. Find the missing value in the table.

n	$6 \times n - 7$
7	35
8	■

A 41
B 36
C 6
D 55

4. Which means "the quotient of t and 6"?

A $6 \div t$
B $\frac{t}{6}$
C $6t$
D $6 + t$

5. What is the value of $3^3 - (15 \div 3) \times 2$?

A 8
B 44
C 1
D 17

6. Which number is *greatest*?

A 12,301,542
B 12,381,536
C 12,311,518
D 12,385,501

7. Nicole is 15 years old. She is 3 years younger than her sister Jackie. Solve the equation $j - 3 = 15$ to find Jackie's age.

A 18
B 17
C 12
D 5

TIP!

TEST TAKING TIP!

Substitute the given value in each equation. See which value gives a correct equation.

8. Which of the following has a solution of 7?

A $p + 14 = 20$
B $7p = 42$
C $\frac{p}{4} = 7$
D $p - 2 = 5$

9. Michael plans to enter a biking and running competition. The total race course is 19 miles long. The map shows the course and the distance Michael will run. How many miles will Michael have to bike?

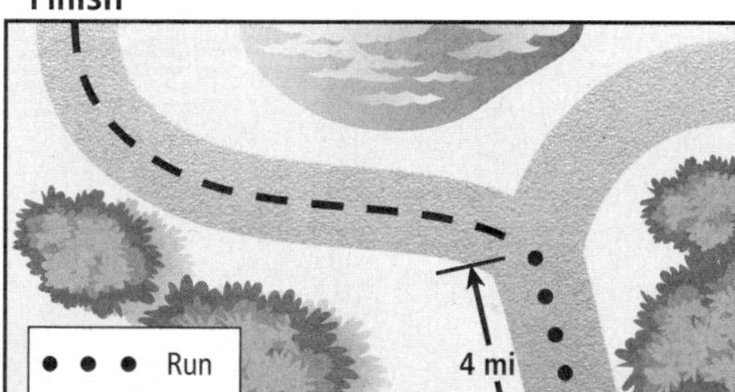

A 4 miles
B 15 miles
C 19 miles
D 24 miles

10. ***SHORT RESPONSE*** Show the steps to solve the equation $x + 5 = 11$. Explain why the only solution is $x = 6$.

11. ***SHORT RESPONSE*** A student said the solution of the equation $\frac{x}{7} = 14$ is $x = 2$. Explain the student's error and show the steps to solve the equation correctly.

Chapter 3

Decimals

TABLE OF CONTENTS

internet connect
Chapter Opener Online
go.hrw.com
KEYWORD: MR4 Ch3

Winning Olympic Performances

Year	Women's 100 Meters (s)	Women's Discus (m)	Men's 100 Meters (s)	Men's Discus (m)
1900	–	36.04	12.0	–
1928	12.2	39.62	10.8	47.32
1952	11.5	51.4	10.4	55.02
1988	10.54	72.3	9.92	68.81
2000	10.75	68.4	9.87	69.29

Career *Sports Historian*

Are people breaking records by running faster and jumping farther and higher? Records are kept for both professional and amateur sports. Many schools keep records of their individual athletes' and teams' performances. Keeping track of sports records is the job of sports historians. One of the most complete records is that of the Olympic games. The table shows the changes in the last century of the winning performances in some men's and women's Olympic sports.

ARE YOU READY?

Choose the best term from the list to complete each sentence.

1. In the metric system, the base unit for measuring length is the ___?___, and the base unit for measuring volume is the ___?___.
2. In the expression 72 ÷ 9, 72 is the ___?___, and 9 is the ___?___.
3. The answer to a subtraction expression is the ___?___.
4. A(n) ___?___ is a mathematical statement that says two quantities are equal.

difference
dividend
divisor
equation
liter
meter
quotient

Complete these exercises to review skills you will need for this chapter.

✔ Place Value of Whole Numbers

Identify the place value of each underlined digit.

5. $1\underline{5}2$ **6.** $\underline{7},903$ **7.** $\underline{1}45,072$

8. $4,8\underline{9}3,025$ **9.** $1\underline{3},796,020$ **10.** $1\underline{4}5,683,032$

✔ Add and Subtract Whole Numbers

Find each sum or difference.

11. \$425 − \$75 **12.** 532 + 145 **13.** 160 − 82

✔ Multiply and Divide Whole Numbers

Find each product or quotient.

14. \$320 × 5 **15.** 125 ÷ 5 **16.** 54 × 3

✔ Exponents

Find each value.

17. 10^3 **18.** 3^6 **19.** 10^5

20. 4^5 **21.** 8^3 **22.** 2^7

✔ Solve Whole Number Equations

Solve each equation.

23. $y + 382 = 743$ **24.** $n - 150 = 322$ **25.** $9x = 108$

Model Decimals

Use with Lesson 3-1

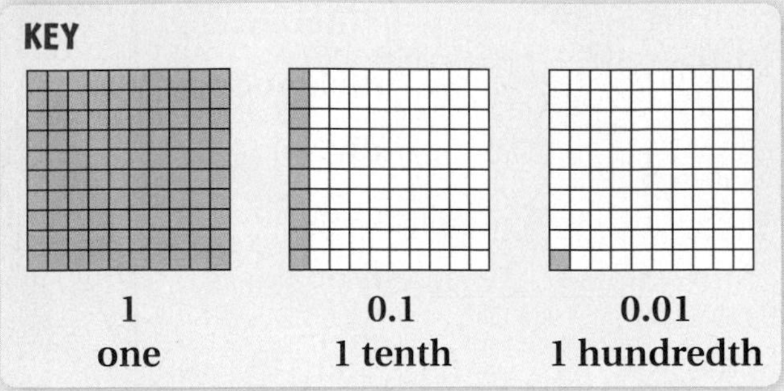

internet connect
Lab Resources Online
go.hrw.com
KEYWORD: MR4 Lab3A

You can use decimal grids to represent decimals. The grid is divided into 100 small squares. One square represents 1 hundredth, or 0.01. Ten squares form a column, which represents 1 tenth, or 0.1. Ten columns make up the grid, which represents one whole, or 1. By shading hundredths, tenths, or whole grids, you can represent decimal numbers.

Activity 1

1 Write the decimal that is represented.

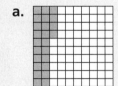

a. *24 hundredths squares are shaded.*

So 0.24 is shaded.

b. *1 whole grid and 8 columns are shaded.*

So 1.8 is shaded.

c. 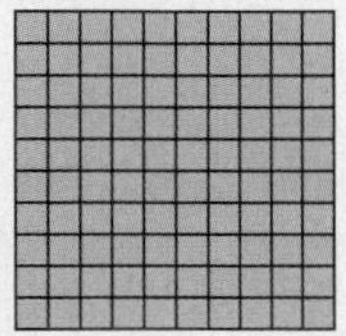 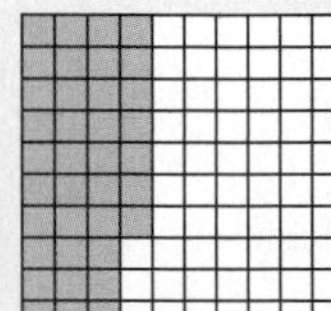

2 whole grids and 37 hundredths are shaded.

So 2.37 is shaded.

Think and Discuss

1. Tell how decimal grids show that 0.30 = 0.3.

Try This

Write the decimal that is represented.

1. 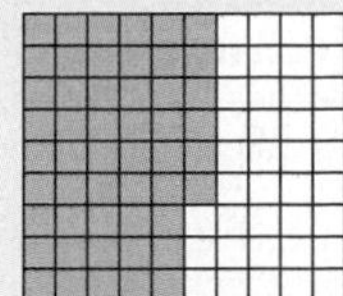

2. 

3. 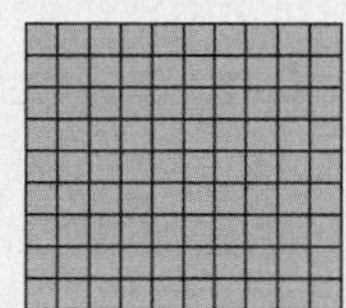 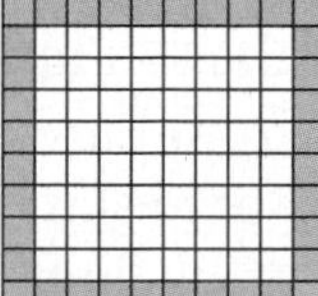

Activity 2

1 Use a decimal grid to represent each decimal.

a. 0.42

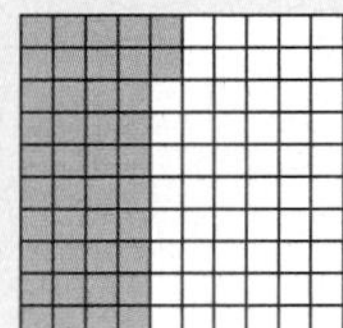

Shade 42 hundredths squares.

b. 1.88

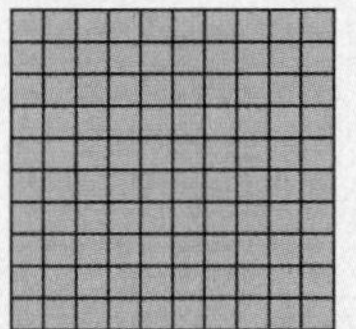 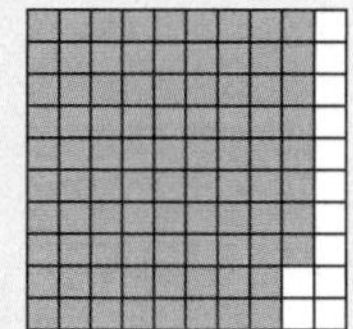

Shade 1 whole grid, 8 columns, and 8 small squares.

c. 2.75

  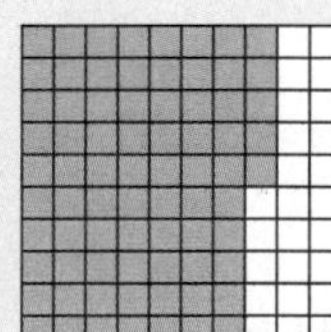

Shade 2 whole grids, 7 columns, and 5 small squares.

Think and Discuss

1. Explain how to represent 0.46 by shading only 10 sections on the grid. (*Hint:* A section is a grid, column, or small square.)

Try This

Use a decimal grid to represent each decimal.

1. 1.02 **2.** 0.04 **3.** 0.4 **4.** 2.14 **5.** 0.53

3-1 Representing, Comparing, and Ordering Decimals

Learn to write, compare, and order decimals using place value and number lines.

The smaller the apparent magnitude of a star, the brighter the star appears when viewed from Earth. The magnitudes of some stars are listed in the table as decimal numbers.

Apparent Magnitudes of Stars	
Star	**Magnitude**
Procyon	0.38
Proxima Centauri	11.0
Wolf 359	13.5
Vega	0.03

Decimal numbers represent combinations of whole numbers and numbers between whole numbers.

Place value can help you understand and write decimal numbers.

Place Value

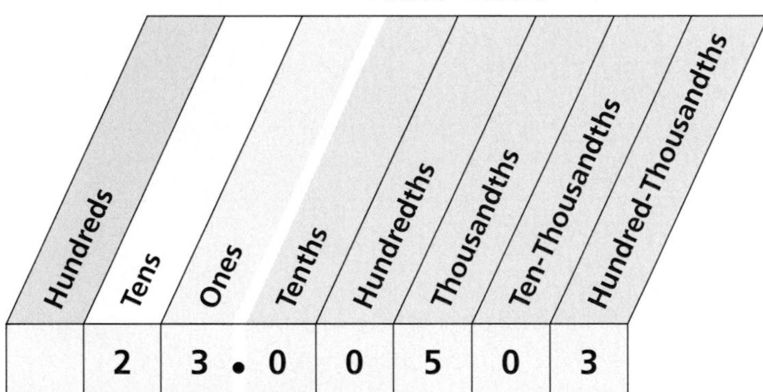

EXAMPLE 1 Reading and Writing Decimals

Write each decimal in standard form, expanded form, and words.

A 1.05

Expanded form: 1 + 0.05
Word form: one *and* five hundredths

Reading Math

Read the decimal point as "and."

B 0.05 + 0.001 + 0.0007

Standard form: 0.0517
Word form: five hundred seventeen ten-thousandths

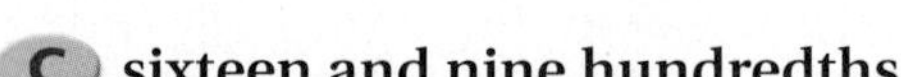

C sixteen and nine hundredths

Standard form: 16.09
Expanded form: 10 + 6 + 0.09

You can use place value to compare decimal numbers.

EXAMPLE 2 *Earth Science Application*

Rigel and Betelgeuse are two stars in the constellation Orion. The apparent magnitude of Rigel is 0.12. The apparent magnitude of Betelgeuse is 0.50. Which star has the smaller magnitude? Which star appears brighter?

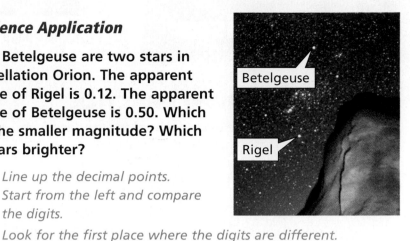

0.12	*Line up the decimal points.* *Start from the left and compare the digits.*
0.50	*Look for the first place where the digits are different.*

1 is less than 5.
0.12 < 0.50

Rigel has a smaller apparent magnitude than Betelgeuse.
The star with the smaller magnitude appears brighter. When seen from Earth, Rigel appears brighter than Betelgeuse.

EXAMPLE 3 Comparing and Ordering Decimals

Helpful Hint

Writing zeros at the end of a decimal does not change the value of the decimal.

0.3 = 0.30 = 0.300

Order the decimals from least to greatest.
14.35, 14.3, 14.05

14.35 14.30	14.30 < 14.35	*Compare two of the numbers at a time. Write 14.3 as "14.30."*
14.35 14.05	14.05 < 14.35	*Start at the left and compare the digits.*
14.30 14.05	14.05 < 14.30	*Look for the first place where the digits are different.*

Graph the numbers on a number line.

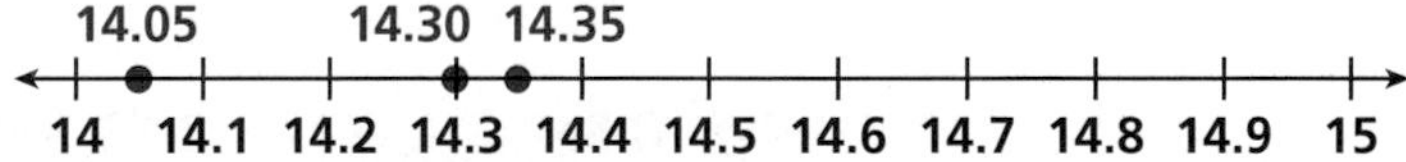

The numbers are ordered when you read the number line from left to right. The numbers in order from least to greatest are 14.05, 14.3, and 14.35.

Think and Discuss

1. **Explain** why 0.5 is greater than 0.29 even though 29 is greater than 5.
2. **Name** the decimal with the least value.
   0.29, 2.09, 2.009, 0.029
3. **Name** three numbers between 1.5 and 1.6.

3-1 Exercises

FOR EOG PRACTICE

see page 644

internet connect
Homework Help Online
go.hrw.com Keyword: MR4 3-1

1.03

GUIDED PRACTICE

See Example 1 **Write each decimal in standard form, expanded form, and words.**

1. 1.98

2. ten and forty-one thousandths

3. 0.07 + 0.006 + 0.0005

4. 0.0472

See Example 2 **5.** Osmium and iridium are precious metals. The density of osmium is 22.58 g/cm^3, and the density of iridium is 22.56 g/cm^3. Which metal is denser?

See Example 3 **Order the decimals from least to greatest.**

6. 9.5, 9.35, 9.65

7. 4.18, 4.1, 4.09

8. 12.39, 12.09, 12.92

INDEPENDENT PRACTICE

See Example 1 **Write each decimal in standard form, expanded form, and words.**

9. 7.0893

10. 12 + 0.2 + 0.005

11. seven and fifteen hundredths

12. 3 + 0.1 + 0.006

See Example 2 **13.** Two meteorites landed in Mexico. The one found in Bacuberito weighed 24.3 tons, and the one found in Chupaderos weighed 26.7 tons. Which meteorite weighed more?

See Example 3 **Order the decimals from least to greatest.**

14. 15.25, 15.2, 15.5

15. 1.56, 1.62, 1.5

16. 6.7, 6.07, 6.23

PRACTICE AND PROBLEM SOLVING

Write each number in words.

17. 9.007

18. 5 + 0.08 + 0.004

19. 10.022

Compare. Write <, >, or =.

20. 8.04 ▢ 8.403

21. 0.907 ▢ 0.6801

22. 1.246 ▢ 1.29

23. one and fifty-two ten-thousandths ▢ 1.0052

Write the value of the red digit in each number.

24. 3.026

25. 17.53703

26. 0.000598

27. 425.1055

Order the numbers from greatest to least.

28. 32.525, 32.5254, 31.6257

29. 0.34, 1.43, 4.034, 1.043, 1.424

Astronomy LINK

Proxima Centauri, the closest star to Earth other than the Sun, was discovered in 1913. It would take about 115,000 years for a spaceship traveling from Earth at 25,000 mi/h to reach Proxima Centauri.

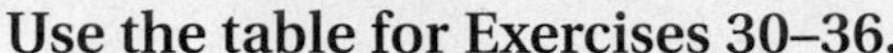

Use the table for Exercises 30–36.

30. Order the stars Sirius, Luyten 726-8, and Lalande 21185 from closest to farthest from Earth.

31. Which star in the table is farthest from Earth?

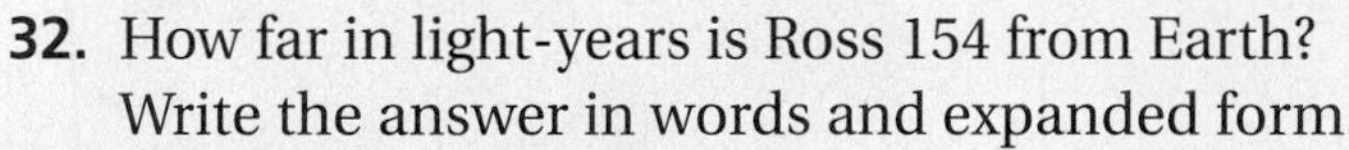

32. How far in light-years is Ross 154 from Earth? Write the answer in words and expanded form.

33. List the stars that are less than 5 light-years from Earth.

34. ***WHAT'S THE ERROR?*** A student wrote the distance of Proxima Centauri from Earth as "four hundred and twenty-two hundredths." Explain the error. Write the correct answer.

35. ***WRITE ABOUT IT*** Which star is closer to Earth, Sirius or Lalande 21185? Explain how you can compare the distances of these stars. Then answer the question.

Distance of Stars from Earth

Star	Distance (light-years)
Alpha Centauri	4.35
Barnard's Star	5.98
Lalande 21185	8.22
Luyten 726-8	8.43
Proxima Centauri	4.22
Ross 154	9.45
Sirius	8.65

36. ***CHALLENGE*** Wolf 359 is located 7.75 light-years from Earth. If the stars in the table were listed in order from closest to farthest from Earth, between which two stars would Wolf 359 be located?

Spiral Review

Compare. Write $<$, $>$, or $=$. (Lesson 1-1)

37. 4,897,204 ■ 4,895,190 **38.** 133,099,588 ■ 133,099,600

Write each expression in exponential form. (Lesson 1-3)

39. $3 \times 3 \times 3 \times 3 \times 3$ **40.** $10 \times 10 \times 10 \times 10$ **41.** $13 \times 13 \times 13$

42. **EOG PREP** For which of the following equations is 8 a solution? (Lesson 2-6)

A $7n = 63$ B $24 + p = 32$ C $t - 5 = 13$ D $\frac{a}{7} = 8$

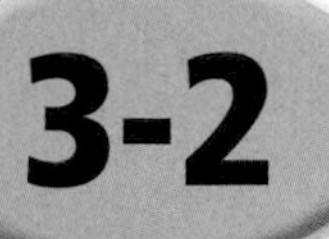

3-2 Estimating Decimals

Learn to estimate decimal sums, differences, products, and quotients.

Vocabulary
clustering
front-end estimation

Beth's health class is learning about fitness and nutrition. The table shows the approximate number of calories burned by someone who weighs 90 pounds.

Activity (45 min)	Calories Burned (App.)
Cycling	198.45
Playing ice hockey	210.6
Rowing	324
Water skiing	194.4

When numbers are about the same value, you can use *clustering* to estimate. **Clustering** means rounding the numbers to the same value.

EXAMPLE  1 ***Health Application***

Beth wants to cycle, play ice hockey, and water ski. If Beth weighs 90 pounds and spends 45 minutes doing each activity, *about* how many calories will she burn in all?

198.45	→	200	*The addends cluster around 200.*
210.6	→	200	*To estimate the total number of calories,*
+ 194.4	→	+ 200	*round each addend to 200.*
		600	*Add.*

Beth burns about 600 calories.

EXAMPLE  2 **Rounding Decimals to Estimate Sums and Differences**

Remember!

When rounding, look at the digit to the right of the place to which you are rounding.
- If it is *5 or greater,* round *up.*
- If it is *less than 5,* round *down.*

Estimate by rounding to the indicated place value.

A **3.92 + 6.48; ones**

3.92 + 6.48 — *Round to the nearest whole number.*

4 + 6 = 10 — *The sum is about 10.*

B **8.6355 − 5.039; hundredths**

8.6355	8.64	*Round to the hundredths.*
− 5.039	− 5.04	*Align the decimals.*
	3.60	*Subtract.*

EXAMPLE 3 Using Compatible Numbers to Estimate Products and Quotients

Remember!

Compatible numbers are close to the numbers that are in the problem and are helpful when you are solving the problem mentally.

Estimate each product or quotient.

A 26.76×2.93

$25 \times 3 = 75$ *25 and 3 are compatible.*

So 26.76×2.93 is about 75.

B $42.64 \div 16.51$

$45 \div 15 = 3$ *45 and 15 are compatible.*

So $42.64 \div 16.51$ is about 3.

You can also use *front-end estimation* to estimate with decimals. **Front-end estimation** means to use only the whole-number part of the decimal.

EXAMPLE 4 Using Front-End Estimation

Estimate a range for the sum.

$9.99 + 22.89 + 8.3$

Use front-end estimation.

9.99	→	9	*Add the whole numbers only.*
22.89	→	22	*The whole-number values of the decimals*
+ 8.30	→	+ 8	*are less than the actual numbers, so the*
		at least 39	*answer is an underestimate.*

The exact answer of $9.99 + 22.89 + 8.3$ is 39 or greater.

You can estimate a range for the sum by adjusting the decimal part of the numbers. Round the decimals to 0, 0.5, or 1.

0.99	→	1.00	*Add the decimal part of the numbers.*
0.89	→	1.00	*Add the whole-number estimate and the*
+ 0.30	→	+ 0.50	*adjusted estimate.*
		2.50	*The adjusted decimals are greater than the*
39.00 + 2.50 = 41.50			*actual decimals, so 41.50 is an overestimate.*

The estimated range for the sum is from 39.00 to 41.50.

Think and Discuss

1. **Tell** what number the following decimals cluster around: 34.5, 36.78, and 35.234.
2. **Determine** whether a front-end estimation without adjustment is always an overestimation or an underestimation.

3-2 Exercises

FOR EOG PRACTICE
see page 644

internet connect
Homework Help Online
go.hrw.com Keyword: MR4 3-2

1.04c, 1.07

GUIDED PRACTICE

See Example 1

1. Elba runs every Monday, Wednesday, and Friday. Last week she ran 3.62 miles on Monday, 3.8 miles on Wednesday, and 4.3 miles on Friday. About how many miles did she run last week?

See Example 2

Estimate by rounding to the indicated place value.

2. 2.746 − 0.866; tenths
3. 6.735 + 4.9528; ones
4. 10.8071 + 5.392; hundredths
5. 5.982 − 0.4832; tenths

See Example 3

Estimate each product or quotient.

6. 38.92 ÷ 4.06
7. 14.51 × 7.89
8. 22.47 ÷ 3.22

See Example 4

Estimate a range for each sum.

9. 7.8 + 31.39 + 6.95
10. 14.27 + 5.4 + 21.86

INDEPENDENT PRACTICE

See Example 1

11. Before Mike's trip, the odometer in his car read 146.8 miles. He drove 167.5 miles to a friend's house and 153.9 miles to the beach. About how many miles did the odometer read when he arrived at the beach?

12. The rainfall in July, August, and September was 16.76 cm, 13.97 cm, and 15.24 cm, respectively. About how many total centimeters of rain fell during those three months?

See Example 2

Estimate by rounding to the indicated place value.

13. 2.0993 + 1.256; tenths
14. 7.504 − 2.3792; hundredths
15. 0.6271 + 4.53027; thousandths
16. 13.274 − 8.5590; tenths

See Example 3

Estimate each product or quotient.

17. 9.64 × 1.769
18. 11.509 ÷ 4.258
19. 19.03 ÷ 2.705

See Example 4

Estimate a range for each sum.

20. 17.563 + 4.5 + 2.31
21. 1.620 + 10.8 + 3.71

PRACTICE AND PROBLEM SOLVING

Estimate by rounding to the nearest whole number.

22. 8.456 + 7.903
23. 12.43 × 3.72
24. 1,576.2 − 150.50
25. Estimate the quotient of 67.25 and 3.83.
26. Estimate $79.45 divided by 17.

Use the table for Exercises 27–31.

Long-Distance Costs for Callers in the United States	
Country	Cost per Minute (¢)
Venezuela	22
Russia	9.9
Japan	7.9
United States	3.7

27. ***MONEY*** Round each cost in the table to the nearest cent. Write your answer using a dollar sign and decimal point.

28. About how much does it cost to phone someone in Russia and talk for 8 minutes?

29. About how much more does it cost to make a 12-minute call to Japan than to make an 18-minute call within the United States?

30. Will the cost of a 30-minute call to someone within the United States be greater or less than $1.20? Explain.

31. Kim is in New York. She calls her grandmother in Venezuela and speaks for 20 minutes, then calls a friend in Japan and talks for 15 minutes, and finally calls her mother in San Francisco and talks for 30 minutes. Estimate the total cost of all her calls.

32. ***HEALTH*** The recommended daily allowance (RDA) for iron is 15 mg/day for teenage girls. Julie eats a hamburger that contains 3.88 mg of iron. About how many more milligrams of iron does she need to meet the RDA? (Round to the nearest whole number.)

33. ***WRITE A PROBLEM*** Write a problem with three decimal numbers that have a total sum between 30 and 32.5.

34. ***WRITE ABOUT IT*** How do you adjust a front-end estimation? Why is this done?

35. ***CHALLENGE*** Place a decimal point in each number so that the sum of the numbers is between 124 and 127.
1059 + 725 + 815 + 1263

Spiral Review

Estimate each sum or difference by rounding to the place value indicated. (Lesson 1-2)

36. 6,319 + 13,804; thousands

37. 25,680 − 18,502; ten thousands

Evaluate each expression. (Lesson 1-4)

38. $15 + 18 \div 6$

39. $4^2 + 19 \times 2 - 30$

40. $(26 - 14) \times 2^3 - 14 \div 2$

41. **EOG PREP** Which of the following is an example of the Distributive Property? (Lesson 1-5)

A $(8 + 2) + 6 = 8 + (2 + 6)$

B $6 \times 3 \times 7 = 7 \times 6 \times 3$

C $3 \times (5 + 4) = (3 \times 5) + (3 \times 4)$

D $6 \times 1 = 1 \times 6$

Explore Decimal Addition and Subtraction

Use with Lesson 3-3

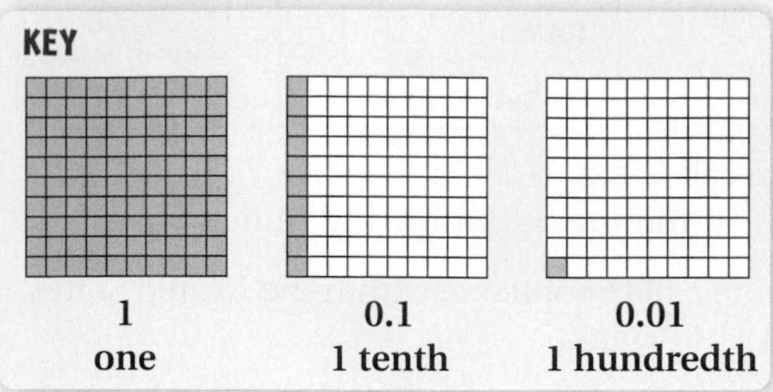

You can model addition and subtraction of decimals with decimal grids.

Activity 1

1. Use decimal grids to find each sum.

a. 0.24 + 0.32

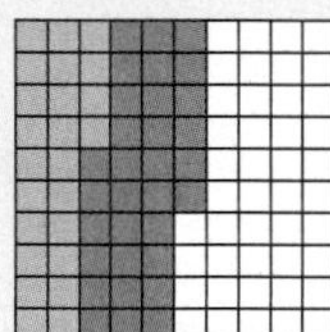

To represent 0.24, shade 24 squares.

To represent 0.32, shade 32 squares in another color.

There are 56 shaded squares representing 0.56.

0.24 + 0.32 = 0.56

b. 1.56 + 0.4

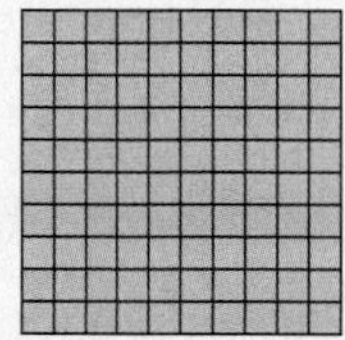

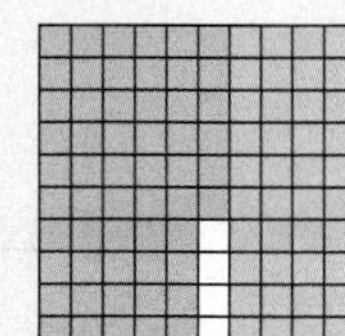

To represent 1.56, shade a whole grid and 56 squares of another.

To represent 0.4, shade 4 columns in another color.

One whole grid and 96 squares are shaded.

1.56 + 0.4 = 1.96

c. 0.75 + 0.68

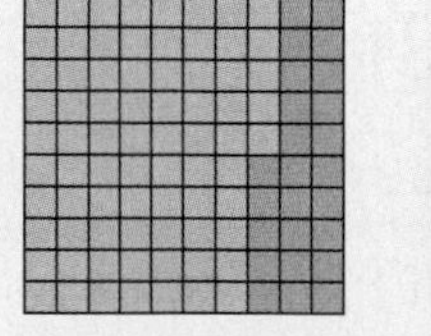

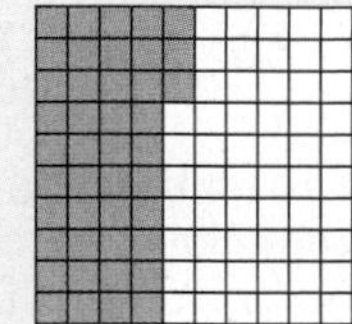

To represent 0.75, shade 75 squares.

To represent 0.68, shade 68 squares in another color. You will need to use another grid.

One whole grid and 43 squares are shaded.

0.75 + 0.68 = 1.43

Think and Discuss

1. How would you shade a decimal grid to represent 0.2 + 0.18?

Try This

Use decimal grids to find each sum.

1. 0.2 + 0.6	**2.** 1.07 + 0.03	**3.** 1.62 + 0.08
4. 0.45 + 0.29	**5.** 0.88 + 0.12	**6.** 1.29 + 0.67
7. 0.07 + 0.41	**8.** 0.51 + 0.51	**9.** 1.01 + 0.23

Activity 2

1 Use a decimal grid to find each difference.

a. 0.6 − 0.38

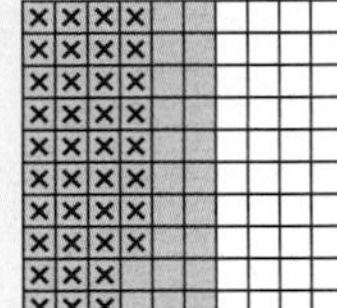

To represent 0.6, shade 6 columns.

Subtract 0.38 by removing 38 squares.

There are 22 remaining squares.

0.6 − 0.38 = 0.22

b. 1.22 − 0.41

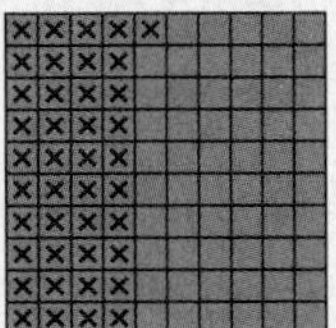

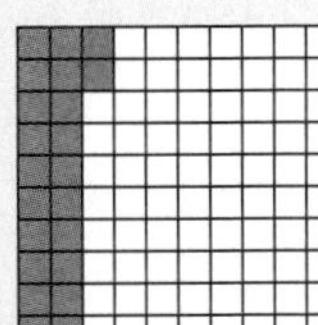

To represent 1.22, shade an entire decimal grid and 22 squares of another.

Subtract 0.41 by removing 41 squares.

There are 81 remaining squares.

1.22 − 0.41 = 0.81

Think and Discuss

1. How would you shade a decimal grid to represent 1.3 − 0.6?

Try This

Use decimal grids to find each difference.

1. 0.9 − 0.3	**2.** 1.2 − 0.98	**3.** 0.6 − 0.41
4. 1.6 − 0.07	**5.** 0.35 − 0.03	**6.** 2.12 − 0.23
7. 2.0− 0.86	**8.** 0.78 − 0.76	**9.** 1.06 − 0.55

3-3 Adding and Subtracting Decimals

Learn to add and subtract decimals.

American gymnast Elise Ray won the 2000 U.S. Championships in the all-around, uneven bars, and floor-exercise events.

Elise Ray's Scores

Event	Points
Floor exercise	9.8
Balance beam	9.7
Vault	9.425
Uneven bars	9.85

Elise Ray competed in the 2000 Summer Olympics in Sydney, Australia.

To find the total number of points, you can add all of the scores.

EXAMPLE 1 *Sports Application*

A What was Elise Ray's total all-around score in the 2000 U.S. Championships?

Find the sum of 9.8, 9.7, 9.425, and 9.85.

$$\begin{array}{ccccccc} 9.8 & + & 9.7 & + & 9.425 & + & 9.85 \\ \downarrow & & \downarrow & & \downarrow & & \downarrow \\ 10 & + & 10 & + & 9 & + & 10 = 39 \end{array}$$

Estimate by rounding to the nearest whole number. The total is about 39 points.

Add.

$$\begin{array}{r} 9.800 \\ 9.700 \\ 9.425 \\ +\ 9.850 \\ \hline 38.775 \end{array}$$

Align the decimal points.

Use zeros as placeholders.

Add. Then place the decimal point.

Since 38.775 is close to the estimate of 39, the answer is reasonable. Elise Ray's total all-around score was 38.775 points.

B How many more points did Elise need on the uneven bars to have a perfect score of 10?

Find the difference between 10 and 9.85.

$$\begin{array}{r} 10.00 \\ -\ 9.85 \\ \hline 0.15 \end{array}$$

Align the decimal points.
Use zeros as placeholders.
Subtract. Then place the decimal point.

Elise needed another 0.15 points to have a perfect score.

Helpful Hint

Estimating before you add or subtract will help you check whether your answer is reasonable.

EXAMPLE 2 Using Mental Math to Add and Subtract Decimals

Find each sum or difference.

A $1.6 + 0.4$

$1.6 + 0.4$ *Think: 0.6 + 0.4 = 1*

$1.6 + 0.4 = 2$

B $3 - 0.8$

$3 - 0.8$ *Think: What number added to*

$3 - 0.8 = 2.2$ *0.8 is 1? 0.8 + 0.2 = 1*

So 1 − 0.8 = 0.2.

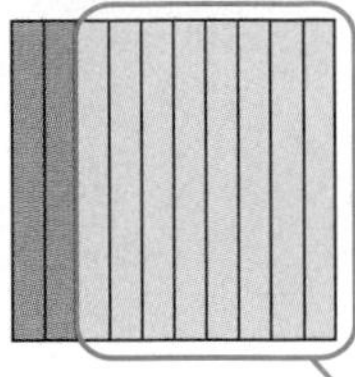

EXAMPLE 3 Evaluating Decimal Expressions

Remember!

You can place any number of zeros at the end of a decimal number without changing its value.

Evaluate 7.52 − *s* for each value of *s*.

A $s = 2.9$

$7.52 - s$

$7.52 - \mathbf{2.9}$ *Substitute 2.9 for s.*

$$\begin{array}{r} 7.52 \\ -\ 2.90 \\ \hline 4.62 \end{array}$$

Align the decimal points.
Use a zero as a placeholder.
Subtract.
Place the decimal point.

B $s = 4.5367$

$7.52 - s$

$7.52 - \mathbf{4.5367}$ *Substitute 4.5367 for s.*

$$\begin{array}{r} 7.5200 \\ -\ 4.5367 \\ \hline 2.9833 \end{array}$$

Align the decimal points.
Use zeros as placeholders.
Subtract.
Place the decimal point.

Think and Discuss

1. **Show** how you would write 2.678 + 124.5 to find the sum.
2. **Tell** why it is a good idea to estimate the answer before you add and subtract.
3. **Explain** how you can use mental math to find how many more points Elise Ray would have needed to have scored a perfect 10 on the floor exercise.

3-3 Exercises

1.04a, c, d, 5.02

FOR EOG PRACTICE
see page 645

internet connect
Homework Help Online
go.hrw.com Keyword: MR4 3-3

GUIDED PRACTICE

See Example 1 **Use the table for Exercises 1 and 2.**

1. How many miles in all is Rea's triathlon training?

2. a. How many miles did Rea run and swim in all?

b. How much farther did Rea cycle than swim?

Rea's Triathlon Training	
Sport	**Distance (mi)**
Cycling	14.25
Running	4.35
Swimming	1.6

See Example 2 **Find each sum or difference.**

3. 2.7 + 0.3 **4.** 6 − 0.4 **5.** 5.2 + 2.8 **6.** 8.9 − 4

See Example 3 **Evaluate 5.35 − *m* for each value of *m*.**

7. $m = 2.37$ **8.** $m = 1.8$ **9.** $m = 4.7612$ **10.** $m = 0.402$

INDEPENDENT PRACTICE

See Example 1

11. During a diving competition, Phil performed two reverse dives and two dives from a handstand position. He received the following scores: 8.765, 9.45, 9.875, and 8.025. What was Phil's total score?

12. Brad works after school at a local grocery store. How much did he earn in all for the month of October?

Brad's Earnings for October				
Week	1	2	3	4
Earnings	$123.48	$165.18	$137.80	$140.92

See Example 2 **Find each sum or difference.**

13. 7.2 + 1.8 **14.** 8.5 − 7 **15.** 3.3 + 0.7 **16.** 15.9 + 2.1

17. 7 − 0.6 **18.** 7.55 − 3.25 **19.** 21.4 + 3.6 **20.** 5 − 2.7

See Example 3 **Evaluate 9.67 − *x* for each value of *x*.**

21. $x = 1.52$ **22.** $x = 3.8$ **23.** $x = 7.21$ **24.** $x = 0.635$

PRACTICE AND PROBLEM SOLVING

Add or subtract.

25. 5.62 + 4.19 **26.** 10.508 − 6.73 **27.** 13.009 + 12.83

28. Find the sum of 0.0679 and 3.75. **29.** Subtract 3.0042 from 7.435.

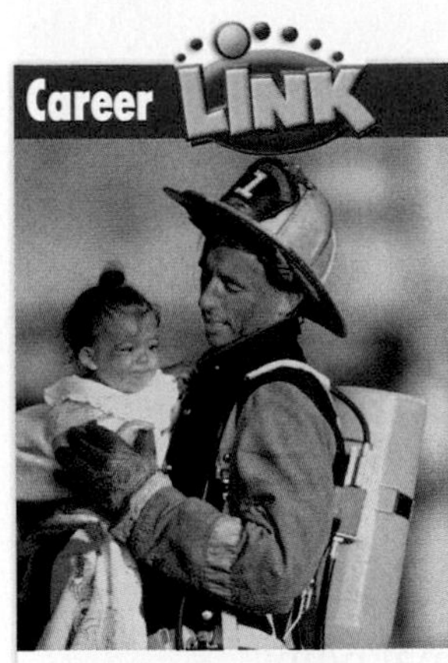

The risk of death in a house fire can be reduced by up to 50% if the home has a working smoke alarm.

Evaluate each expression.

30. $8.09 - a$ for $a = 4.5$

31. $7.03 + 33.8 + n$ for $n = 12.006$

32. $b + (5.68 - 3.007)$ for $b = 6.134$

33. $(2 \times 14) - a + 1.438$ for $a = 0.062$

34. $5^2 - w$ for $w = 3.5$

35. $100 - p$ for $p = 15.034$

36. ***CAREERS*** A fire helmet must be sturdy enough to protect the firefighter's head from dangerous objects and extremely hot temperatures while still being as lightweight as possible. One fire helmet weighs 1.616 kg, and another fire helmet weighs 1.403 kg. What is the difference in weights?

37. ***MONEY*** Logan wants to buy a new bike that costs $135.00. He started with $14.83 in his savings account. Last week, he deposited $15.35 into his account. Today, he deposited $32.40. How much more money does he need to buy the bike?

38. ***SPORTS*** With a time of 60.35 seconds, Martina Moracova broke Jennifer Thompson's world record time in the women's 100-meter medley. How much faster was Thompson than Moracova when, in the next heat, she reclaimed the record with a time of 59.30 seconds?

39. ***SPORTS*** The highest career batting average ever achieved by a professional baseball player is 0.366. Bill Bergen finished with a career 0.170 average. How much lower is Bergen's career average than the highest career average?

40. ***WHAT'S THE QUESTION?*** A cup of rice contains 0.8 mg of iron, and a cup of lima beans contains 4.4 mg of iron. If the answer is 6 mg, what is the question?

41. ***WRITE ABOUT IT*** Why is it important to align the decimal points before adding or subtracting decimal numbers?

42. ***CHALLENGE*** Evaluate $(5.7 + a) \times (9.75 - b)$ for $a = 2.3$ and $b = 7.25$.

Spiral Review

Find each value. (Lesson 1-3)

43. 7^3 **44.** 4^3 **45.** 10^5 **46.** 12^2 **47.** $10^2 + 2^3$

Find an expression for each table. (Lesson 2-1)

48.

x	2	4	9
■	7	17	42

49.

m	3	5	8
■	14	22	34

50. **EOG PREP** Which algebraic expression is the same as "the quotient of n divided by 6"? (Lesson 2-2)

A $6n$ B $\frac{n}{6}$ C $n - 6$ D $6 + n$

3-4 Decimals and Metric Measurement

Learn to multiply and divide decimals by powers of ten and to convert metric measurements.

You know that in a place-value chart each place value is ten times greater than the place value to its right.

The number of zeros in the power of ten, or the exponent in the power of ten, tells you how many places to move the decimal point.

EXAMPLE 1 Multiplying and Dividing by Powers of Ten

Multiply or divide.

A $4,325 \times 1,000$

4,325.000 — *There are 3 zeros in 1,000.*
To multiply, move the decimal point 3 places right.
= 4,325,000 — *Write 3 placeholder zeros.*

B $4,325 \div 1,000$

4,325. — *There are 3 zeros in 1,000.*
To divide, move the decimal point 3 places left.
= 4.325

C $79.95 \div 10^4$

0079.95 — *The power of 10 is 4.*
Move the decimal point 4 places left.
= 0.007995 — *Write placeholder zeros.*

Reading Math

Prefixes:
kilo: thousand
centi: hundredth
milli: thousandth

The metric system uses powers of ten. The base unit for length is the *meter,* for mass the *gram,* and for capacity the *liter*. Prefixes are used to describe units that are greater or smaller than the base unit.

	Unit	Abbreviation	Approximate Comparison
Length	**Kilo**meter	km	Length of 10 football fields
	Meter	m	Width of a door
	Centimeter	cm	Width of your little finger
	Millimeter	mm	Thickness of a dime
Mass	**Kilo**gram	kg	Mass of a textbook
	Gram	g	Mass of a small paperclip
Capacity	Liter	L	Filled bottle of sparkling water
	Milliliter	mL	Half-filled eyedropper

EXAMPLE 2 Choosing Appropriate Units

Use the abbreviation for the most appropriate metric unit.

A **A pencil is about 15 __?__ long.**

Think: A pencil is the length of about 15 little-finger widths.

A pencil is about 15 cm long.

B **The mass of an average man is about 75 __?__.**

Think: A man has a mass of about 75 textbooks.

The mass of a man is about 75 kg.

C **A pail holds about 20 __?__.**

Think: A pail could hold about 20 bottles of water.

A pail holds about 20 L.

In the metric system, each unit of measure is ten times greater than the unit to its right in a place-value chart.

1,000	100	10	1	0.1	0.01	0.001
Thousands	Hundreds	Tens	Ones	Tenths	Hundredths	Thousandths
Kilo-	Hecto-	Deca-	Base unit	Deci-	Centi-	Milli-

To convert units within the metric system, multiply or divide by powers of ten.

EXAMPLE 3 Converting Within the Metric System

Helpful Hint

To convert to smaller units, *multiply.*
To convert to larger units, *divide.*

Convert each measure.

A **The mass of a backpack is about 6,500 g. 6,500 g = __?__ kg**

6,500 g = (6,500 ÷ **1,000**) kg — *1 kg = 1,000 g, so divide by 1,000.*

6,500 g = 6.5 kg — *Move the decimal point 3 places left.*

B **A glass holds about 0.3 L of milk. 0.3 L = __?__ mL**

0.3 L = (0.3 × **1,000**) mL — *1 L = 1,000 mL, so multiply by 1,000.*

0.3 L = 300 mL — *Move the decimal point 3 places right.*

Think and Discuss

1. **Determine** whether any measure in meters can be converted to millimeters.
2. **Tell** how you know whether to multiply or divide when converting units of measure.

3-4 Exercises

FOR EOG PRACTICE

see page 645

internet connect

Homework Help Online
go.hrw.com Keyword: MR4 3-4

1.07, 2.01

GUIDED PRACTICE

See Example 1 **Multiply or divide.**

1. 5,937 × 100
2. $719.25 \div 10^3$
3. 6.0912×10^5

See Example 2 **Use the abbreviation for the most appropriate metric unit.**

4. A piece of paper is about 28 __?__ long.
5. A carton of orange juice holds about 1 __?__.
6. A pencil weighs about 5 __?__.
7. A small glass holds about 250 __?__.

See Example 3 **Convert each measure.**

8. The mass of a cat is about 7 kg. 7 kg = __?__ g
9. A race is about 5,000 m long. 5,000 m = __?__ km
10. A container holds about 0.5 L of solution. 0.5 L = __?__ mL
11. A picture frame is about 18 cm long. 18 cm = __?__ m

INDEPENDENT PRACTICE

See Example 1 **Multiply or divide.**

12. 278 × 1,000
13. 15.09×10^3
14. 810.381 ÷ 100
15. 74.1×10^4
16. $381.8 \div 10^5$
17. 42,516 ÷ 10,000

See Example 2 **Use the abbreviation for the most appropriate metric unit.**

18. A television screen is about 68 __?__ wide.
19. A boy is about 1.5 __?__ tall.
20. A box of cereal weighs about 540 __?__.
21. A spoon holds about 2 __?__.
22. A quarter is about 2.5 __?__ wide.

See Example 3 **Convert each measure.**

23. A sneaker is about 25 cm long. 25 cm = __?__ m
24. A container holds about 2 L of milk. 2 L = __?__ mL
25. A pen is about 0.18 m long. 0.18 m = __?__ cm
26. A dumbbell weighs about 4.5 kg. 4.5 kg = __?__ g
27. Jeremy walks about 10 km. 10 km = __?__ m

PRACTICE AND PROBLEM SOLVING

Find the value of each expression.

28. $2.39 \times 10{,}000$

29. $60.87 \div 10^3$

30. 0.0863×10^3

31. $11.6 - 42 \div 10^2$

32. $(2.3 + 5.67) \times 10^5$

Convert each measure.

33. 6,000 cm = __?__ km

34. 0.75 L = __?__ mL

35. 7.54 kg = __?__ g

36. 17.89 m = __?__ mm

37. Table A is 91.4 centimeters long. Table B is 18.3 decimeters long. Which table is longer? How much longer?

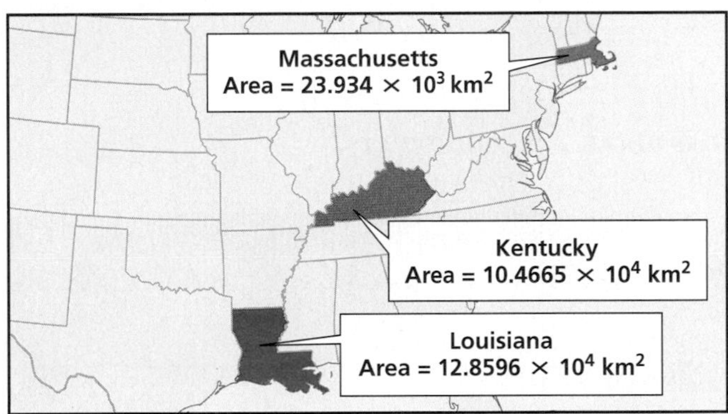

Use the map for Exercises 38 and 39.

38. ***GEOGRAPHY*** Which state has the greater area, Kentucky or Massachusetts? How much greater?

39. ***WHAT'S THE ERROR?*** A student wrote the following equation to find the area of Louisiana: $12.8596 \times 10^4 = 0.00128596$ km^2. Describe the error. Then write the correct answer.

40. ***WRITE ABOUT IT*** When you multiply or divide by a power of ten, how do you know how many places and in which direction to move the decimal point?

41. ***CHALLENGE*** About 3.281 feet equal 1 meter. About how many feet are equal to 2.5 meters?

Spiral Review

Solve each equation. (Lesson 2-6)

42. $b + 53 = 95$

43. $10a = 340$

44. $n - 24 = 188$

45. $w + 20 = 95$

46. $\frac{c}{5} = 60$

47. $73 + p = 82$

48. $r \times 10^2 = 300$

49. $t \div 10^2 = 800$

50. **EOG PREP** Which of the following is the standard form of "four and five hundred six thousandths?" (Lesson 3-1)

A 4.0506 B 4.560 C 4.506 D 0.456

Estimate Measurements

Use with Lesson 3-4

Scientists usually give measurements in metric units, but most people in the United States still use the customary measurement system. To convert from one measurement system to the other, you can estimate units that are approximately equal.

Activity 1

You can see that 1 inch is about 2.5 centimeters.

1 An American football field is 100 yards long. How many meters are in 100 yards?

$100 \text{ yd} = (100 \times 3) \text{ ft} = 300 \text{ ft}$

$300 \text{ ft} = (300 \times 12) \text{ in.} = 3{,}600 \text{ in.}$ *There are 12 inches in a foot.*

$3{,}600 \text{ in.} \approx (3{,}600 \times 2.5) \text{ cm} \approx 9{,}000 \text{ cm}$ *There are about 2.5 centimeters in 1 inch.*

$9{,}000 \text{ cm} = (9{,}000 \div 100) \text{ m} \approx 90 \text{ m}$ *There are 100 centimeters in 1 meter.*

There are about 90 meters in 100 yards.

Since 90 meters is close to 100 meters, a meter is about a yard.

2 There are about 5,000 feet in 1 mile. How many kilometers are in 1 mile?

$1 \text{ mi} \approx 5{,}000 \text{ ft}$

$5{,}000 \text{ ft} \approx (5{,}000 \div 3) \text{ yd} \approx 1{,}666 \text{ yd}$ *There are 3 feet in 1 yard.*

$1{,}666 \text{ yd} \approx 1{,}666 \text{ m}$ *One yard is about 1 meter.*

$1{,}666 \text{ m} \approx (1{,}666 \div 1{,}000) \text{ km} \approx 1.6 \text{ km}$ *There are 1,000 m in 1 km.*

There are about 1.6 kilometers in 1 mile.

Think and Discuss

1. How could you find the number of inches in 1 meter?

Try This

1. Estimate your height in meters.

Activity 2

Fill a one-quart container with water. Pour the water into a one-liter container. A quart is a little less than a liter.

1 There are 4 quarts in a gallon. How many liters are in 1 gallon?

1 gal = 4 qt
4 qt ≈ 4 L *A quart is a little less than a liter.*

There are a little less than 4 liters in 1 gallon.

Think and Discuss

1. How could you find how many milliliters are in 1 gallon?

Try This

1. Estimate the volume in milliliters of a 20-gallon aquarium.

Activity 3

Place a 1-kilogram mass on a scale that gives weights in pounds. One kilogram weighs about 2.2 pounds.

1 There are 16 ounces in 1 pound. How many ounces are in 1 kilogram?

16 oz = 1 lb
1 kg ≈ 2.2 lb *One kilogram weighs about 2.2 pounds.*
2.2 lb ≈ (2.2 × 16) oz ≈ 35.2 oz *There are 16 ounces in 1 pound.*

There are about 35.2 ounces in 1 kilogram.

Think and Discuss

1. How could you find the number of grams in 1 ounce?

Try This

1. The mass of a medium-sized apple is about 100 grams. Estimate the number of apples that would weigh one pound.

2. The average weight of a newborn baby is 7 lb. Estimate this in kilograms.

Chapter 3

Mid-Chapter Quiz

LESSON 3-1 (pp. 92–95)

Write each decimal in standard form, expanded form, and words.

1. 4.012
2. ten and fifty-four thousandths
3. On Monday Jamie ran 3.54 miles. On Wednesday he ran 3.6 miles. On which day did he run farther?

Order the decimals from least to greatest.

4. 3.406, 30.08, 3.6
5. 10.10, 10.01, 101.1, 10.001
6. 16.782, 16.59, 16.79

LESSON 3-2 (pp. 96–99)

7. Matt drove 106.8 miles on Monday, 98.3 miles on Tuesday, and 103.5 miles on Wednesday. About how many miles did he drive in all?

Estimate.

8. $8.345 - 0.6051$; round to the hundredths
9. $16.492 - 2.613$; round to the tenths
10. 18.79×4.68
11. $71.378 \div 8.13$
12. 52.055×7.18

LESSON 3-3 (pp. 102–105)

13. Greg's scores at four gymnastic meets were 9.65, 8.758, 9.884, and 9.500. What was his total score for all four meets?

Find each sum or difference.

14. $0.47 + 0.03$
15. $8 - 0.6$
16. $2.2 + 1.8$

Evaluate $8.67 - s$ for each value of s.

17. $s = 3.4$
18. $s = 2.0871$
19. $s = 7.205$

LESSON 3-4 (pp. 106–109)

Multiply or divide.

20. $516 \times 10{,}000$
21. 16.82×10^3
22. $521.7 \div 10^5$

Use the abbreviation for the most appropriate metric unit.

23. A fork is about 16 __?__ long.

Convert each measure.

24. 5,320 m = __?__ km
25. 1.6 L = __?__ mL

Focus on Problem Solving

Solve

• **Write an equation**

Read the whole problem before you try to solve it. Sometimes you need to solve the problem in more than one step.

Read the problem. Determine the steps needed to solve the problem.

Brian buys erasers and pens for himself and 4 students in his class. The erasers cost \$0.79 each, and the pens cost \$2.95 each. What is the total amount that Brian spends on the erasers and pens?

Here is one way to solve the problem.

5 erasers cost	5 pens cost
$5 \cdot \$0.79$	$5 \cdot \$2.95$

$$(5 \cdot \$0.79) + (5 \cdot \$2.95)$$

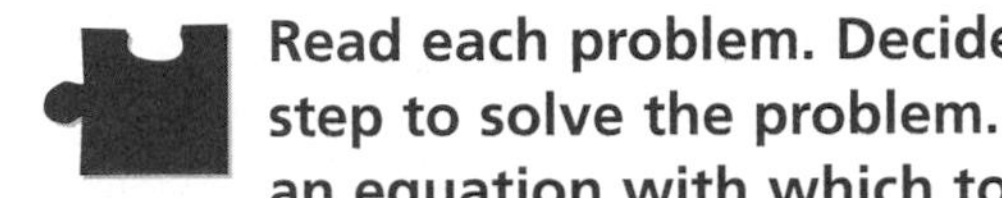

Read each problem. Decide whether you need more than one step to solve the problem. List the possible steps. Then choose an equation with with which to solve the problem.

1. Joan is making some costumes. She cuts 3 pieces of fabric, each 3.5 m long. She has 5 m of fabric left. Which equation can you use to find f, the amount of fabric she had to start with?

   A $(3 \cdot 3.5) + 5 = f$
   B $3 + 3.5 + 5 = f$
   C $(5 \times 3.5) \div 3 = f$
   D $5 - (3 \cdot 3.5) = f$

2. Mario buys 4 chairs and a table. He spends \$245.99 in all. If each chair costs \$38.95, which equation can you use to find T, the cost of the table?

   F $4 + \$245.99 + \$38.95 = T$
   G $(4 \cdot \$38.95) + \$245.99 = T$
   H $\$245.99 - (4 \cdot \$38.95) = T$
   J $\$245.99 \div (4 \cdot \$38.95) = T$

3. Mya skis down Ego Bowl three times and down Fantastic twice. Ego Bowl is 5.85 km long, and Fantastic is 8.35 km long. Which equation can you use to estimate d, the distance Mya skis in all?

   A $(6 \cdot 3) + (8 \cdot 2) = d$
   B $(6 + 8) + (3 + 2) = d$
   C $3(6 + 8) = d$
   D $(6 \div 3) + (8 \div 2) = d$

3-5 Scientific Notation

Learn to write large numbers in scientific notation.

Vocabulary

scientific notation

Pointillism is an art technique in which many small dots are placed close together to form a picture. The famous painting *A Sunday in the Park,* by Georges Seurat, is an example of pointillism. It is made of approximately 3,456,000 dots.

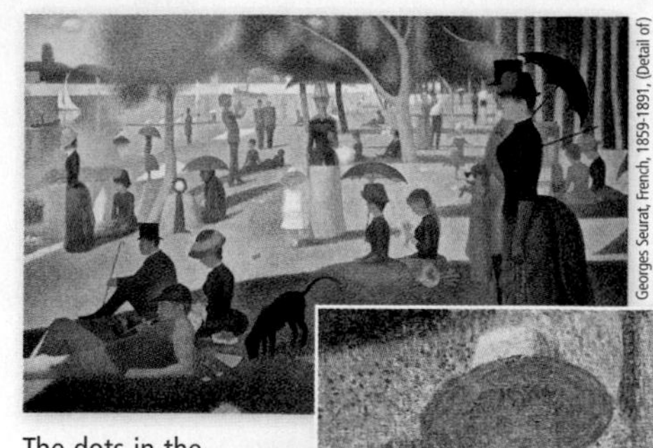

The dots in the painting are as small as 1⁄16 inch. It took Seurat about two years to complete the painting.

Scientific notation is a shorthand method for writing large numbers like 3,456,000.

Remember!

The number of zeros in the power of ten, or the exponent in the power of ten, tells you how many places to move the decimal point.

To write the number 3,456,000 in scientific notation, do the following:

3,456,000 — *Move the decimal point left to form a number that is greater than 1 and less than 10.*

3,456,000 — *Multiply that number by a power of ten.*

3.456×10^6 — *The power of 10 is 6, because the decimal point is moved 6 places left.*

The number of dots written in scientific notation is 3.456×10^6.

A number written in scientific notation has two parts that are multiplied.

$$3.456 \times 10^6$$

The first part is a number that is greater than 1 and less than 10.

The second part is a power of 10.

EXAMPLE 1 Writing Numbers in Scientific Notation

Write each number in scientific notation.

A **700,000**

700,000 — *Move the decimal point 5 places left. The power of 10 is 5.*

$700{,}000 = 7 \times 10^5$

B **8,296,000**

8,296,000 — *Move the decimal point 6 places left. The power of 10 is 6.*

$8{,}296{,}000 = 8.296 \times 10^6$

Write each number in scientific notation.

C 58,000

58,000 — *Move the decimal 4 places left. The power of 10 is 4.*

$58{,}000 = 5.8 \times 10^4$

You can write a large number written in scientific notation in standard form. Look at the power of 10 and move the decimal point that number of places to the right.

EXAMPLE 2 Writing Numbers in Standard Form

Write each number in standard form.

A 8.753×10^2

8.753×10^2 — *The power of 10 is 2.*

8.753 — *Move the decimal point 2 places right.*

$8.753 \times 10^2 = 875.3$

B 3.2×10^7

3.2×10^7 — *The power of 10 is 7.*

3.2000000 — *Move the decimal point 7 places right. Use zeros as placeholders.*

$3.2 \times 10^7 = 32{,}000{,}000$

C 2.001×10^1

2.001×10^1 — *The power of 10 is 1.*

2.001 — *Move the decimal point 1 place right.*

$2.001 \times 10^1 = 20.01$

Think and Discuss

1. **Explain** how you can check whether a number is written correctly in scientific notation.
2. **Tell** why 782.5×10^8 is not correctly written in scientific notation.
3. **Tell** the advantages of writing a number in scientific notation over writing it in standard form. Explain any disadvantages.

3-5 Exercises

FOR EOG PRACTICE

see page 646

internet connect
Homework Help Online
go.hrw.com Keyword: MR4 3-5

1.06

GUIDED PRACTICE

See Example 1 **Write each number in scientific notation.**

1. 62,000 **2.** 500,000 **3.** 6,913,000

4. 130,000 **5.** 7,015,000 **6.** 20,000

See Example 2 **Write each number in standard form.**

7. 6.793×10^6 **8.** 1.4×10^4 **9.** 3.82×10^5

10. 9.401×10^7 **11.** 3.3×10^3 **12.** 1.885×10^4

INDEPENDENT PRACTICE

See Example 1 **Write each number in scientific notation.**

13. 90,000 **14.** 186,000 **15.** 1,607,000

16. 240,000 **17.** 6,000,000 **18.** 16,900,000

19. 1,800 **20.** 12,865,000 **21.** 50,400,000

See Example 2 **Write each number in standard form.**

22. 3.211×10^5 **23.** 1.63×10^6 **24.** 7.7×10^3

25. 2.14×10^4 **26.** 4.03×10^6 **27.** 8.1164×10^8

28. 6.33×10^5 **29.** 9.106×10^7 **30.** 5.5×10^2

PRACTICE AND PROBLEM SOLVING

Write each number in standard form.

31. 7.21×10^3 **32.** 1.234×10^5 **33.** 7.200×10^2

34. 2.08×10^5 **35.** 6.954×10^3 **36.** 5.43×10^1

Write each number in scientific notation.

37. 112,050 **38.** 150,000 **39.** 4,562

40. 1,000 **41.** 65,342 **42.** 95

Write each measurement using scientific notation.

43. 4 km = ___?___ m **44.** 3.78 km = ___?___ cm

45. 18 L = ___?___ mL **46.** 75 kg = ___?___ mg

47. 19.5 kg = ___?___ g **48.** 2 L = ___?___ mL

This F/A-18 Hornet makes a vapor cloud by flying at Mach 0.98, just under the speed of sound.

go.hrw.com
KEYWORD: MR4 SOUND

49. ***EARTH SCIENCE*** The speed of light is about 300,000 km/s. The speed of sound in air that has a temperature of 20°C is 1,125 ft/s. Write both of these values in scientific notation.

50. ***LIFE SCIENCE*** Genes carry the codes used for making proteins that are necessary for life. No one knows yet how many human genes there are. Estimates range from 3.8×10^4 to 1.2×10^5. Write a number in standard form that is within this range.

Use the pictograph for Exercises 51 and 52.

51. Write the capacity of Rungnado Stadium in scientific notation.

52. Estimate the capacity of the largest stadium. Write the estimate in scientific notation.

53. ***TECHNOLOGY*** In the year 2000, there were about 579 million computers in the world. Write this number in standard form and scientific notation.

54. ***SOCIAL STUDIES*** The Library of Congress, in Washington, D.C., is the largest library in the world. It was founded in 1800 and has 24,616,867 books. Round the number of books to the nearest hundred thousand, and write that number in scientific notation.

55. ***WHAT'S THE ERROR?*** A student said the number 56,320,000 written in scientific notation is 56.32×10^6. Describe the error. Then write the correct answer.

56. ***WRITE ABOUT IT*** How does writing numbers in scientific notation make it easier to compare and order the numbers?

57. ***CHALLENGE*** What is 5.32 written in scientific notation?

Spiral Review

Identify a pattern in each sequence, and name the next two terms. (Lesson 1-7)

58. 3, 6, 12, 24, 48, ▒, ▒, ...

59. 1, 3, 8, 24, 29, 87, 92, ▒, ▒, ...

Solve each equation. (Lesson 2-5)

60. $a - 23 = 18$

61. $y - 7 = 45$

62. $x + 16 = 71$

63. **EOG PREP** Which is the product of $30.62 \times 10{,}000$? (Lesson 3-4)

A 0.003062 B 0.3062 C 30,620 D 306,200

Explore Decimal Multiplication and Division

Use with Lessons 3-6 and 3-7

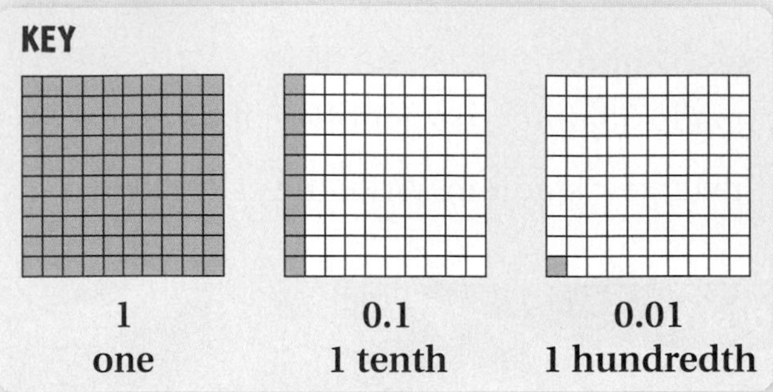

internet connect
Lab Resources Online
go.hrw.com
KEYWORD: MR4 Lab3D

You can use decimal grids to model multiplication and division of decimals.

Activity 1

1. Use decimal grids to find each product.

a. 3 • 0.32

3 • 0.32 = 0.96

To represent 3 • 0.32, shade 32 small squares three times.

Use a different color to shade a different group of 32 small squares each time.

There are 96 shaded squares.

b. 0.3 • 0.5

0.3 • 0.5 = 0.15

To represent 0.3, shade 3 columns.

To represent 0.5, shade 5 rows in another color.

There are 15 squares in the area where the shading overlaps.

Think and Discuss

1. How is multiplying a decimal by a decimal different from multiplying a decimal by a whole number?

2. Why can you shade 5 rows to represent 0.5?

Try This

Use decimal grids to find each product.

1. 3 • 0.14 **2.** 5 • 0.18 **3.** 0.7 • 0.5 **4.** 0.6 • 0.4

Activity 2

1 Use decimal grids to find each quotient.

a. 3.66 ÷ 3

Shade 3 grids and 66 small squares of a fourth grid to represent 3.66.

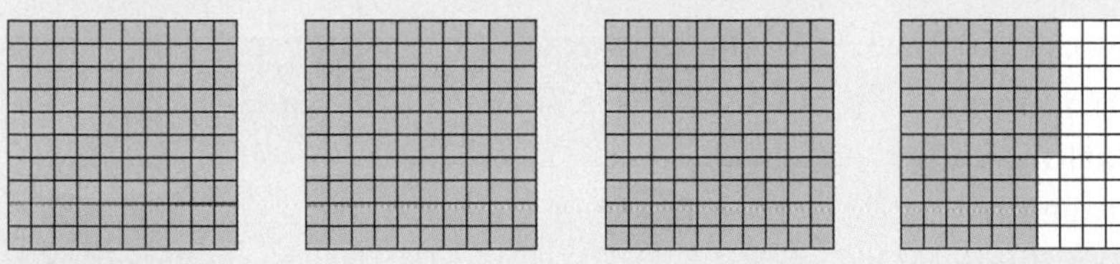

Divide the shaded wholes into 3 equal groups. Use scissors to divide the 66 hundredths into 3 equal groups.

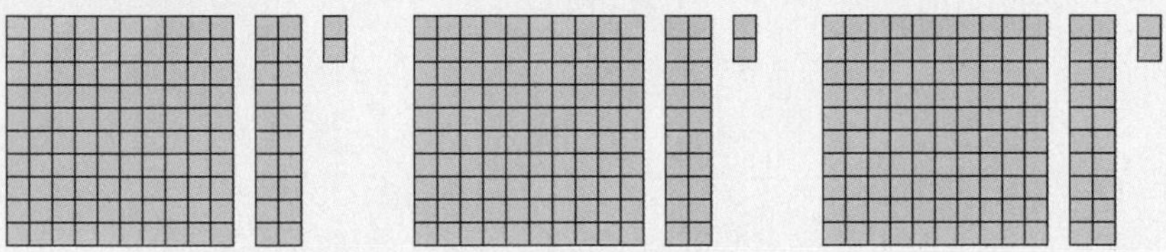

One whole grid and 22 small squares are in each group.

3.66 ÷ 3 = 1.22

b. 3.6 ÷ 1.2

Shade 3 grids and 6 columns of a fourth grid to represent 3.6. Cut apart the 6 tenths.

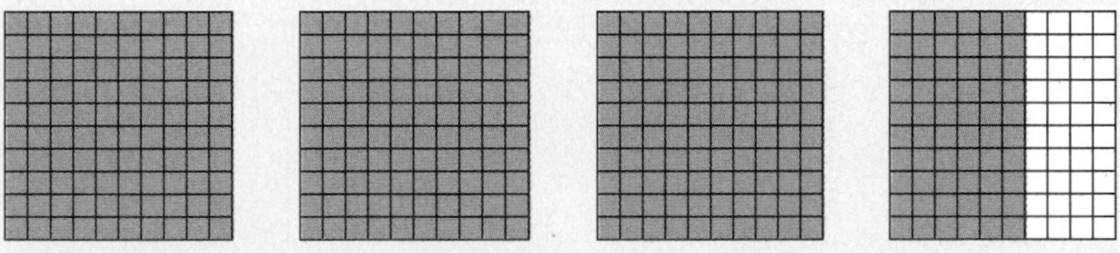

Divide the grids and tenths into equal groups of 1.2.

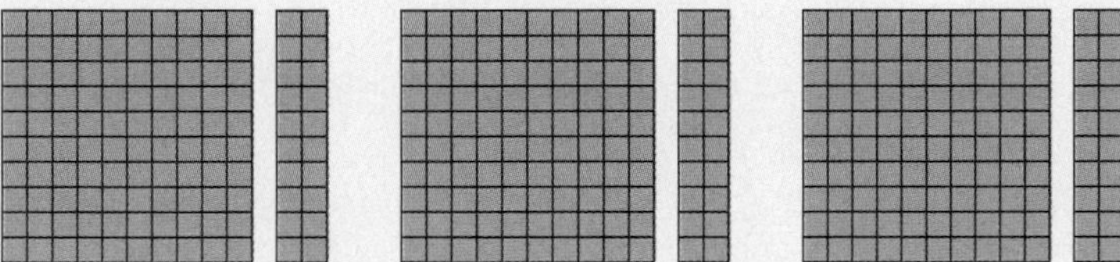

There are 3 equal groups of 1.2.

3.6 ÷ 1.2 = 3

Think and Discuss

1. Find 36 ÷ 12. How does this problem and its quotient compare to 3.6 ÷ 1.2?

Try This

Use decimal grids to find each quotient.

1. 4.04 ÷ 4 **2.** 3.25 ÷ 5 **3.** 7.8 ÷ 1.3 **4.** 5.6 ÷ 0.8

3-6 Multiplying Decimals

Learn to multiply decimals by whole numbers and by decimals.

Because the Moon has less mass than Earth, it has a smaller gravitational effect. An object that weighs 1 pound on Earth weighs only 0.17 pound on the Moon. How much does a 3-pound flag weigh on the Moon?

To find the weight of a 3-pound flag on the Moon, you can add decimals or multiply a decimal by a whole number.

Gravity on Earth is about six times the gravity on the surface of the Moon.

EXAMPLE 1 *Science Application*

Something that weighs 1 lb on Earth weighs 0.17 lb on the Moon. How much does a 3 lb flag weigh on the Moon?

3 × 0.17

$$\begin{array}{r} 0.17 \\ 0.17 \\ +\ 0.17 \\ \hline \mathbf{0.51} \end{array}$$

You can think of multiplication by a whole number as a repeated addition.

You can also multiply as you would with whole numbers.

Place the decimal point by adding the number of decimal places in the numbers multiplied.

$$\begin{array}{r} 0.17 \\ \times\ \ 3 \\ \hline \mathbf{0.51} \end{array}$$

2 decimal places
+ 0 decimal places
2 decimal places

A 3 lb flag on Earth weighs 0.51 lb on the Moon.

EXAMPLE 2 Multiplying a Decimal by a Decimal

Find each product.

A 0.2 × 0.6

Multiply. Then place the decimal point.

$$\begin{array}{r} 0.2 \\ \times\ 0.6 \\ \hline 0.12 \end{array}$$

1 decimal place
+ 1 decimal place
2 decimal places

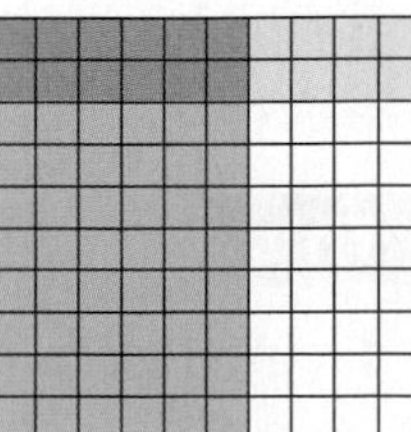

Find each product.

B 0.05×0.9

$0.05 \times 1 = 0.05$ *Estimate the product. 0.9 is close to 1.*

Multiply. Then place the decimal point.

0.05	*2 decimal places*
× 0.9	*+ 1 decimal place*
0.045	*3 decimal places; use a placeholder zero.*

0.045 is close to the estimate of 0.05. The answer is reasonable.

C 3.25×4.8

$3 \times 5 = 15$ *Estimate the product. Round each factor to the nearest whole number.*

Multiply. Then place the decimal point.

3.25	
× 4.8	
2600	*2 decimal places*
13000	*+ 1 decimal place*
15.600	*3 decimal places*

15.600 is close to the estimate of 15. The answer is reasonable.

EXAMPLE 3 Evaluating Decimal Expressions

Remember!

These notations all mean multiply 3 times *x*.

$3 \cdot x$ $3x$ $3(x)$

Evaluate 3*x* for each value of *x*.

A $x = 4.047$

$3x = 3(\mathbf{4.047})$ *Substitute 4.047 for x.*

4.047	*3 decimal places*
× 3	*+ 0 decimal places*
12.141	*3 decimal places*

B $x = 2.95$

$3x = 3(\mathbf{2.95})$ *Substitute 2.95 for x.*

2.95	*2 decimal places*
× 3	*+ 0 decimal places*
8.85	*2 decimal places*

Think and Discuss

1. **Tell** how many decimal places are in the product of 235.2 and 0.24.
2. **Tell** which is greater, 4×0.6 or 4×0.006.
3. **Describe** how the products of 0.3×0.5 and 3×5 are similar. How are they different?

3-6 Exercises

FOR EOG PRACTICE
see page 646

internet connect
Homework Help Online
go.hrw.com Keyword: MR4 3-6

1.04a, c, d, 1.07, 5.02

GUIDED PRACTICE

See Example 1

1. Each can of cat food costs \$0.28. How much will 6 cans of cat food cost?
2. Jorge buys 8 baseballs for \$9.29 each. How much does he spend in all?

See Example 2 **Find each product.**

3. $\begin{array}{r} 0.6 \\ \times\ 0.4 \\ \hline \end{array}$ 4. $\begin{array}{r} 0.008 \\ \times\ 0.5 \\ \hline \end{array}$ 5. $\begin{array}{r} 3.0 \\ \times\ 0.07 \\ \hline \end{array}$ 6. $\begin{array}{r} 0.12 \\ \times\ 0.6 \\ \hline \end{array}$

See Example 3 **Evaluate $5x$ for each value of x.**

7. $x = 3.304$ 8. $x = 4.58$ 9. $x = 7.126$

INDEPENDENT PRACTICE

See Example 1

10. Gwenyth walks her dog each morning. If she walks 0.37 kilometers each morning, how many kilometers will she have walked in 7 days?
11. Apples are on sale for \$0.49 per pound. What is the price for 4 pounds of apples?

See Example 2 **Find each product.**

12. $\begin{array}{r} 0.9 \\ \times\ 0.03 \\ \hline \end{array}$ 13. $\begin{array}{r} 4.5 \\ \times\ 0.5 \\ \hline \end{array}$ 14. $\begin{array}{r} 0.31 \\ \times\ 0.7 \\ \hline \end{array}$ 15. $\begin{array}{r} 1.6 \\ \times\ 0.08 \\ \hline \end{array}$

16. 0.007×0.06 17. 0.04×3.0 18. 2.0×0.006 19. 0.005×0.003

See Example 3 **Evaluate $7x$ for each value of x.**

20. $x = 1.903$ 21. $x = 2.461$ 22. $x = 3.72$
23. $x = 0.164$ 24. $x = 5.89$ 25. $x = 0.3702$

PRACTICE AND PROBLEM SOLVING

Multiply.

26. 0.3×0.03 27. 1.4×0.21 28. 0.06×1.02
29. 12.6×2.1 30. 3.04×0.6 31. 0.66×2.52
32. $0.2 \times 0.94 \times 1.3$ 33. $1.54 \times 3.05 \times 2.6$

Evaluate.

34. $6n$ for $n = 6.23$ 35. $5t + 0.462$ for $t = 3.04$
36. $8^2 - 2b$ for $b = 0.95$ 37. $4^3 + 5c$ for $c = 1.9$

Physical Science LINK

Saturn is the second-largest planet in the solar system. Saturn is covered by thick clouds, and it is thought that there is no Earth-like surface beneath the clouds. Saturn's density is very low. Suppose you weigh 180 pounds on Earth. If you were able to stand on top of Saturn's clouds, you would weigh only 165 pounds. To find the weight of an object on another planet, multiply its weight on Earth by the gravitational pull listed in the table.

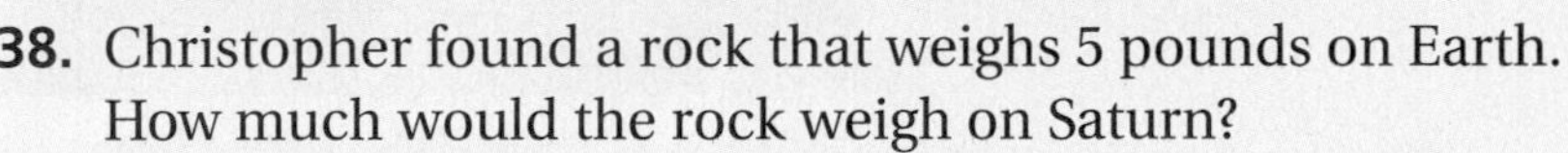

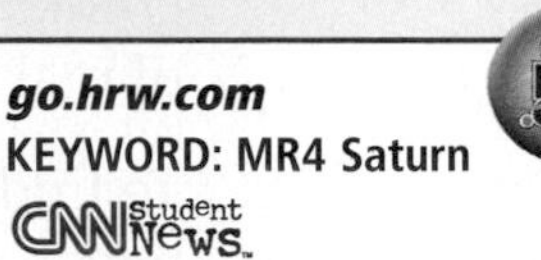

go.hrw.com
KEYWORD: MR4 Saturn
CNN Student News

38. Christopher found a rock that weighs 5 pounds on Earth. How much would the rock weigh on Saturn?

39. On which two planets would the weight of an object be the same?

40. A cat weighs 14 pounds on Earth. Would the cat weigh more or less on Venus? Explain.

41. An object weighs 9 pounds on Earth. How much would this object weigh on Mars?

42. ***WRITE A PROBLEM*** Use the data in the table to write a word problem that can be answered by evaluating an expression with multiplication. Solve your problem.

43. ***WHAT'S THE ERROR?*** A student said that his new baby brother, who weighs 10 pounds, would weigh 120 pounds on Neptune. What is the error? Write the correct answer.

44. ***CHALLENGE*** An object weighs between 2.79 lb and 5.58 lb on Saturn. Give a range for the object's weight on Earth.

Gravitational Pull of Planets (Compared with Earth)

Planet	Gravitational Pull
Mercury	0.38
Venus	0.91
Mars	0.38
Jupiter	2.54
Saturn	0.93
Neptune	1.2

Galileo Galilei was the first person to look at Saturn through a telescope. He thought there were groups of stars on each side of the planet, but it was later determined that he had seen Saturn's rings.

Spiral Review

Find a pattern in each sequence. (Lesson 1-7)

45. 1, 7, 3, 21, 17, 119, 115, …

46. 1, 3^2, 25, 49, 9^2, 121, 169, …

Write each phrase as a numerical or algebraic expression. (Lesson 2-2)

47. the sum of 163 and 24

48. the product of 15 and *c*

49. *y* divided by 8

50. **EOG PREP** What is the sum of 5.6004 + 3.458? (Lesson 3-3)

A 9.0584 B 5.9462 C 9.584 D 8.6462

3-7 Dividing Decimals by Whole Numbers

Learn to divide decimals by whole numbers.

Ethan and two of his friends are going to share equally the cost of making a sculpture for the art fair.

To find how much each person should pay for the materials, you will need to divide a decimal by a whole number.

EXAMPLE 1 Dividing a Decimal by a Whole Number

Find each quotient.

A $0.75 \div 5$

$$\begin{array}{r} 0.15 \\ 5\overline{)0.75} \\ -5 \\ \hline 25 \\ -25 \\ \hline 0 \end{array}$$

Place a decimal point in the quotient directly above the decimal point in the dividend.

Divide as you would with whole numbers.

B $2.52 \div 3$

$$\begin{array}{r} 0.84 \\ 3\overline{)2.52} \\ -2\ 4 \\ \hline 12 \\ -12 \\ \hline 0 \end{array}$$

Place a decimal point in the quotient directly above the decimal point in the dividend.

Divide as you would with whole numbers.

EXAMPLE 2 Evaluating Decimal Expressions

Evaluate $0.435 \div x$ for each given value of x.

A $x = 3$

$0.435 \div x$

$0.435 \div 3$ *Substitute 3 for x.*

$$\begin{array}{r} 0.145 \\ 3\overline{)0.435} \\ -3 \\ \hline 13 \\ -12 \\ \hline 15 \\ -15 \\ \hline 0 \end{array}$$

Divide as you would with whole numbers.

B $x = 15$

$0.435 \div x$

$0.435 \div 15$ *Substitute 15 for x.*

$$\begin{array}{r} 0.029 \\ 15\overline{)0.435} \\ -0 \\ \hline 43 \\ -30 \\ \hline 135 \\ -135 \\ \hline 0 \end{array}$$

Sometimes you need to use a zero as a placeholder.

15 > 4, so place a zero in the quotient and divide 15 into 43.

EXAMPLE 3 *Consumer Application*

Ethan and two of his friends are making a papier-mâché sculpture using balloons, strips of paper, and paint. The materials cost $11.61. If they share the cost equally, how much should each person pay?

> **Remember!**
>
> *quotient*
> *divisor)dividend*
>
> Multiplication can "undo" division. To check your answer to a division problem, multiply the divisor by the quotient.

$11.61 should be divided into three equal groups.
Divide $11.61 by 3.

$$\begin{array}{r} \$3.87 \\ 3\overline{)\$11.61} \\ -\ 9 \\ \hline 2\ 6 \\ -\ 2\ 4 \\ \hline 21 \\ -\ 21 \\ \hline 0 \end{array}$$

Place a decimal point in the quotient directly above the decimal point in the dividend.

Divide as you would with whole numbers.

Check

$3.87 \times 3 = 11.61$

Each person should pay $3.87.

Think and Discuss

1. Tell how you know where to place the decimal point in the quotient.

2. Explain why you can use multiplication to check your answer to a division problem.

3-7 Exercises

FOR EOG PRACTICE
see page 646

internet connect
Homework Help Online
go.hrw.com Keyword: MR4 3-7

1.04a, 5.02

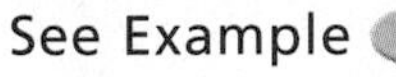

GUIDED PRACTICE

See Example 1

Find each quotient.

1. $1.38 \div 6$ **2.** $0.96 \div 8$ **3.** $1.75 \div 5$ **4.** $0.72 \div 4$

See Example 2

Evaluate $0.312 \div x$ for each given value of x.

5. $x = 4$ **6.** $x = 6$ **7.** $x = 3$ **8.** $x = 12$

See Example 3

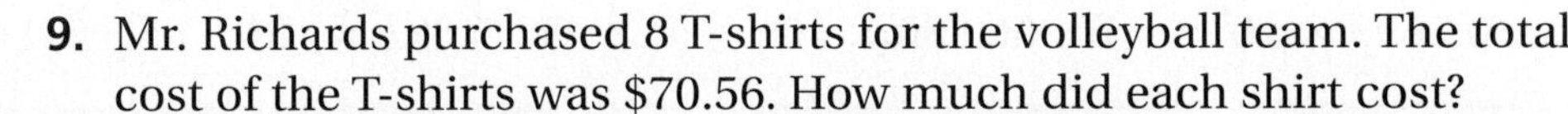

9. Mr. Richards purchased 8 T-shirts for the volleyball team. The total cost of the T-shirts was $70.56. How much did each shirt cost?

INDEPENDENT PRACTICE

See Example 1 — **Find each quotient.**

10. 0.91 ÷ 7 **11.** 1.32 ÷ 6 **12.** 4.68 ÷ 9 **13.** 0.81 ÷ 3

See Example 2 — **Evaluate 0.684 ÷ x for each given value of x.**

14. $x = 3$ **15.** $x = 4$ **16.** $x = 18$ **17.** $x = 9$

See Example 3 — **18.** Charles, Kate, and Kim eat lunch in a restaurant. The bill is $27.12. If they share the bill equally, how much will each person pay?

PRACTICE AND PROBLEM SOLVING

Divide.

19. 3.6 ÷ 4 **20.** 15.35 ÷ 5 **21.** 12.8592 ÷ 6 **22.** 0.729 ÷ 3

Find the value of each expression.

23. (0.49 + 0.0045) ÷ 5 **24.** (4.9 − 3.125) ÷ 5

Evaluate the expression x ÷ 4 for each value of x.

25. $x = 0.504$ **26.** $x = 0.944$ **27.** $x = 57.484$ **28.** $x = 1.648$

29. Dance lessons cost $198.75 for 15 classes. What is the fee for one class?

30. At the grocery store, a 6 lb bag of oranges costs $2.04. Is this more or less expensive than the price shown at the farmers' market?

31. ***CHOOSE A STRATEGY*** Sarah had $1.19 in coins. Jeff asked her for change for a dollar, but she did not have the correct change. What coins did she have?

32. ***WRITE ABOUT IT*** When do you use a placeholder zero in the quotient?

33. ***CHALLENGE*** Evaluate the expression x ÷ 2 for the following values of x: 520, 52, and 5.2. Try to predict the value of the same expression for $x = 0.52$.

Spiral Review

Order the numbers from least to greatest. (Lesson 1-1)

34. 3,673,809; 3,708,211; 3,671,935 **35.** 2,004,801; 225,971; 298,500,004

Compare. Write $<$, $>$, or $=$. (Lesson 3-1)

36. 7.0893 ▢ 7.0798 **37.** 0.0312 ▢ 0.211 **38.** 0.9571 ▢ 1.308

39. **EOG PREP** What is 56,930 expressed in scientific notation? (Lesson 3-5)

A 0.5693×10^5 B 56.930×10^3 C 5.693×10^4 D 5.6930×10^3

3-8 Dividing by Decimals

Learn to divide whole numbers and decimals by decimals.

Julie and her family traveled to the Grand Canyon. They stopped to refill their gas tank with 13.4 gallons of gasoline after they had driven 368.5 miles.

To find the miles that they drove per gallon, you will need to divide a decimal by a decimal.

EXAMPLE 1 Dividing a Decimal by a Decimal

Helpful Hint

Multiplying the divisor and the dividend by the same number does not change the quotient.

$$42 \div 6 = 7$$
$$\times 10 \downarrow \quad \times 10 \downarrow$$
$$420 \div 60 = 7$$

$$42 \div 6 = 7$$
$$\times 100 \downarrow \quad \times 100 \downarrow$$
$$4{,}200 \div 600 = 7$$

Find each quotient.

A $3.6 \div 1.2$

$1.2\overline{)3.6}$ *Multiply the divisor and dividend by the same power of ten.*

$$\begin{array}{r} 3 \\ 12\overline{)36} \\ -\,36 \\ \hline 0 \end{array}$$

There is one decimal place in the divisor. Multiply by 10^1, or 10.

Think: $1.2 \times 10 = 12$ $\quad 3.6 \times 10 = 36$

Divide.

B $41.6 \div 0.39$

$0.39\overline{)41.6}$ *Make the divisor a whole number by multiplying the divisor and dividend by 10^2, or 100.*

Think: $0.39 \times 100 = 39$ $\quad 41.6 \times 100 = 4{,}160$

$$\begin{array}{r} 106.66 \\ 39\overline{)4160.00} \\ -\,39 \\ \hline 26 \\ -\,0 \\ \hline 260 \\ -\,234 \\ \hline 26\,0 \\ -\,23\,4 \\ \hline 2\,60 \\ -\,2\,34 \\ \hline 26 \end{array}$$

Place the decimal point in the quotient. Divide.

When there is a remainder, place a zero after the decimal point in the dividend and continue to divide.

$106.66\ldots = 106.\overline{6}$ *When a repeating pattern occurs, show three dots or draw a bar over the repeating part of the quotient.*

EXAMPLE  2 PROBLEM SOLVING APPLICATION

After driving 368.5 miles, Julie and her family refilled the tank of their car with 13.4 gallons of gasoline. On average, how many miles did they drive per gallon of gas?

1 Understand the Problem

The **answer** will be the average number of miles per gallon.

List the **important information:**

- They drove 368.5 miles.
- They used 13.4 gallons of gas.

2 Make a Plan

Solve a simpler problem by replacing the decimals in the problem with whole numbers.

If they drove 10 miles using 2 gallons of gas, they averaged 5 miles per gallon. You need to divide miles by gallons to solve the problem.

3 Solve

$13.4\overline{)368.5}$ *Multiply the divisor and the dividend by 10.*
Think: $13.4 \times 10 = 134$ $368.5 \times 10 = 3{,}685$

```
       27.5      Place the decimal point in the quotient.
134)3685.0      Divide.
   − 268
     1005
   − 938
      67 0
    − 67 0
         0
```

Julie and her family averaged 27.5 miles per gallon.

4 Look Back

Use compatible numbers to estimate the quotient.

$368.5 \div 13.4 \rightarrow 360 \div 12 = 30$

The answer is reasonable, since 27.5 is close to the estimate of 30.

Think and Discuss

1. **Tell** how the quotient of $48 \div 12$ is similar to the quotient of $4.8 \div 1.2$. How is it different?

3-8 Exercises

FOR EOG PRACTICE

see page 647

internet connect
Homework Help Online
go.hrw.com Keyword: MR4 3-8

1.04a, c, d, 1.07

GUIDED PRACTICE

See Example 1 **Find each quotient.**

1. $6.5 \div 1.3$ **2.** $20.7 \div 0.6$ **3.** $25.5 \div 1.5$

4. $5.4 \div 0.9$ **5.** $13.2 \div 2.2$ **6.** $63.39 \div 0.24$

See Example 2

7. Marcus drove 354.9 miles in 6.5 hours. On average, how many miles per hour did he drive?

8. Anthony spends $87.75 on shrimp. The shrimp cost $9.75 per pound. How many pounds of shrimp does Anthony buy?

INDEPENDENT PRACTICE

See Example 1 **Find each quotient.**

9. $3.6 \div 0.6$ **10.** $8.2 \div 0.5$ **11.** $18.4 \div 2.3$

12. $4.8 \div 1.2$ **13.** $51.2 \div 0.24$ **14.** $32.5 \div 2.6$

15. $50.9 \div 4.5$ **16.** $91.6 \div 0.45$ **17.** $6.5 \div 1.3$

See Example 2

18. Jen spends $5.98 on ribbon. Ribbon costs $0.92 per meter. How many meters of ribbon does Jen buy?

19. Kyle's family drove 329.44 miles. Kyle calculated that the car averaged 28.4 miles per gallon of gas. How many gallons of gas did the car use?

20. Peter is saving $4.95 each week to buy a DVD that costs $24.75, including tax. For how many weeks will he have to save?

PRACTICE AND PROBLEM SOLVING

Divide.

21. $2.52 \div 0.4$ **22.** $12.586 \div 0.35$ **23.** $0.5733 \div 0.003$

24. $10.875 \div 1.2$ **25.** $92.37 \div 0.5$ **26.** $8.43 \div 0.12$

Find the value of each expression.

27. $6.35 \times 10^2 \div 0.5$ **28.** $8.1 \times 10^2 \div 0.9$

29. $20.1 \times 10^3 \div 0.1$ **30.** $2.76 \times 10^2 \div 0.3$

Evaluate.

31. $0.732 \div n$ for $n = 0.06$ **32.** $73.814 \div c$ for $c = 1.3$

33. $b \div 0.52$ for $b = 6.344$ **34.** $r \div 4.17$ for $r = 10.5918$

35. $r \div 3.7$ for $r = 34.928$ **36.** $45.05 \div a$ for $a = 2.5$

The U.S. Mint was established by the Coinage Act in 1792. The first coins were copper and were made in Philadelphia.

37. ***HISTORY*** The U.S. Treasury first printed paper money in 1862. The paper money we use today is 0.0043 inch thick. Estimate the number of bills you would need to stack to make a pile that is 1 inch thick. If you stacked \$20 bills, what would be the total value of the money in the pile?

38. ***EARTH SCIENCE*** A planet's year is the time it takes that planet to revolve around the Sun. A Mars year is 1.88 Earth years. If you are 13 years old in Earth years, about how old would you be in Mars years?

Use the map for Exercises 39–41.

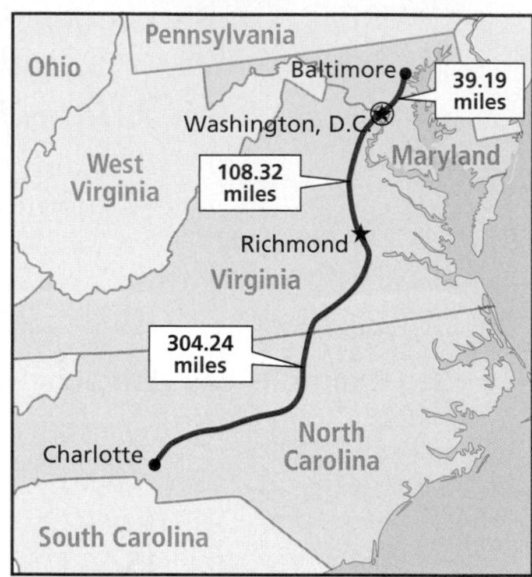

39. Bill drove from Washington, D.C., to Charlotte in 6.5 hours. What was his average speed in miles per hour?

40. Betty drove a truck from Richmond to Washington, D.C., without stopping. It took her about 2.5 hours. Estimate the average speed she was driving.

41. How far is it in miles from Washington, D.C., to Baltimore and back?

42. ***WHAT'S THE ERROR?*** A student incorrectly answered the division problem below. Explain the error and write the correct quotient.

$$0.004\overline{)53.824}\quad\text{quotient: } 13.456$$

43. ***WRITE ABOUT IT*** Explain how you know where to place the decimal point in the quotient when you divide by a decimal number.

44. ***CHALLENGE*** Find the value of a in the division problem.

$$0.4a3\overline{)0.41713}\quad\text{quotient: } 1.01$$

Spiral Review

Use mental math to find each sum or product. (Lesson 1-5)

45. $8 \times 5 \times 9$
46. $49 + 26 + 11 + 14$
47. $4 \times 15 \times 6$

Solve for *y*. (Lesson 2-6)

48. $y - 23 = 40$
49. $14y = 168$
50. $36 + y = 53$

Order the decimals from greatest to least. (Lesson 3-1)

51. 8.304, 8.009, 8.05
52. 15.34, 1.589, 5.62
53. 30.75, 30.211, 30.709

54. **EOG PREP** Find the product of 1.8×0.541. (Lesson 3-6)

A 0.9738 B 9.738 C 97.3800 D 9.73×10^4

3-9 Interpret the Quotient

 Problem Solving Skill

Learn to solve problems by interpreting the quotient.

In science lab, Kim learned to make slime from corn starch, water, and food coloring. She has 0.87 kg of corn starch, and the recipe for one bag of slime calls for 0.15 kg. To find the number of bags of slime Kim can make, you need to divide.

EXAMPLE 1 ***Measurement Application***

Kim will use 0.87 kg of corn starch to make gift bags of slime for her friends. If each bag requires 0.15 kg of corn starch, how many bags of slime can she make?

> **Remember!**
> To divide decimals, first write the divisor as a whole number. Multiply the divisor and dividend by the same power of ten.

The question asks how many whole bags of slime can be made when the corn starch is divided into groups of 0.15 kg.

$0.87 \div 0.15 = ?$

$87 \div 15 = 5.8$

    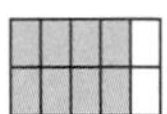

Think: The quotient shows that there is not enough to make 6 bags of slime that are 0.15 kg each. There is only enough for 5 bags. The decimal part of the quotient will not be used in the answer.

Kim can make **5** gift bags of slime.

EXAMPLE 2 ***Photography Application***

There are 246 students in the sixth grade. If Ms. Lee buys rolls of film with 24 exposures each, how many rolls will she need to take every student's picture?

The question asks how many whole rolls are needed to take a picture of every one of the students.

$246 \div 24 = 10.25$

 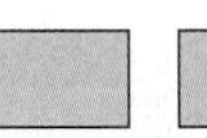 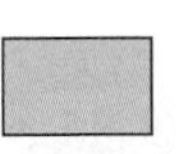 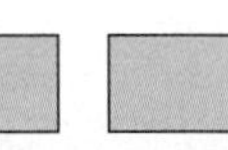 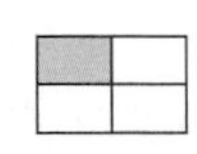

Think: Ten rolls of film will not be enough to take every student's picture. Ms. Lee will need to buy another roll of film. The quotient must be rounded up to the next highest whole number.

Ms. Lee will need 11 rolls of film.

EXAMPLE 3 *Social Studies Application*

Marissa is drawing a time line of the Stone Age. She plans for 6 equal sections, two each for the Paleolithic, Mesolithic, and Neolithic periods. If she has 16.5 meters of paper, how long is each section?

The question asks exactly how long each section will be when the paper is divided into 6 sections.

$16.5 \div 6 = 2.75$ *Think: The question asks for an exact answer, so do not estimate. Use the entire quotient.*

Each section will be **2.75** meters long.

When the question asks	→ You should
How many whole groups can be made when you divide?	→ Drop the decimal part of the quotient.
How many whole groups are needed to put all items from the dividend into a group?	→ Round the quotient up to the next highest whole number.
What is the exact number when you divide?	→ Use the entire quotient as the answer.

Think and Discuss

1. Tell how you would interpret the quotient: A group of 27 students will ride in vans that carry 12 students each. How many vans are needed?

3-9 Exercises

FOR EOG PRACTICE
see page 647

internet connect
Homework Help Online
go.hrw.com Keyword: MR4 3-9

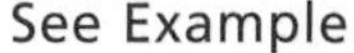

1.04a, d, 1.07

GUIDED PRACTICE

See Example 1
1. Kay is making beaded belts for her friends from 6.5 meters of cord. One belt uses 0.625 meter of cord. How many belts can she make?

See Example 2
2. Julius is supplying cups for a party of 136 people. If cups are sold in packs of 24, how many packs of cups will he need?

See Example 3
3. Miranda is decorating for a party. She has 13 balloons and 29.25 meters of ribbon. She wants to tie the same length of ribbon on each balloon. How long will each ribbon be?

INDEPENDENT PRACTICE

See Example  1

4. There are 0.454 kg of corn starch in a container. How many 0.028 kg portions are in one container?

See Example 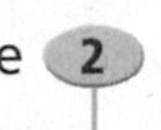 2

5. Tina needs 36 flowers for her next project. The flowers are sold in bunches of 5. How many bunches will she need?

See Example 3

6. Bobby's goal is to run 27 miles a week. If he runs the same distance 6 days a week, how many miles would he have to run each day?

PRACTICE AND PROBLEM SOLVING

7. Nick wants to write thank-you notes to 15 of his friends. The cards are sold in packs of 6. How many packs does Nick need to buy?

8. The science teacher has 7 packs of seeds and 36 students. If the students should each plant the same number of seeds, how many can each student plant?

9. A new parking garage at an apartment complex is 16.8 m high. Each floor is 4.2 m high. How many floors are there?

10. ***WRITE A PROBLEM*** Create a problem that is solved by interpreting the quotient.

11. ***WRITE ABOUT IT*** Explain how a calculator shows the remainder when you divide 145 by 8.

12. ***CHALLENGE*** Leonard wants to place a fence on both sides of a 10-meter walkway. If he puts a post at both ends and at every 2.5 meters in between, how many posts does he use?

Spiral Review

Evaluate the expressions to find the missing values in each table. (Lesson 2-1)

13.

x	*9x*
5	■
6	■
7	■

14.

y	*y* ÷ *11*
121	■
99	■
77	■

Estimate. (Lesson 3-2)

15. 467.32 + 450.64 + 447.9 **16.** 14.87 × 3.78 **17.** 53.67 ÷ 9.18

18.  **EOG PREP** Find the sum of 5.63 + 6.702 + 5.9 + 7.383. (Lesson 3-3)

A 14.707 **B** 256.15 **C** 19.774 **D** 25.615

3-10 Solving Decimal Equations

Learn to solve equations involving decimals.

Felipe has earned \$45.20 by mowing lawns for his neighbors. He wants to buy inline skates that cost \$69.95. Write and solve an equation to find how much more money Felipe must earn to buy the skates.

Let m be the amount of money Felipe needs. $\$45.20 + m = \$69.95$

You can solve equations with decimals using inverse operations just as you solved equations with whole numbers.

$$\begin{aligned} \$45.20 + m &= \$69.95 \\ \underline{-\ \$45.20} \quad &\quad \underline{-\ \$45.20} \\ m &= \$24.75 \end{aligned}$$

Felipe needs \$24.75 more to buy the inline skates.

EXAMPLE 1 Solving One-Step Equations with Decimals

Remember!

Use inverse operations to get the variable alone on one side of the equation.

Solve each equation. Check your answer.

A $g - 3.1 = 4.5$

$g - 3.1 = 4.5$ — *3.1 is subtracted from g.*

$\underline{+\ 3.1} \quad \underline{+\ 3.1}$ — *Add 3.1 to both sides to undo the subtraction.*

$g = 7.6$

Check

$g - 3.1 = 4.5$

$7.6 - 3.1 \stackrel{?}{=} 4.5$ — *Substitute 7.6 for g in the equation.*

$4.5 \stackrel{?}{=} 4.5$ ✔ — *7.6 is the solution.*

B $3k = 8.1$

$3k = 8.1$ — *k is multiplied by 3.*

$\frac{3k}{3} = \frac{8.1}{3}$ — *Divide both sides by 3 to undo the multiplication.*

$k = 2.7$

Check

$3k = 8.1$

$3(2.7) \stackrel{?}{=} 8.1$ — *Substitute 2.7 for k in the equation.*

$8.1 \stackrel{?}{=} 8.1$ ✔ — *2.7 is the solution.*

Solve each equation. Check your answer.

C $\frac{m}{5} = 1.5$

$\frac{m}{5} = 1.5$ *m is divided by 5.*

$\frac{m}{5} \cdot 5 = 1.5 \cdot 5$ *Multiply both sides by 5 to undo the division.*

$m = 7.5$

Check

$\frac{m}{5} = 1.5$

$\frac{7.5}{5} \stackrel{?}{=} 1.5$ *Substitute 7.5 for m in the equation.*

$1.5 \stackrel{?}{=} 1.5$ ✔ *7.5 is the solution.*

EXAMPLE 2 *Measurement Application*

Remember!

The area of a rectangle is its length times its width.

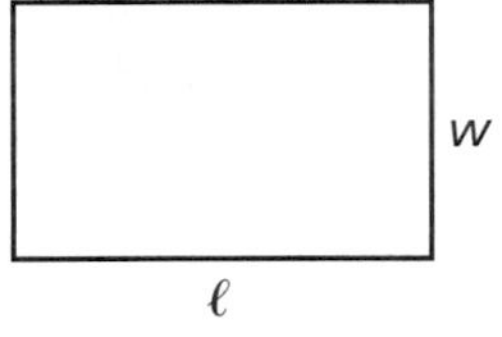

$A = \ell w$

A **The area of the floor in Jonah's bedroom is 28 square meters. If its length is 3.5 meters, what is the width of the bedroom?**

area = length · width

$28 = 3.5 \cdot w$ *Write the equation for the problem.*

$28 = 3.5w$ *Let w be the width of the room.*

$\frac{28}{3.5} = \frac{3.5w}{3.5}$ *w is multiplied by 3.5. Divide both sides by 3.5 to undo the multiplication.*

$8 = w$

The width of Jonah's bedroom is 8 meters.

B **Jonah puts wall-to-wall carpet in his bedroom. The price of the carpet is \$22.50 per square meter. What is the total cost to carpet the bedroom?**

total cost = area · cost of carpet per square meter

$C = 28 \cdot 22.50$ *Let C be the total cost. Write the equation for the problem.*

$C = 630$ *Multiply.*

The cost of carpeting the bedroom is \$630.

Think and Discuss

1. Explain whether the value of m will be less than or greater than 1 when you solve $5m = 4.5$.

2. Tell how you can check the answer in Example 2A.

3-10 Exercises

FOR EOG PRACTICE

see page 647

internet connect

Homework Help Online

go.hrw.com Keyword: MR4 3-10

2.02, 5.03

GUIDED PRACTICE

See Example 1

Solve each equation. Check your answer.

1. $a - 2.3 = 4.8$ **2.** $6n = 8.4$ **3.** $\frac{c}{4} = 3.2$

4. $8.5 = 2.49 + x$ **5.** $\frac{d}{3.2} = 1.09$ **6.** $1.6 = m \cdot 4$

See Example 2

7. The length of a window is 10.5 m, and the width is 5.75 m. Solve the equation $a \div 10.5 = 5.75$ to find the area of the window.

8. The distance around a square picture is 64.8 cm. What is the length of each side?

INDEPENDENT PRACTICE

See Example 1

Solve each equation. Check your answer.

9. $b - 5.6 = 3.7$ **10.** $1.6 = \frac{p}{7}$ **11.** $3r = 62.4$

12. $9.5 = 5x$ **13.** $a - 4.8 = 5.9$ **14.** $\frac{n}{8} = 0.8$

15. $8 + f = 14.56$ **16.** $5.2s = 10.4$ **17.** $1.95 = z - 2.05$

See Example 2

18. The area of a rectangle is 65.8 square units. The length is 7 units. Solve the equation $7 \cdot w = 65.8$ to find the width of the rectangle.

19. Irene bought 1.75 kg of grapes for $5.25. What is the price per kilogram of grapes?

20. Ken placed a fence around his square garden. He used 6.4 meters of fence to enclose all four sides of the garden. How long is each side of his garden?

PRACTICE AND PROBLEM SOLVING

Solve each equation and check your answer.

21. $9.8 = t - 42.1$ **22.** $q \div 2.6 = 9.5$ **23.** $45.36 = 5.6 \cdot m$

24. $1.3b = 5.46$ **25.** $4.93 = 0.563 + m$ **26.** $\frac{a}{5} = 2.78$

27. $w - 64.99 = 13.044$ **28.** $6.205z = 80.665$ **29.** $74.2 = 38.06 + c$

30. $3.7(1.8) = t + 2.9$ **31.** $a - 1.5 = \frac{6.2}{2}$ **32.** $b \cdot 4.4 = 9 + 7.5$

33. The shortest side of the triangle is 10 units long.

a. What are the lengths of the other two sides of the triangle?

b. What is the perimeter of the triangle?

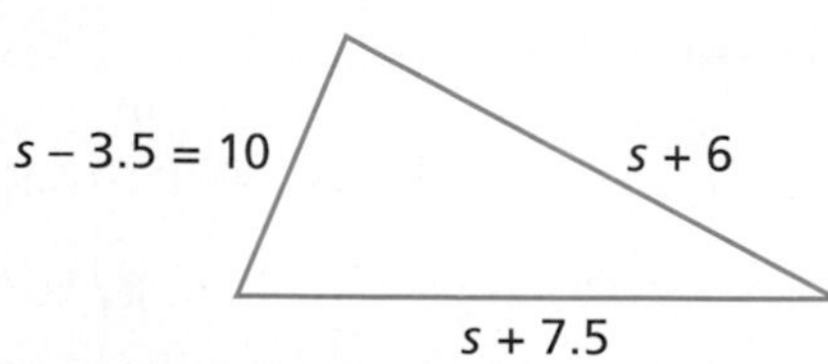

The London Eye is the world's largest Ferris wheel. Use the table for Exercises 34–38.

Weight of wheel	1,900 metric tons
Time to revolve	30 minutes
Height of wheel	135 meters

34. Write the height of the wheel in kilometers.

35. There are 1,000 kilograms in a metric ton.

a. What is the weight of the wheel in kilograms?

b. Write the weight in kilograms in scientific notation.

36. a. How many seconds does it take for the wheel to make one revolution?

b. The wheel moves at a rate of 0.26 meters per second. Use the equation $d \div 0.26 = 1{,}800$ to find the distance of one revolution.

37. Each capsule can hold 25 passengers. How many capsules are needed to hold 210 passengers?

38. Fifteen tickets for the London Eye cost $112.50. What is the cost for one ticket?

39. ***WHAT'S THE ERROR?*** When solving the equation $b - 12.98 = 5.03$, a student said that $b = 7.95$. Describe the error. What is the correct value for b?

40. ***WRITE ABOUT IT*** Explain how you solve for the variable in a multiplication equation such as $2.3a = 4.6$.

41. ***CHALLENGE*** Solve $1.45n \times 3.2 = 23.942 + 4.13$.

Spiral Review

Evaluate each expression. (Lesson 1-4)

42. $6 \times (21 - 15) \div 12$

43. $72 \div 8 + 2^3 \times 5 - 19$

Solve each equation. (Lessons 2-5, 2-6, 2-7)

44. $n - 39 = 20$

45. $3b = 54$

46. $\frac{c}{3} = 12$

47. **EOG PREP** Which is the standard form of 1.7565×10^8? (Lesson 3-5)

A 175,650,000 B 17,565,000 C 1,756,500,000 D 17,565,000,000

EXTENSION Significant Figures

Learn to round measurements to an appropriate number of significant figures.

Vocabulary
significant figures

Measurements are approximations due to the limitations of the tools that are used. For example, a meterstick with tick marks in centimeters cannot give a precise measurement in millimeters.

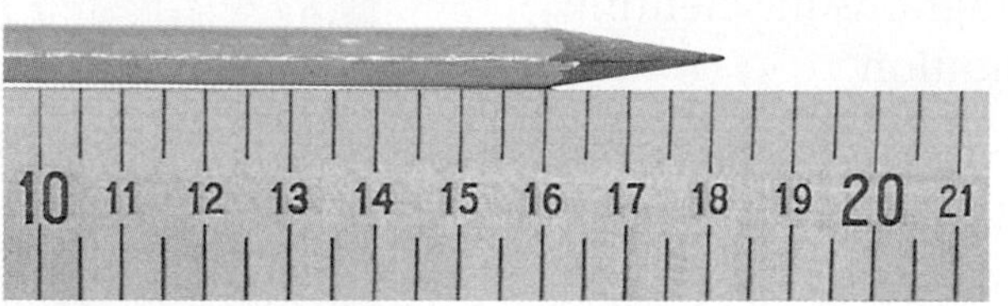

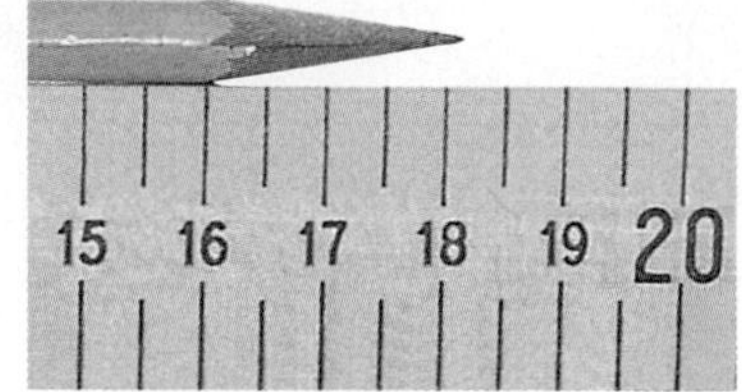

The ruler is marked in centimeters and half centimeters. A measurement in millimeters is only an estimate.

To find out how close your measurement is to an actual measurement, you can use *significant figures.* **Significant figures** are the digits in a measurement that are known with certainty.

Identifying Significant Figures
• Nonzero digits
• Zeros at the end of a number and to the right of the decimal point
• Zeros between significant figures

EXAMPLE 1 Counting Significant Figures

Determine the number of significant figures in each decimal.

A 4.00

4 is a nonzero digit.

The zeros are to the right of the decimal in a number greater than 1.

3 significant figures

B 0.0209

2 and 9 are nonzero digits.

One zero is between nonzero digits.

*Other zeros are **not** significant.*

3 significant figures

When adding or subtracting, find the number in the problem with the least number of *digits* to the right of the decimal point.

EXAMPLE 2 Using Significant Figures in Addition and Subtraction

Write the answer with the appropriate number of significant figures.

35.4 − 7.08

35.4	*1 digit after decimal*
− 7.08	*2 digits after decimal*
28.32	
28.3	*Round the difference to have one digit to the right of the decimal.*

When multiplying and dividing, find the number in the problem with the least number of *significant figures.*

EXAMPLE 3 Using Significant Figures in Multiplication and Division

Write the answer with the appropriate number of significant figures.

0.60 × 1.09

0.60	*2 significant figures*
× 1.09	*3 significant figures*
0.654	
0.65	*Round the product to two significant figures.*

EXTENSION Exercises

Determine the number of significant figures in each decimal.

1. 300.5 **2.** 19.050 **3.** 0.006 **4.** 0.05 **5.** 112.15

Write each answer with the appropriate number of significant figures.

6. 2.5 × 0.6 **7.** 11.54 + 2.8 **8.** 1.20 ÷ 0.8 **9.** 22.57 − 4.85

Use the table for Exercises 10 and 11.

10. How many significant figures are in each score?

11. Using significant figures, tell the difference between the gold medal score and the silver medal score.

Men's 3 m Synchronized Diving Scores (2000 Summer Olympics)

Medal	Country	Score
Gold	China	365.58
Silver	Russia	329.97
Bronze	Australia	322.86

Problem Solving on Location

NORTH CAROLINA

Wild Boars

Wild boars are just one of the many unique wild animals that live in North Carolina. The boar was introduced to Graham County, North Carolina, by a British manufacturing company in the 1900s. Since the 1920s, wild boars have thrived. They now populate other counties, as well as Great Smoky Mountains National Park.

Wild boars are bigger and heavier in the shoulders than in the hips. A boar's mane is formed by bristles, which may reach 5 inches in length. The tusks of a wild boar grow continuously throughout its lifetime. These tusks can be very sharp and can reach a length of 4.75 inches.

Adult males over 2 years old weigh an average of 180 pounds. Females average 155 pounds. Maximum measurements (to the nearest tenth of a foot) are given in the table.

Maximum Measurements of a Wild Boar (ft)	
Body length	5.5
Chest girth	4.3
Ear length	0.5
Hip height	2.8
Shoulder height	2.8
Skull length	1.5
Tail length	0.8
Tusk length	0.4

For 1–5, use the table.

1. Use decimal grids to model the number that represents a boar's maximum shoulder height.
2. What is the difference between a boar's maximum body length and its maximum chest girth?
3. What is the maximum length of a boar's body in inches? (Remember that there are 12 inches in a foot.)
4. Add the tail length, body length, and skull length to find the maximum length of a wild boar.
5. According to the table, which is longer, a boar's ear or a boar's tusk?

Endemic Fish

North Carolina is home to six species of fish that are not found anywhere else in the world. For this reason, the fishes are considered *endemic*, which means they are found only in a particular area. The six endemic fishes in North Carolina are the Waccamaw killifish, Waccamaw silverside, Waccamaw darter, Cape Fear shiner, Pinewoods shiner, and Carolina madtom. The graph shows the maximum lengths for these six species of fish.

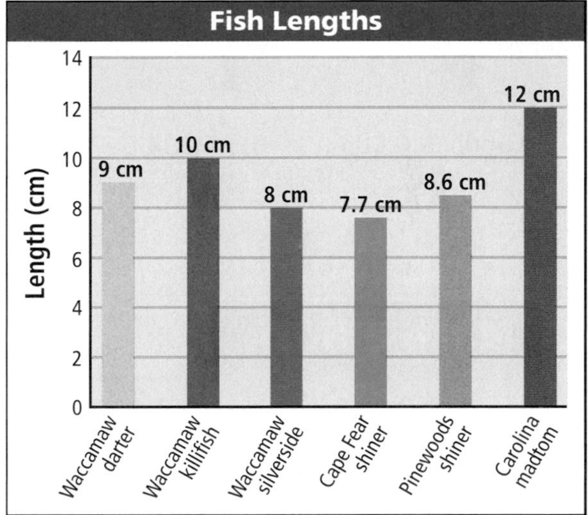

For 1–6, use the graph.

1. Write the length of a Pinewoods shiner in expanded form and in word form.
2. Which endemic fish is the smallest? Which is the largest?
3. How much longer is a Waccamaw darter than a Cape Fear shiner?
4. Find the lengths of the Waccamaw darter, the Waccamaw killifish, and the Waccamaw silverside in millimeters.
5. Find the lengths of the Cape Fear shiner, the Pinewoods shiner, and the Carolina madtom in meters.
6. Write the names of the fish in order from longest fish to shortest fish.

Carolina madtom

Pinewoods shiner

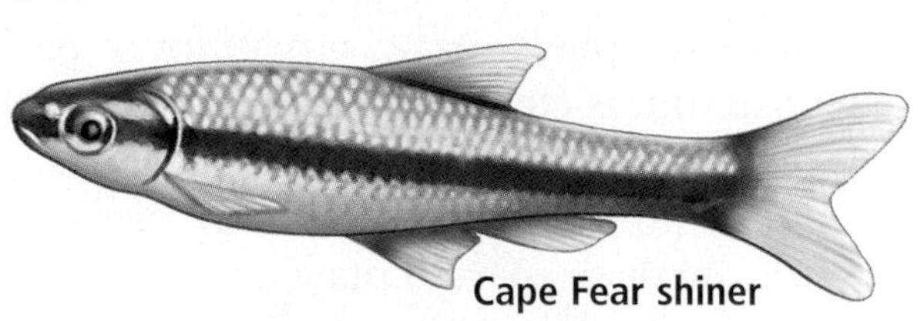

Cape Fear shiner

MATH-ABLES

Jumbles

Do you know what eleven plus two equals?

Use your calculator to evaluate each expression. Keep the letters under the expressions with the answers you get. Then order the answers from least to greatest, and write down the letters in that order. You will spell the answer to the riddle.

$4 - 1.893$	$0.21 \div 0.3$	$0.443 - 0.0042$	$4.509 - 3.526$	$3.14 \cdot 2.44$	$1.56 \cdot 3.678$
E	**L**	**E**	**V**	**E**	**N**

$6.34 \div 2.56$	$1.19 + 1.293$	$8.25 \div 2.5$	$7.4 - 2.356$
P	**L**	**U**	**S**

$0.0003 + 0.003$	$0.3 \cdot 0.04$	$2.17 + 3.42$
T	**W**	**O**

Make a Buck

The object of the game is to win the most points by adding decimal numbers to make a sum close to but not over \$1.00.

Most cards have a decimal number on them representing an amount of money. Others are wild cards: The person who receives a wild card decides its value.

The dealer gives each player four cards. Taking turns, players add the numbers in their hand. If the sum is less than \$1.00, a player can either draw a card from the top of the deck or pass.

When each player has taken a turn or passed, the player whose sum is closest to but not over \$1.00 scores a point. If players tie for the closest sum, each of those players scores a point. All cards are then discarded and four new cards are dealt to each player.

When all of the cards have been dealt, the player with the most points wins.

internet connect go.hrw.com

Go to ***go.hrw.com*** for a complete set of rules and game pieces.
KEYWORD: MR4 Game3

Scientific Notation

internet connect
Lab Resources Online
go.hrw.com
KEYWORD: MR4 TechLab3

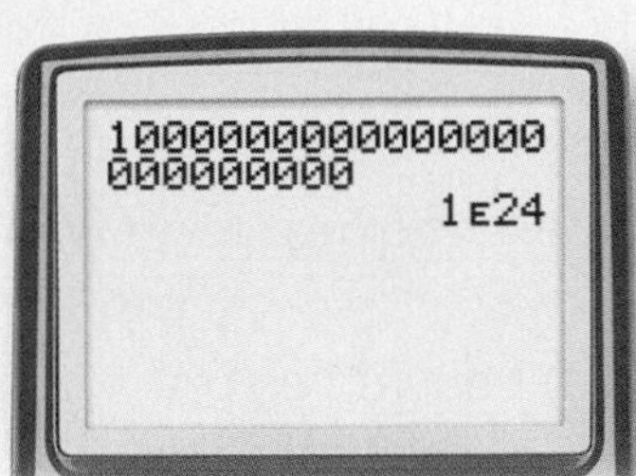

One **septillion** is a very large number. It is 1 followed by 24 zeros, or in exponential notation, 10^{24}.

Enter 1 septillion on your graphing calculator by pressing 1 and then pressing 0 twenty-four times. Press ENTER. What happens?

The calculator displays **1E24**, which means $\mathbf{1 \times 10^{24}}$. A calculator displays very large numbers in **scientific notation.**

Activity

1 Evaluate 248,925 · 259,871 on a calculator. Write the product in standard form.

The product is **6.468838868E10**.
$6.468838868 \times 10^{10} = 64{,}688{,}388{,}680$

The display does not show every digit of the product. The answer is rounded to the nearest ten.

2 Use the **EE** function (the second function of the *comma* key) to enter 2.57×10^7, 5.895×10^{12}, and $3{,}452 \cdot 6.45 \times 10^8$ into a calculator using scientific notation.

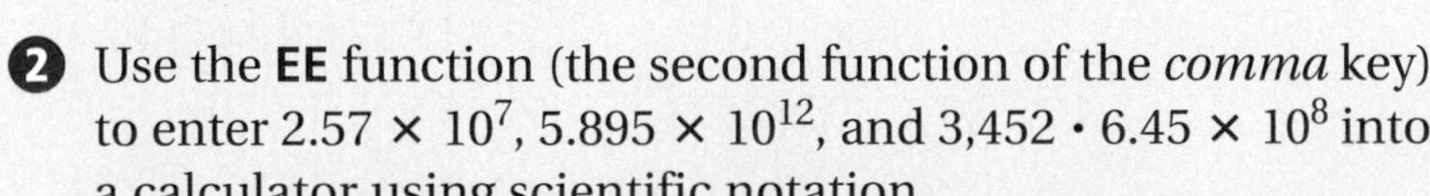
To enter 2.57×10^7, press 2 . 5 7 2nd EE , 7 ENTER.

Notice that, when possible, the calculator displays the results in standard form.

Think and Discuss

1. Tell what happens if you enter a googol into your calculator. A googol is 10^{100}.

Try This

Use your calculator to evaluate each expression. Write the results in standard form.

1. 25,468 · 689,234

2. $5.4 \times 10^4 \cdot 5.6821 \times 10^5$

3. 1.25 · 2 septillion

Chapter 3

Study Guide and Review

Study Guide and Review

Vocabulary

clustering 96

front-end estimation 97

scientific notation 114

Choose the best term from the list above. Words may be used more than once.

1. When you estimate a sum by using only the whole-number part of the decimals, you are using __?__.
2. __?__ is a shorthand method for writing large numbers.
3. __?__ means rounding all the numbers to the same value.

3-1 Representing, Comparing, and Ordering Decimals (pp. 92–95)

EXAMPLES

■ **Write 4.025 in expanded form and words.**

Expanded form: 4 + 0.02 + 0.005

Word form: four and twenty-five thousandths

■ **Order the decimals from least to greatest. 7.8, 7.83, 7.08**

7.08 < 7.80 < 7.83 *Compare the numbers.*
7.08, 7.8, 7.83 *Then order the numbers.*

EXERCISES

Write each in expanded form and words.

4. 5.68
5. 1.0076
6. 1.203
7. 23.005

Order the decimals from least to greatest.

8. 1.2, 1.3, 1.12
9. 11.17, 11.7, 11.07
10. 0.3, 0.303, 0.033
11. 5.009, 5.950, 5.5

3-2 Estimating Decimals (pp. 96–99)

EXAMPLES

■ **Estimate.**

5.35 − 0.7904; round to tenths

$$\begin{array}{r} 5.4 \\ -\,0.8 \\ \hline 4.6 \end{array}$$

Align the decimals.
Subtract.

■ **Estimate each product or quotient.**

49.67 × 2.88

50 × 3 = 150

EXERCISES

Estimate.

12. 8.0954 + 3.218; round to the hundredths
13. 6.8356 − 4.507; round to the tenths

Estimate each product or quotient.

14. 21.19 × 4.23
15. 53.98 ÷ 5.97

3-3 Adding and Subtracting Decimals (pp. 102–105)

EXAMPLE

■ **Find the sum or difference.**

7.62 + 0.563

7.620	*Align the decimal points.*
+ 0.563	*Use zeros as placeholders.*
8.183	*Add. Place the decimal point.*

EXERCISES

Find each sum or difference.

16. 7.08 + 4.5 + 13.27 **17.** 6 − 0.7

Evaluate 6.48 − *s* for each value of *s*.

18. $s = 3.9$ **19.** $s = 3.6082$

3-4 Decimals and Metric Measurement (pp. 106–109)

EXAMPLES

■ **Multiply or divide.**

326 × 10,000 *Move the decimal point 4 places right.*

= 3,260,000 *Write 4 placeholder zeros.*

■ **Convert the measure.**

3,200 g = ___?___ kg *(1,000 g = 1 kg)*

3,200 g = 3.2 kg

EXERCISES

Multiply or divide.

20. 12.6×10^4 **21.** $546 \div 10^3$

22. 67×10^5 **23.** $180.6 \div 10^2$

Convert each measure.

24. 8.9 L = ___?___ mL

25. 18 cm = ___?___ m

3-5 Scientific Notation (pp. 114–117)

EXAMPLES

■ **Write the number in scientific notation.**

60,000 *Move the decimal point 4 places to the left.*

$= 6.0 \times 10^4$

■ **Write each number in standard form.**

7.18×10^5

= 718,000 *Move the decimal point 5 places right.*

EXERCISES

Write each number in scientific notation.

26. 550,000 **27.** 7,230

28. 1,300,000 **29.** 14.8

Write each number in standard form.

30. 3.02×10^4 **31.** 4.293×10^5

32. 1.7×10^6 **33.** 5.39×10^3

3-6 Multiplying Decimals (pp. 120–123)

EXAMPLE

■ **Find the product.**

0.3	*1 decimal place*
× 0.08	*+ 2 decimal places*
0.024	*3 decimal places*

EXERCISES

Find each product.

34. 4 × 2.36 **35.** 0.5 × 1.73

36. 0.6 × 0.012 **37.** 8 × 3.052

Study Guide and Review

3-7 Dividing Decimals by Whole Numbers (pp. 124–126)

EXAMPLE

- **Find the quotient.**

Place a decimal point directly above the decimal point in the dividend. Then divide.

$$5\overline{)0.95} = 0.19$$

EXERCISES

Find each quotient.

38. $6.18 \div 6$

39. $2.16 \div 3$

40. $34.65 \div 9$

41. $20.72 \div 8$

42. If four people equally share a bill for \$14.56, how much should each person pay?

3-8 Dividing by Decimals (pp. 127–130)

EXAMPLE

- **Find the quotient.**

$9.65 \div 0.5$

Make the divisor a whole number.
Place the decimal point in the quotient.

$$5\overline{)96.5} = 19.3$$

EXERCISES

Find each quotient.

43. $4.86 \div 0.6$

44. $1.85 \div 0.3$

45. $34.89 \div 9$

46. $62.73 \div 1.2$

47. Ana cuts some wood that is 3.75 meters long into 5 pieces of equal length. How long is each piece?

3-9 Interpret the Quotient (pp. 131–133)

EXAMPLE

- **Ms. Ald needs 26 stickers for her preschool class. Stickers are sold in packs of 8. How many packs should she buy?**

$26 \div 8 = 3.25$

3.25 is between 3 and 4.
3 packs will not be enough.

Ms. Ald should buy 4 packs of stickers.

EXERCISES

48. Billy has 3.6 liters of juice. How many 0.25 L containers can he fill?

49. There are 34 people going on a field trip. If each car holds 4 people, how many cars will they need for the field trip?

3-10 Solving Decimal Equations (pp. 134–137)

EXAMPLE

- **Solve the equation.**

$4x = 20.8$ — *x is multiplied by 4.*

$\frac{4x}{4} = \frac{20.8}{4}$ — *Divide both sides by 4.*

$x = 5.2$

EXERCISES

Solve each equation.

50. $a - 6.2 = 7.18$

51. $3y = 7.86$

52. $n + 4.09 = 6.38$

53. $\frac{p}{7} = 8.6$

54. Jasmine buys 2.25 kg of apples for \$11.25. How much does 1 kg of apples cost?

Chapter Test

Chapter 3

Write each decimal in standard form, expanded form, and words.

1. 3.107

2. Eight and forty-nine thousandths

Order the decimals from least to greatest.

3. 12.6, 12.07, 12.67

4. 3.5, 3.25, 3.08

5. 23.84, 23.59, 2.899

Estimate by rounding to the indicated place value.

6. 6.178 − 0.2805; hundredths

7. 7.528 + 6.075; ones

Estimate.

8. 21.35×3.18

9. $98.547 \div 4.93$

10. 11.855×8.45

Estimate a range for the sum.

11. 3.89 + 42.71 + 12.32

12. 20.751 + 2.55 + 17.4

13. 4.987 + 28.27 + 0.098

Evaluate.

14. 0.76 + 2.24

15. 7 − 0.4

16. 0.12×0.006

17. $76 \times 10{,}000$

18. $4.57 \div 10^3$

19. $3.44 \div 4$

Convert each measure.

20. 4,700 mL = ___?___ L

21. 3.2 km = ___?___ m

22. 22.8 cm = ___?___ m

Write each number in scientific notation.

23. 16,900

24. 180,500

25. 3,190,000

Write each number in standard form.

26. 3.08×10^5

27. 1.472×10^6

28. 2.973×10^4

Solve each equation.

29. $b - 4.7 = 2.1$

30. $5a = 4.75$

31. $\frac{y}{6} = 7.2$

32. The school band is going to a local competition. There are 165 students in the band. If each bus holds 25 students, how many buses will be needed?

33. Maria and five friends went shopping. All sweaters at the store were on sale for the same price. Each girl chose a sweater. The total bill was $126.24. How much did each sweater cost?

Chapter 3

Performance Assessment

Show What You Know

Create a portfolio of your work from this chapter. Complete this page and include it with your four best pieces of work from Chapter 3. Choose from your homework or lab assignments, mid-chapter quiz, or any journal entries you have done. Put them together using any design you want. Make your portfolio represent what you consider to be your best work.

Short Response

1. Jocelyn and her 2 sisters went shopping for school supplies. They purchased a box of markers for \$2.99, a pack of pencils for \$2.38, 3 notebooks for \$1.59 each, and 3 videos for \$14.94 each. Find the total cost of their purchases. If the cost is divided equally among the girls, how much should each girl pay? Show your work.

2. Emily wants to rent 6 DVD movies to watch over the weekend. On Friday, she can rent 3 movies from the movie store for \$11.25. On Saturday, she can rent 2 movies for \$6.50. On which day should Emily rent movies? Explain your answer.

3. Michael is making costumes for his school play. Each costume requires 1.25 yards of fabric. If Michael has a total of 12 yards of fabric, how many costumes can he make? Explain how you found your answer.

Extended Problem Solving

4. The large rectangular plot of land in the diagram is being made into a park. The total area of the park is 4.56 square kilometers, and the length of the park is 2.4 kilometers.
   a. Let w represent the width of the park. Write an equation that could be used to find the value of w.
   b. Find the width of the park in kilometers. Show your work.
   c. Find the width of the park in meters, and write it in scientific notation.

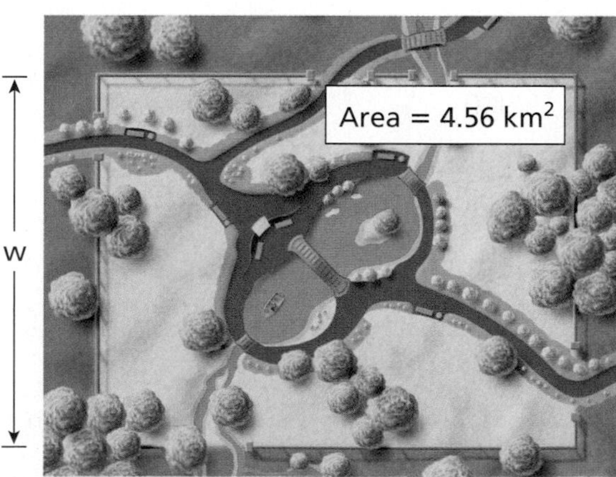

Performance Assessment

internet connect
EOG Test Practice Online
go.hrw.com Keyword: MR4 TestPrep

Getting Ready for EOG

Chapter 3

Cumulative Assessment, Chapters 1–3

1. Which of the following is the standard form for four and seven hundred eighteen ten-thousandths?

A 4.718 C 4.0718

B 4.7018 D 4.00718

2. What is the value of $2^2 \times 11 - 7 + 2^3$?

A 45 C 43

B 17 D 22

3. What is the product of 0.4×3.25?

A 1,300 C 13.00

B 130 D 1.30

TEST TAKING TIP!

To compare decimal numbers, start from the left and compare digits in the same place-value position. Look for the first place value the digits are different.

4. Which number is the *greatest*?

A 18.095 C 18.907

B 18.9 D 18.75

5. Which equation has 8 as the solution?

A $y - 3 = 6$

B $5y = 40$

C $\frac{y}{4} = 8$

D $y + 8 = 32$

6. Which is the quotient of $7.89 \div 3$?

A 263 C 0.263

B 26.30 D 2.63

7. Which measurement is equivalent to 780 cm?

A 78 mm C 7.8 m

B 0.78 km D 7,800 km

8. The area of a rectangle is 83.125 cm^2. Its length is 9.5 cm. Solve $9.5w = 83.125$ to find the width.

A 7.89 cm

B 8.75 cm

C 9 cm

D 0.875 cm

9. Mr. Myers is placing 6 lights along his driveway. If the length of the driveway is 14.4 meters, how far apart should he place the lights to have them equally spaced along the driveway?

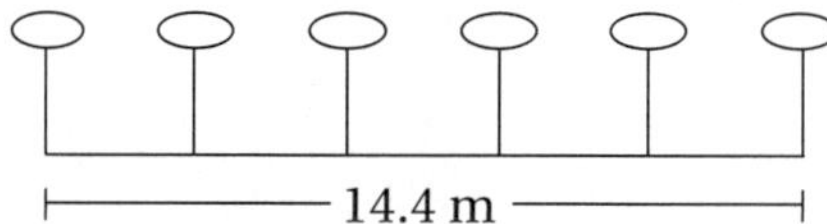

A 86.4 meters

B 2.4 meters

C 20.4 meters

D 2.88 meters

10. ***SHORT RESPONSE*** Explain why 300 cm is the same as 3 m.

11. ***SHORT RESPONSE*** Explain why 54.9×10^6 is not correctly written in scientific notation. What is the correct way to write 54.9×10^6 in scientific notation?

Getting Ready for EOG

Chapter 4

Number Theory and Fractions

TABLE OF CONTENTS

internet connect

Chapter Opener Online
go.hrw.com
KEYWORD: MR4 Ch4

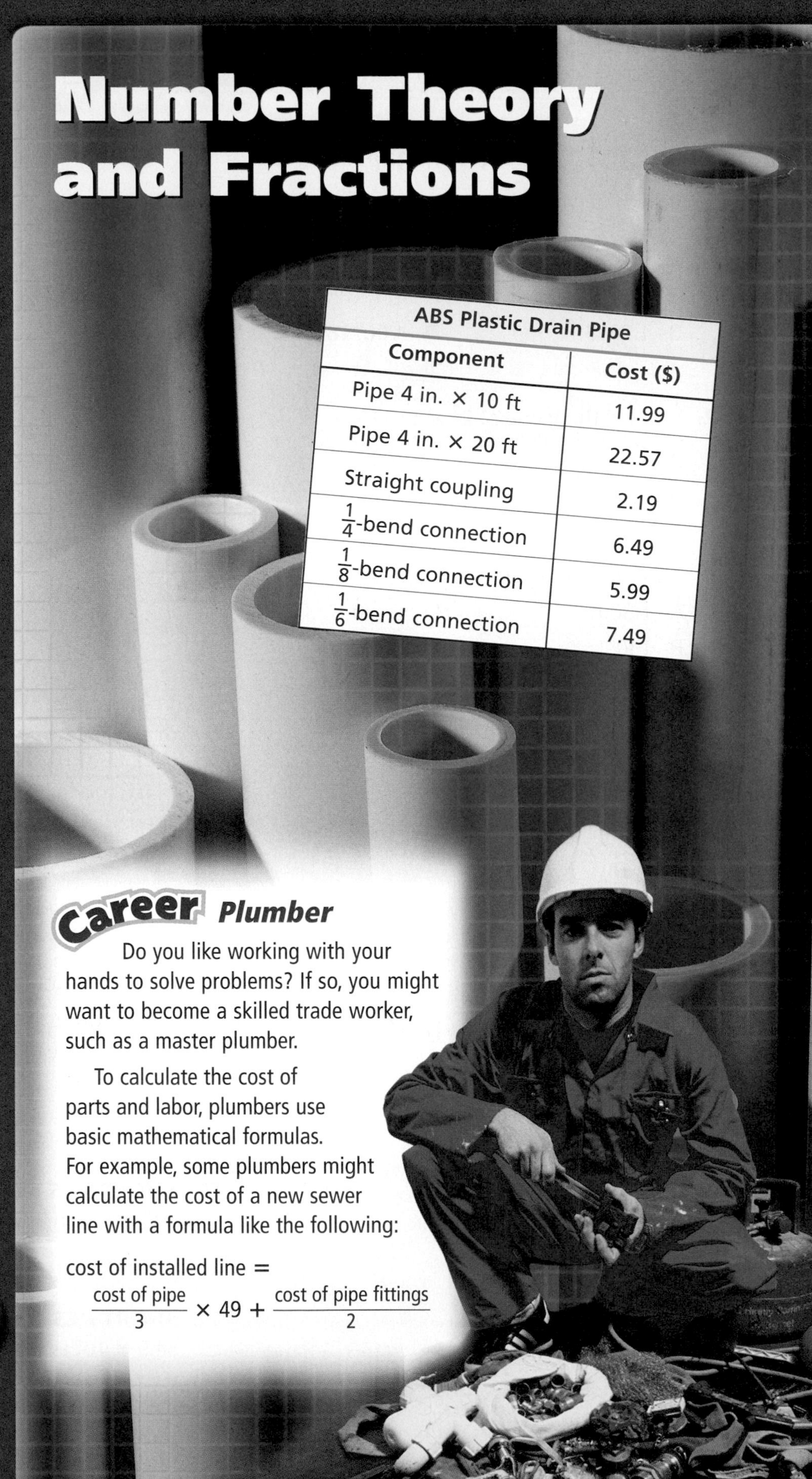

ABS Plastic Drain Pipe	
Component	**Cost ($)**
Pipe 4 in. × 10 ft	11.99
Pipe 4 in. × 20 ft	22.57
Straight coupling	2.19
$\frac{1}{4}$-bend connection	6.49
$\frac{1}{8}$-bend connection	5.99
$\frac{1}{6}$-bend connection	7.49

Career *Plumber*

Do you like working with your hands to solve problems? If so, you might want to become a skilled trade worker, such as a master plumber.

To calculate the cost of parts and labor, plumbers use basic mathematical formulas. For example, some plumbers might calculate the cost of a new sewer line with a formula like the following:

cost of installed line =

$$\frac{\text{cost of pipe}}{3} \times 49 + \frac{\text{cost of pipe fittings}}{2}$$

ARE YOU READY?

Choose the best term from the list to complete each sentence.

denominator
difference
multiples
numerator
product
quotient
add

1. To find the sum of two numbers, you should __?__.
2. Fractions are written as a __?__ over a __?__.
3. In the equation $4 \cdot 3 = 12$, 12 is the __?__.
4. The __?__ of 18 and 10 is 8.
5. The numbers 18, 27, and 72 are __?__ of 9.

Complete these exercises to review skills you will need for this chapter.

✔ Write and Read Decimals

Write each decimal in word form.

6. 0.5 **7.** 2.78 **8.** 0.125

9. 12.8 **10.** 125.49 **11.** 8.024

✔ Identify Sets of Numbers

Determine whether each number is even or odd.

12. 125 **13.** 29 **14.** 24 **15.** 127

16. 213 **17.** 98 **18.** 2 **19.** 17

✔ Multiples

List the first four multiples of each number.

20. 6 **21.** 8 **22.** 5 **23.** 12

24. 7 **25.** 20 **26.** 14 **27.** 9

✔ Evaluate Expressions

Evaluate each expression for the given value of the variable.

28. $y + 4.3$ for $y = 3.2$ **29.** $3c$ for $c = 0.75$

30. $27.8 - d$ for $d = 9.25$ **31.** $\frac{x}{5}$ for $x = 6.4$

32. $a + 4 \div 8$ for $a = 3.75$ **33.** $2.5b$ for $b = 8.4$

4-1 Divisibility

Learn to use divisibility rules.

Vocabulary
divisible
composite number
prime number

This year, 42 girls signed up to play basketball for the Junior Girls League, which has 6 teams. To find whether each team can have the same number of girls, decide if 42 is divisible by 6.

A number is **divisible** by another number if the quotient is a whole number with no remainder.

$42 \div 6 = 7$ ← **Quotient**

Since there is no remainder, 42 is divisible by 6. The Junior Girls League can have 6 teams with 7 girls each.

Divisibility Rules		
A number is divisible by...	**Divisible**	**Not Divisible**
2 if the last digit is even (0, 2, 4, 6, or 8).	3,978	4,975
3 if the sum of the digits is divisible by 3.	315	139
4 if the last two digits form a number divisible by 4.	8,512	7,518
5 if the last digit is 0 or 5.	14,975	10,978
6 if the number is divisible by both 2 and 3.	48	20
9 if the sum of the digits is divisible by 9.	711	93
10 if the last digit is 0.	15,990	10,536

EXAMPLE 1 Checking Divisibility

A **Tell whether 610 is divisible by 2, 3, 4, and 5.**

2	*The last digit, 0, is even.*	Divisible
3	*The sum of the digits is $6 + 1 + 0 = 7$. 7 is not divisible by 3.*	Not divisible
4	*The last two digits form the number 10. 10 is not divisible by 4.*	Not divisible
5	*The last digit is 0.*	Divisible

So 610 is divisible by 2 and 5.

B **Tell whether 387 is divisible by 6, 9, and 10.**

6	*The last digit, 7, is odd, so 387 is not divisible by 2.*	Not divisible
9	*The sum of the digits is 3 + 8 + 7 = 18. 18 is divisible by 9.*	Divisible
10	*The last digit is 7, not 0.*	Not divisible

So 387 is divisible by 9.

Any number greater than 1 is divisible by at least two numbers—1 and the number itself. Numbers that are divisible by more than two numbers are called **composite numbers** .

A **prime number** is divisible by only the numbers 1 and itself. For example, 11 is a prime number because it is divisible by only 1 and 11. The numbers 0 and 1 are neither prime nor composite.

EXAMPLE 2 Identifying Prime and Composite Numbers

Tell whether each number is prime or composite.

A **45**
divisible by 1, 3, 5, 9, 15, 45
composite

B **13**
divisible by 1, 13
prime

C **19**
divisible by 1, 19
prime

D **49**
divisible by 1, 7, 49
composite

The prime numbers from 1 through 50 are highlighted below.

1	2	3	4	5	6	7	8	9	10
11	12	13	14	15	16	17	18	19	20
21	22	23	24	25	26	27	28	29	30
31	32	33	34	35	36	37	38	39	40
41	42	43	44	45	46	47	48	49	50

Think and Discuss

1. Tell which whole numbers are divisible by 1.

2. Explain how you know that 87 is a composite number.

3. Tell how the divisibility rules help you identify composite numbers.

4-1 Exercises

FOR EOG PRACTICE

see page 648

internet connect

Homework Help Online

go.hrw.com Keyword: MR4 4-1

1.04a, 1.05

GUIDED PRACTICE

See Example 1 **Tell whether each number is divisible by 2, 3, 4, 5, 6, 9, and 10.**

1. 508 **2.** 432 **3.** 247 **4.** 189

See Example 2 **Tell whether each number is prime or composite.**

5. 75 **6.** 17 **7.** 27 **8.** 63

INDEPENDENT PRACTICE

See Example 1 **Tell whether each number is divisible by 2, 3, 4, 5, 6, 9, and 10.**

9. 741 **10.** 810 **11.** 675 **12.** 480

13. 908 **14.** 146 **15.** 514 **16.** 405

See Example 2 **Tell whether each number is prime or composite.**

17. 34 **18.** 29 **19.** 61 **20.** 81

21. 51 **22.** 23 **23.** 97 **24.** 93

25. 77 **26.** 41 **27.** 67 **28.** 39

PRACTICE AND PROBLEM SOLVING

Copy and complete the table. Write *yes* if the number is divisible by the given number. Write *no* if it is not.

		2	3	4	5	6	9	10
29.	677	*no*	■	■	*no*	■	■	*no*
30.	290	*yes*	■	■	■	■	■	■
31.	1,744	■	■	■	■	■	■	■
32.	12,180	■	■	■	■	■	■	■

Replace each box with a digit that will make the number divisible by 3.

33. 74■ **34.** 8,10■ **35.** 3,■41

36. ■,335 **37.** 67,■11 **38.** 10,0■1

39. Make a table that shows the prime numbers from 50 to 100.

40. Tell whether each statement is true or false. Explain your answers.

a. All even numbers are divisible by 2.

b. All odd numbers are divisible by 3.

c. Some even numbers are divisible by 5.

41. ***ASTRONOMY*** Earth has a diameter of 7,926 miles. Tell whether this number is divisible by 2, 3, 4, 5, 6, 9, and 10.

42. On which of the bridges in the table could a light fixture be placed every 6 meters so that the first light is at the beginning of the bridge and the last light is at the end of the bridge?

Golden Gate Bridge

Longest Bridges in the U.S.	
Name and State	**Length (m)**
Verrazano Narrows, NY	1,298
Golden Gate, CA	1,280
Mackinac Straits, MI	1,158
George Washington, NY	1,067

43. A number is between 80 and 100 and is divisible by both 5 and 6. What is the number?

44. ***CHOOSE A STRATEGY*** Find the greatest four-digit number that is divisible by 1, 2, 3, and 4.

45. ***WHAT'S THE ERROR?*** To find whether 3,463 is divisible by 4, a student added the digits. The sum, 16, is divisible by 4, so the student stated that 3,463 is divisible by 4. Explain the error.

46. ***WRITE ABOUT IT*** If a number is divisible by both 4 and 9, by what other numbers is it divisible? Explain.

47. ***CHALLENGE*** Find a number that is divisible by 2, 3, 4, 5, 6, and 10, but not 9.

Spiral Review

Compare. Write $<$ or $>$. (Lesson 1-1)

48. 10,976 ▢ 100,100 **49.** 32,107,120 ▢ 32,170,021

50. 60,842,250 ▢ 60,847,205 **51.** 136,422,190 ▢ 136,242,910

Write each phrase as a numerical or algebraic expression. (Lesson 2-2)

52. 562 plus t **53.** the product of n and 16 **54.** the quotient of p and 7

Solve each equation. Check your answer. (Lesson 2-4)

55. $17 + b = 44$ **56.** $x + 31 = 72$ **57.** $28 + y = 57$

58. **EOG PREP** Which decimal is *greatest*? (Lesson 3-1)

A 7.081 B 17.8 C 7.18 D 17.081

4-2 Factors and Prime Factorization

Learn to write prime factorizations of composite numbers.

Vocabulary
factor
prime factorization

Whole numbers that are multiplied to find a product are called **factors** of that product. A number is divisible by its factors.

$2 \cdot 3 = 6$ (2 and 3: Factors; 6: Product)

$6 \div 3 = 2$

$6 \div 2 = 3$

6 is divisible by 3 and 2.

EXAMPLE 1 Finding Factors

List all of the factors of each number.

A 18

Begin listing factors in pairs.

$18 = 1 \cdot 18$ *1 is a factor.*

$18 = 2 \cdot 9$ *2 is a factor.*

$18 = 3 \cdot 6$ *3 is a factor.*

4 is not a factor.

5 is not a factor.

$18 = 6 \cdot 3$ *6 and 3 have already been listed, so stop here.*

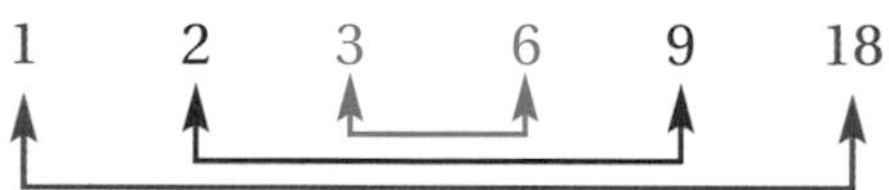

You can draw a diagram to illustrate the factor pairs.

The factors of 18 are 1, 2, 3, 6, 9, and 18.

B 13

$13 = 1 \cdot 13$ *Begin listing factors in pairs. 13 is not divisible by any other whole numbers.*

The factors of 13 are 1 and 13.

> **Helpful Hint**
> When the pairs of factors begin to repeat, then you have found all of the factors of the number you are factoring.

You can use factors to write a number in different ways.

Factorization of 12			
$1 \cdot 12$	$2 \cdot 6$	$3 \cdot 4$	$3 \cdot 2 \cdot 2$

Notice that these factors are all prime.

The **prime factorization** of a number is the number written as the product of its prime factors.

EXAMPLE 2 Writing Prime Factorizations

Write the prime factorization of each number.

36

Method 1: Use a factor tree.

Choose any two factors of 36 to begin. Keep finding factors until each branch ends at a prime factor.

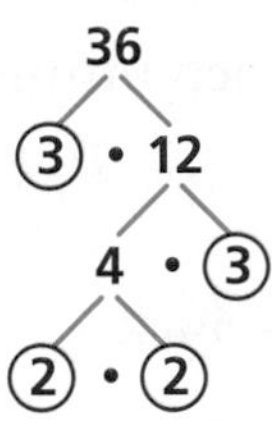

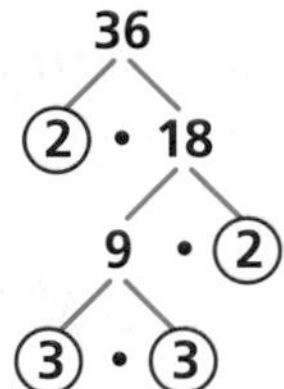

$36 = 3 \cdot 2 \cdot 2 \cdot 3$ $\qquad$ $36 = 2 \cdot 3 \cdot 3 \cdot 2$

The prime factorization of 36 is $2 \cdot 2 \cdot 3 \cdot 3$, or $2^2 \cdot 3^2$.

B 54

Method 2: Use a ladder diagram.

Choose a prime factor of 54 to begin. Keep dividing by prime factors until the quotient is 1.

```
2 | 54          3 | 54
 3 | 27          3 | 18
  3 | 9           2 | 6
   3 | 3           3 | 3
       1               1
```

$54 = 2 \cdot 3 \cdot 3 \cdot 3$ $\qquad$ $54 = 3 \cdot 3 \cdot 2 \cdot 3$

The prime factorization of 54 is $2 \cdot 3 \cdot 3 \cdot 3$, or $2 \cdot 3^3$.

Helpful Hint

You can use exponents to write prime factorizations. Remember that an exponent tells you how many times the base is a factor.

In Example 2, notice that the prime factors may be written in a different order, but they are still the same factors. Except for changes in the order, there is only one way to write the prime factorization of a number.

Think and Discuss

1. **Tell** how you know when you have found all of the factors of a number.
2. **Tell** how you know when you have found the prime factorization of a number.
3. **Explain** the difference between factors of a number and prime factors of a number.

4-2 Exercises

FOR EOG PRACTICE

see page 648

internet connect

Homework Help Online
go.hrw.com Keyword: MR4 4-2

1.04a, 1.05

GUIDED PRACTICE

See Example 1 List all of the factors of each number.

1. 12 **2.** 21 **3.** 52 **4.** 75

See Example 2 Write the prime factorization of each number.

5. 48 **6.** 20 **7.** 66 **8.** 34

INDEPENDENT PRACTICE

See Example 1 List all of the factors of each number.

9. 24 **10.** 37 **11.** 42 **12.** 56

13. 67 **14.** 72 **15.** 85 **16.** 92

See Example 2 Write the prime factorization of each number.

17. 49 **18.** 38 **19.** 76 **20.** 60

21. 81 **22.** 132 **23.** 140 **24.** 87

PRACTICE AND PROBLEM SOLVING

Write each number as a product in two different ways.

25. 34 **26.** 82 **27.** 88 **28.** 50

29. 15 **30.** 78 **31.** 94 **32.** 35

Find the prime factorization of each number.

33. 99 **34.** 249 **35.** 284 **36.** 620

37. 840 **38.** 150 **39.** 740 **40.** 402

41. The prime factorization of 50 is $2 \cdot 5^2$. Without dividing or using a diagram, find the prime factorization of 100.

42. ***GEOMETRY*** The area of a rectangle is the product of its length and width. Suppose the area of a rectangle is 24 in^2. What are the possible whole number measurements of its length and width?

43. ***PHYSICAL SCIENCE*** The speed of sound at sea level at 20°C is 343 meters per second. Write the prime factorization of 343.

44. ***SPORTS*** Little League Baseball began in 1939 in Pennsylvania. When it first started, there were 45 boys on 3 teams.

a. If the teams were equally sized, how many boys were on each team?

b. Name another way the boys could have been divided into equally sized teams. (Remember that a baseball team must have at least 9 players.)

Life Science 

Climate changes, habitat destruction, and overhunting can cause animals and plants to die in large numbers. When the entire population of a species begins to die out, the species is considered endangered.

The graph shows the number of endangered species in each category of animal.

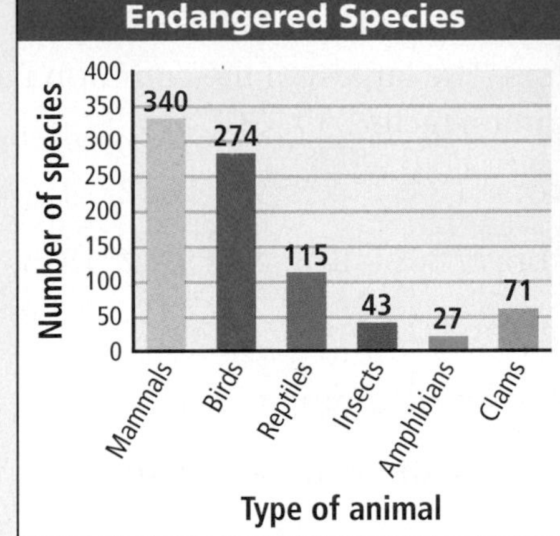

45. How many species of mammals are endangered? Write this number as the product of prime factors.

46. Which categories of animals have a prime number of endangered species?

47. How many species of reptiles and amphibians combined are endangered? Write the answer as the product of prime factors.

48. ***WHAT'S THE ERROR?*** When asked to write the prime factorization of the number of endangered amphibian species, a student wrote 3×9. Explain the error and write the correct answer.

49. ***WRITE ABOUT IT*** A team of five scientists is going to study endangered insect species. The scientists want to divide the species evenly among them. Will they be able to do this? Why or why not?

50. ***CHALLENGE*** Add the number of endangered mammal species to the number of endangered bird species. Find the prime factorization of this number.

Laysan albatross chicks often die from eating plastic that pollutes the oceans and beaches. Clean-up efforts may prevent the albatross from becoming endangered.

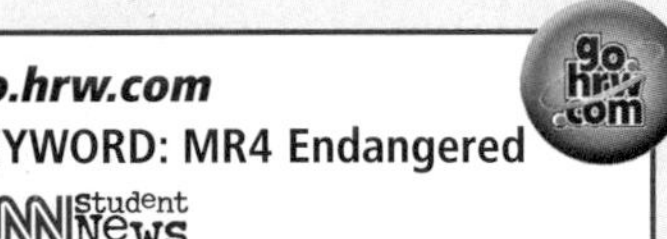

Spiral Review

Estimate each sum or difference by rounding to the place value indicated. (Lesson 1-2)

51. 4,798 + 2,118; thousands

52. 6,293 − 3,192; thousands

53. 23,978 + 18,164; ten thousands

54. 49,169 − 13,919; ten thousands

Find each value. (Lesson 1-3)

55. 5^4

56. 8^3

57. 9^2

58. 10^6

Solve each equation. (Lessons 2-4 and 2-5)

59. $18 + p = 26$

60. $n - 34 = 177$

61. $7 + b = 84$

62. **EOG PREP** Which number is 7,050,000 expressed in scientific notation? (Lesson 3-5)

A 7.05×10^5

B 70.5×10^5

C 0.705×10^7

D 7.05×10^6

4-3 Greatest Common Factor

Learn to find the greatest common factor (GCF) of a set of numbers.

Vocabulary

greatest common factor (GCF)

Factors shared by two or more whole numbers are called common factors. The largest of the common factors is called the **greatest common factor**, or **GCF**.

Factors of 24: **1**, **2**, **3**, **4**, **6**, 8, **12**, 24

Factors of 36: **1**, **2**, **3**, **4**, **6**, 9, **12**, 18, 36

Common factors: 1, 2, 3, 4, 6, (12)

The greatest common factor (GCF) of 24 and 36 is 12.

Example 1 shows three different methods for finding the GCF.

EXAMPLE 1 Finding the GCF

Find the GCF of each set of numbers.

A **16 and 24**

Method 1: List the factors.

factors of 16: **1**, **2**, **4**, (8), 16 — *List all the factors.*

factors of 24: **1**, **2**, 3, **4**, 6, (8), 12, 24 — *Circle the GCF.*

The GCF of 16 and 24 is 8.

B **12, 24, and 32**

Method 2: Use prime factorization.

$12 = 2 \cdot 2 \cdot 3$ — *Write the prime factorization of each number.*

$24 = 2 \cdot 2 \cdot 2 \cdot 3$

$32 = 2 \cdot 2 \cdot 2 \cdot 2 \cdot 2$ — *Find the common prime factors.*

$2 \cdot 2 = 4$ — *Find the product of the common prime factors.*

The GCF of 12, 24, and 32 is 4.

C **12, 18, and 60**

Method 3: Use a ladder diagram.

2	12	18	60
3	6	9	30
	2	3	10

Begin with a factor that divides into each number. Keep dividing until the three numbers have no common factors.

$2 \cdot 3 = 6$ — *Find the product of the numbers you divided by.*

The GCF is 6.

EXAMPLE  PROBLEM SOLVING APPLICATION

There are 12 boys and 18 girls in Mr. Ruiz's science class. The students must form lab groups. Each group must have the same number of boys and the same number of girls. What is the greatest number of groups Mr. Ruiz can make if every student must be in a group?

1 Understand the Problem

The **answer** will be the *greatest* number of groups 12 boys and 18 girls can form so that each group has the same number of boys, and each group has the same number of girls.

2 Make a Plan

You can make an organized list of the possible groups.

3 Solve

Helpful Hint

If more students are put in each group, there will be fewer groups. You need the most groups possible, so put the smallest possible number of students in each team. Start with 1 boy in each group.

There are more girls than boys in the class, so there will be more girls than boys in each group.

Boys	Girls	Groups
1	2	(B GG) (B GG) (B GG) (B GG) (B GG) (B GG) (B GG) (B GG) (B GG) 9 boys, 18 girls: There are 3 boys not in groups. ✗
2	3	(BB GGG) (BB GGG) (BB GGG) (BB GGG) (BB GGG) (BB GGG) 12 boys, 18 girls: Every student is in a group. ✓

The greatest number of groups is 6.

4 Look Back

The number of groups will be a common factor of the number of boys and the number of girls. To form the largest number of groups, find the GCF of 12 and 18.

factors of 12: **1**, **2**, **3**, 4, ⑥, 12 factors of 18: **1**, **2**, **3**, ⑥, 9, 18

The GCF of 12 and 18 is 6.

Think and Discuss

1. **Explain** what the GCF of two prime numbers is.
2. **Tell** what the least common factor of a group of numbers would be.

4-3 Exercises

FOR EOG PRACTICE
see page 649

internet connect
Homework Help Online
go.hrw.com Keyword: MR4 4-3

1.04a, 1.05

GUIDED PRACTICE

See Example 1 **Find the GCF of each set of numbers.**

1. 18 and 27
2. 32 and 72
3. 21, 42, and 56
4. 15, 30, and 60
5. 18, 24, and 36
6. 9, 36, and 81

See Example 2

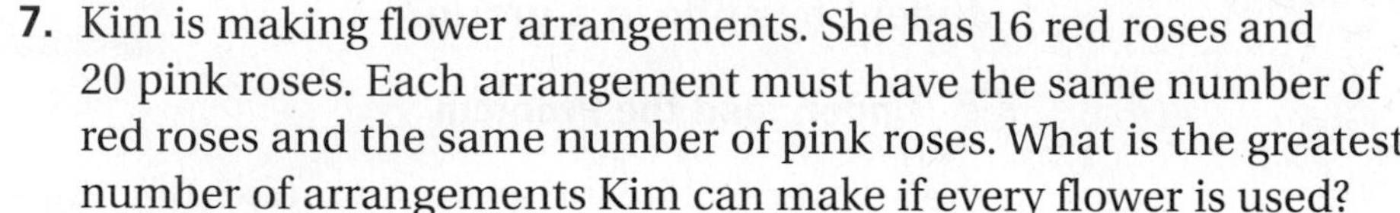

7. Kim is making flower arrangements. She has 16 red roses and 20 pink roses. Each arrangement must have the same number of red roses and the same number of pink roses. What is the greatest number of arrangements Kim can make if every flower is used?

INDEPENDENT PRACTICE

See Example 1 **Find the GCF of each set of numbers.**

8. 10 and 35
9. 28 and 70
10. 36 and 72
11. 26, 48, and 62
12. 16, 40, and 88
13. 12, 60, and 68
14. 30, 45, and 75
15. 24, 48, and 84
16. 16, 48, and 72

See Example 2

17. The local recreation center held a scavenger hunt. There were 15 boys and 9 girls at the event. The group was divided into the greatest number of teams possible with the same number of boys on each team and the same number of girls on each team. How many teams were made if each person was on a team?

18. Ms. Kline makes balloon arrangements. She has 32 blue balloons, 24 yellow balloons, and 16 white balloons. Each arrangement must have the same number of each color. What is the greatest number of arrangements that Ms. Kline can make if every balloon is used?

PRACTICE AND PROBLEM SOLVING

Write the GCF of each set of numbers.

19. 60 and 84
20. 14 and 17
21. 10, 35, and 110
22. 21 and 306
23. 630 and 712
24. 16, 24, and 40
25. 75, 225, and 150
26. 42, 112, and 105
27. 12, 16, 20, and 24
28. 16, 24, 30, and 42
29. 25, 90, 45, and 100
30. 27, 90, 135, and 72
31. $2 \times 2 \times 3$ and 2×2
32. $2 \times 3^2 \times 7$ and $2^2 \times 3$
33. $3^2 \times 7$ and $2 \times 3 \times 5^2$

34. Jared has 12 jars of grape jam, 16 jars of strawberry jam, and 24 jars of raspberry jam. He wants to place the jam into the greatest possible number of boxes so that each box has the same number of jars of each kind of jam. How many boxes does he need?

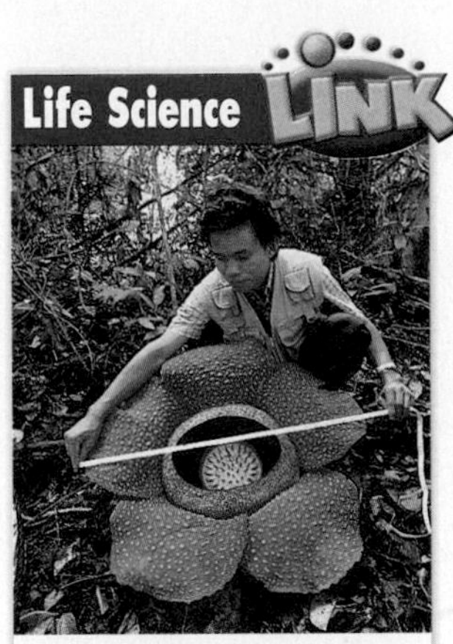

The biggest flower in the world is the Rafflesia. It grows in the Indonesian rain forest. Blossoms can measure four feet across, weigh about 15 pounds, and contain about 1.5 gallons of water.

35. Mr. Rodriguez is planting 4 types of flowers in his garden. He wants each row to contain the same number of each type of flower. What is the greatest number of rows Mr. Rodriguez can plant if every bulb is used?

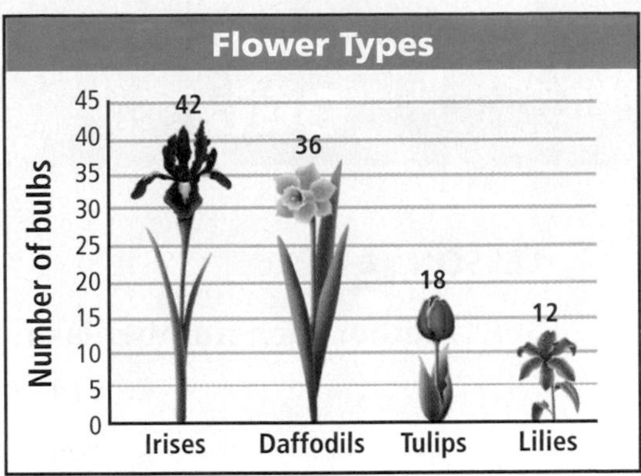

36. Pam is making fruit baskets. She has 30 apples, 24 bananas, and 12 oranges. What is the greatest number of baskets she can make if each type of fruit is distributed equally among the baskets?

37. In a parade, one school band will march directly behind another school band. All rows must have the same number of students. The first band has 36 students, and the second band has 60 students. What is the greatest number of students who can be in each row?

38. ***SOCIAL STUDIES*** Branches of the U.S. Mint in Denver and Philadelphia make all U.S. coins for circulation. A tiny *D* or *P* on the coin tells you where the coin was minted. Suppose you have 32 *D* quarters and 36 *P* quarters. What is the greatest number of groups you can make with the same number of *D* quarters in each group and the same number of *P* quarters in each group so that every quarter is placed in a group?

39. ***WRITE ABOUT IT*** What method do you like best for finding the GCF? Why?

40. ***CHALLENGE*** The GCF of three numbers is 9. The sum of the numbers is 90. Find the three numbers.

Spiral Review

Evaluate each expression. (Lesson 1-4)

41. $5 \times 2^3 + 17 - 3 \times 2$

42. $85 - (44 + 33) \div 7 + (62 - 12)$

43. $14 + 4^3 \div 8 \times 2 - 7$

44. $(76 - 13) \div 3^2 \times 8 + (91 - 47)$

Solve each equation. (Lesson 2-6)

45. $15n = 45$

46. $7t = 147$

47. $6a = 78$

48. $12b = 216$

49. **EOG PREP** ___?___ numbers are divisible by more than two numbers. (Lesson 4-1)

A Prime
B Whole
C Composite
D Equivalent

50. **EOG PREP** The numbers 2, 3, 5, and 6 are all factors of which number? (Lesson 4-2)

A 6
B 10
C 30
D 36

Chapter 4

Mid-Chapter Quiz

Mid-Chapter Quiz

LESSON 4-1 (pp. 152–155)

Tell whether each number is divisible by 2, 3, 4, 5, 6, 9, and 10.

1. 708 **2.** 514 **3.** 470 **4.** 338

5. 200 **6.** 798 **7.** 518 **8.** 309

Tell whether each number is prime or composite.

9. 76 **10.** 59 **11.** 69 **12.** 33

13. 36 **14.** 78 **15.** 93 **16.** 89

LESSON 4-2 (pp. 156–159)

List all of the factors of each number.

17. 26 **18.** 32 **19.** 39 **20.** 84

21. 54 **22.** 85 **23.** 27 **24.** 29

Write the prime factorization of each number.

25. 96 **26.** 50 **27.** 104 **28.** 63

29. 49 **30.** 156 **31.** 62 **32.** 95

LESSON 4-3 (pp. 160–163)

Find the GCF of each set of numbers.

33. 16 and 36 **34.** 22 and 88 **35.** 65 and 91

36. 42, 70, and 84 **37.** 16, 24, and 48 **38.** 20, 55, and 85

39. There are 36 sixth-graders and 40 seventh-graders. What is the greatest number of teams that the students can form if each team has the same number of sixth-graders and the same number of seventh-graders and every student must be on a team?

40. There are 14 girls and 21 boys in Mrs. Sutter's gym class. To play a certain game, the students must form teams. Each team must have the same number of girls and the same number of boys. What is the greatest number of teams Mrs. Sutter can make if every student is on a team?

41. Mrs. Young, an art teacher, is organizing the art supplies. She has 76 red markers, 52 blue markers, and 80 black markers. She wants to divide the markers into boxes with the same number of red, the same number of blue, and the same number of black markers in each box. What is the greatest number of boxes she can have if every marker is placed in a box?

Focus on Problem Solving

Understand the Problem

- **Interpret unfamiliar words**

You must understand the words in a problem in order to solve it. If there is a word you do not know, try to use context clues to figure out its meaning. Suppose there is a problem about red, green, blue, and chartreuse fabric. You may not know the word *chartreuse*, but you can guess that it is probably a color. To make the problem easier to understand, you could replace *chartreuse* with the name of a familiar color, like *white*.

In some problems, the name of a person, place, or thing might be difficult to pronounce, such as *Mr. Joubert*. When you see a proper noun that you do not know how to pronounce, you can use another proper noun or a pronoun in its place. You could replace *Mr. Joubert* with *he*. You could replace *Koenisburg Street* with *K Street*.

Copy each problem. Underline any words that you do not understand. Then replace each word with a more familiar word.

1. Grace is making flower bouquets. She has 18 chrysanthemums and 42 roses. She wants to arrange them in groups that each have the same number of chrysanthemums and the same number of roses. What is the fewest number of flowers that Grace can have in each group? How many chrysanthemums and how many roses will be in each group?

2. Most marbles are made from glass. The glass is liquefied in a furnace and poured. It is then cut into cylinders that are rounded off and cooled. Suppose 1,200 cooled marbles are put into packages of 8. How many packages could be made? Would there be any marbles left over?

3. In ancient times, many civilizations used calendars that divided the year into months of 30 days. A year has 365 days. How many whole months were in these ancient calendars? Were there any days left over? If so, how many?

4. Mrs. LeFeubre is tiling her garden walkway. It is a rectangle that is 4 feet wide and 20 feet long. Mrs. LeFeubre wants to use square tiles, and she does not want to have to cut any tiles. What is the size of the largest square tile that Mrs. LeFeubre can use?

Explore Decimals and Fractions

Use with Lesson 4-4

KEY

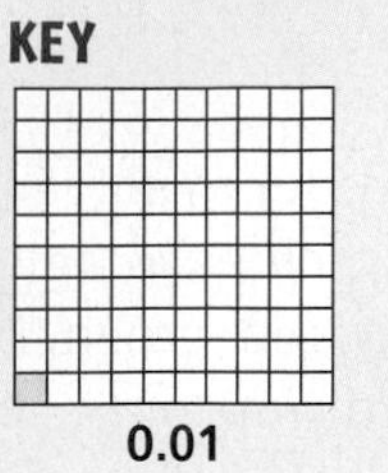
0.01

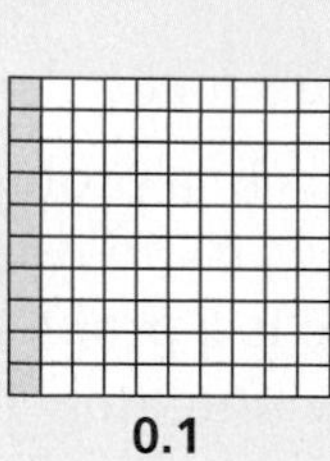
0.1

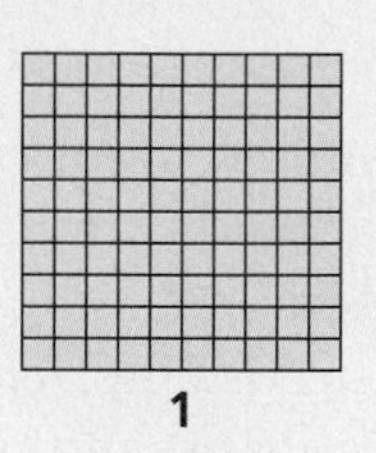
1

internet connect
Lab Resources Online
go.hrw.com
KEYWORD: MR4 Lab4A

You can use decimal grids to show the relationship between fractions and decimals.

Activity

Write the number represented on each grid as a fraction and as a decimal.

1

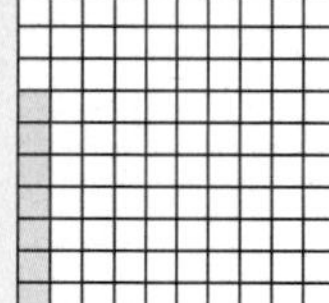

Seven hundredths squares are shaded → 0.07

How many squares are shaded? $\frac{7}{100}$ ← numerator

How many squares are in the whole? ← denominator

$0.07 = \frac{7}{100}$

2

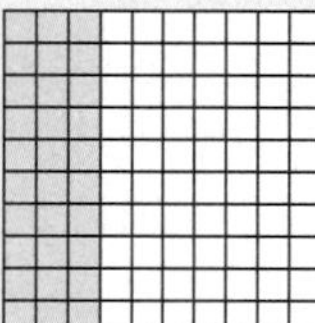

Three tenths columns are shaded → 0.3

How many complete columns are shaded? $\frac{3}{10}$

How many columns are in the whole?

$0.3 = \frac{3}{10}$

Think and Discuss

1. Is 0.09 the same as $\frac{9}{10}$? Use decimal grids to support your answer.

Try This

Use decimal grids to represent each number.

1. 0.8 **2.** $\frac{37}{100}$ **3.** 0.53 **4.** $\frac{1}{10}$ **5.** $\frac{67}{100}$

6. For 1–5, write each decimal as a fraction and each fraction as a decimal.

4-4 Decimals and Fractions

Learn to convert between decimals and fractions.

Vocabulary

mixed number

terminating decimal

repeating decimal

Decimals and fractions can often be used to represent the same number.

For example, a baseball player's batting average can be represented as a fraction:

$$\frac{\text{number of hits}}{\text{number of times at bat}}$$

Ty Cobb, at right, holds the record for the highest career batting average in professional baseball. He had 4,189 hits, and he was at bat 11,434 times. His batting average was $\frac{4{,}189}{11{,}434}$.

Batting averages are usually written as decimals rounded to three decimal places. To find this decimal, use the fact that the fraction bar means "divided by."

$$4{,}189 \div 11{,}434 = 0.3663634773\ldots$$

Ty Cobb's career batting average is reported as .366.

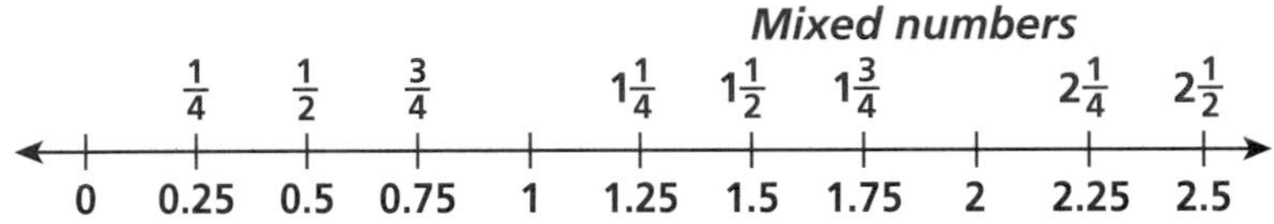

A number that contains both a whole number greater than 0 and a fraction, such as $1\frac{3}{4}$, is called a **mixed number**.

EXAMPLE 1 Writing Decimals as Fractions or Mixed Numbers

Write each decimal as a fraction or mixed number.

A 0.23

0.23 *Identify the place value of the digit farthest to the right.*

$\frac{23}{100}$ *The 3 is in the **hundred**ths place, so use **100** as the denominator.*

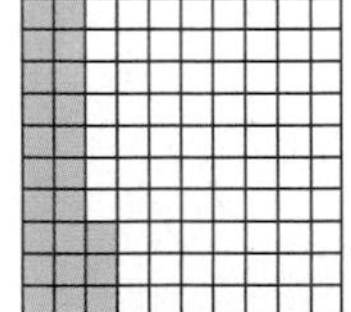

B 1.7

1.7 *Identify the place value of the digit farthest to the right.*

$1\frac{7}{10}$ *Write the whole number, 1. The 7 is in the **ten**ths place, so use **10** as the denominator.*

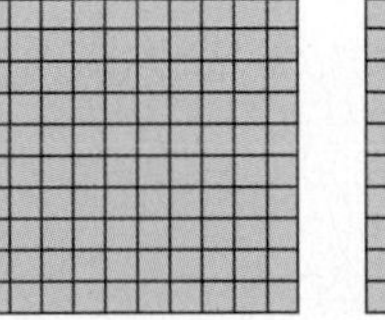

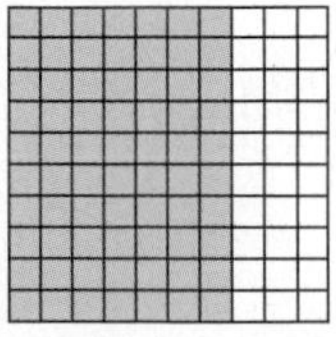

Remember!

Place Value			
Ones	Tenths	Hundredths	Thousandths

EXAMPLE 2 Writing Fractions as Decimals

Write each fraction or mixed number as a decimal.

A $\frac{3}{4}$

$$\begin{array}{r} 0.75 \\ 4\overline{)3.00} \\ -\underline{28} \\ 20 \\ -\underline{20} \\ 0 \end{array}$$

Divide 3 by 4. Add zeros after the decimal point. The remainder is 0.

$\frac{3}{4} = 0.75$

B $5\frac{2}{3}$

$$\begin{array}{r} 0.666 \\ 3\overline{)2.000} \\ -\underline{18} \\ 20 \\ -\underline{18} \\ 20 \\ -\underline{18} \\ 2 \end{array}$$

Divide 2 by 3. Add zeros after the decimal point. The 6 repeats in the quotient.

$5\frac{2}{3} = 5.666... = 5.\overline{6}$

Writing Math

To write a repeating decimal, you can show three dots or draw a bar over the repeating part: $0.666... = 0.\overline{6}$

A **terminating decimal**, such as 0.75, has a finite number of decimal places. A **repeating decimal**, such as 0.666..., has a block of one or more digits that repeat continuously.

Common Fractions and Equivalent Decimals								
$\frac{1}{5}$	$\frac{1}{4}$	$\frac{1}{3}$	$\frac{2}{5}$	$\frac{1}{2}$	$\frac{3}{5}$	$\frac{2}{3}$	$\frac{3}{4}$	$\frac{4}{5}$
0.2	0.25	$0.\overline{3}$	0.4	0.5	0.6	$0.\overline{6}$	0.75	0.8

EXAMPLE 3 Comparing and Ordering Fractions and Decimals

Order the fractions and decimals from least to greatest.

0.5, $\frac{1}{5}$, 0.37

First rewrite the fraction as a decimal. $\frac{1}{5} = 0.2$

Order the three decimals.

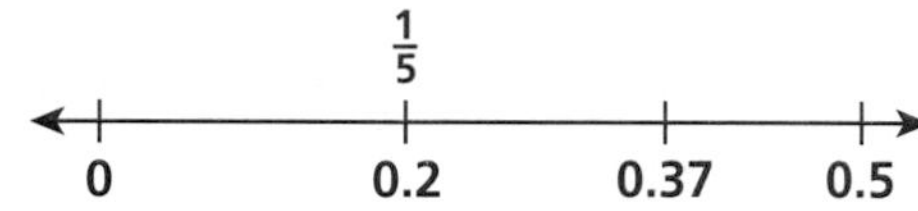

The numbers in order from least to greatest are $\frac{1}{5}$, 0.37, and 0.5.

Think and Discuss

1. **Tell** how reading the decimal 6.9 as "six and nine tenths" helps you to write 6.9 as a mixed number.
2. **Look** at the decimal 0.121122111222.... If the pattern continues, is this a repeating decimal? Why or why not?

4-4 Exercises

FOR EOG PRACTICE

see page 650

internet connect
Homework Help Online
go.hrw.com Keyword: MR4 4-4

1.03

GUIDED PRACTICE

See Example 1

Write each decimal as a fraction or mixed number.

1. 0.15 **2.** 1.25 **3.** 0.43 **4.** 2.6

See Example 2

Write each fraction or mixed number as a decimal.

5. $\frac{2}{5}$ **6.** $2\frac{7}{8}$ **7.** $\frac{1}{8}$ **8.** $4\frac{1}{10}$

See Example 3

Order the fractions and decimals from least to greatest.

9. $\frac{2}{3}$, 0.78, 0.21 **10.** $\frac{5}{16}$, 0.67, $\frac{1}{6}$ **11.** 0.52, $\frac{1}{9}$, 0.3

INDEPENDENT PRACTICE

See Example 1

Write each decimal as a fraction or mixed number.

12. 0.31 **13.** 5.71 **14.** 0.13 **15.** 3.23

16. 0.5 **17.** 2.7 **18.** 0.19 **19.** 6.3

See Example 2

Write each fraction or mixed number as a decimal.

20. $\frac{1}{9}$ **21.** $1\frac{3}{5}$ **22.** $\frac{8}{9}$ **23.** $3\frac{11}{40}$

24. $2\frac{5}{6}$ **25.** $\frac{3}{8}$ **26.** $4\frac{4}{5}$ **27.** $\frac{5}{8}$

See Example 3

Order the fractions and decimals from least to greatest.

28. 0.49, 0.82, $\frac{1}{2}$ **29.** $\frac{3}{8}$, 0.29, $\frac{1}{9}$ **30.** 0.94, $\frac{4}{5}$, 0.6

31. 0.11, $\frac{1}{10}$, 0.13 **32.** $\frac{2}{3}$, 0.42, $\frac{2}{5}$ **33.** $\frac{3}{7}$, 0.76, 0.31

PRACTICE AND PROBLEM SOLVING

Write each decimal in expanded form and use a whole number or fraction for each place value.

34. 0.81 **35.** 92.3 **36.** 13.29 **37.** 107.17

Write each fraction as a decimal. Tell whether the decimal terminates or repeats.

38. $\frac{7}{9}$ **39.** $\frac{1}{6}$ **40.** $\frac{17}{20}$ **41.** $\frac{5}{12}$

Compare. Write $<$, $>$, or $=$.

42. 0.75 ■ $\frac{3}{4}$ **43.** $\frac{5}{8}$ ■ 0.5 **44.** 0.78 ■ $\frac{7}{9}$

Order the mixed numbers and decimals from greatest to least.

45. 4.48, 3.92, $4\frac{1}{2}$

46. $10\frac{5}{9}$, 10.5, $10\frac{1}{5}$

47. 125.205, 125.25, $125\frac{1}{5}$

The table shows batting averages for two baseball seasons. Use the table for Exercises 48 and 49.

Player	Season 1	Season 2
Pedro	0.360	$\frac{3}{10}$
Jill	0.380	$\frac{3}{8}$
Lamar	0.290	$\frac{1}{3}$
Britney	0.190	$\frac{3}{20}$

48. Which players had higher batting averages in season 1 than they had in season 2?

49. Who had the highest batting average in either season?

50. ***LIFE SCIENCE*** Most people with color deficiency (often called color blindness) have trouble distinguishing shades of red and green. About 0.05 of men in the world have color deficiency. What fraction of men have color deficiency?

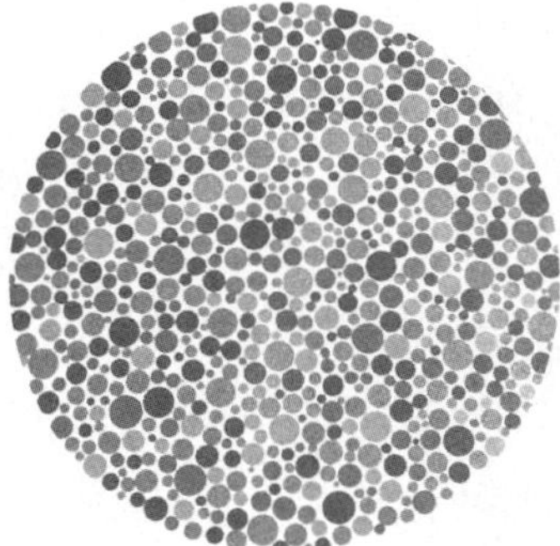
People with normal color vision will see "7" in this color-blindness test.

51. ***WHAT'S THE ERROR?*** A student found the decimal equivalent of $\frac{7}{18}$ to be $0.\overline{38}$. Explain the error. What is the correct answer?

52. ***WRITE ABOUT IT*** The decimal for $\frac{1}{25}$ is 0.04, and the decimal for $\frac{2}{25}$ is 0.08. Without dividing, find the decimal for $\frac{6}{25}$. Explain how you found your answer.

53. ***CHALLENGE*** Write $\frac{1}{999}$ as a decimal.

Spiral Review

Identify the property illustrated by each equation. (Lesson 1-5)

54. $4 + 5 = 5 + 4$

55. $3(4 - 1) = 3(4) - 3(1)$

56. $(91 + 80) + 72 = 91 + (80 + 72)$

Find the number of decimal places in each product. Then multiply. (Lesson 3-6)

57. $2.4 \cdot 1.8$

58. $19 \cdot 0.5$

59. $7.04 \cdot 2.38$

60. $0.4 \cdot 0.1$

61. **EOG PREP** What is the greatest common factor of 12, 18, and 30? (Lesson 4-3)

A 2 B 3 C 6 D 9

Model Equivalent Fractions

Use with Lesson 4-5

KEY

 = 1 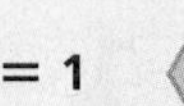 = $\frac{1}{2}$ 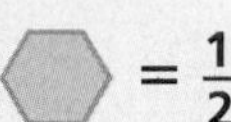  = $\frac{1}{4}$  = $\frac{1}{6}$  = $\frac{1}{12}$

internet connect
Lab Resources Online
go.hrw.com
KEYWORD: MR4 Lab4B

Pattern blocks can be used to model equivalent fractions. To find a fraction that is equivalent to $\frac{1}{2}$, first choose the pattern block that represents $\frac{1}{2}$. Then find all the pieces of one color that will fit evenly on the $\frac{1}{2}$ block. Count these pieces to find the equivalent fraction. You may be able to find more than one equivalent fraction.

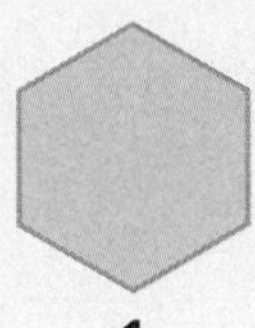 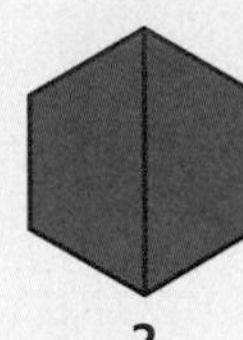 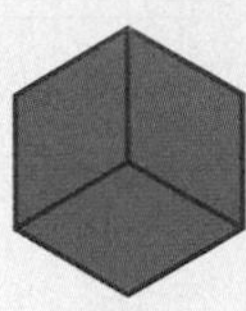 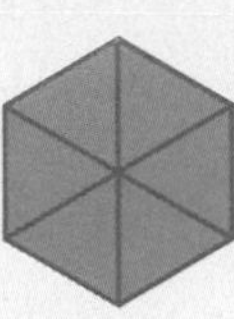

$\frac{1}{2} = \frac{2}{4} = \frac{3}{6} = \frac{6}{12}$

Activity

1 Use pattern blocks to find an equivalent fraction for $\frac{8}{12}$.

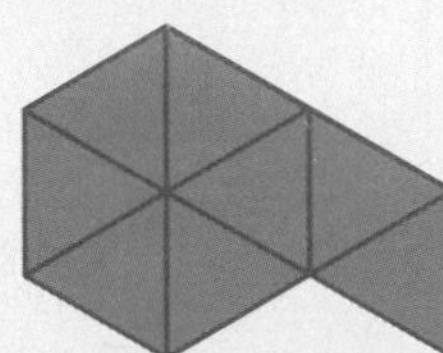

First show $\frac{8}{12}$.

$\frac{8}{12} = \frac{4}{6}$

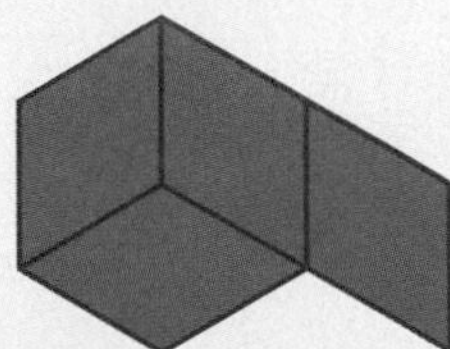

You can cover $\frac{8}{12}$ with four of the $\frac{1}{6}$ pieces.

Think and Discuss

1. Can you find a combination of pattern blocks for $\frac{1}{3}$? Find an equivalent fraction for $\frac{1}{3}$.
2. Are $\frac{9}{12}$ and $\frac{3}{6}$ equivalent? Use pattern blocks to support your answer.

Try This

Write the fraction that is modeled. Then find an equivalent fraction.

1.

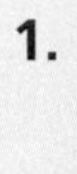
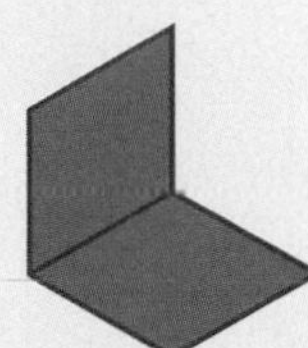

2.

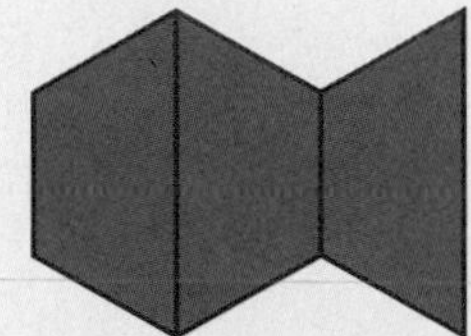

4-5 Equivalent Fractions

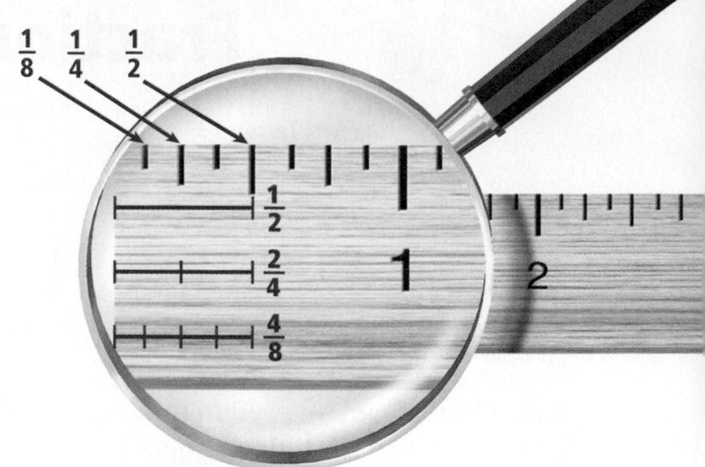

Learn to write equivalent fractions.

Vocabulary

equivalent fractions

simplest form

Rulers often have marks for inches, $\frac{1}{2}$, $\frac{1}{4}$, and $\frac{1}{8}$ inches.

Notice that $\frac{1}{2}$ in., $\frac{2}{4}$ in., and $\frac{4}{8}$ in. all name the same length. Fractions that represent the same value are **equivalent fractions**. So $\frac{1}{2}$, $\frac{2}{4}$, and $\frac{4}{8}$ are equivalent fractions.

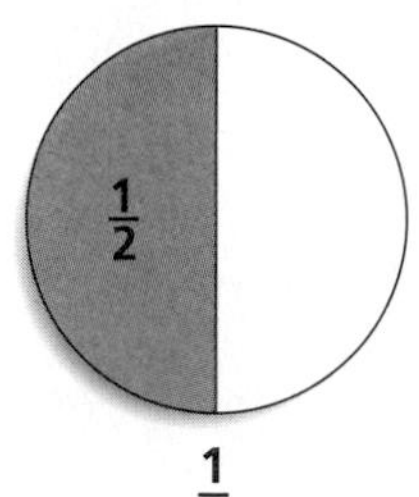

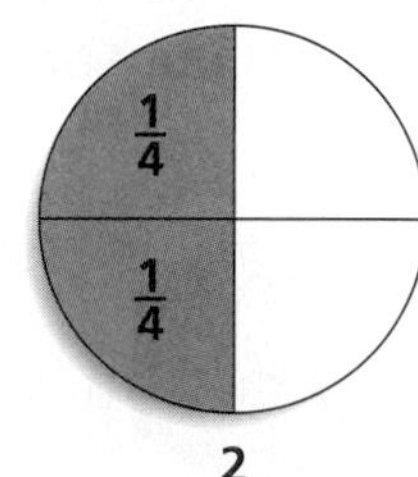

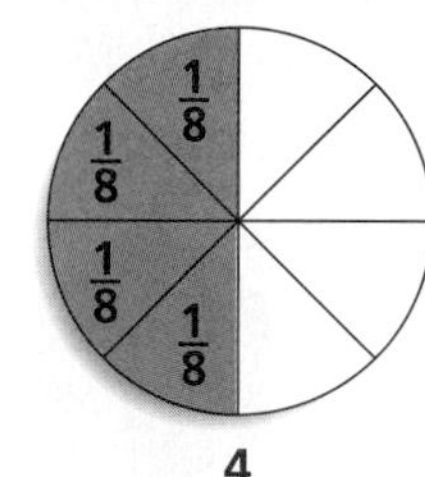

$$\frac{1}{2} = \frac{2}{4} = \frac{4}{8}$$

EXAMPLE 1 Finding Equivalent Fractions

Find two equivalent fractions for $\frac{6}{8}$.

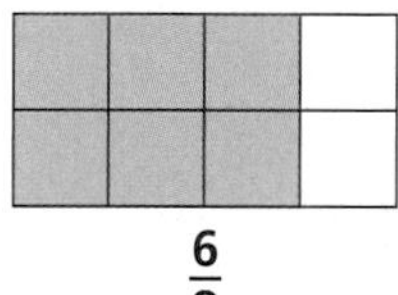

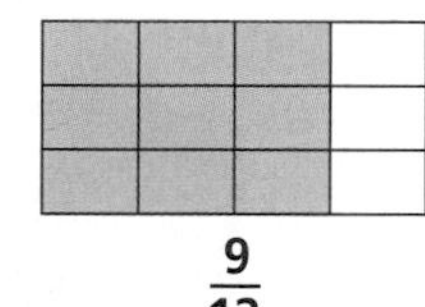

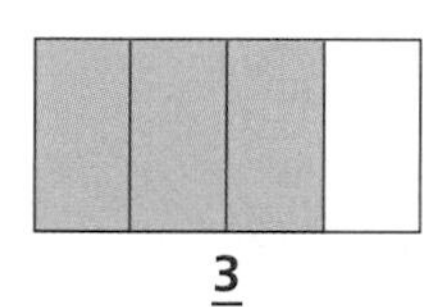

$$\frac{6}{8} = \frac{9}{12} = \frac{3}{4}$$

So $\frac{6}{8}$, $\frac{9}{12}$, and $\frac{3}{4}$ are all equivalent fractions.

EXAMPLE 2 Multiplying and Dividing to Find Equivalent Fractions

Find the missing number that makes the fractions equivalent.

A $\frac{2}{3} = \frac{\square}{18}$

$\frac{2 \cdot 6}{3 \cdot 6} = \frac{12}{18}$ *In the denominator, 3 is multiplied by 6 to get 18. Multiply the numerator, 2, by the same number, 6.*

So $\frac{2}{3}$ is equivalent to $\frac{12}{18}$.

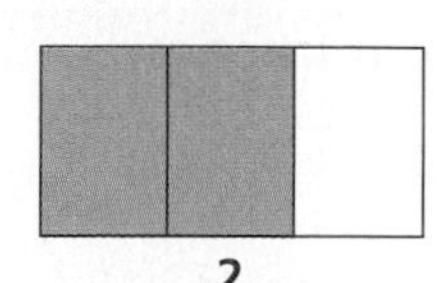

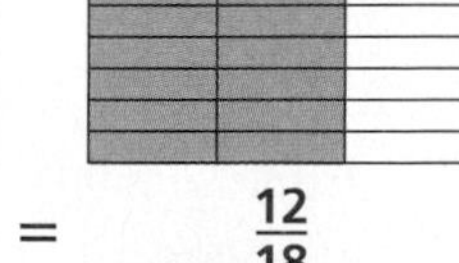

$$\frac{2}{3} = \frac{12}{18}$$

Find the missing number that makes the fractions equivalent.

B $\frac{70}{100} = \frac{7}{\square}$

$\frac{70 \div \mathbf{10}}{100 \div \mathbf{10}} = \frac{7}{10}$ *In the numerator, 70 is divided by 10 to get 7. Divide the denominator by the same number, 10.*

So $\frac{70}{100}$ is equivalent to $\frac{7}{10}$.

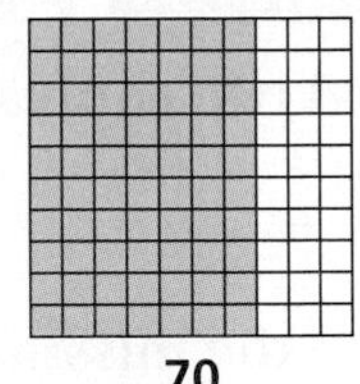

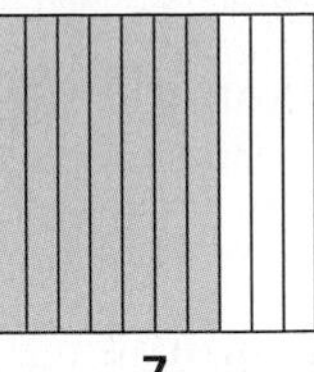

$\frac{70}{100} = \frac{7}{10}$

Every fraction has one equivalent fraction that is called the simplest form of the fraction. A fraction is in **simplest form** when the GCF of the numerator and the denominator is 1.

Example 3 shows two methods for writing a fraction in simplest form.

EXAMPLE 3 Writing Fractions in Simplest Form

Write each fraction in simplest form.

A $\mathbf{\frac{18}{24}}$

The GCF of 18 and 24 is 6, so $\frac{18}{24}$ is not in simplest form.

Method 1: Use the GCF.

$\frac{18 \div \mathbf{6}}{24 \div \mathbf{6}} = \frac{3}{4}$ *Divide 18 and 24 by their GCF, 6.*

Method 2: Use a ladder diagram.

```
2 | 18/24
3 |  9/12
     (3/4)
```

Use a ladder. Divide 18 and 24 by any common factor (except 1) until you cannot divide anymore.

So $\frac{18}{24}$ written in simplest form is $\frac{3}{4}$.

B $\mathbf{\frac{8}{9}}$

The GCF of 8 and 9 is 1, so $\frac{8}{9}$ is already in simplest form.

Helpful Hint

Method 2 is useful when you know that the numerator and denominator have common factors, but you are not sure what the GCF is.

Think and Discuss

1. **Explain** whether a fraction is equivalent to itself.
2. **Tell** which of the following fractions are in simplest form: $\frac{9}{21}$, $\frac{20}{25}$, and $\frac{5}{13}$. Explain.
3. **Explain** how you know that $\frac{7}{16}$ is in simplest form.

4-5 Exercises

1.03

FOR EOG PRACTICE

see page 650

internet connect

Homework Help Online

go.hrw.com Keyword: MR4 4-5

GUIDED PRACTICE

See Example 1 Find two equivalent fractions for each fraction.

1. $\frac{4}{6}$ **2.** $\frac{3}{12}$ **3.** $\frac{3}{6}$ **4.** $\frac{6}{16}$

See Example 2 Find the missing numbers that make the fractions equivalent.

5. $\frac{2}{5} = \frac{10}{■}$ **6.** $\frac{7}{21} = \frac{1}{■}$ **7.** $\frac{3}{4} = \frac{■}{28}$

See Example 3 Write each fraction in simplest form.

8. $\frac{2}{10}$ **9.** $\frac{6}{18}$ **10.** $\frac{4}{16}$ **11.** $\frac{9}{15}$

INDEPENDENT PRACTICE

See Example 1 Find two equivalent fractions for each fraction.

12. $\frac{3}{9}$ **13.** $\frac{2}{10}$ **14.** $\frac{3}{21}$ **15.** $\frac{3}{18}$

16. $\frac{12}{15}$ **17.** $\frac{4}{10}$ **18.** $\frac{10}{12}$ **19.** $\frac{6}{10}$

See Example 2 Find the missing numbers that make the fractions equivalent.

20. $\frac{3}{7} = \frac{■}{35}$ **21.** $\frac{6}{48} = \frac{1}{■}$ **22.** $\frac{2}{5} = \frac{28}{■}$

23. $\frac{2}{7} = \frac{■}{21}$ **24.** $\frac{8}{32} = \frac{■}{4}$ **25.** $\frac{2}{7} = \frac{40}{■}$

See Example 3 Write each fraction in simplest form.

26. $\frac{2}{8}$ **27.** $\frac{10}{15}$ **28.** $\frac{6}{30}$ **29.** $\frac{6}{14}$

30. $\frac{12}{16}$ **31.** $\frac{4}{28}$ **32.** $\frac{4}{8}$ **33.** $\frac{10}{35}$

PRACTICE AND PROBLEM SOLVING

Write the equivalent fractions represented by each picture.

34.

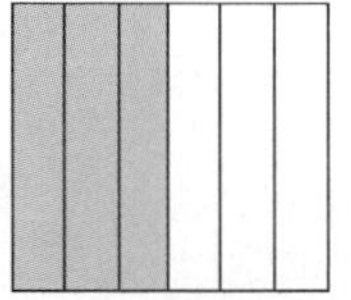

=

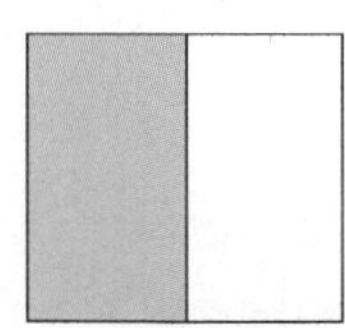

35.

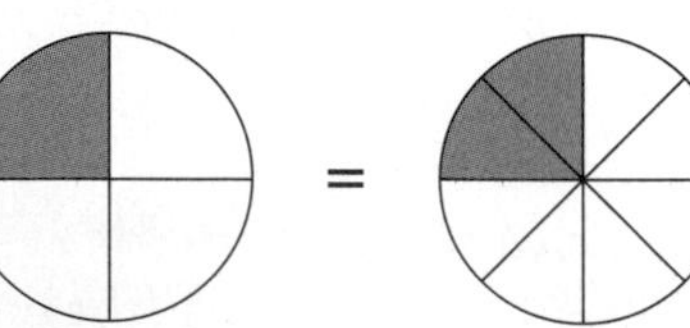

36.

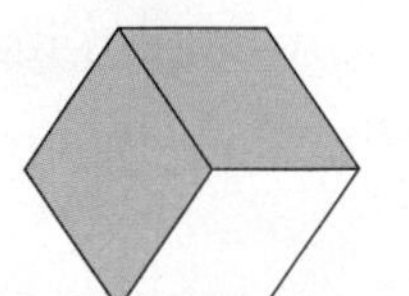

37.

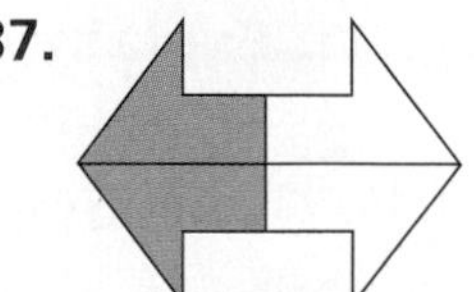

Social Studies 

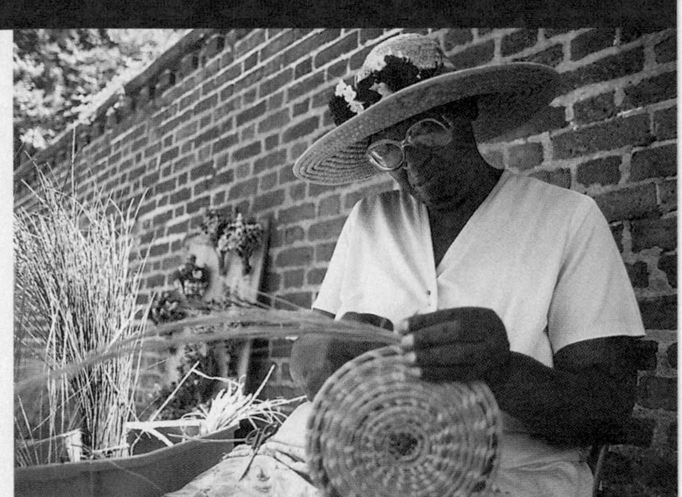

The Old City Market is a public market in Charleston, South Carolina. Local artists, craftspeople, and vendors display and sell their goods in open-sided booths.

38. You can buy food, such as southern sesame seed cookies, at $\frac{1}{10}$ of the booths. Write two equivalent fractions for $\frac{1}{10}$.

39. Handwoven sweetgrass baskets are a regional specialty. About 8 out of every 10 baskets sold are woven at the market. Write a fraction for "8 out of 10." Then write this fraction in simplest form.

40. Suppose the circle graph shows the number of each kind of craft booth at the Old City Market. For each type of booth, tell what fraction it represents of the total number of craft booths. Write these fractions in simplest form.

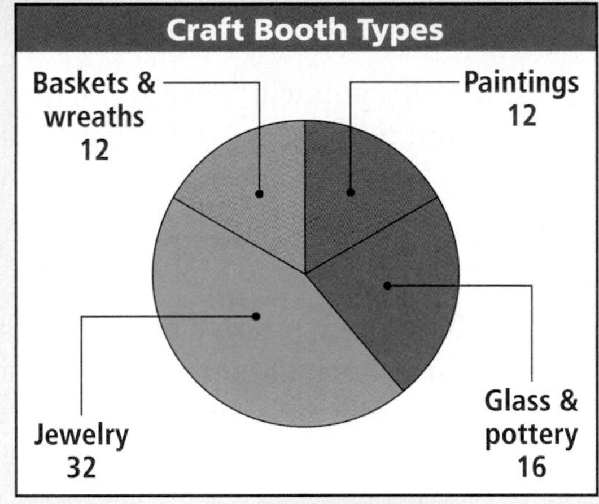

41. Customers can buy packages of dried rice and black-eyed peas, which can be made into black-eyed pea soup. One recipe for black-eyed pea soup calls for $\frac{1}{2}$ tsp of basil. How could you measure the basil if you had only a $\frac{1}{4}$ tsp measuring spoon? What if you had only a $\frac{1}{8}$ tsp measuring spoon?

42. ***WRITE ABOUT IT*** The recipe for soup also calls for $\frac{1}{4}$ tsp of pepper. How many fractions are equivalent to $\frac{1}{4}$? Explain.

43. ***CHALLENGE*** Silver jewelry is a popular item at the market. Suppose there are 28 bracelets at one jeweler's booth and that $\frac{3}{7}$ of these bracelets have red stones. How many bracelets have red stones?

Spiral Review

Order the numbers in each set from greatest to least. (Lesson 1-1)

44. 740, 680, 749, 168 **45.** 204, 271, 640, 644 **46.** 4,192; 4,286; 4,181; 4,287

Compare. Write <, >, or =. (Lesson 3-1)

47. 4.23 ▇ 4.28 **48.** 12.05 ▇ 8.79 **49.** 0.45 ▇ 0.8

50. 50 ▇ 0.5 **51.** 14.006 ▇ 14.3003 **52.** 23.1945 ▇ 23.1928

53. **EOG PREP** Which is the prime factorization of 189? (Lesson 4-2)

A $9 \cdot 3 \cdot 7$ B $7^2 \cdot 3^2$ C $3^3 \cdot 7$ D $3 \cdot 63$

Explore Fraction Measurement

Use with Lesson 4-5

Look at a standard ruler. It probably has marks for inches, half inches, quarter inches, eighth inches, and sixteenth inches.

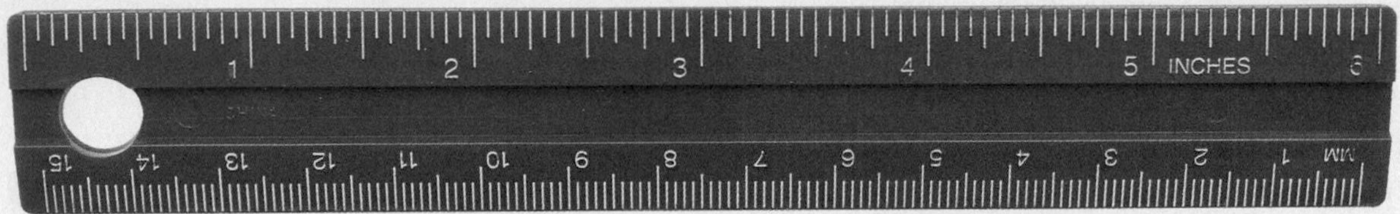

In this activity, you will make some of your own rulers and use them to help you find and understand equivalent fractions.

Activity

1. You will need four strips of paper. On one strip, use your ruler to make a mark for every half inch. Number each mark, beginning with 1. Label this strip "half-inch ruler."

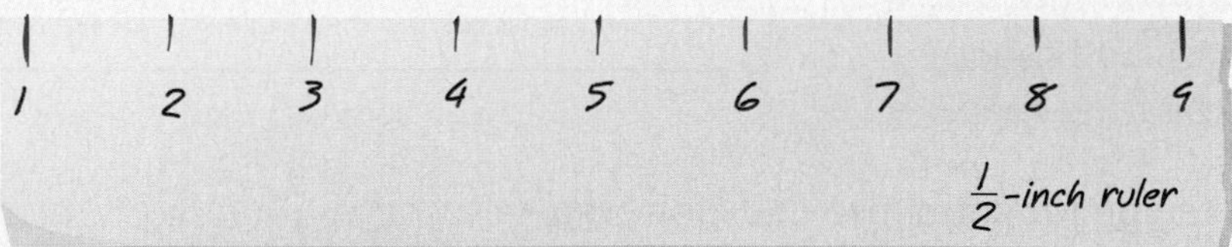

On a second strip, make a mark for every quarter inch. Again, number each mark, beginning with 1. Label this strip "quarter-inch ruler."

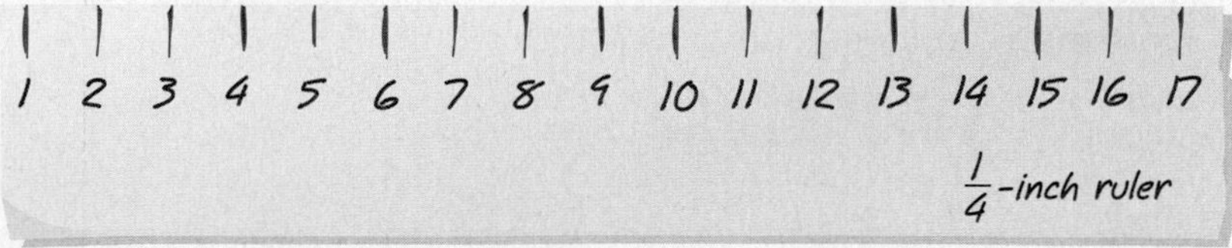

Do the same thing for eighth inches and sixteenth inches.

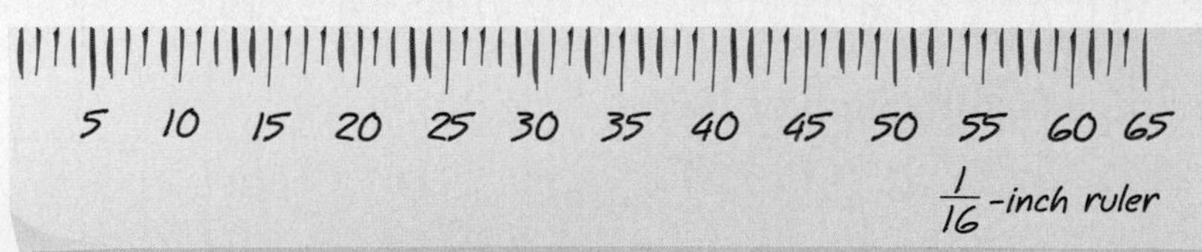

2. Now use the half-inch ruler you made to measure the line segment at right. How many half inches long is the segment?

Use your quarter-inch ruler to measure the line segment again. How many quarter inches long is the segment?

How many eighth inches long is the segment?

How many sixteenth inches?

Fill in the blanks: $\frac{1}{2} = \frac{\square}{4} = \frac{\square}{8} = \frac{\square}{16}$.

3. Use your quarter-inch ruler to measure the line segment below.

How long is the segment?

Now use your eighth-inch ruler to measure the line segment again. How many eighth inches long is the segment?

How many sixteenth inches?

Fill in the blanks: $\frac{3}{4} = \frac{\square}{8} = \frac{\square}{16}$.

Think and Discuss

1. How does a ruler show that equivalent fractions have the same value?
2. Look at your lists of equivalent fractions from 2 and 3. Do you notice any patterns? Describe them.
3. Use your rulers to measure an object longer than 1 inch. Use your measurements to write equivalent fractions. What do you notice about these fractions?

Try This

1. Use your rulers to measure the items below. Use your measurements to write equivalent fractions.

2. Use your rulers to measure several items in your classroom. Use your measurements to write equivalent fractions.

4-6 Comparing and Ordering Fractions

Learn to use pictures and number lines to compare and order fractions.

Vocabulary

like fractions

unlike fractions

common denominator

Rachel and Hannah are making a kind of cookie called *hamantaschen.* They have $\frac{1}{2}$ cup of strawberry jam, but the recipe requires $\frac{1}{3}$ cup.

To determine if they have enough for the recipe, they need to compare the fractions $\frac{1}{2}$ and $\frac{1}{3}$.

Hamantaschen
1/2 cup butter
2 egg yolks
1 1/2 cups flour
2 tablespoons sugar
3 tablespoons ice water
1/3 cup strawberry jam

When you are comparing fractions, first check their denominators. When fractions have the same denominator, they are called **like fractions**. For example, $\frac{7}{10}$ and $\frac{3}{10}$ are like fractions.

EXAMPLE 1 Comparing Like Fractions

Helpful Hint

When two fractions have the same denominator, the one with the larger numerator is greater.

$\frac{2}{5} < \frac{3}{5}$ $\frac{3}{8} > \frac{1}{8}$

Compare. Write <, >, or =.

A $\frac{7}{10}$ ▪ $\frac{3}{10}$

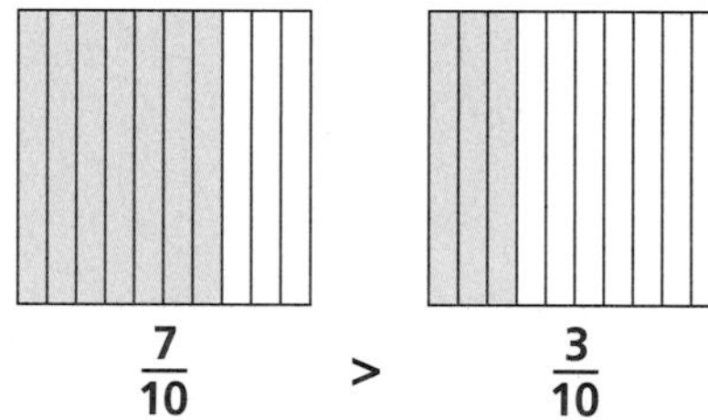

From the model, $\frac{7}{10} > \frac{3}{10}$.

B $\frac{1}{8}$ ▪ $\frac{5}{8}$

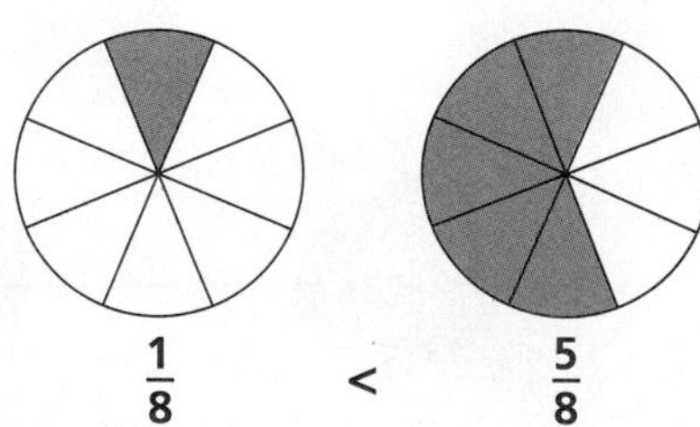

From the model, $\frac{1}{8} < \frac{5}{8}$.

When two fractions have different denominators, they are called **unlike fractions** . To compare unlike fractions, first rename the fractions so they have the same denominator. This is called finding a **common denominator** .

EXAMPLE 2 *Cooking Application*

Rachel and Hannah have $\frac{1}{2}$ cup of strawberry jam. They need $\frac{1}{3}$ cup to make hamantaschen. Do they have enough strawberry jam for the recipe?

Compare $\frac{1}{2}$ and $\frac{1}{3}$.

Find a common denominator by multiplying the denominators.
$2 \cdot 3 = 6$

Find equivalent fractions with 6 as the denominator.

$$\frac{1}{2} = \frac{\square}{6} \qquad \frac{1}{3} = \frac{\square}{6}$$

$$\frac{1 \cdot 3}{2 \cdot 3} = \frac{3}{6} \qquad \frac{1 \cdot 2}{3 \cdot 2} = \frac{2}{6}$$

$$\frac{1}{2} = \frac{3}{6} \qquad \frac{1}{3} = \frac{2}{6}$$

Compare the like fractions.

$$\frac{3}{6} > \frac{2}{6}, \text{ so } \frac{1}{2} > \frac{1}{3}.$$

Since $\frac{1}{2}$ cup is more than $\frac{1}{3}$ cup, they have enough strawberry jam.

EXAMPLE 3 Ordering Fractions

Remember!

Numbers increase in value as you move from left to right on a number line.

Order $\frac{3}{7}$, $\frac{3}{4}$, and $\frac{1}{4}$ from least to greatest.

$\frac{3 \cdot 4}{7 \cdot 4} = \frac{12}{28}$ $\quad \frac{3 \cdot 7}{4 \cdot 7} = \frac{21}{28}$ $\quad \frac{1 \cdot 7}{4 \cdot 7} = \frac{7}{28}$ *Rename with like denominators.*

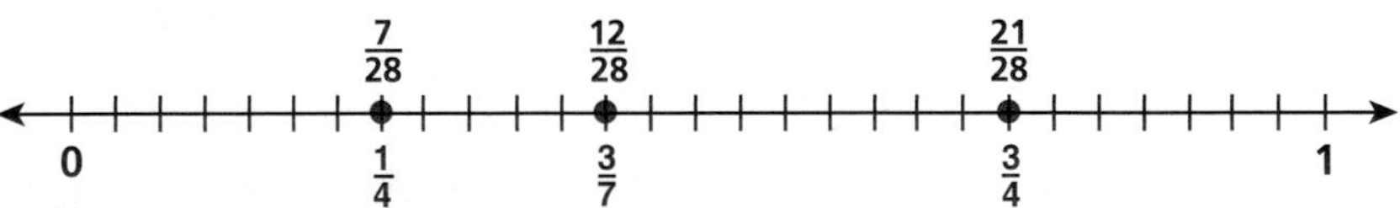

The fractions in order from least to greatest are $\frac{1}{4}, \frac{3}{7}, \frac{3}{4}$.

Think and Discuss

1. **Tell** whether the values of the fractions change when you rename two fractions so that they have common denominators.
2. **Explain** how to compare $\frac{2}{5}$ and $\frac{4}{5}$.

4-6 Exercises

FOR EOG PRACTICE

see page 651

internet connect
Homework Help Online
go.hrw.com Keyword: MR4 4-6

1.03

GUIDED PRACTICE

See Example 1

Compare. Write <, >, or =.

1. $\frac{3}{5} \square \frac{2}{5}$
2. $\frac{1}{9} \square \frac{2}{9}$
3. $\frac{6}{8} \square \frac{3}{4}$

See Example 2

4. Arsenio has $\frac{2}{3}$ cup of brown sugar. The recipe he is using requires $\frac{1}{4}$ cup of brown sugar. Does he have enough brown sugar for the recipe? Explain.

See Example 3

Order the fractions from least to greatest.

5. $\frac{3}{8}, \frac{1}{5}, \frac{2}{3}$
6. $\frac{1}{4}, \frac{2}{5}, \frac{1}{3}$
7. $\frac{5}{9}, \frac{1}{8}, \frac{2}{7}$

INDEPENDENT PRACTICE

See Example 1

Compare. Write <, >, or =.

8. $\frac{2}{5} \square \frac{4}{5}$
9. $\frac{1}{10} \square \frac{3}{10}$
10. $\frac{3}{4} \square \frac{15}{20}$
11. $\frac{4}{5} \square \frac{5}{5}$
12. $\frac{2}{4} \square \frac{1}{2}$
13. $\frac{4}{6} \square \frac{16}{24}$

See Example 2

14. Kelly needs $\frac{2}{3}$ gallon of paint to finish painting her deck. She has $\frac{5}{8}$ gallon of paint. Does she have enough paint to finish her deck? Explain.

See Example 3

Order the fractions from least to greatest.

15. $\frac{1}{2}, \frac{3}{5}, \frac{3}{7}$
16. $\frac{1}{6}, \frac{2}{5}, \frac{1}{4}$
17. $\frac{4}{9}, \frac{3}{8}, \frac{1}{3}$
18. $\frac{3}{4}, \frac{7}{10}, \frac{2}{3}$
19. $\frac{13}{18}, \frac{5}{9}, \frac{5}{6}$
20. $\frac{3}{8}, \frac{1}{4}, \frac{2}{3}$

PRACTICE AND PROBLEM SOLVING

Compare. Write <, >, or =.

21. $\frac{4}{15} \square \frac{3}{10}$
22. $\frac{7}{12} \square \frac{13}{30}$
23. $\frac{5}{9} \square \frac{4}{11}$
24. $\frac{3}{5} \square \frac{26}{65}$
25. $\frac{3}{5} \square \frac{2}{21}$
26. $\frac{24}{41} \square \frac{2}{7}$

Order the fractions from least to greatest.

27. $\frac{2}{5}, \frac{1}{2}, \frac{3}{10}$
28. $\frac{3}{4}, \frac{3}{5}, \frac{7}{10}$
29. $\frac{7}{15}, \frac{2}{3}, \frac{1}{5}$
30. $\frac{2}{5}, \frac{4}{9}, \frac{11}{15}$
31. $\frac{7}{12}, \frac{5}{8}, \frac{1}{2}$
32. $\frac{5}{8}, \frac{3}{4}, \frac{5}{12}$

There are over 3,500 different uses for corn products, from ethanol fuel and industrial products to the household items you see above.

33. Kyle operates a hot dog cart in a large city. He spends $\frac{2}{5}$ of his budget on supplies, $\frac{1}{12}$ on advertising, and $\frac{2}{25}$ on taxes and fees. Does Kyle spend more on advertising or more on taxes and fees?

34. The Dixon Dragons must win at least $\frac{3}{7}$ of their remaining games to qualify for their district playoffs. If they have 15 games left and they win 7 of them, will the Dragons compete in the playoffs? Explain.

35. ***AGRICULTURE*** The table shows the fraction of the world's total corn each country produces. List the countries in order from the country that produces the most corn to the country that produces the least corn.

World's Corn Production	
United States	$\frac{41}{100}$
China	$\frac{1}{5}$
Brazil	$\frac{1}{20}$

36. ***WRITE A PROBLEM*** Write a problem that involves comparing two fractions with different denominators.

37. ***WRITE ABOUT IT*** Compare the following fractions.

$\frac{1}{2}$ ■ $\frac{1}{4}$ $\quad$ $\frac{2}{3}$ ■ $\frac{2}{5}$ $\quad$ $\frac{3}{4}$ ■ $\frac{3}{7}$ $\quad$ $\frac{4}{5}$ ■ $\frac{4}{9}$

What do you notice about two fractions that have the same numerator but different denominators? Which one is greater?

38. ***CHALLENGE*** Name a fraction that would make the inequality true.

$$\frac{1}{4} > \blacksquare > \frac{1}{5}$$

Spiral Review

Evaluate each expression when $a = 4$, $b = 2.8$, and $c = 0.9$. (Lesson 3-3)

39. $a + b$ **40.** $b - c$ **41.** $a + c$ **42.** $a - b$

43. $b + c$ **44.** $a + b + c$ **45.** $a - c$ **46.** $a + b - c$

Write each number in scientific notation. (Lesson 3-5)

47. 45 **48.** 820 **49.** 319

50. 36,000 **51.** 405,000 **52.** 23,000,000

53. **EOG PREP** Nadine divided a bag of pretzels among 6 friends. Each friend received the same number of pretzels, and 4 extra pretzels remained in the bag. Which number could be the number of pretzels that were in the bag before Nadine shared them? (Lesson 3-9)

A 24 B 36 C 44 D 64

4-7 Mixed Numbers and Improper Fractions

Learn to convert between mixed numbers and improper fractions.

Vocabulary

improper fraction

proper fraction

Have you ever witnessed a total eclipse of the sun? It occurs when the sun's light is completely blocked out. A total eclipse is rare—only three have been visible in the continental United States since 1963.

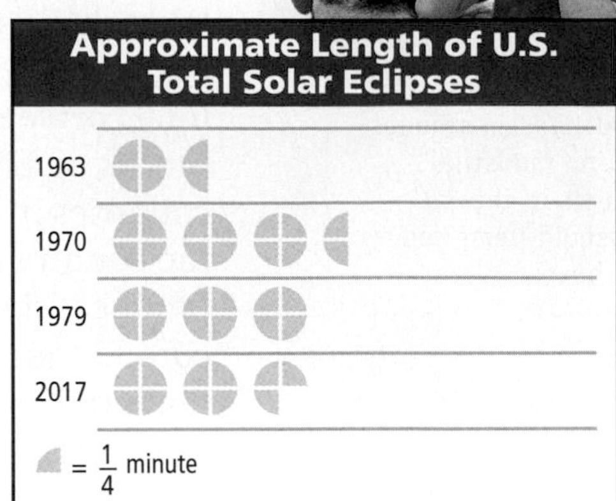

The graph shows that the eclipse in 2017 will last $2\frac{3}{4}$ minutes. There are eleven $\frac{1}{4}$-minute sections, so $2\frac{3}{4} = \frac{11}{4}$.

Reading Math

$\frac{11}{4}$ is read as "eleven-fourths."

An **improper fraction** is a fraction in which the numerator is greater than or equal to the denominator, such as $\frac{11}{4}$.

Whole numbers can be written as improper fractions. The whole number is the numerator, and the denominator is 1. For example, $7 = \frac{7}{1}$.

When the numerator is less than the denominator, the fraction is called a **proper fraction** .

Improper and Proper Fractions		
Improper Fractions		
• Numerator equals denominator ➤ fraction is equal to 1	$\frac{3}{3} = 1$	$\frac{102}{102} = 1$
• Numerator greater than denominator ➤ fraction is greater than 1	$\frac{9}{5} > 1$	$\frac{13}{1} > 1$
Proper Fractions		
• Numerator less than denominator ➤ fraction is less than 1	$\frac{2}{5} < 1$	$\frac{102}{351} < 1$

You can write an improper fraction as a mixed number.

EXAMPLE 1 *Astronomy Application*

The longest total solar eclipse in the next 200 years will take place in 2186. It will last about $\frac{15}{2}$ minutes. Write $\frac{15}{2}$ as a mixed number.

Method 1: Use a model.

Draw squares divided into half sections. Shade 15 of the half sections.

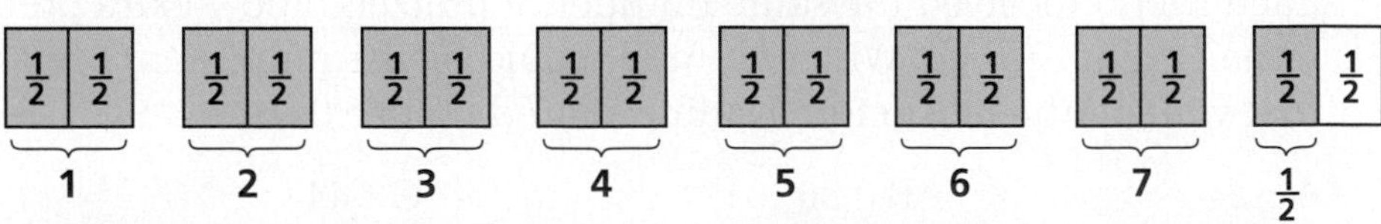

There are 7 whole squares and 1 half square, or $7\frac{1}{2}$ squares, shaded.

Method 2: Use division.

$$\begin{array}{r} 7\frac{1}{2} \\ 2\overline{)15} \\ -14 \\ \hline 1 \end{array}$$

Divide the numerator by the denominator.

To form the fraction part of the quotient, use the remainder as the numerator and the divisor as the denominator.

The 2186 eclipse will last about $7\frac{1}{2}$ minutes.

EXAMPLE 2 Writing Mixed Numbers as Improper Fractions

Write $2\frac{1}{5}$ as an improper fraction.

Method 1: Use a model.

You can draw a diagram to illustrate the whole and fractional parts.

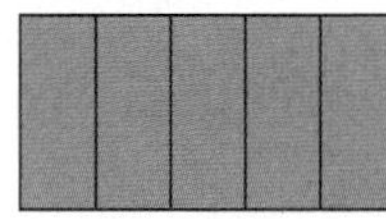
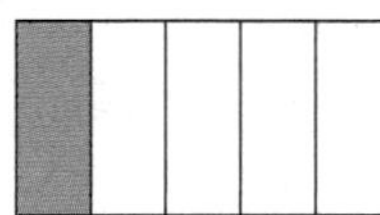

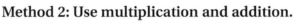

There are 11 fifths, or $\frac{11}{5}$. *Count the fifths in the diagram.*

Method 2: Use multiplication and addition.

When you are changing a mixed number to an improper fraction, spiral clockwise as shown in the picture. The order of operations will help you remember to multiply before you add.

$$2\frac{1}{5} = \frac{(5 \cdot 2) + 1}{5}$$ *Multiply the whole number by the denominator and add the numerator. Keep the same denominator.*

$$= \frac{10 + 1}{5}$$

$$= \frac{11}{5}$$

Think and Discuss

1. **Read** each improper fraction: $\frac{10}{7}, \frac{25}{9}, \frac{31}{16}$.
2. **Tell** whether each fraction is less than 1, equal to 1, or greater than 1: $\frac{21}{21}, \frac{54}{103}, \frac{9}{11}, \frac{7}{3}$.
3. **Explain** why any mixed number written as a fraction will be improper.

4-7 Exercises

FOR EOG PRACTICE

see page 651

internet connect

Homework Help Online
go.hrw.com Keyword: MR4 4-7

1.03

GUIDED PRACTICE

See Example 1

1. The fifth largest meteorite found in the United States is named the Navajo. The Navajo weighs $\frac{12}{5}$ tons. Write $\frac{12}{5}$ as a mixed number.

See Example 2

Write each mixed number as an improper fraction.

2. $1\frac{1}{4}$ **3.** $2\frac{2}{3}$ **4.** $1\frac{2}{7}$ **5.** $2\frac{2}{5}$

INDEPENDENT PRACTICE

See Example 1

6. Saturn is the sixth planet from the Sun. It takes Saturn $\frac{59}{2}$ years to revolve around the Sun. Write $\frac{59}{2}$ as a mixed number.

7. Pluto has the lowest surface gravity of all the planets in the solar system. A person who weighs 143 pounds on Earth weighs $\frac{43}{5}$ pounds on Pluto. Write $\frac{43}{5}$ as a mixed number.

See Example 2

Write each mixed number as an improper fraction.

8. $1\frac{3}{5}$ **9.** $2\frac{2}{9}$ **10.** $3\frac{1}{7}$ **11.** $4\frac{1}{3}$

12. $2\frac{3}{8}$ **13.** $4\frac{1}{6}$ **14.** $1\frac{4}{9}$ **15.** $3\frac{4}{5}$

PRACTICE AND PROBLEM SOLVING

Write each improper fraction as a mixed number or whole number. Tell whether your answer is a mixed number or whole number.

16. $\frac{21}{4}$ **17.** $\frac{32}{8}$ **18.** $\frac{20}{3}$ **19.** $\frac{43}{5}$

20. $\frac{108}{9}$ **21.** $\frac{87}{10}$ **22.** $\frac{98}{11}$ **23.** $\frac{105}{7}$

Write each mixed number as an improper fraction.

24. $9\frac{1}{4}$ **25.** $4\frac{9}{11}$ **26.** $11\frac{4}{9}$ **27.** $18\frac{3}{5}$

Replace each shape with a number that will make the equation correct.

28. $\blacksquare\frac{2}{5} = \frac{17}{\bullet}$ **29.** $\blacksquare\frac{6}{11} = \frac{83}{\bullet}$ **30.** $\blacksquare\frac{1}{9} = \frac{118}{\bullet}$

31. $\blacksquare\frac{6}{7} = \frac{55}{\bullet}$ **32.** $\blacksquare\frac{9}{10} = \frac{29}{\bullet}$ **33.** $\blacksquare\frac{1}{3} = \frac{55}{\bullet}$

34. $21\frac{\blacksquare}{3} = \frac{65}{\bullet}$ **35.** $5\frac{\blacksquare}{15} = \frac{77}{\bullet}$ **36.** $31\frac{\blacksquare}{19} = \frac{607}{\bullet}$

37. $14\frac{\blacksquare}{3} = \frac{45}{\bullet}$ **38.** $12\frac{\blacksquare}{10} = \frac{129}{\bullet}$ **39.** $9\frac{\blacksquare}{17} = \frac{154}{\bullet}$

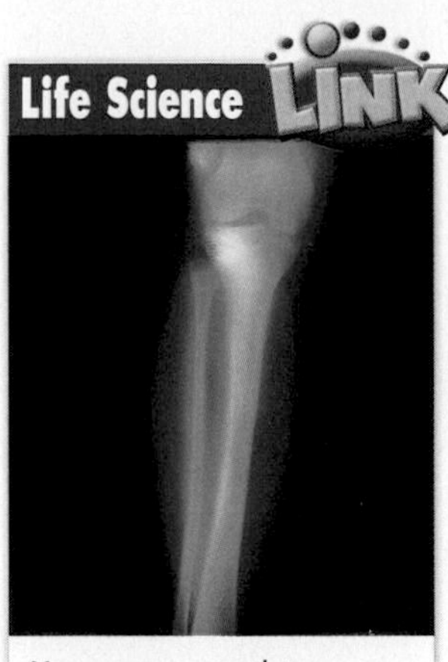

X-rays are used to produce images of bones. Bones absorb different amounts of x-rays than other body tissues. This is what makes them appear on the film.

go.hrw.com
KEYWORD: MR4 X-ray

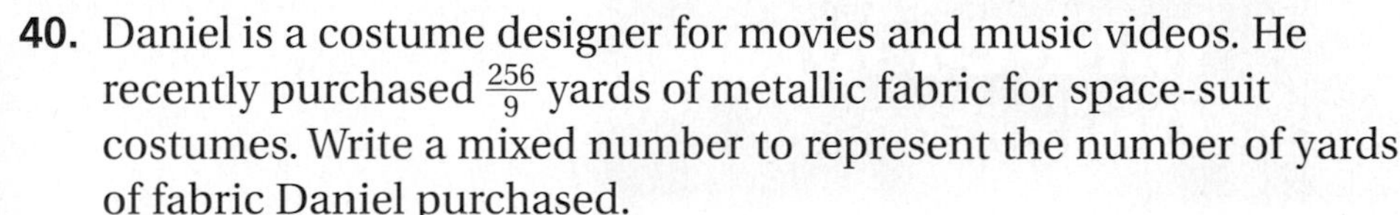

40. Daniel is a costume designer for movies and music videos. He recently purchased $\frac{256}{9}$ yards of metallic fabric for space-suit costumes. Write a mixed number to represent the number of yards of fabric Daniel purchased.

Use the table for Exercises 41–43.

41. ***LIFE SCIENCE*** Write the length of the ulna as an improper fraction. Then do the same for the length of the humerus.

42. ***LIFE SCIENCE*** Write the length of the fibula as a mixed number. Then do the same for the length of the femur.

43. ***LIFE SCIENCE*** Write the bones in order from longest to shortest.

44. ***SOCIAL STUDIES*** The European country of Monaco, with an area of only $1\frac{4}{5}$ km^2, is one of the smallest countries in the world. Write $1\frac{4}{5}$ as an improper fraction.

45. ***WHAT'S THE QUESTION?*** The lengths of Victor's three favorite movies are $\frac{11}{4}$ hours, $\frac{9}{4}$ hours, and $\frac{7}{4}$ hours. The answer is $2\frac{1}{4}$ hours. What is the question?

46. ***WRITE ABOUT IT*** Draw models representing $\frac{4}{4}$, $\frac{5}{5}$, and $\frac{9}{9}$. Use your models to explain why a fraction whose numerator is the same as its denominator is equal to 1.

47. ***CHALLENGE*** Write $\frac{65}{12}$ as a decimal.

Longest Human Bones	
Fibula (outer lower leg)	$\frac{81}{2}$ cm
Ulna (inner lower arm)	$28\frac{1}{5}$ cm
Femur (upper leg)	$\frac{101}{2}$ cm
Humerus (upper arm)	$36\frac{1}{2}$ cm
Tibia (inner lower leg)	43 cm

Spiral Review

Find each value. (Lesson 1-3)

48. 3^3 **49.** 9^2 **50.** 2^6 **51.** 4^4

Solve each equation. (Lesson 2-4)

52. $12 + y = 23$ **53.** $38 + y = 80$ **54.** $y + 76 = 230$

55. **EOG PREP** Which number is *not* divisible by 3? (Lesson 4-1)

A 240 C 522
B 413 D 735

56. **EOG PREP** Which is the prime factorization of 50? (Lesson 4-2)

A 2×5^2 C 10^5
B 2×5^{10} D 5×10

Chapter 4

Mid-Chapter Quiz

Mid-Chapter Quiz

LESSONS 4-1 AND 4-2 (pp. 152–159)

1. List all of the factors of 38. Is 38 prime or composite?
2. Write the prime factorization of 84.

LESSON 4-3 (pp. 160–163)

3. Find the GCF of 25 and 45.
4. Find the GCF of 26, 65, and 91.

LESSON 4-4 (pp. 167–170)

Write each decimal as a fraction.

5. 0.67
6. 0.9
7. 0.43
8. 0.17

Write each fraction as a decimal.

9. $\frac{2}{5}$
10. $\frac{1}{6}$
11. $\frac{3}{4}$
12. $\frac{5}{8}$

LESSON 4-5 (pp. 172–175)

Write two equivalent fractions for each fraction.

13. $\frac{9}{12}$
14. $\frac{18}{42}$
15. $\frac{25}{30}$
16. $\frac{8}{20}$

Write each fraction in simplest form.

17. $\frac{20}{24}$
18. $\frac{14}{49}$
19. $\frac{12}{28}$
20. $\frac{16}{36}$

LESSON 4-6 (pp. 178–181)

Compare. Write <, >, or =.

21. $\frac{3}{4}$ ■ $\frac{2}{3}$
22. $\frac{7}{9}$ ■ $\frac{5}{6}$
23. $\frac{4}{9}$ ■ $\frac{4}{7}$
24. $\frac{5}{11}$ ■ $\frac{3}{5}$

Order the fractions from least to greatest.

25. $\frac{5}{8}, \frac{1}{2}, \frac{3}{4}$
26. $\frac{3}{4}, \frac{3}{5}, \frac{7}{10}$
27. $\frac{1}{3}, \frac{3}{8}, \frac{1}{4}$
28. $\frac{2}{5}, \frac{4}{9}, \frac{11}{15}$

LESSON 4-7 (pp. 182–185)

29. The proboscis bat, with a length of $\frac{19}{5}$ cm, is one of the smallest bats. Write $\frac{19}{5}$ as a mixed number.
30. Write $\frac{15}{2}$ as a mixed number.

Focus on Problem Solving

Understand the Problem

• Write the problem in your own words

One way to understand a problem better is to write it in your own words. Before you do this, you may need to read it over several times, perhaps aloud so that you can hear yourself say the words.

When you write a problem in your own words, try to make the problem simpler. Use smaller words and shorter sentences. Leave out any extra information, but make sure to include all the information you need to answer the question.

Write each problem in your own words. Check that you have included all the information you need to answer the question.

1. Martin is making muffins for his class bake sale. The recipe calls for $2\frac{1}{3}$ cups of flour, but Martin's only measuring cup holds $\frac{1}{3}$ cup. How many of his measuring cups should he use?

2. Mariko sold an old book to a used bookstore. She had hoped to sell it for \$0.80, but the store gave her $\frac{3}{4}$ of a dollar. What is the difference between the two amounts?

3. Koalas of eastern Australia feed mostly on eucalyptus leaves. They select certain trees over others to find the $1\frac{1}{4}$ pounds of food they need each day. Suppose a koala has eaten $1\frac{1}{8}$ pounds of food. Has the koala eaten enough food for the day?

4. The first day of the Tour de France is called the prologue. Each of the days after that is called a stage, and each stage covers a different distance. The total distance covered in the race is about 3,600 km. If a cyclist has completed $\frac{1}{3}$ of the race, how many kilometers has he ridden?

4-8 Adding and Subtracting with Like Denominators

Learn to add and subtract fractions with like denominators.

You can estimate the age of an oak tree by measuring around the trunk at four feet above the ground.

The distance around a young oak tree's trunk increases at a rate of approximately $\frac{1}{8}$ inch per month.

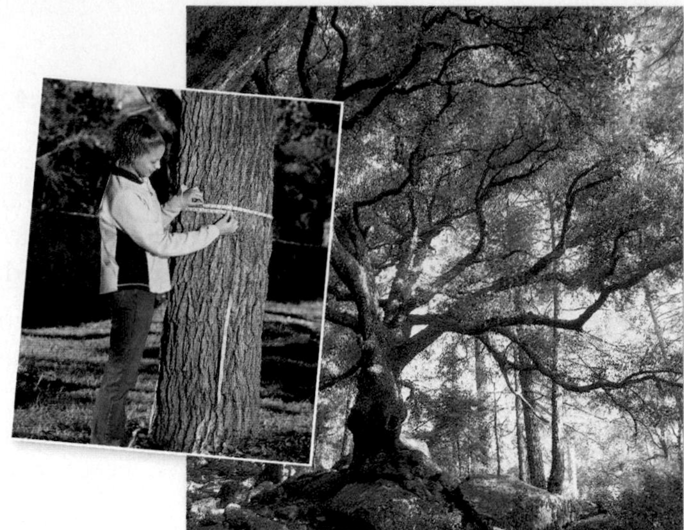

EXAMPLE 1 *Life Science Application*

Sophie plants a young oak tree in her backyard. The distance around the trunk grows at a rate of $\frac{1}{8}$ inch per month. Find how much this distance will increase in two months. Write your answer in simplest form.

$\frac{1}{8} + \frac{1}{8}$

$\frac{1}{8} + \frac{1}{8} = \frac{2}{8}$ *Add the numerators. Keep the same denominator.*

$= \frac{1}{4}$ *Write your answer in simplest form.*

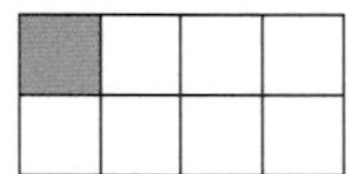

The distance around the trunk will increase by $\frac{1}{4}$ inch.

EXAMPLE 2 Subtracting Like Fractions and Mixed Numbers

Remember!

When the numerator equals the denominator, the fraction is equal to 1.

$\frac{3}{3} = 1$ $\frac{173}{173} = 1$

Subtract. Write each answer in simplest form.

A $1 - \frac{2}{3}$

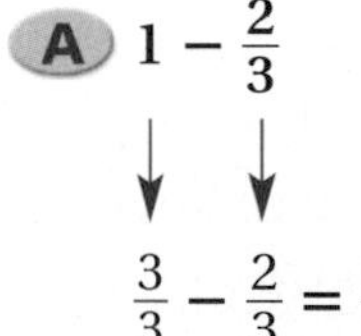

$\frac{3}{3} - \frac{2}{3} = \frac{1}{3}$

To get a common denominator, rewrite 1 as a fraction with a denominator of 3.

Subtract the numerators. Keep the same denominator.

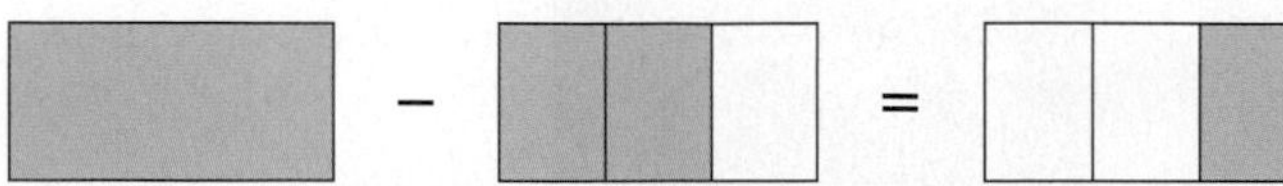

Subtract. Write each answer in simplest form.

B $3\frac{7}{12} - 1\frac{1}{12}$

$3\frac{7}{12} - 1\frac{1}{12}$ *Subtract the fractions.*

$2\frac{6}{12}$ *Then subtract the whole numbers.*

$2\frac{1}{2}$ *Write your answer in simplest form.*

EXAMPLE 3 Evaluating Expressions with Fractions

Evaluate each expression for $x = \frac{3}{8}$. Write each answer in simplest form.

A $\frac{5}{8} - x$

$\frac{5}{8} - x$ *Write the expression.*

$\frac{5}{8} - \frac{3}{8} = \frac{2}{8}$ *Substitute $\frac{3}{8}$ for x and subtract the numerators. Keep the same denominator.*

$= \frac{1}{4}$ *Write your answer in simplest form.*

Helpful Hint

When adding a fraction to a mixed number, you can think of the fraction as having a whole number of 0.

$\frac{3}{8} = 0\frac{3}{8}$

B $x + 1\frac{1}{8}$

$x + 1\frac{1}{8}$ *Write the expression.*

$\frac{3}{8} + 1\frac{1}{8} = 1\frac{4}{8}$ *Substitute $\frac{3}{8}$ for x. Add the fractions. Then add the whole numbers.*

$= 1\frac{1}{2}$ *Write your answer in simplest form.*

C $x + \frac{7}{8}$

$x + \frac{7}{8}$ *Write the expression.*

$\frac{3}{8} + \frac{7}{8} = \frac{10}{8}$ *Substitute $\frac{3}{8}$ for x and add the numerators. Keep the same denominator.*

$= \frac{5}{4}$ or $1\frac{1}{4}$ *Write your answer in simplest form.*

Think and Discuss

1. **Explain** how to add or subtract like fractions.
2. **Tell** why the sum of $\frac{1}{5}$ and $\frac{3}{5}$ is not $\frac{4}{10}$. Give the correct sum.
3. **Describe** how you would add $2\frac{3}{8}$ and $1\frac{1}{8}$. How would you subtract $1\frac{1}{8}$ from $2\frac{3}{8}$?

4-8 Exercises

1.04a, 5.02

FOR EOG PRACTICE

see page 652

internet connect
Homework Help Online
go.hrw.com Keyword: MR4 4-8

GUIDED PRACTICE

See Example 1

1. Marta is filling a bucket with water. The height of the water is increasing $\frac{1}{6}$ foot each minute. Find how much the height of the water will change in three minutes. Write your answer in simplest form.

See Example 2

Subtract. Write each answer in simplest form.

2. $2 - \frac{3}{5}$
3. $8 - \frac{6}{7}$
4. $4\frac{2}{3} - 1\frac{1}{3}$
5. $8\frac{7}{12} - 3\frac{5}{12}$

See Example 3

Evaluate each expression for $x = \frac{3}{10}$. Write each answer in simplest form.

6. $\frac{9}{10} - x$
7. $x + \frac{1}{10}$
8. $x + \frac{9}{10}$
9. $x - \frac{1}{10}$

INDEPENDENT PRACTICE

See Example 1

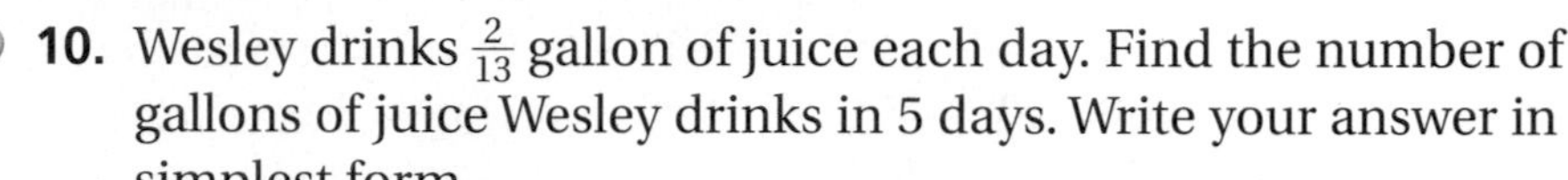

10. Wesley drinks $\frac{2}{13}$ gallon of juice each day. Find the number of gallons of juice Wesley drinks in 5 days. Write your answer in simplest form.

See Example 2

Subtract. Write each answer in simplest form.

11. $1 - \frac{5}{7}$
12. $1 - \frac{3}{8}$
13. $2\frac{4}{5} - 1\frac{1}{5}$
14. $9\frac{9}{14} - 5\frac{3}{14}$

See Example 3

Evaluate each expression for $x = \frac{11}{20}$. Write each answer in simplest form.

15. $x + \frac{13}{20}$
16. $x - \frac{3}{20}$
17. $x - \frac{9}{20}$
18. $x + \frac{17}{20}$

PRACTICE AND PROBLEM SOLVING

Write each sum or difference in simplest form.

19. $\frac{1}{16} + \frac{9}{16}$
20. $\frac{15}{26} - \frac{11}{26}$
21. $\frac{10}{33} + \frac{4}{33}$
22. $1 - \frac{9}{10}$
23. $\frac{26}{75} + \frac{24}{75}$
24. $\frac{100}{999} + \frac{899}{999}$
25. $37\frac{13}{18} - 24\frac{7}{18}$
26. $\frac{1}{20} + \frac{7}{20} + \frac{3}{20}$
27. $\frac{11}{24} + \frac{1}{24} + \frac{5}{24}$

Evaluate. Write each answer in simplest form.

28. $a + \frac{7}{18}$ for $a = \frac{1}{18}$
29. $\frac{6}{13} - j$ for $j = \frac{4}{13}$
30. $c + c$ for $c = \frac{5}{14}$
31. $m - \frac{6}{17}$ for $m = 1$
32. $8\frac{14}{15} - z$ for $z = \frac{4}{15}$
33. $13\frac{1}{24} + y$ for $y = 2\frac{5}{24}$

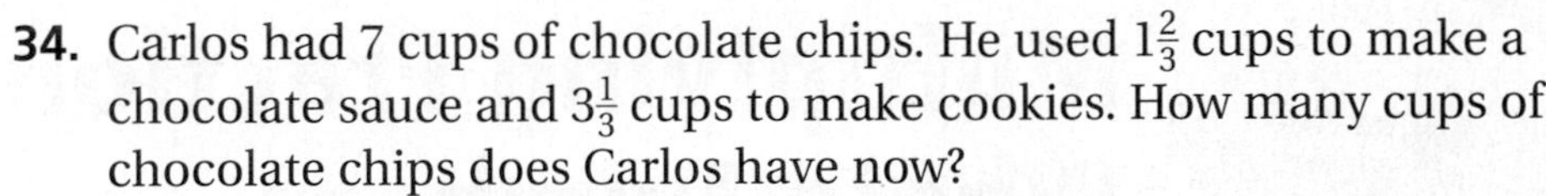

34. Carlos had 7 cups of chocolate chips. He used $1\frac{2}{3}$ cups to make a chocolate sauce and $3\frac{1}{3}$ cups to make cookies. How many cups of chocolate chips does Carlos have now?

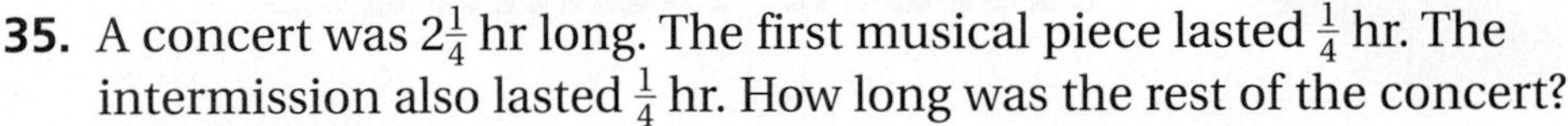

35. A concert was $2\frac{1}{4}$ hr long. The first musical piece lasted $\frac{1}{4}$ hr. The intermission also lasted $\frac{1}{4}$ hr. How long was the rest of the concert?

Life Science LINK

The Venus' flytrap, like other carnivorous plants, has adapted to nutrient-poor soil by capturing insects. Their traps spring shut in a fraction of a second, and then enzymes dissolve the insect.

36. A flight from Washington, D.C., stops in San Francisco and then continues to Seattle. The trip to San Francisco takes $4\frac{5}{8}$ hr. The trip to Seattle takes $1\frac{1}{8}$ hr. What is the total flight time?

Use the graph for Exercises 37–39.

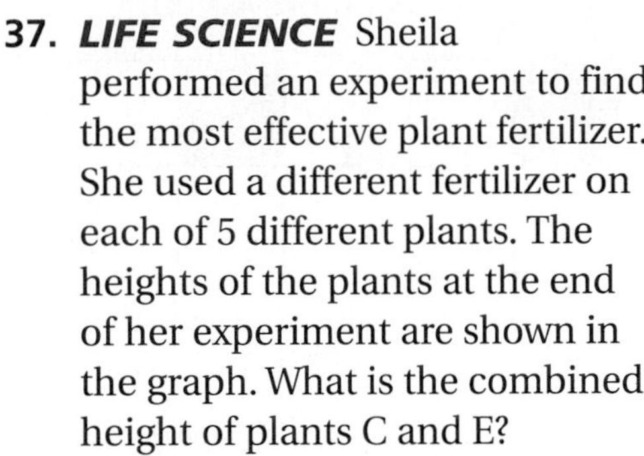

37. ***LIFE SCIENCE*** Sheila performed an experiment to find the most effective plant fertilizer. She used a different fertilizer on each of 5 different plants. The heights of the plants at the end of her experiment are shown in the graph. What is the combined height of plants C and E?

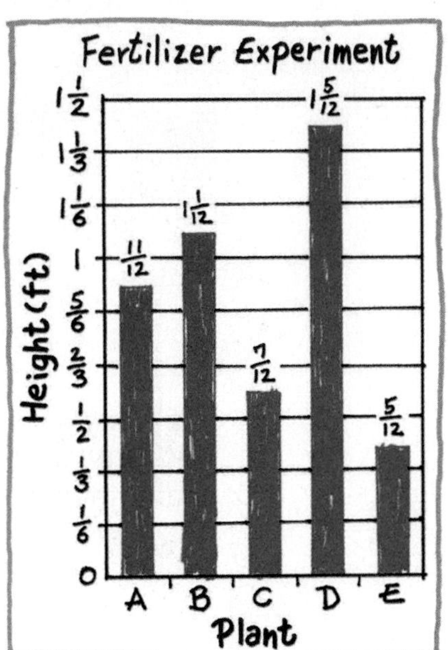

38. ***LIFE SCIENCE*** What is the difference in height between the tallest plant and the shortest plant?

39. ***WHAT'S THE ERROR?*** Sheila found the combined heights of plants B and E to be $1\frac{6}{24}$ feet. Explain the error and give the correct answer in simplest form.

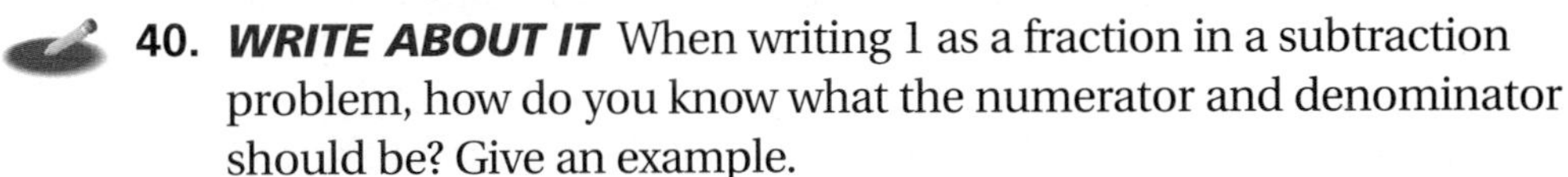

40. ***WRITE ABOUT IT*** When writing 1 as a fraction in a subtraction problem, how do you know what the numerator and denominator should be? Give an example.

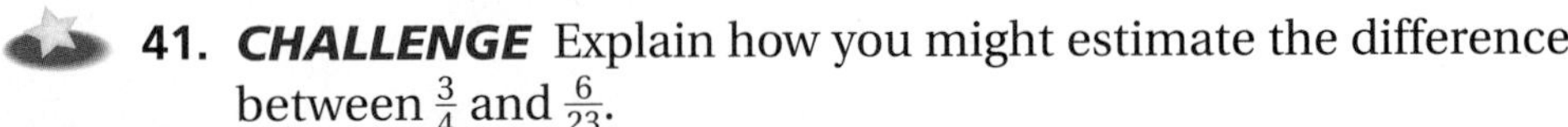

41. ***CHALLENGE*** Explain how you might estimate the difference between $\frac{3}{4}$ and $\frac{6}{23}$.

Spiral Review

Evaluate. (Lesson 1-4)

42. $2 + 3 \cdot 4 - 5$ 43. $(9 + 4) \div (3 + 10)$ 44. $23 - 16 + 28$

Identify a pattern and find the next term. (Lesson 1-7)

45. 3, 10, 17, 24, … 46. 5, 10, 15, 20, … 47. 1, 4, 2, 5, 3, …

48. **EOG PREP** Paul is seven years younger than his friend Rhonda. If r stands for Rhonda's age, which expression can be used to represent Paul's age? (Lesson 2-2)

A $7 - r$ B $r - 7$ C $7r$ D $r + 7$

4-9 Multiplying Fractions by Whole Numbers

Learn to multiply fractions by whole numbers.

Recall that multiplication by a whole number can be represented as repeated addition. For example, $4 \cdot 5 = 5 + 5 + 5 + 5$. You can multiply a whole number by a fraction using the same method.

$$3 \cdot \frac{1}{4} = \frac{1}{4} + \frac{1}{4} + \frac{1}{4} = \frac{3}{4}$$

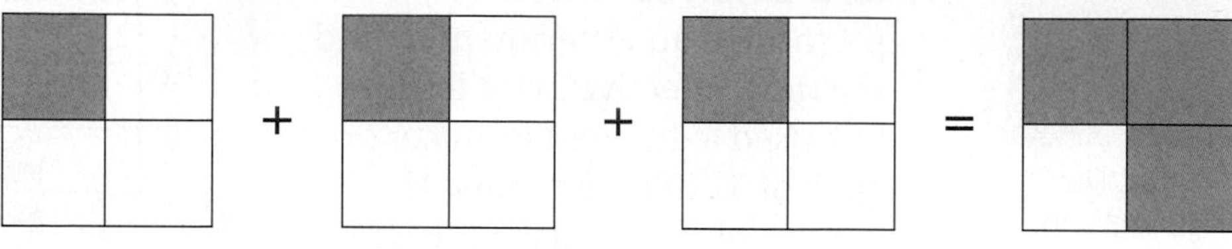

$$3 \cdot \frac{1}{4} = \frac{3}{4}$$

There is another way to multiply with fractions. Remember that a whole number can be written as an improper fraction with 1 in the denominator. So $3 = \frac{3}{1}$.

$$\frac{3}{1} \cdot \frac{1}{4} = \frac{3 \cdot 1}{1 \cdot 4} = \frac{3}{4}$$

Multiply numerators.
Multiply denominators.

EXAMPLE 1 **Multiplying Fractions and Whole Numbers**

Multiply. Write each answer in simplest form.

A $5 \cdot \frac{1}{8}$

$\frac{5}{1} \cdot \frac{1}{8} = \frac{5 \cdot 1}{1 \cdot 8}$ *Write 5 as a fraction. Multiply numerators and denominators.*

$= \frac{5}{8}$

B $3 \cdot \frac{1}{9}$

$\frac{3}{1} \cdot \frac{1}{9} = \frac{3 \cdot 1}{1 \cdot 9}$ *Write 3 as a fraction. Multiply numerators and denominators.*

$= \frac{3}{9}$

$= \frac{1}{3}$ *Write your answer in simplest form.*

C $4 \cdot \frac{7}{8}$

$\frac{4}{1} \cdot \frac{7}{8} = \frac{4 \cdot 7}{1 \cdot 8}$ *Write 4 as a fraction. Multiply numerators and denominators.*

$= \frac{28}{8}$

$= \frac{7}{2}$ or $3\frac{1}{2}$ *Write your answer in simplest form.*

EXAMPLE 2 Evaluating Fraction Expressions

Evaluate 6*x* for each value of *x*. Write each answer in simplest form.

A $x = \frac{1}{8}$

$6x$	*Write the expression.*
$6 \cdot \frac{1}{8}$	*Substitute $\frac{1}{8}$ for x.*
$\frac{6}{1} \cdot \frac{1}{8} = \frac{6}{8}$	*Multiply.*
$= \frac{3}{4}$	*Write your answer in simplest form.*

B $x = \frac{2}{3}$

$6x$	*Write the expression.*
$6 \cdot \frac{2}{3}$	*Substitute $\frac{2}{3}$ for x.*
$\frac{6}{1} \cdot \frac{2}{3} = \frac{12}{3}$	*Multiply.*
$= \frac{4}{1}$	
$= 4$	

Sometimes the denominator of an improper fraction will divide evenly into the numerator, as in Example 2B. When this happens, the improper fraction is equivalent to a whole number, not a mixed number.

$$\frac{12}{3} = 4$$

This makes sense if you remember that the fraction bar means "divided by."

EXAMPLE 3 *Social Studies Application*

Any proposed amendment to the U.S. Constitution must be ratified, or approved, by $\frac{3}{4}$ of the states. When the 13th Amendment abolishing slavery was proposed in 1865, there were 36 states. How many states needed to ratify this amendment in order for it to pass?

To find $\frac{3}{4}$ of 36, multiply.

$$\frac{3}{4} \cdot 36 = \frac{3}{4} \cdot \frac{36}{1}$$
$$= \frac{108}{4}$$
$$= 27$$

For the 13th Amendment to pass, 27 states had to ratify it.

Think and Discuss

1. Describe a model you could use to show the product of $4 \cdot \frac{1}{5}$.

2. Choose the expression that is correctly multiplied.

$$2 \cdot \frac{3}{7} = \frac{6}{7} \qquad 2 \cdot \frac{3}{7} = \frac{6}{14}$$

4-9 Exercises

FOR EOG PRACTICE

see page 653

internet connect
Homework Help Online
go.hrw.com Keyword: MR4 4-9

1.04a

GUIDED PRACTICE

See Example 1 **Multiply. Write each answer in simplest form.**

1. $8 \cdot \frac{1}{9}$
2. $2 \cdot \frac{1}{5}$
3. $12 \cdot \frac{1}{4}$
4. $7 \cdot \frac{4}{9}$
5. $3 \cdot \frac{1}{7}$
6. $4 \cdot \frac{2}{11}$
7. $8 \cdot \frac{3}{4}$
8. $18 \cdot \frac{1}{3}$

See Example 2 **Evaluate $12x$ for each value of x. Write the answer in simplest form.**

9. $x = \frac{2}{3}$
10. $x = \frac{1}{2}$
11. $x = \frac{3}{4}$
12. $x = \frac{5}{6}$

See Example 3

13. The school Community Service Club has 45 members. Of these 45 members, $\frac{3}{5}$ are boys. How many boys are members of the Community Service Club?

INDEPENDENT PRACTICE

See Example 1 **Multiply. Write each answer in simplest form.**

14. $4 \cdot \frac{1}{10}$
15. $6 \cdot \frac{1}{8}$
16. $3 \cdot \frac{1}{12}$
17. $2 \cdot \frac{2}{5}$
18. $6 \cdot \frac{10}{11}$
19. $2 \cdot \frac{3}{11}$
20. $15 \cdot \frac{2}{15}$
21. $20 \cdot \frac{1}{2}$

See Example 2 **Evaluate $8x$ for each value of x. Write the answer in simplest form.**

22. $x = \frac{1}{2}$
23. $x = \frac{3}{4}$
24. $x = \frac{1}{8}$
25. $x = \frac{1}{4}$

See Example 3

26. Kiesha spent 120 minutes completing her homework last night. Of those minutes, $\frac{1}{6}$ were spent on Spanish. How many minutes did Kiesha spend on her Spanish homework?

PRACTICE AND PROBLEM SOLVING

Evaluate each expression. Write each answer in simplest form.

27. $12b$ for $b = \frac{7}{12}$
28. $20m$ for $m = \frac{1}{20}$
29. $33z$ for $z = \frac{5}{11}$
30. $\frac{2}{3}y$ for $y = 18$
31. $\frac{1}{4}x$ for $x = 20$
32. $\frac{3}{5}a$ for $a = 30$
33. $\frac{4}{5}c$ for $c = 12$
34. $14x$ for $x = \frac{3}{8}$
35. $\frac{9}{10}n$ for $n = 50$

Compare. Write $<$, $>$, or $=$.

36. $9 \cdot \frac{1}{16} \blacksquare \frac{1}{2}$
37. $15 \cdot \frac{2}{5} \blacksquare 5$
38. $\frac{8}{13} \blacksquare 4 \cdot \frac{2}{13}$
39. $3 \cdot \frac{2}{9} \blacksquare \frac{2}{3}$
40. $6 \cdot \frac{4}{15} \blacksquare \frac{11}{24}$
41. $5 \blacksquare 12 \cdot \frac{3}{4}$
42. $3 \cdot \frac{1}{7} \blacksquare 3 \cdot \frac{1}{5}$
43. $7 \cdot \frac{3}{4} \blacksquare 6 \cdot \frac{3}{7}$
44. $2 \cdot \frac{5}{6} \blacksquare 6 \cdot \frac{2}{5}$

Life Science LINK

The General Sherman, a giant sequoia tree in California's Sequoia National Park, is one of the largest trees in the world. Its weight is estimated to be equal to the combined weights of 740 elephants.

California also has the nation's tallest grand fir, ponderosa pine, and sugar pine. The table below shows how the heights of these three trees compare with the height of the General Sherman. For example, the height of the grand fir is $\frac{23}{25}$ the height of the General Sherman.

45. The General Sherman tree is 275 ft tall. Find the heights of the three trees in the table. Write your answers as whole numbers or as mixed numbers in simplest form.

46. The world's tallest bluegum eucalyptus tree is $\frac{3}{5}$ of the height of the General Sherman tree. How tall is this bluegum eucalyptus?

Tree Heights Compared with the General Sherman	
Tallest Grand Fir	$\frac{23}{25}$
Tallest Ponderosa Pine	$\frac{41}{50}$
Tallest Sugar Pine	$\frac{21}{25}$

Source: The Top 10 of Everything 2000

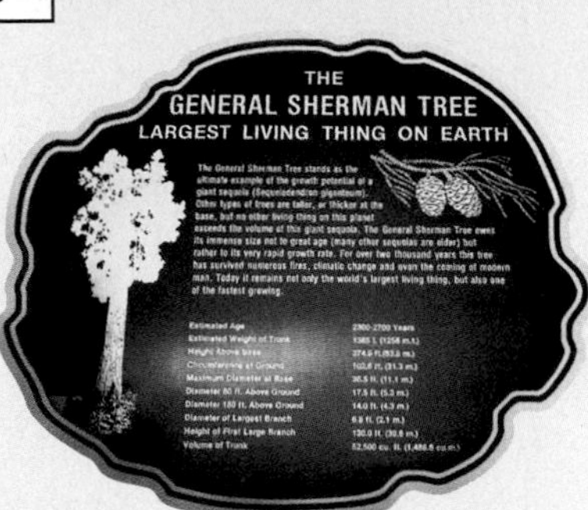

47. ***WHAT'S THE QUESTION?*** California is also the location of Joshua Tree National Park. Joshua trees can grow to be 40 ft tall. The answer is $\frac{8}{55}$. What is the question?

48. ***WRITE ABOUT IT*** Find $\frac{1}{5}$ the height of the General Sherman. Then divide the height of the General Sherman by 5. What do you notice? Why does this make sense?

49. ***CHALLENGE*** The world's tallest incense cedar tree is 152 ft tall. What is $\frac{1}{5}$ of $\frac{1}{2}$ of $\frac{1}{4}$ of 152?

Spiral Review

Identify the base and the exponent. (Lesson 1-3)

50. 5^3 **51.** 4^8 **52.** 9^2 **53.** 12

Solve each equation. (Lesson 2-5)

54. $x - 25 = 40$ **55.** $c - 18 = 20$ **56.** $56 = d - 0$ **57.** $e - 64 = 64$

58. **EOG PREP** Which fraction is *not* equivalent to $\frac{1}{6}$? (Lesson 4-5)

A $\frac{6}{1}$ B $\frac{2}{12}$ C $\frac{3}{18}$ D $\frac{6}{36}$

EXTENSION

Sets of Numbers

Learn to make Venn diagrams to describe number sets.

Vocabulary

set
empty set
element
subset
Venn diagram
intersection
union

A group of items is called a **set**. The items in a set are called **elements**. In this chapter, you saw several sets of numbers, such as prime numbers, composite numbers, and factors.

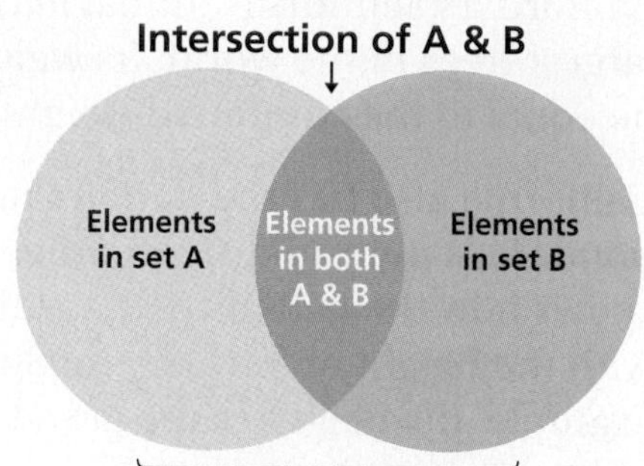

In a **Venn diagram**, circles are used to show relationships between sets. The overlapped region represents elements that are in both set A *and* set B. This set is called the **intersection** of A and B. Elements that are in set A *or* set B make up the **union** of A and B.

EXAMPLE 1 Identifying Elements and Drawing Venn Diagrams

Identify the elements in each set. Then draw a Venn diagram. What is the intersection? What is the union?

A **Set A: prime numbers** **Set B: composite numbers**

Elements of A: 2, 3, 5, 7, … Elements of B: 4, 6, 8, 9, …

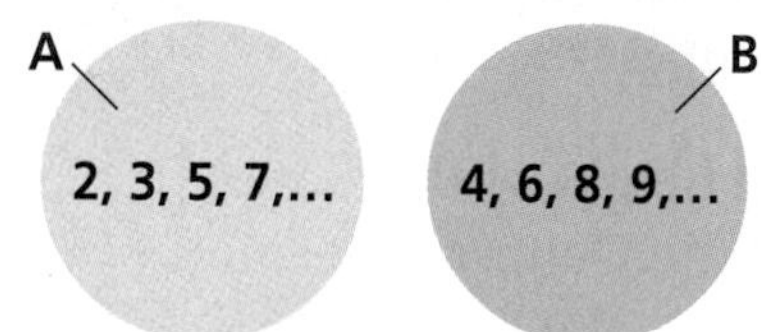

The circles do not overlap because no number is both prime and composite.

Intersection: none. When a set has no elements, it is called an **empty set**. The intersection of A and B is empty.

Union: all numbers that are prime *or* composite—all whole numbers except 0 and 1.

B **Set A: factors of 36** **Set B: factors of 24**

Elements of A: 1, 2, 3, 4, 6, 9, 12, 18, 36

Elements of B: 1, 2, 3, 4, 6, 8, 12, 24

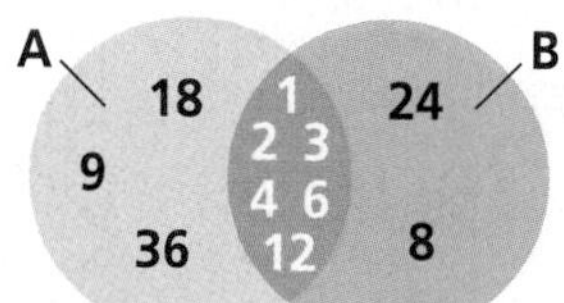

The circles overlap because some factors of 36 are also factors of 24.

Intersection: 1, 2, 3, 4, 6, 12 *factors of 36* ***and*** *24*

Union: 1, 2, 3, 4, 6, 8, 9, 12, 18, 24, 36 *factors of 36* ***or*** *24*

Identify the elements in each set. Then draw a Venn diagram. What is the intersection? What is the union?

C **Set A: factors of 36** **Set B: factors of 12**

Elements of A: 1, 2, 3, 4, 6, 9, 12, 18, 36
Elements of B: 1, 2, 3, 4, 6, 12

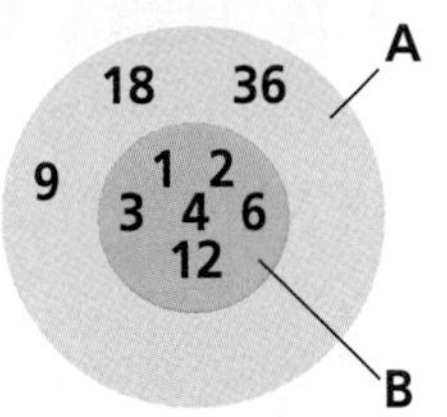

The circle for set B is entirely inside the circle for set A because all factors of 12 are also factors of 36.

Intersection: 1, 2, 3, 4, 6, 12 — *factors of 36* ***and*** *12*
Union: 1, 2, 3, 4, 6, 9, 12, 18, 36 — *factors of 36* ***or*** *12*

Helpful Hint

To decide whether set B is a subset of set A, ask yourself, "Is every element of B also in A?" If the answer is yes, then B is a subset of A.

Look at Example 1C. When one set is entirely contained in another set, we say the first set is a **subset** of the second set.

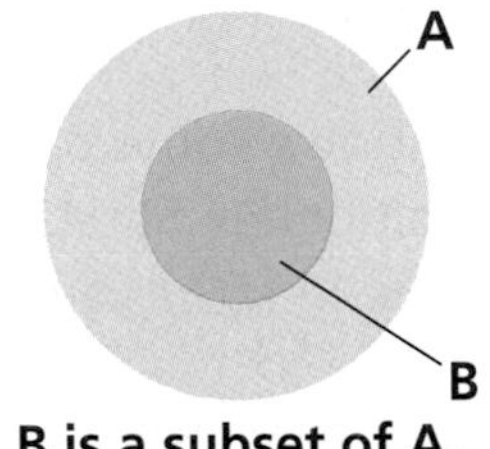

B is a subset of A.

EXTENSION Exercises

Identify the elements in each set. Then draw a Venn diagram. What is the intersection? What is the union?

1. Set A: even numbers
   Set B: odd numbers
2. Set A: factors of 18
   Set B: factors of 40
3. Set A: factors of 72
   Set B: factors of 36
4. Set A: even numbers
   Set B: composite numbers

Tell whether set A is a subset of set B.

5. Set A: whole numbers less than 10
   Set B: whole numbers less than 12
6. Set A: whole numbers less than 8
   Set B: whole numbers greater than 9
7. Set A: prime numbers
   Set B: odd numbers
8. Set A: numbers divisible by 6
   Set B: numbers divisible by 3

9. ***WRITE ABOUT IT*** How could you use a Venn diagram to help find the greatest common factor of two numbers? Give an example.
10. ***CHALLENGE*** How could you use a Venn diagram to help find the greatest common factor of three numbers? Give an example.

Problem Solving on Location

NORTH CAROLINA

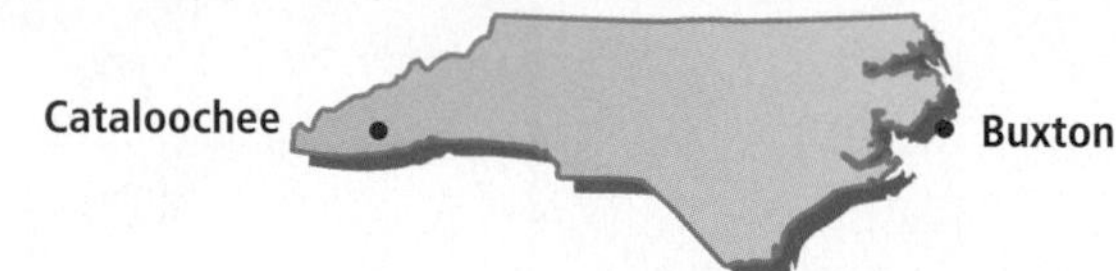

Snow Skiing

Cataloochee Ski Area, in Maggie Valley, North Carolina, has ten ski slopes. Of those ten slopes, $\frac{1}{4}$ of them are for beginners, $\frac{1}{2}$ are for intermediate-level skiers, and $\frac{1}{4}$ are reserved for advanced skiers.

The table shows some of the slopes of the Cataloochee Ski Area and their approximate lengths in miles.

Cataloochee Ski Area Trails

Slope	Length (mi)
1 Upper Omigosh	$\frac{1}{5}$
2 Lower Omigosh	$\frac{1}{3}$
3 Snowbird Trail	$\frac{2}{9}$
9 Easy Way	$\frac{1}{9}$
5 Alley Cat	$\frac{1}{6}$
6 Rabbit Hill	$\frac{1}{3}$
4 Rock Island Run	$\frac{1}{9}$

For 1–6, use the table.

1. Which is longer, Lower Omigosh or Upper Omigosh?
2. If you skied Snowbird Trail, Easy Way, and Rock Island Run, how many total miles would you have skied? Write your answer in simplest form.
3. How many Easy Way slopes would it take to equal the length of one Snowbird Trail slope?
4. If you skied Alley Cat twice, you would have skied a distance equal to the lengths of what other trails?
5. Write the ski slopes in order from shortest to longest.
6. Raul skied the Snowbird Trail slope, and April skied the Lower Omigosh slope. Who skied farther?

Lighthouses

The $63\frac{2}{5}$-meter-tall Cape Hatteras Lighthouse is the tallest lighthouse in the United States. Located in Buxton, North Carolina, the tower was built of 1.2 million bricks, and its walls are 13 feet thick.

The Cape Hatteras Lighthouse sits over the dangerous shallow sandbars of Diamond Shoals. It projects light more than 50 miles into the Atlantic Ocean, warning travelers away from the area known as the graveyard of the Atlantic.

The following table lists the approximate heights in meters of six operating lighthouses in North Carolina.

For 1–4, use the table.

North Carolina Lighthouses	
Lighthouse	**Height (m)**
Bodie Island (Pea Island)	$27\frac{11}{25}$
Cape Hatteras	$63\frac{2}{5}$
Cape Lookout	$51\frac{14}{125}$
Currituck Beach	$49\frac{1}{3}$
Oak Island	$51\frac{14}{125}$
Ocracoke Island	$23\frac{4}{25}$

1. Write the heights of the Bodie Island lighthouse, the Cape Hatteras lighthouse, and the Currituck Beach lighthouse as decimals.
2. Which two lighthouses are the same height?
3. Which lighthouse is the shortest?
4. Write the lighthouses in order from tallest to shortest.
5. Suppose the six lighthouses were evenly spaced along the coastline with one lighthouse at one end and one at the other end. To find the distance between each lighthouse, would you divide the 300 miles of coastline by 6? Explain your answer.

MATH-ABLES

Riddle Me This

"When you go from there to here,
you'll find I disappear.
Go from here to there, and then
you'll see me again.
What am I?"

To solve this riddle, copy the square below. If a number is divisible by 3, color that box red. Remember the divisibility rule for 3. If a number is not divisible by 3, color that box blue.

102	981	210	6,015	72
79	1,204	576	10,019	1,771
548	3,416	12,300	904	1,330
217	2,662	1,746	3,506	15,025
34,351	725	2,352	5,675	6,001

On a Roll

The object is to be the first person to fill in all the squares on your game board.

On your turn, roll a number cube and record the number rolled in any blank square on your game board. Once you have placed a number in a square, you cannot move that number. If you cannot place the number in a square, then your turn is over. The winner is the first player to complete their game board correctly.

internet connect

Go to ***go.hrw.com*** for a complete set of rules and game pieces.
KEYWORD: MR4 Game4

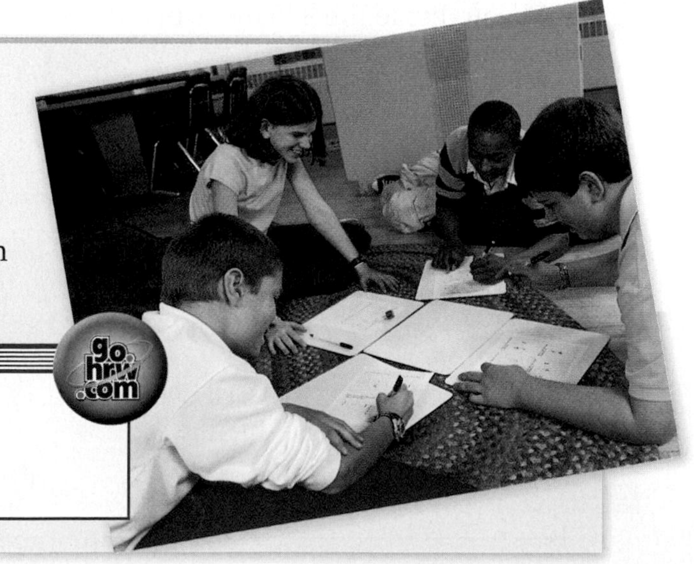

Greatest Common Factor

internet connect
Lab Resources Online
go.hrw.com
KEYWORD: MR4 TechLab4

You can use a graphing calculator to quickly find the greatest common factor (GCF) of two or more numbers. A calculator is particularly useful when you need to find the GCF of large numbers.

Activity

1 Find the GCF of 504 and 3,150.

The GCF is also known as the *greatest common divisor,* or GCD. The GCD function is found on the **MATH** menu.

To find the GCD on a graphing calculator, press MATH. Press ▶ to highlight NUM, and then use ▼ to scroll down and highlight 9:.

Press ENTER 504 , 3150) ENTER.

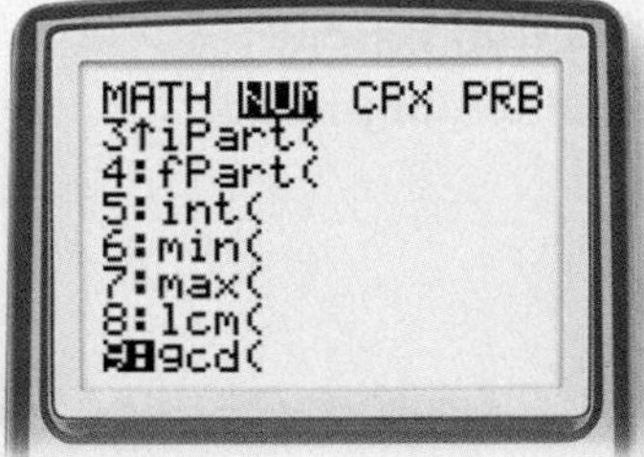

The greatest common factor of 504 and 3,150 is 126.

Think and Discuss

1. Suppose your calculator will not allow you to enter three numbers into the GCD function. How could you still use your calculator to find the GCF of the three following numbers: 4,896; 2,364; and 656? Explain your strategy and why it works.
2. Would you use your calculator to find the GCF of 6 and 18? Why or why not?

Try This

Find the GCF of each set of numbers.

1. 14, 48	**2.** 18, 54	**3.** 99, 121	**4.** 144, 196
5. 200, 136	**6.** 246, 137	**7.** 72, 860	**8.** 55, 141, 91

Chapter 4

Study Guide and Review

Vocabulary

Complete the sentences below with vocabulary words from the list above. Words may be used more than once.

1. The number $\frac{11}{9}$ is an example of a(n) ___?___, and $3\frac{1}{6}$ is an example of a(n) ___?___.
2. A(n) ___?___, such as 0.3333..., has a block of one or more digits that repeat without end. A(n) ___?___, such as 0.25, has a finite number of decimal places.
3. A(n) ___?___ is divisible by only two numbers, 1 and itself. A(n) ___?___ is divisible by more than two numbers.

4-1 Divisibility (pp. 152–155)

EXAMPLE

■ **Tell whether 210 is divisible by 2, 3, 4, and 6.**

2	*The last digit, 0, is even.*	Divisible
3	*The sum of the digits is divisible by 3.*	Divisible
4	*The number formed by the last two digits is not divisible by 4.*	Not divisible
6	*210 is divisible by 2 and 3.*	Divisible

■ **Tell whether each number is prime or composite.**

17 *only divisible by 1 and 17* prime
25 *divisible by 1, 5, and 25* composite

EXERCISES

Tell whether each number is divisible by 2, 3, 4, 5, 6, 9, and 10.

4. 118
5. 90
6. 342
7. 284
8. 170
9. 393

Tell whether each number is prime or composite.

10. 121
11. 77
12. 13
13. 118
14. 67
15. 93
16. 39
17. 97
18. 85
19. 61

4-2 Factors and Prime Factorization (pp. 156–159)

EXAMPLES

- **List all the factors of 10.**

  $10 = 1 \cdot 10$ $\quad$ $10 = 2 \cdot 5$

  The factors of 10 are 1, 2, 5, and 10.

- **Write the prime factorization of 30.**

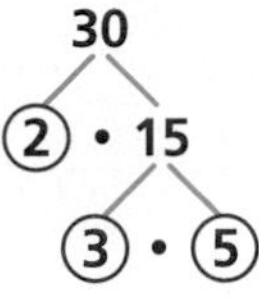

  $30 = 2 \cdot 3 \cdot 5$

EXERCISES

List all the factors of each number.

20. 60 **21.** 72
22. 29 **23.** 56
24. 85 **25.** 71

Write the prime factorization of each number.

26. 65 **27.** 94 **28.** 110
29. 81 **30.** 99 **31.** 76
32. 97 **33.** 55 **34.** 46

4-3 Greatest Common Factor (pp. 160–163)

EXAMPLE

- **Find the GCF of 35 and 50.**

  factors of 35: 1, ⑤, 7, 35
  factors of 50: 1, 2, ⑤, 10, 25, 50

  The GCF of 35 and 50 is 5.

EXERCISES

Find the GCF of each set of numbers.

35. 36 and 60
36. 50, 75, and 125
37. 45, 81, and 99

4-4 Decimals and Fractions (pp. 167–170)

EXAMPLES

- **Write 1.29 as a mixed number.**

  $1.29 = 1\frac{29}{100}$

- **Write $\frac{3}{5}$ as a decimal.**

  $5\overline{)3.0}$ = 0.6 $\quad$ $\frac{3}{5} = 0.6$

EXERCISES

Write as a fraction or mixed number.

38. 0.37 **39.** 1.8 **40.** 0.4

Write as a decimal.

41. $\frac{7}{8}$ **42.** $\frac{2}{5}$ **43.** $\frac{7}{9}$

4-5 Equivalent Fractions (pp. 172–175)

EXAMPLES

- **Find an equivalent fraction for $\frac{4}{5}$.**

  $\frac{4}{5} = \frac{\blacksquare}{15}$ $\quad$ $\frac{4 \cdot 3}{5 \cdot 3} = \frac{12}{15}$

- **Write $\frac{8}{12}$ in simplest form.**

  $\frac{8 \div 4}{12 \div 4} = \frac{2}{3}$

EXERCISES

Find two equivalent fractions.

44. $\frac{4}{6}$ **45.** $\frac{4}{5}$ **46.** $\frac{3}{12}$

Write each fraction in simplest form.

47. $\frac{14}{16}$ **48.** $\frac{9}{30}$ **49.** $\frac{7}{10}$

4-6 Comparing and Ordering Fractions (pp. 178–181)

EXAMPLE

- Order from least to greatest.

$\frac{3}{5}, \frac{2}{3}, \frac{1}{3}$ *Rename with like denominators.*

$\frac{3 \cdot 3}{5 \cdot 3} = \frac{9}{15}$ $\frac{2 \cdot 5}{3 \cdot 5} = \frac{10}{15}$ $\frac{1 \cdot 5}{3 \cdot 5} = \frac{5}{15}$

$\frac{1}{3}, \frac{3}{5}, \frac{2}{3}$

EXERCISES

Compare. Write <, >, or =.

50. $\frac{6}{8}$ ■ $\frac{3}{8}$ **51.** $\frac{7}{9}$ ■ $\frac{2}{3}$

Order from least to greatest.

52. $\frac{3}{8}, \frac{2}{3}, \frac{7}{8}$ **53.** $\frac{4}{6}, \frac{3}{12}, \frac{1}{3}$

4-7 Mixed Numbers and Improper Fractions (pp. 182–185)

EXAMPLE

- Write $3\frac{5}{6}$ as an improper fraction.

$3\frac{5}{6} = \frac{(3 \cdot 6) + 5}{6} = \frac{18 + 5}{6} = \frac{23}{6}$

- Write $\frac{19}{4}$ as a mixed number.

$4\overline{)19}$ = 4R3 $\frac{19}{4} = 4\frac{3}{4}$

EXERCISES

Write as an improper fraction.

54. $3\frac{7}{9}$ **55.** $2\frac{5}{12}$ **56.** $5\frac{2}{7}$

Write as a mixed number.

57. $\frac{23}{6}$ **58.** $\frac{17}{5}$ **59.** $\frac{41}{8}$

4-8 Adding and Subtracting with Like Denominators (pp. 188–191)

EXAMPLE

- Subtract $4\frac{5}{6} - 2\frac{1}{6}$. Write your answer in simplest form.

$4\frac{5}{6} - 2\frac{1}{6} = 2\frac{4}{6} = 2\frac{2}{3}$

EXERCISES

Add or subtract. Write each answer in simplest form.

60. $\frac{1}{5} + \frac{4}{5}$ **61.** $1 - \frac{3}{12}$

62. $\frac{9}{10} - \frac{3}{10}$ **63.** $4\frac{2}{7} + 2\frac{3}{7}$

4-9 Multiplying Fractions by Whole Numbers (pp. 192–195)

EXAMPLE

- Multiply $3 \cdot \frac{3}{5}$. Write your answer in simplest form.

$3 \cdot \frac{3}{5} = \frac{3}{1} \cdot \frac{3}{5} = \frac{3 \cdot 3}{1 \cdot 5} = \frac{9}{5}$ or $1\frac{4}{5}$

EXERCISES

Multiply. Write each answer in simplest form.

64. $5 \cdot \frac{1}{7}$ **65.** $2 \cdot \frac{3}{8}$

66. $3 \cdot \frac{6}{7}$ **67.** $4 \cdot \frac{2}{9}$

Chapter 4

Chapter Test

Tell whether each number is divisible by 2, 3, 4, 5, 6, 9, and 10.

1. 384 **2.** 815 **3.** 724 **4.** 624

List all the factors of each number. Then tell whether each number is prime or composite.

5. 98 **6.** 40 **7.** 45 **8.** 41

Write the prime factorization of each number.

9. 64 **10.** 130 **11.** 49 **12.** 28

Find the GCF of each set of numbers.

13. 24 and 108 **14.** 45, 18, and 39 **15.** 49, 77, and 84

Write each decimal as a fraction or mixed number.

16. 0.37 **17.** 1.9 **18.** 0.92 **19.** 5.03

Write each fraction or mixed number as a decimal.

20. $\frac{3}{8}$ **21.** $9\frac{3}{5}$ **22.** $\frac{2}{3}$ **23.** $2\frac{1}{8}$

Write each fraction in simplest form.

24. $\frac{4}{12}$ **25.** $\frac{6}{9}$ **26.** $\frac{3}{15}$ **27.** $\frac{7}{8}$

Write each mixed number as an improper fraction.

28. $4\frac{7}{8}$ **29.** $7\frac{5}{12}$ **30.** $3\frac{5}{7}$ **31.** $1\frac{8}{11}$

Compare. Write <, >, or =.

32. $\frac{5}{6}$ ■ $\frac{3}{6}$ **33.** $\frac{3}{4}$ ■ $\frac{7}{8}$ **34.** $\frac{4}{5}$ ■ $\frac{7}{10}$

Order the fractions and decimals from least to greatest.

35. 2.17, 2.3, $2\frac{1}{9}$ **36.** 0.1, $\frac{3}{8}$, 0.3 **37.** 0.9, $\frac{2}{8}$, 0.35

38. On Monday, it snowed $2\frac{1}{4}$ inches. On Tuesday, an additional $3\frac{3}{4}$ inches of snow fell. How much snow fell altogether on Monday and Tuesday?

Multiply. Write each answer in simplest form.

39. $4 \cdot \frac{1}{3}$ **40.** $2 \cdot \frac{3}{8}$ **41.** $\frac{1}{4} \cdot 14$

Chapter 4

Performance Assessment

Show What You Know

Create a portfolio of your work from this chapter. Complete this page and include it with your four best pieces of work from Chapter 4. Choose from your homework or lab assignments, mid-chapter quizzes, or any journal entries you have done. Put them together using any design you want. Make your portfolio represent what you consider your best work.

Short Response

1. In Mrs. Matika's class, there are 9 girls and 15 boys. Mrs. Matika wants to divide the class into groups for a project. Each group should have the same number of boys and the same number of girls. What is the greatest number of groups she can make if every student is put in a group? Explain how you determined your answer.

2. Kerry, Janice, Marcos, and Carl ordered a pizza for dinner. Kerry ate $\frac{1}{8}$ of the pizza, Janice ate $\frac{3}{8}$ of the pizza, and Carl ate $\frac{2}{8}$ of the pizza. If there were no leftovers, how much pizza did Marcos eat? Show your work.

3. Find the value of the expression $1\frac{3}{5} + 2\frac{4}{5}$. Write the answer as a mixed number, an improper fraction, and a decimal. Show your work.

Extended Problem Solving

4. Trent plans to purchase the rug in the photograph to place in his den. The rug is 3 yards wide and $5\frac{1}{2}$ yards long.
   a. Find the perimeter of the rug by adding the four side measures. Show your work.
   b. Find the area of the rug by multiplying the length and width. Write your answer in simplest form.
   c. To determine if the rug will fit, Trent measures his den. It is 144 inches wide and 168 inches long. He calculates the area of the den as $18\frac{2}{3}$ square yards. Since the area of the den is greater than the area of the rug, Trent decides to purchase the rug. Do you agree with Trent's decision? Explain your answer.

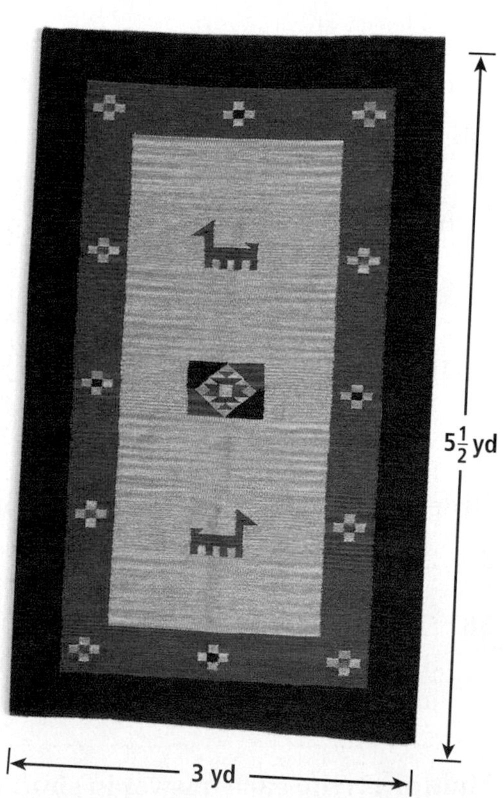

Performance Assessment

internet connect
EOG Test Practice Online
go.hrw.com Keyword: MR4 TestPrep

Getting Ready for EOG

Chapter 4

Cumulative Assessment, Chapters 1–4

1. Which is a prime number?
   A 9 C 47
   B 39 D 51

2. Which number is *greatest*?
   A 7.056 C $7\frac{3}{5}$
   B 7.06 D $7\frac{1}{2}$

3. What is the value of $5^2 \times 3 + 6 - 2^2$?
   A 77 C 125
   B 221 D 19

4. Which fraction is *not* equivalent to $\frac{4}{6}$?
   A $\frac{2}{3}$ C $\frac{8}{12}$
   B $\frac{10}{15}$ D $\frac{16}{18}$

5. The bar graph shows the four most common kinds of insects and the approximate number of known species of each. Which is the number of true flies written in scientific notation?

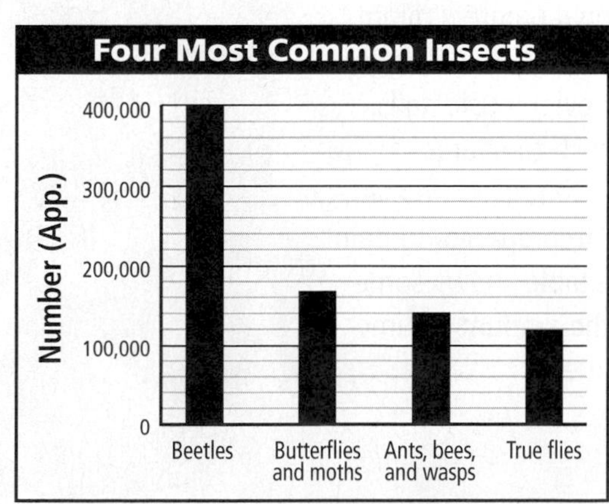

   A 16.5×10^4 C 1.2×10^5
   B 1.4×10^5 D 12.0×10^4

6. What is the product of 4 and $\frac{1}{3}$?
   A $\frac{1}{12}$ C $\frac{3}{4}$
   B $1\frac{1}{3}$ D $\frac{5}{3}$

7. Which is 2.04 written as a mixed number?
   A $2\frac{1}{4}$ C $2\frac{2}{10}$
   B $2\frac{2}{5}$ D $2\frac{1}{25}$

8. Which measure is equivalent to 12 meters?
   A 120 cm C 0.12 km
   B 1.2 km D 1.2×10^3 cm

TEST TAKING TIP!
Estimate the answer before solving. Use your estimate to check the reasonableness of your solution.

9. ***SHORT RESPONSE*** At the store, Ryan bought three items that cost \$4.65, \$3.99, and \$2.50. He gave the cashier \$15.00. How much change should Ryan receive? Explain how you found your answer.

10. ***SHORT RESPONSE*** Use prime factorization to find the GCF of 24, 56, and 72. Show your work.

Getting Ready for EOG

Chapter 5

Fraction Operations

TABLE OF CONTENTS

internet connect

Chapter Opener Online
go.hrw.com
KEYWORD: MR4 Ch5

Painting Times

Object	Paint	Time (hr)
Wall (100 ft^2)	Oil-based	$\frac{3}{10}$
Wall (100 ft^2)	Latex	$\frac{2}{5}$
Chair rail (100 ft)	Latex	$\frac{3}{4}$
Chair rail (100 ft)	Stain	$\frac{3}{5}$
Door	Oil-based	$\frac{1}{2}$
Window	Oil-based	$\frac{3}{4}$

Career ***Painter***

Have you ever wondered how painters estimate how much to charge for a job? Professional painters might paint houses, schools, office buildings, sports stadiums, or even music halls. To estimate how much to charge, many painters use a table that lists the average time it should take to prepare and paint certain objects. The table shows some painting jobs and the amount of time they take to complete.

ARE YOU READY?

Choose the best term from the list to complete each sentence.

1. The first five ___?___ of 6 are 6, 12, 18, 24, and 30. The ___?___ of 6 are 1, 2, 3, and 6.
2. Fractions with the same denominator are called ___?___.
3. A fraction is in ___?___ when the GCF of the numerator and the denominator is 1.
4. The fraction $\frac{13}{9}$ is a(n) ___?___ because the ___?___ is greater than the ___?___.

denominator
factors
improper fraction
like fractions
multiples
numerator
proper fraction
simplest form
unlike fractions

Complete these exercises to review skills you will need for this chapter.

✔ Simplify Fractions

Write each fraction in simplest form.

5. $\frac{6}{10}$ **6.** $\frac{5}{15}$ **7.** $\frac{14}{8}$ **8.** $\frac{8}{12}$

9. $\frac{10}{100}$ **10.** $\frac{12}{144}$ **11.** $\frac{33}{121}$ **12.** $\frac{15}{17}$

✔ Write Mixed Numbers as Fractions

Write each mixed number as an improper fraction.

13. $1\frac{1}{8}$ **14.** $2\frac{3}{4}$ **15.** $2\frac{4}{5}$ **16.** $1\frac{7}{9}$

17. $3\frac{1}{5}$ **18.** $5\frac{2}{3}$ **19.** $4\frac{4}{7}$ **20.** $3\frac{11}{12}$

✔ Add and Subtract Like Fractions

Add or subtract. Write each answer in simplest form.

21. $\frac{5}{8} + \frac{1}{8}$ **22.** $\frac{3}{7} + \frac{5}{7}$ **23.** $\frac{9}{10} - \frac{3}{10}$ **24.** $\frac{5}{9} - \frac{2}{9}$

✔ Multiplication Facts

Multiply.

25. 8×11 **26.** 7×8 **27.** 4×12 **28.** 12×7

29. 10×13 **30.** 9×7 **31.** 6×8 **32.** 11×12

Model Fraction Multiplication

Use with Lessons 5-1 and 5-2

internet connect
Lab Resources Online
go.hrw.com
KEYWORD: MR4 Lab5A

You can use grids to help you understand fraction multiplication.

Activity 1

1 Think of $\frac{1}{2} \cdot \frac{1}{3}$ as $\frac{1}{2}$ of $\frac{1}{3}$.

Shade $\frac{1}{3}$ of a square.

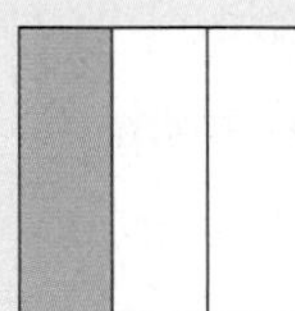

Divide the square into halves.

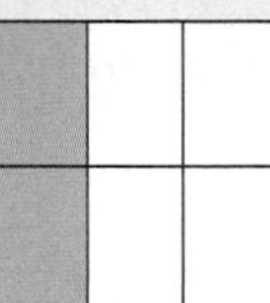

Look at $\frac{1}{2}$ of the part you shaded.

What fraction of the whole is this?

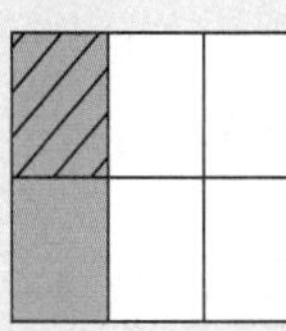

$\frac{1}{2}$ of $\frac{1}{3}$ is $\frac{1}{6}$.

2 Think of $\frac{2}{3} \cdot \frac{1}{2}$ as $\frac{2}{3}$ of $\frac{1}{2}$.

Shade $\frac{1}{2}$ of a square.

Divide the square into thirds.

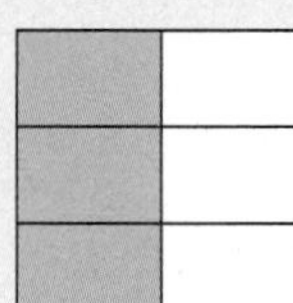

$\frac{2}{3}$ of $\frac{1}{2}$ is $\frac{2}{6}$, or $\frac{1}{3}$.

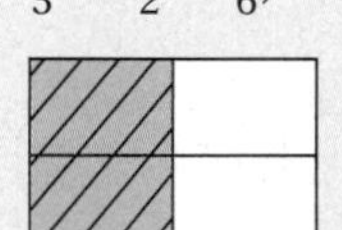

Think and Discuss

1. Tell whether the product is greater than or less than the fractions you started with.

Try This

Write the multiplication expression modeled on each grid.

1.

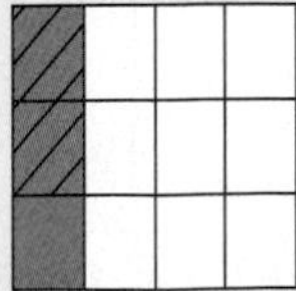

2.

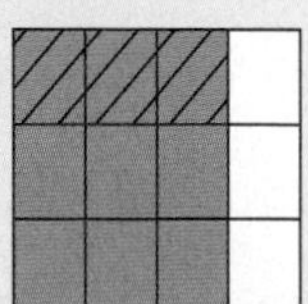

3.

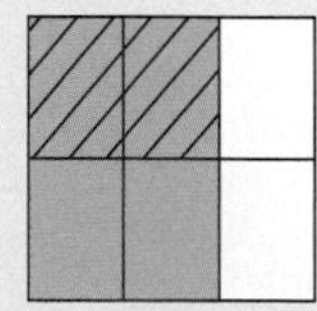

Use a grid to model each multiplication expression.

4. $\frac{1}{3} \cdot \frac{1}{2}$ **5.** $\frac{2}{3} \cdot \frac{1}{3}$ **6.** $\frac{1}{4} \cdot \frac{2}{3}$ **7.** $\frac{1}{3} \cdot \frac{3}{4}$

You can also use grids to model multiplication of mixed numbers.

Activity 2

1 Think of $\frac{1}{2} \cdot 2\frac{1}{2}$ as $\frac{1}{2}$ of $2\frac{1}{2}$.

Shade $2\frac{1}{2}$ squares.

Divide the squares into halves.

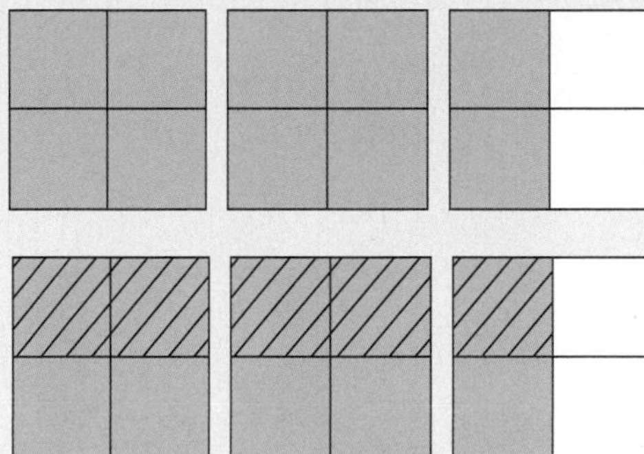

Look at $\frac{1}{2}$ of the part you shaded.

What fraction of the model is this?

$\frac{1}{2}$ of $2\frac{1}{2}$ is $1\frac{1}{4}$.

Think and Discuss

1. Describe how modeling multiplication of mixed numbers is like modeling multiplication of fractions.

Try This

Write the multiplication expression modeled on each grid.

1.

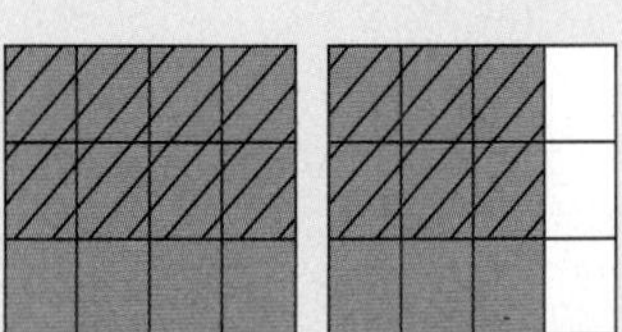

2.

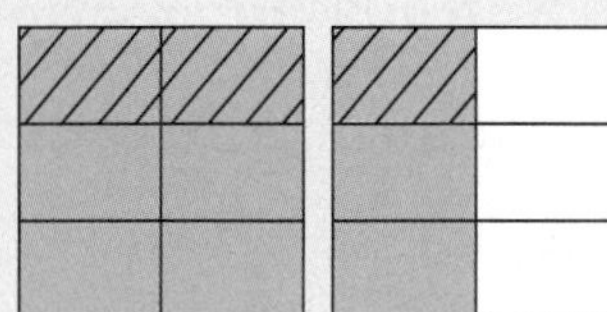

3.

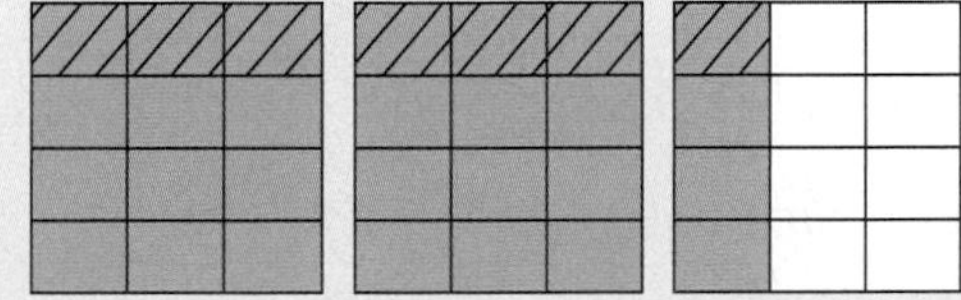

Use a grid to model each multiplication expression.

4. $\frac{1}{3} \cdot 1\frac{1}{2}$ **5.** $\frac{2}{3} \cdot 2\frac{1}{3}$ **6.** $\frac{1}{4} \cdot 2\frac{2}{3}$ **7.** $\frac{1}{3} \cdot 1\frac{3}{4}$

5-1 Multiplying Fractions

Learn to multiply fractions.

Your brain keeps working even when you're asleep. It makes sure that you keep breathing and that your heart keeps beating.

On average, people spend $\frac{1}{3}$ of their lives asleep. About $\frac{1}{4}$ of the time they sleep, they dream. What fraction of a lifetime does a person typically spend dreaming?

One way to find $\frac{1}{4}$ *of* $\frac{1}{3}$ is to make a model.

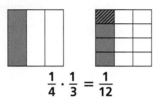

Find $\frac{1}{4}$ of $\frac{1}{3}$.

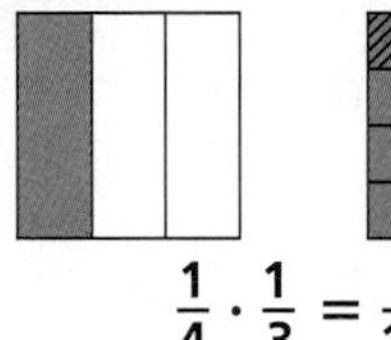

$$\frac{1}{4} \cdot \frac{1}{3} = \frac{1}{12}$$

You can also multiply fractions without making a model.

$\frac{1}{4} \cdot \frac{1}{3} = \frac{1 \cdot 1}{4 \cdot 3}$ ← *Multiply the numerators.* ← *Multiply the denominators.*

$= \frac{1}{12}$ *The answer is in simplest form.*

A person typically spends $\frac{1}{12}$ of his or her lifetime dreaming.

EXAMPLE 1 Multiplying Fractions

Multiply. Write each answer in simplest form.

A $\frac{1}{3} \cdot \frac{3}{5}$

$\frac{1}{3} \cdot \frac{3}{5} = \frac{1 \cdot 3}{3 \cdot 5}$ *Multiply numerators. Multiply denominators.*

$= \frac{3}{15}$ *The GCF of 3 and 15 is 3.*

$= \frac{1}{5}$ *The answer is in simplest form.*

B $\frac{6}{7} \cdot \frac{2}{3}$

$\frac{\overset{2}{\cancel{6}}}{7} \cdot \frac{2}{\underset{1}{\cancel{3}}} = \frac{2}{7} \cdot \frac{2}{1}$ *Use the GCF to simplify the fractions before multiplying. The GCF of 6 and 3 is 3.*

$= \frac{2 \cdot 2}{7 \cdot 1}$ *Multiply numerators. Multiply denominators.*

$= \frac{4}{7}$ *The answer is in simplest form.*

Multiply. Write each answer in simplest form.

C $\frac{3}{8} \cdot \frac{2}{9}$

$\frac{3}{8} \cdot \frac{2}{9} = \frac{3 \cdot 2}{8 \cdot 9}$ — *Multiply numerators. Multiply denominators.*

$= \frac{6}{72}$ — *The GCF of 6 and 72 is 6.*

$= \frac{1}{12}$ — *The answer is in simplest form.*

EXAMPLE 2 Evaluating Fraction Expressions

Evaluate the expression $a \cdot \frac{1}{3}$ for each value of a. Write the answer in simplest form.

A $a = \frac{5}{8}$

$a \cdot \frac{1}{3}$

$\frac{5}{8} \cdot \frac{1}{3}$ — *Substitute $\frac{5}{8}$ for a.*

$\frac{5 \cdot 1}{8 \cdot 3}$ — *Multiply.*

$\frac{5}{24}$ — *The answer is in simplest form.*

B $a = \frac{9}{10}$

$a \cdot \frac{1}{3}$

$\frac{9}{10} \cdot \frac{1}{3}$ — *Substitute $\frac{9}{10}$ for a.*

$\frac{\overset{3}{\cancel{9}}}{10} \cdot \frac{1}{\underset{1}{\cancel{3}}}$ — *Use the GCF to simplify.*

$\frac{3 \cdot 1}{10 \cdot 1}$ — *Multiply.*

$\frac{3}{10}$ — *The answer is in simplest form.*

C $a = \frac{3}{4}$

$a \cdot \frac{1}{3}$

$\frac{3}{4} \cdot \frac{1}{3}$ — *Substitute $\frac{3}{4}$ for a.*

$\frac{3 \cdot 1}{4 \cdot 3}$ — *Multiply numerators. Multiply denominators.*

$\frac{3}{12}$ — *The GCF of 3 and 12 is 3.*

$\frac{1}{4}$ — *The answer is in simplest form.*

Helpful Hint

You can look for a common factor in a numerator and a denominator to determine whether you can simplify before multiplying.

Think and Discuss

1. **Determine** whether the product of two proper fractions is greater than or less than each factor.
2. **Name** the missing denominator in the equation $\frac{1}{\blacksquare} \cdot \frac{2}{3} = \frac{2}{21}$.
3. **Tell** how to find the product of $\frac{4}{21} \cdot \frac{6}{10}$ in two different ways.

5-1 Exercises

FOR EOG PRACTICE
see page 654

internet connect
Homework Help Online
go.hrw.com Keyword: MR4 5-1

1.04a, 5.02

GUIDED PRACTICE

See Example 1 Multiply. Write each answer in simplest form.

1. $\frac{1}{2} \cdot \frac{1}{3}$
2. $\frac{2}{5} \cdot \frac{1}{4}$
3. $\frac{4}{7} \cdot \frac{3}{4}$
4. $\frac{5}{6} \cdot \frac{3}{5}$
5. $\frac{4}{9} \cdot \frac{3}{8}$
6. $\frac{2}{11} \cdot \frac{2}{3}$

See Example 2 Evaluate the expression $b \cdot \frac{1}{5}$ for each value of b. Write the answer in simplest form.

7. $b = \frac{2}{3}$
8. $b = \frac{5}{8}$
9. $b = \frac{3}{5}$

INDEPENDENT PRACTICE

See Example 1 Multiply. Write each answer in simplest form.

10. $\frac{1}{3} \cdot \frac{2}{7}$
11. $\frac{1}{3} \cdot \frac{1}{5}$
12. $\frac{5}{6} \cdot \frac{2}{3}$
13. $\frac{1}{3} \cdot \frac{6}{7}$
14. $\frac{3}{10} \cdot \frac{5}{6}$
15. $\frac{7}{9} \cdot \frac{3}{5}$
16. $\frac{1}{2} \cdot \frac{10}{11}$
17. $\frac{3}{5} \cdot \frac{3}{4}$
18. $\frac{8}{9} \cdot \frac{3}{4}$

See Example 2 Evaluate the expression $x \cdot \frac{1}{6}$ for each value of x. Write the answer in simplest form.

19. $x = \frac{4}{5}$
20. $x = \frac{6}{7}$
21. $x = \frac{3}{4}$
22. $x = \frac{8}{9}$
23. $x = \frac{9}{10}$
24. $x = \frac{5}{8}$

PRACTICE AND PROBLEM SOLVING

Find each product. Simplify the answer.

25. $\frac{3}{5} \cdot \frac{4}{9}$
26. $\frac{5}{12} \cdot \frac{9}{10}$
27. $\frac{2}{5} \cdot \frac{2}{7} \cdot \frac{5}{8}$

Compare. Write $<$, $>$, or $=$.

28. $\frac{2}{3} \cdot \frac{1}{4}$ ■ $\frac{1}{3} \cdot \frac{3}{4}$
29. $\frac{3}{5} \cdot \frac{3}{4}$ ■ $\frac{1}{2} \cdot \frac{9}{10}$
30. $\frac{5}{6} \cdot \frac{2}{3}$ ■ $\frac{1}{3} \cdot \frac{2}{3}$

31. A walnut muffin recipe calls for $\frac{3}{4}$ cup walnuts. Mrs. Hooper wants to make $\frac{1}{3}$ of the recipe. What fraction of a cup of walnuts will she need?

32. Jim spent $\frac{5}{6}$ of an hour doing chores. He spent $\frac{2}{5}$ of that time washing dishes. What fraction of an hour did he spend washing dishes?

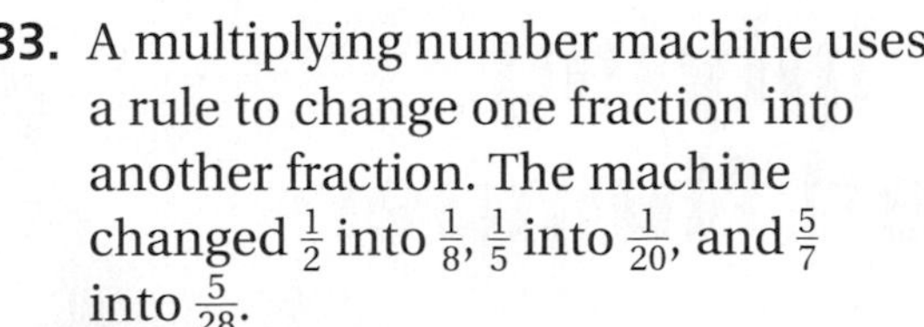

33. A multiplying number machine uses a rule to change one fraction into another fraction. The machine changed $\frac{1}{2}$ into $\frac{1}{8}$, $\frac{1}{5}$ into $\frac{1}{20}$, and $\frac{5}{7}$ into $\frac{5}{28}$.

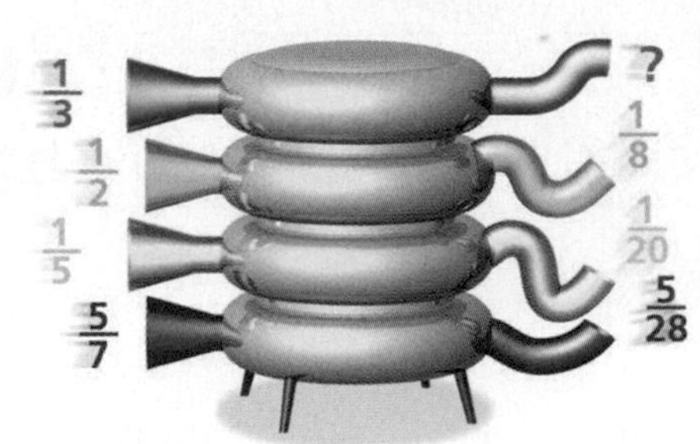

a. What is the rule?

b. Into what fraction will the machine change $\frac{1}{3}$?

Life Science LINK

There are about 1,000 species of bats in the world. Bats make up about $\frac{1}{4}$ of the world's mammals.

go.hrw.com
KEYWORD: MR4 Bats

34. ***LIFE SCIENCE*** A bat can eat half its weight in insects in one night. If a bat weighing $\frac{3}{4}$ lb eats half its weight in insects, how much do the insects weigh?

35. The seating plan shows Oak School's theater. The front section has $\frac{3}{4}$ of the seats, and the rear section has $\frac{1}{4}$ of the seats. The school has reserved $\frac{1}{2}$ of the seats in the front section for students. What fraction of the seating is reserved for students?

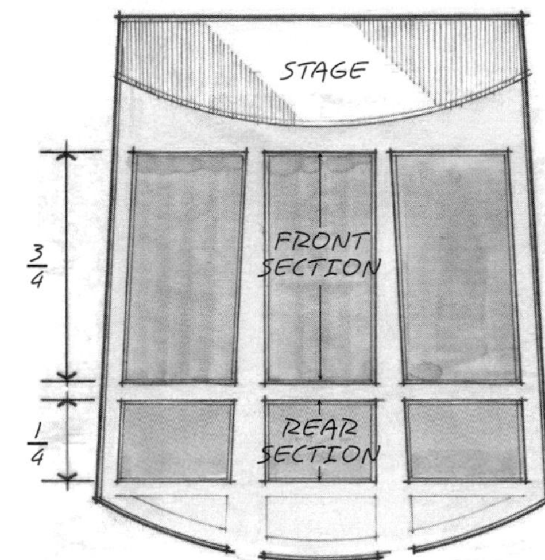

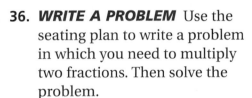

36. ***WRITE A PROBLEM*** Use the seating plan to write a problem in which you need to multiply two fractions. Then solve the problem.

37. ***WRITE ABOUT IT*** Explain how you can use the GCF before multiplying so that the product of two fractions is in simplest form.

38. ***CHALLENGE*** Evaluate the expression. Then simplify your answer.

$$\frac{(2+6)}{5} \cdot \frac{1}{4} \cdot 6$$

Spiral Review

Evaluate each expression. (Lesson 1-4)

39. $3^3 + 12 \div 6 - (5 + 3)$

40. $30 \div (3 + 2) \times 3 + 12 - 6$

Compare. Write <, >, or =. (Lesson 3-1)

41. 0.303 ▢ 0.033 **42.** 10.17 ▢ 1.701 **43.** 3.104 ▢ 3.91

44. **EOG PREP** A dog's mass is about 9 kg. What is the dog's mass in grams? (Lesson 3-4)

A 9,000 grams B 900 grams C 90 grams D 0.9 grams

45. **EOG PREP** Find the quotient of 3.45 ÷ 0.3. (Lesson 3-8)

A 0.115 B 1.15 C 10.15 D 11.5

5-2 Multiplying Mixed Numbers

Learn to multiply mixed numbers.

Janice and Carlos are making homemade pasta from a recipe that calls for $1\frac{1}{2}$ cups of flour. They want to make $\frac{1}{3}$ of the recipe.

You can find $\frac{1}{3}$ *of* $1\frac{1}{2}$, or multiply $\frac{1}{3}$ by $1\frac{1}{2}$, to find how much flour Janice and Carlos need.

EXAMPLE 1 Multiplying Fractions and Mixed Numbers

Remember!

To write a mixed number as an improper fraction, start with the whole number, multiply by the denominator, and add the numerator. Use the same denominator.

$1\frac{1}{5} = \frac{1 \cdot 5 + 1}{5} = \frac{6}{5}$

Multiply. Write each answer in simplest form.

A $\frac{1}{3} \cdot 1\frac{1}{2}$

$\frac{1}{3} \cdot \frac{3}{2}$ *Write $1\frac{1}{2}$ as an improper fraction. $1\frac{1}{2} = \frac{3}{2}$*

$\frac{1 \cdot 3}{3 \cdot 2}$ *Multiply numerators. Multiply denominators.*

$\frac{3}{6}$

$\frac{1}{2}$ *Write the answer in simplest form.*

B $1\frac{1}{5} \cdot \frac{2}{3}$

$\frac{6}{5} \cdot \frac{2}{3}$ *Write $1\frac{1}{5}$ as an improper fraction. $1\frac{1}{5} = \frac{6}{5}$*

$\frac{6 \cdot 2}{5 \cdot 3}$ *Multiply numerators. Multiply denominators.*

$\frac{12}{15}$

$\frac{4}{5}$ *Write the answer in simplest form.*

C $\frac{3}{4} \cdot 2\frac{1}{3}$

$\frac{3}{4} \cdot \frac{7}{3}$ *Write $2\frac{1}{3}$ as an improper fraction. $2\frac{1}{3} = \frac{7}{3}$*

$\frac{\cancel{3}^{1}}{4} \cdot \frac{7}{\cancel{3}_{1}}$ *Use the GCF to simplify before multiplying.*

$\frac{1 \cdot 7}{4 \cdot 1}$

$\frac{7}{4} = 1\frac{3}{4}$ *You can write the answer as a mixed number.*

EXAMPLE 2 Multiplying Mixed Numbers

Find each product. Write the answer in simplest form.

A $2\frac{1}{2} \cdot 1\frac{1}{3}$

$\frac{5}{2} \cdot \frac{4}{3}$ — *Write the mixed numbers as improper fractions.* $2\frac{1}{2} = \frac{5}{2}$ $1\frac{1}{3} = \frac{4}{3}$

$\frac{5 \cdot 4}{2 \cdot 3}$

$\frac{20}{6}$ — *Multiply numerators. Multiply denominators.*

$3\frac{2}{6}$ — *You can write the improper fraction as a mixed number.*

$3\frac{1}{3}$ — *Simplify.*

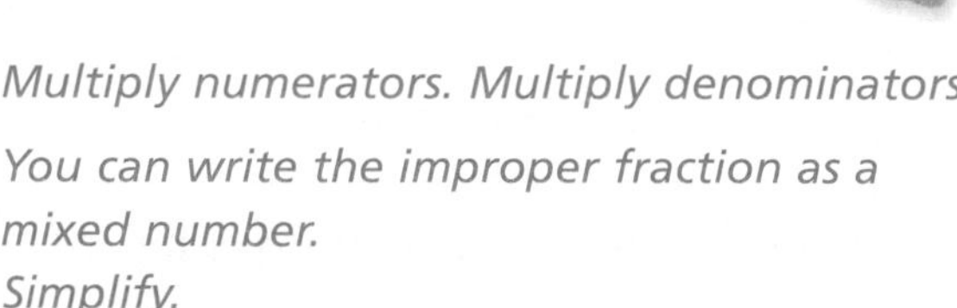

B $1\frac{1}{4} \cdot 1\frac{1}{3}$

$\frac{5}{4} \cdot \frac{4}{3}$ — *Write the mixed numbers as improper fractions.* $1\frac{1}{4} = \frac{5}{4}$ $1\frac{1}{3} = \frac{4}{3}$

$\frac{5}{\cancel{4}_1} \cdot \frac{\cancel{4}^1}{3}$ — *Use the GCF to simplify before multiplying.*

$\frac{5 \cdot 1}{1 \cdot 3}$ — *Multiply numerators. Multiply denominators.*

$\frac{5}{3}$

$1\frac{2}{3}$ — *You can write the answer as a mixed number.*

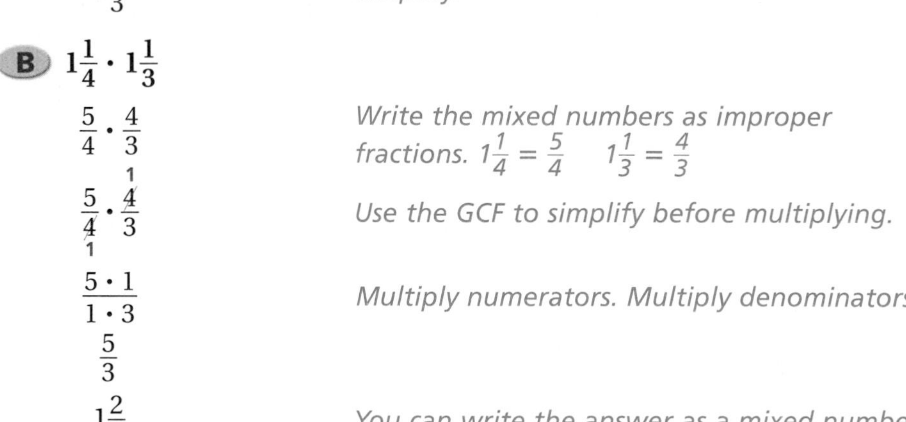

C $5 \cdot 3\frac{2}{11}$

$5 \cdot 3\frac{2}{11}$

$5 \cdot \left(3 + \frac{2}{11}\right)$

$(5 \cdot 3) + \left(5 \cdot \frac{2}{11}\right)$ — *Use the Distributive Property.*

$(5 \cdot 3) + \left(\frac{5}{1} \cdot \frac{2}{11}\right)$

$15 + \frac{10}{11}$ — *Multiply.*

$15\frac{10}{11}$ — *Add.*

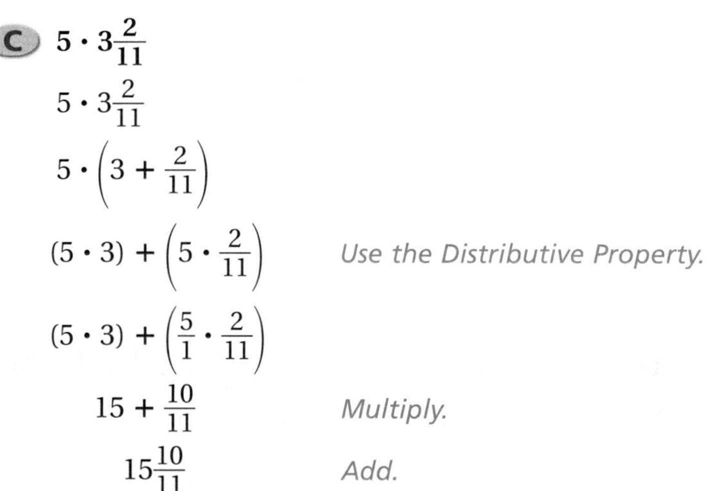

Think and Discuss

1. **Tell** how you multiply a mixed number by a mixed number.
2. **Explain** two ways you would multiply a mixed number by a whole number.

5-2 Exercises

FOR EOG PRACTICE
see page 654

internet connect
Homework Help Online
go.hrw.com Keyword: MR4 5-2

1.04a

GUIDED PRACTICE

See Example 1 **Multiply. Write each answer in simplest form.**

1. $1\frac{1}{4} \cdot \frac{2}{3}$
2. $2\frac{2}{3} \cdot \frac{1}{4}$
3. $\frac{3}{7} \cdot 1\frac{5}{6}$
4. $1\frac{1}{3} \cdot \frac{6}{7}$
5. $\frac{2}{3} \cdot 1\frac{3}{10}$
6. $2\frac{6}{11} \cdot \frac{2}{7}$

See Example 2 **Find each product. Write the answer in simplest form.**

7. $1\frac{5}{6} \cdot 1\frac{1}{8}$
8. $2\frac{2}{5} \cdot 1\frac{1}{12}$
9. $4 \cdot 5\frac{3}{7}$

INDEPENDENT PRACTICE

See Example 1 **Multiply. Write each answer in simplest form.**

10. $1\frac{1}{4} \cdot \frac{3}{4}$
11. $\frac{4}{7} \cdot 1\frac{1}{4}$
12. $1\frac{1}{6} \cdot \frac{2}{5}$
13. $2\frac{1}{6} \cdot \frac{3}{7}$
14. $\frac{5}{9} \cdot 1\frac{9}{10}$
15. $2\frac{2}{9} \cdot \frac{3}{5}$
16. $1\frac{3}{10} \cdot \frac{5}{7}$
17. $\frac{3}{5} \cdot 2\frac{2}{9}$
18. $2\frac{8}{11} \cdot \frac{3}{10}$

See Example 2 **Find each product. Write the answer in simplest form.**

19. $1\frac{1}{3} \cdot 1\frac{5}{7}$
20. $1\frac{2}{3} \cdot 2\frac{3}{10}$
21. $4 \cdot 3\frac{7}{8}$
22. $6 \cdot 2\frac{1}{3}$
23. $5 \cdot 4\frac{7}{10}$
24. $2\frac{2}{3} \cdot 3\frac{5}{8}$
25. $1\frac{1}{2} \cdot 2\frac{2}{5}$
26. $3\frac{5}{6} \cdot 2\frac{3}{4}$
27. $2\frac{1}{4} \cdot 1\frac{2}{9}$

PRACTICE AND PROBLEM SOLVING

Write each product in simplest form.

28. $1\frac{2}{3} \cdot \frac{2}{9}$
29. $3\frac{1}{3} \cdot \frac{7}{10}$
30. $2 \cdot \frac{5}{8}$
31. $\frac{3}{8} \cdot \frac{4}{9}$
32. $2\frac{1}{12} \cdot 1\frac{3}{5}$
33. $3\frac{3}{10} \cdot 4\frac{1}{6}$
34. $2 \cdot \frac{4}{5} \cdot 1\frac{2}{3}$
35. $3\frac{5}{6} \cdot \frac{9}{10} \cdot 4\frac{2}{3}$
36. $1\frac{7}{8} \cdot 2\frac{1}{3} \cdot 4$

Evaluate each expression.

37. $\frac{1}{2} \cdot c$ for $c = 4\frac{2}{5}$
38. $1\frac{5}{7} \cdot x$ for $x = \frac{5}{6}$
39. $1\frac{3}{4} \cdot b$ for $b = 1\frac{1}{7}$
40. $1\frac{5}{9} \cdot n$ for $n = 18$
41. $2\frac{5}{9} \cdot t$ for $t = 4$
42. $3\frac{3}{4} \cdot p$ for $p = \frac{1}{2}$
43. $\frac{4}{5} \cdot m$ for $m = 2\frac{2}{3}$
44. $6y$ for $y = 3\frac{5}{8}$

Computer Science 

In a survey, 240 people were asked how many hours per week they spend using the Internet. The circle graph shows which fractions of the people use the Internet for which amounts of time.

Use the graph for Exercises 45–51.

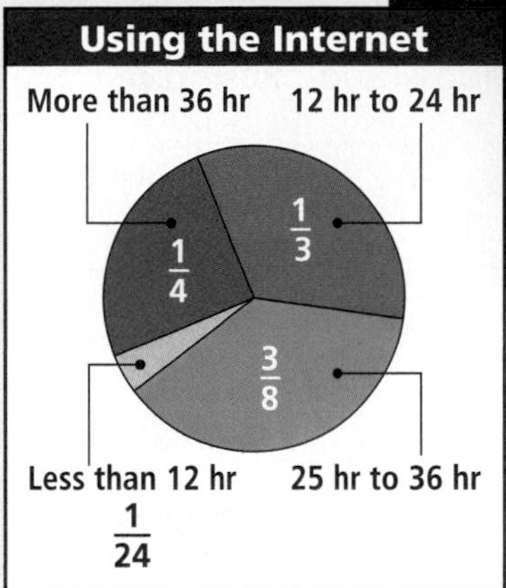

45. How many people in all were surveyed?

46. Find the number of people who said they use the Internet for 12 hours to 24 hours a week.

47. Find the number of people who said they use the Internet for 25 hours to 36 hours a week.

The World Wide Web was developed by Tim Berners-Lee at CERN, in Switzerland. It was designed to help physicists from different parts of the world share information.

go.hrw.com
KEYWORD: MR4 Internet
CNN Student News.

48. Toni's grandfather uses the Internet for $1\frac{1}{2}$ hours each day.

a. How many hours does he use the Internet in one week? (Write the answer as a mixed number.)

b. If Toni's grandfather were included in the survey, in which time section of the circle graph would his data be?

49. ***CHOOSE A STRATEGY*** Which set of tallies could represent the number of people who use the Internet for fewer than 12 hours a week?

A 卌 卌
B 卌 卌 ||
C 卌 卌 卌 卌
D 卌 卌 卌 卌 ||||

50. ***WRITE ABOUT IT*** Explain how you can find the number of people surveyed who use the Internet for more than 36 hours a week.

51. ***CHALLENGE*** Five-sixths of the people who use the Internet for 25 hours to 36 hours said they use it for 30 hours each week. Find the number of people who use the Internet for 30 hours each week.

Spiral Review

Solve each equation. (Lesson 2-4)

52. $n + 52 = 71$ **53.** $30 = k - 15$ **54.** $22 - 18 + c = 30$

55. **EOG PREP** Which number is *not* between 7.5 and 8.25? (Lesson 3-1)

A 8.039 B 8.219 C 7.501 D 7.051

Model Fraction Division

Use with Lesson 5-3

internet connect
Lab Resources Online
go.hrw.com
KEYWORD: MR4 Lab5B

You can use grids to help you understand division of fractions.

Activity 1

1 Think of $4\frac{1}{2} \div 3$ as dividing $4\frac{1}{2}$ into 3 equal groups.

Shade $4\frac{1}{2}$ squares.

Divide the squares into 3 equal groups.

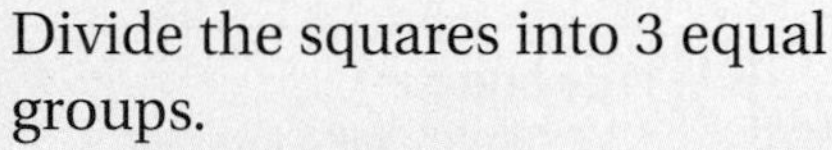

Look at one of the shaded groups.

What fraction is this?

$4\frac{1}{2} \div 3$ is $1\frac{1}{2}$.

Think and Discuss

1. Explain how you know the number of groups into which you must divide the squares.

Try This

Write the division expression modeled on each grid.

1.

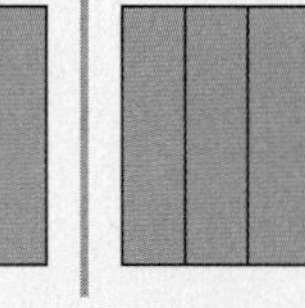

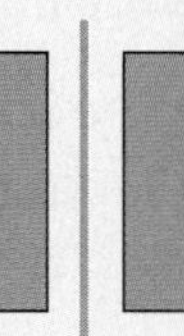

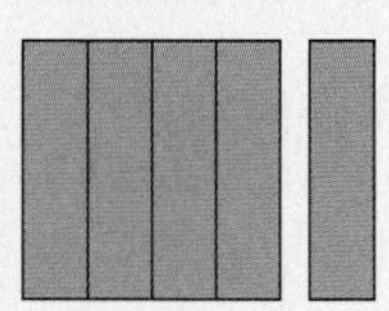

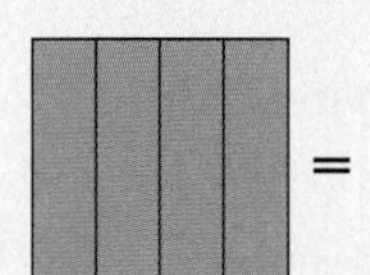

= 1

Use grids to model each division expression. Then find the value of the expression.

2. $9\frac{1}{3} \div 4$ **3.** $3\frac{3}{4} \div 5$ **4.** $4\frac{2}{3} \div 2$ **5.** $4\frac{1}{5} \div 3$

Activity 2

1 To find $2\frac{2}{3} \div \frac{2}{3}$, think, "How many groups of $\frac{2}{3}$ are in $2\frac{2}{3}$?"

Shade $2\frac{2}{3}$ squares.

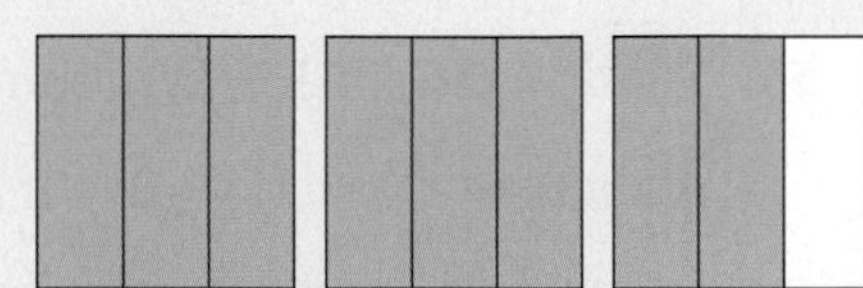

Divide the shaded squares and shaded thirds into equal groups of $\frac{2}{3}$.

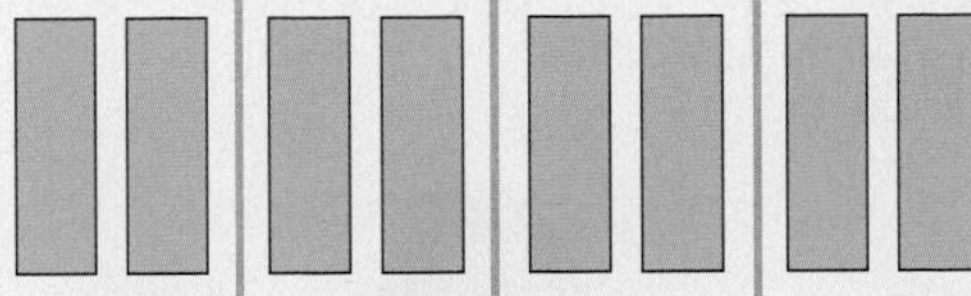

There are 4 groups of $\frac{2}{3}$ in $2\frac{2}{3}$. $2\frac{2}{3} \div \frac{2}{3} = 4$.

2 To find $3 \div \frac{3}{4}$, think, "How many groups of $\frac{3}{4}$ are in 3?"

Shade 3 squares. Then divide the squares into fourths because the denominator of $\frac{3}{4}$ is 4.

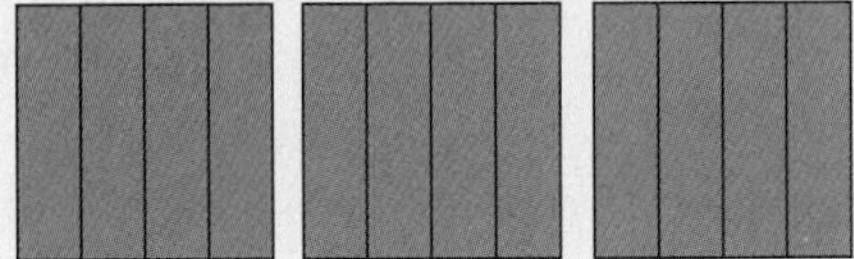

Divide the shaded squares into equal groups of $\frac{3}{4}$.

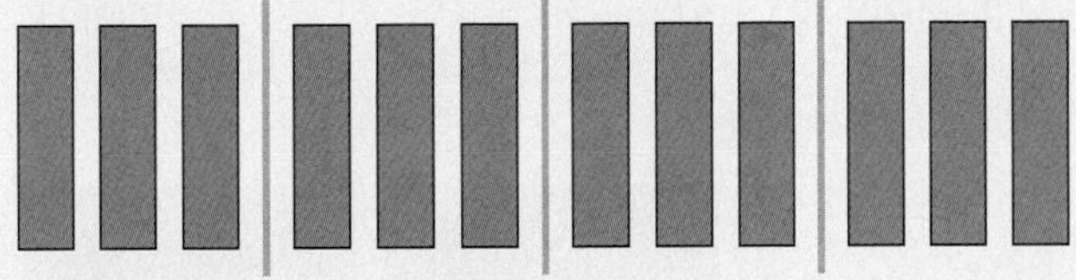

There are 4 groups of $\frac{3}{4}$ in 3. $3 \div \frac{3}{4} = 4$.

Think and Discuss

1. Explain what prediction you can make about the value of $6 \div \frac{3}{4}$ if you know that $3 \div \frac{3}{4}$ is 4.

Try This

Write the division expression modeled by each grid.

1. = 1

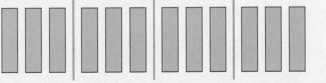

2.

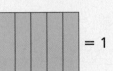

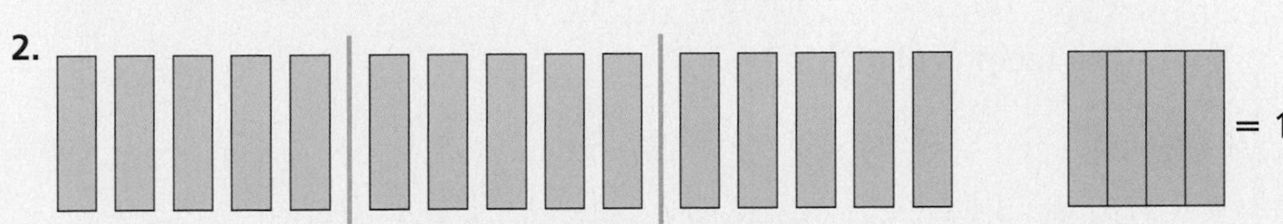

Use grids to model each division expression. Then find the value of the expression.

3. $4 \div 1\frac{1}{3}$

4. $3\frac{3}{4} \div \frac{3}{4}$

5. $5\frac{1}{3} \div \frac{2}{3}$

6. $6\frac{2}{3} \div 1\frac{2}{3}$

5-3 Dividing Fractions and Mixed Numbers

Learn to divide fractions and mixed numbers.

Vocabulary
reciprocal

Curtis is making sushi rolls. First, he will place a sheet of seaweed, called *nori,* on the sushi rolling mat. Then, he will use the mat to roll up rice, cucumber, avocado, and crabmeat. Finally, he will slice the roll into smaller pieces.

Curtis has 2 cups of rice and will use $\frac{1}{3}$ cup for each sushi roll. How many sushi rolls can he make?

Think: How many $\frac{1}{3}$ pieces equal 2 wholes?

There are six $\frac{1}{3}$ pieces in 2 wholes.

Curtis can make 6 sushi rolls.

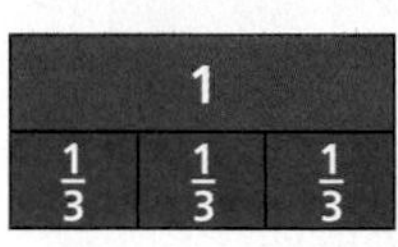

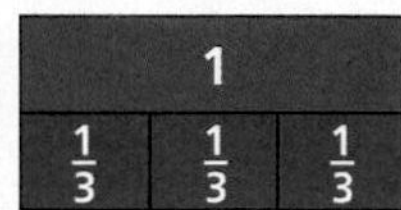

Reciprocals can help you divide by fractions. Two numbers are **reciprocals** if their product is 1.

EXAMPLE 1 Finding Reciprocals

Find the reciprocal.

A $\frac{1}{5}$

$\frac{1}{5} \cdot \blacksquare = 1$ — *Think: $\frac{1}{5}$ of what number is 1?*

$\frac{1}{5} \cdot 5 = 1$ — *$\frac{1}{5}$ of $\frac{5}{1}$ is 1.*

The reciprocal of $\frac{1}{5}$ is 5.

B $\frac{3}{4}$

$\frac{3}{4} \cdot \blacksquare = 1$ — *Think: $\frac{3}{4}$ of what number is 1?*

$\frac{3}{4} \cdot \frac{4}{3} = \frac{12}{12} = 1$ — *$\frac{3}{4}$ of $\frac{4}{3}$ is 1.*

The reciprocal of $\frac{3}{4}$ is $\frac{4}{3}$.

C $2\frac{1}{3}$

$\frac{7}{3} \cdot \blacksquare = 1$ — *Write $2\frac{1}{3}$ as $\frac{7}{3}$.*

$\frac{7}{3} \cdot \frac{3}{7} = \frac{21}{21} = 1$ — *$\frac{7}{3}$ of $\frac{3}{7}$ is 1.*

The reciprocal of $\frac{7}{3}$ is $\frac{3}{7}$.

Look at the relationship between the fractions $\frac{3}{4}$ and $\frac{4}{3}$. If you switch the numerator and denominator of a fraction, you will find its reciprocal. Dividing by a number is the same as multiplying by its reciprocal.

$$24 \div 4 = 6 \qquad 24 \cdot \frac{1}{4} = 6$$

EXAMPLE 2 Using Reciprocals to Divide Fractions and Mixed Numbers

Divide. Write each answer in simplest form.

A $\frac{4}{5} \div 5$

$\frac{4}{5} \div 5 = \frac{4}{5} \cdot \frac{1}{5}$ — *Rewrite as multiplication using the reciprocal of 5, $\frac{1}{5}$.*

$= \frac{4 \cdot 1}{5 \cdot 5}$ — *Multiply by the reciprocal.*

$= \frac{4}{25}$ — *The answer is in simplest form.*

B $\frac{3}{4} \div \frac{1}{2}$

$\frac{3}{4} \div \frac{1}{2} = \frac{3}{4} \cdot \frac{2}{1}$ — *Rewrite as multiplication using the reciprocal of $\frac{1}{2}$, $\frac{2}{1}$.*

$= \frac{3 \cdot \cancel{2}^{1}}{\cancel{4}_{2} \cdot 1}$ — *Simplify before multiplying.*

$= \frac{3}{2}$ — *Multiply.*

$= 1\frac{1}{2}$ — *You can write the answer as a mixed number.*

C $2\frac{2}{3} \div 1\frac{1}{6}$

$2\frac{2}{3} \div 1\frac{1}{6} = \frac{8}{3} \div \frac{7}{6}$ — *Write the mixed numbers as improper fractions. $2\frac{2}{3} = \frac{8}{3}$ and $1\frac{1}{6} = \frac{7}{6}$*

$= \frac{8}{3} \cdot \frac{6}{7}$ — *Rewrite as multiplication.*

$= \frac{8 \cdot \cancel{6}^{2}}{\cancel{3}_{1} \cdot 7}$ — *Simplify before multiplying.*

$= \frac{16}{7}$ — *Multiply.*

$= 2\frac{2}{7}$ — *You can write the answer as a mixed number.*

Think and Discuss

1. **Explain** how you can use mental math to find the value of n in the equation $\frac{5}{8} \cdot n = 1$.
2. **Explain** how to find the reciprocal of $3\frac{6}{11}$.

5-3 Exercises

FOR EOG PRACTICE
see page 655

internet connect
Homework Help Online
go.hrw.com Keyword: MR4 5-3

1.04a

GUIDED PRACTICE

See Example 1

Find the reciprocal.

1. $\frac{2}{7}$
2. $\frac{5}{9}$
3. $\frac{1}{9}$
4. $\frac{3}{11}$

See Example 2

Divide. Write each answer in simplest form.

5. $\frac{5}{6} \div 3$
6. $2\frac{1}{7} \div 1\frac{1}{4}$
7. $\frac{5}{12} \div 5$
8. $\frac{2}{3} \div \frac{1}{6}$
9. $\frac{3}{10} \div 1\frac{2}{3}$
10. $\frac{4}{7} \div 1\frac{1}{7}$

INDEPENDENT PRACTICE

See Example 1

Find the reciprocal.

11. $\frac{7}{8}$
12. $\frac{1}{10}$
13. $\frac{3}{8}$
14. $\frac{11}{12}$
15. $\frac{8}{11}$
16. $\frac{5}{6}$
17. $\frac{6}{7}$
18. $\frac{2}{9}$

See Example 2

Divide. Write each answer in simplest form.

19. $\frac{7}{8} \div 4$
20. $2\frac{3}{8} \div 1\frac{3}{4}$
21. $\frac{8}{9} \div 12$
22. $3\frac{5}{6} \div 1\frac{5}{9}$
23. $\frac{9}{10} \div 3$
24. $2\frac{4}{5} \div 1\frac{5}{7}$
25. $\frac{5}{8} \div \frac{1}{2}$
26. $1\frac{1}{2} \div 2\frac{1}{4}$
27. $\frac{7}{12} \div 2\frac{5}{8}$

PRACTICE AND PROBLEM SOLVING

Multiply or divide. Write each answer in simplest form.

28. $2\frac{3}{4} \div 2\frac{1}{5}$
29. $4\frac{4}{5} \div 2\frac{6}{7}$
30. $\frac{3}{8} \cdot \frac{5}{12}$
31. $6 \cdot \frac{7}{9}$
32. $3\frac{1}{7} \div 5$
33. $\frac{9}{14} \div \frac{1}{6}$
34. $5\frac{3}{5} \div \frac{4}{7}$
35. $\frac{9}{11} \cdot 2\frac{2}{3}$
36. $2\frac{7}{10} \div 3\frac{3}{5}$
37. $\frac{11}{12} \cdot \frac{9}{10} \div 1\frac{1}{4}$
38. $2\frac{3}{4} \cdot 1\frac{2}{3} \div 5$
39. $1\frac{1}{2} \div \frac{3}{4} \cdot \frac{2}{5}$
40. $\frac{3}{4} \cdot \left(\frac{5}{7} \div \frac{1}{2}\right)$
41. $4\frac{2}{3} \div \left(6 \cdot \frac{3}{5}\right)$
42. $5\frac{1}{5} \cdot \left(3\frac{2}{5} \cdot 2\frac{1}{3}\right)$

Decide whether the fractions in each pair are reciprocals. If not, write the reciprocal of each fraction.

43. $\frac{1}{2}, 2$
44. $\frac{3}{8}, \frac{16}{6}$
45. $\frac{7}{9}, \frac{21}{27}$
46. $\frac{5}{6}, \frac{12}{10}$
47. $1\frac{1}{2}, \frac{2}{3}$
48. $\frac{2}{5}, \frac{4}{25}$
49. $\frac{3}{7}, 2\frac{1}{3}$
50. $5, \frac{5}{1}$

LIFE SCIENCE The bar graph shows the lengths of some species of snakes found in the United States.

Use the bar graph for Exercises 51–53.

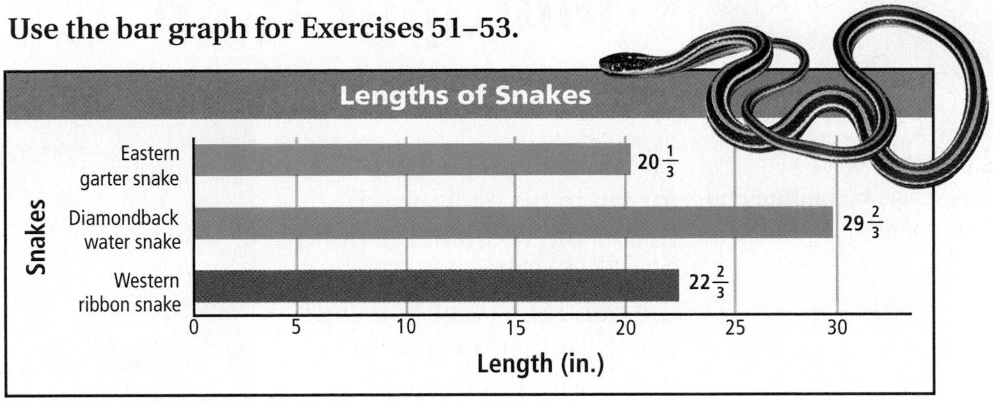

Lengths of Snakes

Snakes	Length (in.)
Eastern garter snake	$20\frac{1}{3}$
Diamondback water snake	$29\frac{2}{3}$
Western ribbon snake	$22\frac{2}{3}$

51. Is the length of the eastern garter snake greater than or less than $\frac{1}{2}$ yd? Explain.

52. What is the average length of all the snakes?

53. Jim measured the length of a rough green snake. It was $27\frac{1}{3}$ in. long. What would the average length of the snakes be if Jim's measure of a rough green snake were added?

54. At Lina's restaurant, one serving of chili is $1\frac{1}{2}$ cups. The chef makes 48 cups of chili each night. How many servings of chili are in 48 cups?

55. Rhula bought 12 lb of raisins. She packed them into freezer bags so that each bag weighs $\frac{3}{4}$ lb. How many freezer bags did she pack?

56. Lisa had some wood that was $12\frac{1}{2}$ feet long. She cut it into 5 pieces that are equal in length. How long is each piece of wood?

57. ***WHAT'S THE ERROR?*** A student said the reciprocal of $6\frac{2}{3}$ is $6\frac{3}{2}$. Explain the error. Then write the correct reciprocal.

58. ***WRITE ABOUT IT*** Explain how you can divide fractions to find the quotient of $\frac{3}{4} \div 2\frac{1}{3}$.

59. ***CHALLENGE*** Evaluate the expression $\frac{(6-3)}{4} \div \frac{1}{8} \cdot 5$.

Spiral Review

Identify a pattern in each sequence and name the missing term. (Lesson 1-7)

60. 85, 80, 75, 70, 65, ▇, …

61. 1, 4, 7, 10, 13, ▇, …

62. 2, 6, 5, 9, 8, ▇, …

63. **EOG PREP** Which is the most reasonable measure for the length of a pencil? (Lesson 3-4)

A 15 m B 15 cm C 15 yd D 15 oz

64. **EOG PREP** Of which set of numbers is 16 the GCF? (Lesson 4-3)

A 16, 32, 48 B 12, 24, 32 C 24, 48, 60 D 8, 80, 100

5-4 Solving Fraction Equations: Multiplication and Division

Learn to solve equations by multiplying and dividing fractions.

Josef is building a fish pond for koi in his backyard. He makes the width of the pond $\frac{2}{3}$ of the length. The width of the pond is 14 feet. You can use the equation $\frac{2}{3}\ell = 14$ to find the length of the pond.

Small koi in a backyard pond usually grow 2 to 4 inches per year.

EXAMPLE 1 Solving Equations by Multiplying and Dividing

Solve each equation. Write the answer in simplest form.

A $\frac{2}{3}\ell = 14$

$$\frac{2}{3}\ell = 14$$

$$\frac{2}{3}\ell \div \frac{2}{3} = 14 \div \frac{2}{3}$$ *Divide both sides of the equation by $\frac{2}{3}$.*

$$\frac{2}{3}\ell \cdot \frac{3}{2} = 14 \cdot \frac{3}{2}$$ *Multiply by $\frac{3}{2}$, the reciprocal of $\frac{2}{3}$.*

$$\ell = 14 \cdot \frac{3}{2}$$

$$\ell = \frac{14 \cdot 3}{1 \cdot 2}$$

$$\ell = \frac{42}{2}, \text{ or } 21$$

B $2x = \frac{1}{3}$

$$2x = \frac{1}{3}$$

$$\frac{2x}{1} \cdot \frac{1}{2} = \frac{1}{3} \cdot \frac{1}{2}$$ *Multiply both sides by the reciprocal of 2.*

$$x = \frac{1 \cdot 1}{3 \cdot 2}$$

$$x = \frac{1}{6}$$ *The answer is in simplest form.*

C $\frac{5x}{6} = 4$

$$\frac{5x}{6} = 4$$

$$\frac{5x}{6} \div \frac{5}{6} = \frac{4}{1} \div \frac{5}{6}$$ *Divide both sides by $\frac{5}{6}$.*

$$\frac{5x}{6} \cdot \frac{6}{5} = \frac{4}{1} \cdot \frac{6}{5}$$ *Multiply by the reciprocal of $\frac{5}{6}$.*

$$x = \frac{24}{5}, \text{ or } 4\frac{4}{5}$$

Remember!

Dividing by a number is the same as multiplying by its reciprocal.

EXAMPLE  2 PROBLEM SOLVING APPLICATION

Dexter makes dog biscuits for the animal shelter. He makes $\frac{3}{4}$ of a recipe and uses 15 cups of powdered milk. How many cups of powdered milk are in the recipe?

1 Understand the Problem

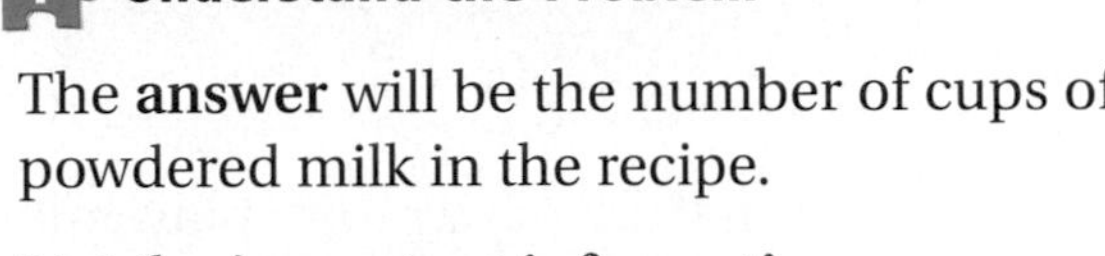

The **answer** will be the number of cups of powdered milk in the recipe.

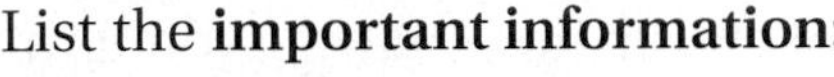

List the **important information:**

- He makes $\frac{3}{4}$ of the recipe.
- He uses 15 cups of powdered milk.

2 Make a Plan

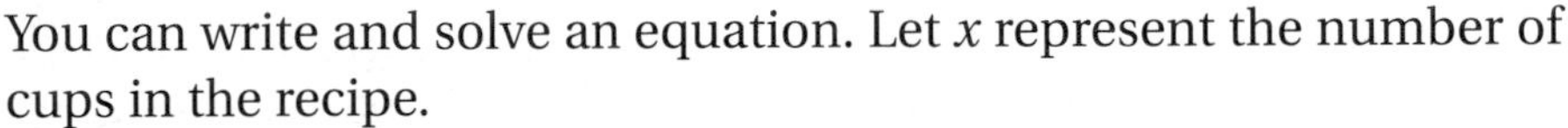

You can write and solve an equation. Let x represent the number of cups in the recipe.

He uses 15 cups, which is three-fourths of the amount in the recipe.
$15 = \frac{3}{4}x$

3 Solve

$15 = \frac{3}{4}x$

$15 \cdot \frac{4}{3} = \frac{3}{4}x \cdot \frac{4}{3}$ *Multiply both sides by $\frac{4}{3}$, the reciprocal of $\frac{3}{4}$.*

$\frac{\overset{5}{\cancel{15}}}{1} \cdot \frac{4}{\underset{1}{\cancel{3}}} = x$ *Simplify. Then multiply.*

$20 = x$

There are 20 cups of powdered milk in the recipe.

4 Look Back

Check $15 = \frac{3}{4}x$

$15 \overset{?}{=} \frac{3}{4}(20)$ *Substitute 20 for x.*

$15 \overset{?}{=} \frac{\overset{15}{\cancel{60}}}{\underset{1}{\cancel{4}}}$ *Multiply and simplify.*

$15 \overset{?}{=} 15$ ✔ *20 is the solution.*

Life Science LINK

No more than $\frac{1}{10}$ of a dog's diet should consist of treats and biscuits.

Think and Discuss

1. **Explain** whether $\frac{2}{3}x = 4$ is the same as $\frac{2}{3} = 4x$.
2. **Tell** how you know which numbers to divide by in the following equations: $\frac{2}{3}x = 4$ and $\frac{4}{5} = 8x$.

5-4 Exercises

FOR EOG PRACTICE
see page 655

internet connect
Homework Help Online
go.hrw.com Keyword: MR4 5-4

5.03

GUIDED PRACTICE

See Example 1

Solve each equation. Write the answer in simplest form.

1. $\frac{3}{4}z = 12$ **2.** $4n = \frac{3}{5}$ **3.** $\frac{2x}{3} = 5$

See Example 2

4. In PE class, $\frac{3}{8}$ of the students want to play volleyball. If 9 students want to play volleyball, how many students are in the class?

INDEPENDENT PRACTICE

See Example 1

Solve each equation. Write the answer in simplest form.

5. $3t = \frac{2}{7}$ **6.** $\frac{1}{3}x = 3$ **7.** $\frac{3r}{5} = 9$

8. $\frac{4}{5}a = 1$ **9.** $\frac{y}{4} = 5$ **10.** $2b = \frac{6}{7}$

See Example 2

11. Jason uses 2 cans of paint to paint $\frac{1}{2}$ of his room. How many cans of paint will he use to paint the whole room?

12. Cassandra baby-sits for $\frac{4}{5}$ of an hour and earns \$8. What is her hourly rate?

PRACTICE AND PROBLEM SOLVING

Solve each equation. Write the answer in simplest form.

13. $m = \frac{3}{8} \cdot 4$ **14.** $\frac{3y}{5} = 6$ **15.** $4z = \frac{7}{10}$

16. $\frac{3}{5}a = \frac{3}{5}$ **17.** $\frac{1}{6}b = 2\frac{1}{3}$ **18.** $5c = \frac{2}{3} \div \frac{2}{3}$

19. $\frac{1}{2} = \frac{w}{4}$ **20.** $8 = \frac{2n}{3}$ **21.** $\frac{1}{4} \cdot \frac{1}{2} = 4d$

Write each equation. Then solve, and check the solution.

22. A number n is multiplied by $\frac{1}{3}$ and the product is 12.

23. A number n is divided by 4 and the quotient is $\frac{1}{2}$.

24. A number n is multiplied by $1\frac{1}{2}$ and the product is 9.

25. A recipe for a loaf of bread calls for $\frac{3}{4}$ cup of oatmeal.

a. How much oatmeal do you need if you make half the recipe?

b. How much oatmeal do you need if you double the recipe?

26. ***ENTERTAINMENT*** Connie rode the roller coaster at the amusement park. After 3 minutes, the ride was $\frac{3}{4}$ complete. How long did the entire ride take?

The northwest corner of Madagascar is home to black lemurs. These primates live in groups of 7–10, and they have an average life span of 20–25 years. Much of their habitat is being destroyed by human agricultural activity.

27. ***LIFE SCIENCE*** Sasha's book report is about animals in Madagascar. She writes 10 pages, which represents $\frac{1}{3}$ of her report, about lemurs. How many more pages does Sasha have to write to complete her book report?

28. Alder cut 3 pieces of fabric from a roll. Each piece of fabric she cut is $1\frac{1}{2}$ yd long. She has 2 yards of fabric left on the roll. How much fabric was on the roll before she cut it?

Use the circle graph for Exercises 29 and 30.

29. The circle graph shows the results of a survey of people who were asked to choose their favorite kind of bagel.
    a. One hundred people chose plain bagels as their favorite kind of bagel. How many people were surveyed in all?
    b. One-fifth of the people who chose sesame bagels also chose plain cream cheese as their favorite spread. How many people chose plain cream cheese? (*Hint*: Use the answer to part **a** to help you solve this problem.)

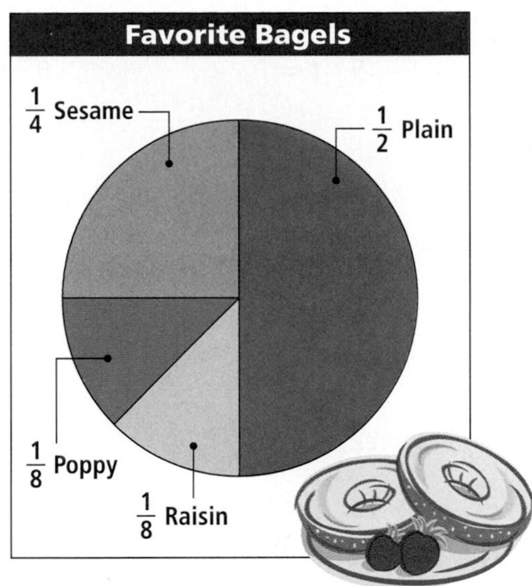

30. ***WHAT'S THE QUESTION?*** If the answer is 25 people, what is the question?

31. ***WRITE ABOUT IT*** Explain how to solve $\frac{3}{5}x = 4$.

32. ***CHALLENGE*** Solve.

$$2\frac{3}{4}n = \frac{11}{12}$$

Spiral Review

Estimate each sum or difference by rounding to the thousands place. (Lesson 1-2)

33. 9,074 + 2,123 **34.** 7,232 − 4,245 **35.** 9,903 + 5,367

36. 5,234 − 1,901 **37.** 8,523 + 2,459 **38.** 15,050 − 10,810

Move the decimal point to multiply or divide by powers of 10. (Lesson 3-4)

39. $718 \times 1{,}000$ **40.** 23.06×10^3 **41.** $420.34 \div 100$

42. $1{,}000 \times 1{,}500$ **43.** 5.03×10^2 **44.** $88.5 \div 1{,}000$

45. **EOG PREP** Which fraction is *not* equivalent to $\frac{1}{4}$? (Lesson 4-6)

A $\frac{3}{12}$ B $\frac{4}{16}$ C $\frac{5}{24}$ D $\frac{12}{48}$

Chapter 5

Mid-Chapter Quiz

LESSON 5-1 (pp. 212–215)

Multiply. Write each answer in simplest form.

1. $\frac{2}{7} \cdot \frac{3}{4}$
2. $\frac{3}{5} \cdot \frac{2}{3}$
3. $\frac{7}{12} \cdot \frac{4}{5}$
4. $\frac{5}{8} \cdot \frac{9}{10}$

Evaluate the expression $t \cdot \frac{1}{8}$ for each value of t. Write the answer in simplest form.

5. $t = \frac{4}{9}$
6. $t = \frac{4}{5}$
7. $t = \frac{2}{3}$
8. $t = \frac{6}{7}$

LESSON 5-2 (pp. 216–219)

Multiply. Write each answer in simplest form.

9. $\frac{1}{4} \cdot 2\frac{1}{3}$
10. $1\frac{1}{6} \cdot \frac{2}{3}$
11. $\frac{7}{8} \cdot 2\frac{2}{3}$

Find each product. Write the answer in simplest form.

12. $2\frac{1}{4} \cdot 1\frac{1}{6}$
13. $1\frac{2}{3} \cdot 2\frac{1}{5}$
14. $3 \cdot 4\frac{2}{7}$
15. $\frac{5}{6}$ of $1\frac{3}{5}$
16. $\frac{1}{5}$ of $2\frac{1}{3}$
17. $\frac{3}{4}$ of $1\frac{1}{2}$

LESSON 5-3 (pp. 222–225)

Find the reciprocal.

18. $\frac{2}{7}$
19. $\frac{5}{12}$
20. $\frac{3}{5}$
21. $\frac{1}{10}$

Divide. Write each answer in simplest form.

22. $\frac{3}{5} \div 4$
23. $1\frac{3}{10} \div 3\frac{1}{4}$
24. $1\frac{1}{5} \div 2\frac{1}{3}$
25. $10 \div 2\frac{1}{2}$
26. $1\frac{5}{11} \div 1\frac{5}{11}$
27. $\frac{3}{10} \div \frac{3}{100}$

LESSON 5-4 (pp. 226–229)

Solve each equation.

28. $\frac{2y}{3} = 10$
29. $6p = \frac{3}{4}$
30. $\frac{2x}{3} = 9$
31. Michael has a black cat and a gray kitten. The black cat weighs 12 pounds. The gray kitten weighs $\frac{3}{5}$ the weight of the black cat. How much does the gray kitten weigh?
32. Ronald rode his bike $7\frac{1}{5}$ miles in an hour. At that speed, how far will he travel in the next $\frac{3}{4}$ hour?
33. Amy has some beads that are each $\frac{1}{2}$ inch long. How many beads will she need to make a necklace that is 18 inches long?

Focus on Problem Solving

Solve

Choose the operation: multiplication or division

Read the whole problem before you try to solve it. Determine what action is taking place in the problem. Then decide whether you need to multiply or divide in order to solve the problem.

If you are asked to combine equal groups, you need to multiply. If you are asked to share something equally or to separate something into equal groups, you need to divide.

Action	Operation	
Combining equal groups	Multiplication	
Sharing things equally or separating into equal groups	Division	

Read each problem, and determine the action taking place. Choose an operation to solve the problem. Then solve, and write the answer in simplest form.

1. Jason picked 30 cups of raspberries. He put them in freezer bags with $\frac{3}{4}$ cup in each bag. How many bags does he have?

2. When the cranberry flowers start to open in June, cranberry growers usually bring in about $1\frac{1}{2}$ beehives per acre of cranberries to pollinate the flowers. A grower has 36 acres of cranberries. About how many beehives does she need?

3. A recipe that makes 3 cranberry banana loaves calls for 4 cups of cranberries. Linh wants to make only 1 loaf. How many cups of cranberries does she need?

4. Clay wants to double a recipe for blueberry muffins that calls for $\frac{3}{4}$ cup of blueberries. How many blueberries will he need?

5-5 Least Common Multiple

Learn to find the least common multiple (LCM) of a group of numbers.

Vocabulary

least common multiple (LCM)

After games in Lydia's soccer league, one player's family brings snacks for both teams to share. This week Lydia's family will provide juice boxes and granola bars for 24 players.

You can make a model to help you find the least number of juice and granola packs Lydia's family should buy. Use colored counters, drawings, or pictures to illustrate the problem.

EXAMPLE  1 *Consumer Application*

Juice comes in packs of 6, and granola bars in packs of 8. If there are 24 players, what is the least number of packs needed so that each player has a drink and granola bar and there are none left over?

Draw juice boxes in groups of 6. Draw granola bars in groups of 8. Stop when you have drawn the same number of each.

There are 24 juice boxes and 24 granola bars.

Lydia's family should buy 4 packs of juice and 3 packs of granola bars.

The smallest number that is a multiple of two or more numbers is the **least common multiple (LCM)**. In Example 1, the LCM of 6 and 8 is 24.

EXAMPLE 2 Using Multiples to Find the LCM

Find the least common multiple (LCM).

Method 1: Use a number line.

A **6 and 9**

Use a number line to skip count by 6 and 9.

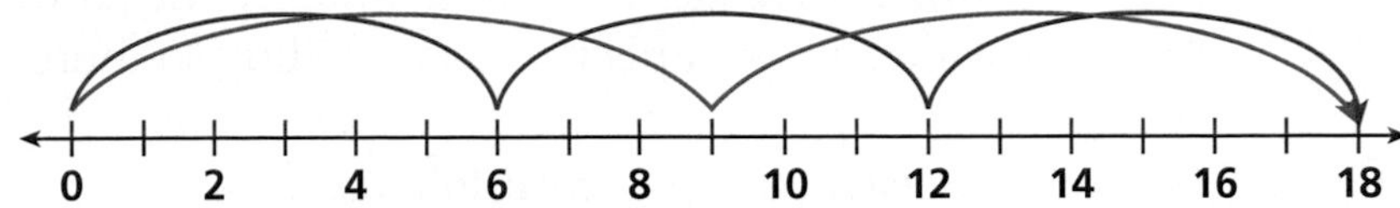

The least common multiple (LCM) of 6 and 9 is 18.

Method 2: Use a list.

B **3, 5, and 6**

3: 3, 6, 9, 12, 15, 18, 21, 24, 27, **30**, 33, . . .
5: 5, 10, 15, 20, 25, **30**, 35, . . .
6: 6, 12, 18, 24, **30**, 36, . . .

LCM: 30

List multiples of 3, 5, and 6. Find the smallest number that is in all the lists.

Method 3: Use prime factorization.

Remember!

The prime factorization of a number is the number written as a product of its prime factors.

C **8 and 12**

$8 = 2 \cdot 2 \cdot 2$

$12 = 2 \cdot 2 \cdot \quad 3$ *Write the prime factorization of each number.*

$\downarrow \; \downarrow \; \downarrow \; \downarrow$ *Line up the common factors.*

$2 \cdot 2 \cdot 2 \cdot 3$ *To find the LCM, multiply one*

$2 \cdot 2 \cdot 2 \cdot 3 = 24$ *number from each column.*

LCM: 24

D **12, 10, and 15**

$12 = 2^2 \cdot 3$

$10 = 2 \quad \cdot \quad 5$

$15 = \quad 3 \cdot 5$

Write the prime factorization of each number in exponential form.

$\downarrow \; \downarrow \; \downarrow$

$2^2 \cdot 3 \cdot 5$

$2^2 \cdot 3 \cdot 5 = 60$

To find the LCM, multiply each prime factor once with the greatest exponent used in any of the prime factorizations.

LCM: 60

Think and Discuss

1. **Explain** why you cannot find a greatest common multiple for a group of numbers.
2. **Tell** whether the LCM of a set of numbers can ever be smaller than any of the numbers in the set.

5-5 Exercises

1.05

FOR EOG PRACTICE
see page 656

internet connect
Homework Help Online
go.hrw.com Keyword: MR4 5-5

GUIDED PRACTICE

See Example 1

1. Pencils are sold in packs of 12, and erasers in packs of 9. Mr. Joplin wants to give each of 36 students a pencil and an eraser. What is the least number of packs he should buy so there are none left over?

See Example 2

Find the least common multiple (LCM).

2. 3 and 5	**3.** 4 and 9	**4.** 2, 3, and 6	**5.** 2, 4, and 5
6. 4 and 12	**7.** 6 and 16	**8.** 4, 6, and 8	**9.** 2, 5, and 8
10. 6 and 10	**11.** 21 and 63	**12.** 3, 5, and 9	**13.** 5, 6, and 25

INDEPENDENT PRACTICE

See Example 1

14. String-cheese sticks are sold in packs of 10, and celery sticks in packs of 15. Ms. Sobrino wants to give each of 30 students one string-cheese stick and one celery stick. What is the least number of packs she should buy so there are none left over?

See Example 2

Find the least common multiple (LCM).

15. 2 and 8	**16.** 3 and 7	**17.** 4 and 10	**18.** 3 and 9
19. 3, 6, and 9	**20.** 4, 8, and 10	**21.** 4, 6, and 12	**22.** 4, 6, and 7
23. 3, 8, and 12	**24.** 3, 7, and 10	**25.** 2, 6, and 11	**26.** 2, 3, 6, and 9
27. 2, 4, 5, and 6	**28.** 10 and 11	**29.** 4, 5, and 7	**30.** 2, 3, 6, and 8

PRACTICE AND PROBLEM SOLVING

31. What is the LCM of 6 and 12?

32. What is the LCM of 5 and 11?

33. **Find the missing numbers in the diagram.**
   a. a two-digit multiple of 4 that is not a multiple of 6
   b. a two-digit multiple of 6 that is not a multiple of 4
   c. the LCM of 4 and 6
   d. a three-digit common multiple of 4 and 6

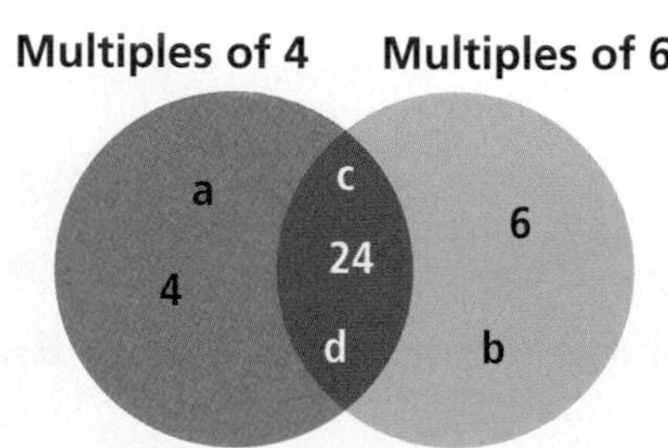

34. During its grand opening weekend, a restaurant gave every eighth customer a complimentary appetizer, every twelfth customer a complimentary beverage, and every fifteenth customer a complimentary dish of frozen yogurt.

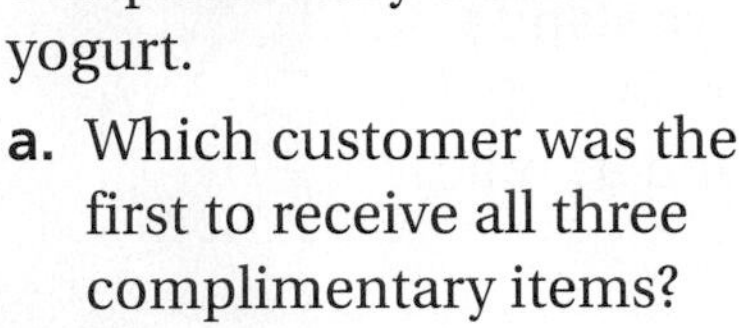

a. Which customer was the first to receive all three complimentary items?

b. Which customer was the first to receive a complimentary appetizer and frozen yogurt?

c. If the restaurant served 500 customers that weekend, how many of those customers received all three complimentary items?

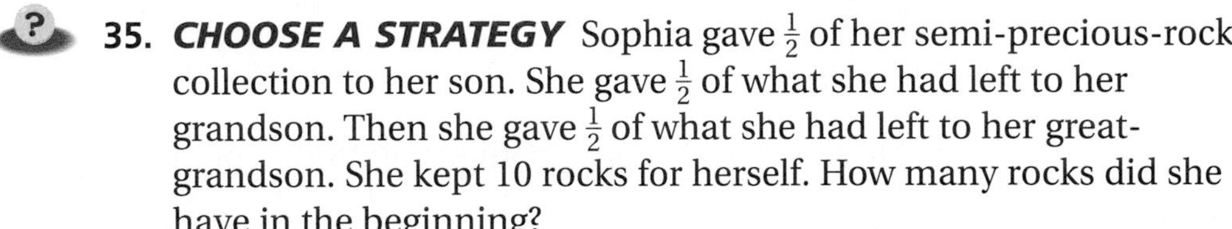

35. ***CHOOSE A STRATEGY*** Sophia gave $\frac{1}{2}$ of her semi-precious-rock collection to her son. She gave $\frac{1}{2}$ of what she had left to her grandson. Then she gave $\frac{1}{2}$ of what she had left to her great-grandson. She kept 10 rocks for herself. How many rocks did she have in the beginning?

A 40 B 80 C 100 D 160

36. ***WRITE ABOUT IT*** Explain the steps you can use to find the LCM of two numbers. Choose two numbers to show an example of your method.

37. ***CHALLENGE*** Find the LCM of each pair of numbers.

a. 4 and 6 b. 8 and 9 c. 5 and 7 d. 8 and 10

When is the LCM of two numbers equal to the product of the two numbers?

Spiral Review

Order the numbers from least to greatest. (Lesson 1-1)

38. 1,235; 354; 846 **39.** 978; 679; 879 **40.** 1,264; 1,098; 1,104

Determine whether the given value of each variable is a solution. (Lesson 2-3)

41. $y + 37 = 64$ for $y = 27$ **42.** $43 - c = 19$ for $c = 24$ **43.** $72 \div z = 9$ for $z = 7$

Write each number in standard form. (Lesson 3-5)

44. 6.479×10^3 **45.** 0.208×10^2 **46.** 13.507×10^4 **47.** 7.1×10^5

48. **EOG PREP** The prime factorization of which number is $2 \times 2 \times 3 \times 5$? (Lesson 4-2)

A 100 B 60 C 30 D 20

5-6 Estimating Fraction Sums and Differences

Learn to estimate sums and differences of fractions and mixed numbers.

Members of the Nature Club went mountain biking in Canyonlands National Park, Utah. They biked $10\frac{3}{10}$ miles on Monday and $9\frac{3}{4}$ miles on Tuesday.

Canyonlands National Park, Utah, is a 337,570-acre park that has many canyons, mesas, arches, and spires.

You can estimate fractions by rounding to 0, $\frac{1}{2}$, or 1.

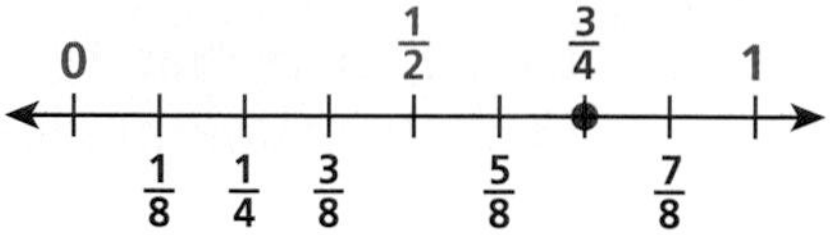

0
1/2
3/4
1
1/8
1/4
3/8
5/8
7/8

The fraction $\frac{3}{4}$ rounds to 1.

You can round fractions by comparing the numerator and denominator.

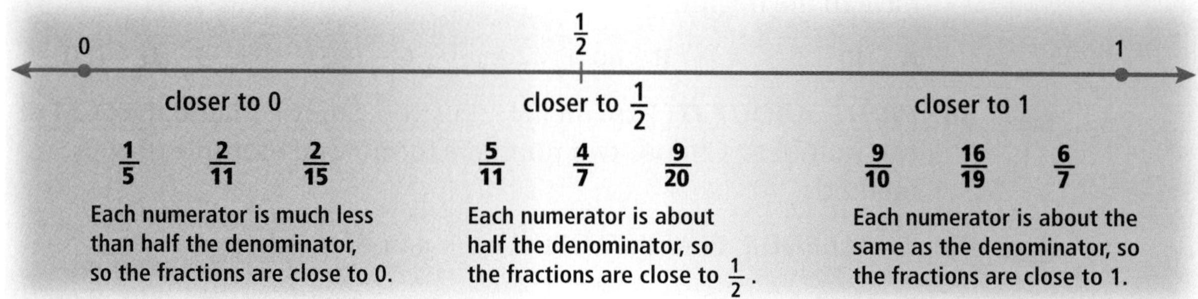

0
1/2
1
closer to 0
1/5 2/11 2/15
Each numerator is much less than half the denominator, so the fractions are close to 0.
closer to 1/2
5/11 4/7 9/20
Each numerator is about half the denominator, so the fractions are close to 1/2.
closer to 1
9/10 16/19 6/7
Each numerator is about the same as the denominator, so the fractions are close to 1.

EXAMPLE 1 Estimating Fractions

Estimate each sum or difference by rounding to 0, $\frac{1}{2}$, or 1.

A $\frac{8}{9} + \frac{2}{11}$

$\frac{8}{9} + \frac{2}{11}$ *Think: $\frac{8}{9}$ rounds to 1 and $\frac{2}{11}$ rounds to 0.*

$1 + 0 = 1$

$\frac{8}{9} + \frac{2}{11}$ is **about** 1.

B $\frac{7}{12} - \frac{8}{15}$

$\frac{7}{12} - \frac{8}{15}$ *Think: $\frac{7}{12}$ rounds to $\frac{1}{2}$ and $\frac{8}{15}$ rounds to $\frac{1}{2}$.*

$\frac{1}{2} - \frac{1}{2} = 0$

$\frac{7}{12} - \frac{8}{15}$ is **about** 0.

You can also estimate by rounding mixed numbers. You compare each mixed number to the two nearest whole numbers and the nearest $\frac{1}{2}$.

Does $10\frac{3}{10}$ round to 10, $10\frac{1}{2}$, or 11?

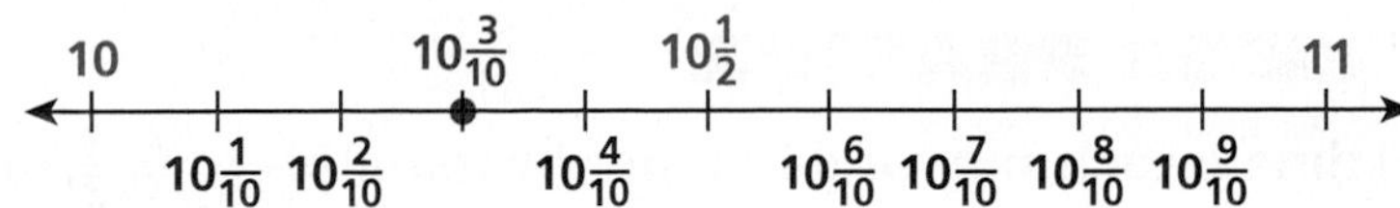

The mixed number $10\frac{3}{10}$ rounds to $10\frac{1}{2}$.

EXAMPLE 2 *Sports Application*

The table shows the distances the Nature Club biked in Utah.

Nature Club's Biking Distances

Day	Distances (mi)
Monday	$10\frac{3}{10}$
Tuesday	$9\frac{3}{4}$
Wednesday	$12\frac{1}{4}$
Thursday	$4\frac{7}{10}$

A **About how far did the Nature Club ride on Monday and Tuesday?**

$10\frac{3}{10} + 9\frac{3}{4}$

$10\frac{1}{2} + 10 = 20\frac{1}{2}$

They rode **about** $20\frac{1}{2}$ miles.

B **About how much farther did the Nature Club ride on Wednesday than on Thursday?**

$12\frac{1}{4} - 4\frac{7}{10}$

$12\frac{1}{2} - 4\frac{1}{2} = 8$

They rode **about** 8 miles farther on Wednesday than on Thursday.

C **Estimate the total distance that the Nature Club rode on Monday, Tuesday, and Wednesday.**

$10\frac{3}{10} + 9\frac{3}{4} + 12\frac{1}{4}$

$10\frac{1}{2} + 10 + 12\frac{1}{2} = 33$

They rode **about** 33 miles.

Think and Discuss

1. **Tell** whether each fraction rounds to 0, $\frac{1}{2}$, or 1: $\frac{5}{6}, \frac{2}{15}, \frac{7}{13}$.
2. **Explain** how to round mixed numbers to the nearest whole number.
3. **Determine** whether the Nature Club met their goal to ride at least 35 total miles.

FOR EOG PRACTICE
see page 656

internet connect
Homework Help Online
go.hrw.com Keyword: MR4 5-6

1.04c, 1.07

GUIDED PRACTICE

See Example 1 **Estimate each sum or difference by rounding to 0, $\frac{1}{2}$, or 1.**

1. $\frac{8}{9} + \frac{1}{6}$
2. $\frac{11}{12} - \frac{4}{9}$
3. $\frac{3}{7} + \frac{1}{12}$
4. $\frac{6}{13} - \frac{2}{5}$

See Example 2 **Use the table for Exercises 5 and 6.**

5. About how far did Mark run during week 1 and week 2?
6. About how much farther did Mark run during week 2 than during week 3?

Mark's Running Distances	
Week	Distance (mi)
1	$8\frac{3}{4}$
2	$7\frac{1}{5}$
3	$5\frac{5}{6}$

INDEPENDENT PRACTICE

See Example 1 **Estimate each sum or difference by rounding to 0, $\frac{1}{2}$, or 1.**

7. $\frac{7}{8} - \frac{3}{8}$
8. $\frac{3}{10} + \frac{3}{4}$
9. $\frac{5}{6} - \frac{7}{8}$
10. $\frac{7}{10} + \frac{1}{6}$

See Example 2 **Use the table for Exercises 11–13.**

11. About how much do the meteorites in Brenham and Goose Lake weigh together?
12. About how much more does the meteorite in Willamette weigh than the meteorite in Norton County?
13. About how much do the two meteorites in Kansas weigh together?

Meteorites in the United States	
Location	Weight (tons)
Willamette, AZ	$16\frac{1}{2}$
Brenham, KS	$2\frac{3}{5}$
Goose Lake, CA	$1\frac{3}{10}$
Norton County, KS	$1\frac{1}{10}$

PRACTICE AND PROBLEM SOLVING

Estimate each sum or difference to compare. Write $<$ or $>$.

14. $\frac{5}{6} + \frac{7}{9}$ ■ 3
15. $2\frac{8}{15} - 1\frac{1}{11}$ ■ 1
16. $1\frac{2}{21} + \frac{3}{7}$ ■ 2
17. $1\frac{7}{13} - \frac{8}{9}$ ■ 1
18. $3\frac{2}{10} + 2\frac{2}{5}$ ■ 6
19. $4\frac{6}{9} - 2\frac{3}{19}$ ■ 2

Estimate.

20. $\frac{7}{8} + \frac{4}{7} + \frac{7}{13}$
21. $\frac{6}{11} + \frac{9}{17} + \frac{3}{5}$
22. $\frac{8}{9} + \frac{3}{4} + \frac{9}{10}$
23. $1\frac{5}{8} + 2\frac{1}{15} + 2\frac{12}{13}$
24. $4\frac{11}{12} + 3\frac{1}{19} + 5\frac{4}{7}$
25. $10\frac{1}{9} + 8\frac{5}{9} + 11\frac{13}{14}$

Use an inch ruler for Exercises 26–29. Measure to the nearest $\frac{1}{4}$ inch.

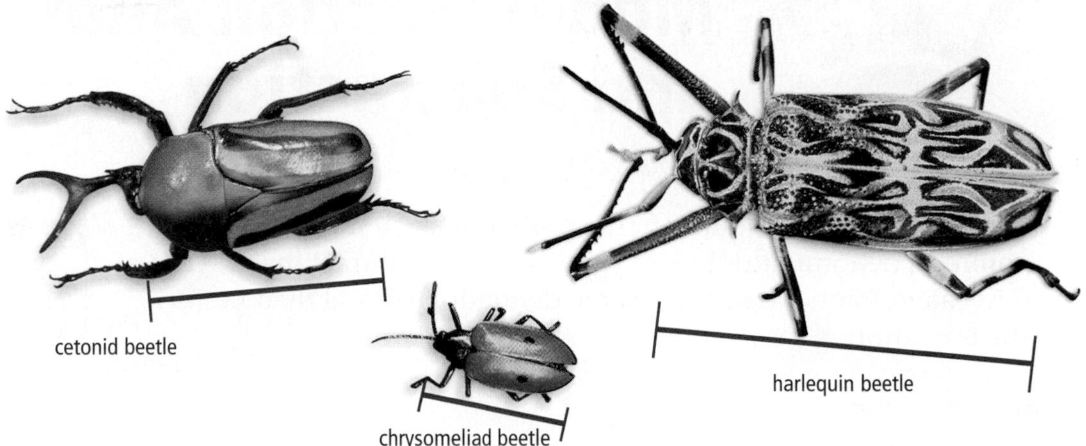

26. About how long is the chrysomeliad beetle?

27. About how long is the cetonid beetle?

28. About how much longer is the harlequin beetle than the cetonid beetle?

29. About how much longer is the harlequin beetle than the chrysomeliad beetle?

30. ***WRITE A PROBLEM*** Write a problem about a trip that can be solved by estimating fractions. Exchange with a classmate and solve.

31. ***WRITE ABOUT IT*** Explain how to estimate the sum of two mixed numbers. Give an example to explain your answer.

32. ***CHALLENGE*** Estimate.

$$\left[5\frac{7}{8} - 2\frac{3}{20}\right] + 1\frac{4}{7}$$

Spiral Review

Write each expression in exponential form. (Lesson 1-3)

33. $8 \times 8 \times 8 \times 8 \times 8$

34. $4 \times 4 \times 4 \times 4$

35. $7 \times 7 \times 7 \times 7 \times 7 \times 7$

Find the missing values in each table. (Lesson 2-1)

36.

x	6	7	9
$x^2 - 5$	■	■	■

37.

a	12	10	8
■	28	24	20

Evaluate 4*y* for each value of *y*. (Lesson 3-6)

38. $y = 2.13$

39. $y = 4.015$

40. $y = 3.6$

41. $y = 0.78$

42. **EOG PREP** Which number is *not* a factor of 42? (Lesson 4-2)

A 2

B 6

C 4

D 21

Model Fraction Addition and Subtraction

Use with Lessons 5-7 and 5-8

When fractions have different denominators, you need to find a common denominator before you can add or subtract them. Write equivalent fractions with the same denominator, and then perform the operation.

Activity 1

1 Find $\frac{1}{8} + \frac{1}{4}$.

Use fraction bars to represent both fractions.

$\frac{1}{8}$	$\frac{1}{4}$

Which fractions fit exactly across $\frac{1}{8}$ and $\frac{1}{4}$?

$\frac{1}{8}$	$\frac{1}{4}$	
$\frac{1}{8}$	$\frac{1}{8}$	$\frac{1}{8}$

$\frac{1}{4} = \frac{2}{8}$ $\qquad$ $\frac{1}{8} + \frac{2}{8} = \frac{3}{8}$

2 Find $\frac{2}{3} + \frac{1}{2}$.

Use fraction bars to represent both fractions.

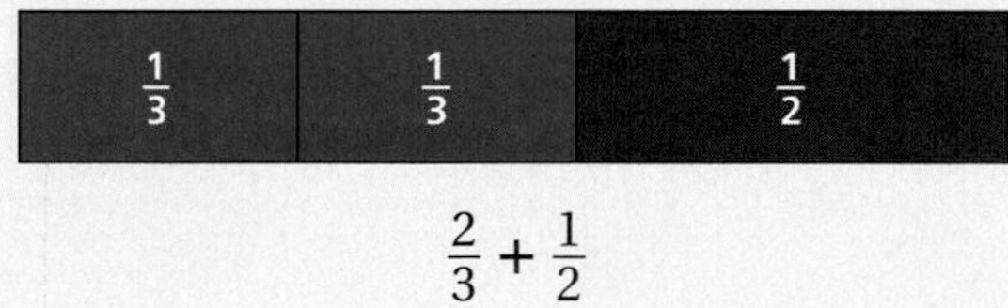

$\frac{2}{3} + \frac{1}{2}$

Which fractions fit exactly across $\frac{2}{3}$ and $\frac{1}{2}$?

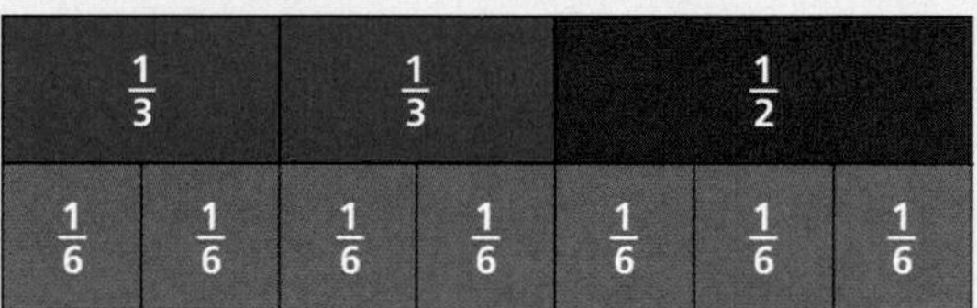

$\frac{2}{3} = \frac{4}{6}$ $\qquad$ $\frac{1}{2} = \frac{3}{6}$

$$\frac{4}{6} + \frac{3}{6} = \frac{7}{6} = 1\frac{1}{6}$$

Think and Discuss

1. Explain what the denominators of $\frac{1}{6}$, $\frac{1}{4}$, $\frac{2}{3}$, and $\frac{1}{2}$ have in common. (*Hint:* Think of common multiples.)

Try This

Model each addition expression with fraction bars, and find the sum.

1. $\frac{1}{4} + \frac{1}{2}$ $\qquad$ **2.** $\frac{3}{8} + \frac{1}{4}$ $\qquad$ **3.** $\frac{1}{2} + \frac{2}{5}$ $\qquad$ **4.** $\frac{3}{4} + \frac{1}{6}$

5. $\frac{1}{3} + \frac{1}{6}$ $\qquad$ **6.** $\frac{7}{8} + \frac{3}{4}$ $\qquad$ **7.** $\frac{2}{3} + \frac{1}{4}$ $\qquad$ **8.** $\frac{5}{8} + \frac{1}{4}$

Activity 2

1 Find $\frac{1}{3} - \frac{1}{6}$.

Use fraction bars to represent both fractions.

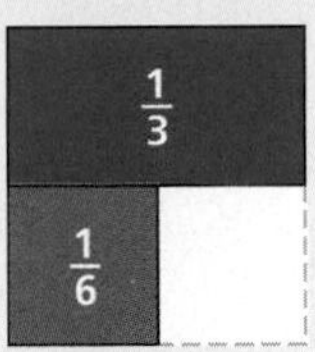

Which fractions fit exactly across $\frac{1}{3}$ and $\frac{1}{6}$?

$\frac{1}{3} = \frac{2}{6}$

Subtract $\frac{1}{6}$ from $\frac{2}{6}$.

$\frac{2}{6} - \frac{1}{6} = \frac{1}{6}$

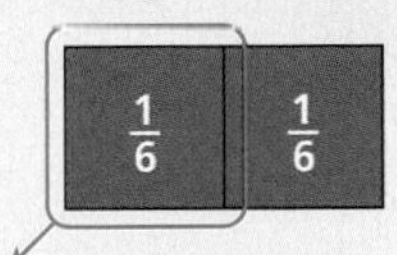

2 Find $\frac{1}{2} - \frac{1}{3}$.

Use fraction bars to represent both fractions.

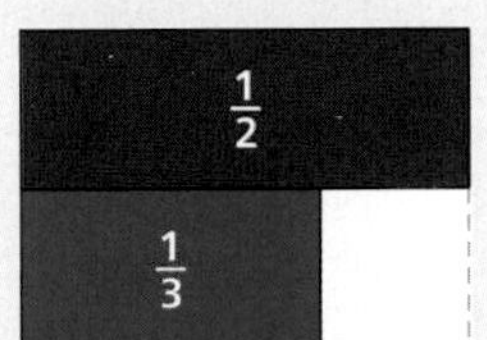

Which fractions fit exactly across $\frac{1}{2}$ and $\frac{1}{3}$?

$\frac{1}{2} = \frac{3}{6}$

$\frac{1}{3} = \frac{2}{6}$

Subtract $\frac{2}{6}$ from $\frac{3}{6}$.

$\frac{3}{6} - \frac{2}{6} = \frac{1}{6}$

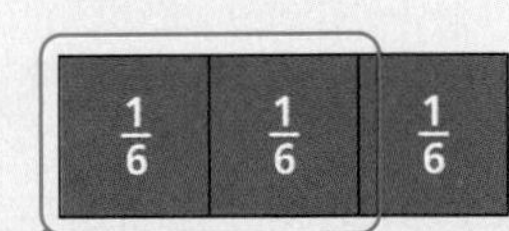

Think and Discuss

1. Explain what the area surrounded by a dashed line represents.

Try This

Model each subtraction expression with fraction bars, and find the difference.

1. $\frac{3}{4} - \frac{1}{3}$ **2.** $\frac{1}{3} - \frac{1}{4}$ **3.** $\frac{1}{2} - \frac{2}{5}$ **4.** $\frac{5}{6} - \frac{1}{3}$

5. $\frac{1}{2} - \frac{5}{12}$ **6.** $\frac{7}{8} - \frac{3}{4}$ **7.** $\frac{1}{4} - \frac{1}{8}$ **8.** $\frac{1}{4} - \frac{1}{6}$

5-7 Adding and Subtracting with Unlike Denominators

Learn to add and subtract fractions with unlike denominators.

Vocabulary

least common denominator (LCD)

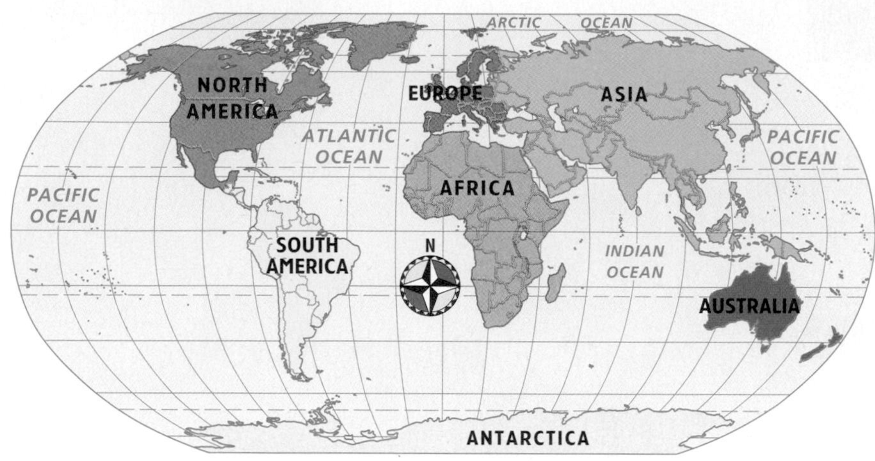

The Pacific Ocean covers $\frac{1}{3}$ of Earth's surface. The Atlantic Ocean covers $\frac{1}{5}$ of Earth's surface. To find the fraction of Earth's surface that is covered by both oceans, you can add $\frac{1}{3}$ and $\frac{1}{5}$, which are unlike fractions.

Remember!

Fractions that represent the same value are equivalent.

To add or subtract unlike fractions, first rewrite them as equivalent fractions with a common denominator.

EXAMPLE 1 *Social Studies Application*

What fraction of Earth's surface is covered by the Atlantic and Pacific Oceans?

Add $\frac{1}{3}$ and $\frac{1}{5}$.

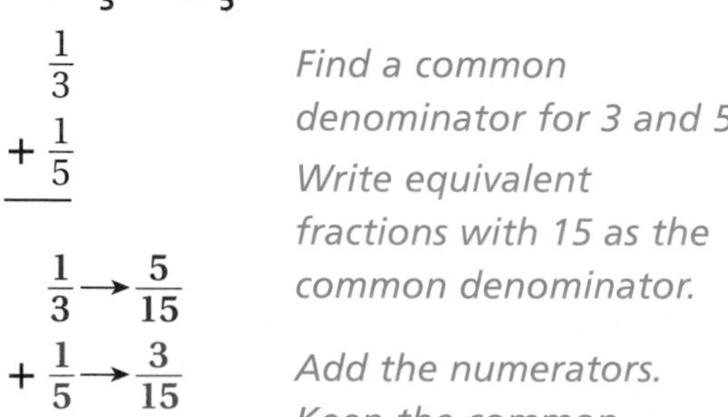

$$\begin{array}{r} \frac{1}{3} \\ +\frac{1}{5} \\ \hline \end{array}$$

Find a common denominator for 3 and 5.

Write equivalent fractions with 15 as the common denominator.

$$\begin{array}{r} \frac{1}{3} \rightarrow \frac{5}{15} \\ +\frac{1}{5} \rightarrow \frac{3}{15} \\ \hline \frac{8}{15} \end{array}$$

Add the numerators. Keep the common denominator.

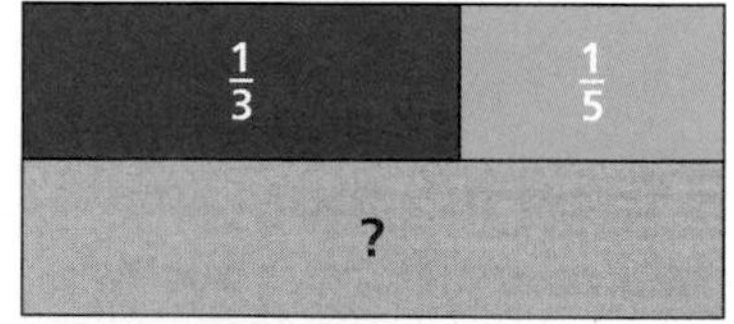

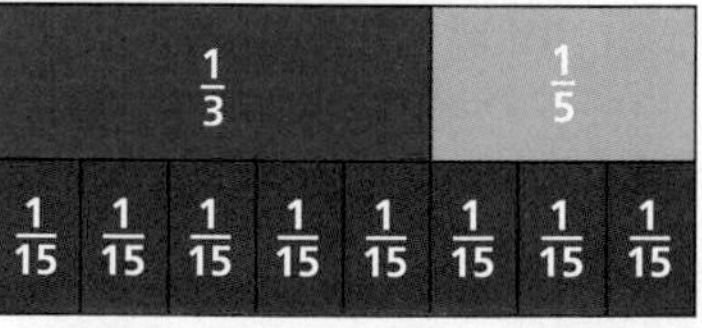

The Pacific and Atlantic Oceans cover $\frac{8}{15}$ of Earth's surface.

You can use *any* common denominator or the *least common denominator* to add and subtract unlike fractions. The **least common denominator (LCD)** is the least common multiple of the denominators.

EXAMPLE 2 Adding and Subtracting Unlike Fractions

Add or subtract. Write each answer in simplest form.

Method 1: Multiply denominators.

A $\frac{9}{10} - \frac{7}{8}$

$\frac{9}{10} - \frac{7}{8}$ *Multiply the denominators. 10 · 8 = 80*

$\frac{72}{80} - \frac{70}{80}$ *Write equivalent fractions.*

$\frac{2}{80}$ *Subtract.*

$\frac{1}{40}$ *Write the answer in simplest form.*

Method 2: Use the LCD.

B $\frac{9}{10} - \frac{7}{8}$

$\frac{9}{10} - \frac{7}{8}$ *The LCD is 40.*

$\frac{36}{40} - \frac{35}{40}$ *Write equivalent fractions.*

$\frac{1}{40}$ *Subtract.*

Method 3: Use mental math.

C $\frac{5}{12} + \frac{1}{6}$

$\frac{5}{12} + \frac{1}{6}$ *Think: 12 is a multiple of 6, so the LCD is 12.*

$\frac{5}{12} + \frac{2}{12}$ *Rewrite $\frac{1}{6}$ with a denominator of 12.*

$\frac{7}{12}$ *Add.*

D $\frac{5}{12} - \frac{1}{6}$

$\frac{5}{12} - \frac{1}{6}$ *Think: 12 is a multiple of 6, so the LCD is 12.*

$\frac{5}{12} - \frac{2}{12}$ *Rewrite $\frac{1}{6}$ with a denominator of 12.*

$\frac{3}{12}$ *Subtract.*

$\frac{1}{4}$ *Write the answer in simplest form.*

Think and Discuss

1. **Explain** an advantage of using the least common denominator (LCD) when adding unlike fractions.
2. **Tell** when the least common denominator (LCD) of two fractions is the product of their denominators.
3. **Explain** how you can use mental math to subtract $\frac{1}{12}$ from $\frac{3}{4}$.

5-7 Exercises

FOR EOG PRACTICE

see page 656

internet connect

Homework Help Online

go.hrw.com Keyword: MR4 5-7

1.04a

GUIDED PRACTICE

See Example 1

1. A trailer hauling wood weighs $\frac{2}{3}$ ton. The trailer weighs $\frac{1}{4}$ ton without the wood. What is the weight of the wood?

See Example 2

Add or subtract. Write each answer in simplest form.

2. $\frac{1}{3} + \frac{1}{9}$
3. $\frac{7}{10} - \frac{2}{5}$
4. $\frac{2}{3} - \frac{2}{5}$
5. $\frac{1}{2} + \frac{3}{7}$

INDEPENDENT PRACTICE

See Example 1

6. Approximately $\frac{1}{5}$ of the world's population lives in China. The people of India make up about $\frac{1}{6}$ of the world's population. What fraction of the world's people live in either China or India?

7. Cedric is making an Italian dish using a recipe that calls for $\frac{2}{3}$ cup of grated mozarella cheese. If Cedric has grated $\frac{1}{2}$ cup of mozarella cheese, how much more does he need to grate?

See Example 2

Add or subtract. Write each answer in simplest form.

8. $\frac{3}{4} - \frac{1}{2}$
9. $\frac{1}{6} + \frac{5}{12}$
10. $\frac{5}{6} - \frac{3}{4}$
11. $\frac{1}{5} + \frac{1}{4}$
12. $\frac{7}{10} + \frac{1}{8}$
13. $\frac{1}{3} + \frac{4}{5}$
14. $\frac{8}{9} - \frac{2}{3}$
15. $\frac{5}{8} + \frac{1}{2}$

PRACTICE AND PROBLEM SOLVING

Find each sum or difference. Write your answer in simplest form.

16. $\frac{3}{10} + \frac{1}{2}$
17. $\frac{4}{5} - \frac{1}{3}$
18. $\frac{5}{8} - \frac{1}{6}$
19. $\frac{1}{6} + \frac{2}{9}$
20. $\frac{2}{7} + \frac{2}{5}$
21. $\frac{7}{12} - \frac{1}{4}$
22. $\frac{7}{8} - \frac{2}{3}$
23. $\frac{2}{11} + \frac{2}{3}$

Evaluate each expression for $b = \frac{1}{2}$. Write your answer in simplest form.

24. $b + \frac{1}{3}$
25. $\frac{8}{9} - b$
26. $b - \frac{2}{11}$
27. $\frac{2}{7} + b$
28. $b + b$
29. $b - b$

Evaluate. Write each answer in simplest form.

30. $\frac{1}{3} + \frac{1}{9} + \frac{1}{3}$
31. $\frac{9}{10} - \frac{2}{10} - \frac{1}{5}$
32. $\frac{1}{2} + \frac{1}{4} - \frac{1}{8}$
33. $\frac{5}{6} - \frac{2}{3} + \frac{7}{12}$
34. $\frac{2}{3} + \frac{1}{4} - \frac{1}{6}$
35. $\frac{2}{9} + \frac{1}{6} + \frac{1}{3}$

36. Bailey spent $\frac{2}{3}$ of his monthly allowance at the movies and $\frac{1}{5}$ of it on baseball cards. What fraction of Bailey's allowance is left?

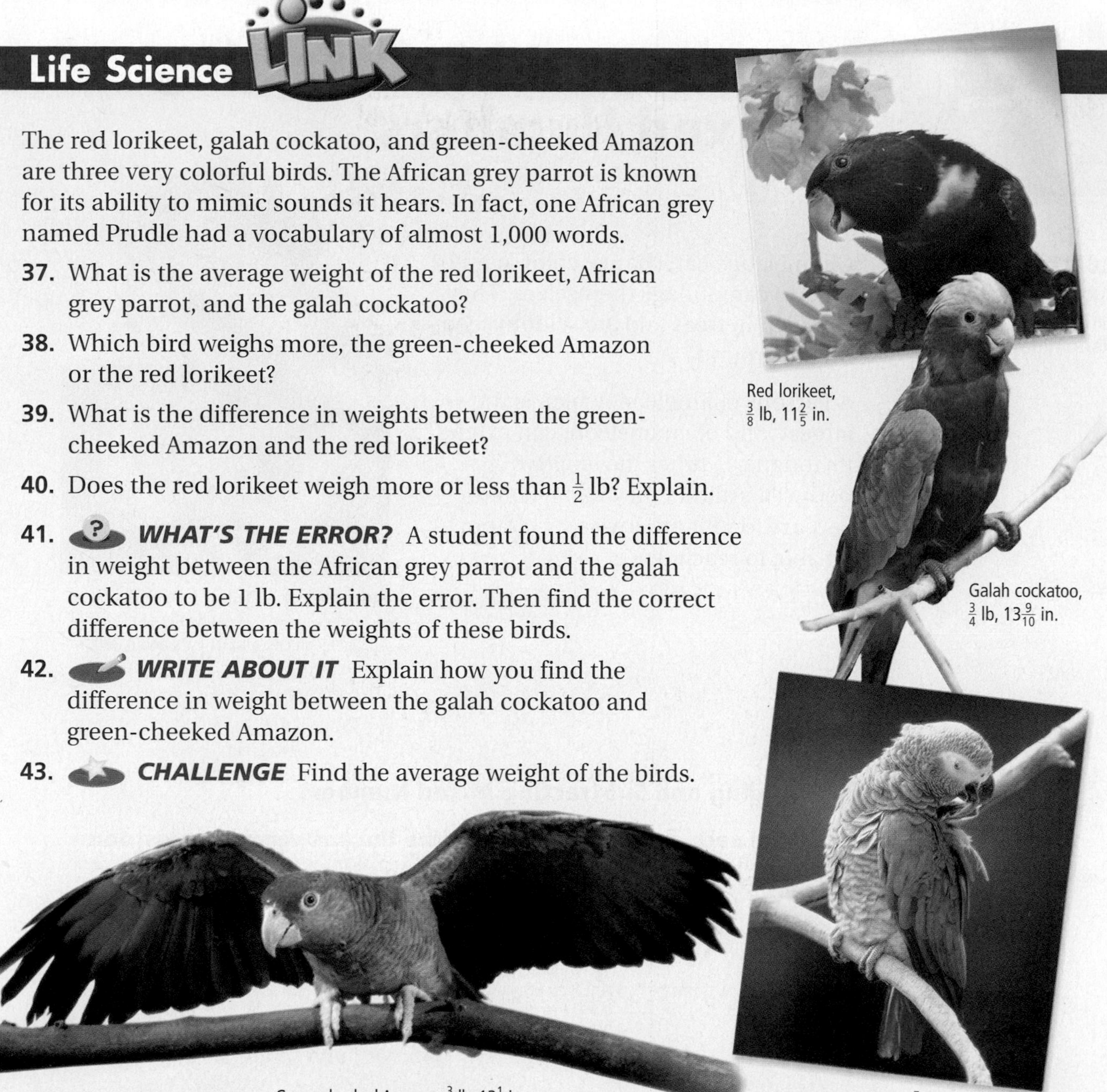

Life Science LINK

The red lorikeet, galah cockatoo, and green-cheeked Amazon are three very colorful birds. The African grey parrot is known for its ability to mimic sounds it hears. In fact, one African grey named Prudle had a vocabulary of almost 1,000 words.

37. What is the average weight of the red lorikeet, African grey parrot, and the galah cockatoo?

38. Which bird weighs more, the green-cheeked Amazon or the red lorikeet?

39. What is the difference in weights between the green-cheeked Amazon and the red lorikeet?

40. Does the red lorikeet weigh more or less than $\frac{1}{2}$ lb? Explain.

41. ***WHAT'S THE ERROR?*** A student found the difference in weight between the African grey parrot and the galah cockatoo to be 1 lb. Explain the error. Then find the correct difference between the weights of these birds.

42. ***WRITE ABOUT IT*** Explain how you find the difference in weight between the galah cockatoo and green-cheeked Amazon.

43. ***CHALLENGE*** Find the average weight of the birds.

Red lorikeet, $\frac{3}{8}$ lb, $11\frac{2}{5}$ in.

Galah cockatoo, $\frac{3}{4}$ lb, $13\frac{9}{10}$ in.

Green-cheeked Amazon, $\frac{3}{5}$ lb, $13\frac{1}{5}$ in.

African grey parrot, $\frac{7}{8}$ lb, 13 in.

Spiral Review

Find each value. (Lesson 1-3)

44. 4^2 **45.** 2^4 **46.** 10^5 **47.** 7^3

48. **EOG PREP** Which list is ordered from *least* to *greatest*? (Lesson 3-1)

A 75.4, 75.09, 75.28
B 75.28, 75.09, 75.4
C 75.09, 75.4, 75.28
D 75.09, 75.28, 75.4

49. **EOG PREP** Of which pair of numbers is 1 the only common factor? (Lesson 4-3)

A 12 and 9 B 7 and 16 C 21 and 15 D 3 and 6

5-8 Adding and Subtracting Mixed Numbers

Learn to add and subtract mixed numbers with unlike denominators.

Chameleons can change color at any time to camouflage themselves. They live high in trees and are seldom seen on the ground.

A Parsons chameleon, which is the largest kind of chameleon, can extend its tongue $1\frac{1}{2}$ times the length of its body. This allows the chameleon to capture food it otherwise would not be able to reach.

The chameleon is the only animal capable of moving each eye independently of the other. A chameleon can turn its eyes about 360°.

EXAMPLE 1 Adding and Subtracting Mixed Numbers

Find each sum or difference. Write the answer in simplest form.

A $2\frac{3}{4} + 1\frac{1}{6}$

$$2\frac{3}{4} \rightarrow 2\frac{18}{24}$$

Multiply the denominators. 4 • 6 = 24

$$+\,1\frac{1}{6} \rightarrow +\,1\frac{4}{24}$$

Write equivalent fractions with a denominator of 24.

$$3\frac{22}{24} = 3\frac{11}{12}$$

Add the fractions and then the whole numbers, and simplify.

B $4\frac{5}{6} - 2\frac{2}{9}$

The LCD is 18.

$$4\frac{5}{6} \rightarrow 4\frac{15}{18}$$

$$-\,2\frac{2}{9} \rightarrow -\,2\frac{4}{18}$$

Write equivalent fractions with a denominator of 18.

$$2\frac{11}{18}$$

Subtract the fractions and then the whole numbers.

C $2\frac{2}{3} + 1\frac{3}{4}$

The LCD is 12.

$$2\frac{2}{3} \rightarrow 2\frac{8}{12}$$

$$+\,1\frac{3}{4} \rightarrow +\,1\frac{9}{12}$$

Write equivalent fractions with a denominator of 12.

$$3\frac{17}{12} = 4\frac{5}{12}$$

Add the fractions and then the whole numbers. $3\frac{17}{12} = 3 + 1\frac{5}{12}$

Find each sum or difference. Write the answer in simplest form.

D $8\frac{2}{5} - 6\frac{3}{10}$

$$\begin{array}{rcr} 8\frac{2}{5} & \longrightarrow & 8\frac{4}{10} \\ -\,6\frac{3}{10} & \longrightarrow & -\,6\frac{3}{10} \\ \hline & & 2\frac{1}{10} \end{array}$$

Think: 10 is a multiple of 5, so 10 is the LCD.

Write equivalent fractions with a denominator of 10.

Subtract the fractions and then the whole numbers.

EXAMPLE 2 *Measurement Application*

The length of a Parsons chameleon's body is $23\frac{1}{2}$ inches. The chameleon can extend its tongue $35\frac{1}{4}$ inches. What is the total length of its body and its tongue?

Add $23\frac{1}{2}$ and $35\frac{1}{4}$.

$$\begin{array}{rcr} 23\frac{1}{2} & \longrightarrow & 23\frac{2}{4} \\ +\,35\frac{1}{4} & \longrightarrow & +\,35\frac{1}{4} \\ \hline & & 58\frac{3}{4} \end{array}$$

Find a common denominator. Write equivalent fractions with the LCD, 4, as the denominator.

Add the fractions and then the whole numbers.

The total length of the chameleon's body and tongue is $58\frac{3}{4}$ inches.

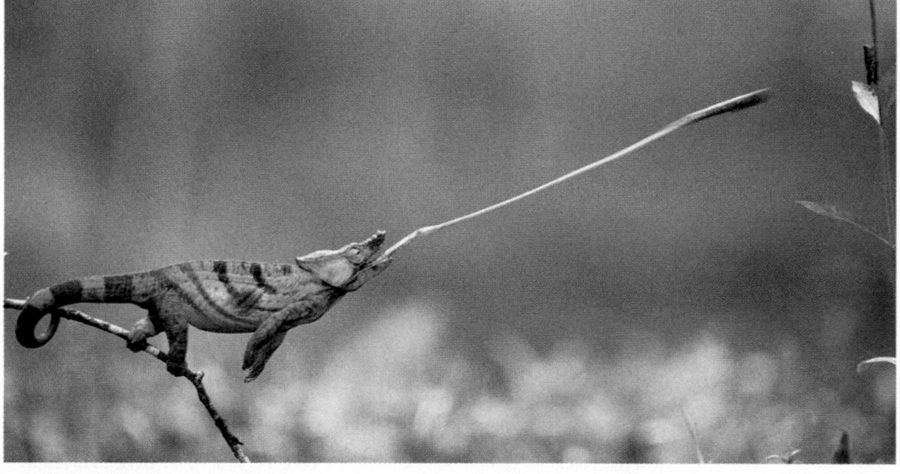

Helpful Hint

You can use mental math to find an LCD. *Think:* 4 is a multiple of 2 and 4.

Think and Discuss

1. **Tell** what mistake was made when subtracting $2\frac{1}{2}$ from $5\frac{3}{5}$ gave the following result: $5\frac{3}{5} - 2\frac{1}{2} = 3\frac{2}{3}$.
2. **Explain** why you first find equivalent fractions when adding $1\frac{1}{5}$ and $1\frac{1}{2}$.
3. **Tell** how you know that $5\frac{1}{2} - 3\frac{1}{4}$ is more than 2.

5-8 Exercises

FOR EOG PRACTICE
see page 657

internet connect
Homework Help Online
go.hrw.com Keyword: MR4 5-8

1.04a

GUIDED PRACTICE

See Example 1

Find each sum or difference. Write the answer in simplest form.

1. $7\frac{1}{12} + 3\frac{1}{3}$
2. $2\frac{1}{6} + 2\frac{3}{8}$
3. $8\frac{5}{6} - 2\frac{3}{4}$
4. $6\frac{6}{7} - 1\frac{1}{2}$

See Example 2

5. A sea turtle traveled $7\frac{3}{4}$ hours in two days. It traveled $3\frac{1}{2}$ hours on the first day. How many hours did it travel on the second day?

INDEPENDENT PRACTICE

See Example 1

Find each sum or difference. Write the answer in simplest form.

6. $3\frac{9}{10} - 1\frac{2}{5}$
7. $2\frac{1}{6} + 4\frac{5}{12}$
8. $5\frac{9}{11} + 5\frac{1}{3}$
9. $9\frac{3}{4} - 3\frac{1}{2}$

See Example 2

10. The drama club rehearsed $1\frac{3}{4}$ hours Friday and $3\frac{1}{6}$ hours Saturday. How many total hours did the students rehearse?

PRACTICE AND PROBLEM SOLVING

Add or subtract. Write each answer in simplest form.

11. $6\frac{3}{10} + 3\frac{2}{5}$
12. $10\frac{2}{3} - 2\frac{1}{12}$
13. $14\frac{3}{4} - 6\frac{5}{12}$
14. $19\frac{1}{10} + 10\frac{1}{2}$
15. $15\frac{5}{6} + 18\frac{2}{3}$
16. $17\frac{1}{6} + 12\frac{1}{4}$
17. $23\frac{9}{10} - 20\frac{3}{9}$
18. $32\frac{5}{7} - 13\frac{2}{5}$
19. $28\frac{11}{12} - 8\frac{5}{9}$
20. $12\frac{2}{11} + 20\frac{2}{3}$
21. $36\frac{5}{8} - 24\frac{5}{12}$
22. $48\frac{9}{11} + 2\frac{1}{4}$

Evaluate. Write each answer as a fraction in simplest form.

23. $0.3 + \frac{2}{5}$
24. $\frac{4}{5} + 0.9$
25. $5\frac{4}{5} - 3.2$
26. $6.3 + \frac{4}{5}$
27. $23\frac{3}{4} - 10.5$
28. $18.9 - 6\frac{1}{2}$

Evaluate each expression for $n = 2\frac{1}{3}$. Write your answer in simplest form.

29. $2\frac{2}{3} + n$
30. $5 - \left(1\frac{2}{3} + n\right)$
31. $n - 1\frac{1}{4}$
32. $n + 5\frac{7}{9}$
33. $6\left(3\frac{4}{9} + n\right)$
34. $2n - n$

35. ***MEASUREMENT*** Kyle's backpack weighs $14\frac{7}{20}$ lb. Kirsten's backpack weighs $12\frac{1}{4}$ lb.
    a. How much do the backpacks weigh together?
    b. How much more does Kyle's backpack weigh than Kirsten's backpack?
    c. Kyle takes his $3\frac{1}{4}$ lb math book out of his backpack. How much does his backpack weigh now?

Life Science LINK

Research shows that elephants hear through their feet. The feet act as drums, sending vibrations through the bones to the elephant's ears.

36. ***LIFE SCIENCE*** Elephants can communicate through low-frequency infrasonic rumbles. Such sounds can travel from $\frac{1}{8}$ km to $9\frac{1}{2}$ km. Find the difference between these two distances.

37. The route Jo usually takes to work is $4\frac{2}{5}$ mi. After heavy rains, when that road is flooded, she must take a different route that is $4\frac{9}{10}$ mi. How much longer is Jo's alternate route?

38. Mr. Hansley used $1\frac{2}{3}$ c of flour to make muffins and $4\frac{1}{2}$ c to make bread. If he has $3\frac{5}{6}$ c left, how much flour did Mr. Hansley have before making the muffins and bread?

Use the drawing for Exercises 39–42.

39. Sarah is a landscape architect designing a garden. Based on her drawing, how much longer is the south side of the building than the west side?

40. Sarah needs to determine how many azalea bushes she can plant along both sides of the path. What is the sum of the lengths of the two sides of the path?

41. How wide is the path?

42. ***WHAT'S THE QUESTION?*** The answer is $63\frac{2}{3}$ yd. What is the question?

43. ***WRITE ABOUT IT*** Explain how you would use the sum of $\frac{2}{5}$ and $\frac{1}{3}$ to find the sum of $10\frac{2}{5}$ and $6\frac{1}{3}$.

44. ***CHALLENGE*** Find each missing numerator.

a. $3\frac{x}{9} + 4\frac{2}{9} = 7\frac{7}{9}$ **b.** $1\frac{3}{10} + 9\frac{x}{2} = 10\frac{4}{5}$

Spiral Review

Move the decimal point to multiply or divide by powers of 10. (Lesson 3-4)

45. $1{,}465 \times 100$ **46.** $32.09 \div 10^2$ **47.** $209{,}467 \times 1{,}000$ **48.** $1.7 \div 10^4$

Find each product. (Lesson 5-1)

49. $\frac{2}{3} \times \frac{1}{5}$ **50.** $\frac{3}{7} \times \frac{1}{4}$ **51.** $\frac{2}{9} \times \frac{3}{8}$ **52.** $\frac{1}{4} \times \frac{6}{7}$

53. **EOG PREP** What is the least common multiple of 5 and 8? (Lesson 5-5)

A 40 B 20 C 80 D 60

Model Subtraction with Renaming

Use with Lesson 5-9

Sometimes you need to rename a mixed number before you can subtract. To rename a mixed number, divide one or more of the whole numbers into fractional parts.

Activity

1 Find $2\frac{1}{3} - 1\frac{2}{3}$.

Use fraction bars to model $2\frac{1}{3}$.

1	1	$\frac{1}{3}$

You cannot subtract $\frac{2}{3}$ from $\frac{1}{3}$. You need to rename $2\frac{1}{3}$.

You can subtract $1\frac{2}{3}$ from $1\frac{4}{3}$.

1	$\frac{1}{3}$	$\frac{1}{3}$	$\frac{1}{3}$	$\frac{1}{3}$

$$2\frac{1}{3} - 1\frac{2}{3} = 1\frac{4}{3} - 1\frac{2}{3} = \frac{2}{3}$$

2 Find $2\frac{1}{6} - 1\frac{5}{12}$.

Use fraction bars to model $2\frac{1}{6}$.

1	1	$\frac{1}{6}$

Write an equivalent fraction with a denominator of 12.

1	1	$\frac{1}{12}$	$\frac{1}{12}$

Still, you cannot subtract $\frac{5}{12}$ from $\frac{2}{12}$. You need to rename $2\frac{2}{12}$.

You can subtract $1\frac{5}{12}$ from $1\frac{14}{12}$.

1	$\frac{1}{12}$	$\frac{1}{12}$	$\frac{1}{12}$	$\frac{1}{12}$	$\frac{1}{12}$	$\frac{1}{12}$	$\frac{1}{12}$	$\frac{1}{12}$	$\frac{1}{12}$	$\frac{1}{12}$	$\frac{1}{12}$	$\frac{1}{12}$	$\frac{1}{12}$	$\frac{1}{12}$

$$2\frac{1}{6} - 1\frac{5}{12} = 1\frac{14}{12} - 1\frac{5}{12} = \frac{9}{12}, \text{ or } \frac{3}{4}$$

3 Find $2\frac{1}{4} - 1\frac{3}{8}$.

Use fraction bars to model $2\frac{1}{4}$.

Write an equivalent fraction.

Still, you cannot subtract $\frac{3}{8}$ from $\frac{2}{8}$. You need to rename $2\frac{2}{8}$.

You can subtract $1\frac{3}{8}$ from $1\frac{10}{8}$.

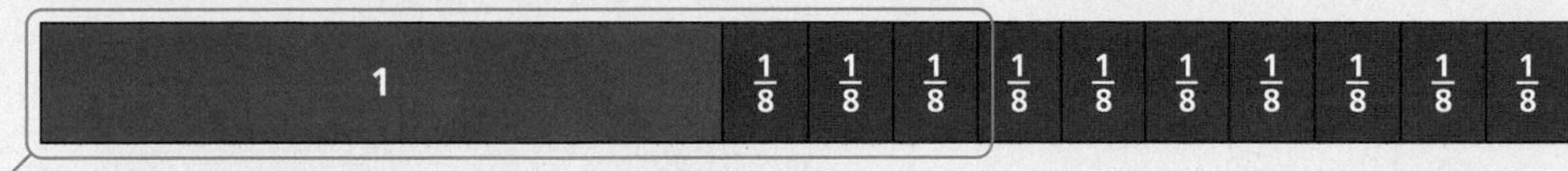

$$2\frac{1}{4} - 1\frac{3}{8} = 1\frac{10}{8} - 1\frac{3}{8} = \frac{7}{8}$$

Think and Discuss

1. Tell whether you need to rename before you subtract $3\frac{3}{8} - 1\frac{1}{8}$.
2. Tell whether you need to rename before you subtract $4\frac{3}{5} - 1\frac{7}{10}$.

Try This

Give the subtraction expression that is modeled.

1.

2.

3.

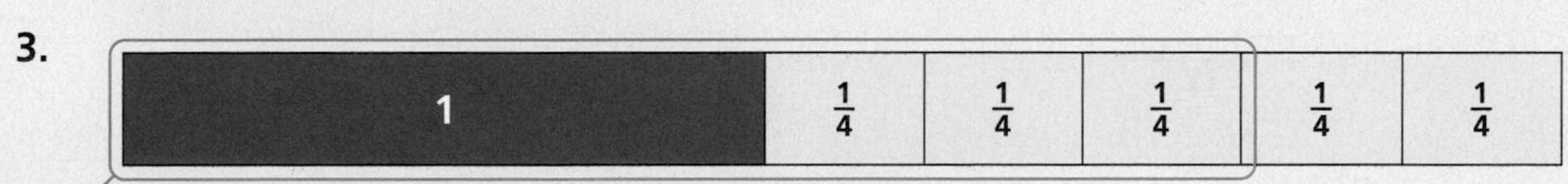

4.

Use fraction bars to subtract. Draw each model.

5. $2\frac{3}{8} - 1\frac{1}{2}$
6. $3\frac{1}{3} - 1\frac{2}{3}$
7. $4\frac{1}{4} - 2\frac{5}{6}$
8. $3\frac{1}{6} - 1\frac{1}{4}$

5-9 Renaming to Subtract Mixed Numbers

Learn to rename mixed numbers to subtract.

Jimmy and his family planted a tree when it was $1\frac{3}{4}$ ft tall. Now the tree is $2\frac{1}{4}$ ft tall. How much has the tree grown since it was planted?

The difference in the heights is $2\frac{1}{4} - 1\frac{3}{4}$.

You will need to rename $2\frac{1}{4}$ because the fraction in $1\frac{3}{4}$ is greater than $\frac{1}{4}$.

Divide *one whole* of $2\frac{1}{4}$ into fourths.

1	1	$\frac{1}{4}$

1	$\frac{1}{4}$	$\frac{1}{4}$	$\frac{1}{4}$	$\frac{1}{4}$	$\frac{1}{4}$

1	$\frac{1}{4}$	$\frac{1}{4}$	$\frac{1}{4}$		$\frac{1}{4}$	$\frac{1}{4}$

Rename $2\frac{1}{4}$ as $1\frac{5}{4}$.

$$\begin{array}{rcr} 2\frac{1}{4} & \rightarrow & 1\frac{5}{4} \\ -\,1\frac{3}{4} & \rightarrow & -\,1\frac{3}{4} \\ \hline & & \frac{2}{4} = \frac{1}{2} \end{array}$$

The tree has grown $\frac{1}{2}$ ft since it was planted.

EXAMPLE 1 Renaming Mixed Numbers

Subtract. Write each answer in simplest form.

A $6\frac{5}{12} - 2\frac{7}{12}$

$$\begin{array}{rcr} 6\frac{5}{12} & \rightarrow & 5\frac{17}{12} \\ -\,2\frac{7}{12} & \rightarrow & -\,2\frac{7}{12} \\ \hline & & 3\frac{10}{12} = 3\frac{5}{6} \end{array}$$

Rename $6\frac{5}{12}$ as $5 + 1\frac{5}{12} = 5 + \frac{12}{12} + \frac{5}{12}$.

Subtract the fractions and then the whole numbers.

Write the answer in simplest form.

B $7\frac{2}{3} - 2\frac{5}{6}$

$$\begin{array}{rcr} 7\frac{4}{6} & \rightarrow & 6\frac{10}{6} \\ -\,2\frac{5}{6} & \rightarrow & -\,2\frac{5}{6} \\ \hline & & 4\frac{5}{6} \end{array}$$

6 is a multiple of 3, so 6 is a common denominator.

Rename $7\frac{4}{6}$ as $6 + 1\frac{4}{6} = 6 + \frac{6}{6} + \frac{4}{6}$.

Subtract the fractions and then the whole numbers.

Subtract. Write each answer in simplest form.

C $8\frac{1}{4} - 5\frac{2}{3}$

$$\begin{array}{rcr} 8\frac{3}{12} & \longrightarrow & 7\frac{15}{12} \\ -\,5\frac{8}{12} & \longrightarrow & -\,5\frac{8}{12} \\ \hline & & 2\frac{7}{12} \end{array}$$

The LCM of 4 and 3 is 12.
Rename $8\frac{3}{12}$ as $7 + 1\frac{3}{12} = 7 + \frac{12}{12} + \frac{3}{12}$.
Subtract the fractions and then the whole numbers.

D $8 - 5\frac{3}{4}$

$$\begin{array}{rcr} 8 & \longrightarrow & 7\frac{4}{4} \\ -\,5\frac{3}{4} & \longrightarrow & -\,5\frac{3}{4} \\ \hline & & 2\frac{1}{4} \end{array}$$

Write 8 as a mixed number with a denominator of 4. Rename 8 as $7 + \frac{4}{4}$.
Subtract the fractions and then the whole numbers.

EXAMPLE 2 *Measurement Application*

Dave is re-covering an old couch and cushions. He determines that he needs 17 yards of fabric for the job.

A **Dave has $1\frac{2}{3}$ yards of fabric. How many more yards does he need?**

$$\begin{array}{rcr} 17 & \longrightarrow & 16\frac{3}{3} \\ -\,1\frac{2}{3} & \longrightarrow & -\,1\frac{2}{3} \\ \hline & & 15\frac{1}{3} \end{array}$$

Write 17 as a mixed number with a denominator of 3. Rename 17 as $16 + \frac{3}{3}$.
Subtract the fractions and then the whole numbers.

Dave needs another $15\frac{1}{3}$ yards of material.

B **If Dave uses $9\frac{5}{6}$ yards of fabric to cover the couch frame, how much of the 17 yards will he have left?**

$$\begin{array}{rcr} 17 & \longrightarrow & 16\frac{6}{6} \\ -\,9\frac{5}{6} & \longrightarrow & -\,9\frac{5}{6} \\ \hline & & 7\frac{1}{6} \end{array}$$

Write 17 as a mixed number with a denominator of 6. Rename 17 as $16 + \frac{6}{6}$.
Subtract the fractions and then the whole numbers.

Dave will have $7\frac{1}{6}$ yards of material left.

Think and Discuss

1. **Explain** why you rename 2 as $1\frac{8}{8}$ instead of $1\frac{3}{3}$ when you find $2 - 1\frac{3}{8}$.
2. **Give an example** of a subtraction expression in which you would need to rename the first mixed number to subtract.

5-9 Exercises

FOR EOG PRACTICE

see page 657

internet connect
Homework Help Online
go.hrw.com Keyword: MR4 5-9

1.04a

GUIDED PRACTICE

See Example 1

Subtract. Write each answer in simplest form.

1. $2\frac{1}{2} - 1\frac{3}{4}$ **2.** $8\frac{2}{9} - 2\frac{7}{9}$ **3.** $3\frac{2}{6} - 1\frac{2}{3}$ **4.** $7\frac{1}{4} - 4\frac{11}{12}$

See Example 2

5. Mr. Jones purchased a 4-pound bag of flour. He used $1\frac{2}{5}$ pounds of flour to make bread. How many pounds of flour are left?

INDEPENDENT PRACTICE

See Example 1

Subtract. Write each answer in simplest form.

6. $6\frac{3}{11} - 3\frac{10}{11}$ **7.** $9\frac{2}{5} - 5\frac{3}{5}$ **8.** $4\frac{3}{10} - 3\frac{3}{5}$ **9.** $10\frac{1}{2} - 2\frac{5}{8}$

10. $11\frac{3}{4} - 9\frac{1}{8}$ **11.** $7\frac{5}{9} - 2\frac{5}{6}$ **12.** $6 - 2\frac{2}{3}$ **13.** $5\frac{7}{10} - 3\frac{1}{2}$

See Example 2

14. A standard piece of notebook paper has a length of 11 inches and a width of $8\frac{1}{2}$ inches. What is the difference between these two measures?

PRACTICE AND PROBLEM SOLVING

Find each difference. Write the answer in simplest form.

15. $8 - 6\frac{4}{7}$ **16.** $13\frac{1}{9} - 11\frac{2}{3}$ **17.** $10\frac{3}{4} - 6\frac{1}{2}$ **18.** $13 - 4\frac{2}{11}$

19. $15\frac{2}{5} - 12\frac{3}{4}$ **20.** $17\frac{5}{9} - 6\frac{1}{3}$ **21.** $18\frac{1}{4} - 14\frac{3}{8}$ **22.** $20\frac{1}{6} - 7\frac{4}{9}$

Evaluate each expression. Write the answer in simplest form.

23. $4\frac{2}{3} + 5\frac{1}{3} - 7\frac{1}{8}$ **24.** $12\frac{5}{9} - 6\frac{2}{3} + 1\frac{4}{9}$

25. $7\frac{4}{11} - 2\frac{8}{11} - \frac{10}{11}$ **26.** $8\frac{1}{3} - 5\frac{8}{9} + 8\frac{1}{2}$

Evaluate each expression for $a = 6\frac{2}{3}$, $b = 8\frac{1}{2}$, and $c = 1\frac{3}{4}$. Write the answer in simplest form.

27. $a - c$ **28.** $b - c$ **29.** $b - a$ **30.** $10 - b$

31. $b - (a + c)$ **32.** $c + (b - a)$ **33.** $(a + b) - c$ **34.** $(10 - c) - a$

35. ***ECONOMICS*** A single share of stock in a company cost $\$23\frac{2}{5}$ on Monday. By Tuesday, the cost of a share in the company had fallen to $\$19\frac{1}{5}$. By how much did the price of a share fall?

36. Octavio used a brand new 6-hour tape to record some television shows. He recorded a movie that is $1\frac{1}{2}$ hours long and a cooking show that is $1\frac{1}{4}$ hours long. How much time is left on the tape?

Use the table for Exercises 37–40.

37. Gustavo is working at a gift wrap center. He has 2 yd^2 of wrapping paper to wrap a small box. How much wrapping paper will be left after he wraps the gift?

38. Gustavo must now wrap two extra-large boxes. If he has 6 yd^2 of wrapping paper, how much more wrapping paper will he need to wrap the two gifts?

39. To wrap a large box, Gustavo used $\frac{3}{4}$ yd^2 less wrapping paper than the amount listed in the table. How many square yards did he use to wrap the gift?

Gustavo's Gift Wrap Table

Gift Size	Paper Needed (yd^2)
Small	$\frac{11}{12}$
Medium	$1\frac{5}{9}$
Large	$2\frac{2}{3}$
X-large	$3\frac{1}{9}$

40. ***WHAT'S THE ERROR?*** Gustavo calculated the difference between the amount needed to wrap an extra-large box and the amount needed to wrap a medium box to be $2\frac{4}{9}$ yd^2. Explain his error and find the correct answer.

41. ***WRITE ABOUT IT*** Explain why you write equivalent fractions before you rename them. Explain why you do not rename them first.

42. ***CHALLENGE*** Fill in the box with a mixed number that makes the inequality true.

$$12\frac{1}{2} - 8\frac{3}{4} > 10 - \blacksquare$$

Spiral Review

Solve. (Lesson 3-9)

43. Gina made 4 dozen enchiladas for a dinner party. If she made 3 enchiladas for each guest, how many guests were at Gina's party?

Find each product. Write your answers in simplest form. (Lesson 5-1)

44. $\frac{2}{3} \times \frac{1}{5}$ 45. $\frac{3}{7} \times \frac{1}{4}$ 46. $\frac{2}{9} \times \frac{3}{8}$ 47. $\frac{1}{4} \times \frac{6}{7}$

48. **EOG PREP** Which is the correct value of $\frac{3}{4}$ of $\frac{3}{4}$? (Lesson 5-1)

A 1 B $\frac{9}{16}$ C $\frac{1}{3}$ D 3

49. **EOG PREP** Solve $\frac{3}{4}x = 9$. (Lesson 5-4)

A 6 B 8 C 10 D 12

5-10 Solving Fraction Equations: Addition and Subtraction

Learn to solve equations by adding and subtracting fractions.

Sugarcane is the main source of the sugar we use to sweeten our foods. It grows in tropical areas, such as Costa Rica and Haiti.

In one year, the average person in Costa Rica consumes $24\frac{1}{4}$ lb less sugar than the average person in the United States consumes.

This painting depicts the landscape of Haiti, a tropical area where sugarcane grows.

EXAMPLE 1 Solving Equations by Adding and Subtracting

Solve each equation. Write the solution in simplest form.

A $x + 6\frac{2}{3} = 11$

$$x + 6\frac{2}{3} = 11$$

$$-6\frac{2}{3} \quad -6\frac{2}{3}$$ *Subtract $6\frac{2}{3}$ from both sides to undo the addition.*

$$x = 10\frac{3}{3} - 6\frac{2}{3}$$ *Rename 11 as $10\frac{3}{3}$.*

$$x = 4\frac{1}{3}$$ *Subtract.*

B $2\frac{1}{4} = x - 3\frac{1}{2}$

$$2\frac{1}{4} = x - 3\frac{1}{2}$$

$$+3\frac{1}{2} \quad +3\frac{1}{2}$$ *Add $3\frac{1}{2}$ to both sides to undo the subtraction.*

$$2\frac{1}{4} + 3\frac{2}{4} = x$$ *Find a common denominator.*

$$5\frac{3}{4} = x$$ *Add.*

C $5\frac{3}{5} = m + \frac{7}{10}$

$$5\frac{3}{5} = m + \frac{7}{10}$$

$$-\frac{7}{10} \quad -\frac{7}{10}$$ *Subtract $\frac{7}{10}$ from both sides to undo the addition.*

$$5\frac{6}{10} - \frac{7}{10} = m$$ *Find a common denominator.*

$$4\frac{16}{10} - \frac{7}{10} = m$$ *Rename $5\frac{6}{10}$ as $4\frac{10}{10} + \frac{6}{10}$.*

$$4\frac{9}{10} = m$$ *Subtract.*

Solve each equation. Write the solution in simplest form.

D $w - \frac{1}{2} = 2\frac{3}{4}$

$$\begin{aligned} w - \tfrac{1}{2} &= 2\tfrac{3}{4} \\ +\tfrac{1}{2} \quad & \quad +\tfrac{1}{2} \end{aligned}$$

Add $\frac{1}{2}$ to both sides to undo the subtraction.

$w = 2\frac{3}{4} + \frac{1}{2}$

$w = 2\frac{3}{4} + \frac{2}{4}$ *Find a common denominator.*

$w = 2\frac{5}{4}$ *Add.*

$w = 3\frac{1}{4}$ *$2\frac{5}{4} = 2 + 1\frac{1}{4}$*

EXAMPLE 2 *Social Studies Application*

On average, a person in Costa Rica consumes $132\frac{1}{4}$ lb of sugar per year. If the average person in Costa Rica consumes $24\frac{1}{4}$ lb less than the average person in the U.S., what is the average sugar consumption per year by a person in the U.S.?

$u - 24\frac{1}{4} = 132\frac{1}{4}$ *Let u represent the average amount of sugar consumed in the U.S.*

$+24\frac{1}{4} \quad +24\frac{1}{4}$ *Add $24\frac{1}{4}$ to both sides to undo the subtraction.*

$u = 156\frac{2}{4} = 156\frac{1}{2}$ *Simplify.*

Check

$u - 24\frac{1}{4} = 132\frac{1}{4}$

$156\frac{1}{2} - 24\frac{1}{4} \stackrel{?}{=} 132\frac{1}{4}$ *Substitute $156\frac{1}{2}$ for u.*

$156\frac{2}{4} - 24\frac{1}{4} \stackrel{?}{=} 132\frac{1}{4}$ *Find a common denominator.*

$132\frac{1}{4} \stackrel{?}{=} 132\frac{1}{4}$ ✔ *$156\frac{1}{2}$ is the solution.*

On average, a person in the U.S. consumes $156\frac{1}{2}$ lb of sugar per year.

Think and Discuss

1. **Explain** how renaming a mixed number when subtracting is similar to regrouping when subtracting whole numbers.
2. **Give an example** of an addition equation with a solution that is a fraction between 3 and 4.

5-10 Exercises

FOR EOG PRACTICE

see page 657

internet connect

Homework Help Online
go.hrw.com Keyword: MR4 5-10

5.03

GUIDED PRACTICE

See Example 1

Solve each equation. Write the solution in simplest form.

1. $x + 2\frac{1}{2} = 7$
2. $3\frac{1}{3} = x - 5\frac{1}{9}$
3. $9\frac{3}{4} = x + 4\frac{1}{8}$
4. $x + 1\frac{1}{5} = 5\frac{3}{10}$

See Example 2

5. A tailor increased the length of a robe by $2\frac{1}{4}$ inches. The new length of the robe is 60 inches. What was the original length?

INDEPENDENT PRACTICE

See Example 1

Solve each equation. Write the solution in simplest form.

6. $x - 4\frac{3}{4} = 1\frac{1}{12}$
7. $x + 5\frac{3}{8} = 9$
8. $3\frac{1}{2} = 1\frac{3}{10} + x$
9. $4\frac{2}{3} = x - \frac{1}{6}$

See Example 2

10. Robert is taking a movie-making class in school. He edited his short video and cut $3\frac{2}{5}$ minutes. The new length of the video is $12\frac{1}{10}$ minutes. How long was his video before he cut it?

PRACTICE AND PROBLEM SOLVING

Find the solution to each equation. Check your answers.

11. $y + 8\frac{2}{4} = 10$
12. $p - 1\frac{2}{5} = 3\frac{7}{10}$
13. $6\frac{2}{3} + n = 7\frac{5}{6}$
14. $5\frac{3}{5} = s - 2\frac{3}{10}$
15. $k - 8\frac{1}{4} = 2\frac{2}{3} - 1\frac{1}{3}$
16. $\frac{5}{6} + \frac{1}{8} = c + \frac{5}{8}$
17. $m + 4 = 6\frac{3}{8} - 1\frac{1}{4}$
18. $12\frac{1}{6} - 10\frac{1}{9} + 2\frac{2}{3} = y$
19. $3\frac{2}{9} - 1\frac{1}{3} = p - 5\frac{1}{2}$
20. $q - 4\frac{1}{4} = 1\frac{1}{6} + 1\frac{1}{2}$
21. $h = 9\frac{3}{11} - 6\frac{2}{3} + 2\frac{1}{11}$
22. $a + 5\frac{1}{4} + 2\frac{1}{2} = 13\frac{1}{6}$
23. $6\frac{2}{9} = n - 2\frac{3}{8} - 1\frac{1}{9}$
24. $d - 20\frac{1}{4} + 2\frac{1}{10} = 12\frac{3}{10}$
25. $4\frac{1}{2} + \frac{1}{5} = z - 5\frac{1}{5}$
26. $11\frac{2}{7} = w + 3\frac{1}{2} - 1\frac{1}{7}$
27. $4\frac{1}{8} + 2\frac{3}{4} + 5\frac{1}{2} = r$
28. $9 - 5\frac{7}{8} = x - 1\frac{1}{8}$
29. The difference between Cristina's and Erin's heights is $\frac{1}{2}$ foot. Erin's height is $4\frac{1}{4}$ feet, and she is shorter than Cristina. How tall is Cristina?
30. ***MEASUREMENT*** Lori used $2\frac{5}{8}$ ounces of shampoo to wash her dog. When she was finished, the bottle contained $13\frac{3}{8}$ ounces of shampoo. How many ounces of shampoo were in the bottle before Lori washed her dog?

31. ***SPORTS*** Jack decreased his best time in the 400-meter race by $1\frac{3}{10}$ seconds. His new best time is $52\frac{3}{5}$ seconds. What was Jack's old time in the 400-meter race?

32. ***CRAFTS*** Juan makes bracelets to sell at his mother's gift shop. He alternates between green and blue beads.

What is the length of the green bead?

$\frac{11}{16}$ in.

$\frac{5}{16}$ in.

33. ***MUSIC*** A string quartet is performing Antonio Vivaldi's *The Four Seasons.* The concert is scheduled to last 45 minutes.

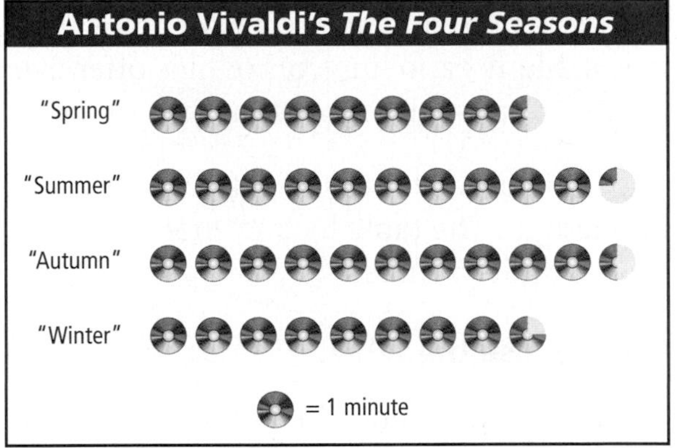

   a. After playing "Spring," "Summer," and "Autumn," how much time will be left in the concert?

   b. Is the concert long enough to play the four movements and another piece that is $6\frac{1}{2}$ minutes long? Explain.

34. ***WRITE A PROBLEM*** Use the pictograph to write a subtraction problem with two mixed numbers.

35. ***CHOOSE A STRATEGY*** How can you draw a line that is 5 inches long using only one sheet of $8\frac{1}{2}$ in. × 11 in. notebook paper?

36. ***WRITE ABOUT IT*** Explain how you know whether to add a number to or subtract a number from both sides of an equation in order to solve the equation.

37. ***CHALLENGE*** Use the numbers 1, 2, 3, 4, 5, and 6 to write a subtraction problem with two mixed numbers that have a difference of $4\frac{13}{20}$.

Spiral Review

Order the numbers from least to greatest. (Lesson 1-1)

38. 1,497; 2,560; 1,038 **39.** 10,462; 9,198; 11,320 **40.** 4,706; 11,765; 1,765

41. **EOG PREP** Which expression contains a variable? (Lesson 2-1)

A $19p$ B $\frac{1}{4}$ C $0.25\overline{5}$ D 8^2

42. **EOG PREP** What is the standard form of 10 + 4 + 0.2 + 0.06 + 0.003? (Lesson 3-1)

A 14.263 B 14,263 C 142.63 D 1.4263

Problem Solving on Location

NORTH CAROLINA

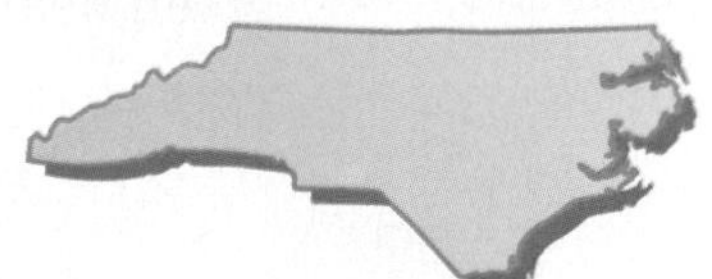

Canoeing

Canoeing is a popular sport that allows you to combine exercise with nature. While canoeing, you can enjoy fishing, bird watching, and scenic views. Many canoeing companies offer overnight canoe trips for those who want to spend more than just one day on the water.

North Carolina has many canoe trails that range in difficulty and length. The table lists several of these trails, along with their lengths in miles.

Canoe Trails	
Trail Name	**Length (mi)**
Ararat River	$8\frac{4}{5}$
Big Swamp	$6\frac{1}{5}$
Contentnea Creek	$7\frac{1}{2}$
Cape Fear River	$15\frac{2}{5}$
Deep River	$12\frac{7}{10}$
Eno River	$1\frac{7}{10}$
Fishing Creek	$9\frac{4}{5}$
Little River	$2\frac{7}{10}$
Trent River	$15\frac{1}{2}$

For 1–5, use the table.

1. Find the difference in length between the Cape Fear River Trail and the Big Swamp Trail.
2. Find the difference in length between the Ararat River Trail and the Contentnea Creek Trail.
3. If Derrick canoed the Eno River Trail and then the Deep River Trail, how much distance did he cover?
4. If Gwen canoes the Little River Trail once each week for five weeks, how many total miles will she canoe?
5. Suppose it takes Karrie 30 minutes to canoe 1 mile. How many hours would it take her to canoe the Trent River Trail?

Hiking Trails

North Carolina hiking trails wind through open meadows, over mountain peaks, and past cascading waterfalls. Along the trails are many places to view wildlife that includes white-tailed deer, raccoons, bobcats, and red and gray foxes. In the late spring, hikers also see many varieties of plant life. The trails range from 0.3 mile to over 30 miles in length, offering many options to hikers of different abilities.

The table lists some of North Carolina's hiking trails, their approximate lengths in miles, and their levels of difficulty.

For 1–5, use the table.

1. Luis hiked the Fox Hunters Paradise Trail, and Susan hiked the Cumberland Knob Trail. Who hiked farther? By how much?
2. Jeremy hiked a total of $12\frac{3}{5}$ miles on the Cedar Ridge Trail. How many times did he hike the trail?
3. If you were to hike the Crabtree Falls Loop Trail three times, how many total miles would you hike?
4. Angela is planning to hike the Fox Hunters Paradise Trail, the Bluff Mountain Trail, and then the Linville Gorge Trail. How many total miles does she plan to cover?
5. Suppose that Jessica and John walk at about the same speed. Jessica hikes the Linville Gorge Trail, while John hikes the Cumberland Knob Trail. Based on the levels of difficulty, who do you think will finish first? Explain.

North Carolina Hiking Trails

Trail Name	Length (mi)	Difficulty
Cumberland Knob Trail	$\frac{1}{2}$	Easy
Fox Hunters Paradise Trail	$\frac{1}{5}$	Easy
Bluff Mountain Trail	$7\frac{1}{2}$	Moderate
Cedar Ridge Trail	$4\frac{1}{5}$	Moderate
Crabtree Falls Loop Trail	$2\frac{3}{5}$	Strenuous
Linville Gorge Trail	$\frac{1}{2}$	Strenuous

MATH-ABLES

Fraction Riddles

1. What is the value of one-half of two-thirds of three-fourths of four-fifths of five-sixths of six-sevenths of seven-eighths of eight-ninths of nine-tenths of one thousand?

2. What is the next fraction in the sequence below?

   $\frac{1}{12}, \frac{1}{6}, \frac{1}{4}, \frac{1}{3}, \ldots$

3. I am a three-digit number. My hundreds digit is one-third of my tens digit. My tens digit is one-third of my ones digit. What number am I?

4. A *splorg* costs three-fourths of a dollar plus three-fourths of a *splorg*. How much does a *splorg* cost?

5. How many cubic inches of dirt are in a hole that measures $\frac{1}{3}$ feet by $\frac{1}{4}$ feet by $\frac{1}{2}$ feet?

Fraction Bingo

The object is to be the first player to cover five squares in a row horizontally, vertically, or diagonally.

One person is the caller. On each of the caller's cards, there is an expression containing fractions. When the caller draws a card, he or she reads the expression aloud for the players.

The players must find the value of the expression. If a square on the player's card has that value or a fraction equivalent to that value, they cover the square.

The first player to cover five squares in a row is the winner. Take turns being the caller. A variation can be played in which the winner is the first person to cover all their squares.

internet connect

Go to ***go.hrw.com*** to print out cards for Fraction Bingo.
KEYWORD: MR4 Game5

go.hrw.com

Fraction Operations

To enter a fraction, like $\frac{1}{8}$, into a graphing calculator, use ÷. When you press ENTER, the fraction is converted to a decimal.
To convert a decimal number, such as 0.625, to a fraction, enter the number on your calculator. Press MATH, and then press ENTER to select Frac.

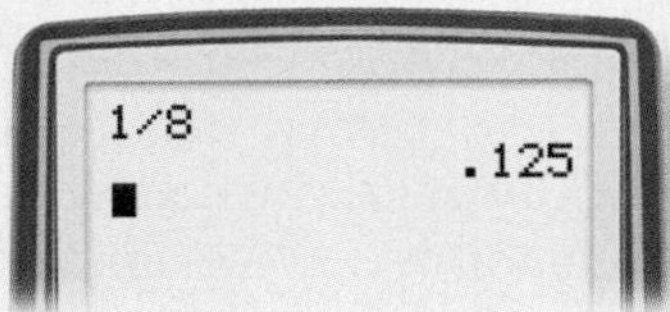

When you press ENTER again, the calculator displays the fraction equivalent.

The decimal 0.625 is equivalent to $\frac{5}{8}$.

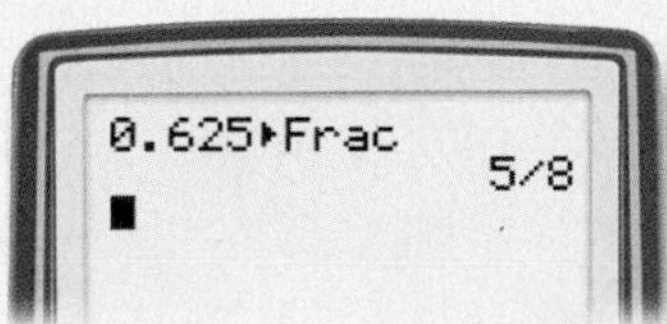

Activity

1. Use your calculator to find the value of $\frac{2}{5} + \frac{3}{7}$. Write the sum as a fraction.

Step 1: To find the sum, enter $\frac{2}{5} + \frac{3}{7}$, and then press ENTER. The calculator displays the sum as a decimal.

Step 2: To convert the decimal to a fraction, press MATH, press ENTER to select Frac, and then press ENTER again.

$$\frac{2}{5} + \frac{3}{7} = \frac{29}{35}$$

Think and Discuss

1. Without using a calculator, write 0.10001 as a fraction. Then use a graphing calculator to convert 0.10001 to a fraction. What happens? Why do you think this happened?

Try This

Add or subtract. Write the answer as a fraction or a mixed number and as a decimal rounded to the nearest hundredth.

1. $\frac{1}{4} + \frac{2}{3}$ **2.** $\frac{2}{3} - \frac{1}{4}$ **3.** $\frac{3}{5} + \frac{5}{3}$ **4.** $\frac{4}{3} - \frac{3}{4}$

Chapter 5

Study Guide and Review

Vocabulary

Complete the sentences below with vocabulary words from the list above. Words may be used more than once.

1. Two numbers are ___?___ if their product is 1.
2. The ___?___ is the smallest number that is a common multiple of two or more denominators.

5-1 Multiplying Fractions (pp. 212–215)

EXAMPLE

- **Multiply. Write the answer in simplest form.**

$\frac{3}{4} \cdot \frac{1}{3}$ *Multiply. Then simplify.*

$\frac{3 \cdot 1}{4 \cdot 3} = \frac{3}{12} = \frac{1}{4}$

EXERCISES

Multiply. Write each answer in simplest form.

3. $\frac{5}{6} \cdot \frac{2}{5}$
4. $\frac{5}{7} \cdot \frac{3}{4}$
5. $\frac{4}{5} \cdot \frac{1}{8}$
6. $\frac{7}{10} \cdot \frac{2}{5}$
7. $\frac{1}{9} \cdot \frac{5}{9}$
8. $\frac{1}{4} \cdot \frac{6}{7}$

5-2 Multiplying Mixed Numbers (pp. 216–219)

EXAMPLE

- **Multiply. Write the answer in simplest form.**

$\frac{2}{5} \cdot 1\frac{2}{3} = \frac{2}{5} \cdot \frac{5}{3} = \frac{10}{15} = \frac{2}{3}$

EXERCISES

Multiply. Write each answer in simplest form.

9. $\frac{2}{5} \cdot 2\frac{1}{4}$
10. $\frac{3}{4} \cdot 1\frac{2}{3}$
11. $3\frac{1}{3} \cdot \frac{3}{5}$

5-3 Dividing Fractions and Mixed Numbers (pp. 222–225)

EXAMPLE

- **Divide. Write the answer in simplest form.**

$\frac{3}{4} \div 6 = \frac{3 \cdot 1}{4 \cdot 6} = \frac{3}{24} = \frac{1}{8}$

EXERCISES

Divide. Write each answer in simplest form.

12. $\frac{4}{7} \div 3$
13. $\frac{3}{10} \div 2$
14. $1\frac{1}{3} \div 2\frac{2}{5}$

5-4 Solving Fraction Equations: Multiplication and Division (pp. 226–229)

EXAMPLE

- Solve the equation.

$\frac{4}{5}n = 12$

$\frac{4}{5}n \div \frac{4}{5} = 12 \div \frac{4}{5}$ *Divide both sides by $\frac{4}{5}$.*

$\frac{4}{5}n \cdot \frac{5}{4} = 12 \cdot \frac{5}{4}$ *Multiply by the reciprocal.*

$n = \frac{60}{4} = 15$

EXERCISES

Solve each equation.

15. $4a = \frac{1}{2}$

16. $\frac{3b}{4} = 1\frac{1}{2}$

17. $\frac{2m}{7} = 5$

18. $6g = \frac{4}{5}$

19. $\frac{5}{6}r = 9$

20. $\frac{s}{8} = 6\frac{1}{4}$

5-5 Least Common Multiple (pp. 232–235)

EXAMPLE

- Find the least common multiple (LCM) of 4, 6, and 8.

4: 4, 8, 12, 16, 20, **24**, 28, …
6: 6, 12, 18, **24**, 30, …
8: 8, 16, **24**, 32, …
LCM: 24

EXERCISES

Find the least common multiple (LCM).

21. 3, 5, and 10

22. 6, 8, and 16

23. 3, 9, and 27

24. 4, 12, and 30

25. 25 and 45

26. 12, 22, and 30

5-6 Estimating Fraction Sums and Differences (pp. 236–239)

EXAMPLE

- Estimate the sum or difference by rounding fractions to 0, $\frac{1}{2}$, or 1.

$\frac{7}{8} + \frac{1}{7}$ *Think: 1 + 0.*

$\frac{7}{8} + \frac{1}{7}$ is about 1.

EXERCISES

Estimate each sum or difference by rounding fractions to 0, $\frac{1}{2}$, or 1.

27. $\frac{3}{5} + \frac{3}{7}$

28. $\frac{6}{7} - \frac{5}{9}$

29. $4\frac{9}{10} + 6\frac{1}{5}$

30. $7\frac{5}{11} - 4\frac{3}{4}$

5-7 Adding and Subtracting with Unlike Denominators (pp. 242–245)

EXAMPLE

- $\frac{7}{9} + \frac{2}{3}$

$\frac{7}{9} + \frac{2}{3}$ *Write equivalent fractions. Add.*

$\frac{7}{9} + \frac{6}{9} = \frac{13}{9} = 1\frac{4}{9}$

EXERCISES

Add or subtract. Write each answer in simplest form.

31. $\frac{1}{5} + \frac{5}{8}$

32. $\frac{1}{6} + \frac{7}{12}$

33. $\frac{13}{15} - \frac{4}{5}$

34. $\frac{7}{8} - \frac{2}{3}$

5-8 Adding and Subtracting Mixed Numbers (pp. 246–249)

EXAMPLE

■ **Find the difference. Write the answer in simplest form.**

$5\frac{5}{8} - 3\frac{1}{6}$

$5\frac{15}{24} - 3\frac{4}{24}$ *Write equivalent fractions.*

$2\frac{11}{24}$ *Subtract.*

EXERCISES

Find each sum or difference. Write the answer in simplest form.

35. $1\frac{3}{10} + 3\frac{2}{5}$

36. $4\frac{5}{9} - 1\frac{1}{2}$

37. $5\frac{5}{12} + 6\frac{3}{10}$

38. $2\frac{1}{4} + 1\frac{5}{6}$

39. $2\frac{9}{10} - 1\frac{1}{4}$

40. $6\frac{3}{4} - 4\frac{3}{8}$

41. Angela had $\frac{7}{10}$ gallon of paint. She used $\frac{1}{3}$ gallon for a project. How much paint did she have left?

5-9 Renaming to Subtract Mixed Numbers (pp. 252–255)

EXAMPLE

■ **Subtract.**

$4\frac{7}{10} - 2\frac{9}{10}$

$3\frac{17}{10} - 2\frac{9}{10}$ *Rename $4\frac{7}{10}$. Subtract.*

$1\frac{8}{10}$

$1\frac{4}{5}$

EXERCISES

Subtract. Write each answer in simplest form.

42. $7\frac{2}{9} - 3\frac{5}{9}$

43. $3\frac{1}{5} - 1\frac{7}{10}$

44. $8\frac{7}{12} - 2\frac{11}{12}$

45. $5\frac{3}{8} - 2\frac{3}{4}$

46. $11\frac{6}{7} - 4\frac{13}{14}$

47. $10 - 8\frac{7}{8}$

48. Georgette needs 8 feet of ribbon to decorate gifts. She has $3\frac{1}{4}$ feet of ribbon. How many more feet of ribbon does Georgette need?

5-10 Solving Fraction Equations: Addition and Subtraction (pp. 256–259)

EXAMPLE

■ **Solve $n + 2\frac{5}{7} = 8$.**

$n + 2\frac{5}{7} - 2\frac{5}{7} = 8 - 2\frac{5}{7}$

$n = 8 - 2\frac{5}{7}$

$n = 7\frac{7}{7} - 2\frac{5}{7}$

$n = 5\frac{2}{7}$

EXERCISES

Solve each equation. Write the solution in simplest form.

49. $x - 12\frac{3}{4} = 17\frac{2}{5}$

50. $t + 6\frac{11}{12} = 21\frac{5}{6}$

51. $3\frac{2}{3} = m - 1\frac{3}{4}$

52. $5\frac{2}{3} = p + 2\frac{2}{9}$

53. $y - 1\frac{2}{3} = 3\frac{4}{5}$

54. Jon poured $1\frac{1}{2}$ oz of lemon juice onto a salad. He has $5\frac{1}{2}$ oz lemon juice left in the bottle. How many ounces of lemon juice were in the bottle before Jon poured some on the salad?

Chapter 5

Chapter Test

Find the reciprocal.

1. $\frac{3}{5}$
2. $\frac{7}{11}$
3. $\frac{5}{9}$
4. $\frac{1}{8}$

Find the least common multiple (LCM).

5. 10 and 15
6. 4, 6, and 18
7. 9, 10, and 12
8. 6, 15, and 20

Estimate each sum or difference by rounding to 0, $\frac{1}{2}$, or 1.

9. $\frac{1}{8} + \frac{4}{7}$
10. $\frac{11}{12} - \frac{4}{9}$
11. $\frac{4}{5} + \frac{1}{9}$

Evaluate each expression. Write the answer in simplest form.

12. $4\frac{1}{9} - 2\frac{4}{9}$
13. $\frac{2}{5} \cdot \frac{5}{6}$
14. $2\frac{1}{3} \div \frac{5}{6}$
15. $1\frac{7}{10} + 3\frac{3}{4}$
16. $\frac{3}{7} \cdot \frac{4}{9}$
17. $\frac{2}{3} - \frac{3}{8}$
18. $1\frac{3}{8} \cdot \frac{6}{11}$
19. $3\frac{1}{8} \div 1\frac{1}{4}$
20. $\frac{7}{8} \div 2$
21. $\frac{3}{8} \cdot \frac{3}{4}$
22. $3\frac{1}{3} \div 1\frac{5}{12}$
23. $4 \cdot 2\frac{2}{7}$
24. $2\frac{1}{4} \cdot 2\frac{2}{3}$
25. $\frac{1}{12} + \frac{5}{6}$
26. $\frac{4}{5} \cdot 1\frac{1}{3}$
27. $\frac{3}{8} \cdot \frac{2}{3}$

Evaluate the expression $n \cdot \frac{1}{4}$ for each value of n. Write the answer in simplest form.

28. $n = \frac{7}{8}$
29. $n = \frac{2}{5}$
30. $n = \frac{8}{9}$
31. $n = \frac{4}{11}$

Solve each equation. Write the solution in simplest form.

32. $3r = \frac{9}{10}$
33. $n + 3\frac{1}{6} = 12$
34. $5\frac{5}{6} = x - 3\frac{1}{4}$
35. $\frac{2}{5}t = 9$
36. $\frac{4}{5}m = 7$
37. $y - 15\frac{3}{5} = 2\frac{1}{3}$
38. Jessica purchased a bag of cat food. She feeds her cat 1 cup of cat food each day. After 7 days, she has fed her cat $\frac{2}{3}$ of the food in the bag. How many cups of food were in the bag of cat food when Jessica bought it?
39. On Saturday, Cecelia ran $3\frac{3}{7}$ miles. On Sunday, she ran $4\frac{5}{6}$ miles. About how much farther did Cecelia run on Sunday than on Saturday?
40. Michael studied social studies for $\frac{3}{4}$ of an hour, Spanish for $1\frac{1}{2}$ hours, and math for $1\frac{1}{4}$ hours. How many hours did Michael spend studying all three subjects?

Chapter 5 Performance Assessment

Show What You Know

Create a portfolio of your best work from this chapter. Complete this page and include it with your four best pieces of work from Chapter 5. Choose from your homework or lab assignments, mid-chapter quiz, or any journal entries you have done. Put them together using any design you want. Make your portfolio represent what you consider your best work.

Short Response

1. Use prime factorization to find the least common multiple of 7, 12, and 15. Show your work.

2. Daphne will distribute cereal samples with pamphlets about good nutrition. The samples come in packages of 15. The pamphlets come in packages of 20. What is the least number of cereal samples and pamphlets that Daphne can get to have the same number of each? How many packages of each will she have? Show your work.

3. Estimate the sum of $\frac{4}{5}$ and $\frac{9}{10}$. Is your answer an overestimate or an underestimate? Explain.

CLOSE TO HOME JOHN McPHERSON

Bobby's excitement about going to summer camp faded as soon as he read the sign.

Extended Problem Solving

4. During the summer, Garrett attends a day camp for 6 hours each day. The circle graph shows what fraction of each day he spends doing different activities.
   a. How long does Garrett spend doing each activity? Write the activities in order from longest to shortest.
   b. Sports activities and playground games are all held on the camp fields. What fraction of the day does Garrett spend on the fields? Write your answer in simplest form.
   c. Lunch and crafts are held in the cafeteria. How many hours does Garrett spend in the cafeteria during a 5-day week at day camp? Write your answer in simplest form, and show the work necessary to determine the correct answer.

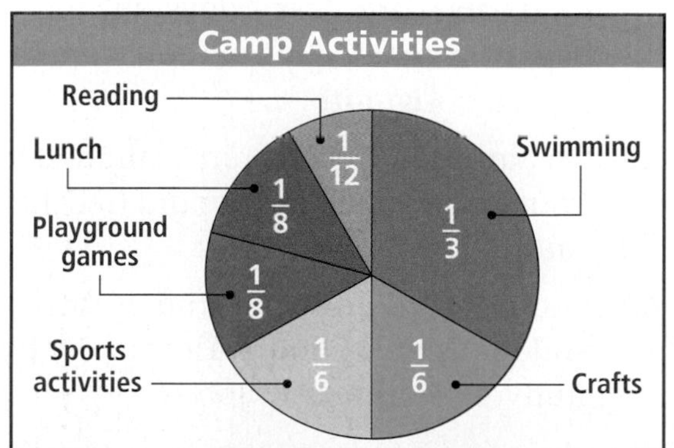

internet connect

EOG Test Practice Online
go.hrw.com Keyword: MR4 TestPrep

Getting Ready for EOG

Chapter 5

Cumulative Assessment, Chapters 1–5

1. Which number is *greater* than $\frac{4}{5}$?

A $\frac{1}{2}$ C $\frac{3}{4}$

B $\frac{5}{6}$ D $\frac{2}{3}$

2. Which number is divisible by 2, 3, 6, and 9 but *not* by 4, 5, or 10?

A 882 C 684

B 768 D 180

3. Which is the standard form of thirty-one and twenty-two thousandths?

A 31.22 C 31.022

B 31,022 D 31,022,000

TEST TAKING TIP!

To multiply mixed numbers, write the mixed number as an improper fraction. Multiply the numerators, and then multiply the denominators.

4. Find the product. $2\frac{2}{3} \cdot 3\frac{1}{2}$

A $6\frac{1}{3}$ C $\frac{1}{3}$

B 2 D $9\frac{1}{3}$

5. For which equation is $n = 5$ a solution?

A $7n = 25 + 3$

B $\frac{n}{8} = 5$

C $47 - n = (8 \times 5)$

D $\frac{n}{6} \cdot \frac{1}{3} = \frac{5}{18}$

6. What is the LCM of 4, 7, and 14?

A 2 C 28

B 14 D 56

7. Which is a common denominator of $\frac{1}{4}$, $\frac{5}{6}$, and $\frac{3}{8}$?

A 2 C 15

B 12 D 24

8. What is the distance around the rectangular picture frame shown?

A $5\frac{1}{4}$ in. C $2\frac{5}{8}$ in.

B 4 in. D $3\frac{7}{8}$ in.

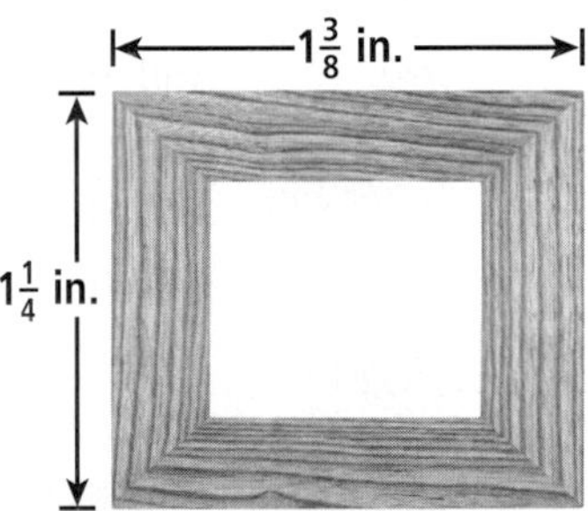

9. ***SHORT RESPONSE*** Lucy has a total of $\frac{5}{6}$ yard of ribbon to wrap gifts for her friends. The bow on each gift requires $\frac{1}{6}$ yard of ribbon. How can you determine how many bows b Lucy can make? Find the correct value for b.

10. ***SHORT RESPONSE*** Mr. Frost is painting his garage, which has a surface area of 500 square feet. He has $1\frac{3}{4}$ gallons of paint. If one gallon of paint covers 150 square feet, how much more paint must Mr. Frost buy? Show your work.

Chapter 6

Collect and Display Data

TABLE OF CONTENTS

Chapter Opener Online
go.hrw.com
KEYWORD: MR4 Ch6

National Park	Average High Temperatures (°C)		
	Jun	Jul	Aug
Badlands, SD	27	33	32
Big Bend, TX	32	31	31
Crater Lake, OR	19	25	24
Everglades, FL	31	32	32

Career *Meteorologist*

Weather affects our daily activities, and weather information is useful and often necessary. Businesses such as farms, ski resorts, and airlines need to know weather conditions.

This information comes from meteorologists, people who study and forecast the weather. They gather data such as temperature, wind speed, and rainfall. They then study this data and make predictions.

The table lists the average daily high temperatures during the summer in some popular national parks.

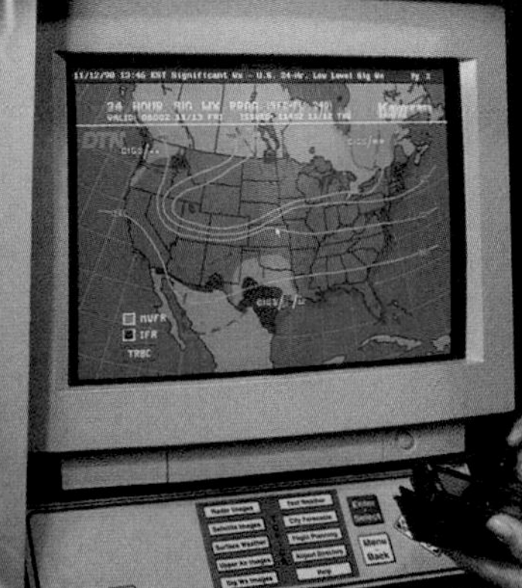

ARE YOU READY?

Choose the best term from the list to complete each sentence.

1. The answer to an addition problem is called the ___?___.
2. The ___?___ of the 6 in 5,672 is hundreds.
3. When you move ___?___, you move left or right.
   When you move ___?___, you move up or down.

place value
horizontally
vertically
sum
quotients

Complete these exercises to review skills you will need for this chapter.

✔ Place Value

Write the digit in the tens place of each number.

4. 718 **5.** 989 **6.** 55 **7.** 7,709

✔ Compare and Order Whole Numbers

Order the numbers from least to greatest.

8. 40, 32, 51, 78, 26, 43, 27

9. 132, 150, 218, 176, 166

10. 92, 91, 84, 92, 87, 90

11. 23, 19, 33, 27, 31, 31, 28, 18

Find the greatest number in each set.

12. 452, 426, 502, 467, 530, 512

13. 711, 765, 723, 778, 704, 781

14. 143, 122, 125, 137, 140, 118, 139

15. 1,053; 1,106; 1,043; 1,210; 1,039; 1,122

✔ Write Fractions as Decimals

Write each fraction as a decimal.

16. $\frac{1}{4}$ **17.** $\frac{5}{8}$ **18.** $\frac{1}{6}$ **19.** $\frac{2}{5}$

✔ Locate Points on a Number Line

Name the point on the number line that corresponds to each given value.

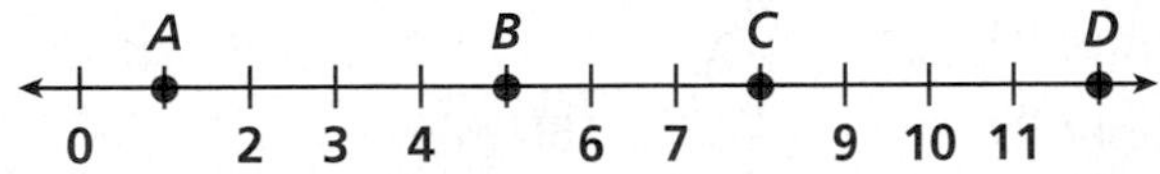

20. 5 **21.** 12 **22.** 8 **23.** 1

6-1 Make a Table

Problem Solving Strategy

Learn to use tables to record and organize data.

Weather forecasters collect data about weather. By organizing and interpreting this data, they can often warn people of severe weather before it happens. This advance warning can save lives.

This satellite image shows a hurricane approaching Florida's coastline.

One way to organize data is to make a table. By looking at a table, you may see patterns and relationships.

EXAMPLE 1 *Weather Application*

Use the data about Hurricane Mitch to make a table. Then use your table to describe how the hurricane's strength changed over time.

On October 24, 1998, Hurricane Mitch's wind speed was 90 mi/h. On October 26, its wind speed was 130 mi/h. On October 27, its wind speed was 150 mi/h. On October 31, its wind speed was 40 mi/h. On November 1, its wind speed was 30 mi/h.

Date (1998)	Wind Speed
October 24	90 mi/h
October 26	130 mi/h
October 27	150 mi/h
October 31	40 mi/h
November 1	30 mi/h

Make a table. Write the dates in order so that you can see how the hurricane's strength changed over time.

From the table, you can see that Hurricane Mitch became stronger from October 24 to October 27 and then weakened from October 27 to November 1.

Weather LINK

The National Weather Service estimated that Mitch's wind speed reached 180 mi/h. This made Mitch a Category 5 hurricane, which is the strongest type.

go.hrw.com
KEYWORD: MR4 Hurricane

EXAMPLE 2 **Organizing Data in a Table**

Use the temperature data to make a table. Then use your table to find a pattern in the data and draw a conclusion.

At 10 A.M., the temperature was 62°F. At noon, it was 65°F. At 2 P.M., it was 68°F. At 4 P.M., it was 70°F. At 6 P.M., it was 66°F.

Time	Temperature (°F)
10 A.M.	62
Noon	65
2 P.M.	68
4 P.M.	70
6 P.M.	66

The temperature rose until 4 P.M., and then it dropped. One conclusion is that the high temperature on this day was at least 70°F.

Think and Discuss

1. **Tell** how a table helps you organize data.
2. **Explain** why the data in Example 2 was arranged from earliest to latest time instead of from lowest to highest temperature.

6-1 Exercises

5.04

FOR EOG PRACTICE
see page 658

internet connect
Homework Help Online
go.hrw.com Keyword: MR4 6-1

GUIDED PRACTICE

See Example 1

1. On Monday, the high temperature was 72°F. On Tuesday, the high was 75°F. On Wednesday, the high was 68°F. On Thursday, the high was 62°F. On Friday, the high was 55°F. Use this data to make a table.

See Example 2

2. Use your table from Exercise 1 to find a pattern in the data and draw a conclusion.

INDEPENDENT PRACTICE

See Example 1

3. On his first math test, Joe made a grade of 70. On the second test, Joe made a grade of 75. On the third test, Joe made a grade of 80. On the fourth test, Joe made a grade of 85. On the fifth test, Joe made a grade of 90. Use this data to make a table.

See Example 2

4. Use your table from Exercise 3 to find a pattern in the data and draw a conclusion.

PRACTICE AND PROBLEM SOLVING

5. For ice-skating on a frozen pond to be safe, the ice should be at least 7 inches thick. Use the data below to make a table, and estimate the date on which it first became safe to ice-skate.

   On December 3, the ice was 1 in. thick. On December 18, the ice was 2 in. thick. On January 3, the ice was 5 in. thick. On January 18, the ice was 11 in. thick. On February 3, the ice was 17 in. thick.

6. ***WHAT'S THE ERROR?*** A student read this table about the populations of large cities and decided that Buenos Aires had the smallest population. Why might the student have made this mistake? Which city does have the smallest population?

City	Population
Buenos Aires	13,430,000
Calcutta	13,400,000
Jakarta	13,300,000
Mexico City	19,430,000
Mumbai	16,630,000

7. ***WRITE ABOUT IT*** The tables below were made using identical data that have been organized differently. When might each table be useful?

Time	Temperature (°F)
6 A.M.	55
10 A.M.	68
2 P.M.	75
6 P.M.	62
10 P.M.	58

Time	Temperature (°F)
2 P.M.	75
10 A.M.	68
6 P.M.	62
10 P.M.	58
6 A.M.	55

8. ***CHALLENGE*** Arthur, Victoria, and Jeffrey are in the sixth, seventh, and eighth grades, although not necessarily in that order. Victoria is not in eighth grade. The sixth-grader is in choir with Arthur and in band with Victoria. Which student is in which grade? Use a yes/no table like the one at right to help you answer this question.

	Arthur	Victoria	Jeffrey
6th			
7th			
8th		No	

Spiral Review

Write each number in scientific notation. (Lesson 3-5)

9. 5,234,000 10. 23 11. 12.078

Multiply. (Lesson 3-6)

12. $0.3 \cdot 0.1$ 13. $0.16 \cdot 0.5$ 14. $1.2 \cdot 0.2$

15. **EOG PREP** Find the quotient: $0.64 \div 8$. (Lesson 3-7)

A 80 B 8 C 0.8 D 0.08

6-2 Range, Mean, Median, and Mode

Learn **to find the range, mean, median, and mode of a data set.**

Vocabulary

range

mean

median

mode

Players on a volleyball team measured how high they could jump. The results in inches are recorded in the table.

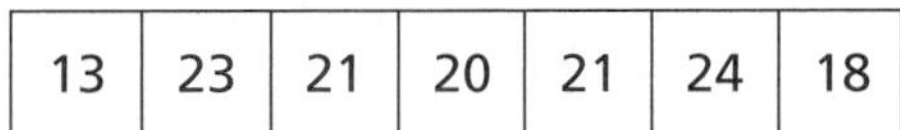

13	23	21	20	21	24	18

Some descriptions of a set of data are called the *range, mean, median,* and *mode.*

- The **range** is the difference between the least and greatest values in the set.
- The **mean** is the sum of all the items, divided by the number of items in the set. (The mean is sometimes called the *average.*)
- The **median** is the middle value when the data are in numerical order, or the mean of the two middle values if there are an even number of items.
- The **mode** is the value or values that occur most often. There may be more than one mode for a data set. When all values occur an equal number of times, the data set has no mode.

EXAMPLE 1 Finding the Range, Mean, Median, and Mode of a Data Set

Find the range, mean, median, and mode of each data set.

A

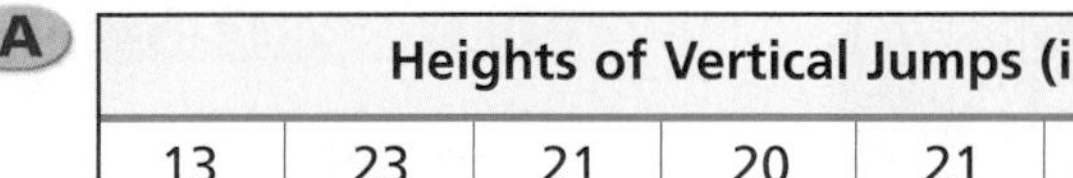

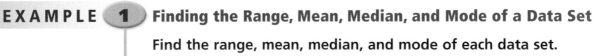

Heights of Vertical Jumps (in.)						
13	23	21	20	21	24	18

Start by writing the data in numerical order.

13, 18, 20, 21, 21, 23, 24

range: $24 - 13 = 11$ — *Subtract least value from greatest value.*

mean: $13 + 18 + 20 + 21 + 21 + 23 + 24 = 140$ — *Add all values.*

$140 \div 7 = 20$ — *Divide the sum by the number of items.*

median: 21 — *There are an odd number of items, so find the middle value.*

mode: 21 — *21 occurs most often.*

The range is 11 in.; the mean is 20 in.; the median is 21 in.; and the mode is 21 in.

Find the range, mean, median, and mode of each data set.

B

NFL Career Touchdowns			
Marcus Allen	145	Franco Harris	100
Jim Brown	126	Walter Payton	125

Write the data in numerical order: 100, 125, 126, 145

range: $145 - 100 = 45$

mean: $\frac{145 + 126 + 100 + 125}{4}$

$= 124$

median: 100, (125, 126), 145

$\frac{125 + 126}{2} = 125.5$

There are an even number of items, so find the mean of the two middle values.

mode: none

The range is 45 touchdowns; the mean is 124 touchdowns; the median is 125.5 touchdowns; and there is no mode.

Think and Discuss

1. **Describe** what you can say about the values in a data set if the set has a small range.
2. **Tell** how many modes are in the following data set. Explain your answer. 15, 12, 13, 15, 12, 11

6-2 Exercises

FOR EOG PRACTICE
see page 658

internet connect
Homework Help Online
go.hrw.com Keyword: MR4 6-2

See Example 1 Find the range, mean, median, and mode of the data set.

1.

Heights of Students (in.)	51	67	63	52	49	48	48

INDEPENDENT PRACTICE

See Example 1 Find the range, mean, median, and mode of each data set.

2.

Ages of Students (yr)	14	16	15	17	16	12

Find the range, mean, median, and mode of each data set.

3.

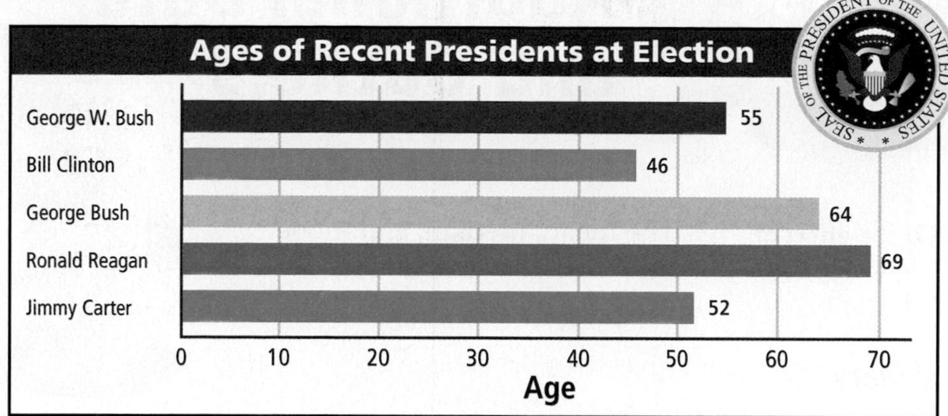

PRACTICE AND PROBLEM SOLVING

4. Frank has 3 nickels, 5 dimes, and 2 quarters. Find the range, mean, median, and mode of the values of Frank's coins.

5. ***EDUCATION*** For the six New England states, the mean scores on the math section of the SAT one year were as follows: Connecticut, 509; Maine, 500; Massachusetts, 513; New Hampshire, 519; Rhode Island, 500; and Vermont, 508. Create a table using this data. Then find the range, mean, median, and mode.

6. ***WHAT'S THE QUESTION?*** On an exam, three students scored 75, four students scored 82, three students scored 88, four students scored 93, and one student scored 99. If the answer is 88, what is the question?

7. ***CHALLENGE*** In the Super Bowls from 1997 to 2002, the winning team won by a mean of $12\frac{1}{6}$ points. By how many points did the Green Bay Packers win in 1997?

Year	Super Bowl Champion	Points Won By
2002	New England Patriots	3
2001	Baltimore Ravens	27
2000	St. Louis Rams	7
1999	Denver Broncos	15
1998	Denver Broncos	7
1997	Green Bay Packers	■

Spiral Review

Order the fractions from least to greatest. (Lesson 4-6)

8. $\frac{3}{7}, \frac{5}{4}, \frac{2}{6}$ 9. $\frac{2}{3}, \frac{4}{11}, \frac{5}{8}$ 10. $\frac{3}{10}, \frac{3}{8}, \frac{1}{3}$

Multiply. Write your answers in simplest form. (Lesson 5-1)

11. $\frac{3}{5} \cdot \frac{6}{7}$ 12. $\frac{2}{3} \cdot \frac{4}{5}$ 13. $\frac{7}{9} \cdot \frac{3}{4}$ 14. $\frac{1}{7} \cdot \frac{1}{2}$

15. **EOG PREP** Write the product in simplest form: $3\frac{6}{7} \cdot \frac{1}{3}$. (Lesson 5-2)

A $1\frac{27}{21}$ B $1\frac{2}{7}$ C $1\frac{5}{7}$ D $1\frac{3}{4}$

6-3 Additional Data and Outliers

USA's Jim Shea in Men's Skeleton at the 2002 Winter Olympics

Learn the effect of additional data and outliers.

The mean, median, and mode may change when you add data to a data set.

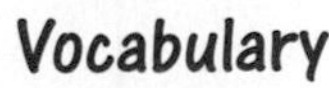

Vocabulary

outlier

EXAMPLE 1 *Sports Application*

A **Find the mean, median, and mode of the data in the table.**

U.S. Winter Olympic Medals Won								
Year	2002	1998	1994	1992	1988	1984	1980	1976
Medals	34	13	13	11	6	8	12	10

mean = 13.375 mode = 13 median = 11.5

B **The United States also won 8 medals in 1972 and 5 medals in 1968. Add this data to the data in the table and find the mean, median, and mode.**

mean = 12 modes = 8, 13 median = 10.5

The mean decreased by 1.375, there is an additional mode, and the median decreased by 1.

An **outlier** is a value in a set that is very different from the other values.

EXAMPLE 2 *Social Studies Application*

In 2001, 64-year-old Sherman Bull became the oldest person to reach the top of Mount Everest. Other climbers to reach the summit that day were 33, 31, 31, 32, 33, and 28 years old. Find the mean, median, and mode without and with Bull's age.

Data without Bull's age: mean ≈ 31.3 modes = 31, 33 median = 31.5

Data with Bull's age: mean = 36 modes = 31, 33 median = 32

When you add Bull's age, the mean increases by 4.7, the modes stay the same, and the median increases by 0.5. The mean is the most affected by the outlier—notice that it is greater than every age in the set except Bull's. The median is closer to most of the climbers' ages.

Helpful Hint

Sherman Bull's age is an outlier because he is much older than the others in the group.

Sometimes one or two data values can greatly affect the mean, median, or mode. When one of these values is affected like this, you should choose a different value to best describe the data set.

EXAMPLE 3 Describing a Data Set

The Seawells are shopping for a DVD player. They found ten DVD players with the following prices:

\$175, \$180, \$130, \$150, \$180, \$500, \$160, \$180, \$150, \$160

What are the mean, median, and mode of this data set? Which one best describes the data set?

mean = \$196.50 mode = \$180 median = \$167.50

The median price is the best description of the prices. Most of the DVD players cost *about* \$167.50.

The mean is higher than most of the prices because of the \$500 player, and the mode is higher because of the three players that cost \$180 each.

Some data sets do not contain numbers. For example, the circle graph shows the results of a survey to find people's favorite color.

When it does not contain numbers, the only way to describe the data set is with the mode. You cannot find a mean or a median for a set of colors.

The mode for this data set is blue. Most people in this survey chose blue as their favorite color.

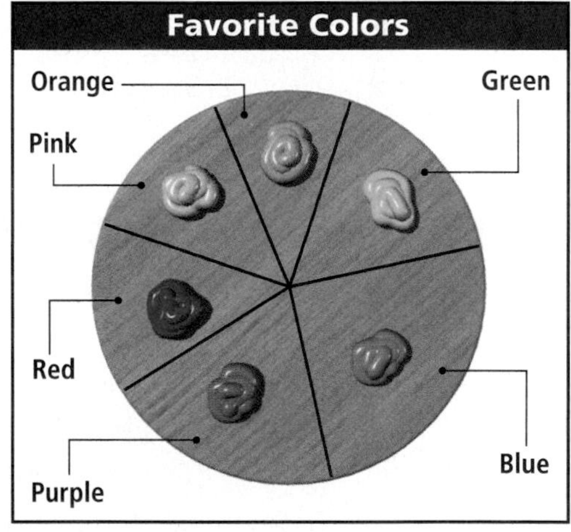

Think and Discuss

1. **Explain** how an outlier with a large value will affect the mean of a data set. What is the effect of a small outlier value?
2. **Explain** why the mean would not be a good description of the following high temperatures that occurred over 7 days: 72°F, 73°F, 70°F, 68°F, 70°F, 71°F, and 39°F.
3. **Give an example** of a data set that could be described only by its mode.

6-3 Exercises

FOR EOG PRACTICE
see page 658

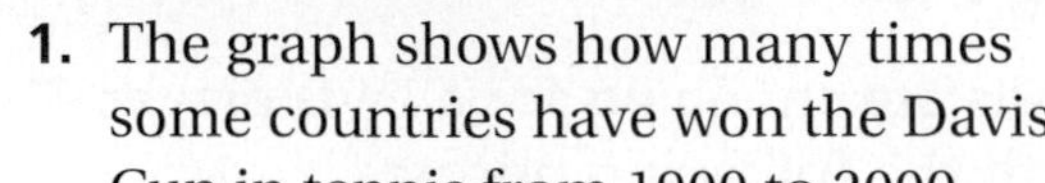

GUIDED PRACTICE

See Example 1

1. The graph shows how many times some countries have won the Davis Cup in tennis from 1900 to 2000.
   a. Find the mean, median, and mode of the data.
   b. The United States won 31 Davis Cups between 1900 and 2000. Add this number to the data in the graph and find the mean, median, and mode.

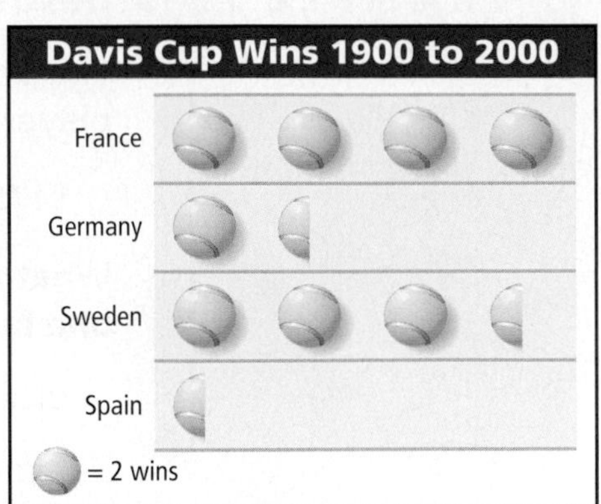

See Example 2

2. In 1998, 77-year-old John Glenn became the oldest person to travel into space. Other astronauts traveling on that same mission were 43, 37, 38, 46, 35, and 42 years old. Find the mean, median, and mode of all their ages with and without Glenn's age.

See Example 3

3. Kate read books that were 240, 450, 180, 160, 195, 170, 240, and 165 pages long. What are the mean, median, and mode of this data set? Which one best describes the data set?

INDEPENDENT PRACTICE

See Example 1

4. The graph shows the ages of the 10 youngest signers of the Declaration of Independence.
   a. Find the mean, median, and mode of the data.
   b. Benjamin Franklin was 70 years old when he signed the Declaration of Independence. Add his age to the data in the graph and find the mean, median, and mode.

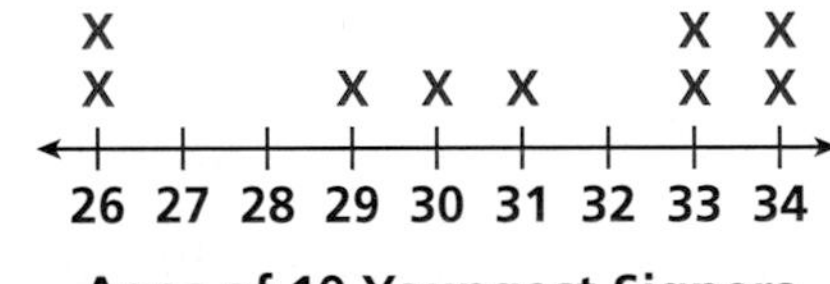

Ages of 10 Youngest Signers of Declaration of Independence

See Example 2

5. The map shows the population densities of several states along the Atlantic coast. Find the mean, median, and mode of the data with and without Maine's population density.

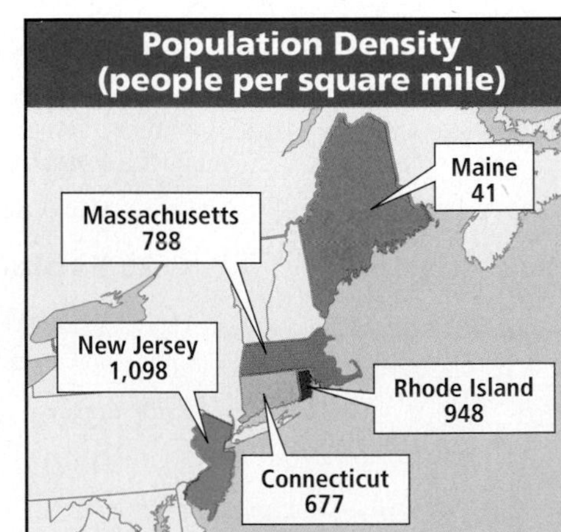

See Example 3

6. The passengers in a van are 16, 19, 17, 18, 15, 14, 32, 32, and 41 years old. What are the mean, median, and mode of this data set? Which one best describes the data set?

Earth Science LINK

On September 13, 1922, the temperature in El Azizia, Libya, reached 136°F, the record high for the planet. (*Source: The World Almanac and Book of Facts*)

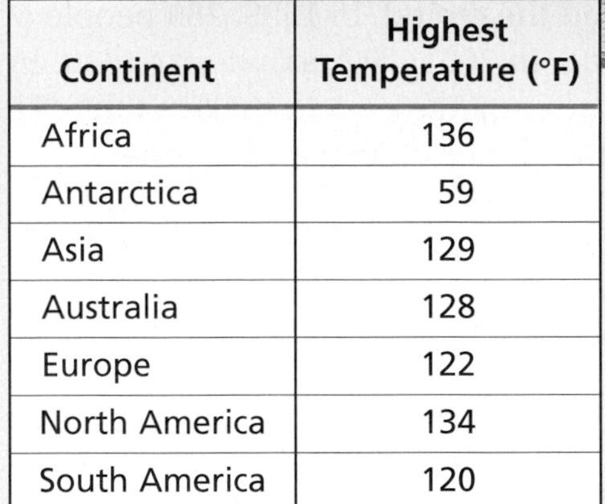

This satellite map shows the world's surface temperature. The dark blue areas are coldest, and the deep red areas are hottest.

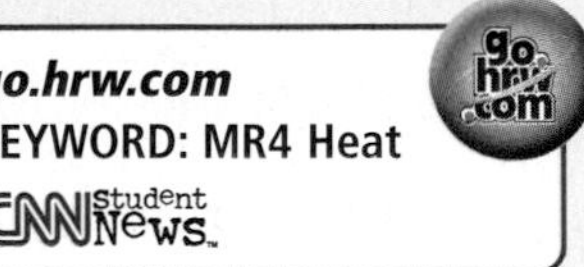

Continent	Highest Temperature (°F)
Africa	136
Antarctica	59
Asia	129
Australia	128
Europe	122
North America	134
South America	120

7. What are the mean, median, and mode of the highest recorded temperatures on each continent?

8. a. Which temperature is an outlier?

b. What are the mean, median, and mode of the temperatures if the outlier is not included?

9. ***WHAT'S THE ERROR?*** A student stated that the median temperature would rise to 120.6°F if a new record high of 75°F were recorded in Antarctica. Explain the error. How would the median temperature actually be affected if a high of 75°F were recorded in Antarctica?

10. ***WRITE ABOUT IT*** Is the data in the table best described by the mean, median, or mode? Explain.

11. ***CHALLENGE*** Suppose a new high temperature were recorded in Europe, and the new mean temperature became 120°F. What is Europe's new high temperature?

Spiral Review

List all the factors of each number. (Lesson 4-2)

12. 57 **13.** 36 **14.** 54

Find the GCF of each set of numbers. (Lesson 4-3)

15. 6 and 15 **16.** 18 and 56 **17.** 12, 16, and 32 **18.** 24, 63, and 81

19. **EOG PREP** What is the least common multiple of 4, 12, and 15? (Lesson 5-5)

A 30 B 60 C 45 D 90

20. **EOG PREP** Over 5 days, Pedro jogged 6 mi, 5 mi, 2 mi, 2 mi, and 4 mi. Find the mean distance that Pedro jogged. (Lesson 6-2)

A 4 mi B 2 mi C 3.8 mi D 4.75 mi

Chapter 6

Mid-Chapter Quiz

LESSON 6-1 (pp. 272–274)

1. The local dance studio holds a spring recital each year. In 1998, 220 people attended the recital. In 1999, 235 people attended. In 2000, 250 people attended. In 2001, 242 people attended. In 2002, 258 people attended. Use the attendance data to make a table. Then use your table to describe how attendance changed over time.

LESSON 6-2 (pp. 275–277)

Find the range, mean, median, and mode of each data set.

2.

Distance (mi)					
5	6	4	7	3	5

3.

Test Scores				
78	80	86	92	90

4.

Ages of Students (yr)							
11	13	12	12	12	13	9	14

5.

Number of Pages in Each Book						
145	119	156	158	125	128	135

LESSON 6-3 (pp. 278–281)

6. The table shows the number of people who attended each monthly meeting from January to May.

Number of People Attending				
Jan	Feb	Mar	Apr	May
27	26	32	30	30

   a. Find the mean, median, and mode of the attendances.

   b. In June, 39 people attended the meeting, and in July, 26 people attended the meeting. Add this data to the table and find the mean, median, and mode with the new data.

7. The four states with the longest coastlines are Alaska, Florida, California, and Hawaii. Alaska's coastline is 6,640 miles. Florida's coastline is 1,350 miles. California's coastline is 840 miles, and Hawaii's coastline is 750 miles. Find the mean, median, and mode of the lengths with and without Alaska's.

8. The daily snowfall amounts for the first ten days of December are listed below.

   2 in., 5 in., 0 in., 0 in., 15 in., 1 in., 0 in., 3 in., 1 in., 4 in.

   What are the mean, median, and mode of this data set? Which one best describes the data set?

Focus on Problem Solving

Make a Plan

- **Prioritize and sequence information**

Some problems give you a lot of information. Read the entire problem carefully to be sure you understand all of the facts. You may need to read it over several times, perhaps aloud so that you can hear yourself say the words.

Then decide which information is most important (prioritize). Is there any information that is absolutely necessary to solve the problem? This information is important.

Finally, put the information in order (sequence). Use comparison words like *before, after, longer, shorter,* and so on to help you. Write the sequence down before you try to solve the problem.

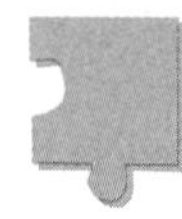

Read the problems below and answer the questions that follow.

1. The compact disc (CD) was invented 273 years after the piano. The tape recorder was invented in 1898. Thomas Edison invented the phonograph 21 years before the tape recorder and 95 years before the compact disc. What is the date of each invention?

   a. Which invention's date can you use to find the dates of all the others?

   b. Can you solve the problem without this date? Explain.

   c. List the inventions in order from earliest invention to latest invention.

2. Jon recorded the heights of his family members. There are 4 people in Jon's family, including Jon. Jon's mother is 2 inches taller than Jon's father. Jon is 56 inches tall. Jon's sister is 4 inches taller than Jon and 5 inches shorter than Jon's father. What are the heights of Jon and his family members?

   a. Whose height can you use to find the heights of all the others?

   b. Can you solve the problem without this height? Explain.

   c. List Jon's family members in order from shortest to tallest.

?

1898

?

?

6-4 Bar Graphs

Learn to display and analyze data in bar graphs.

Vocabulary

bar graph

double-bar graph

A biome is a large region characterized by a specific climate. There are ten land biomes on Earth. Some are pictured at right. Each gets a different amount of rainfall.

A *bar graph* can be used to display and compare data about rainfall. A **bar graph** displays data with vertical or horizontal bars.

EXAMPLE 1 Reading a Bar Graph

Use the bar graph to answer each question.

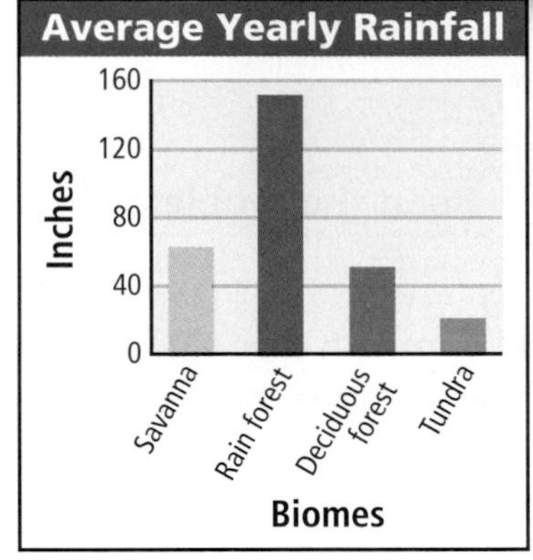

A **Which biome in the graph has the most rainfall?**

Find the highest bar.

The rain forest has the most rainfall.

B **Which biomes in the graph have an average yearly rainfall less than 40 inches?**

Find the bar or bars whose heights measure less than 40.

The tundra has an average yearly rainfall less than 40 inches.

EXAMPLE 2 Making a Bar Graph

Use the given data to make a bar graph.

Coal Reserves (billion metric tons)		
Asia	Europe	Africa
695	404	66

Step 1: Find an appropriate scale and interval. The scale must include all of the data values. The interval separates the scale into equal parts.

Step 2: Use the data to determine the lengths of the bars. Draw bars of equal width. The bars cannot touch.

Step 3: Title the graph and label the axes.

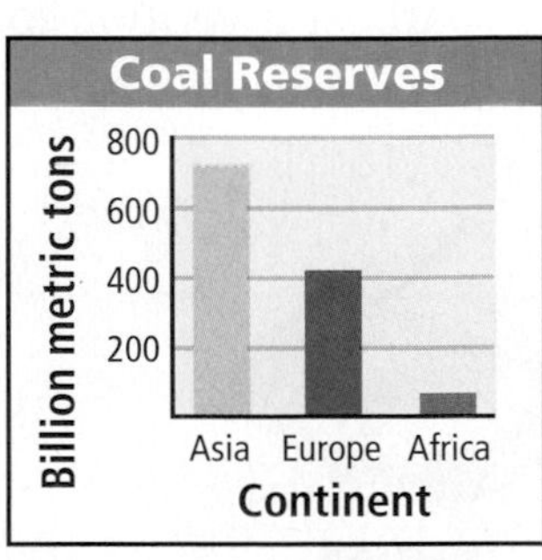

A **double-bar graph** shows two sets of related data.

EXAMPLE 3 PROBLEM SOLVING APPLICATION

PROBLEM SOLVING

Make a double-bar graph to compare the data in the table.

Life Expectancies in Atlantic South America				
	Brazil	**Argentina**	**Uruguay**	**Paraguay**
Male (yr)	59	71	73	70
Female (yr)	69	79	79	74

1 Understand the Problem

You are asked to use a graph to compare the data given in the table. You will need to use all of the information given.

2 Make a Plan

You can make a double-bar graph to display the two sets of data.

3 Solve

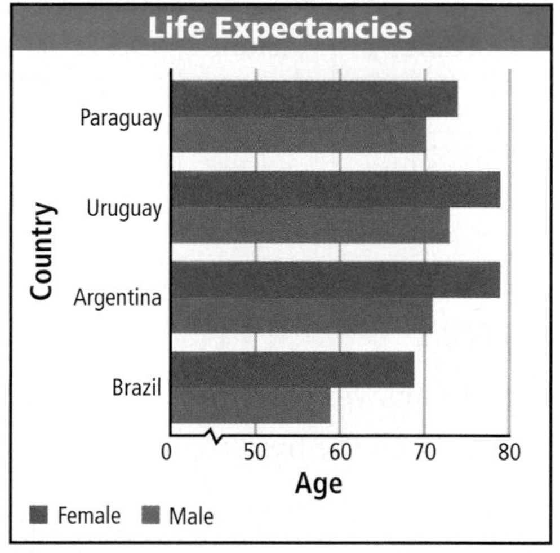

Determine appropriate scales for both sets of data.

Use the data to determine the lengths of the bars. Draw bars of equal width. Bars should be in pairs. Use a different color for male ages and female ages.

Title the graph and label both axes.

Include a key to show what each bar represents.

4 Look Back

You could make two separate graphs, one of male ages and one of female ages. However, it is easier to compare the two data sets when they are on the same graph.

Reading Math

This symbol means there is a break in the scale. Some numbers were left out because they were not needed for the graph.

Think and Discuss

1. Give comparisons you can make by looking at a bar graph.

2. Describe the kind of data you would display in a bar graph.

3. Tell why the graph in Example 3 needs a key.

6-4 Exercises

FOR EOG PRACTICE

see page 660

internet connect

Homework Help Online
go.hrw.com Keyword: MR4 6-4

GUIDED PRACTICE

See Example 1

Use the bar graph to answer each question.

1. Which color was the least common among the cars in the parking lot?
2. Which colors appeared more than ten times in the parking lot?

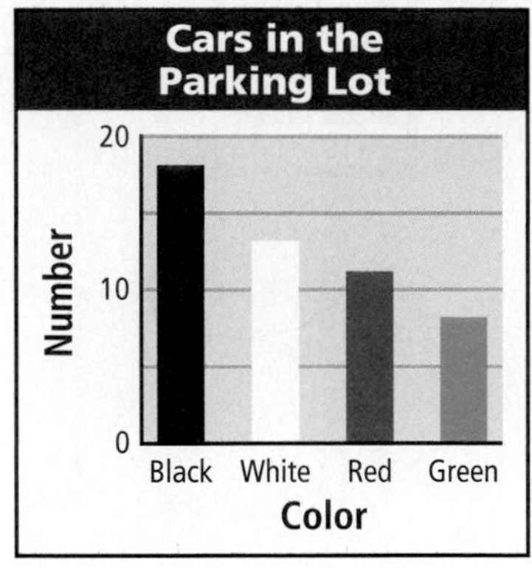

See Example 2

3. Use the given data to make a bar graph.

Students in Mr. Jones's History Classes			
Period 1	28	Period 6	22
Period 2	27	Period 7	7

See Example 3

4. Make a bar graph to compare the data in the table.

Movie Preferences of Men and Women Polled at the Mall						
	Comedy	Action	Sci-Fi	Horror	Drama	Other
Men	16	27	16	23	12	6
Women	21	14	8	18	30	9

INDEPENDENT PRACTICE

See Example 1

Use the bar graph to answer each question.

5. Which fruit was liked the best?
6. Which fruits were liked by equal numbers of people?

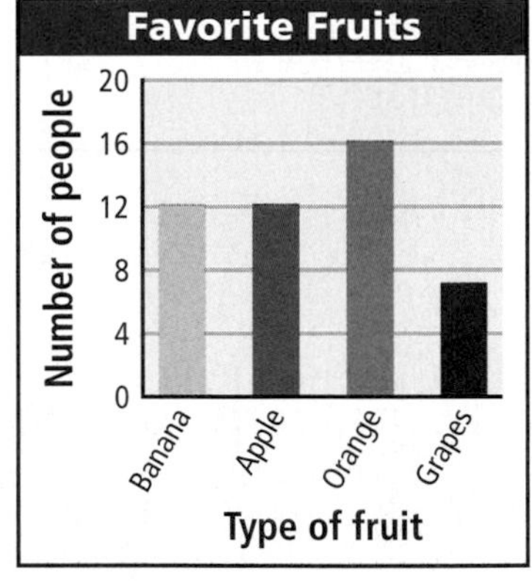

See Example 2

7. Use the given data to make a bar graph.

Days with Rainfall			
January	14	March	16
February	12	April	23

See Example 3

8. Make a bar graph to compare the data in the table.

Heart Rates Before and After Exercise (beats per minute)						
	Jason	Jamal	Ray	Tonya	Peter	Brenda
Before	60	62	61	65	64	65
After	131	140	128	140	135	120

PRACTICE AND PROBLEM SOLVING

Use the bar graph for Exercises 9–12.

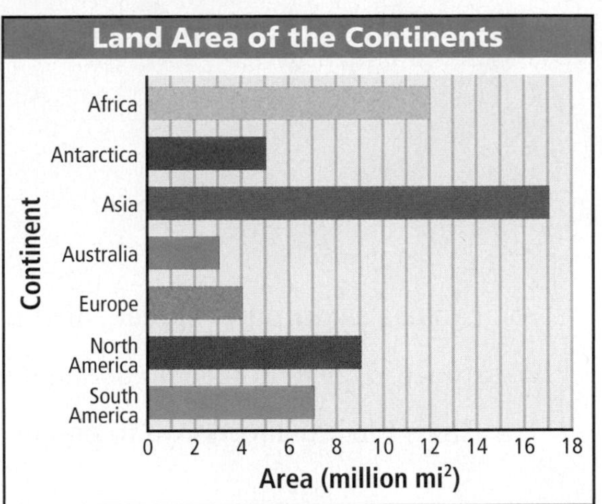

9. What is the range of the land area of the continents?

10. What is the mode of the land area of the continents?

11. What is the mean of the land area of the continents?

12. What is the median of the land area of the continents?

13. ***CHOOSE A STRATEGY*** The heights of Maria, Glenn, Carol, and Luis are shown in the graph, but the labels are missing.

- Maria is neither the tallest nor the shortest.
- Glenn is taller than Carol.
- There is only one person taller than Luis.

Which student's name should go with each bar in the graph?

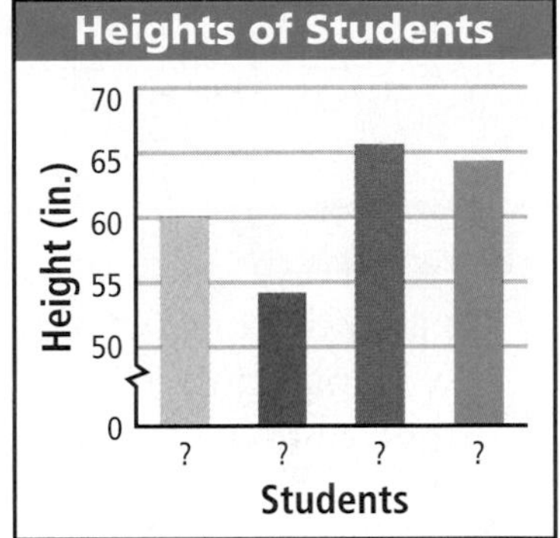

14. ***WRITE ABOUT IT*** Explain how you would make a bar graph of the five most populated cities in the United States.

15. ***CHALLENGE*** Create a bar graph displaying the number of A's, B's, C's, D's, and F's in Ms. Walker's class if the grades were the following: 81, 87, 80, 75, 77, 98, 52, 78, 75, 82, 74, 95, 76, 52, 76, 53, 86, 77, 90, 83, 96, 83, 74, 67, 90, 65, 69, 93, 68, and 76.

Grading System	
A	90–100
B	80–89
C	70–79
D	60–69
F	0–59

Spiral Review

Write each phrase as a numerical or algebraic expression. (Lesson 2-2)

16. 739 minus 103 **17.** the product of 7 and z **18.** the difference of 12 and n

Write each mixed number as an improper fraction. (Lesson 4-7)

19. $2\frac{2}{5}$ **20.** $1\frac{3}{4}$ **21.** $4\frac{1}{7}$ **22.** $3\frac{1}{3}$

23. **EOG PREP** Which of the following is the solution to $4x = \frac{3}{4}$? (Lesson 5-4)

A $x = \frac{3}{16}$ B $x = \frac{3}{4}$ C $x = 3$ D $x = 5\frac{1}{3}$

Create Bar Graphs

Use with Lesson 6-4

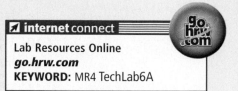

internet connect
Lab Resources Online
go.hrw.com
KEYWORD: MR4 TechLab6A

You can use a computer spreadsheet to draw bar graphs. The Chart Wizard icon, [icon], on a spreadsheet menu looks like a bar graph. The Chart Wizard allows you to create different types of graphs.

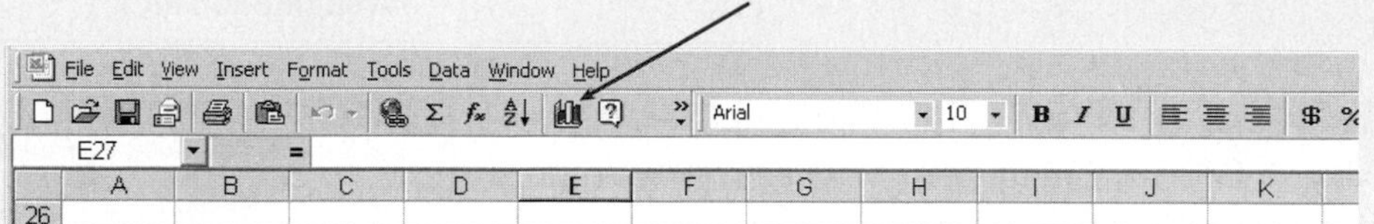

Activity

In a study conducted in December 2001 at Texas A&M University, the population of Texas through 2035 was projected. Make a bar graph of this data.

Texas Population	
Year	**Population**
2000	20,851,820
2005	23,207,929
2010	25,897,018
2015	28,971,283
2020	32,427,282
2025	36,273,829
2030	40,538,290
2035	45,283,746

1. Type the titles *Year* and *Population* into cells A1 and B1. Then type the data into columns A (year) and B (population).
2. Select the cells containing the titles and the data. Do this by placing your pointer in A1, clicking and holding the mouse button, and dragging the pointer down to B9.
3. Click the Chart Wizard icon. Highlight **Column** to make a vertical bar graph. Click **Next**.

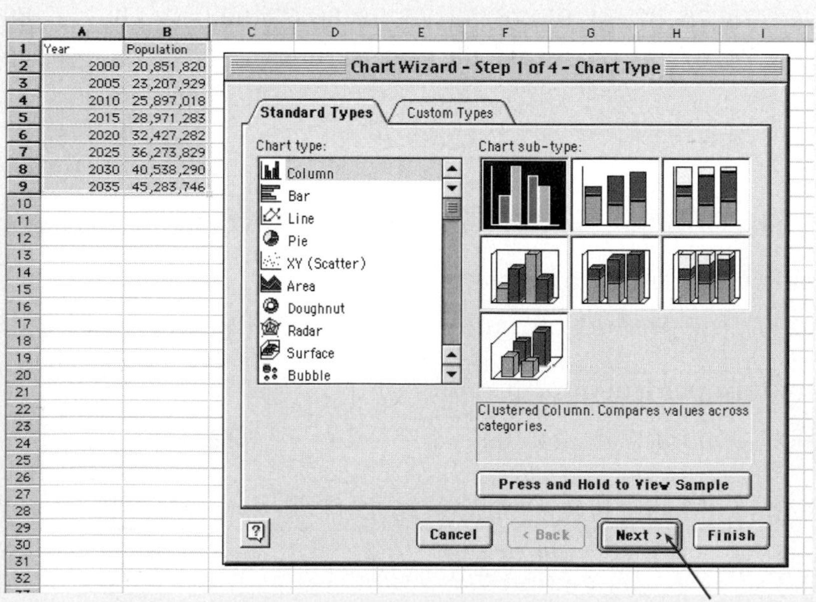

4. The next screen shows where the data from the graph comes from. Click **Next**.

5. Title your graph and both axes. Click the **Legend** tab. Click the box next to **Show Legend** to turn off the key. (You would need a key if you were making a double-bar graph.) Click **Next** when you are finished.

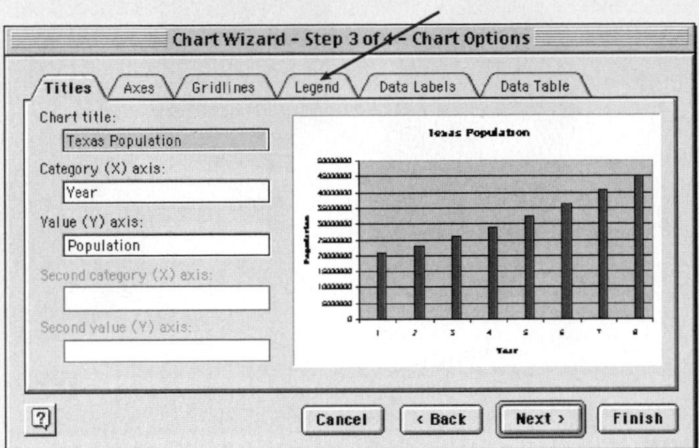

6. The next screen asks you where you want to place your chart. Click **Finish** to place it in your spreadsheet.

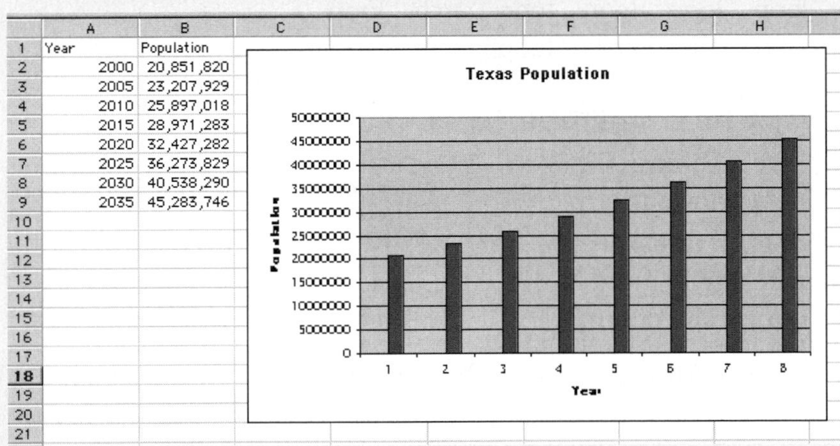

Think and Discuss

1. Do you think the population of Texas will be 32,427,282 in the year 2020 as shown in the graph? Explain.

Try This

1. Redraw the bar graph in the activity to show the population of Texas as 39,000,000 in 2035 and 33,000,000 in 2040.

2. The table shows the number of countries in which some common languages are spoken. Make a bar graph of the data.

Language	English	Arabic	Spanish	French
Number of Countries	54	24	21	33

6-5 Frequency Tables and Histograms

Learn to organize data in frequency tables and histograms.

Vocabulary

frequency table

cumulative frequency

histogram

Your fingerprints are unlike anyone else's. Even identical twins have slightly different fingerprint patterns.

All fingerprints have one of three patterns: whorl, arch, or loop.

EXAMPLE 1 Making a Tally Table

Each student in Mrs. Choe's class recorded their fingerprint pattern. Which type do most students in Mrs. Choe's class have?

whorl	loop	loop	loop	loop	arch	loop
whorl	arch	loop	arch	loop	arch	whorl

Make a *tally table* to organize the data.

Step 1: Make a column for each fingerprint pattern.

Step 2: For each fingerprint, make a tally mark in the appropriate column.

Number of Fingerprint Patterns		
Whorl	Arch	Loop
///	////	~~////~~ //

Most students in Mrs. Choe's class have a loop fingerprint pattern.

Reading Math

A group of four tally marks with a line through it means *five*.

~~////~~ = 5

~~////~~ ~~////~~ = 10

A **frequency table** tells the number of times an event, category, or group occurs. The **cumulative frequency** column shows a running total of all frequencies.

EXAMPLE 2 Making a Cumulative Frequency Table

Use the tally table above to make a cumulative frequency table.

Step 1: Make a row for each pattern.

Step 2: The frequency is how many times each pattern occurred.

Step 3: Find the cumulative frequency for each row by adding all frequency values above or in that row.

Number of Fingerprint Patterns		
Fingerprint Pattern	Frequency	Cumulative Frequency
Whorl	3	3
Arch	4	7
Loop	7	14

EXAMPLE 3 Making a Frequency Table with Intervals

Social Studies LINK

The number of representatives a state has depends on its population. A state can gain or lose representatives after a population count called a census.

Use the data in the table to make a frequency table with intervals.

Number of Representatives per State in the U.S. House of Representatives												
7	1	6	4	52	6	6	1	1	23	11	2	2
20	10	5	4	6	7	2	8	10	16	8	5	9
1	3	2	2	13	3	31	12	1	19	6	5	21
2	6	1	9	30	3	1	11	9	3	9		

Step 1: Choose equal intervals.

Step 2: Find the number of data values in each interval. Write these numbers in the "Frequency" row.

Number of Representatives per State in the U.S. House of Representatives									
Number	0–5	6–11	12–17	18–23	24–29	30–35	36–41	42–47	48–53
Frequency	22	18	3	4	0	2	0	0	1

This table shows that 22 states have between 0 and 5 representatives, 18 states have between 6 and 11 representatives, and so on.

A **histogram** is a bar graph that shows the number of data items that occur within each interval.

EXAMPLE 4 Making a Histogram

Use the frequency table in Example 3 to make a histogram.

Step 1: Choose an appropriate scale and interval.

Step 2: Draw a bar for the number of states in each interval. The bars should touch but not overlap.

Step 3: Title the graph and label the axes.

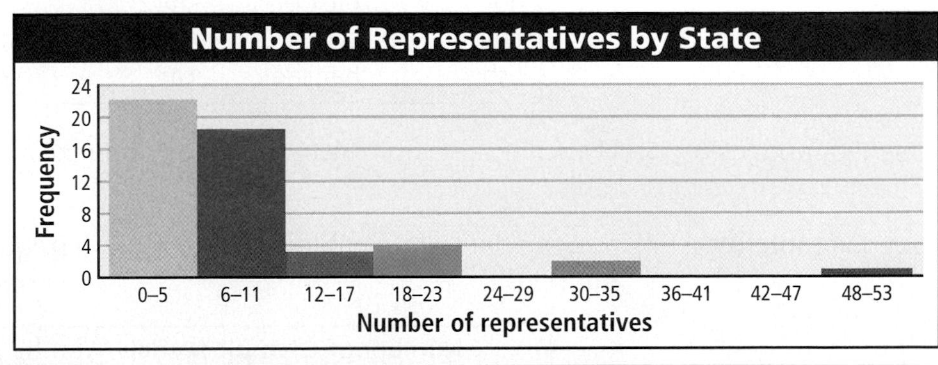

Think and Discuss

1. Tell how to find cumulative frequency.

6-5 Exercises

FOR EOG PRACTICE

see page 660

internet connect

Homework Help Online
go.hrw.com Keyword: MR4 6-5

4.01, 4.06

GUIDED PRACTICE

See Example 1

1. Each student in the band recorded the type of instrument he or she plays. The results are shown in the box. Make a tally table to organize the data. Which instrument do the fewest students play?

trumpet	tuba	French horn	drums	trombone
drums	trombone	trombone	trumpet	trumpet
trumpet	French horn	trumpet	French horn	French horn

See Example 2

2. Use your tally table from Exercise 1 to make a cumulative frequency table.

See Example 3

3. Use the data in the table below to make a frequency table with intervals.

Length of Each U.S. Presidency (yr)																				
8	4	8	8	8	4	8	4	0	4	4	1	3	4	4	4	4	8	4	0	4
4	4	4	4	8	4	8	2	6	4	12	8	8	2	6	5	3	4	8	4	8

See Example 4

4. Use your frequency table from Exercise 3 to make a histogram.

INDEPENDENT PRACTICE

See Example 1

5. Students recorded the type of pet they own. The results are shown in the box. Make a tally table. Which type of pet do most students own?

cat	cat	bird	dog	dog
dog	bird	dog	bird	fish
bird	cat	fish	dog	cat
fish	hamster	cat	hamster	dog

See Example 2

6. Use your tally table from Exercise 5 to make a cumulative frequency table.

See Example 3

7. Use the data in the table below to make a frequency table with intervals.

Number of Olympic Medals Won by 27 Countries													
8	88	59	12	11	57	38	17	14	28	28	26	25	23
18	8	29	34	14	17	13	13	58	12	97	10	9	

See Example 4

8. Use your frequency table from Exercise 7 to make a histogram.

PRACTICE AND PROBLEM SOLVING

9. a. Students in a gym class recorded their favorite sports. Use their data to make a tally table.

basketball football soccer	hockey track and field track and field	hockey football football	soccer football baseball	tennis basketball track and field

b. Use your tally table to make a cumulative frequency table.

10. ***SOCIAL STUDIES*** The map shows the populations of Australia's states and territories. Use the data to make a frequency table with intervals.

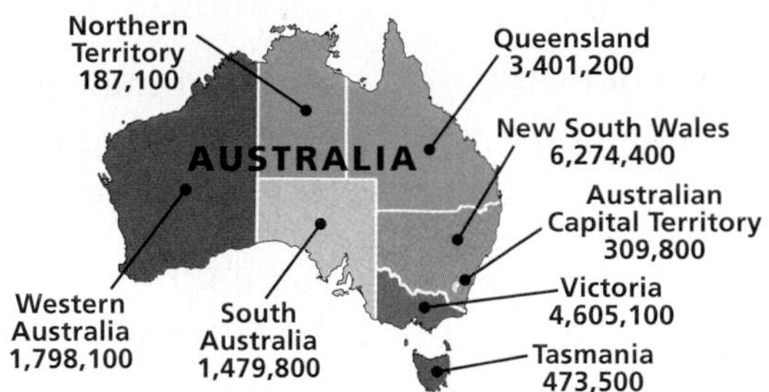

11. ***SOCIAL STUDIES*** Use your frequency table from Exercise 10 to make a histogram.

12. ***WRITE ABOUT IT*** Choose one of the histograms you made for this lesson and redraw it using different intervals. How did the histogram change? Explain.

13. ***WHAT'S THE ERROR?*** Describe the error in the frequency table.

Age of Consumers		
Age	**Frequency**	**Cumulative Frequency**
Child	3	
Teenager	2	15
Adult	10	

14. ***CHALLENGE*** Can you find the mean, median, and mode price using this frequency table? If so, find them. If not, explain why not.

Cost of Video Game Rentals at Different Stores				
Price	\$2.00–\$2.99	\$3.00–\$3.99	\$4.00–\$4.99	\$5.00–\$5.99
Frequency	5	12	8	5

Spiral Review

Write each decimal in expanded form and word form. (Lesson 3-1)

15. 1.23 **16.** 0.45 **17.** 26.07 **18.** 80.002

Find the reciprocal of each number. (Lesson 5-3)

19. 6 **20.** $\frac{4}{7}$ **21.** $\frac{2}{9}$ **22.** $\frac{11}{5}$

23. **EOG PREP** The ___?___ of a data set is always a value in the set. (Lesson 6-2)

A range B median C mode D mean

6-6 Ordered Pairs

San Diego, CA. Image courtesy of spaceimaging.com.

Learn to graph ordered pairs on a coordinate grid.

Vocabulary

coordinate grid

ordered pair

Cities, towns, and neighborhoods are often laid out on a grid. This makes it easier to map and find locations.

A **coordinate grid** is formed by horizontal and vertical lines and is used to locate points.

Each point on a coordinate grid can be located by using an **ordered pair** of numbers, such as (4, 6). The starting point is (0, 0).

- The first number tells how far to move horizontally from (0, 0).
- The second number tells how far to move vertically.

EXAMPLE 1 **Identifying Ordered Pairs**

Name the ordered pair for each location.

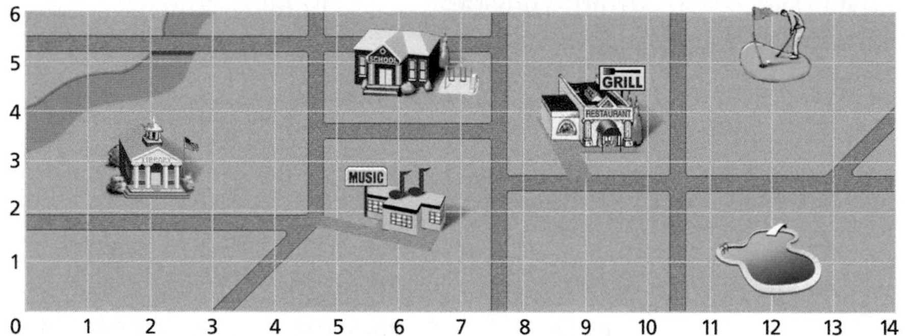

A **Library**

Start at (0, 0). Move right 2 units and then up 3 units.

The library is located at (2, 3).

B **School**

Start at (0, 0). Move right 6 units and then up 5 units.

The school is located at (6, 5).

C **Pool**

Start at (0, 0). Move right 12 units and up 1 unit.

The pool is located at (12, 1).

EXAMPLE 2 Graphing Ordered Pairs

Graph and label each point on a coordinate grid.

A $Q(4, 6)$ *Start at (0, 0).*
Move right 4 units.
Move up 6 units.

B $S(0, 4)$ *Start at (0, 0).*
Move right 0 units.
Move up 4 units.

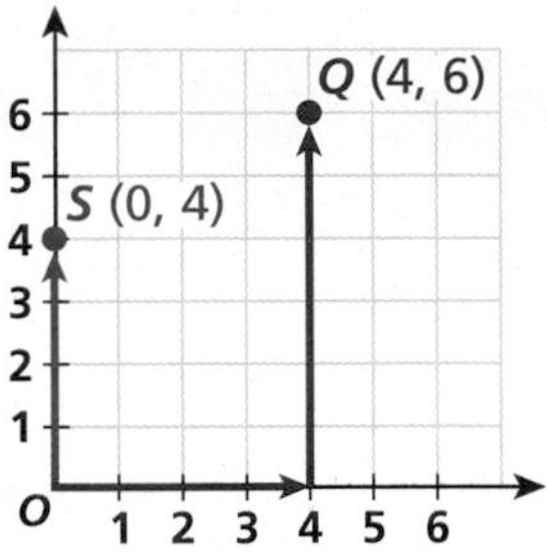

Think and Discuss

1. **Tell** what point is the starting location when you are graphing on a coordinate grid.
2. **Describe** how to graph (2, 8) on a coordinate grid.

6-6 Exercises

3.04

FOR EOG PRACTICE
see page 660

internet connect
Homework Help Online
go.hrw.com Keyword: MR4 6-6

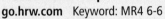

GUIDED PRACTICE

See Example 1 **Name the ordered pair for each location.**

1. school
2. store
3. hospital
4. mall
5. office
6. hotel

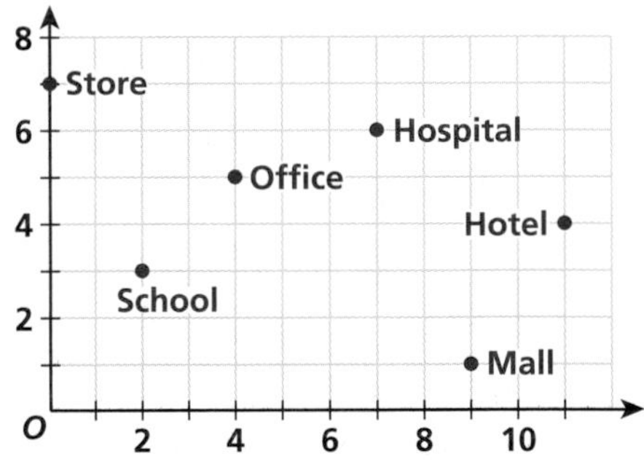

See Example 2 **Graph and label each point on a coordinate grid.**

7. $T(3, 4)$
8. $S(2, 8)$
9. $U(5, 5)$
10. $V(4, 1)$

INDEPENDENT PRACTICE

See Example 1 **Name the ordered pair for each location.**

11. diner
12. library
13. store
14. bank
15. theater
16. town hall

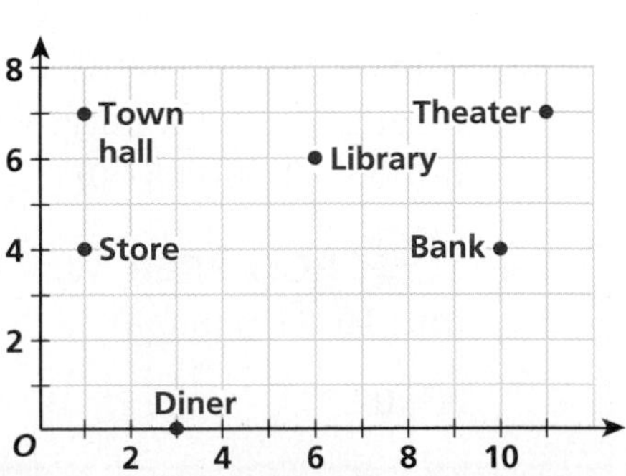

See Example 2 **Graph and label each point on a coordinate grid.**

17. $P(5, 1)$ **18.** $R(2, 4)$ **19.** $Q(3, 2)$

20. $V(6, 5)$ **21.** $X(1, 3)$ **22.** $Y(7, 4)$

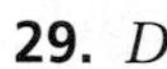

PRACTICE AND PROBLEM SOLVING

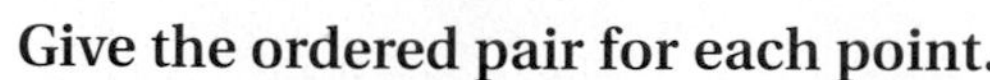

Use the coordinate grid for Exercises 23–34. Name the point found at each location.

23. (1, 7) **24.** (5, 9) **25.** (3, 3)

26. (4, 7) **27.** (7, 4) **28.** (7, 7)

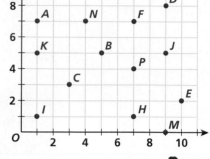

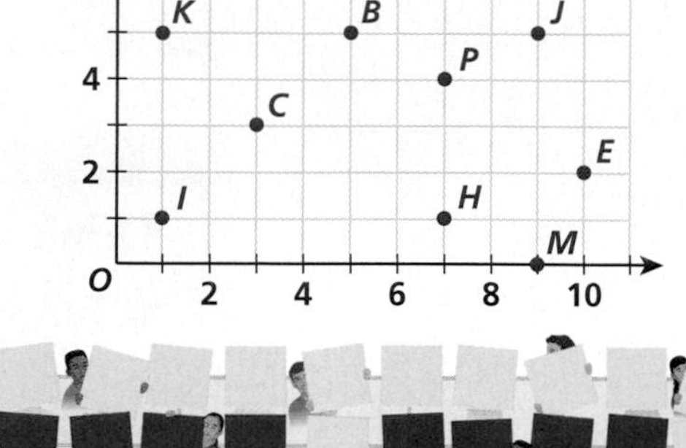

Give the ordered pair for each point.

29. D **30.** H **31.** K

32. Q **33.** M **34.** B

35. The Spirit Club marches in a large grid formation at sporting events. To spell words, some students hold red cards while others hold gold cards. Give five ordered pairs where students are holding red cards in the formation at right.

36. ***WRITE ABOUT IT*** Explain the difference between the points (3, 2) and (2, 3).

37. ***WHAT'S THE QUESTION?*** If the answer is "Start at (0, 0) and move 3 units to the right," what is the question?

38. ***CHALLENGE*** Locate and graph points that can be connected to form your initials. What are the ordered pairs for these points?

Spiral Review

Find the prime factorization of each number. (Lesson 4-2)

39. 18 **40.** 20 **41.** 33 **42.** 50

43. The marching band's halftime show was $10\frac{5}{6}$ minutes long. Then the director added a song. The new length of the show is $12\frac{1}{3}$ minutes. How long is the song that was added? (Lesson 5-10)

44. **EOG PREP** Which number, if any, is an outlier in the set 0, 1, 4, 0, 3, 4, 2, and 1? (Lesson 6-3)

A 0 B 1 C 4 D No outlier

6-7 Line Graphs

Learn to display and analyze data in line graphs.

Vocabulary

line graph

double-line graph

The first permanent English settlement in the New World was founded in 1607. It contained 104 colonists. Population increased quickly as more and more immigrants left Europe for North America.

A New England Dame School, 1713

The table shows the estimated population of English American colonies from 1650 to 1700.

Population of American Colonies				
Year	1650	1670	1690	1700
Population	50,400	111,900	210,400	250,900

Data that shows change over time is best displayed in a *line graph.* A **line graph** displays a set of data using line segments.

EXAMPLE 1 Making a Line Graph

Use the data in the table above to make a line graph.

Step 1: Place *years* on the horizontal axis and *population* on the vertical axis. Label the axes.

Step 2: Determine an appropriate scale and interval for each axis.

Step 3: Mark a point for each data value. Connect the points with straight lines.

Step 4: Title the graph.

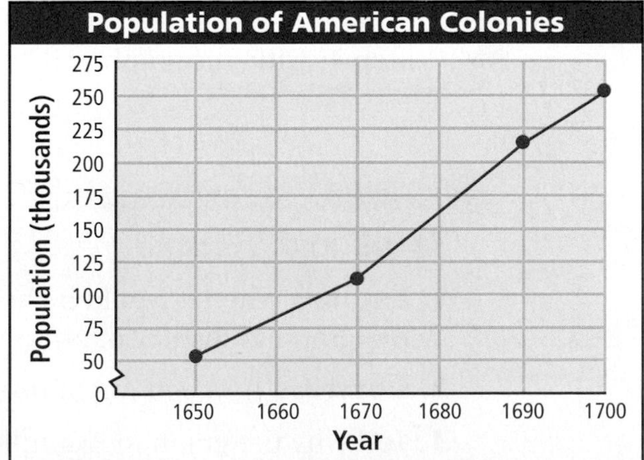

Helpful Hint

Because time passes whether or not the population changes, time is *independent* of population. Always put the independent quantity on the horizontal axis.

EXAMPLE 2 Reading a Line Graph

Use the line graph to answer each question.

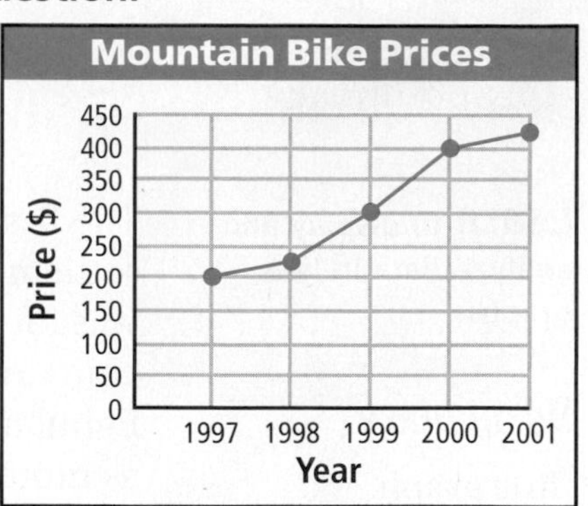

A **In which year did mountain bikes cost the least?** 1997

B **About how much did mountain bikes cost in 1999?** about $300

C **Did mountain bike prices increase or decrease from 1997 through 2001?** They increased.

Line graphs that display two sets of data are called **double-line graphs**.

EXAMPLE 3 Making a Double-Line Graph

Use the data in the table to make a double-line graph.

Life Expectancy in the United States							
	1970	**1975**	**1980**	**1985**	**1990**	**1995**	**2000**
Male (yr)	67	69	70	71	72	73	74
Female (yr)	71	77	77	78	79	79	80

Helpful Hint

Use different colors of lines to connect the male and female values so you will easily be able to tell the data apart.

Step 1: Determine an appropriate scale and interval.

Step 2: Mark a point for each male value and connect the points.

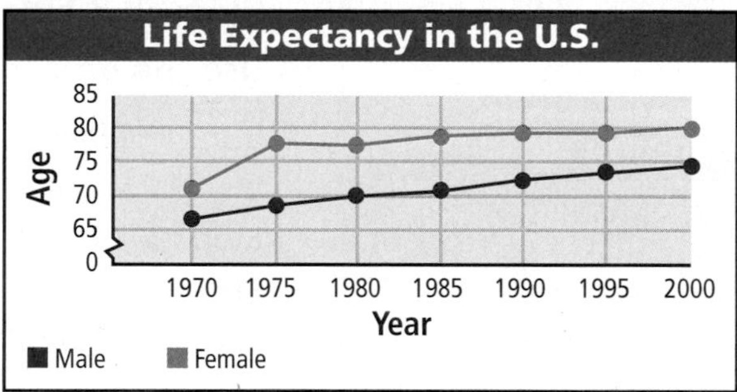

Step 3: Mark a point for each female value and connect the points.

Step 4: Title the graph and label both axes. Include a key.

Think and Discuss

1. **Explain** when it would be helpful to use a line graph instead of a bar graph to display data.
2. **Describe** how you might use a line graph to make predictions.
3. **Tell** why the graph in Example 3 needs a key.

6-7 Exercises

FOR EOG PRACTICE

see page 661

internet connect
Homework Help Online
go.hrw.com Keyword: MR4 6-7

GUIDED PRACTICE

See Example 1

1. Use the data in the table to make a line graph.

School Enrollment				
Year	2000	2001	2002	2003
Students	2,000	2,500	2,750	3,500

See Example 2 **Use the line graph to answer each question.**

2. In which year did the most students participate in the science fair?
3. Did the number of students increase or decrease from 2000 to 2001?

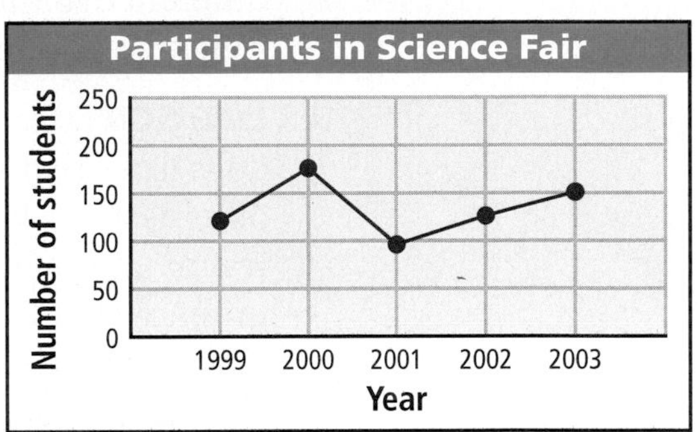

See Example 3

4. Use the data in the table to make a double-line graph.

	January	February	March	April	May
Stock A	$10	$12	$20	$25	$22
Stock B	$8	$8	$12	$20	$30

INDEPENDENT PRACTICE

See Example 1

5. Use the data in the table to make a line graph.

Winning Times in the Iditarod Dog Sled Race							
Year	1995	1996	1997	1998	1999	2000	2001
Time (hr)	219	222	225	222	231	217	236

See Example 2 **Use the line graph to answer each question.**

6. About how many personal computers were in use in the United States in 1996?
7. When was the number of personal computers in use about 105 million?

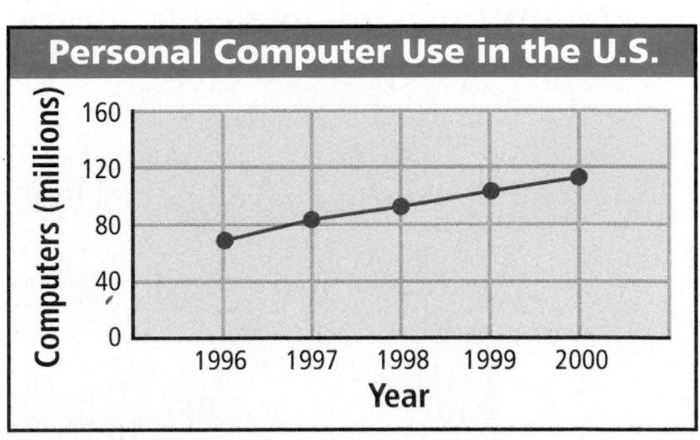

See Example 3 **8.** Use the data in the table to make a double-line graph.

Soccer Team's Total Fund-Raising Sales						
Day	0	1	2	3	4	5
Team A	$0	$100	$225	$300	$370	$450
Team B	$0	$50	$100	$150	$200	$250

PRACTICE AND PROBLEM SOLVING

Use the line graph for Exercises 9 and 10.

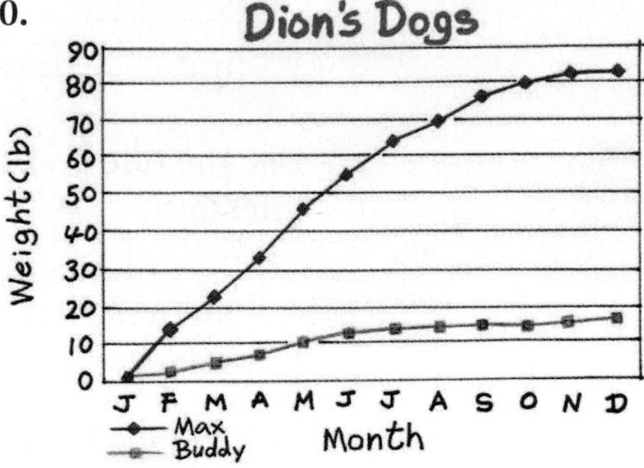

9. ***LIFE SCIENCE*** Estimate the difference in the dogs' weights in March.

10. ***LIFE SCIENCE*** One of Dion's dogs is a Great Dane, and the other is a miniature Dalmatian. Which dog is probably the Great Dane? Justify your answer.

11. ***LIFE SCIENCE*** The table shows the weights in pounds for Sara Beth's two pets. Use the data to make a line graph that is similar to Dion's.

	Jan	Feb	Mar	Apr	May	Jun	Jul	Aug	Sep	Oct	Nov	Dec
Ginger	3	9	15	21	24	25	26	25	26	27	26	28
Toto	4	8	13	17	24	26	27	29	25	26	28	28

12. ***WRITE ABOUT IT*** Suppose you have a bowl of soup with lunch. Draw a line graph that could represent the changes in the soup's temperature during lunch. Explain.

13. ***CHALLENGE*** Describe a situation that this graph could represent.

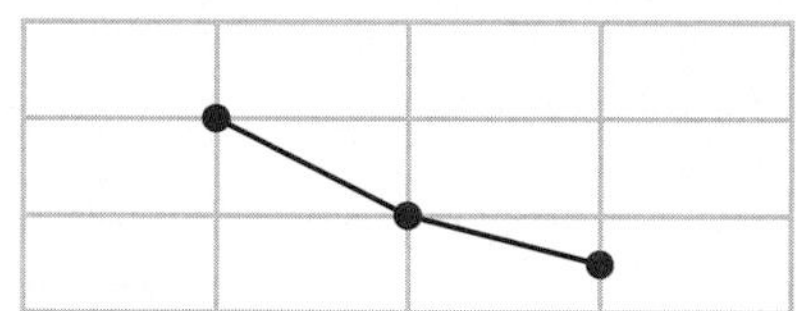

Spiral Review

Identify the property that is illustrated by each equation. (Lesson 1-5)

14. $3 + (4 + 5) = (3 + 4) + 5$ **15.** $19(24) = 19(20) + 19(4)$ **16.** $(2)(13) = (13)(2)$

17. Four friends split the cost of a pizza and four drinks. The pizza cost $12, and each drink cost $2.00. How much did each person pay? (Lesson 3-8)

18. **EOG PREP** Two apples weigh $\frac{1}{4}$ lb and $\frac{3}{16}$ lb. Find the difference in their weights. (Lesson 5-7)

A $\frac{1}{6}$ lb B $\frac{1}{16}$ lb C $\frac{7}{16}$ lb D $\frac{1}{4}$ lb

6-8 Misleading Graphs

Learn to recognize misleading graphs.

Data can be displayed in many different ways. Sometimes people who make graphs choose to display data in a misleading way.

This bar graph was created by a group of students who believe their school should increase support of the football team. How could this bar graph be misleading?

At a glance, you might conclude that about three times as many students prefer football to basketball. But if you look at the values of the bars, you can see that only 20 more students chose football over basketball.

EXAMPLE 1 Misleading Bar Graphs

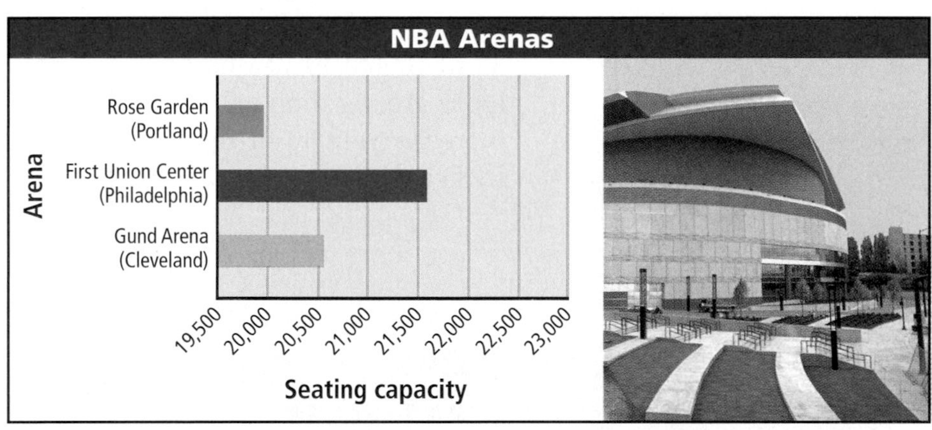

A **Why is this bar graph misleading?**

Because the lower part of the horizontal scale is missing, the differences in seating capacities are exaggerated.

B **What might people believe from the misleading graph?**

People might believe that the First Union Center holds 2–4 times as many people as Gund Arena and the Rose Garden. In reality, the First Union Center holds only one to two thousand more people than the other two arenas.

EXAMPLE 2 Misleading Line Graphs

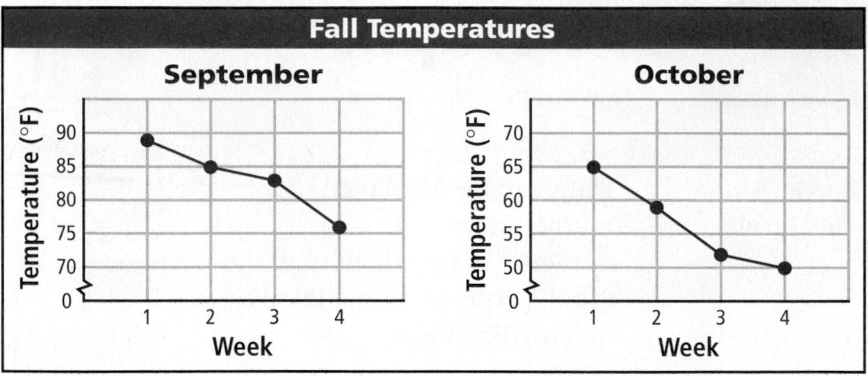

A **Why are these line graphs misleading?**

If you look at the scale for each graph, you will notice that the September graph goes from 75°F to 90°F and the October graph goes from 50°F to 65°F.

B **What might people believe from these misleading graphs?**

People might believe that the temperatures in October were about the same as the temperatures in September. In reality, the temperatures in September were 20–30 degrees higher.

C **Why is this line graph misleading?**

The scale does not have equal intervals. So, for example, an increase from 35 sit-ups to 40 sit-ups looks greater than an increase from 30 sit-ups to 35 sit-ups.

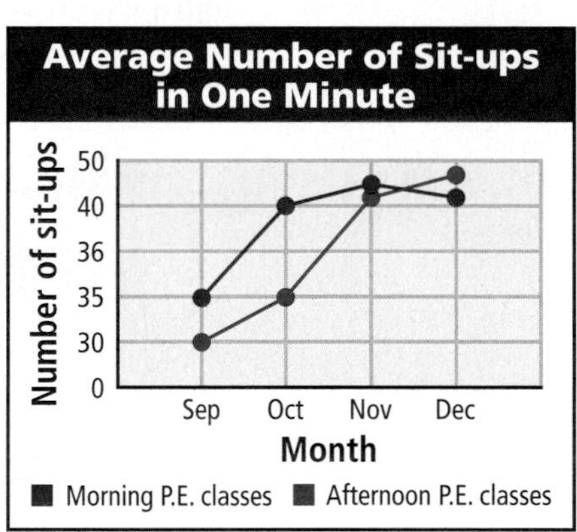

Think and Discuss

1. **Give an example** of a situation in which you think someone would intentionally try to make a graph misleading.
2. **Tell** who might have made the misleading graph in Example 2C.
3. **Tell** how you could change the graph in Example 2C so that it is not misleading.

6-8 Exercises

FOR EOG PRACTICE

see page 661

internet connect

Homework Help Online

go.hrw.com Keyword: MR4 6-8

GUIDED PRACTICE

See Example 1

1. Why is this bar graph misleading?
2. What might people believe from the misleading graph?

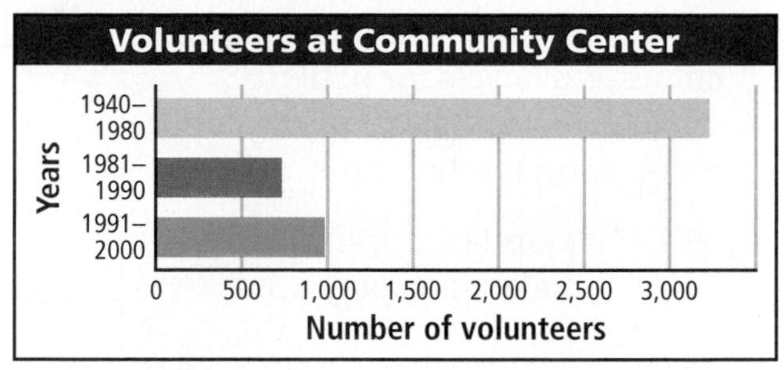

See Example 2

3. Why is this line graph misleading?
4. What might people believe from the misleading graph?

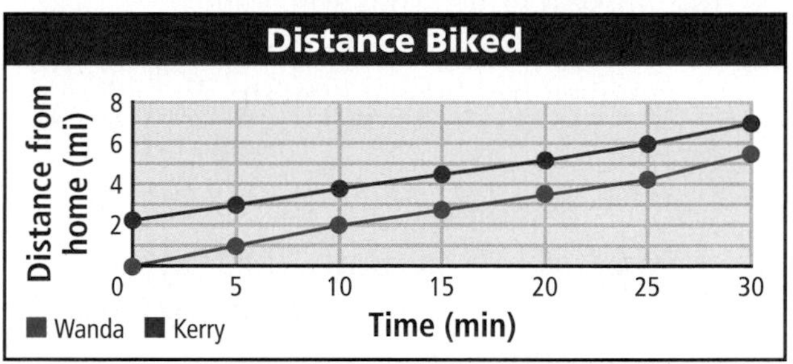

INDEPENDENT PRACTICE

See Example  1

5. Why is this bar graph misleading?
6. What might people believe from the misleading graph?

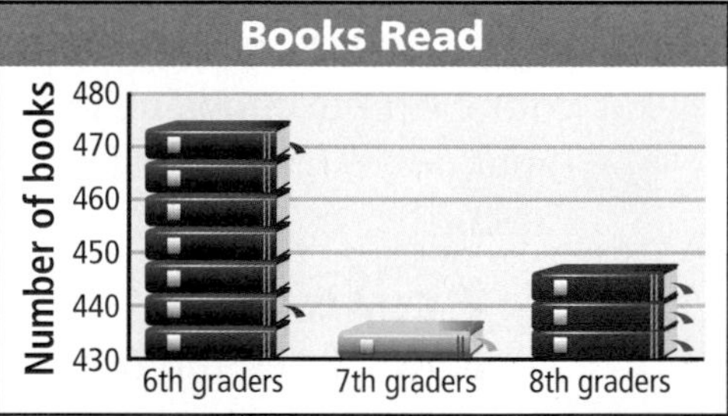

See Example  2

7. Why is this line graph misleading?
8. What might people believe from the misleading graph?

Health LINK

A research company has developed a new medication for lowering cholesterol levels. The table shows the mean monthly cholesterol levels for patients who have been taking the medication for 6 months.

Mean Cholesterol Level of Patients Taking New Medicine	
Month	**Total Cholesterol**
1	300
2	275
3	240
4	230
5	210
6	190

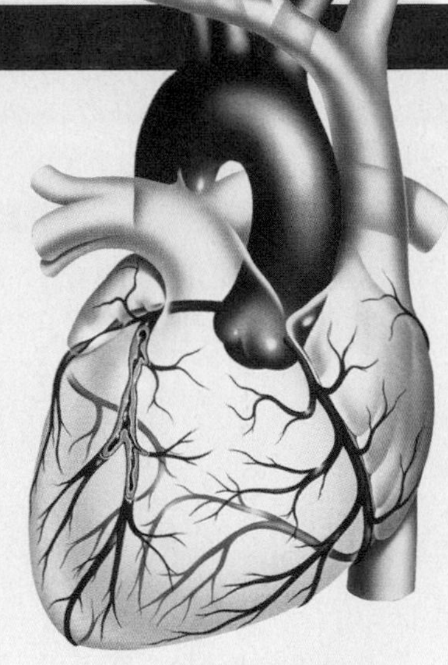

A heart with coronary artery disease, caused by buildup of fatty deposits

9. What kind of graph would you make to display this data? Why?

10. Make a graph that suggests the medication greatly reduces cholesterol levels. Explain how your graph does this.

11. Make a graph that suggests the medication has little effect on cholesterol levels. Explain how your graph does this.

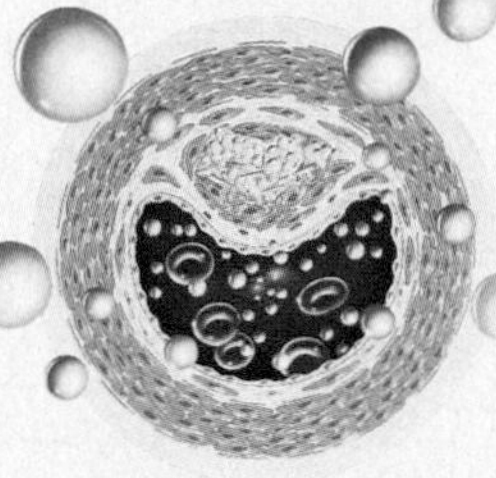

An artery that has been narrowed by high levels of blood cholesterol

12. ***WHAT'S THE QUESTION?*** Look at the entries in the table. If the answer is 110, what is the question?

13. ***WRITE ABOUT IT*** Suppose you saw your graph from Exercise 11 in an advertisement. What do you think it might be an advertisement for? Explain.

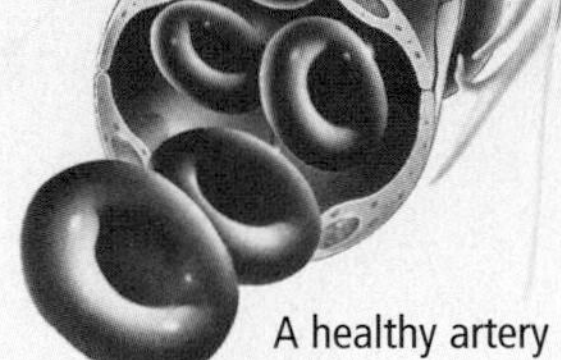

A healthy artery

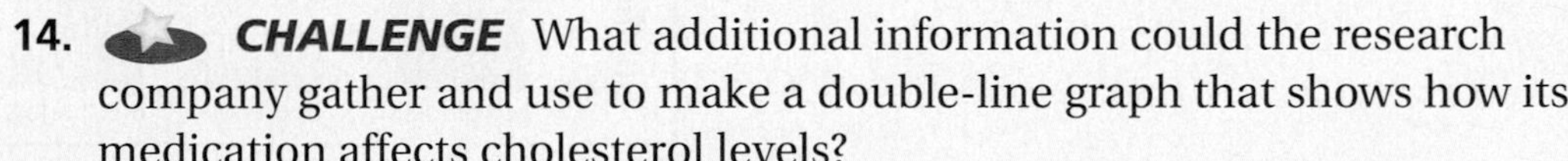

14. ***CHALLENGE*** What additional information could the research company gather and use to make a double-line graph that shows how its medication affects cholesterol levels?

Spiral Review

Find each sum or product. (Lesson 1-5)

15. 13 + 6 + 17 + 24 **16.** $4 \cdot 11 \cdot 3$ **17.** 45 + 12 + 35 + 28

Find each product. Write your answers in simplest form. (Lesson 5-1)

18. $\frac{2}{3} \cdot \frac{1}{5}$ **19.** $\frac{3}{7} \cdot \frac{1}{4}$ **20.** $\frac{2}{9} \cdot \frac{3}{8}$ **21.** $\frac{1}{4} \cdot \frac{6}{7}$

22. **EOG PREP** Which type of graph would you use to display two sets of data that change over time? (Lesson 6-7)

A Bar graph **B** Line graph **C** Tally table **D** Double-line graph

6-9 Stem-and-Leaf Plots

Learn to make and analyze stem-and-leaf plots.

Vocabulary

stem-and-leaf plot

A **stem-and-leaf plot** shows data arranged by place value. You can use a stem-and-leaf plot when you want to display data in an organized way that allows you to see each value.

The Explorer Scouts had a competition to see who could build the highest card tower. The table shows the number of levels reached by each scout.

Bryan Berg and his card model of the Iowa State Capitol

Number of Card-Tower Levels									
12	23	31	50	14	17	25	44	51	20
23	18	35	15	19	15	23	42	21	13

EXAMPLE 1 Creating Stem-and-Leaf Plots

Use the data in the table above to make a stem-and-leaf plot.

Step 1: Group the data by tens digits.

Step 2: Order the data from least to greatest.

Step 3: List the tens digits of the data in order from least to greatest. Write these in the "stems" column.

12 13 14 15 15 17 18 19
20 21 23 23 23 25
31 35
42 44
50 51

Step 4: For each tens digit, record the ones digits of each data value in order from least to greatest. Write these in the "leaves" column.

Step 5: Title the graph and add a key.

Number of Card Tower Levels

Stems	Leaves
1	2 3 4 5 5 7 8 9
2	0 1 3 3 3 5
3	1 5
4	2 4
5	0 1

Key: 1|5 means 15

Helpful Hint

To write 42 in a stem-and-leaf plot, write each digit in a separate column.

4|2

Stem Leaf

EXAMPLE 2 Reading Stem-and-Leaf Plots

Find the least value, greatest value, mean, median, mode, and range of the data.

Stems	Leaves
5	8
6	8 9
7	2 4 8
8	0 4 5 6 8
9	0 0 2 3 6 7 8
10	
11	7

Key: 5|8 means 58

The least stem and least leaf give the least value, 58.

The greatest stem and greatest leaf give the greatest value, 117.

Use the data values to find the mean.

$(58 + \ldots + 117) \div 19 = 85$

The median is the middle value in the table, 86.

To find the mode, look for the number that occurs most often in a row of leaves. Then identify its stem. The mode is 90.

The range is the difference between the greatest and least value.

$117 - 58 = 59$

Helpful Hint

If a stem has no leaves, there are no data points with that stem. In the stem-and-leaf plot in Example 2, there are no data values between 100 and 109.

Think and Discuss

1. Describe how to show 25 on a stem-and-leaf plot.

6-9 Exercises

FOR EOG PRACTICE see page 661

internet connect
Homework Help Online
go.hrw.com Keyword: MR4 6-9

GUIDED PRACTICE

See Example 1

1. Use the data in the table to make a stem-and-leaf plot.

Daily High Temperatures (°F)	45	56	40	39	37	48	51

See Example 2

Find each value of the data.

2. smallest value
3. largest value
4. mean
5. median
6. mode
7. range

Stems	Leaves
1	0 2
2	
3	2
4	1 4

Key: 1|0 means 10

INDEPENDENT PRACTICE

See Example 1

8. Use the data in the table to make a stem-and-leaf plot.

Heights of Plants (cm)	30	12	27	28	15	47	37	28	40	20

See Example 2 **Find each value of the data.**

9. least value **10.** greatest value

11. mean **12.** median

13. mode **14.** range

Stems	Leaves
4	1 2 2
5	1 3
6	7 8

Key: 4|1 means 41

PRACTICE AND PROBLEM SOLVING

For Exercises 15 and 16, write the letter of the stem-and-leaf plot described.

A.

Stems	Leaves
1	0 3 4
2	0 0 1 1 1 3
3	4 5 9
4	8
5	

Key: 1|0 means 10

B.

Stems	Leaves
1	6
2	2 3
3	0 1 4
4	1 4 8
5	8 8 8

Key: 1|6 means 16

C.

Stems	Leaves
1	4
2	
3	
4	3 6 8
5	2 2 4

Key: 1|4 means 14

15. The data set has a mode of 58. **16.** The data set has a median of 48.

Use the table for Exercises 17 and 18.

17. Karla recorded the number of cars with only one passenger that came through a toll booth each day. Use Karla's data to make a stem-and-leaf plot.

Cars with Only One Passenger					
82	103	95	125	88	94
89	92	94	99	87	80
109	101	100	83	124	81

18. ***WHAT'S THE ERROR?*** Karla's classmate looked at the stem-and-leaf plot and said that the mean number of cars with only one passenger is 4. Explain Karla's classmate's error. What is the correct mean?

19. ***CHALLENGE*** Josh is the second youngest of 4 teenage boys, all 2 years apart in age. Josh's mother is 3 times as old as Josh is, and she is 24 years younger than her father. Make a stem-and-leaf plot to show the ages of Josh, his brothers, his mother, and his grandfather.

Spiral Review

20. Holly read 128 pages on Monday, 239 pages on Tuesday, and 152 pages each day on Wednesday through Friday. Estimate the number of pages Holly read to the nearest ten. (Lesson 1-2)

Find each sum or difference. (Lesson 3-3)

21. 12.56 + 8.91 **22.** 19.05 − 2.27 **23.** 5 + 8.25 + 10.2 **24.** 40 − 20.66

25. **EOG PREP** Which value is *not* always a number in the data set it represents? (Lesson 6-2)

A Mode B Lowest value C Highest value D Mean

EXTENSION

Box-and-Whisker Plots

Learn to make and read box-and-whisker plots.

Vocabulary

box-and-whisker plot
lower extreme
lower quartile
upper quartile
upper extreme

A **box-and-whisker plot** shows how data is distributed. To make a box-and-whisker plot of a data set, you need to know the following five values:

- the **lower extreme**, the least value
- the **lower quartile**, the median of the lower half of the data
- the median of the data
- the **upper quartile**, the median of the upper half of the data
- the **upper extreme**, the greatest value

EXAMPLE 1

Making a Box-and-Whisker Plot

Use the data in the table to make a box-and-whisker plot.

Cans Collected (lb)	10	23	15	17	26	27	21	22	19	11	16

10, 11, 15, 16, 17, 19, 21, 22, 23, 26, 27 — *Order from least to greatest.*

10, 11, 15, 16, 17, 19, 21, 22, 23, 26, (27) — *Find the upper extreme.*

(10), 11, 15, 16, 17, 19, 21, 22, 23, 26, 27 — *Find the lower extreme.*

10, 11, 15, 16, 17, (19), 21, 22, 23, 26, 27 — *Find the median.*

10, 11, (15), 16, 17, 19, 21, 22, 23, 26, 27 — *Find the lower quartile.*

10, 11, 15, 16, 17, 19, 21, 22, (23), 26, 27 — *Find the upper quartile.*

Step 1: Make a box using the median and the upper and lower quartiles.

Step 2: Place a dot at the upper and lower extremes.

Step 3: Connect the dots to the box with segments called whiskers.

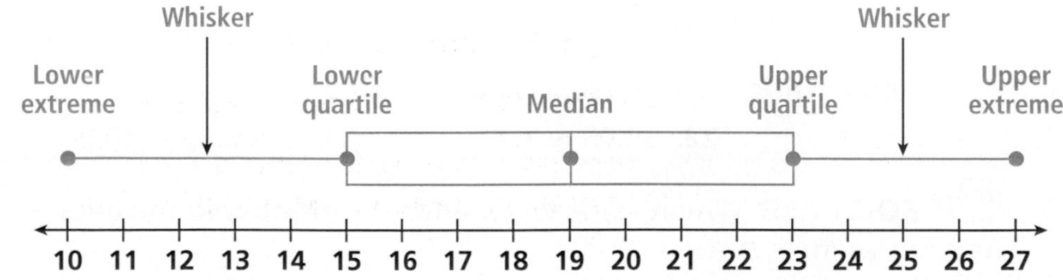

The range is the difference between the upper and lower extremes.
$27 - 10 = 17$

The interquartile range is the difference between the upper and lower quartiles. $23 - 15 = 8$

EXTENSION

Exercises

Use the box-and-whisker plot to find each value.

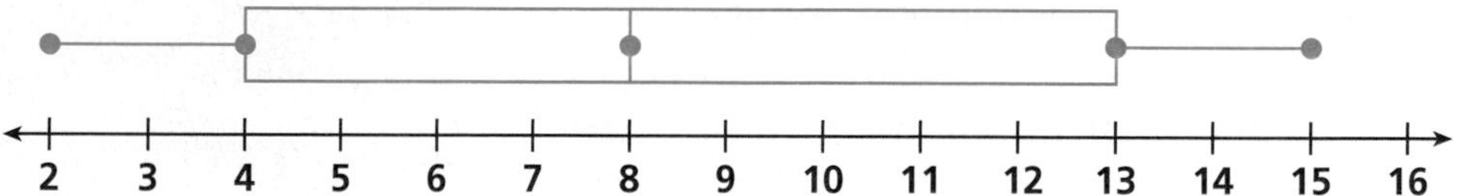

1. lower extreme

2. median

3. upper quartile

4. lower quartile

5. upper extreme

6. interquartile range

7. Use the data in the table to create a box-and-whisker plot.

Waiting Times for Movie Tickets (min)					
0	0	5	5	2	9
4	4	1	8	20	3

The two box-and-whisker plots represent the test scores for two different math classes. Use the box-and-whisker plots for Exercises 8–12.

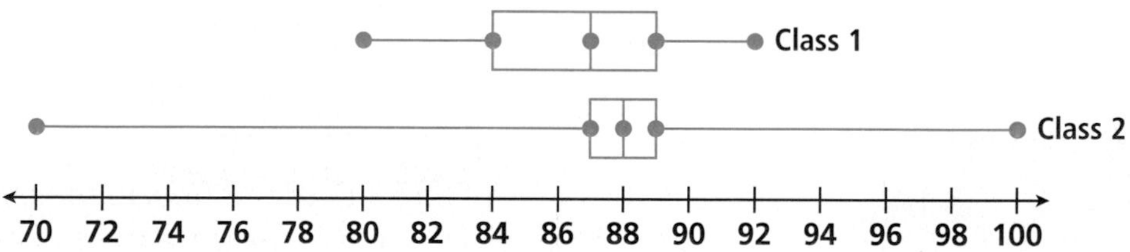

8. What is each class's median score?

9. Which class's interquartile range is the greatest?

10. What is the highest score from either class?

11. What is the difference between the ranges of the two sets of scores?

12. Can you use these plots to determine which class has more students? Explain why or why not.

Problem Solving on Location

NORTH CAROLINA

Rainfall

In the months leading up to September 1999, most of North Carolina was experiencing a drought. Then between September 4 and October 17, Hurricanes Dennis, Floyd, and Irene hit the state. These three storms dropped so much rain that rainfall records that had stood for more than 80 years were broken.

For 1–5, use the table.

The table lists various locations within North Carolina and the amount of rainfall each received during the three hurricanes.

Rainfall During September and October 1999 Hurricanes (in.)			
Location	**Hurricane Dennis**	**Hurricane Floyd**	**Hurricane Irene**
Chapel Hill	12.52	4.67	0.84
Enfield	7.01	11.84	4.30
Greenville	7.03	12.63	3.29
Raleigh	8.46	6.55	1.50
Whiteville	1.52	16.76	5.97
Willamston	7.20	16.28	5.54

1. Find the range of the rainfall amounts in the table during Hurricane Dennis. Which location had the least amount of rainfall during this storm? the greatest?

2. Find the mean rainfall amount for the six locations during Hurricane Floyd. Round your answer to the nearest hundredth.

3. Find the median rainfall amount in the table during Hurricane Irene. Round your answer to the nearest hundredth, if necessary.

4. During which hurricane did the most rain fall in Raleigh?

5. Which of the six locations received the most rainfall during a single hurricane? Which hurricane was it?

Pottery

For almost 300 years, potters in Seagrove, North Carolina, and the nearby area have been turning pots by hand. Today, about 100 potteries in the area continue the tradition.

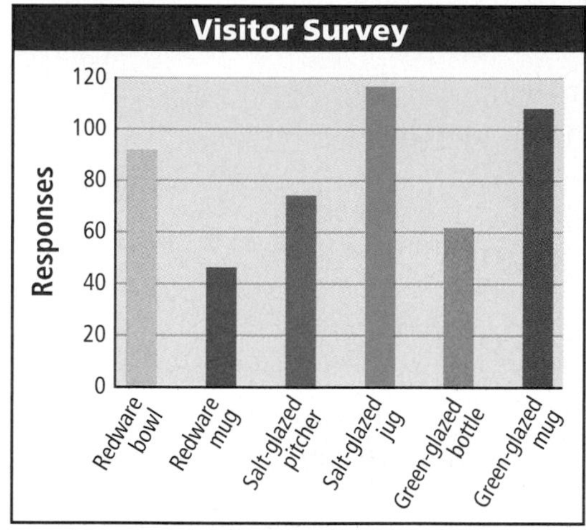

For 1–5, use the graph.

Students in Seagrove surveyed 500 visitors. The visitors were asked to choose a favorite piece of pottery from among six items. The bar graph shows the responses of the people surveyed.

1. Use the bar graph to determine which of the six pieces of pottery was least popular with the 500 visitors.
2. Which three examples of pottery were the most popular pieces in the survey?
3. Estimate how many more visitors chose the redware bowl than the green-glazed bottle.
4. About what fraction of the visitors responding to the survey liked the green-glazed mug the most?
5. Use the results from the survey to place the items in order from the most popular to the least popular.

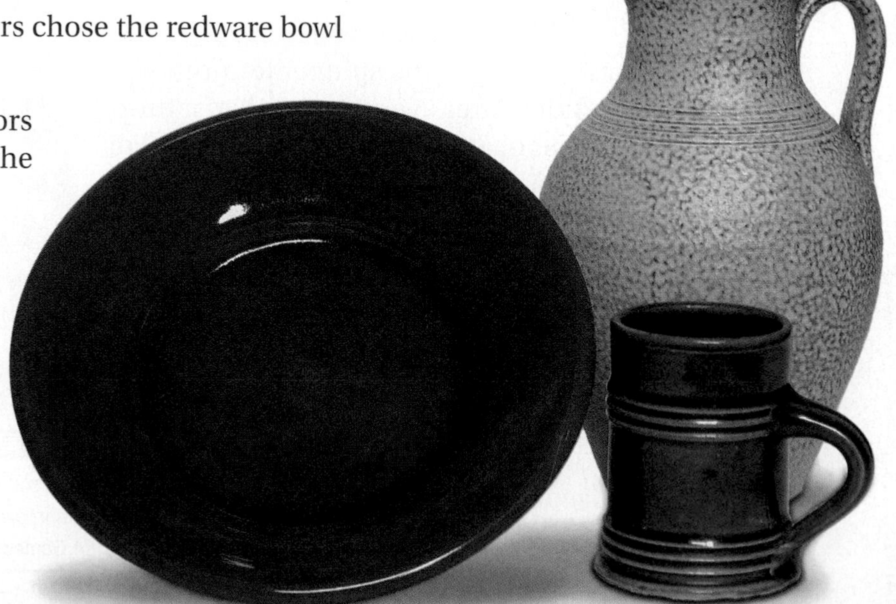

MATH-ABLES

A Thousand Words

Did you ever hear the saying "A picture is worth a thousand words"? A graph can be worth a thousand words too!

Each of the graphs below tells a story about a student's trip to school. Read each story and think about what each graph is showing. Can you match each graph with its story?

Kyla:
I rode my bike to school at a steady pace. I had to stop and wait for the light to change at two intersections.

Tom:
I walked to my bus stop and waited there for the bus. After I boarded the bus, it was driven straight to school.

Megan:
On my way to school, I stopped at my friend's house. She wasn't ready yet, so I waited for her. Then we walked to school.

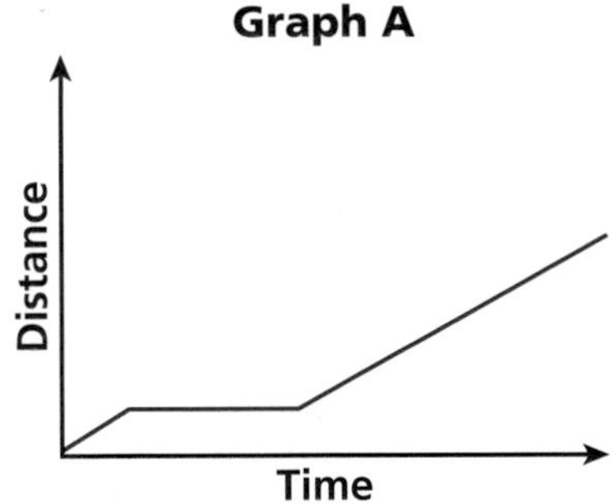

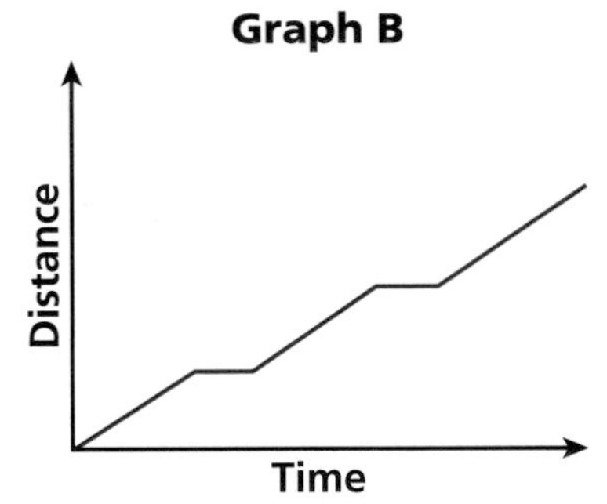

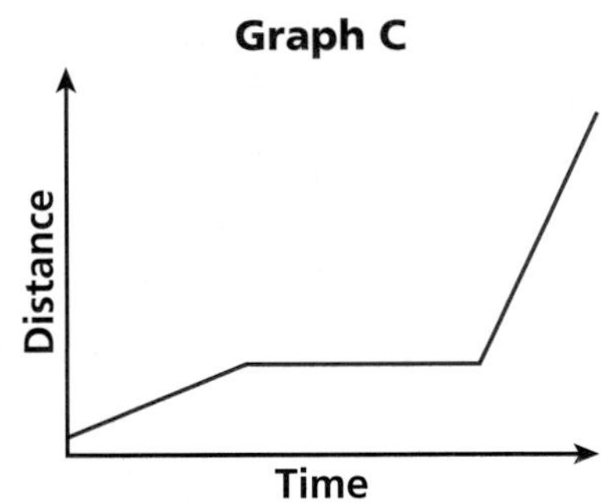

Spinnermeania

Round 1: On your turn, spin the spinner four times and record the results. After everyone has had a turn, find the mean, median, and mode of your results. For every category in which you have the highest number, you get one point. If there is a tie in a category, each player with that number gets a point. If your data set has more than one mode, use the greatest one.

Spin five times in round 2, eight times in round 3, ten times in round 4, and twelve times in round 5. The player with the highest score at the end of five rounds is the winner.

internet connect

Go to ***go.hrw.com*** for a complete set of game pieces.
KEYWORD: MR4 Game6

Create Line Graphs

internet connect
Lab Resources Online
go.hrw.com
KEYWORD: MR4 TechLab6

Activity

The table shows the tide heights at various times. Use the data to make a line graph.

Time (A.M.)	9:00	9:06	9:12	9:18	9:24
Height (ft)	2.41	2.31	2.21	2.10	2.00

1 Enter the data. Put times in the **L1** list and heights in the **L2** list. You cannot enter times, so use 0 for 9:00, 1 for 9:06, and so on.

Times: STAT ENTER 0 ENTER 1 ENTER 2 ENTER 3 ENTER 4 ENTER

Heights: ▶ 2.41 ENTER 2.31 ENTER 2.21 ENTER 2.10 ENTER 2.00 ENTER

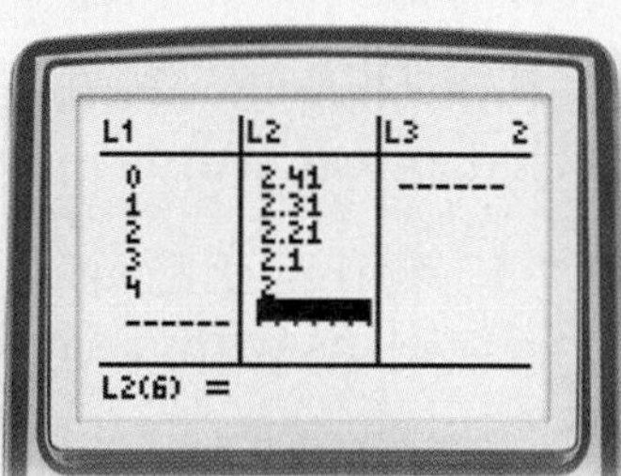

2 Choose a scale. Press WINDOW. The calculator uses **X** for the horizontal axis (time) and **Y** for the vertical axis (height). Set **Xmin** to 0, **Xmax** to 5, **Ymin** to 1, and **Ymax** to 3.

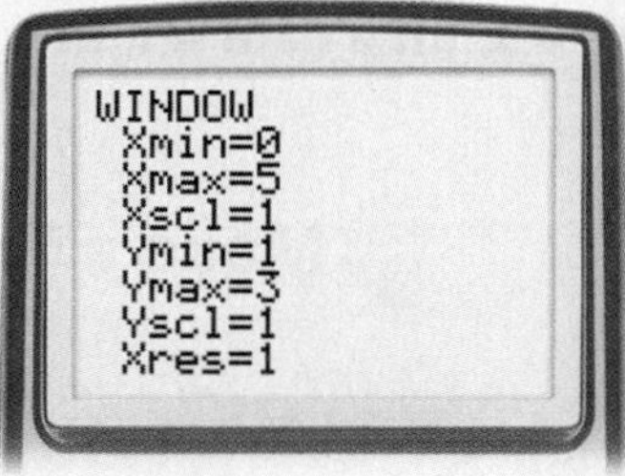

3 Make the graph.

Select Plot 1: 2nd PLOT Y= ENTER ENTER.
Use the arrow keys to highlight the line graph icon and press ENTER. Then press GRAPH.

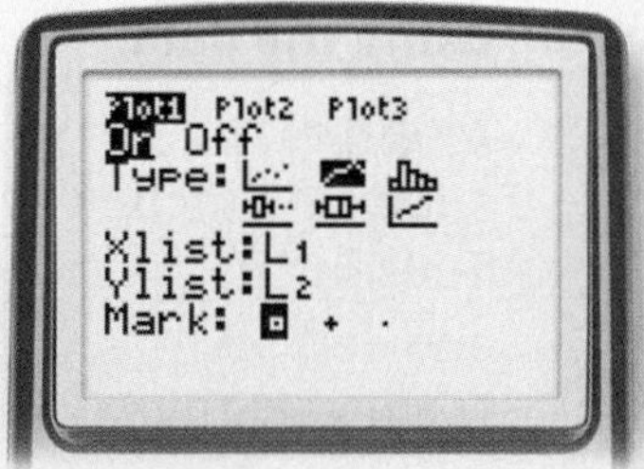

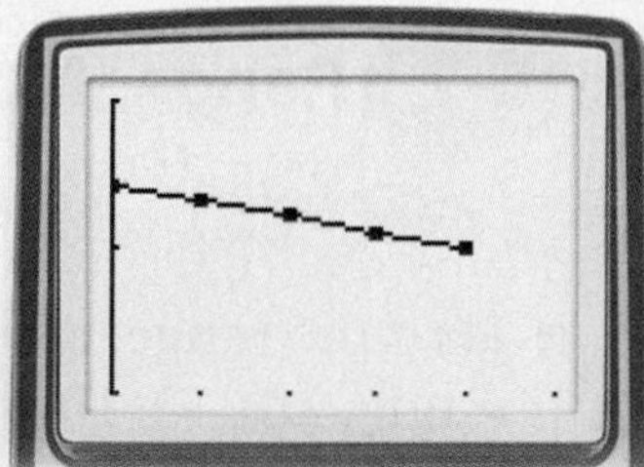

Think and Discuss

1. What does the line graph tell you about the data? What do you think was happening when the measurements were taken?

Try This

1. The table shows amounts of snowfall during one winter. Use the data to make a line graph.

Month	Nov	Dec	Jan	Feb	Mar
Snowfall (in.)	2	6	9	6	2

Chapter 6

Study Guide and Review

Vocabulary

Complete the sentences below with vocabulary words from the list above. Words may be used more than once.

1. A(n) __?__ uses vertical or horizontal bars to show the number of items within each interval.
2. A point can be located by using a(n) __?__ of numbers such as (3, 5).
3. In a data set, the __?__ is the value or values that occur most often.

6-1 Make a Table (pp. 272–274)

EXAMPLES

- **Make a table using the data.** Monday it snowed 2 inches. Tuesday it snowed 3.5 inches. Thursday it snowed 4.25 inches.

Day	Snowfall
Mon	2 in.
Tue	3.5 in.
Thu	4.25 in.

EXERCISES

4. Make a table using the data on snake lengths.

An anaconda can be up to 35 ft long. A diamond python can be up to 21 ft long. A king cobra can be up to 19 ft long. A boa constrictor can be up to 16 ft long.

6-2 Range, Mean, Median, and Mode (pp. 275–277)

EXAMPLE

- **Find the range, mean, median, and mode. 7, 8, 12, 10, 8**

range: $12 - 7 = 5$
mean: $7 + 8 + 8 + 10 + 12 = 45$
$45 \div 5 = 9$
median: 8
mode: 8

EXERCISES

Find the range, mean, median, and mode.

5.

Hours Worked Each Week						
32	39	39	38	36	39	36

6-3 Additional Data and Outliers (pp. 278–281)

EXAMPLE

- **Find the mean, median, and mode with and without the outlier.**

10, 4, 7, 8, 34, 7, 7, 12, 5, 8 *The outlier is 34.*
Without: **mean** = 10.2, **mode** = 7,
median = 7.5
With: **mean** ≈ 7.555, **mode** = 7, **median** = 7

EXERCISES

Find the mean, median, and mode of each data set with and without the outlier.

6. 12, 11, 9, 38, 10, 8, 12
7. 34, 12, 32, 45, 32
8. 16, 12, 15, 52, 10, 13

6-4 Bar Graphs (pp. 284–287)

EXAMPLE

- **Which grades have more than 200 students?** 6th grade and 8th grade

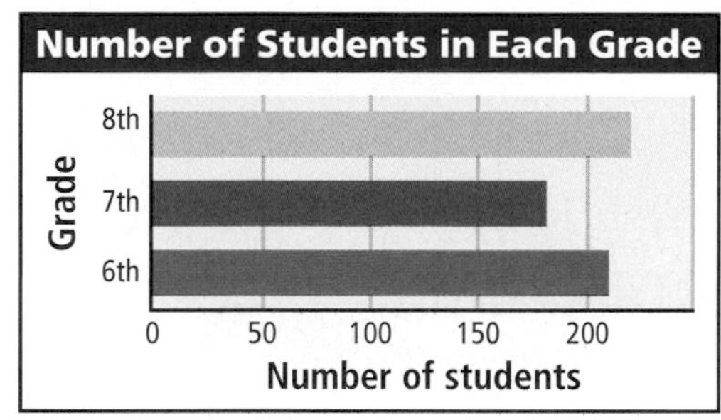

EXERCISES

Use the bar graph at left for Exercise 9.

9. Which grade has the most students?
10. Use the data to make a bar graph.

Test	Math	English	History	Science
Grade	95	85	90	80

6-5 Frequency Tables and Histograms (pp. 290–293)

EXAMPLE

- **Make a frequency table with intervals.**

Ages of people at Irene's birthday party: 37, 39, 18, 15, 13

Ages of People at Irene's Birthday Party				
Ages	13–19	20–26	27–33	34–40
Frequency	3	0	0	2

EXERCISES

11. Make a frequency table with intervals.

Points Scored					
6	4	5	4	7	10

12. Use the frequency table from Exercise 11 to make a histogram.

6-6 Ordered Pairs (pp. 294–296)

EXAMPLE

- **Name the ordered pair for *A*.**

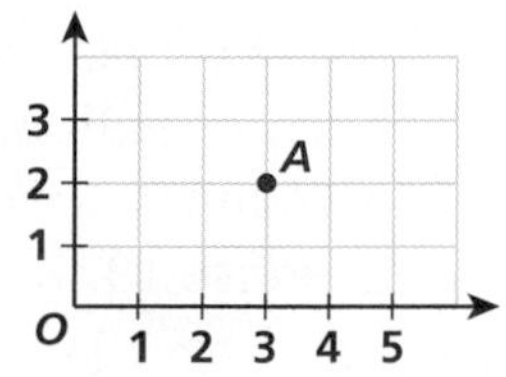

A is at (3, 2).

EXERCISES

Name the ordered pair for each location.

13. Bob's house
14. toy store

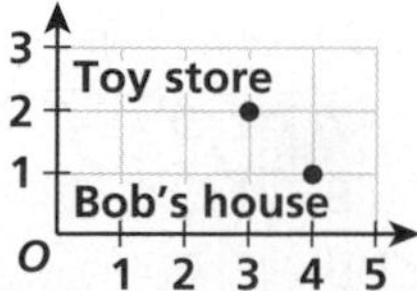

6-7 Line Graphs (pp. 297–300)

EXAMPLE

■ **Use the data to make a line graph.**

Temperature Recording (°F)				
Day 1	Day 2	Day 3	Day 4	Day 5
32	36	38	40	36

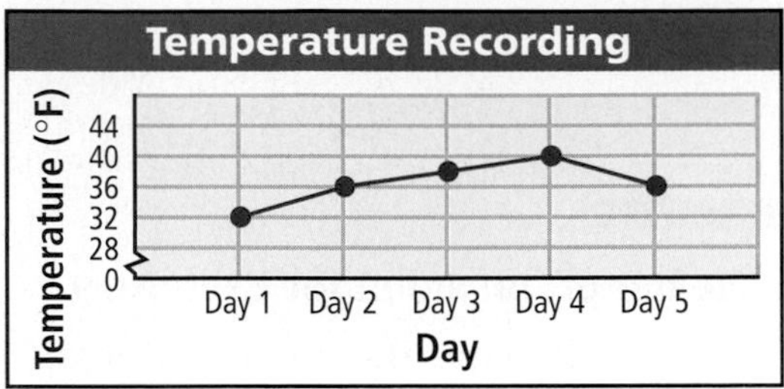

EXERCISES

15. Use the data to make a line graph.

Bookstore Sales			
Jan	Feb	Mar	Apr
$425	$320	$450	$530

Use your line graph from Exercise 15.

16. When were bookstore sales the greatest?

17. Describe the trend in bookstore sales over the four months.

6-8 Misleading Graphs (pp. 301–304)

EXAMPLES

■ **Why is this graph misleading?**

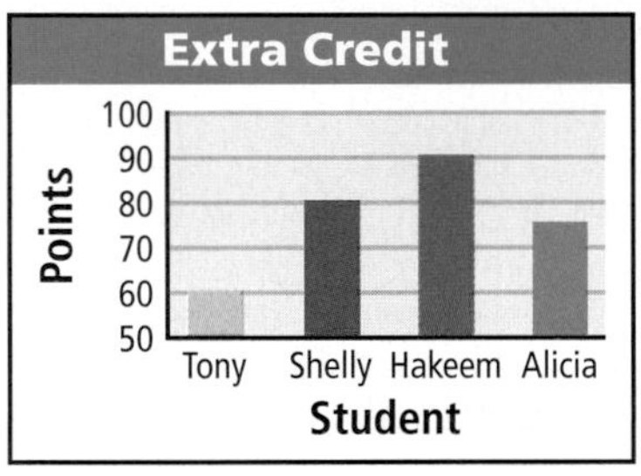

The lower part of the scale is missing.

EXERCISES

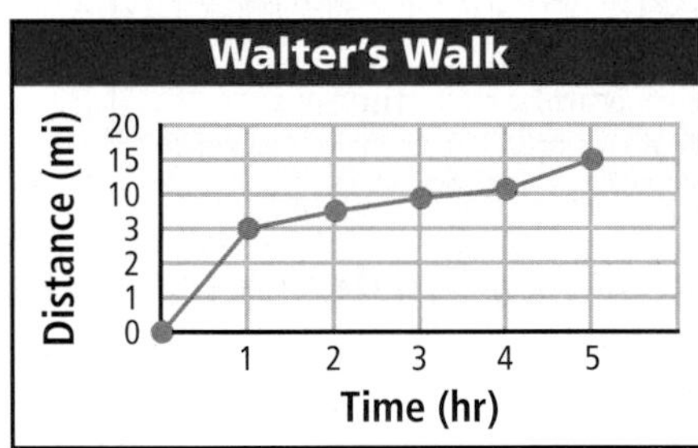

18. Explain why this graph is misleading.

6-9 Stem-and-Leaf Plots (pp. 305–307)

EXAMPLE

■ **Make a stem-and-leaf plot.**

Test Scores							
66	72	80	92	88	86	85	94

Stems	Leaves
6	6
7	2
8	0 5 6 8
9	2 4

Key: 6|6 means 66

EXERCISES

19. Make a stem-and-leaf plot.

Basketball Scores							
22	26	34	46	20	44	40	28

20. List the least value, greatest value, mean, median, mode, and range of the data in the stem-and-leaf plot from Exercise 19.

Chapter Test

Chapter 6

1. Use the data about sound to make a table.

   The loudness of a sound is measured by the size of its vibrations. The unit of measurement is the decibel (dB). A soft whisper is 30 dB. Conversation is 60 dB. A loud shout is 100 dB. The pain threshold for humans is 130 dB. An airplane takeoff at 100 ft is 140 dB.

Use the bar graph for Exercises 2–5.

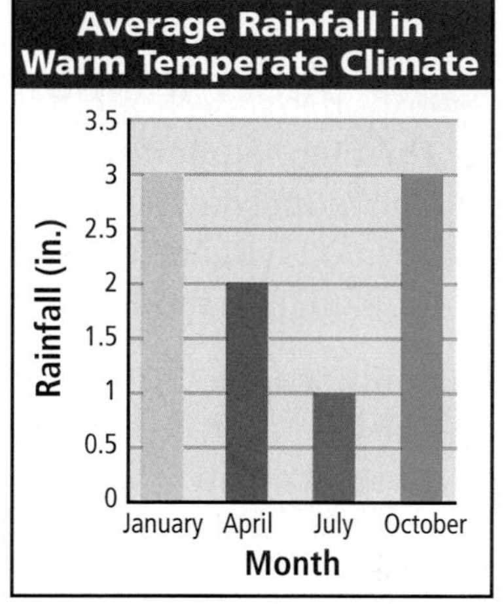

2. Find the mean, median, and mode of the rainfall amounts.
3. Which month had the lowest average rainfall?
4. Which months had rainfall amounts greater than 2 inches?
5. In a tropical climate, the rainfall in January was 0 in. In April, it was 2 in. In July, it was 22 in. In October, it was 7 in. Make a double-bar graph to compare the rainfall in the tropical climate and the warm temperate climate.
6. Use the book report data to make a frequency table. Then use the frequency table to make a histogram.

Scores on Book Reports							
82	88	60	75	95	92	71	82
78	93	87	76	90	80	70	85

Name the ordered pair for each point on the grid.

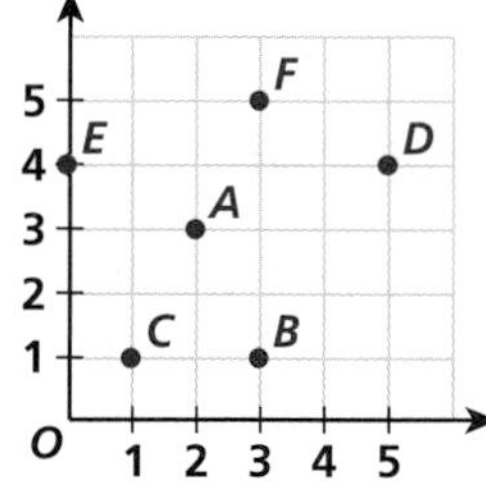

7. *A*
8. *B*
9. *C*
10. *D*
11. *E*
12. *F*

Graph and label the following points on a coordinate grid.

13. *T*(3, 4)
14. *M*(0, 6)
15. *P*(5, 1)
16. *S*(3, 2)

17. Make a double-line graph of the height data at right. How could you alter your graph to magnify the difference between males and females?

Average Heights				
	Birth	**4 yr**	**12 yr**	**18 yr**
Male	50 cm	100 cm	150 cm	180 cm
Female	50 cm	100 cm	152 cm	164 cm

18. Make a stem-and-leaf plot of the push-up data. Use your stem-and-leaf plot to find the mean, median, and mode of the data.

Number of Push-ups Performed						
35	33	25	45	52	21	18
41	27	35	40	53	24	38

Chapter 6

Performance Assessment

Show What You Know

Create a portfolio of your work from this chapter. Complete this page and include it with your four best pieces of work from Chapter 6. Choose from your homework or lab assignments, mid-chapter quiz, or any journal entries you have done. Put them together using any design you want. Make your portfolio represent what you consider your best work.

Short Response

1. The high temperature on Monday was 54°F. On Tuesday, it was 62°F. On Wednesday, it was 65°F. On Thursday, it was 60°F. On Friday, it was 62°F. Organize this data in a table. Find the range, mean, median, and mode of the data. Show your work.

2. Emily scored 75, 85, 35, 85, 70, and 10 on her first six math quizzes. The best score Emily can make on a math quiz is 100. What is the greatest mean score Emily could have after taking the seventh math quiz? Show your work.

3. The batting averages for the players on two softball teams are given below. Assuming each player had the same number of at bats, tell which team has the higher batting average. Show your work.

Team A	.213	.138	.115	.152	.297	.101	.198	.176	.189
Team B	.145	.313	.103	.228	.184	.183	.261	.149	.163

Extended Problem Solving

4. A group of people at a shopping mall were surveyed to determine their favorite frozen fruit bars. The graph at right shows the results of the survey.
   a. Explain why the graph is misleading.
   b. Use the same data to construct a graph that is not misleading.
   c. Create a cumulative frequency table of the data. How many people at the mall participated in the survey?

Performance Assessment

internet connect

EOG Test Practice Online
go.hrw.com Keyword: MR4 TestPrep

Getting Ready for EOG

Chapter 6

Cumulative Assessment, Chapters 1–6

1. What is the mode of the following data set?
17, 13, 14, 13, 21, 18, 16, 19
   A 13
   B 16
   C 16.5
   D 16.375

2. What is the LCM of 6, 8, and 12?
   A 2
   B 12
   C 24
   D 48

3. Which expression has the greatest value?
   A $5.35 \cdot 1.6$
   B $12\frac{2}{3} \div 2$
   C $35.7 \div 6.8$
   D $2\frac{1}{3} \cdot 3\frac{3}{5}$

4. Which is a type of graph that uses bars and intervals to display data?
   A Stem-and-leaf plot
   B Histogram
   C Double-line graph
   D Cumulative frequency

5. $7\frac{5}{10} + 4\frac{3}{4}$
   A $11\frac{4}{7}$
   B $11\frac{1}{4}$
   C $12\frac{3}{4}$
   D $12\frac{1}{4}$

6. Which decimal is equivalent to $4\frac{3}{8}$?
   A 4.125
   B 4.375
   C 4.38
   D 4.8

TIP! TEST TAKING TIP!

Eliminate possibilities. If there are an odd number of items in the data set, the median will be a value from the set.

7. Find the mean and median of the following data set: 25, 30, 27, 26, 32, 32, and 24.
   A Mean: 28; median: 32
   B Mean: 28; median: 27
   C Mean: 27; median: 28
   D Mean: 28; median: 28

8. What is the outlier of the following data set: 55, 62, 71, 64, 28, 64?
   A 64
   B 63
   C 36
   D 28

9. ***SHORT RESPONSE*** Terry and Carl collect baseball cards. Terry has 151 fewer cards than Carl. Let *c* represent the number of cards Carl has and write an expression for the number of cards Terry has. Explain how to find the number of cards Carl has if Terry has 728.

10. ***SHORT RESPONSE*** Name the ordered pair for each point on the coordinate grid. Suppose a new point *F* is two units to the left and three units above point *B*. Explain how to find the ordered pair for point *F*.

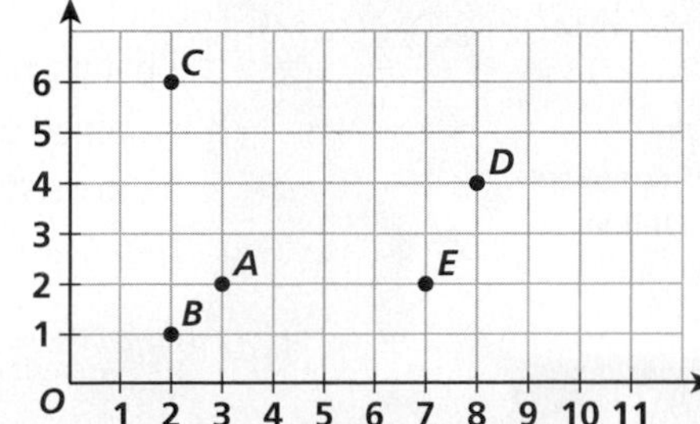

Getting Ready for EOG

Chapter 7

Plane Geometry

TABLE OF CONTENTS

internet connect

Chapter Opener Online
go.hrw.com
KEYWORD: MR4 Ch7

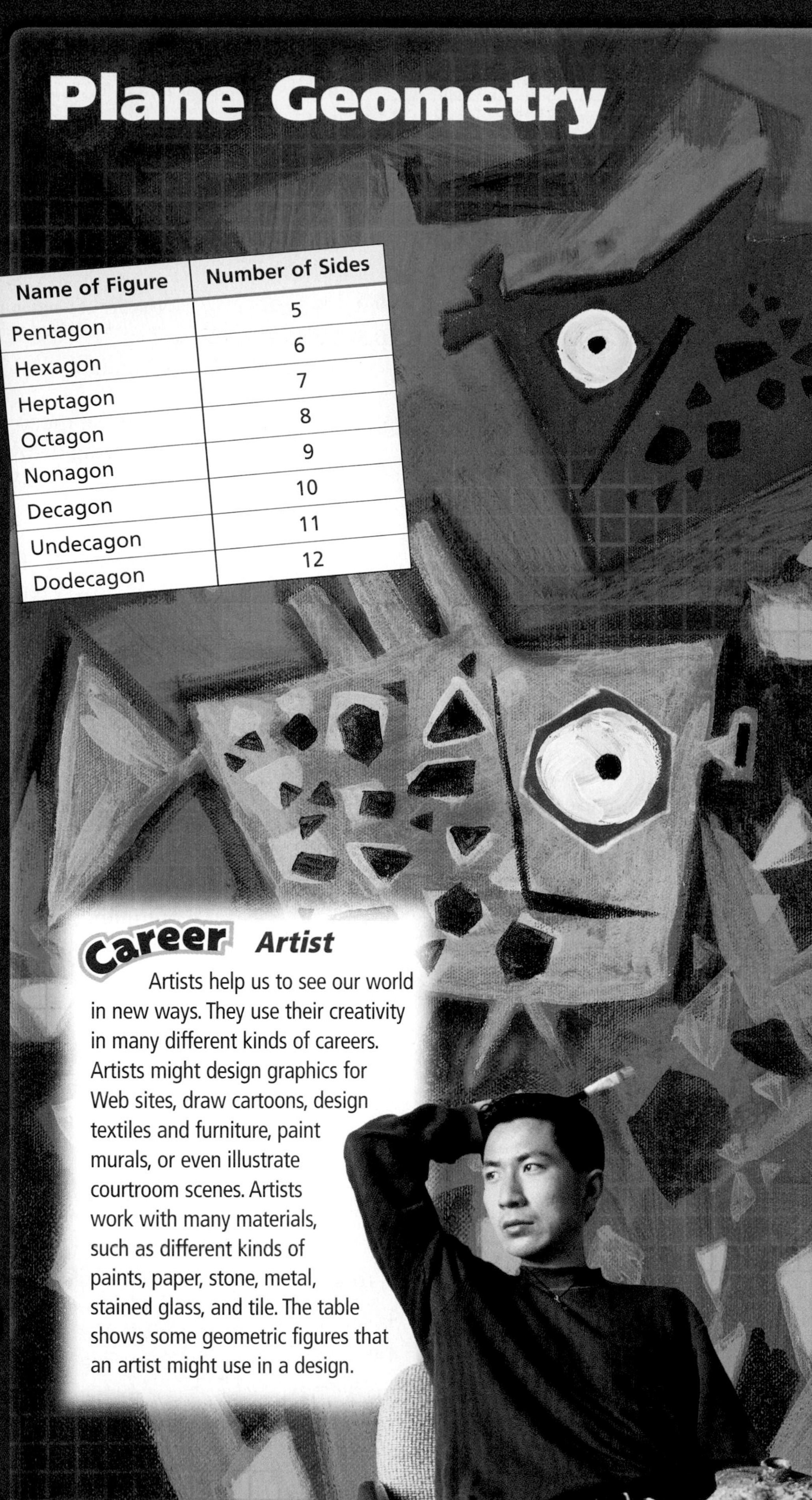

Name of Figure	Number of Sides
Pentagon	5
Hexagon	6
Heptagon	7
Octagon	8
Nonagon	9
Decagon	10
Undecagon	11
Dodecagon	12

Career *Artist*

Artists help us to see our world in new ways. They use their creativity in many different kinds of careers. Artists might design graphics for Web sites, draw cartoons, design textiles and furniture, paint murals, or even illustrate courtroom scenes. Artists work with many materials, such as different kinds of paints, paper, stone, metal, stained glass, and tile. The table shows some geometric figures that an artist might use in a design.

ARE YOU READY?

Choose the best term from the list to complete each sentence.

protractor
ruler
triangle
quadrilateral
horizontal
vertical
clockwise
counterclockwise

1. A closed figure with three sides is a __?__, and a closed figure with four sides is a __?__.
2. A __?__ is used to measure and draw angles.
3. 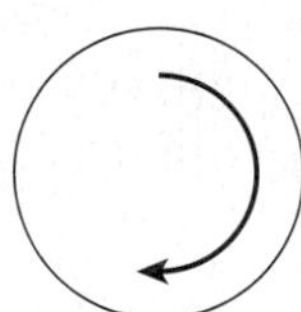

   The arrow inside the circle is moving __?__.
4. A line that extends left to right is __?__.

Complete these exercises to review skills you will need for this chapter.

✔ Measure with Customary and Metric Units

Use an inch ruler to measure each line segment to the nearest $\frac{1}{2}$ in.

5. ______________________
6. ________________

Use a centimeter ruler to measure each line segment to the nearest centimeter.

7. _____________
8. ______

✔ Identify Polygons

Tell how many sides and angles each figure has.

9. 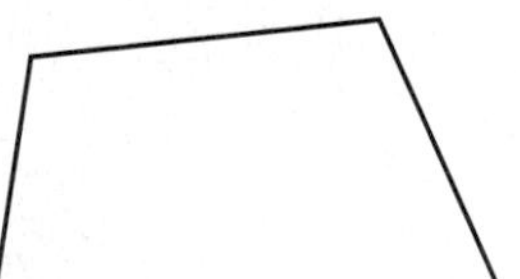
10. 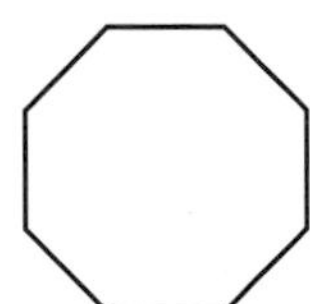
11. 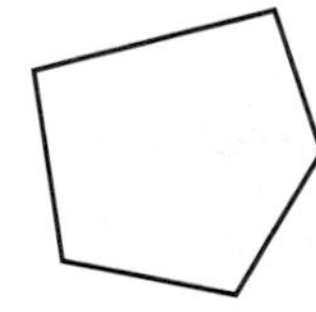

✔ Identify Congruent Figures

Which two figures are exactly the same size and shape but in different positions?

12. A 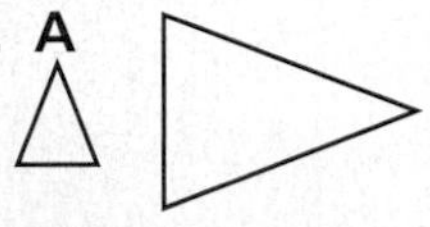 B 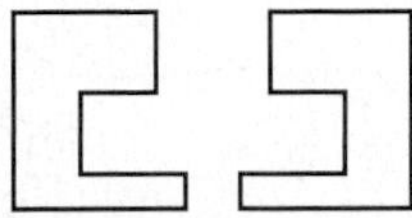 C 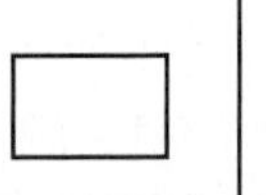 D 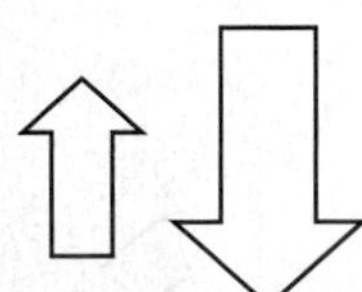

7-1 Points, Lines, and Planes

Learn to describe figures by using the terms of geometry.

Vocabulary

point

line

plane

line segment

ray

The building blocks of geometry are *points, lines,* and *planes.*

Definition	Figure	Name
A **point** is an exact location.	• *P* *A point is named by a capital letter.*	point *P*, *P*
A **line** is a straight path that extends without end in opposite directions.	*A* *B* *A line is named by two points on the line.*	line *AB*, $\overleftrightarrow{AB}$, line *BA*, $\overleftrightarrow{BA}$
A **plane** is a flat surface that extends without end in all directions.	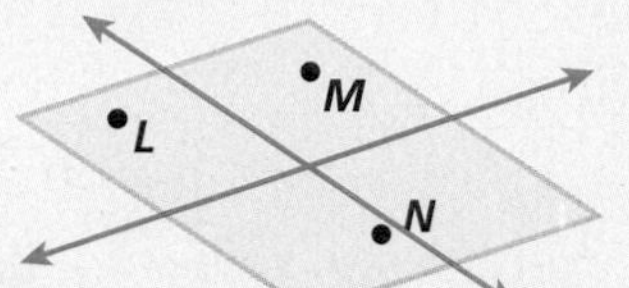 *A plane is named by three points on the plane that are not on the same line.*	plane *LMN*, plane *MLN*, plane *NLM*

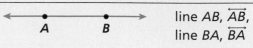

EXAMPLE 1 Identifying Points, Lines, and Planes

Use the diagram to name each geometric figure.

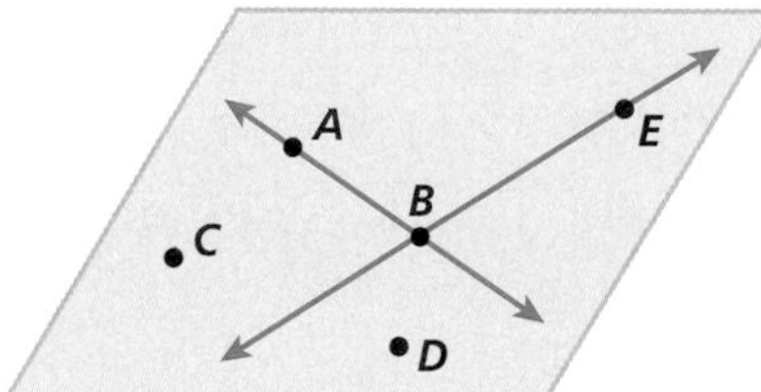

A three points

A, *C*, and *D*

Five points are labeled: points A, B, C, D, and E.

B two lines

$\overleftrightarrow{AB}$ and $\overleftrightarrow{BE}$

You can also write $\overleftrightarrow{BA}$ *and* $\overleftrightarrow{EB}$.

C a point shared by two lines

point *B*

Point B is a point on $\overleftrightarrow{AB}$ *and* $\overleftrightarrow{BE}$.

D a plane

plane *ADC*

Use any three points in the plane that are not on the same line. Write the three points in any order.

Line segments and rays are parts of lines. Use points on a line to name line segments and rays.

A **line segment** is made of two endpoints and all the points between the endpoints.	 A line segment is named by its endpoints.	line segment XY, $\overline{XY}$, line segment YX, $\overline{YX}$
A **ray** has one endpoint. From the endpoint, the ray extends without end in one direction only.	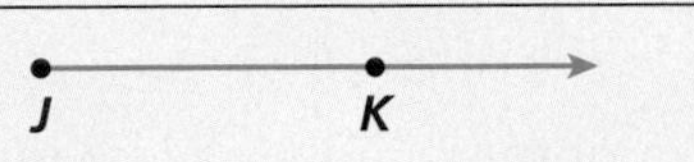 *A ray is named by its endpoint first followed by another point on the ray.*	ray JK, $\overrightarrow{JK}$

EXAMPLE 2 Identifying Line Segments and Rays

Use the diagram to give a possible name to each figure.

A **three different line segments**

$\overline{TU}$, $\overline{UV}$, and $\overline{TV}$

You can also write $\overline{UT}$, $\overline{VU}$, and $\overline{VT}$.

B **three ways to name the line**

$\overleftrightarrow{UT}$, $\overleftrightarrow{VU}$, and $\overleftrightarrow{VT}$

You can also write $\overleftrightarrow{TU}$, $\overleftrightarrow{UV}$, and $\overleftrightarrow{TV}$.

C **six different rays**

$\overrightarrow{TU}$, $\overrightarrow{TV}$, $\overrightarrow{VT}$, $\overrightarrow{VU}$, $\overrightarrow{UV}$, and $\overrightarrow{UT}$

D **another name for ray TU**

$\overrightarrow{TV}$

T is still the endpoint. V is another point on the ray.

Think and Discuss

1. **Name** the geometric figure suggested by each of the following: a page of a book; a dot (also called a *pixel*) on a computer screen; the path of a jet across the sky.
2. **Explain** how $\overrightarrow{XY}$ is different from $\overleftrightarrow{XY}$.
3. **Explain** how $\overline{AB}$ is different from $\overleftrightarrow{AB}$.

7-1 Exercises

FOR EOG PRACTICE

see page 662

internet connect

Homework Help Online

go.hrw.com Keyword: MR4 7-1

GUIDED PRACTICE

See Example 1 **Use the diagram to name each geometric figure.**

1. two points
2. a line
3. a point shared by two lines
4. a plane

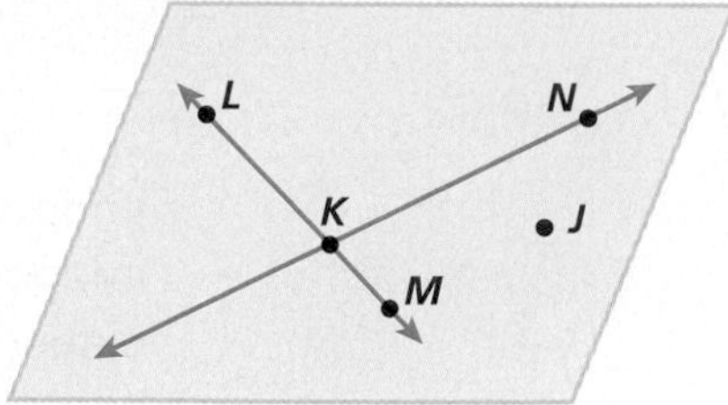

See Example 2 **Use the diagram to give a possible name to each figure.**

5. two different ways to name the line
6. four different names for rays
7. another name for $\overrightarrow{AC}$

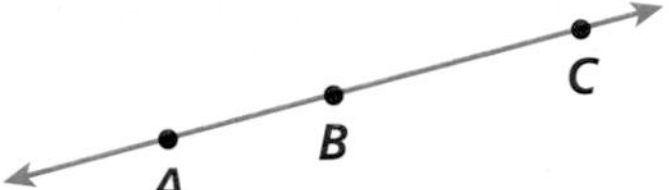

INDEPENDENT PRACTICE

See Example 1 **Use the diagram to name each geometric figure.**

8. three points
9. two lines
10. a point shared by a line and a ray
11. a plane

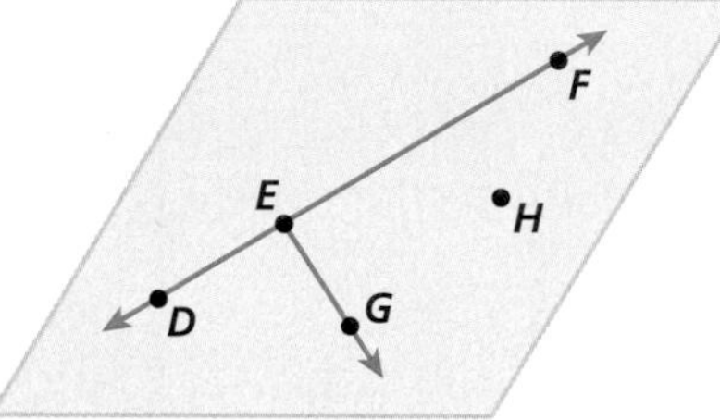

See Example 2 **Use the diagram to give a possible name to each figure.**

12. two different line segments
13. six different names for rays
14. another name for $\overline{YX}$

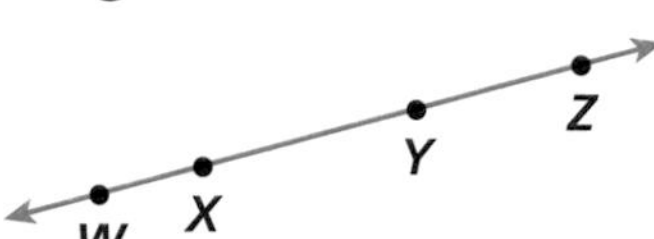

PRACTICE AND PROBLEM SOLVING

Use the diagram to find a name for each geometric figure described.

15. a point shared by three lines
16. two points on the same line
17. two different rays
18. another name for $\overrightarrow{AD}$
19. two different names for the same line

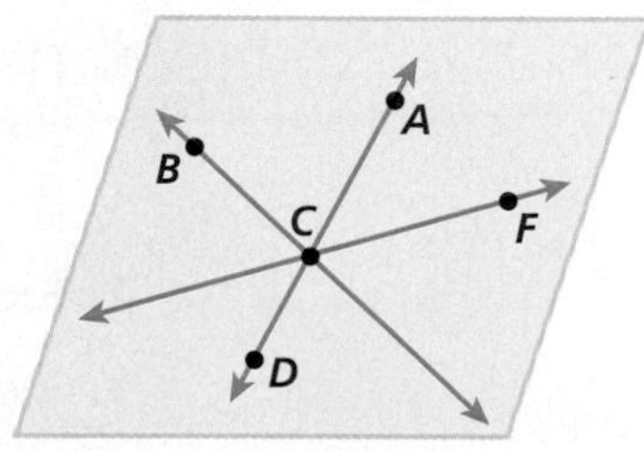

Geography LINK

Mapmakers often must use a kind of code to give information on a map. To aid with interpreting this code, they also usually include a *compass rose,* a *scale,* and a *legend.*

The compass rose has arrows that point north, south, east, and west. The scale tells you the real-world distance that is represented by a distance on the map. The legend explains what each symbol on the map represents.

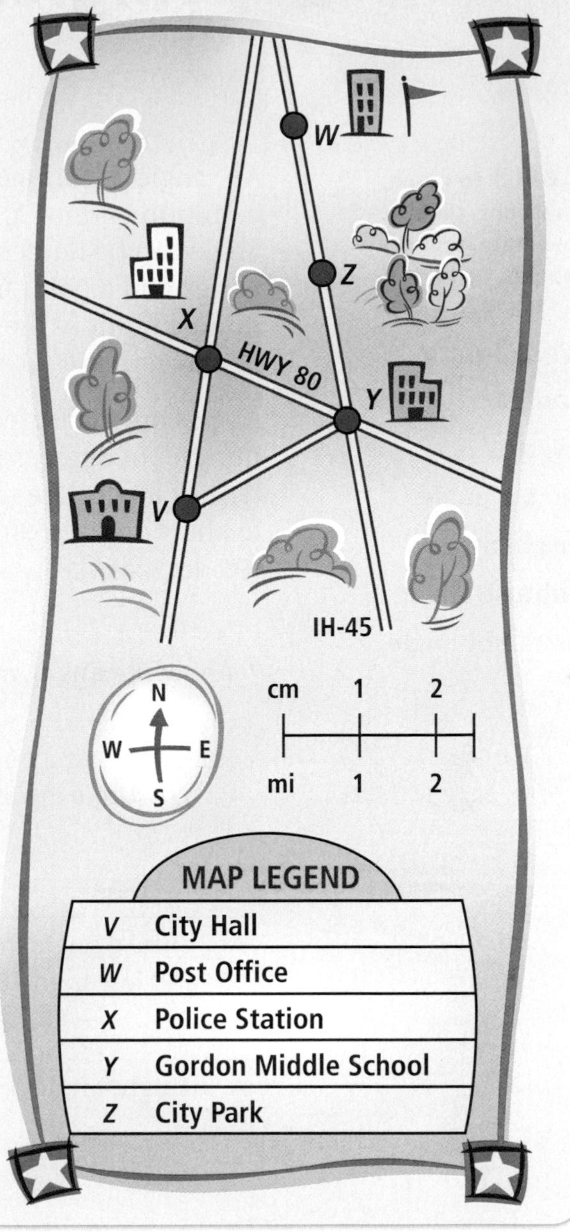

20. Name the geometric figure suggested by each part of the map.

a. City Hall and Gordon Middle School

b. Highway 80

c. the road from the park to the post office

21. Use a centimeter ruler to measure the route from the police station to Gordon Middle School. Use the scale to find the actual distance your measurement represents.

22. ***WHAT'S THE QUESTION?*** The answer is the route from City Hall to Gordon Middle School. What is the question?

23. ***WRITE ABOUT IT*** Explain why the road from City Hall that goes past the police station suggests a ray named $\overrightarrow{VX}$ rather than a ray named $\overrightarrow{XV}$.

24. ***CHALLENGE*** What are all the possible names for the line suggested by IH-45?

Spiral Review

25. A cocker spaniel weighs seven times more than her puppy. Write an algebraic expression for the weight of the mother. Use *p* to represent the weight of the puppy. (Lesson 2-2)

Find the quotient. (Lessons 3-7, 3-8)

26. $45.5 \div 5$ **27.** $103.7 \div 2$ **28.** $35 \div 2.5$ **29.** $4.25 \div 0.25$

30. **EOG PREP** Which sum is the *greatest*? (Lesson 5-7)

A $\frac{1}{3} + \frac{1}{4}$ B $\frac{1}{6} + \frac{5}{12}$ C $\frac{1}{4} + \frac{7}{12}$ D $\frac{1}{12} + \frac{1}{2}$

7-2 Angles

Learn to name, measure, classify, estimate, and draw angles.

Vocabulary

angle
vertex
acute angle
right angle
obtuse angle
straight angle

An **angle** is formed by two rays with a common endpoint, called the **vertex**. An angle can be named by its vertex or by its vertex and a point from each ray. The middle point in the name should always be the vertex.

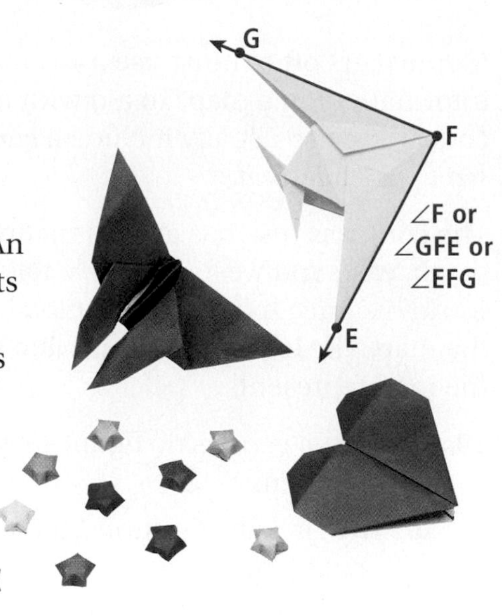

Angles are measured in degrees. The number of degrees determines the type of angle. Use the symbol ° to show degrees: 90° means "90 degrees."

An **acute angle** measures less than 90°.

A **right angle** measures exactly 90°.

An **obtuse angle** measures more than 90° and less than 180°.

A **straight angle** measures exactly 180°.

EXAMPLE 1 Measuring an Angle with a Protractor

Use a protractor to measure the angle. Tell what type of angle it is.

- Place the center point of the protractor on the vertex of the angle.
- Place the protractor so that ray *YZ* passes through the 0° mark.
- Using the scale that starts with 0° along ray *YZ*, read the measure where ray *YX* crosses.
- The measure of $\angle XYZ$ is 75°. Write this as $m\angle XYZ = 75°$.
- Since $75° < 90°$, the angle is acute.

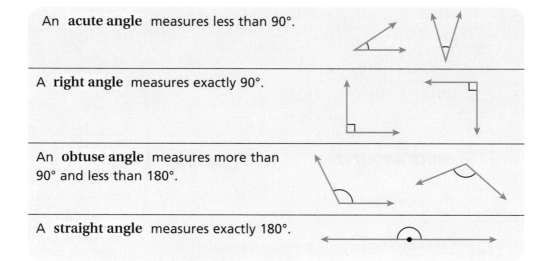

EXAMPLE 2 Drawing an Angle with a Protractor

Use a protractor to draw an angle that measures 150°.

- Draw a ray on a sheet of paper.
- Place the center point of the protractor on the endpoint of the ray.
- Place the protractor so that the ray passes through the 0° mark.
- Make a mark at 150° above the scale on the protractor.
- Use a straightedge to draw a ray from the endpoint of the first ray through the mark you made at 150°.

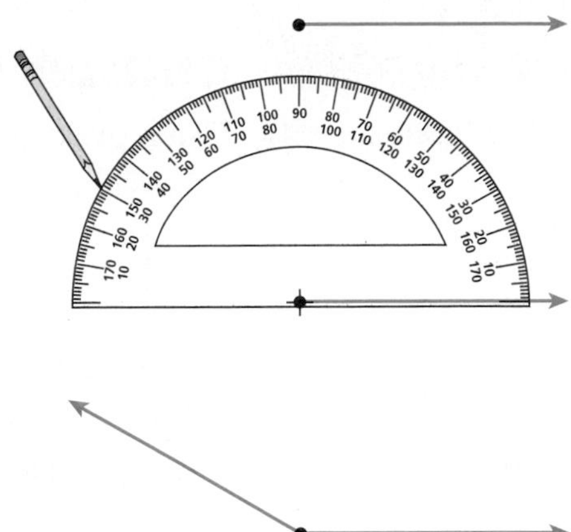

To estimate the measure of an angle, compare it with an angle whose measure you already know. A right angle has half the measure of a straight angle. A 45° angle has half the measure of a right angle.

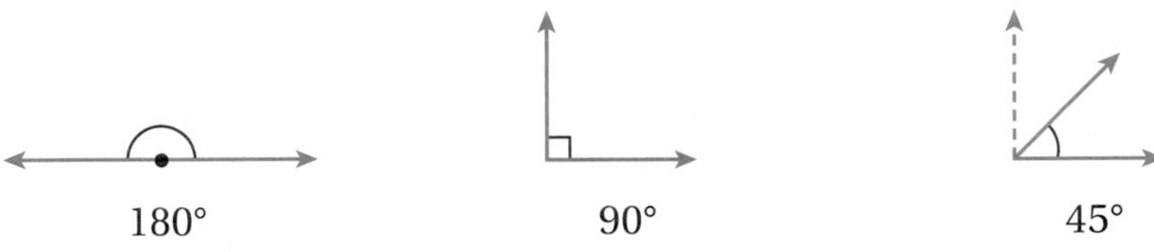

EXAMPLE 3 Estimating Angle Measures

Estimate the measure of the angle, and then use a protractor to check the reasonableness of your estimate.

Think: The measure of the angle is close to 45°, but it is a little less. A good estimate would be about 35°.

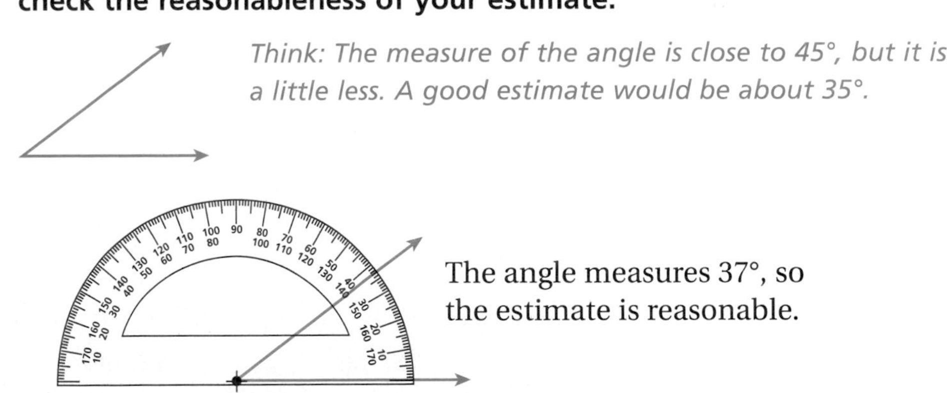

The angle measures 37°, so the estimate is reasonable.

Think and Discuss

1. **Tell** what types of angles are in Examples 2 and 3.
2. **Explain** how you know which scale to read on a protractor.

7-2 Exercises

FOR EOG PRACTICE

see page 662

internet connect

Homework Help Online
go.hrw.com Keyword: MR4 7-2

2.01

GUIDED PRACTICE

See Example 1 Use a protractor to measure each angle. Tell what type of angle it is.

1. 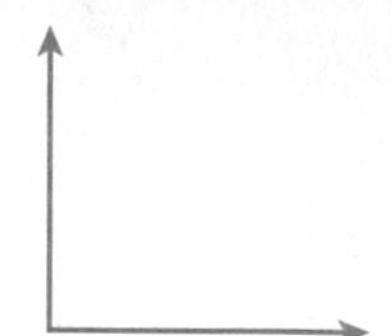

2. 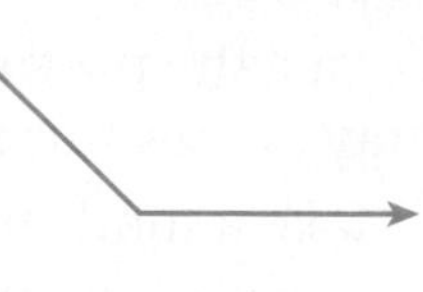

3. 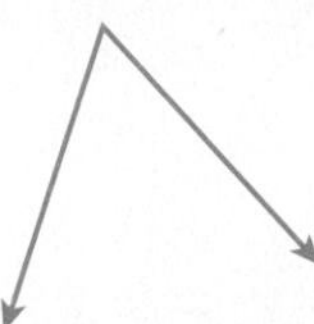

See Example 2 Use a protractor to draw an angle with each given measure.

4. 55° **5.** 135° **6.** 20° **7.** 190°

See Example 3 Estimate the measure of each angle, and then use a protractor to check the reasonableness of your estimate.

8. 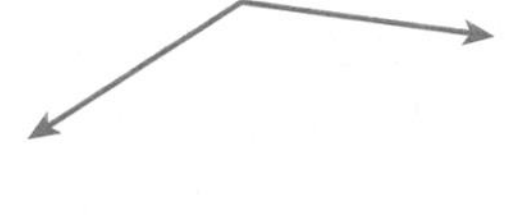

9. 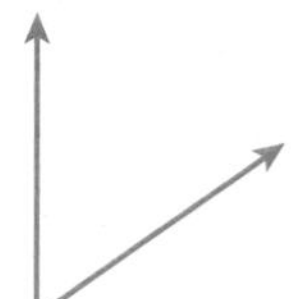

10. 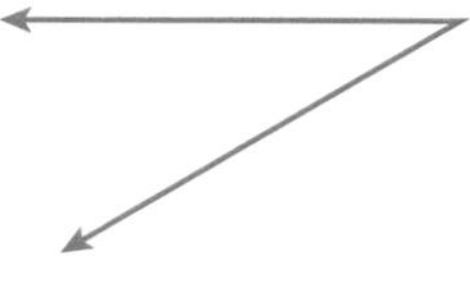

INDEPENDENT PRACTICE

See Example 1 Use a protractor to measure each angle. Tell what type of angle it is.

11. 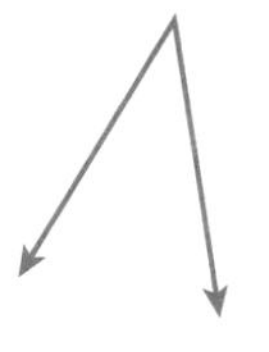

12. 

13. 

14. 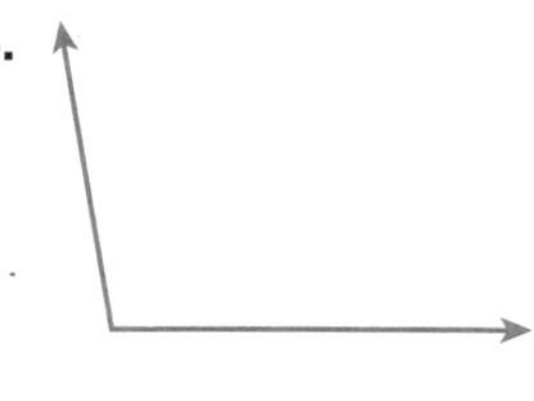

15. 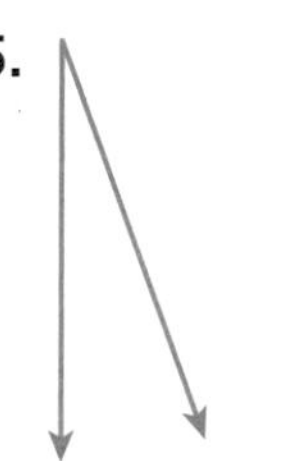

16. 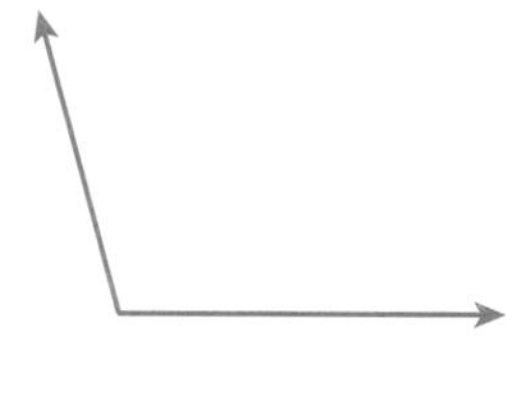

See Example 2 Use a protractor to draw an angle with each given measure.

17. 150° **18.** 38° **19.** 90° **20.** 72°

21. 112° **22.** 180° **23.** 19° **24.** 45°

See Example 3 Estimate the measure of each angle, and then use a protractor to check the reasonableness of your estimate.

25. 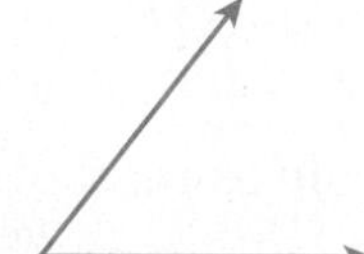

26. 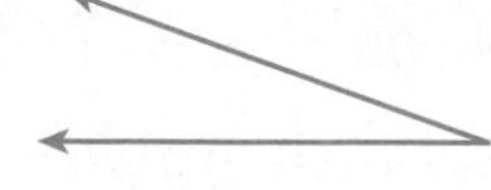

27. 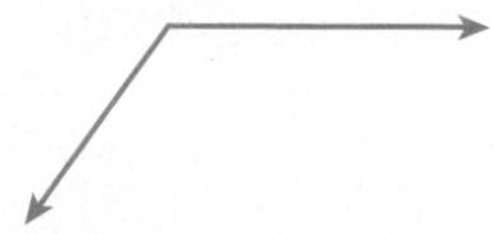

PRACTICE AND PROBLEM SOLVING

Use a protractor to draw each angle.

28. an acute angle whose measure is less than 45°

29. an obtuse angle whose measure is between 100° and 160°

30. a right angle

31. two acute angles that together form a right angle

Name the smallest angle formed by the hands on each clock.

32. 

33. 

34. 

35. ***AVIATION*** Classify the angle the body of an airplane makes with the runway as the airplane takes off.

36. ***SPORTS*** The quarterback of a football team throws a long pass, and the angle the path of the ball makes with the ground is 30°. Draw an angle with this measurement.

37. ***WHAT'S THE ERROR?*** A student wrote that the measure of this angle is 156°. Explain the error the student may have made, and give the correct measure of the angle. How can the student avoid making the same mistake again?

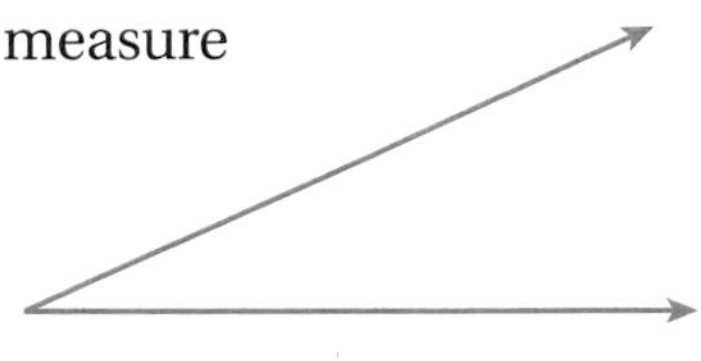

38. ***WRITE ABOUT IT*** Describe how an acute angle and an obtuse angle are different.

39. ***CHALLENGE*** How many times during the day do the hands of a clock form a straight angle?

Spiral Review

Find each value. (Lesson 1-3)

40. 9^2 **41.** 2^6 **42.** 3^3 **43.** 1^{12}

Find each sum or difference. (Lesson 3-3)

44. $2.1 + 0.9$ **45.** $4 - 1.2$ **46.** $1.5 + 3.2$ **47.** $5 - 2.6$

48. **EOG PREP** Which fraction is *not* equivalent to two-thirds? (Lesson 4-6)

A $\frac{8}{12}$ B $\frac{4}{9}$ C $\frac{2}{3}$ D $\frac{10}{15}$

Construct Congruent Segments and Angles

Use with Lesson 7-3

Line segments are congruent if they are the same length. You can use a compass and straightedge to construct congruent line segments.

internet connect
Lab Resources Online
go.hrw.com
KEYWORD: MR4 Lab7A

Activity 1

1 Draw a line segment congruent to line segment AM.

A M

Draw a ray, and label the endpoint B.

B

Place your compass point on point A. Open the compass to the length of $\overline{AM}$. Use the same opening to draw an arc that intersects the ray. Label the intersection point Y.

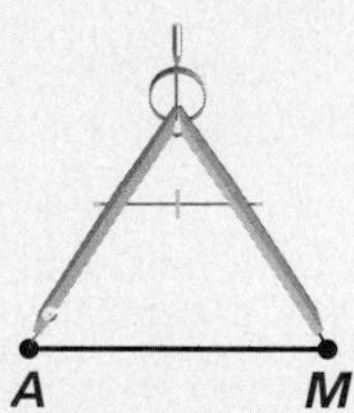

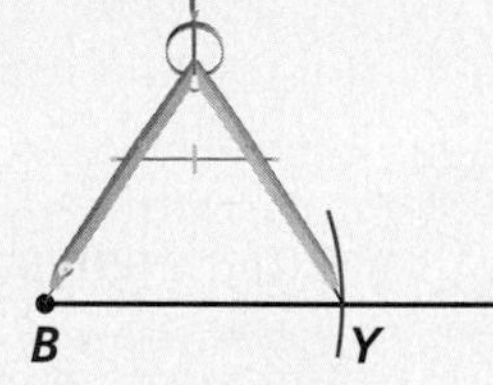

$\overline{AM} \cong \overline{BY}$

Think and Discuss

1. Tell whether it is more accurate to use a ruler or a compass to construct congruent line segments. Explain your answer.

Try This

Construct a line segment that is congruent to the given line segment.

1. 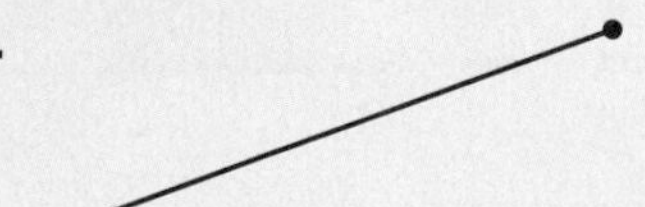

2. 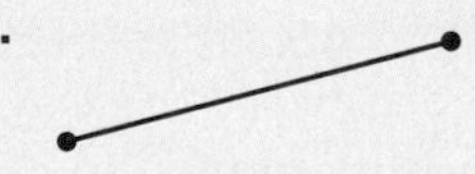

3. 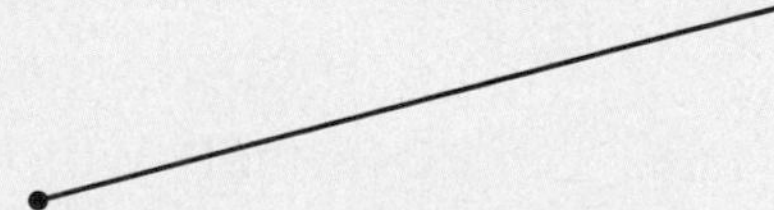

Angles are congruent if they have the same measure in degrees. You can use a compass and straightedge to construct congruent angles.

Activity 2

1 Draw an angle congruent to ∠C.
Draw a ray, and label the endpoint *P*.

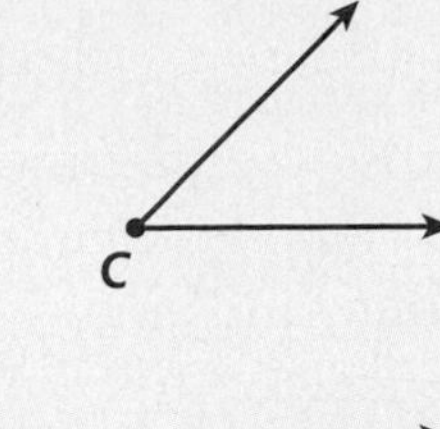

P

Place your compass point on point *C* and draw an arc through ∠*C*. Label the points of intersection *D* and *E*. Place the compass point on *P*, and draw a similar arc through the ray. Label the point of intersection *R*.

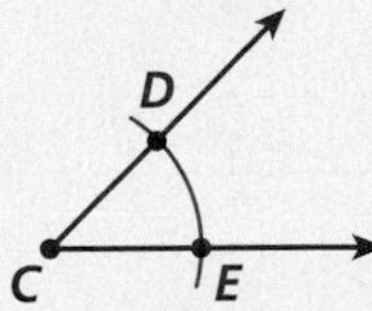

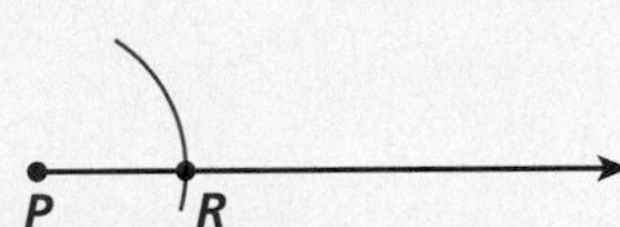

Use the compass to measure the arc in ∠*DCE*. Then place the compass point on *R* and draw an arc that intersects the first one. Label the point of intersection *Q*. Draw $\overrightarrow{PQ}$.

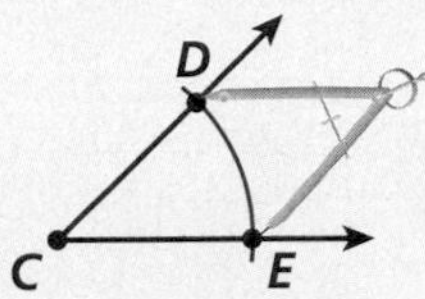

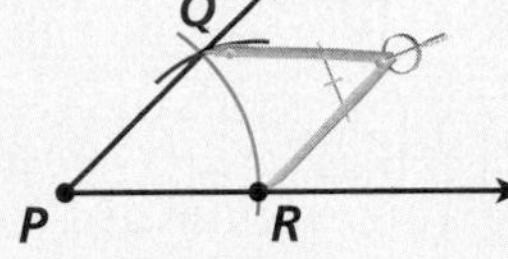

∠*DCE* ≅ ∠*QPR*

Think and Discuss

1. Explain how drawing and constructing are different.

Try This

Construct a congruent angle for each given angle.

1\.

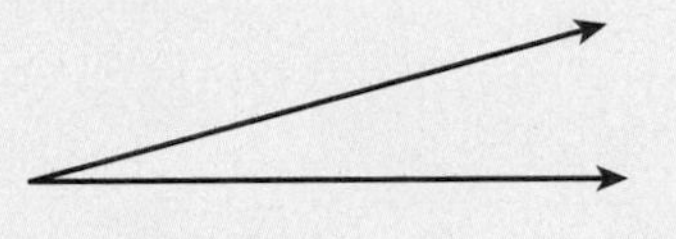

2\.

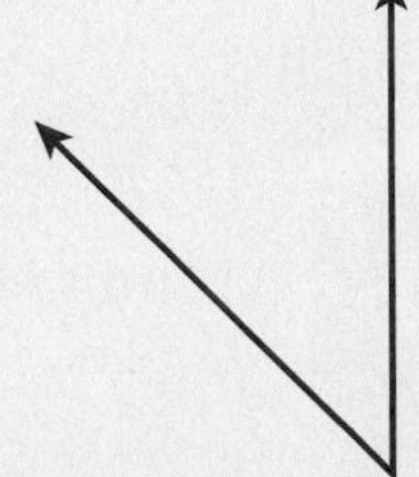

3\.

7-3 Angle Relationships

Learn to understand relationships of angles.

Vocabulary
- congruent
- vertical angles
- adjacent angles
- complementary angles
- supplementary angles

Angle relationships play an important role in many sports and games. Miniature-golf players must understand angles to know where to aim the ball. In the miniature-golf hole shown, $m\angle 1 = m\angle 2$, $m\angle 3 = m\angle 4$, and $m\angle 5 = m\angle 6$.

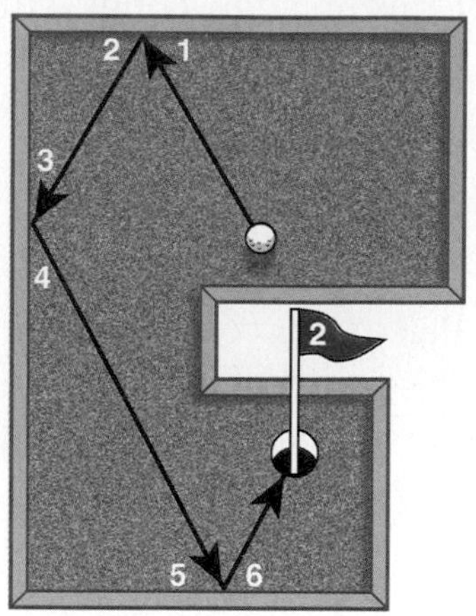

When angles have the same measure, they are said to be **congruent**.

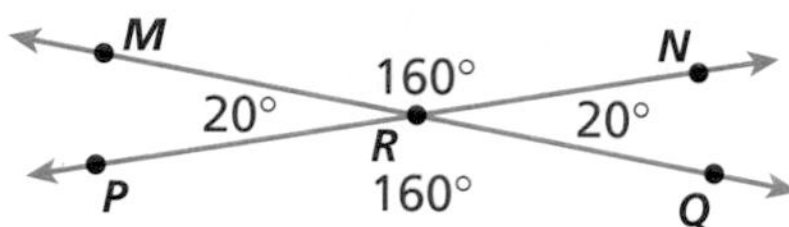

Reading Math

$m\angle 1$ is read "the measure of angle 1."

Vertical angles are formed opposite each other when two lines intersect. Vertical angles have the same measure, so they are always congruent.

$\angle MRP$ and $\angle NRQ$ are vertical angles.
$\angle MRN$ and $\angle PRQ$ are vertical angles.

Adjacent angles are side by side and have a common vertex and ray. Adjacent angles may or may not be congruent.

$\angle MRN$ and $\angle NRQ$ are adjacent angles. They share vertex R and $\overrightarrow{RN}$.
$\angle NRQ$ and $\angle QRP$ are adjacent angles. They share vertex R and $\overrightarrow{RQ}$.

EXAMPLE 1 Identifying Types of Angle Pairs

Identify the type of each angle pair shown.

A

$\angle 1$ and $\angle 2$ are opposite each other and are formed by two intersecting lines.

They are vertical angles.

B

$\angle 3$ and $\angle 4$ are side by side and have a common vertex and ray.

They are adjacent angles.

Complementary angles are two angles whose measures have a sum of 90°.

$65° + 25° = 90°$

$\angle LMN$ and $\angle NMP$ are complementary.

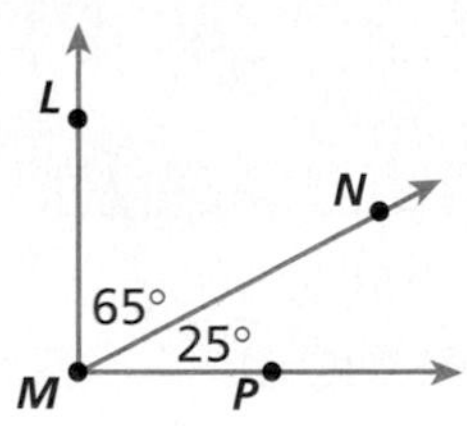

Supplementary angles are two angles whose measures have a sum of 180°.

$65° + 115° = 180°$

$\angle GHK$ and $\angle KHJ$ are supplementary.

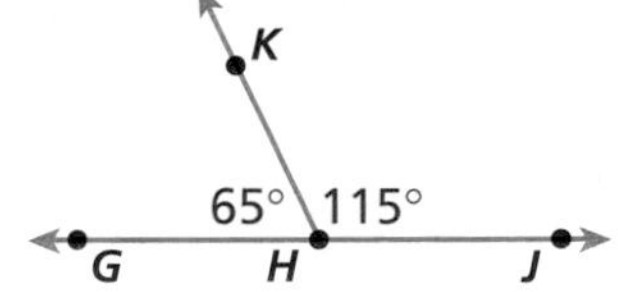

EXAMPLE 2 Identifying an Unknown Angle Measure

Find each unknown angle measure.

A The angles are complementary.

$$\begin{aligned} 55° + a &= 90° \\ -55° \quad &\quad -55° \\ a &= 35° \end{aligned}$$

The sum of the measures is 90°.

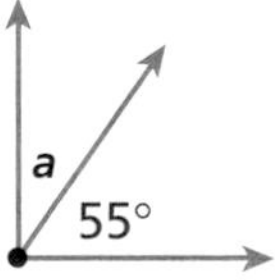

B The angles are supplementary.

$$\begin{aligned} 75° + b &= 180° \\ -75° \quad &\quad -75° \\ b &= 105° \end{aligned}$$

The sum of the measures is 180°.

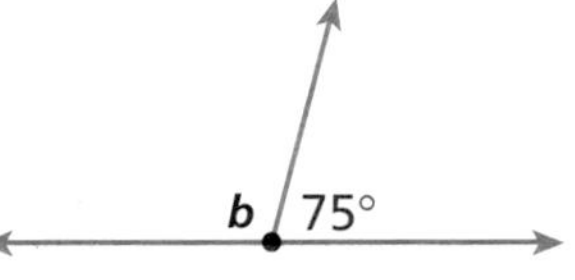

C The angles are vertical angles.

$c = 51°$ *Vertical angles are congruent.*

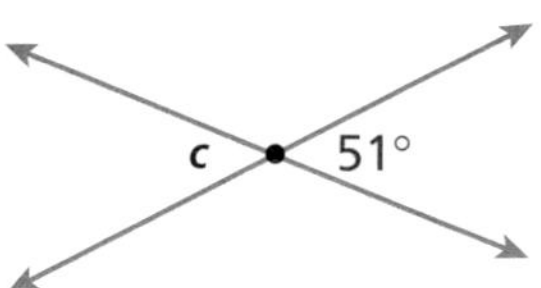

D Angles *JGF* and *KGH* are congruent.

$$\begin{aligned} d + e + 136° &= 180° \\ -136° \quad &\quad -136° \\ d + e &= 44° \end{aligned}$$

$d = 22°$ and $e = 22°$

The sum of the measures is 180°. Each angle measures half of 44°.

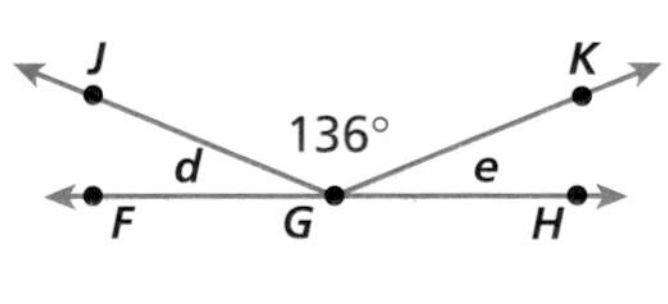

Think and Discuss

1. **Give** the measure of $\angle 2$ if $\angle 1$ and $\angle 2$ are vertical angles and $m\angle 1 = 40°$.
2. **Give** the measure of $\angle 3$ if $\angle 3$ and $\angle 4$ are supplementary and $m\angle 4 = 150°$.
3. **Tell** whether the angles in Example 1B are supplementary or complementary.

7-3 Exercises

3.01

FOR EOG PRACTICE
see page 663

internet connect
Homework Help Online
go.hrw.com Keyword: MR4 7-3

GUIDED PRACTICE

See Example 1 **Identify the type of each angle pair shown.**

1. 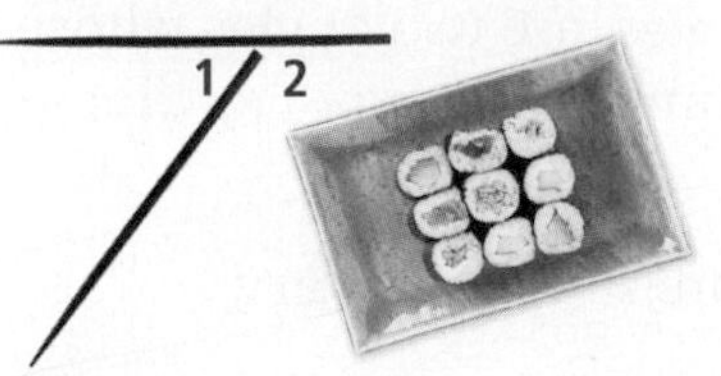

2. 

See Example 2 **Find each unknown angle measure.**

3. The angles are complementary.

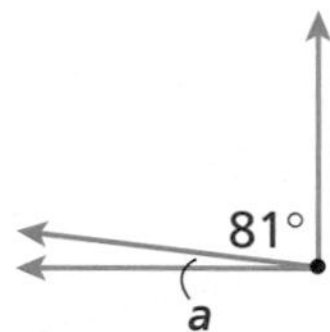

4. The angles are supplementary.

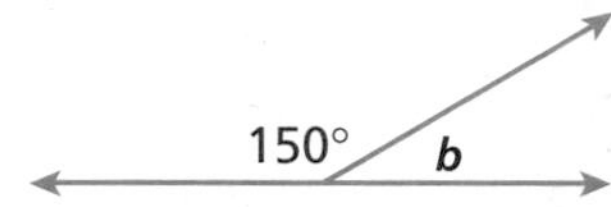

INDEPENDENT PRACTICE

See Example 1 **Identify the type of each angle pair shown.**

5. 

6. 

See Example 2 **Find each unknown angle measure.**

7. The angles are vertical angles.

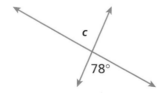

c

78°

8. The angles are supplementary.

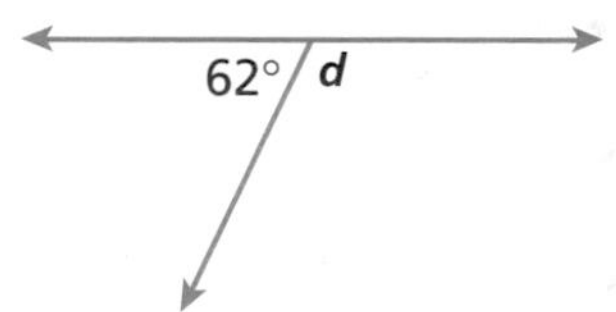

PRACTICE AND PROBLEM SOLVING

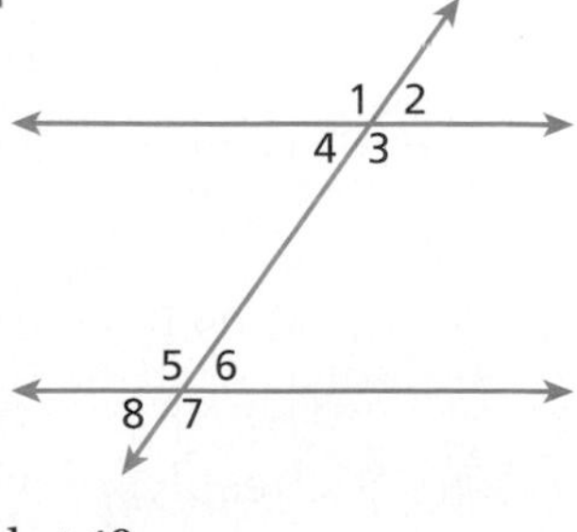

Use the figure for Exercises 9–12.

9. Which angles are not adjacent to ∠3?

10. Name all the pairs of vertical angles that include ∠8.

11. If the m∠6 is 72°, what are the measures of ∠5, ∠7, and ∠8?

12. What is the sum of the measures of ∠1, ∠2, ∠3, and ∠4?

Use the figure for Exercises 13–16.

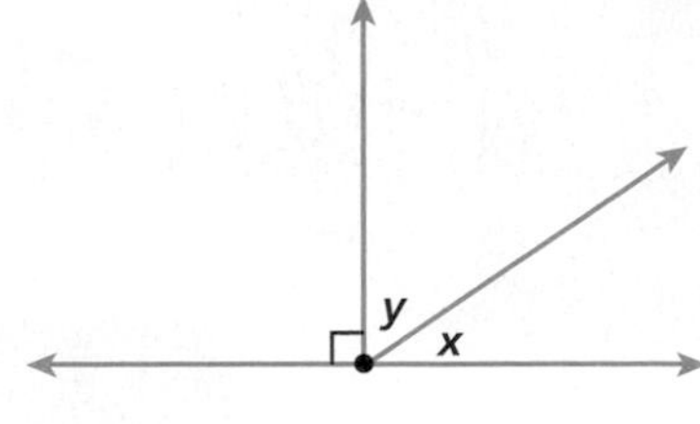

13. If $y = 51°$, what does x equal?

14. If $x = 64°$, what does y equal?

15. If $x = 40.09°$, what does y equal?

16. If $y = 27\frac{1}{3}°$, what does x equal?

Find the measure of the angle that is complementary to each given angle. Use a protractor to draw both angles.

17. 47° **18.** 62° **19.** 55° **20.** 31°

Find the measure of the angle that is supplementary to each given angle. Use a protractor to draw both angles.

21. 75° **22.** 102° **23.** 136° **24.** 81°

25. Angles A and B are complementary. If the measure of angle A equals the measure of angle B, what is the measure of each angle?

26. Angles C and D are each complementary to angle F. How are angle C and angle D related?

27. The measure of angle 1 is 43°. Angle 2 is complementary to angle 1, and angle 3 is supplementary to angle 1.

a. Give $m\angle 2$ and $m\angle 3$.

b. Use a protractor to draw $\angle 1$, $\angle 2$, and $\angle 3$.

28. ***WRITE A PROBLEM*** Draw a pair of adjacent supplementary angles. Write a problem in which the measure of one of the angles must be found.

29. ***WRITE ABOUT IT*** Two angles are supplementary to the same angle. Explain the relationship between the measures of these angles.

30. ***CHALLENGE*** The measure of angle A is 38°. Angle B is complementary to angle A. Angle C is supplementary to angle B. What is the measure of angle C?

Spiral Review

31. Tami worked 4 hours on Saturday at the city pool. She spent $1\frac{3}{4}$ hours cleaning the pool and the remaining time working as a lifeguard. How many hours did Tami spend working as a lifeguard? (Lesson 5-9)

32. **EOG PREP** Which type of data would *not* be appropriate for a line graph? (Lesson 6-7)

A Average rainfall over a ten-year period

B A tennis player's earnings during his career

C The value of a share of stock during the last 14 months

D The results of a survey on favorite television programs

7-4 Classifying Lines

Learn to classify the different types of lines.

Vocabulary
- parallel lines
- perpendicular lines
- skew lines

The photograph of the houses and the table below show some of the ways that lines can relate to each other. The yellow lines are intersecting. The purple lines are parallel. The green lines are perpendicular. The white lines are skew.

Intersecting lines are lines that cross at one common point.	W, Y, X, Z	Line *YZ* intersects line *WX*. $\overleftrightarrow{YZ}$ intersects $\overleftrightarrow{WX}$.
Parallel lines are lines in the same plane that never intersect.	B, A, L, M	Line *AB* is parallel to line *ML*. $\overleftrightarrow{AB} \parallel \overleftrightarrow{ML}$
Perpendicular lines intersect to form 90° angles, or right angles.	R, T, U, S	Line *RS* is perpendicular to line *TU*. $\overleftrightarrow{RS} \perp \overleftrightarrow{TU}$
Skew lines are lines that lie in different planes. They are neither parallel nor intersecting.	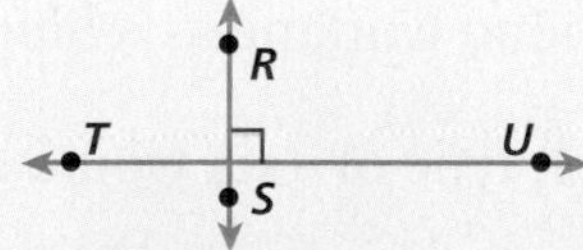	Line *AB* and line *ML* are skew. $\overleftrightarrow{AB}$ and $\overleftrightarrow{ML}$ are skew.

Writing Math

The square inside a right angle shows that the rays of the angle are perpendicular.

EXAMPLE 1 Classifying Pairs of Lines

Classify each pair of lines.

A

The lines are in the same plane. They do not appear to intersect.
They are parallel.

B

The lines cross at one common point.
They are intersecting.

C

The lines intersect to form right angles.
They are perpendicular.

D

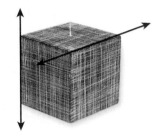

The lines are in different planes and are not parallel or intersecting.
They are skew.

EXAMPLE 2 *Science Application*

The particles in a transverse wave move up and down as the wave travels to the right. What type of line relationship does this represent?

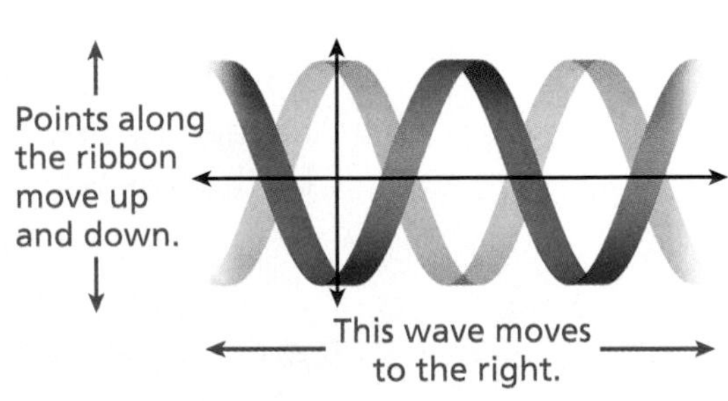

The direction that the particles move forms a right angle with the direction that the wave is traveling. The lines are perpendicular.

Think and Discuss

1. **Give an example** of intersecting, parallel, perpendicular, and skew lines or line segments in your classroom.
2. **Determine** whether two lines must be parallel if they do not intersect. Explain.

7-4 Exercises

3.01

FOR EOG PRACTICE
see page 663

internet connect
Homework Help Online
go.hrw.com Keyword: MR4 7-4

GUIDED PRACTICE

See Example 1 Classify each pair of lines.

1.
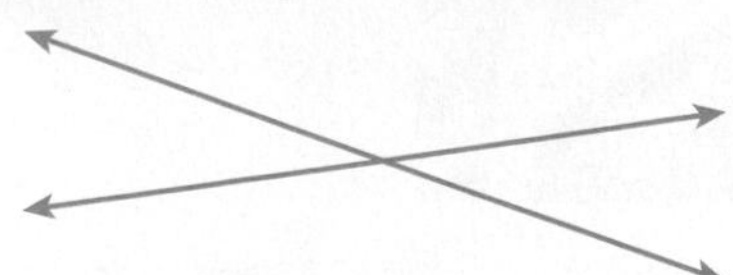

2.
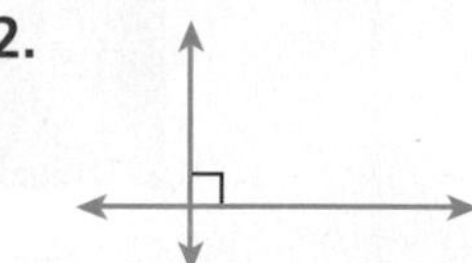

See Example 2

3. Jamal dropped a fishing line from a pier, as shown in the drawing. What type of relationship is formed by the lines?

INDEPENDENT PRACTICE

See Example 1 Classify each pair of lines.

4.

5.
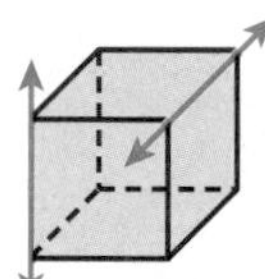

See Example 2

6. The drawing shows where an archaeologist found two fossils. What type of relationship is formed by the lines suggested by the fossils?

PRACTICE AND PROBLEM SOLVING

Describe each pair of lines as parallel, skew, intersecting, or perpendicular.

7.

8.

9.
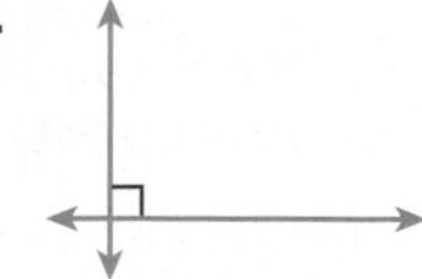

10.
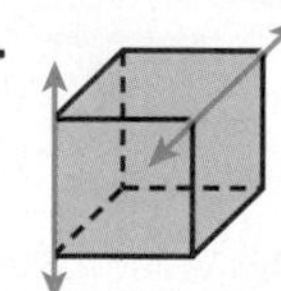

The lines in the figure intersect to form a rectangular box.

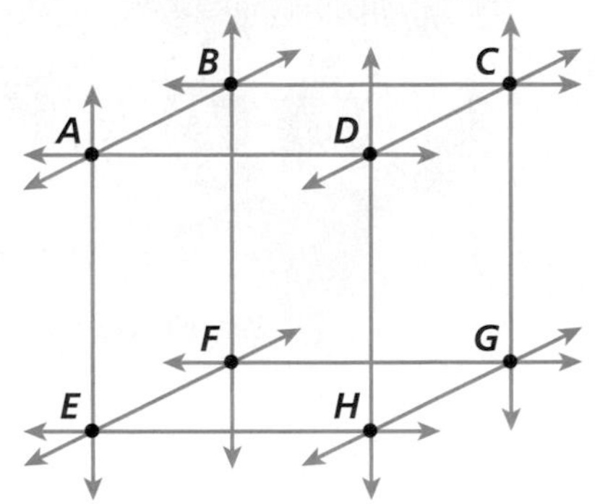

11. Name all the lines that are parallel to $\overleftrightarrow{AD}$.

12. Name all the lines that are perpendicular to $\overleftrightarrow{FG}$.

13. Name a pair of lines that are skew.

14. Name all the lines that are not parallel to and do not intersect $\overleftrightarrow{DH}$.

Tell whether each statement is *always, sometimes,* or *never* true.

15. Intersecting lines are parallel.

16. Intersecting lines are perpendicular.

17. Perpendicular lines are intersecting.

18. Parallel lines are in the same plane.

19. Parallel lines are skew.

20. Capitol Street intersects 1st, 2nd, and 3rd Avenues, which are parallel to each other. West Street and East Street are perpendicular to 2nd Avenue.

a. Draw a map showing the six streets.

b. Suppose East and West Streets were perpendicular to Capitol Street rather than 2nd Avenue. Draw a map showing the streets.

21. ***WHAT'S THE ERROR?*** A student drew two lines and claimed that the lines were both parallel and intersecting. Explain the error.

22. ***WRITE ABOUT IT*** Explain the similarities and differences between perpendicular and intersecting lines.

23. ***CHALLENGE*** Lines x, y, and z are in a plane. If lines x and y are parallel and line z intersects line x, does line z intersect line y? Explain.

Spiral Review

Multiply. Write each answer in simplest form. (Lesson 4-9)

24. $5 \cdot \frac{1}{10}$ **25.** $21 \cdot \frac{1}{3}$ **26.** $\frac{2}{7} \cdot 14$ **27.** $\frac{5}{12} \cdot 2$

Subtract. Write each answer in simplest form. (Lesson 5-9)

28. $5\frac{2}{3} - 4\frac{5}{6}$ **29.** $12\frac{4}{7} - 3\frac{6}{7}$ **30.** $9\frac{7}{12} - 2\frac{1}{3}$ **31.** $11\frac{5}{8} - 5\frac{1}{4}$

32. **EOG PREP** Which is the outlier of the data set 24, 76, 28, 24, 35, 31, 28, 24, 27? (Lesson 6-3)

A 24 B 33 C 28 D 76

Parallel Line Relationships

Use with Lesson 7-4

Parallel lines are in the same plane and never intersect. You can use a straightedge and protractor to draw parallel lines.

internet connect
Lab Resources Online
go.hrw.com
KEYWORD: MR4 Lab7B

Activity

1. Draw a line on your paper. Label two points *A* and *B*.

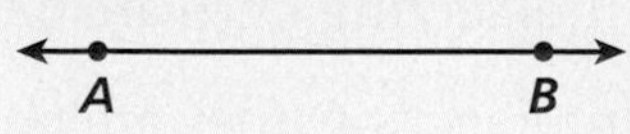

Use your protractor to measure and mark a 90° angle at each point.

Draw rays with endpoints *A* and *B* through the marks you made with the protractor.

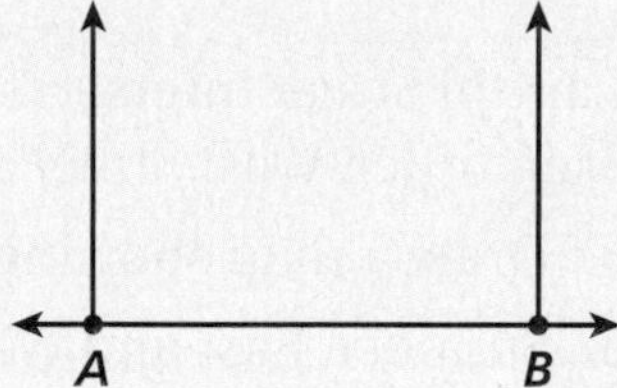

Place the point of your compass on point *A*, and draw an arc through the ray.

Use the same compass opening to draw an arc through the ray at point *B*.

Label the points of intersection *X* and *Y*.

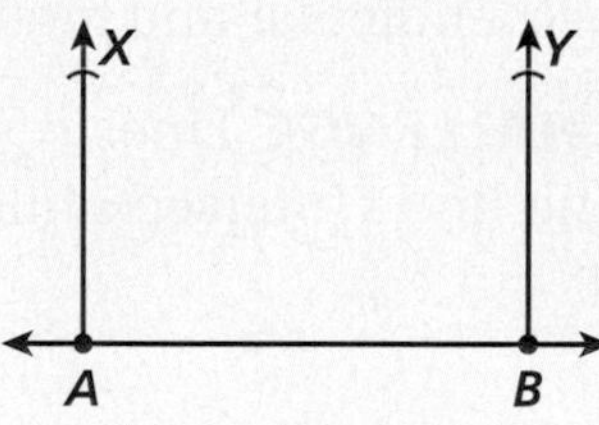

Now use your straightedge to draw a line through *X* and *Y*.

$\overline{AB} \parallel \overline{XY}$

When a pair of parallel lines is intersected by a third line, the angles formed have special relationships.

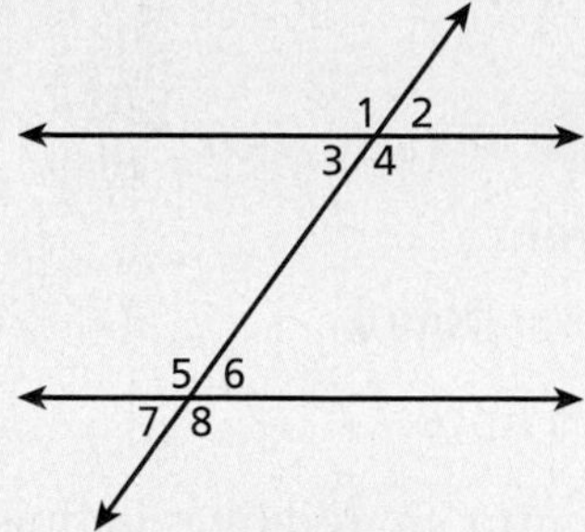

2 Draw a pair of parallel lines and a third line that intersects them. Label the angles 1 through 8, as shown.

Angles inside the parallel lines are called *interior angles*. The interior angles here are angles 3, 4, 5, and 6.

Angles outside the parallel lines are called *exterior angles*. The exterior angles here are angles 1, 2, 7, and 8.

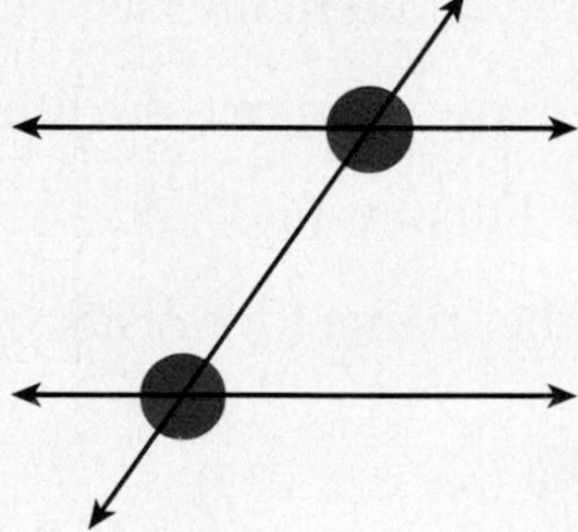

Measure each angle, and write its measurement inside the angle.

Shade angles with the same measure with the same color.

The interior angles with the same measure are called *alternate interior angles*. They are angles 3 and 6 and angles 4 and 5.

The exterior angles with the same measure are called *alternate exterior angles*. They are angles 1 and 8 and angles 2 and 7.

Angles in the same position when the third line intersects the parallel lines are called *corresponding angles.*

Think and Discuss

1. Name three pairs of corresponding angles.
2. Tell the relationship between the measure of interior angles and the measure of exterior angles.

Try This

Follow the steps to construct and label the diagram.

1. Draw a pair of parallel lines, and draw a third line intersecting them where one angle measures 75°.
2. Label each angle on the diagram using the measure you know.

Chapter 7

Mid-Chapter Quiz

LESSON  7-1 (pp. 322–325)

Use the diagram.

1. Name three points.
2. Name two lines.
3. Name a point shared by two lines.
4. Name a plane.

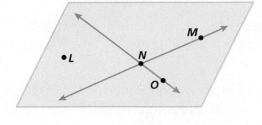

Use the diagram.

5. Name three different line segments.
6. Give three ways to name the line.
7. Name six different rays.
8. Give another name for ray *RS*.

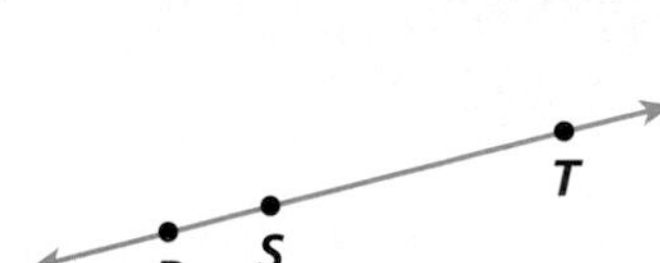

LESSON  7-2 (pp. 326–329)

Estimate the measure of each angle, and use a protractor to check the reasonableness of your estimate. Tell what type each angle is.

9. 

10. 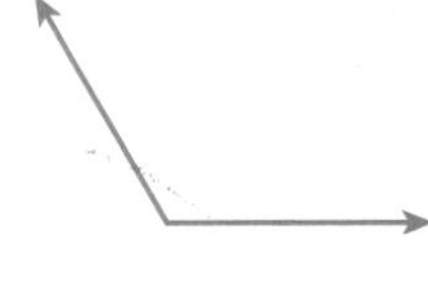

11. 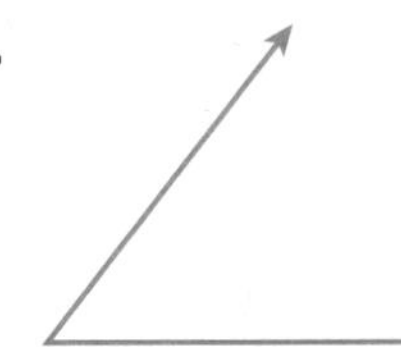

12. 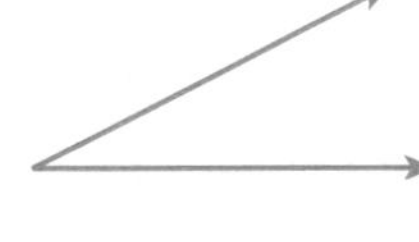

LESSON  7-3 (pp. 332–335)

Find each unknown angle measure.

13. 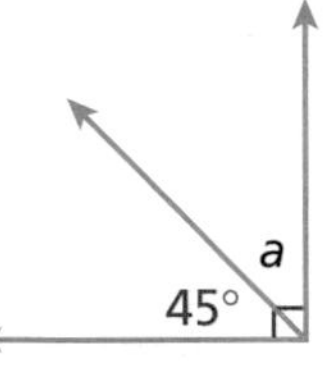

14. 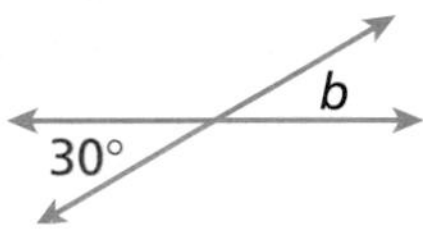

15. 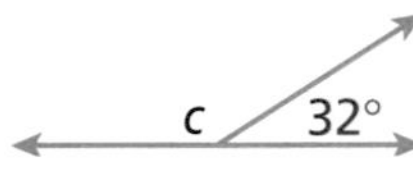

16. 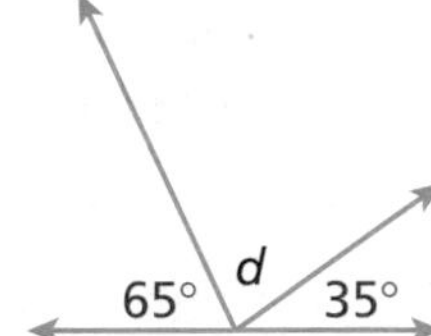

LESSON  7-4 (pp. 336–339)

Classify each pair of lines.

17. 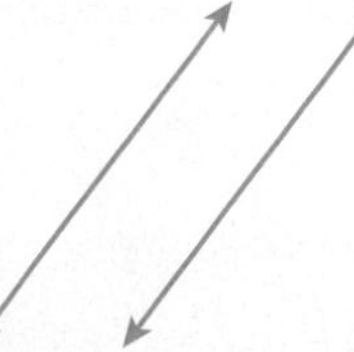

18. 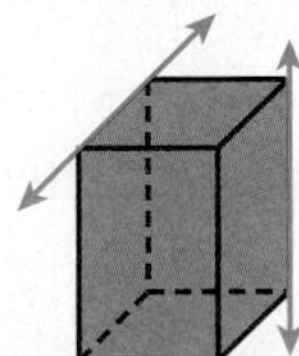

19. 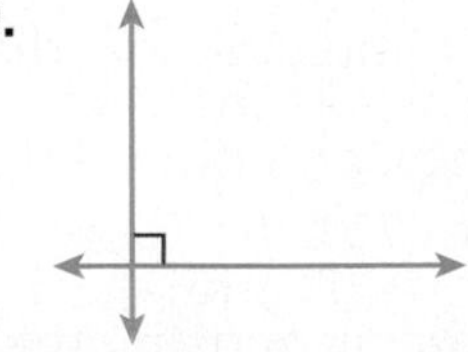

20. 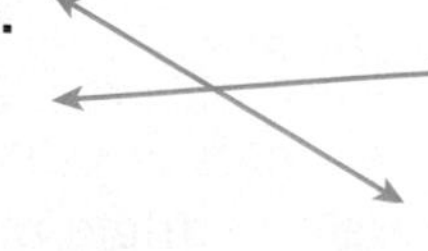

Focus on Problem Solving

Solve

• Eliminate answer choices

Sometimes, when a problem has multiple answer choices, you can eliminate some of the choices to help you solve the problem.

For example, a problem reads, "The missing shape is not a red triangle." If one of the answer choices is a red triangle, you can eliminate that answer choice.

Read each problem, and look at the answer choices. Determine whether you can eliminate any of the answer choices before solving the problem. Then solve.

Smileys are letters and symbols that look like faces if you turn them around. When you write an e-mail to someone, you can use smileys to show how you are feeling.

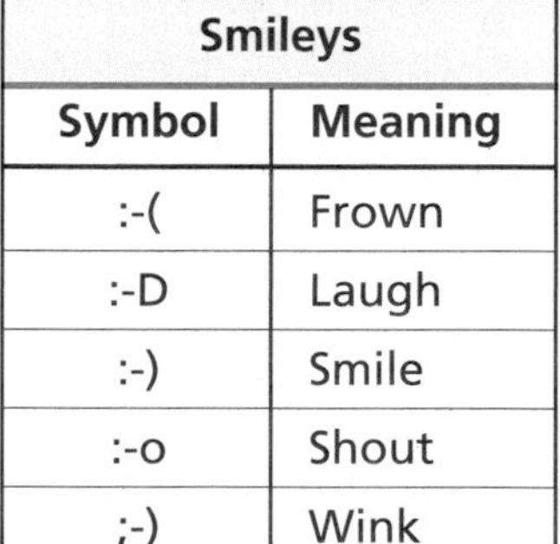

Smileys	
Symbol	**Meaning**
:-(	Frown
:-D	Laugh
:-)	Smile
:-o	Shout
;-)	Wink

For 1–3, use the table.

1 Dora made a pattern with smileys. Which smiley will she probably use next?

:-D :-) :-D :-) :-D :-) :-D :-) :-D ▢

A :-D **C** :-)

B :-) **D** :-D

2 Troy made a pattern with smileys. Identify a pattern. Which smiley is missing?

:-(;-) :-o :-(;-) :-o :-(▢ :-o

F :-(**H** ;-)

G :-o **J** ;-)

3 To end an e-mail, Mya typed four smileys in a row. The shout is first. The wink is between the frown and the smile. The smile is not last. In which order did Mya type the smileys?

A :-o :-(;-) :-) **C** :-) ;-) :-o :-(

B :-o :-) ;-) :-(**D** :-o ;-) :-(:-)

CarlyQ - Instant Message

File Edit View

Mindy2005: My mom says she will take us to the movie Fri. :-)

CarlyQ: Cool. Are Jaime and Rachel going too? :-D

Mindy2005: No, Rachel and Jaime are going to visit their grandmother. :-(

link

CarlyQ: We will have fun anyway. ;-)

Send

7-5 Triangles

Learn to classify triangles and solve problems involving angle and side measures of triangles.

Vocabulary

acute triangle
obtuse triangle
right triangle
scalene triangle
isosceles triangle
equilateral triangle

A triangle is a closed figure with three line segments and three angles. Triangles can be classified by the measures of their angles. An **acute triangle** has only acute angles. An **obtuse triangle** has one obtuse angle. A **right triangle** has one right angle.

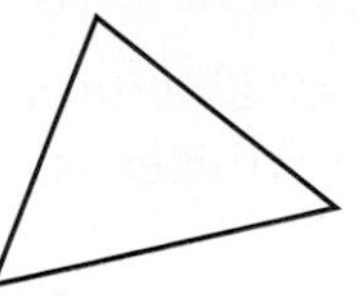

Acute triangle

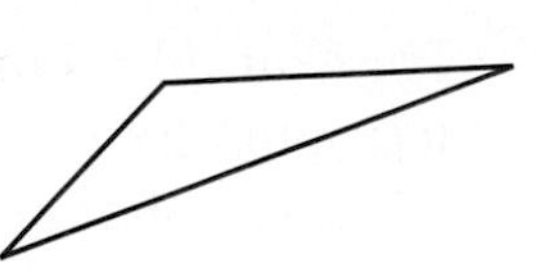

Obtuse triangle

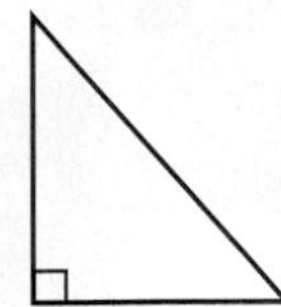

Right triangle

To decide whether a triangle is acute, obtuse, or right, you need to know the measures of its angles.

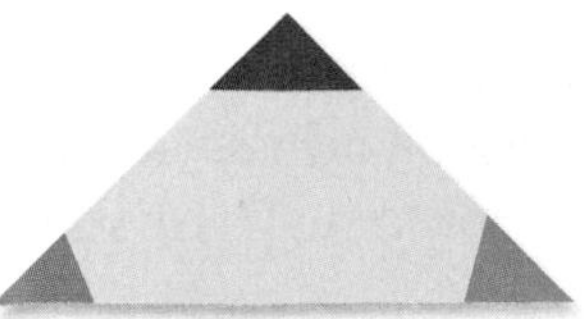

The sum of the measures of the angles in any triangle is 180°. You can see this if you tear the corners from a triangle and arrange them around a point on a line.

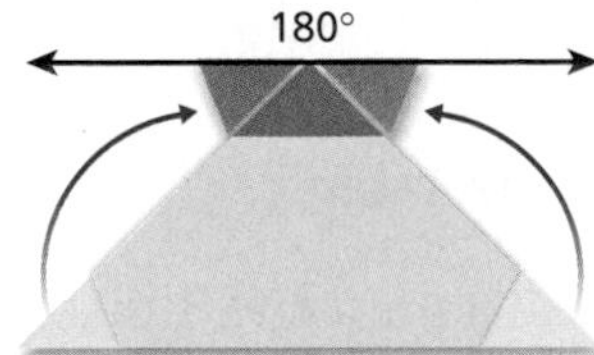

By knowing the sum of the measures of the angles in a triangle, you can find unknown angle measures.

EXAMPLE 1 ***Sports Application***

Boat sails are often shaped like triangles. The measure of ∠*A* is 70°, and the measure of ∠*B* is 45°. Classify the triangle.

To classify the triangle, find the measure of ∠*C* on the sail.

$m\angle C = 180° - (70° + 45°)$
$m\angle C = 180° - 115°$ *Subtract the sum of the known angle measures from 180°.*
$m\angle C = 65°$

So the measure of ∠*C* is 65°. Because △*ABC* has only acute angles, the boat sail is an acute triangle.

You can use what you know about vertical, adjacent, complementary, and supplementary angles to find the missing measures of angles.

EXAMPLE 2 Using Properties of Angles to Label Triangles

Use the diagram to find the measure of each indicated angle.

Remember!

Vertical angles are congruent. The sum of the measures of complementary angles is 90°. The sum of the measures of supplementary angles is 180°.

A $\angle BDE$

$\angle BDE$ and $\angle ADC$ are vertical angles, so $m\angle BDE = m\angle ADC$.

$$\begin{aligned} m\angle ADC &= 180° - (30° + 35°) \\ &= 180° - 65° \\ &= 115° \end{aligned}$$

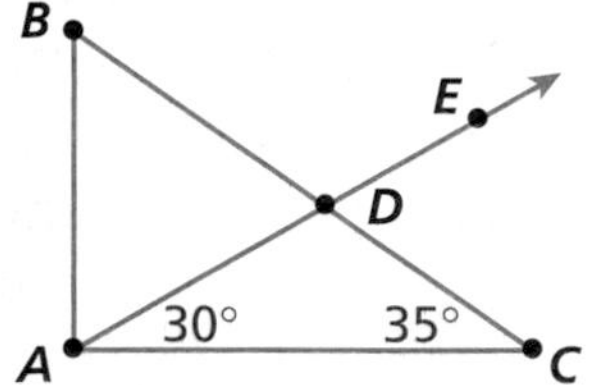

$m\angle BDE = 115°$

B $\angle ADB$

The sum of $m\angle BDE$ and $m\angle ADB$ is 180°.

$$\begin{aligned} m\angle ADB &= 180° - 115° \\ &= 65° \end{aligned}$$

$m\angle ADB = 65°$

Triangles can be classified by the lengths of their sides. A **scalene triangle** has no congruent sides. An **isosceles triangle** has at least two congruent sides. An **equilateral triangle** has three congruent sides.

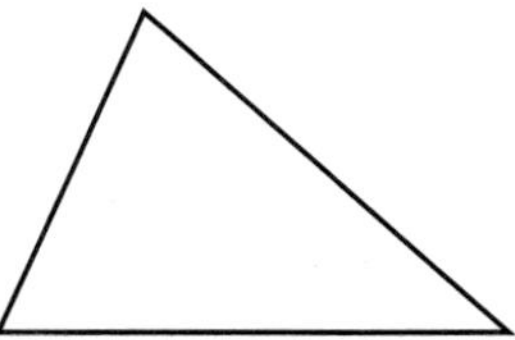

Scalene triangle

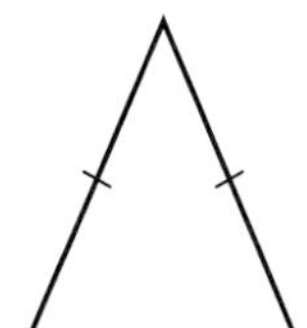

Isosceles triangle

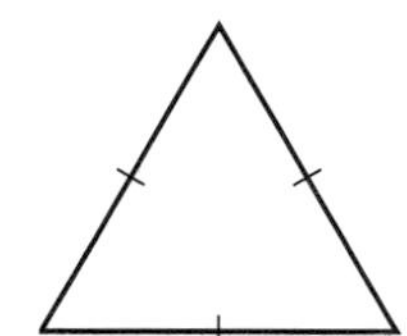

Equilateral triangle

EXAMPLE 3 Classifying Triangles by Lengths of Sides

Classify the triangle. The sum of the lengths of the sides is 7.8 cm.

$$\begin{aligned} a + (3.8 + 2) &= 7.8 \\ a + 5.8 &= 7.8 \\ a + 5.8 - 5.8 &= 7.8 - 5.8 \\ a &= 2 \end{aligned}$$

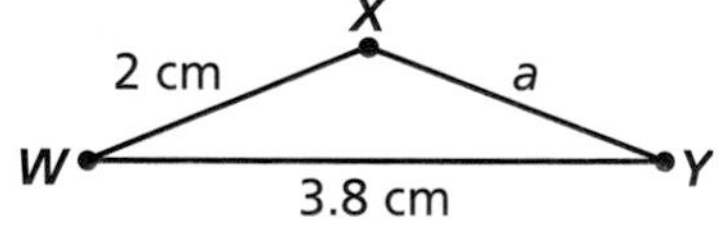

Side a is 2 centimeters long. Because $\triangle WXY$ has at least two sides, but not three, that are the same length, it is an isosceles triangle.

Think and Discuss

1. Explain why a triangle cannot have two obtuse angles.

2. Tell whether a right triangle can also be an acute triangle. Explain.

7-5 Exercises

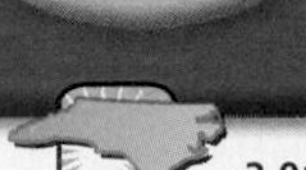

3.01

FOR EOG PRACTICE

see page 664

internet connect
Homework Help Online
go.hrw.com Keyword: MR4 7-5

GUIDED PRACTICE

See Example 1

1. Three stars form a triangular constellation. Two of the angles measure 20° and 50°. Classify the triangle.

See Example 2

Use the diagram to find the measure of each indicated angle.

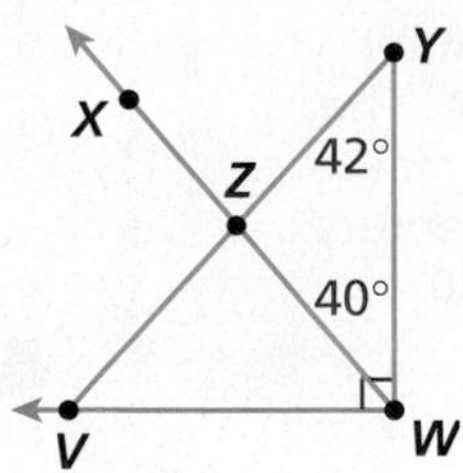

2. $\angle XZV$
3. $\angle VZW$

See Example 3

Classify each triangle using the given information.

4. The sum of the lengths of the three sides is 24 cm.

5. The sum of the lengths of the three sides is 30 ft.

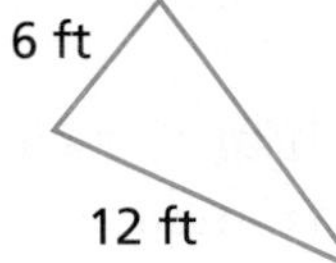

INDEPENDENT PRACTICE

See Example 1

6. Interstate highways connecting towns *R*, *S*, and *T* form a triangle. Two of the angles measure 40° and 42°. Classify the triangle.

See Example 2

Use the diagram to find the measure of each indicated angle.

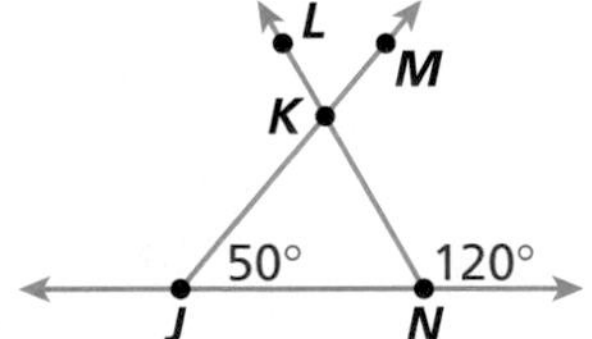

7. $\angle KNJ$
8. $\angle LKM$

See Example 3

Classify each triangle using the given information.

9. The sum of the lengths of the three sides is 10.5 in.

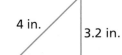

4 in.
3.2 in.

10. The sum of the lengths of the three sides is 231 km.

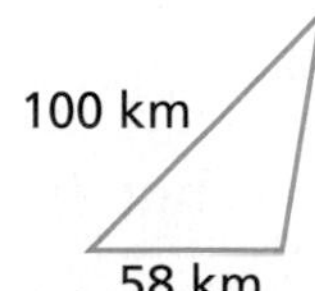

PRACTICE AND PROBLEM SOLVING

If the angles can form a triangle, classify it as acute, obtuse, or right.

11. 45°, 90°, 45°
12. 51°, 88°, 41°
13. 55°, 102°, 33°
14. 37°, 40°, 103°

Social Studies LINK

The Pony Express was established in 1860. This system of horseback messengers carried information across the United States before our modern postal service.

The lengths of two sides are given for $\triangle ABC$. Use the sum of the lengths of the three sides to calculate the length of the third side and classify each triangle.

15. $AB = 7$ cm; $BC = 7$ cm; sum = 15.9 cm

16. $AB = 24$ in.; $BC = 30$ in.; sum = 92 in.

17. $AB = 1\frac{1}{6}$ ft; $BC = 1\frac{1}{6}$ ft; sum = $3\frac{1}{2}$ ft

18. $AB = 9.5$ m; $BC = 9.5$ m; sum = 30 m

Draw an example of each triangle described.

19. a scalene acute triangle

20. an isosceles right triangle

21. an isosceles obtuse triangle

22. a scalene right triangle

23. ***SOCIAL STUDIES*** Some triangular stamps are made by dividing a rectangle into two parts. Classify the triangle that is made by cutting on a line that connects one corner of a rectangle to the opposite corner.

24. ***MEASUREMENT*** Use a centimeter ruler to measure each side of triangle A. Add the lengths of any two sides and compare the sum to the length of the third side. Add a different pair of lengths and compare the sum to the third side. Do the same for triangles B and C. What do you notice?

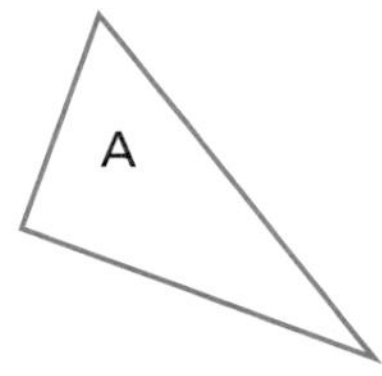

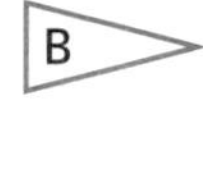

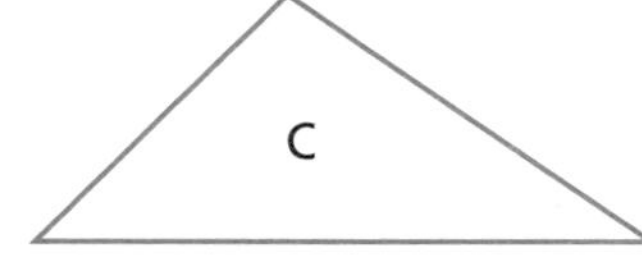

25. ***CHOOSE A STRATEGY*** How many triangles are in the figure at right?

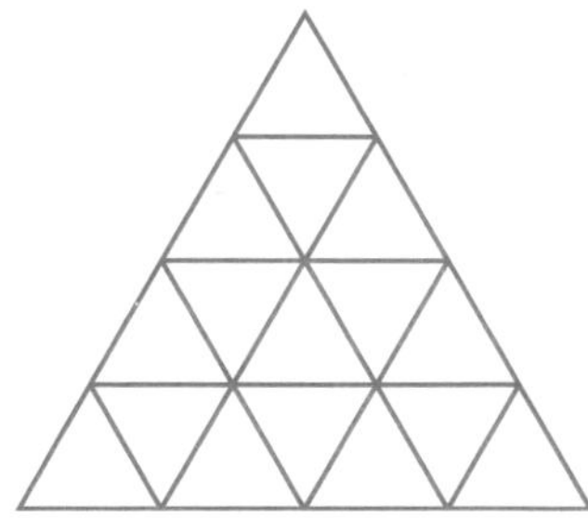

26. ***WRITE ABOUT IT*** Explain why a triangle cannot have two right angles.

27. ***CHALLENGE*** Find the sum of the angles of a square. (*Hint:* Divide the square into two triangles.)

Spiral Review

Find the greatest common factor of each set of numbers. (Lesson 4-3)

28. 12, 36

29. 15, 24

30. 18, 24, 42

31. 5, 14, 17

Plot each point on a coordinate plane. (Lesson 6-6)

32. A (3, 5)

33. B (6, 2)

34. C (0, 4)

35. D (1, 0)

36. **EOG PREP** Which number could represent the measure of an acute angle? (Lesson 7-2)

A 71°

B 90°

C 112°

D 180°

7-6 Quadrilaterals

Learn to identify, classify, and compare quadrilaterals.

Vocabulary
- quadrilateral
- parallelogram
- rectangle
- rhombus
- square
- trapezoid

A **quadrilateral** is a plane figure with four sides and four angles.

Five special types of quadrilaterals and their properties are shown in the table below. The same mark on two or more sides of a figure indicates that the sides are congruent.

Parallelogram	Opposite sides are parallel and congruent. Opposite angles are congruent.
Rectangle	Parallelogram with four right angles
Rhombus	Parallelogram with four congruent sides
Square	Rectangle with four congruent sides
Trapezoid	Quadrilateral with exactly two parallel sides May have two right angles

EXAMPLE 1 Naming Quadrilaterals

Give the most descriptive name for each figure.

A 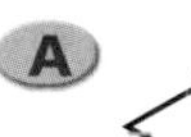 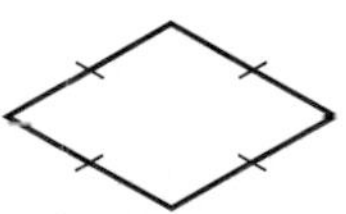

The figure is a quadrilateral, a parallelogram, and a rhombus.

Rhombus is the most descriptive name.

B 

The figure is a quadrilateral and a trapezoid.

Trapezoid is the most descriptive name.

Give the most descriptive name for each figure.

C

The figure is a quadrilateral, parallelogram, rectangle, rhombus, and square.

Square is the most descriptive name.

D

This figure is a plane figure, but it has more than 4 sides.

The figure is not a quadrilateral.

You can draw a diagram to classify quadrilaterals based on their properties.

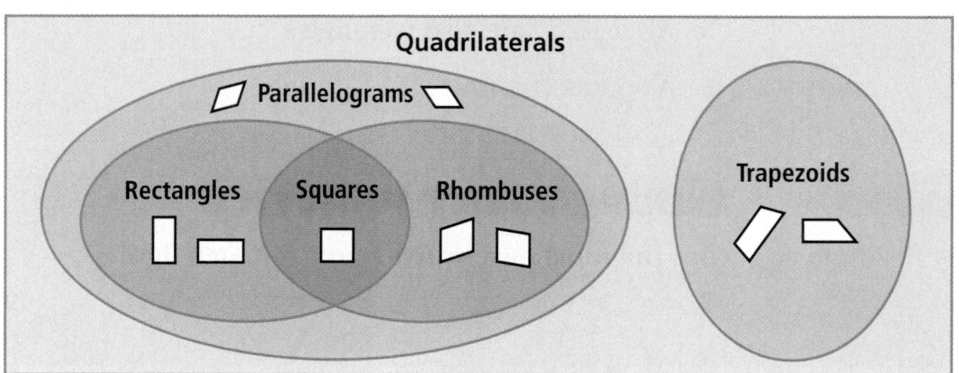

EXAMPLE 2 Classifying Quadrilaterals

Complete each statement.

A **A rhombus that is a rectangle is also a ___?___.**

A rhombus has four congruent sides, and the opposite sides are parallel. If it is a rectangle, it has four right angles, which makes it a **square**.

B **A square can also be called a ___?___, ___?___, and ___?___.**

A square has opposite sides that are parallel; it can be called a **parallelogram**.

A square has four congruent sides; it can be called a **rhombus**.

A square has four right angles; it can be called a **rectangle**.

Think and Discuss

1. **Tell** whether all squares are rhombuses and whether all rhombuses are squares.
2. **Compare** a trapezoid with a rectangle.

7-6 Exercises

FOR EOG PRACTICE

see page 664

internet connect

Homework Help Online
go.hrw.com Keyword: MR4 7-6

GUIDED PRACTICE

See Example 1 **Give the most descriptive name for each figure.**

1. 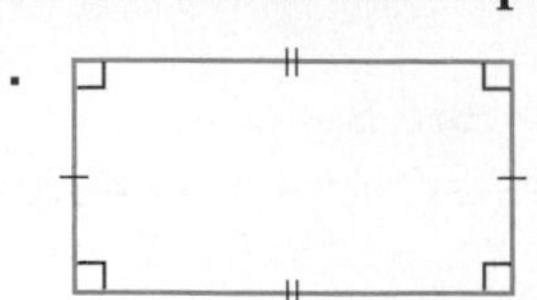

2.

3. 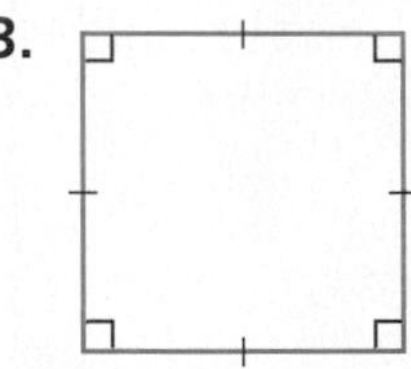

See Example 2 **Complete each statement.**

4. A trapezoid is also a __?__.

5. All __?__ are also rectangles.

6. A square has four __?__ angles.

INDEPENDENT PRACTICE

See Example 1 **Give the most descriptive name for each figure.**

7. 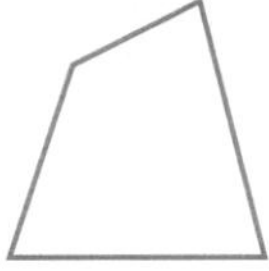

8. 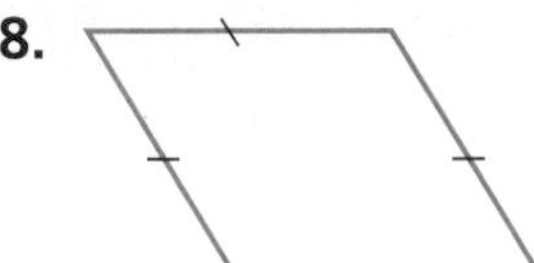

9. 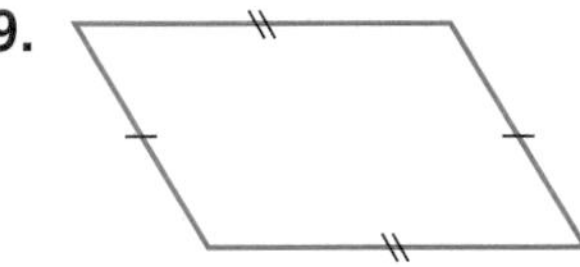

See Example 2 **Complete each statement.**

10. A rhombus with four right angles is a __?__.

11. A parallelogram cannot be a __?__.

12. A quadrilateral with four congruent sides and no right angles can be called a __?__ and a __?__.

PRACTICE AND PROBLEM SOLVING

Give all of the possible names for each figure. Circle the most exact name.

13. 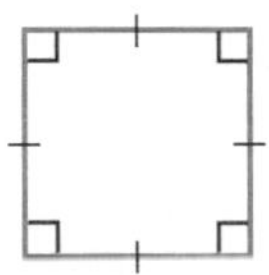

14. 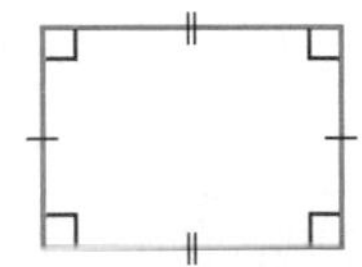

15. 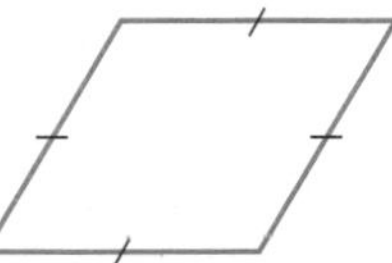

Determine if the given statements are *sometimes, always,* or *never* true.

16. A square is a rectangle.

17. A trapezoid is a parallelogram.

18. A rhombus is a square.

19. A parallelogram is a quadrilateral.

20. A rectangle is a rhombus.

21. Four-sided figures are parallelograms.

22. A rectangle is a square.

23. A trapezoid has one right angle.

The first baseball game played at Shea Stadium was on April 17, 1964. In September 1974, the Mets played a 25-inning game against the Cardinals, almost three times the length of a regulation game.

go.hrw.com
KEYWORD: MR4 Baseball

CNN Student News

Draw each quadrilateral as described. If it is not possible to draw, explain why.

24. a rectangle that is also a square

25. a rhombus that is also a trapezoid

26. a parallelogram that is not a rectangle

27. a square that is not a rhombus

28. ***SPORTS*** A baseball diamond is in the shape of a square. The distance from home plate to first base is 90 ft. What is the distance around the baseball diamond?

29. A rectangular picture frame is 3 in. wider than it is tall. The total length of the four sides of the frame is 38 in.

a. The dimensions of the frame could be 10 in. by 13 in. because one dimension is 3 in. longer than the other. Explain how you know the frame is not 10 in. by 13 in.

b. How can you use your answer from part **a** to find the dimensions of the frame?

c. Using parts **a** and **b,** what are the dimensions of the frame?

30. Anika drew a quadrilateral. Then she drew a line segment connecting one pair of opposite corners. She saw that she had divided the original quadrilateral into two right isosceles triangles. Classify the quadrilateral she began with.

31. ***WHAT'S THE ERROR?*** A student said that any quadrilateral with two right angles and a pair of parallel sides is a rectangle. What is the error in the statement?

32. ***WRITE ABOUT IT*** Explain why a square is also a rectangle and a rhombus.

33. ***CHALLENGE*** Part of a quadrilateral is hidden. What are the possible types of quadrilaterals that the figure could be?

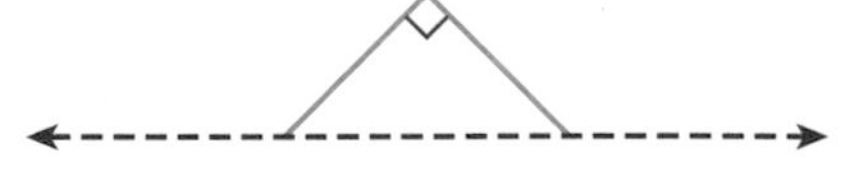

Spiral Review

34. A reporter interviewed 100 drivers and asked them how many times each had received a speeding ticket. Create a bar graph that displays the data. (Lesson 6-4)

Tickets	Number of Drivers
0	48
1	34
2	10
3	5
4+	3

35. Angles M and N are supplementary. If the measure of angle M is 33°, what is the measure of angle N? (Lesson 7-3)

36. **EOG PREP** ___?___ are lines in the same plane that never intersect. (Lesson 7-4)

A Skew lines

B Intersecting lines

C Parallel lines

D Perpendicular lines

7-7 Polygons

Learn to identify regular and not regular polygons and to find the angle measures of regular polygons.

Vocabulary

polygon

regular polygon

Triangles and quadrilaterals are examples of polygons. A **polygon** is a closed plane figure formed by three or more line segments. A **regular polygon** is a polygon in which all sides are congruent and all angles are congruent.

Polygons are named by the number of their sides and angles.

	Triangle	Quadrilateral	Pentagon	Hexagon	Octagon
Sides and Angles	3	4	5	6	8
Regular					
Not Regular					

EXAMPLE 1 Identifying Polygons

Tell whether each shape is a polygon. If so, give its name and tell whether it appears to be regular or not regular.

There are 4 sides and 4 angles.

quadrilateral

The sides and angles appear to be congruent.

regular

There are 4 sides and 4 angles.

quadrilateral

All 4 sides do not appear to be congruent.

not regular

The sum of the interior angle measures in a triangle is 180°, so the sum of the interior angle measures in a quadrilateral is 360°.

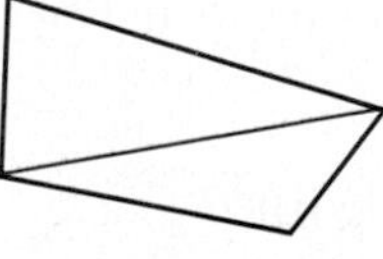

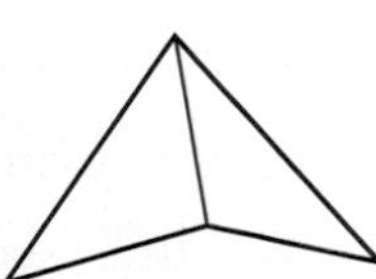

EXAMPLE 2 PROBLEM SOLVING APPLICATION

A stop sign is in the shape of a regular octagon. What is the measure of each angle of the stop sign?

1 Understand the Problem

The **answer** will be the measure of each angle in a regular octagon. List the **important information:**

- A regular octagon has 8 congruent sides and 8 congruent angles.

2 Make a Plan

Make a table to look for a pattern using regular polygons.

3 Solve

Draw some regular polygons and divide each into triangles.

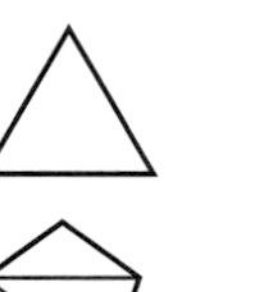

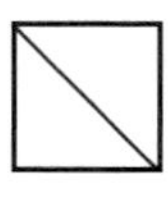

Reading Math

The prefixes in the names of the polygons tell you how many sides and angles there are.

tri- = three
quad- = four
penta- = five
hexa- = six
octa- = eight

Polygon	Sides	Triangles	Sum of Angle Measures
Triangle	3	1	180°
Quadrilateral	4	2	$2 \times 180° = 360°$
Pentagon	5	3	$3 \times 180° = 540°$
Hexagon	6	4	$4 \times 180° = 720°$

The number of triangles is always 2 fewer than the number of sides.
An octagon can be divided into $8 - 2 = 6$ triangles.
The sum of the interior angle measures in an octagon is $6 \times 180° = 1{,}080°$.
So the measure of each angle is $1{,}080° \div 8 = 135°$.

4 Look Back

Each angle in a regular octagon is obtuse. 135° is a reasonable answer, because an obtuse angle is between 90° and 180°.

Think and Discuss

1. **Classify** the angles in each figure: a regular triangle, a regular hexagon, and a rectangle.
2. **Name** an object that is in the shape of a pentagon and an object that is in the shape of an octagon.

7-7 Exercises

FOR EOG PRACTICE
see page 665

internet connect
Homework Help Online
go.hrw.com Keyword: MR4 7-7

GUIDED PRACTICE

See Example 1 Tell whether each shape is a polygon. If so, give its name and tell whether it appears to be regular or not regular.

1. 

2. 

3. 

See Example 2

4. A flower bed in the park is in the shape of a rhombus. The distance around the flower bed is 160 meters. What is the length of one side of the flower bed?

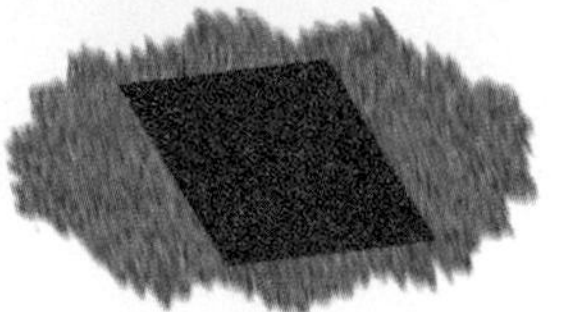

INDEPENDENT PRACTICE

See Example 1 Tell whether each shape is a polygon. If so, give its name and tell whether it appears to be regular or not regular.

5. 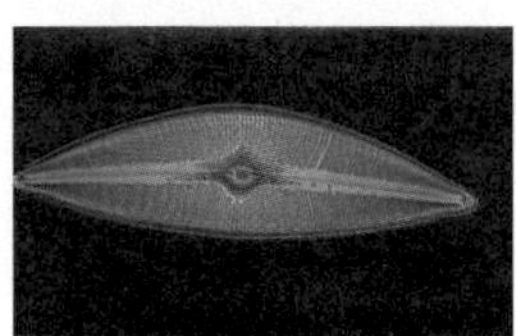

6. 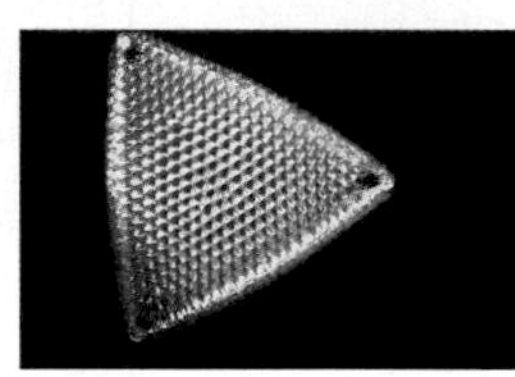

7. 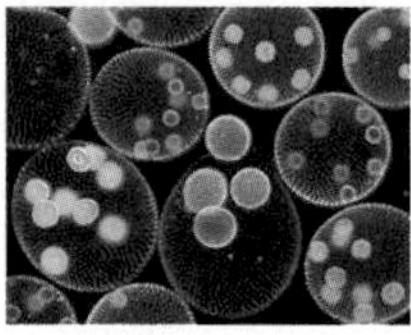

See Example 

8. Janet made a sign for her room in the shape of a regular pentagon. What is the measure of each angle of the pentagon?

PRACTICE AND PROBLEM SOLVING

Explain why each shape is *not* a polygon.

9. 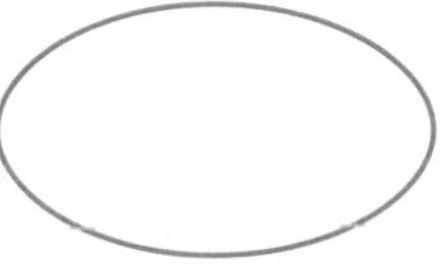

10. 

11. 

Name each polygon.

12. 

13. 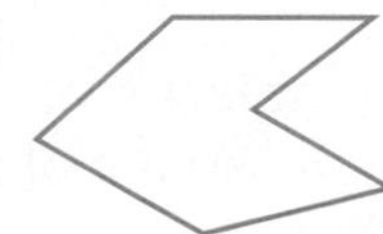

14. 

Classify each of the following polygons as either *always* regular, *sometimes* regular, or *never* regular.

		Always	Sometimes	Never
15.	Equilateral triangle	?	?	?
16.	Trapezoid	?	?	?
17.	Right triangle	?	?	?
18.	Parallelogram	?	?	?

MEASUREMENT A *diagonal* is a line segment that connects two nonadjacent vertices of a polygon. One diagonal is shown in each figure.

19. a. How many diagonals does a rectangle have?

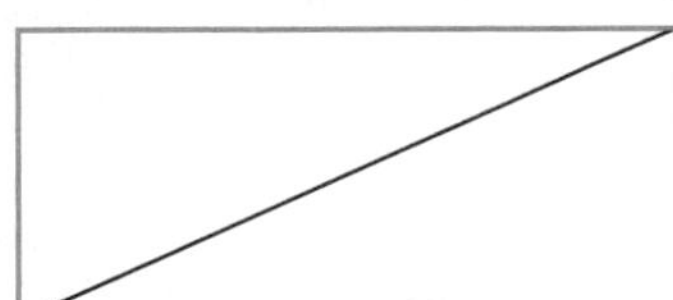

b. How many diagonals does a pentagon have?

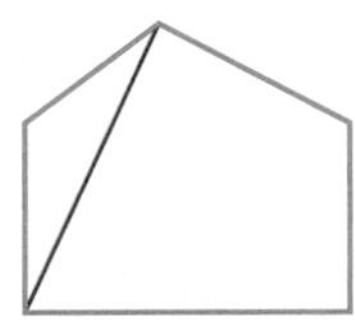

20. *WHAT'S THE ERROR?* A student said a rectangle is never a regular polygon because the lengths of all the sides are not congruent. What error did the student make? Explain why a rectangle is sometimes a regular polygon.

21. *WRITE ABOUT IT* What polygon is formed when two equilateral triangles are placed side by side, with one upside down? Draw examples, and explain whether the polygon formed by the two triangles is regular.

22. *CHALLENGE* A figure is formed by placing 6 equilateral triangle tiles around a regular hexagon tile. The distance around the regular hexagon is 60 cm. A snail moves along the sides of the figure. How far will the snail travel until it gets back to its starting point?

Spiral Review

Use the pattern to write the first five terms of the sequence. (Lesson 1-7)

23. Start with 10; add 3.

24. Start with 4; multiply by 4.

Write each decimal as a fraction in simplest form. (Lesson 4-4)

25. 0.9 **26.** 0.71 **27.** 0.20 **28.** 0.88

29. **EOG PREP** Which is the least common multiple (LCM) of 4, 6, and 12? (Lesson 5-5)

A 2 B 3 C 12 D 24

7-8 Geometric Patterns

Learn to recognize, describe, and extend geometric patterns.

Native American art often involves geometric patterns. The patterns are based on the shape, color, size, position, or number of geometric figures.

This blanket has a geometric pattern. The first row with a complete figure has a parallelogram with a horse in its center. The next row has two parallelograms with cows in the centers. This pattern continues. If the weaver wanted to make a longer blanket, the next row would be two parallelograms with pictures of cows.

This Navajo blanket was made in the late seventeenth century.

EXAMPLE  1 Extending Geometric Patterns

Identify a possible pattern. Use the pattern to draw the next figure.

A

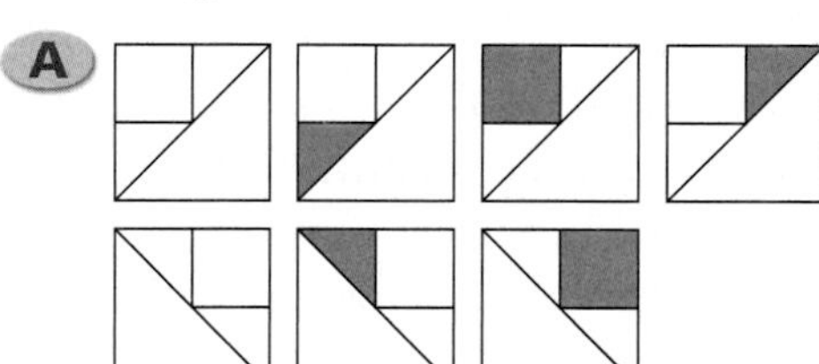

The small shapes within the figure are shaded one at a time from bottom to top. Then the figure is rotated and the top triangle is shaded.

So the next figure might look like this:

B

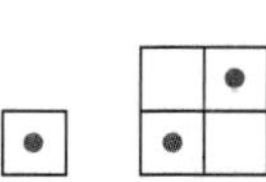

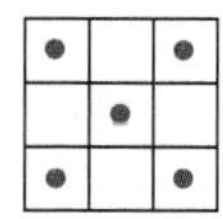

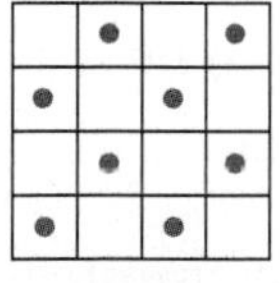

The figures from left to right are a 1 × 1 square, a 2 × 2 square, a 3 × 3 square, and a 4 × 4 square.

So the next figure might look like this:

Remember!

Perfect squares, such as 2^2, 3^2, and 4^2, are also called "square numbers" because they can be modeled as a square array.

EXAMPLE 2 Completing Geometric Patterns

Identify a possible pattern. Use the pattern to draw the missing figure.

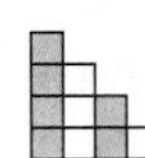 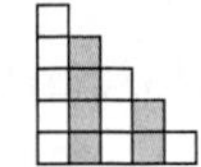 ? 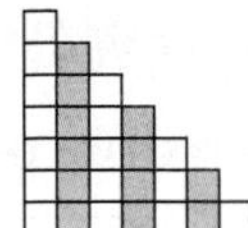

The first figure from the bottom row to the top has 4 squares and then 3, 2, and 1 square. The next figure has 5 squares in the bottom row and then 4, 3, 2, and 1.

So the missing figure might look like this:

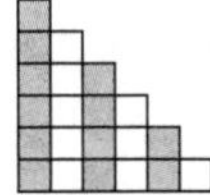

B

 ?  

Each figure is an equilateral triangle. The first figure has 3 red triangles along the base. The third figure has 5 red triangles, and the last figure has 6.

So the missing figure might look like this:

EXAMPLE 3 *Art Application*

Dan is painting a clay pot. Identify a pattern in Dan's design and tell what the finished pot might look like.

The pattern from bottom to top is narrow stripe, wide stripe, narrow stripe, wide stripe. The color pattern from bottom to top is blue, green, yellow, blue, green.

If this pattern is followed, the finished pot might look like the pot at left.

Think and Discuss

1. **Explain** how you can use a pattern to find the number of squares in the next, or fifth, figure in Example 2A.
2. **Tell** how you can use a pattern to find the number of small red triangles in the sixth figure in Example 2B.

7-8 Exercises

FOR EOG PRACTICE

see page 665

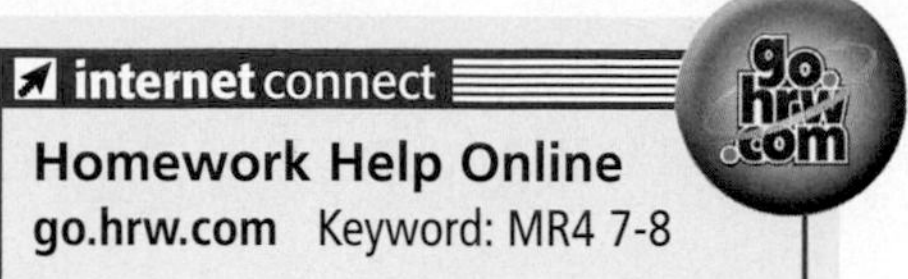

GUIDED PRACTICE

See Example 1 **Identify a possible pattern. Use the pattern to draw the next figure.**

1. 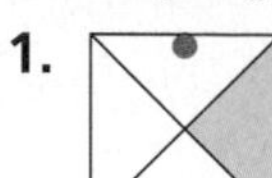 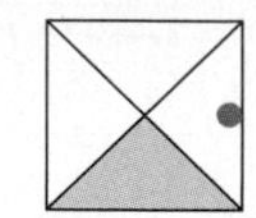 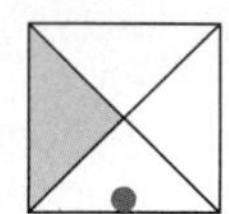 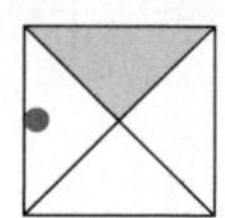

See Example 2 **Identify a possible pattern. Use the pattern to draw the missing figure.**

2.  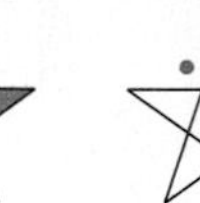  

?

See Example 3

3. Oscar is making a beaded necklace. Identify a pattern in Oscar's design. Then tell which five beads Oscar will probably use next.

INDEPENDENT PRACTICE

See Example 1 **Identify a possible pattern. Use the pattern to draw the next figure.**

4. 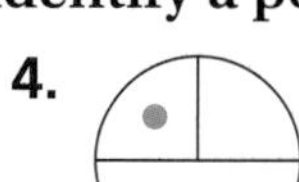 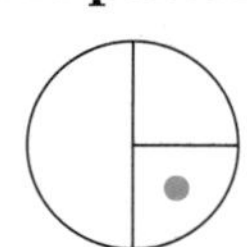 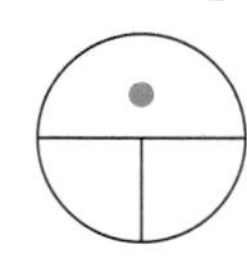 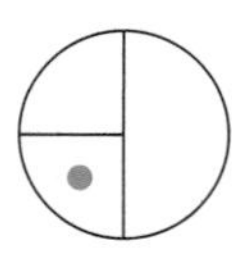 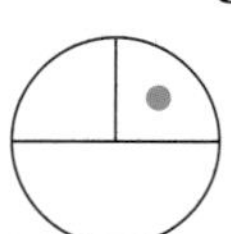

See Example 2 **Identify a possible pattern. Use the pattern to draw the missing figure.**

5.  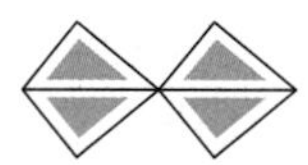  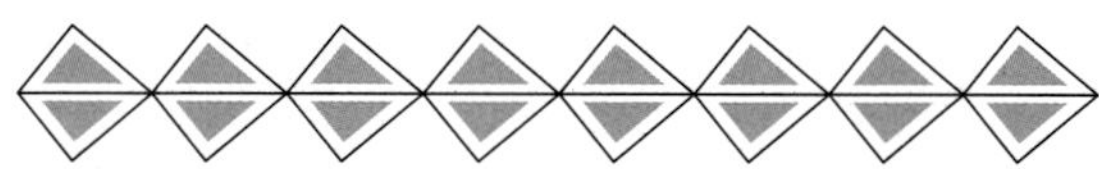

?

See Example 3

6. Tamara is planting flowers in her garden. She makes groups of purple flowers and groups of pink flowers.

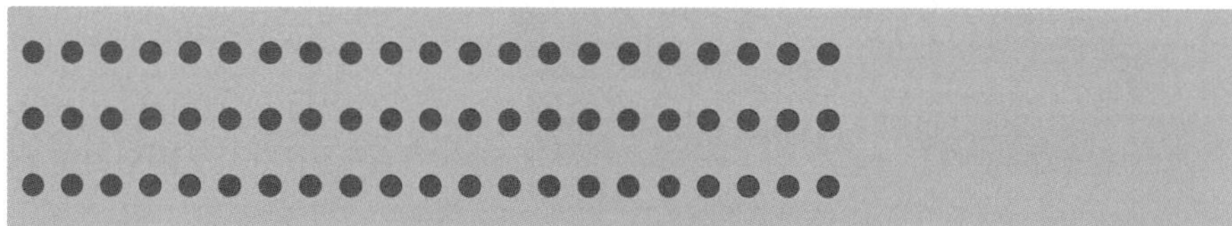

If she continues this pattern, how many flowers might Tamara plant in the next group of purple flowers? How many flowers might she plant in the next group of pink flowers?

PRACTICE AND PROBLEM SOLVING

Draw the next figure in the pattern.

7.    

Social Studies

The art of decorating houses is practiced throughout Africa. In South Africa, Ndebele and Basotho people paint their houses with brightly colored patterns made up of geometric shapes.

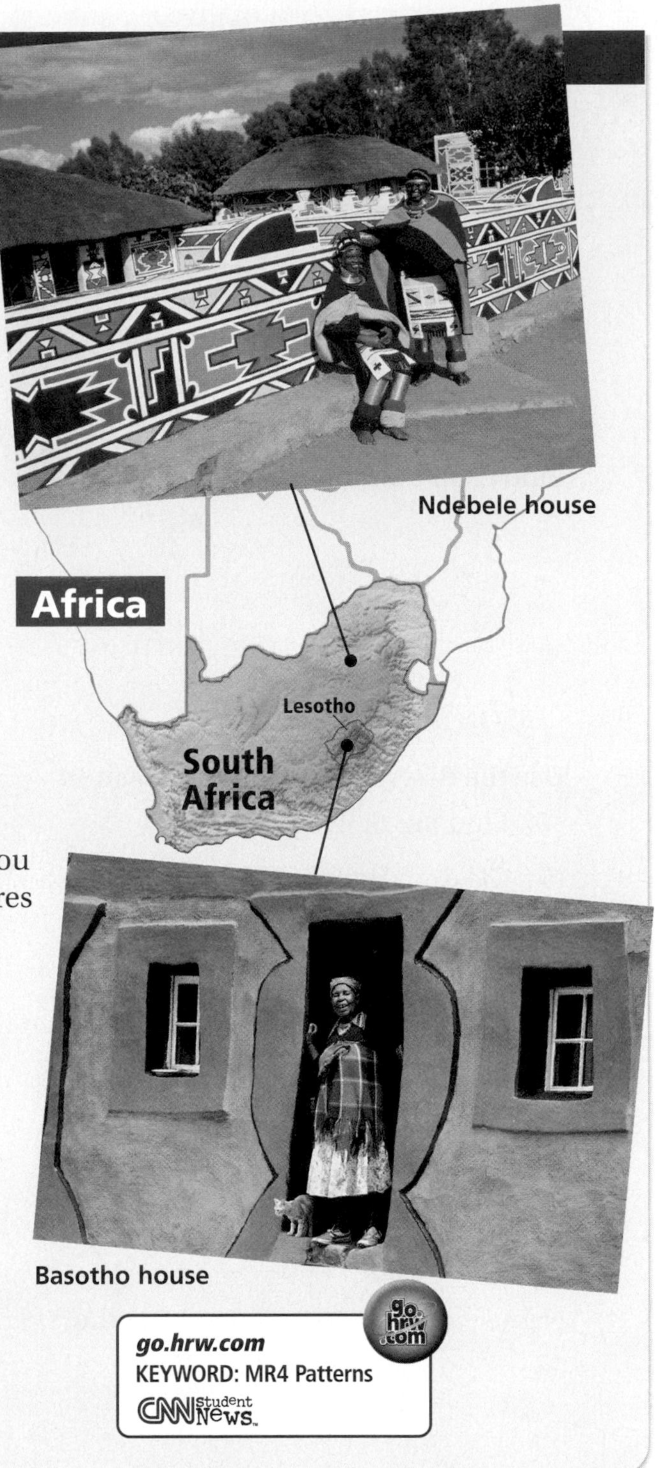

Ndebele house

Basotho house

8. Look at the shapes found on the wall surrounding the Ndebele house. Identify a possible pattern that was used to paint the top band of the wall. Use the pattern to draw the shapes hidden by the Ndebele people. (You do not need to make color part of the pattern.)

9. Look at the design on the Basotho house. Identify a possible pattern. Use the pattern to draw a picture of what the house might look like.

10. ***WRITE ABOUT IT*** Look closely at the Ndebele house. Draw four geometric figures you see painted on the house. Then use those figures to make a pattern. Describe your pattern.

11. ***CHALLENGE*** Look at the designs below, which were made using an African motif. Identify a possible pattern. If the pattern continues, how many motifs will be in the sixth design? If there are 45 motifs, what will the design number be?

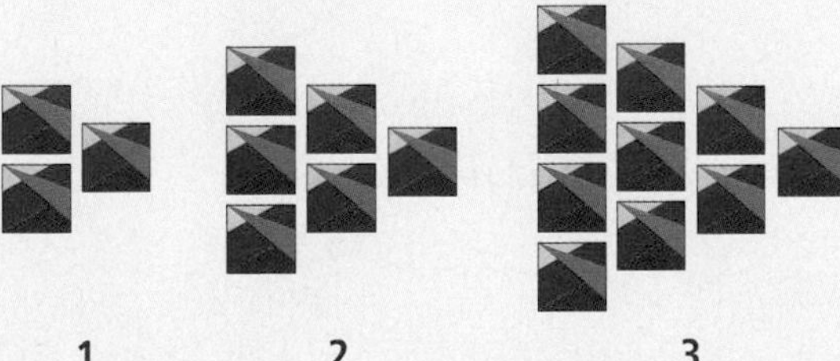

go.hrw.com
KEYWORD: MR4 Patterns
CNN Student News

Spiral Review

Write each expression in exponential form. (Lesson 4-2)

12. $3 \times 3 \times 3 \times 5 \times 5$ **13.** $7 \times 7 \times 4 \times 4$ **14.** $2 \times 2 \times 2 \times 2 \times 3 \times 3 \times 5$

15. **EOG PREP** Multiply $\frac{4}{5} \cdot \frac{3}{5}$. Write the answer in simplest form. (Lesson 5-1)

A $1\frac{2}{5}$ B $\frac{3}{5}$ C $\frac{12}{25}$ D $1\frac{1}{3}$

Chapter 7

Mid-Chapter Quiz

LESSON 7-1 (pp. 322–325)

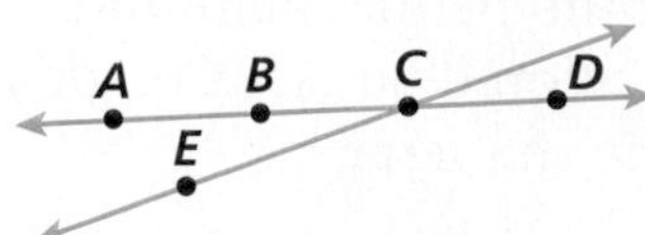

Use the diagram for problems 1 and 2.

1. Name two lines.
2. Name three different rays.

LESSON  7-2 and 7-3 and  7-4 (pp. 326–339)

Find each unknown angle measure, and tell what type of angle it is. Classify lines ℓ and m.

3. 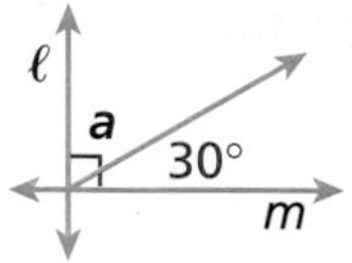

4. 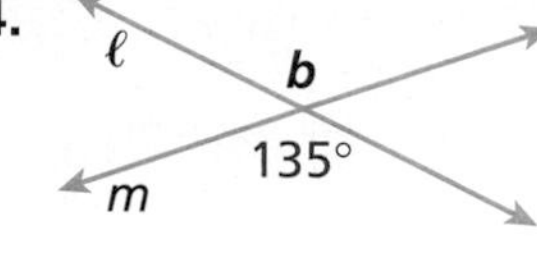

5. 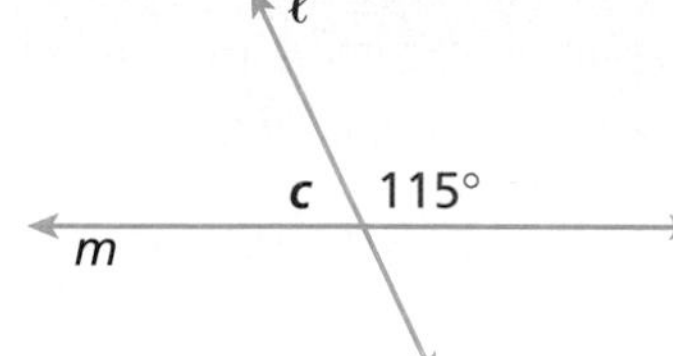

LESSON  7-5 (pp. 344–347)

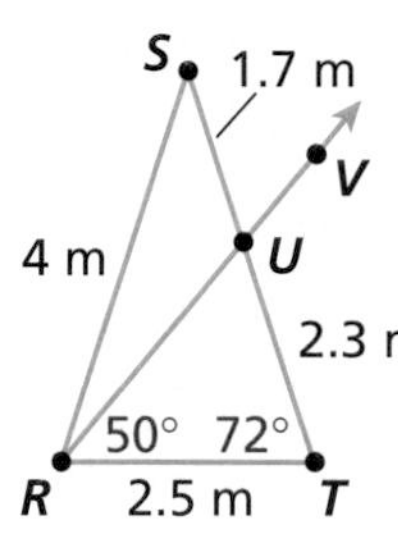

Use the diagram for problems 6 and 7.

6. Find m∠*SUV*.
7. Classify triangle *STR* by its angles and by its sides.

LESSON 7-6 (pp. 348–351)

Give the most descriptive name for each figure.

8. 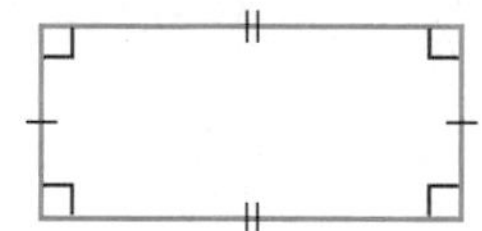

9.

10. 

11. 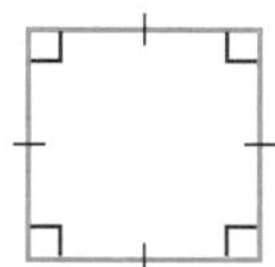

LESSON  7-7 (pp. 352–355)

Name each polygon, and tell whether it appears to be regular or not regular.

12. 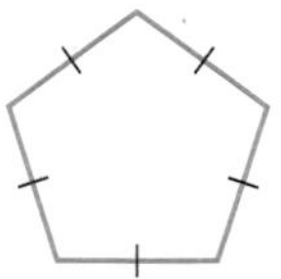

13. 

14. 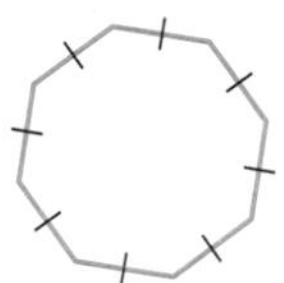

15. 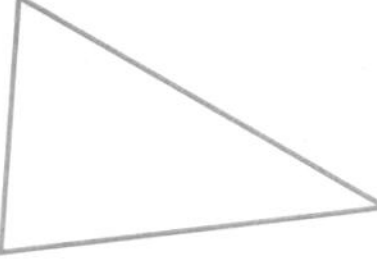

LESSON  7-8 (pp. 356–359)

Identify a possible pattern. Use the pattern to draw the missing figure.

16. 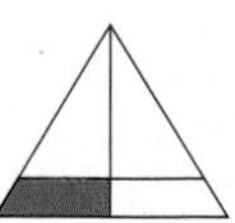 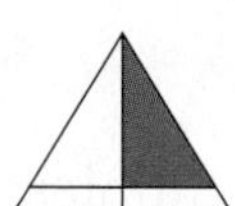 ?

17. 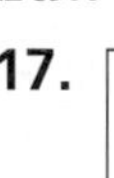  ? 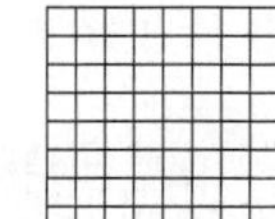

Focus on Problem Solving

Make a Plan

• **Draw a diagram**

Sometimes a problem seems difficult because it is described in words only. You can draw a diagram to help you picture the problem. Try to label all the information you are given on your diagram. Then use the diagram to solve the problem.

Read each problem. Draw a diagram to help you solve the problem. Then solve.

1. Bob used a ruler to draw a quadrilateral. First he drew a line 3 in. long and labeled it $\overline{AB}$. From B, he drew a line 2 in. long and labeled the endpoint C. From A, he drew a line $2\frac{1}{2}$ in. long and labeled the endpoint D. What is the length of $\overline{CD}$ if the perimeter of Bob's quadrilateral is $12\frac{1}{2}$ in?

2. Karen has a vegetable garden that is 12 feet long and 10 feet wide. She plans to plant tomatoes in one-half of the garden. She will divide the other half of the garden equally into three beds, where she'll grow cabbage, pumpkin, and radishes.
   - **a.** What are the possible whole number dimensions of the tomato bed?
   - **b.** What fraction of the garden will Karen use to grow cabbage?

3. Pam draws three parallel lines that are an equal distance apart. The two outside lines are 8 cm apart. How far apart is the middle line from the outside lines?

4. Jan connected the following points on a coordinate grid: (2, 4), (4, 6), (6, 6), (6, 2), (3, 2), and (2, 4).
   - **a.** What figure did Jan draw?
   - **b.** How many right angles does the figure have?

5. Triangle ABC is isoceles. The measure of angle B is equal to the measure of angle C. The measure of angle B equals 50°. What is the measure of angle A?

7-9 Congruence

Learn **to identify congruent figures and to use congruence to solve problems.**

You know that angles that have the same measure are congruent. Figures that have the same shape and same size are also congruent.

You can use stencils to decorate pages of a scrapbook. The stencil helps you draw congruent figures.

EXAMPLE 1 Identifying Congruent Figures

Decide whether the figures in each pair are congruent. If not, explain.

A

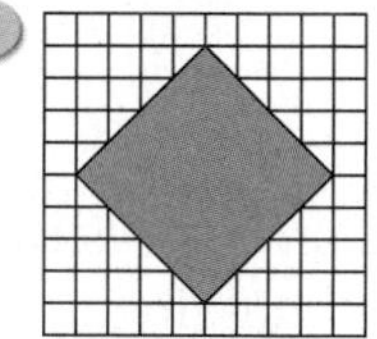

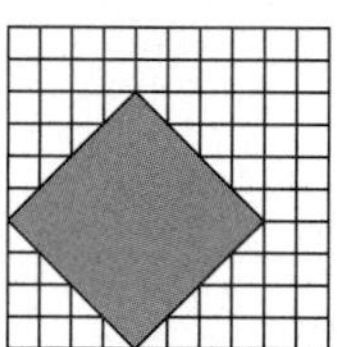

The figures are congruent.

These figures have the same shape and size.

B

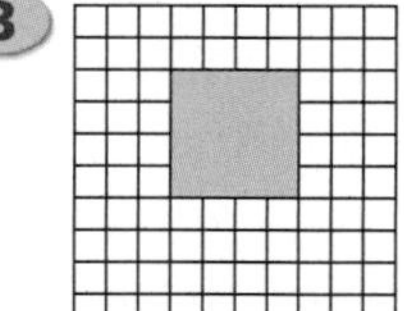

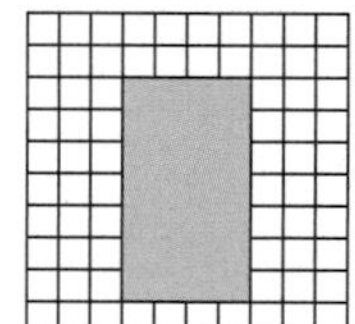

The figures are not congruent.

These figures are both quadrilaterals. But they are neither the same size nor the same shape.

C

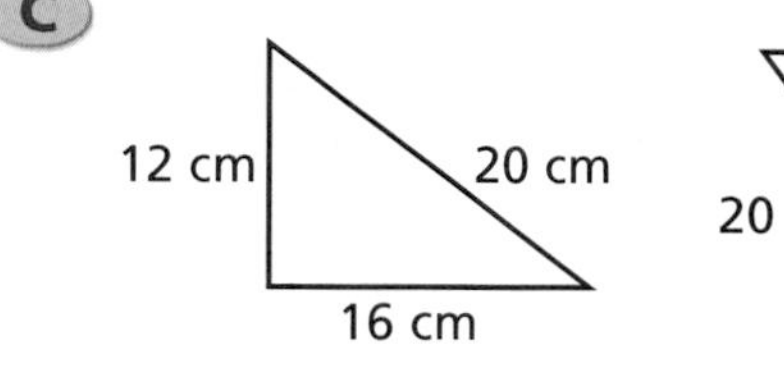

The triangles are congruent.

Each triangle has a 12 cm side, a 16 cm side, and a 20 cm side.

D

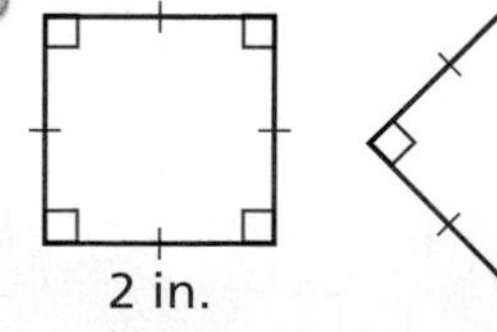

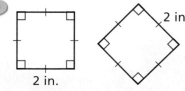

The figures are congruent.

Each figure is a square. Each side of each square measures 2 inches.

EXAMPLE 2 Consumer Application

Landra needs a ground cloth that is congruent to the tent floor. Which ground cloth should she buy?

Tent floor

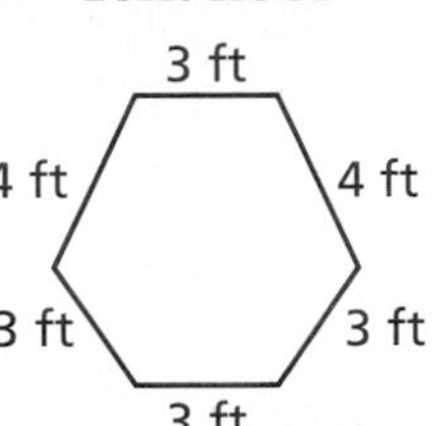

Ground cloth A

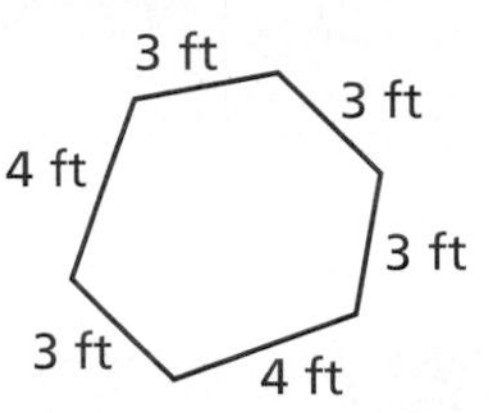

Ground cloth B

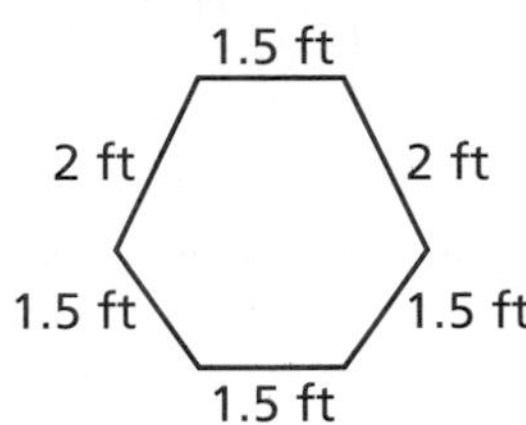

Which ground cloth is the same size and shape as the tent floor?
Both cloths are hexagons. Only Cloth A is the same size as the floor.

Cloth A is congruent to the tent floor.

Think and Discuss

1. **Explain** whether you can determine that figures are congruent just by looking at them.
2. **Tell** what information you would need to know about two rectangles to determine whether they are congruent.

7-9 Exercises

3.04

FOR EOG PRACTICE
see page 666

internet connect
Homework Help Online
go.hrw.com Keyword: MR4 7-9

GUIDED PRACTICE

See Example 1 **Decide whether the figures in each pair are congruent. If not, explain.**

1. 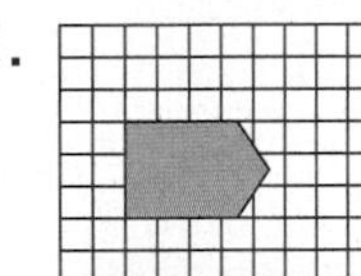 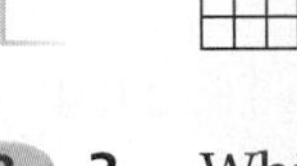

2. 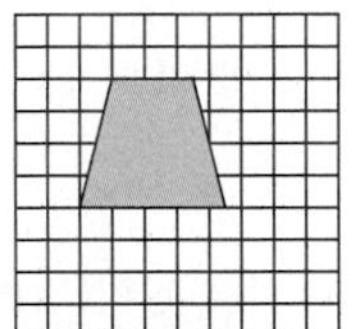

See Example 2 3. Which quadrilateral is congruent to the bottom of the box?

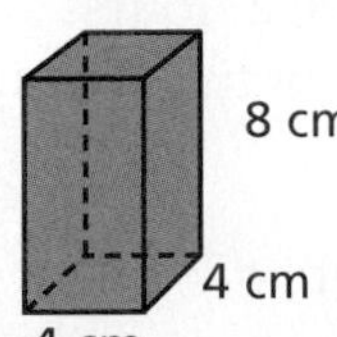

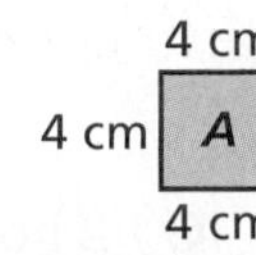

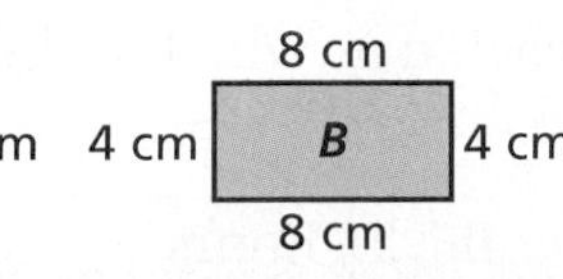

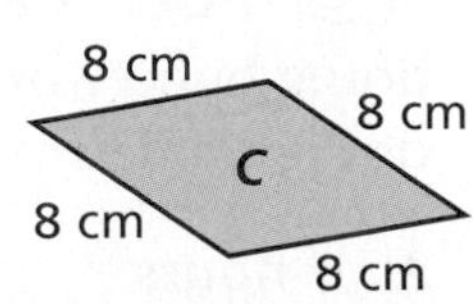

INDEPENDENT PRACTICE

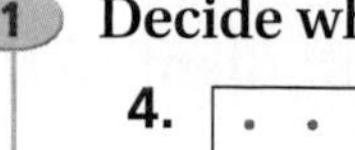

See Example 1 Decide whether the figures in each pair are congruent. If not, explain.

4. 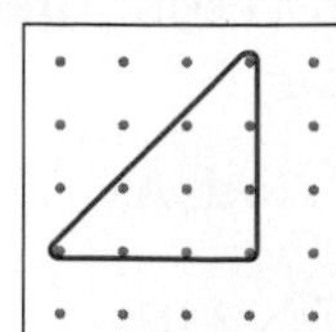 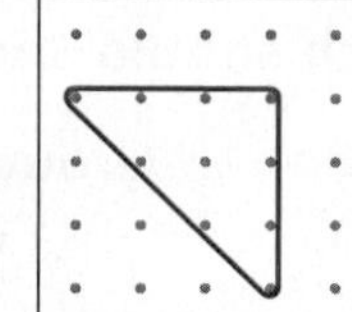

5. 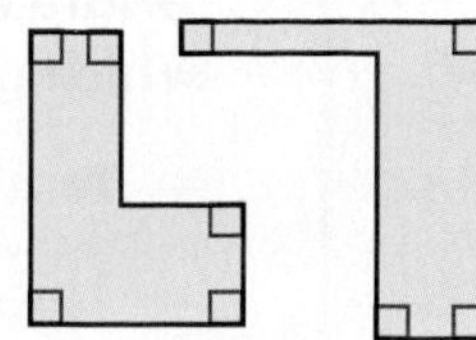

See Example 2

6. Which puzzle piece will fit into the empty space?

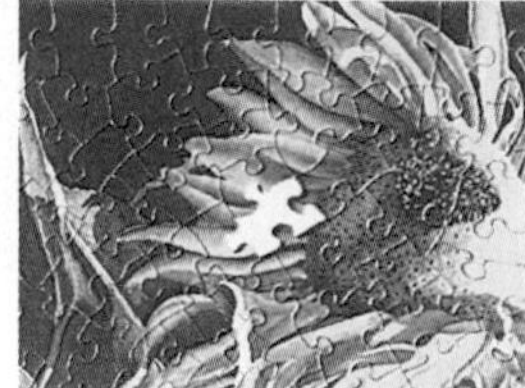

a. 

b. 

c.

PRACTICE AND PROBLEM SOLVING

7. Copy the dot grid. Then draw three figures congruent to the given figure. The figures can have common sides but should not overlap.

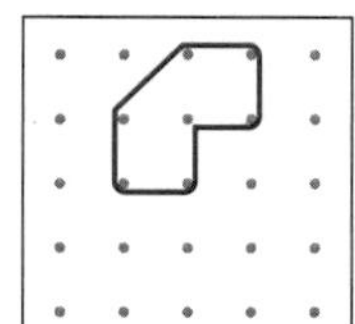

8. ***MEASUREMENT*** Use an inch ruler to draw two congruent rectangles with side lengths that are longer than 2 in. and shorter than 6 in. Label each side length.

9. ***WRITE ABOUT IT*** Explain how to tell whether two polygons are congruent.

10. ***CHALLENGE*** Two quadrilaterals have side lengths 2 cm, 2 cm, 5 cm, and 5 cm. Are the two quadrilaterals congruent? Explain.

Spiral Review

Solve each equation. (Lessons 2-6 and 2-7)

11. $5t = 45$ **12.** $72 = 3n$ **13.** $\frac{s}{6} = 8$

Find the GCF of each set of numbers. (Lesson 4-3)

14. 12, 18, 24 **15.** 15, 18, 30 **16.** 16, 24, 42

17. **EOG PREP** Last week Leo spent 2 hours, 3 hours, 2 hours, and 1 hour doing homework each day. Which is the mean amount of time he spent doing homework? (Lesson 6-2)

A 8 hours B $2\frac{1}{2}$ hours C 2 hours D $1\frac{1}{2}$ hours

7-10 Transformations

Learn to use translations, reflections, and rotations to transform geometric shapes.

Vocabulary

transformation

translation

rotation

reflection

line of reflection

A rigid **transformation** moves a figure without changing its size or shape. So the original figure and the transformed figure are always congruent.

The illustrations of the alien show three transformations: a translation, a rotation, and a reflection. Notice the transformed alien does not change in size or shape.

A **translation** is the movement of a figure along a straight line.

Only the location of the figure changes with a translation.

A **rotation** is the movement of a figure around a point. A point of rotation can be on or outside a figure.

The location and position of a figure can change with a rotation.

When a figure flips over a line, creating a mirror image, it is called a **reflection**. The line the figure is flipped over is called the **line of reflection**.

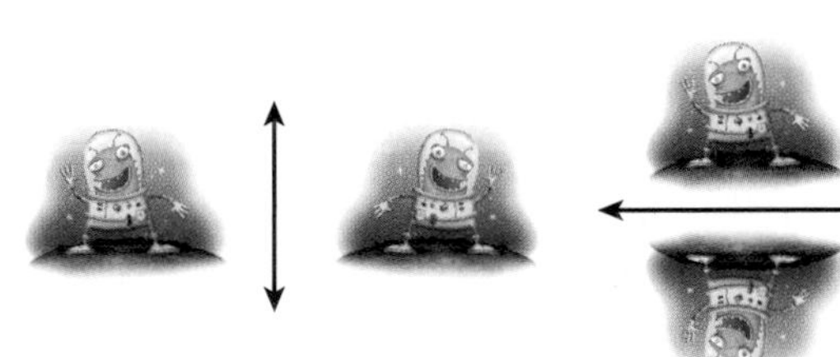

The location and position of a figure change with a reflection.

EXAMPLE  1 Identifying Transformations

Tell whether each is a translation, rotation, or reflection.

A

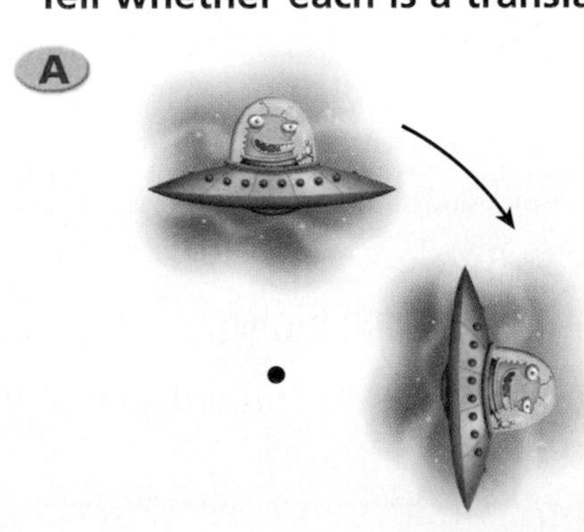

The figure moves around a point.

It is a rotation.

Tell whether each is a translation, rotation, or reflection.

B

The figure is flipped over a line.

It is a reflection.

C

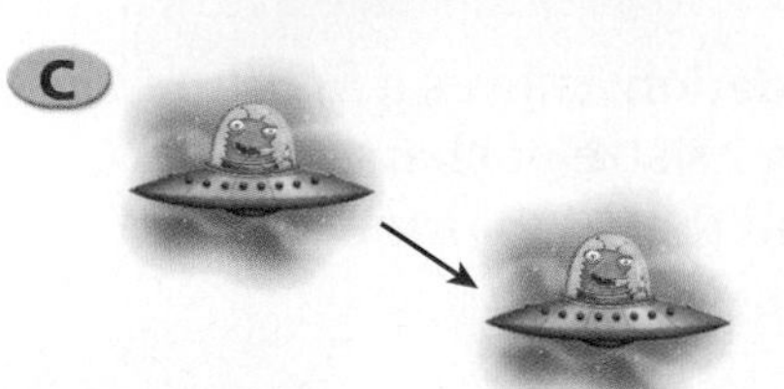

The figure is moved along a line.

It is a translation.

A full turn is a 360° rotation. So a $\frac{1}{4}$ turn is 90°, and a $\frac{1}{2}$ turn is 180°.

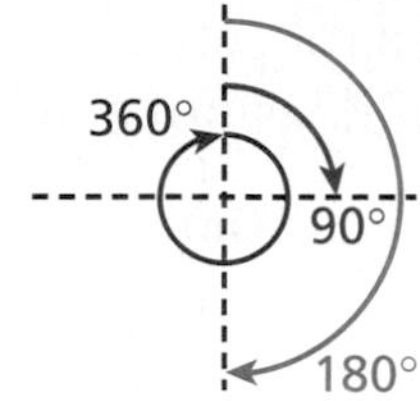

EXAMPLE 2 Drawing Transformations

Draw each transformation.

A

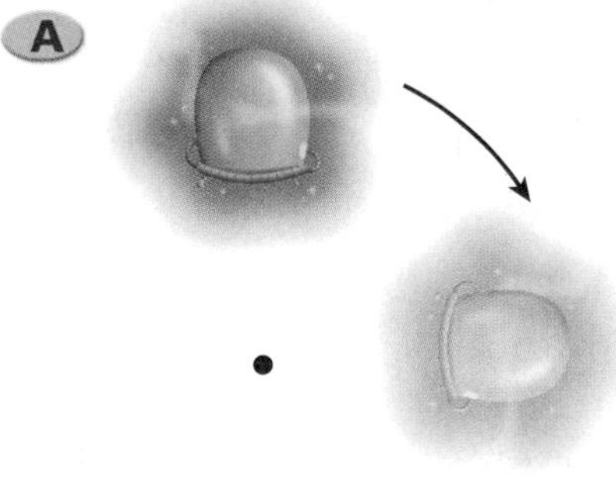

Draw a 90° clockwise rotation about the point shown.

Trace the figure and the point of rotation.

Place your pencil on the point of rotation.

Rotate the figure clockwise 90°.

Trace the figure in its new location.

B

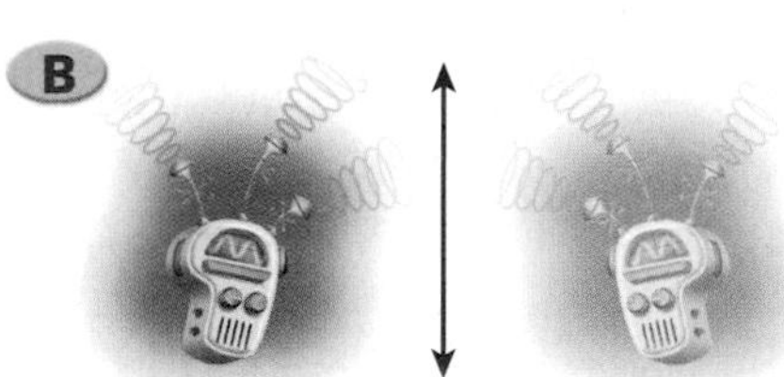

Draw a horizontal reflection.

Trace the figure and the line of reflection.

Fold along the line of reflection.

Trace the figure in its new location.

Think and Discuss

1. **Give examples** of reflections that occur in the real world.
2. **Name** a figure that can be rotated so that it will land on top of itself.

7-10 Exercises

FOR EOG PRACTICE
see page 666

internet connect
Homework Help Online
go.hrw.com Keyword: MR4 7-10

3.03

GUIDED PRACTICE

See Example 1

Tell whether each is a translation, rotation, or reflection.

1.

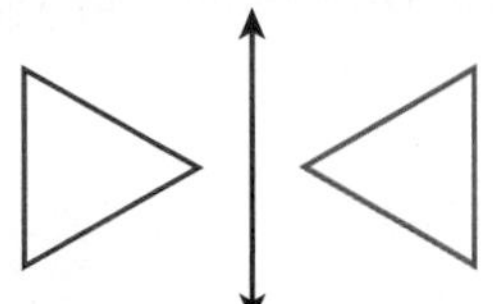

2.

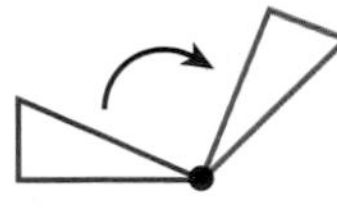

3.

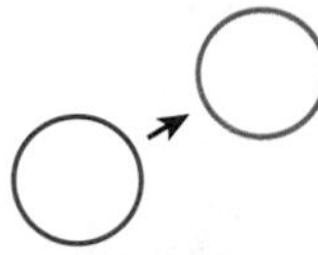

See Example 2

Draw each transformation.

4. Draw a 180° clockwise rotation about the point shown.

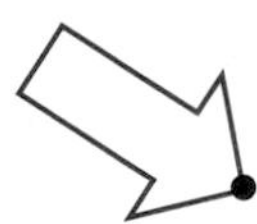

5. Draw a horizontal reflection across the line.

INDEPENDENT PRACTICE

See Example 1

Tell whether each is a translation, rotation, or reflection.

6.

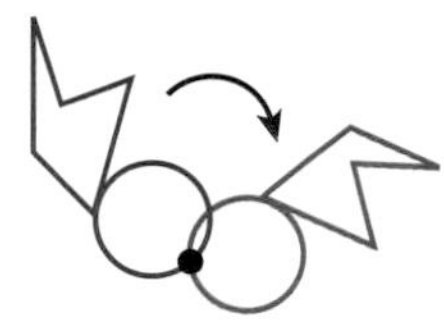

7.

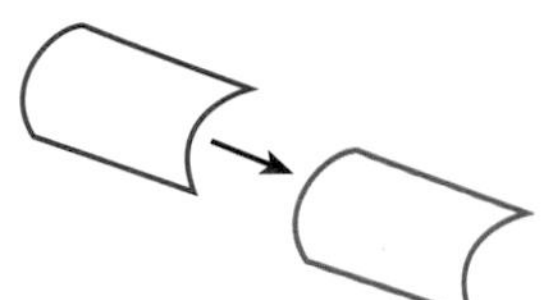

8.

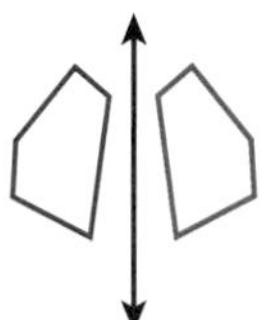

See Example 2

Draw each transformation.

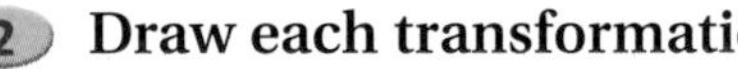

9. Draw a vertical reflection across the line.

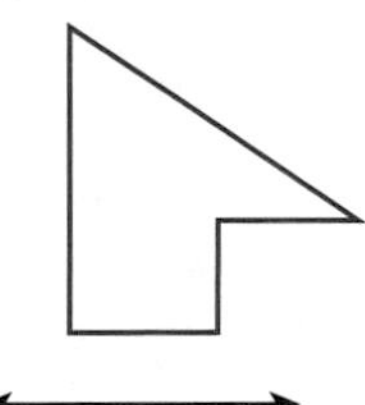

10. Draw a 90° counterclockwise rotation about the point.

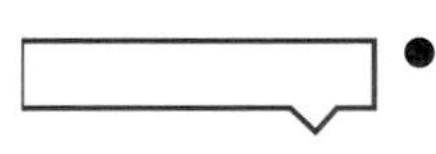

11. Draw a translation.

12. Draw a translation.

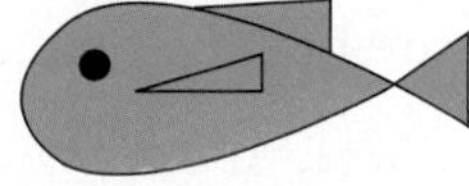

13. Draw a 90° clockwise rotation about the point.

14. Draw a horizontal reflection across the line.

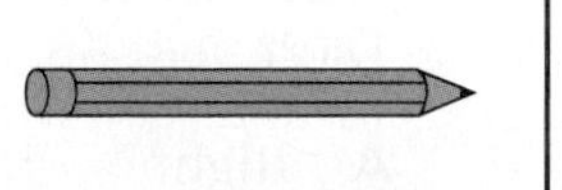

PRACTICE AND PROBLEM SOLVING

In a game of chess, each player has 318,979,564,000 possible ways to make the first four moves.

15. Which is a horizontal reflection of this red arrow?

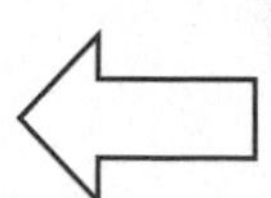

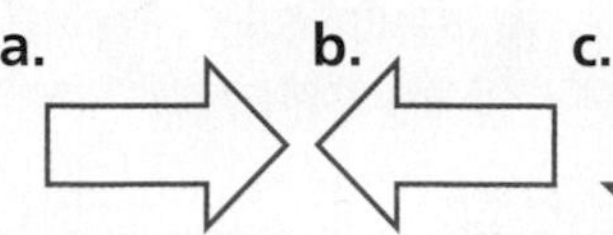

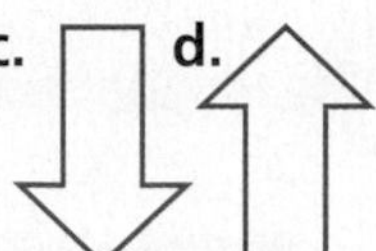

16. ***LANGUAGE ARTS*** Which letters in the alphabet can be horizontally reflected and still look the same? Which letters can be vertically reflected and still look the same?

Use the chessboard for Exercises 17–20.

HOBBIES Chess is a game of skill that is played on a board divided into 64 squares. Each chess piece is moved differently.

17. Copy the lower left corner of the chessboard. Then show the indicated knight moving in a translation of two forward and one right.

18. ***CHOOSE A STRATEGY*** If the knight, king, and pawn are placed in a straight line, how many ways can they be arranged?

A 3 **B** 4 **C** 6 **D** 12

19. ***WRITE ABOUT IT*** Draw one of the chess pieces. Then draw a translation, rotation, and reflection of that piece. Describe each transformation.

20. ***CHALLENGE*** Draw one of the chess pieces rotated 90° clockwise around the vertex of a square and then horizontally reflected.

Spiral Review

Find the least common multiple (LCM). (Lesson 5-5)

21. 4 and 12 **22.** 7, 14, and 21 **23.** 6, 9, and 24

Use the stem-and-leaf plot. (Lessons 6-2, 6-9)

24. Find the median.

25. Find the mode.

26. Find the range.

Stems	Leaves
2	0 1 3
3	2 5 6 6 6 7
4	5 8 9
5	2 3 3

Key: 2|3 means 23

27. **EOG PREP** Which type of angle has a measure between 90° and 180°? (Lesson 7-2)

A Right **B** Straight **C** Acute **D** Obtuse

7-11 Symmetry

Learn to identify line symmetry.

Vocabulary

line symmetry

line of symmetry

A figure has **line symmetry** if it can be folded or reflected so that the two parts of the figure match, or are congruent. The line of reflection is called the **line of symmetry**.

You could draw a line of symmetry on this windmill. The shape of the building and the position of the blades are symmetrical.

EXAMPLE 1 Identifying Lines of Symmetry

Determine whether each dashed line appears to be a line of symmetry.

A

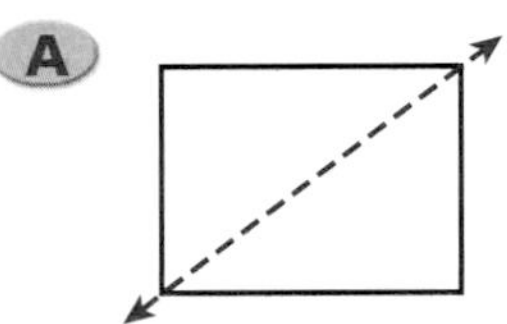

The two parts of the figure are congruent, but they do not match exactly when folded or reflected across the line.

The line does not appear to be a line of symmetry.

B

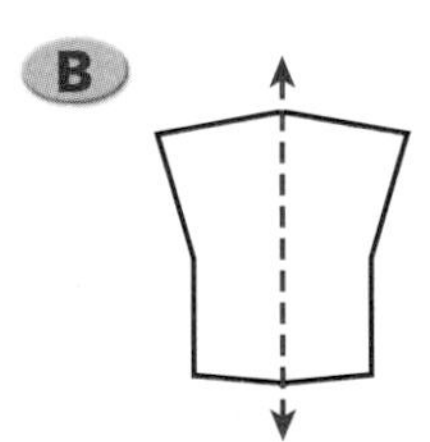

The two parts of the figure appear to match exactly when folded or reflected across the line.

The line appears to be a line of symmetry.

Some figures have more than one line of symmetry.

EXAMPLE 2 Finding Multiple Lines of Symmetry

Find all of the lines of symmetry in each regular polygon.

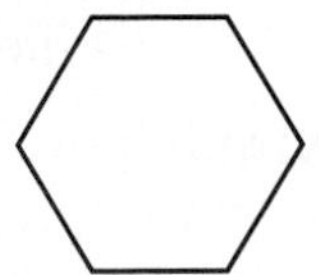

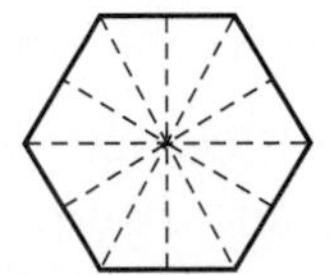

Trace each figure and cut it out. Fold the figure in half in different ways. Count the lines of symmetry.

6 lines of symmetry

Find all of the lines of symmetry in each regular polygon.

B

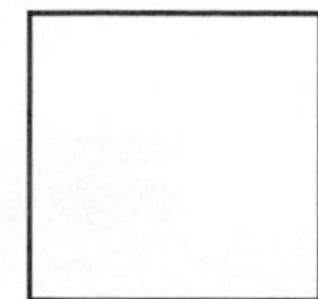

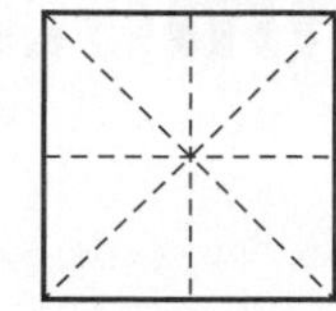

4 lines of symmetry

Count the lines of symmetry.

C

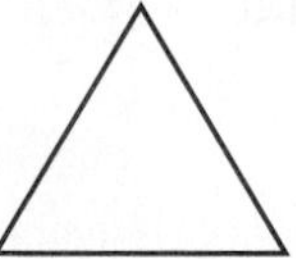

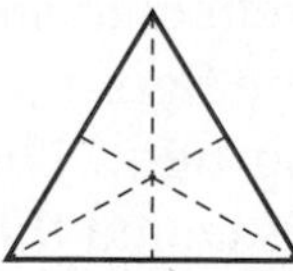

3 lines of symmetry

Count the lines of symmetry.

EXAMPLE 3 *Social Studies Application*

Find all of the lines of symmetry in each flag design.

A Antigua and Barbuda

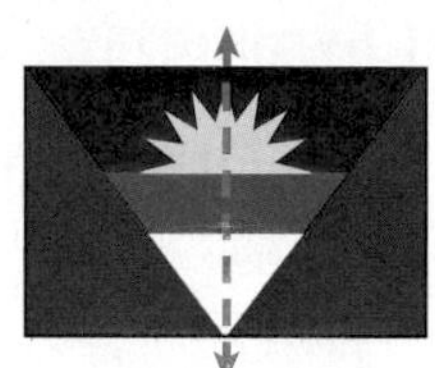

1 line of symmetry

B Macedonia

2 lines of symmetry

C Norway

1 line of symmetry

D Lesotho

There are no lines of symmetry.

Think and Discuss

1. **Explain** how you can use your knowledge of reflection to create a figure that has a line of symmetry.
2. **Determine** whether all hexagons have six lines of symmetry.
3. **Name** objects with line symmetry in your classroom. Tell how many lines of symmetry each of these objects has.

7-11 Exercises

FOR EOG PRACTICE
see page 667

internet connect
Homework Help Online
go.hrw.com Keyword: MR4 7-11

GUIDED PRACTICE

See Example  Determine whether each dashed line appears to be a line of symmetry.

1. 

2. 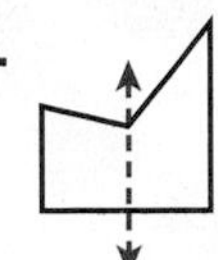

3. 

See Example 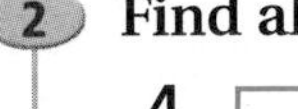 Find all of the lines of symmetry in each regular polygon.

4. 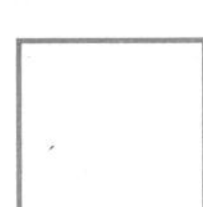

5. 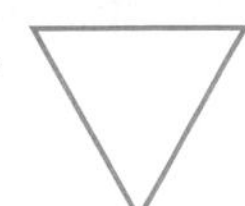

6. 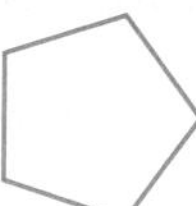

See Example  Find all of the lines of symmetry in each design.

7. 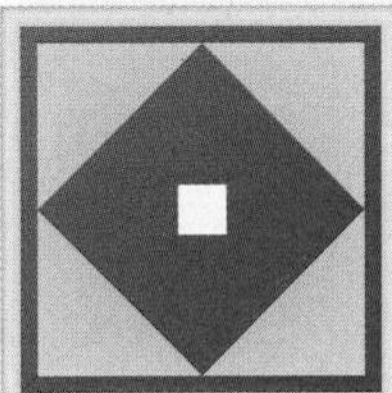

8. 

INDEPENDENT PRACTICE

See Example 1 Determine whether each dashed line appears to be a line of symmetry.

9. 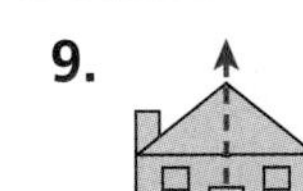

10. 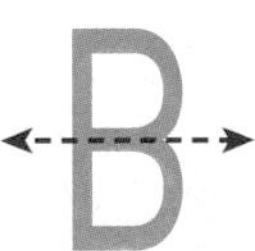

11. 

See Example  Find all of the lines of symmetry in each regular polygon.

12. 

13. 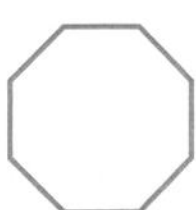

14. 

See Example  Find all of the lines of symmetry in each object.

15. 

16. 

Music LINK

Music is an art form enjoyed by many cultures. Some cultures play music on unique instruments. You might hear the sun drum or turtle drum in Native American music. In music made by people from the Appalachian Mountains, you might hear the strains of a dulcimer. The photo shows young musicians playing sitars, instruments heard in north Indian classical music.

17. Determine whether the dashed line in each drawing is a line of symmetry.

a. 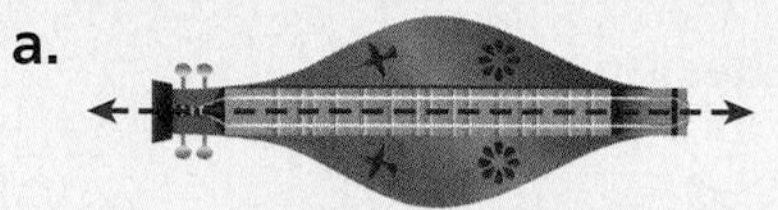

b. 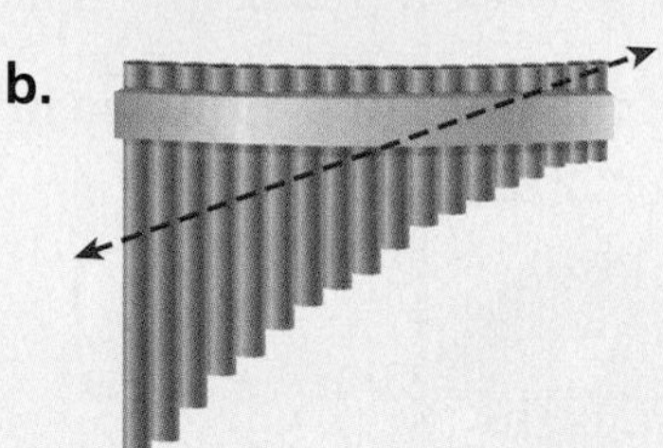

18. The triangle is a percussion instrument formed by a rod of steel or chrome that is bent into the shape of an equilateral triangle. It is open at one corner to allow the notes to resonate. How many lines of symmetry can you find in an equilateral triangle?

19. ***WRITE ABOUT IT*** The turtle drum is a regular octagon. How can you find all of the lines of symmetry in a regular polygon?

20. ***CHALLENGE*** A student drew a drum in the shape of an octagon on a grid. What are the coordinates of the vertices of the unfolded half of the drum drawing if the fold shown is a line of symmetry?

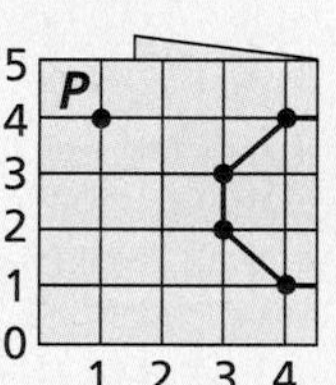

Spiral Review

Write each number in word form. (Lessons 1-1 and 3-1)

21. 101.25 **22.** 3,004,506 **23.** 12,030,921,000 **24.** 47.0305

25. **EOG PREP** Which fraction is equivalent to $10\frac{3}{4}$? (Lesson 4-7)

A $\frac{166}{8}$ B $\frac{83}{8}$ C $\frac{86}{8}$ D $\frac{33}{4}$

26. **EOG PREP** Look at the coordinate grid in Exercise 20. Which ordered pair best represents point P? (Lesson 6-6)

A (4, 1) B (1, 4) C (0, 3) D (3, 0)

7-12 Tessellations

Learn to identify tessellations and shapes that can tessellate.

Vocabulary
tessellation

A **tessellation** is a repeating arrangement of one or more shapes that completely covers a plane with no gaps and no overlaps.

In the design shown, the shape of a fish is used to make a tessellation.

The shape of the fish was copied over and over until the design completely covered an area with no gaps or overlaps.

Although most tessellations are made by humans, a few occur in nature. Honeycombs are naturally occuring tessellations of regular hexagons.

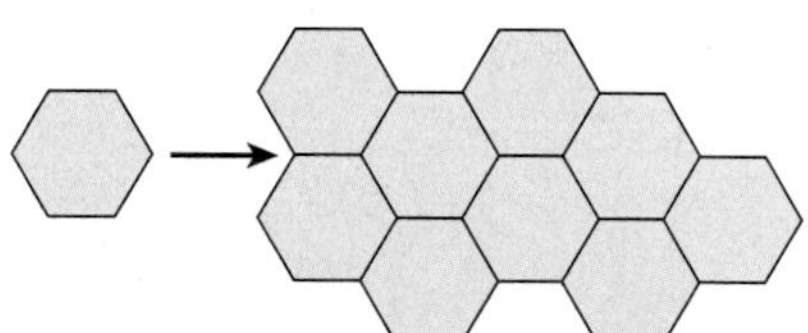

EXAMPLE 1 Identifying Polygons That Tessellate the Plane

Identify whether each polygon can tessellate the plane. Make a drawing to show your answer.

A

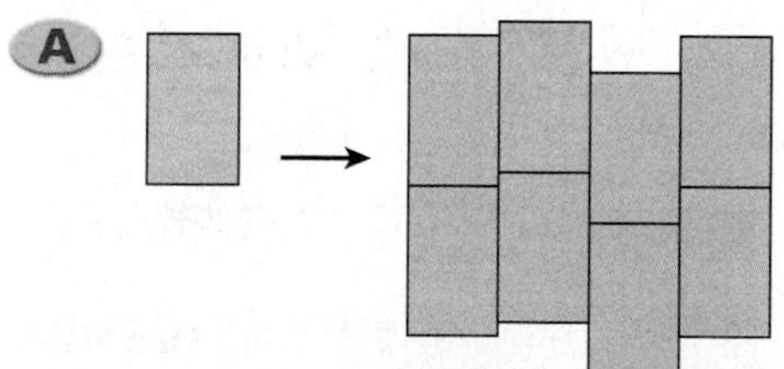

The rectangles cover the plane without any gaps or overlaps.

The rectangle can tessellate the plane.

B

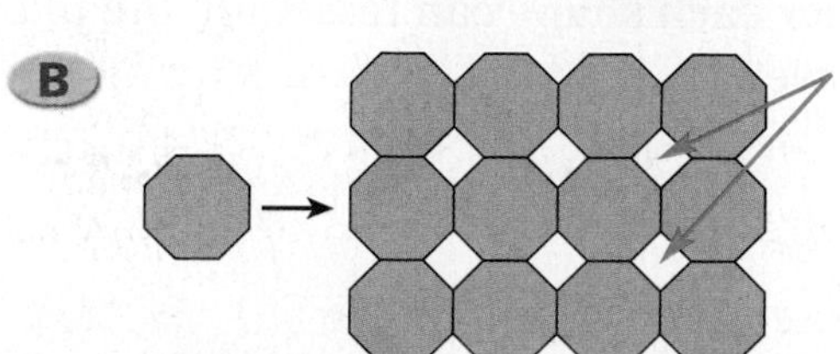

There are gaps between the octagons.

The octagon cannot tessellate the plane.

EXAMPLE 2 Identifying Nonpolygons That Tessellate the Plane

Identify whether each shape can tessellate the plane. Make a drawing to show your answer.

A

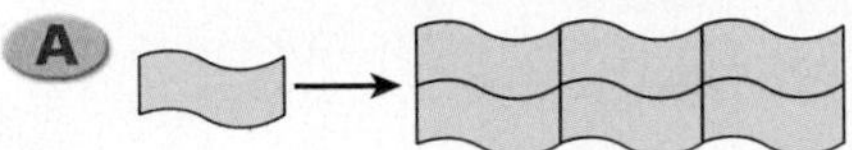

The shapes cover the plane without any gaps or overlaps.

This shape can tessellate the plane.

B

There are gaps between the shapes.

This shape cannot tessellate the plane.

Think and Discuss

1. **Explain** how you know when a pattern of shapes forms a tessellation.
2. **Give an example** of a shape that cannot form a tessellation.

7-12 Exercises

3.04

FOR EOG PRACTICE

see page 667

internet connect

Homework Help Online

go.hrw.com Keyword: MR4 7-12

GUIDED PRACTICE

See Example 1

Identify whether each polygon can tessellate the plane. Make a drawing to show your answer.

1. 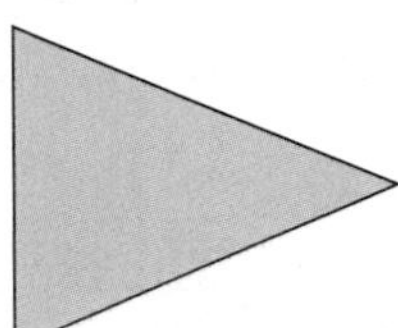

2. 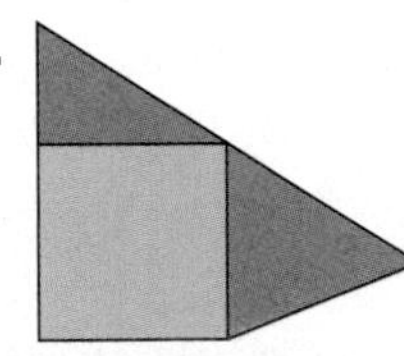

3. 

See Example 2

Identify whether each shape can tessellate the plane. Make a drawing to show your answer.

4. 

5. 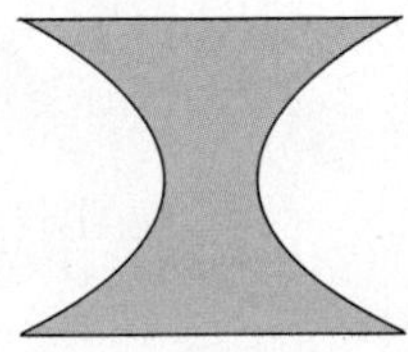

6. 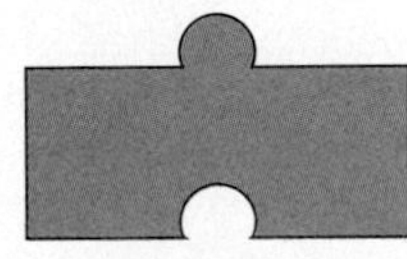

INDEPENDENT PRACTICE

See Example 1

Identify whether each polygon can tessellate the plane. Make a drawing to show your answer.

7. 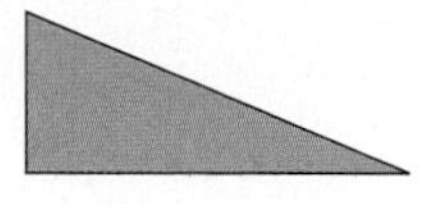

8. 

9. 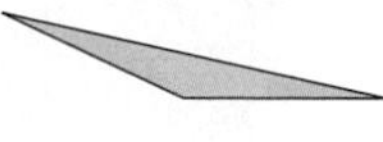

See Example 2

Identify whether each shape can tessellate the plane. Make a drawing to show your answer.

10. 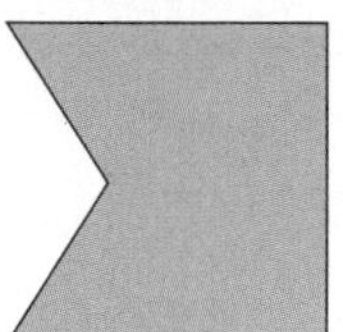

11. 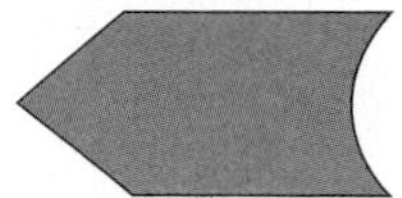

12. 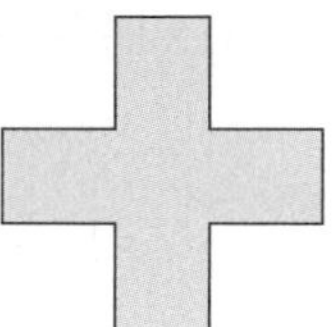

PRACTICE AND PROBLEM SOLVING

13. ***CRAFTS*** This quilt design is a tessellation. Triangular pieces of fabric have been sewn together. Which piece of fabric is missing from the design?

A 

B 

C 

14. ***CRAFTS*** Determine whether the design on the Japanese wallpaper at right is a tessellation. If not, explain.

15. ***WRITE ABOUT IT*** What does it mean to tessellate a plane?

16. ***CHALLENGE*** Draw three figures that can be used together to make a tessellation.

Spiral Review

Write two word phrases for each expression. (Lesson 2-2)

17. $b + 13$ **18.** $(2)(12)$ **19.** $26 - c$ **20.** $m \div 3$

21. **EOG PREP** Which is the reciprocal of $2\frac{2}{3}$? (Lesson 5-3)

A $\frac{3}{2}$ B $2\frac{3}{2}$ C $\frac{8}{3}$ D $\frac{3}{8}$

Create Tessellations

Use with Lesson 7-12

A repeating arrangement of one or more shapes that completely covers a plane, with no gaps or overlaps, is called a *tessellation*. You can make your own tessellations using paper, scissors, and tape.

Activity

1 Start with a square.

Use scissors to cut out a shape from one side of the square.

Translate the shape you cut out to the opposite side of the square and tape the two pieces together.

Trace this new shape to form at least two rows of a tessellation. You will need to translate, rotate, or reflect the shape.

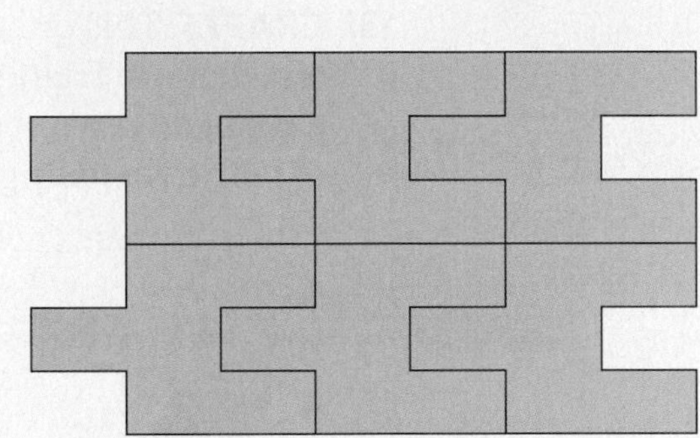

2 Start again with a square.

Use scissors to cut out shapes and move them to the opposite sides of the square.

Trace this new shape to form at least two rows of a tessellation. You will need to translate, rotate, or reflect the shape.

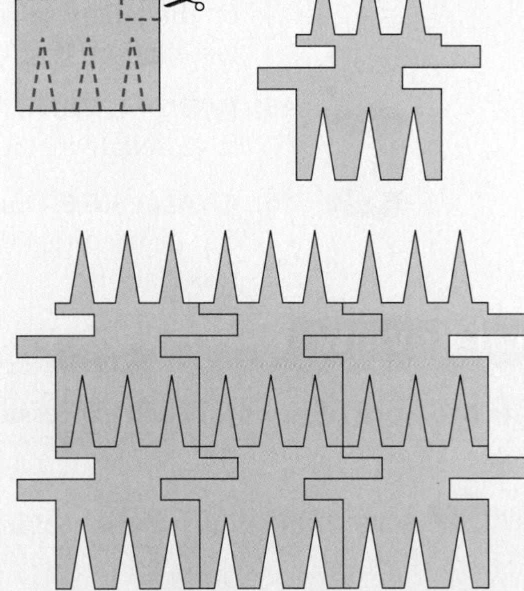

3 You can base a tessellating shape on other polygons.

Try starting with a hexagon.

Use scissors to cut out a shape from one side of the hexagon. Translate the shape to the opposite side of the hexagon.

Try repeating these steps on other sides of the hexagon.

Trace the new shape to form a tessellation. You will need to translate, rotate, or reflect the shape.

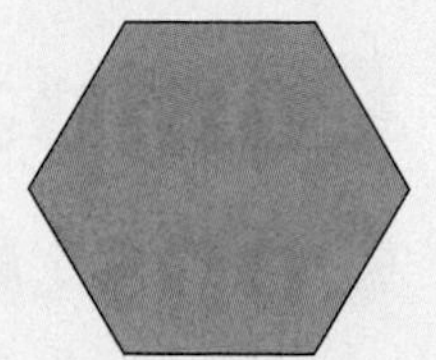

Think and Discuss

1. Tell whether you can make a tessellation out of circles.
2. Tell whether any polygon can make a tessellation.

Try This

Make each tessellation shape described. Then form two rows of a tessellation.

1. 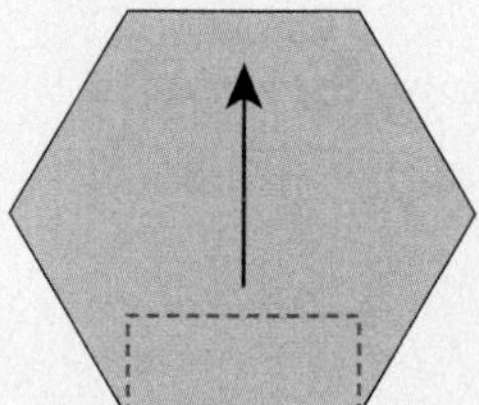

2. 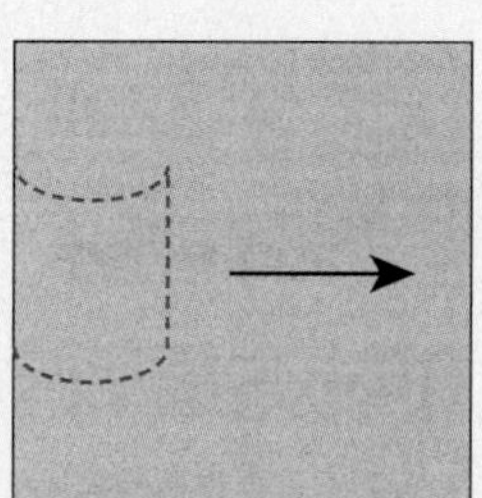

3. 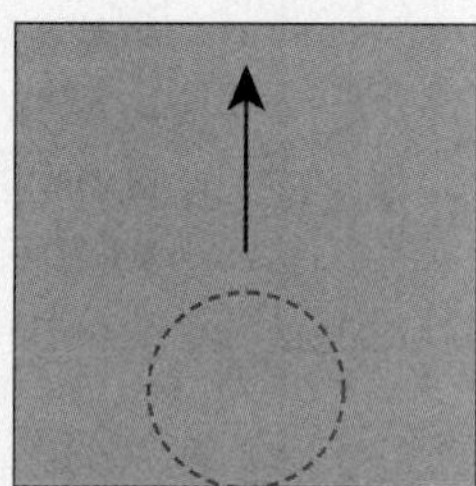

Tell whether each shape can be used to form a tessellation.

4. 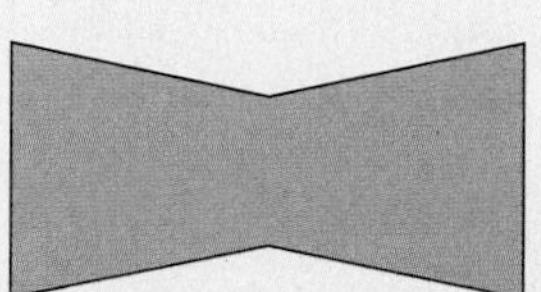

5. 

6. 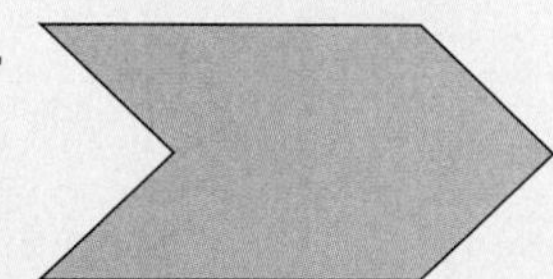

7. Cut out a polygon, and then change it by cutting out a part of one side. Translate the cut-out part to the opposite side. Can your shape form a tessellation? Make a drawing to show your answer.

EXTENSION Compass and Straightedge Constructions

Learn to construct perpendicular bisectors and angle bisectors.

Vocabulary

bisect

To **bisect** something means to divide it into two congruent parts. The perpendicular bisector of a line segment divides it into two line segments of equal length and is perpendicular to the original segment. The midpoint of a segment is the point halfway between the endpoints of the segment. The midpoint bisects the line segment.

You can use a compass and straightedge to bisect a line segment.

EXAMPLE 

Constructing a Perpendicular Bisector

Given $\overline{AC}$, construct its perpendicular bisector.

Step 1

Open your compass to a distance greater than half of $\overline{AC}$ and do not adjust it throughout the construction.

Step 2

Place your compass point on A and draw an arc as shown.

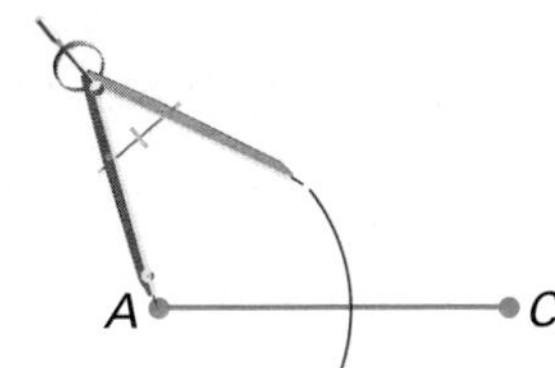

Step 3

Place your compass point on C and draw a second arc as shown. Label the intersections of the two arcs B and D.

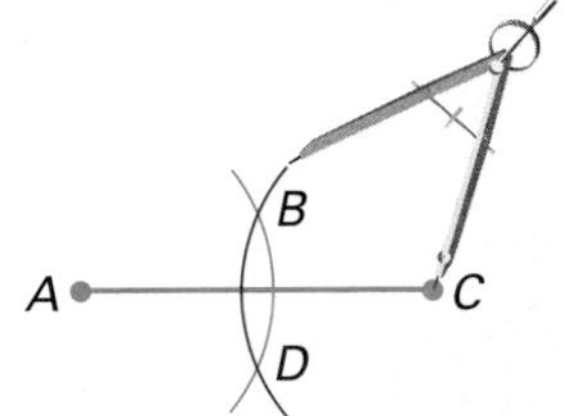

Step 4

Use a straightedge to draw $\overleftrightarrow{BD}$, the perpendicular bisector of $\overline{AC}$.

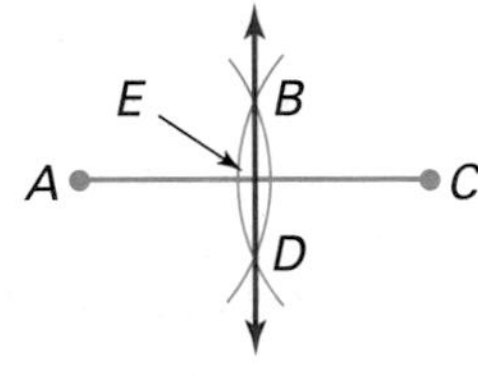

Step 5 ***Check***

Use a ruler to measure the length of $\overline{AE}$ and $\overline{EC}$. $\overline{AE} \cong \overline{EC}$.

Use a protractor to measure $\angle AEB$ and $\angle BEC$.

$m\angle AEB = 90°$ and $m\angle BEC = 90°$.

$\overline{AC} \perp \overleftrightarrow{BD}$

The bisector of an angle divides the angle into two angles of equal measure.

EXAMPLE 2 Constructing an Angle Bisector

Given ∠A, construct its bisector.

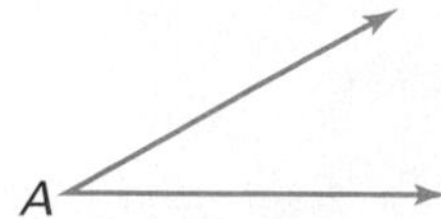

Step 1

Place your compass point on A and draw an arc through the rays of the angle as shown. Label the intersection points B and C.

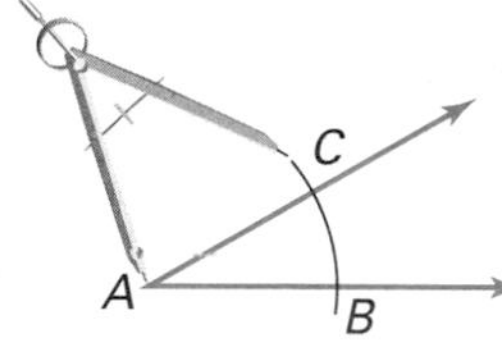

Step 2

Place your compass point on B and draw an arc inside of ∠A. Then, without adjusting your compass, place the compass point on C and draw an arc that intersects the first arc. Label the intersection of the arcs F.

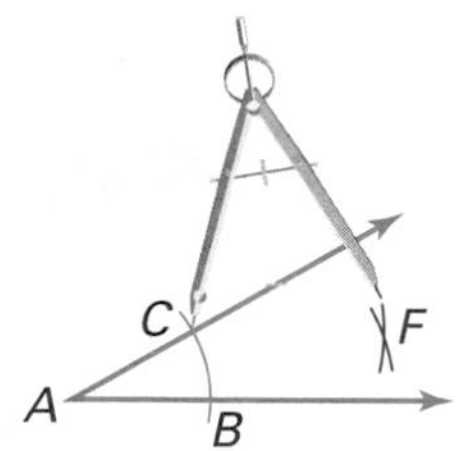

Step 3

Draw a ray from A through F. $\overrightarrow{AF}$ is the bisector of ∠BAC.

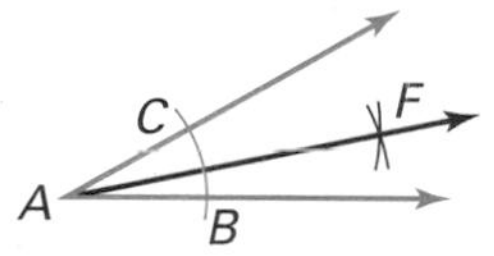

EXTENSION Exercises

1. Draw line segment DH 4 centimeters long. Then construct $\overleftrightarrow{XY}$, a perpendicular bisector of $\overline{DH}$.
2. Draw $\angle L$, and then construct a bisector $\overrightarrow{LG}$. $m\angle L = 120°$.
3. Draw $\angle S$, and then construct a bisector $\overrightarrow{SF}$. $m\angle S = 60°$.

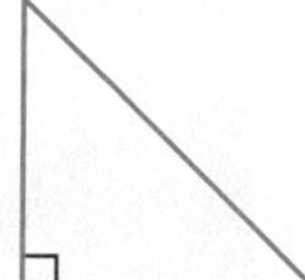

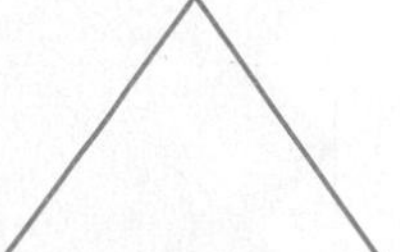

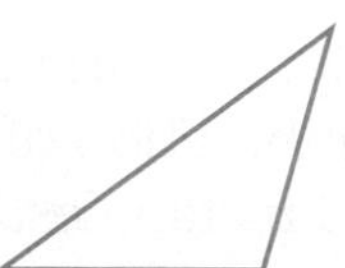

4. Trace each of the triangles above. For each triangle, construct the perpendicular bisector of each side. What do you notice?
5. Trace each of the triangles above. For each triangle, construct the bisector of each angle. What do you notice?

Problem Solving on Location

NORTH CAROLINA

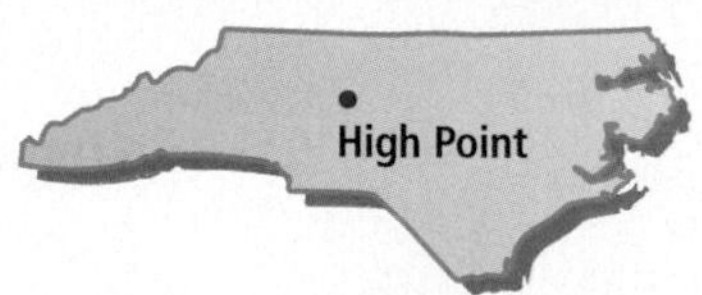

Quilts

Quilters in North Carolina have used the properties of plane geometry to produce warm blankets for the winter as well as beautiful works of art. Most quilt designs are made of blocks that, when used in repeated patterns, result in the skillful combination of basic geometric shapes.

1. Name and classify three polygons used to create the churn dash design.
2. How many lines of symmetry does this quilt block have?

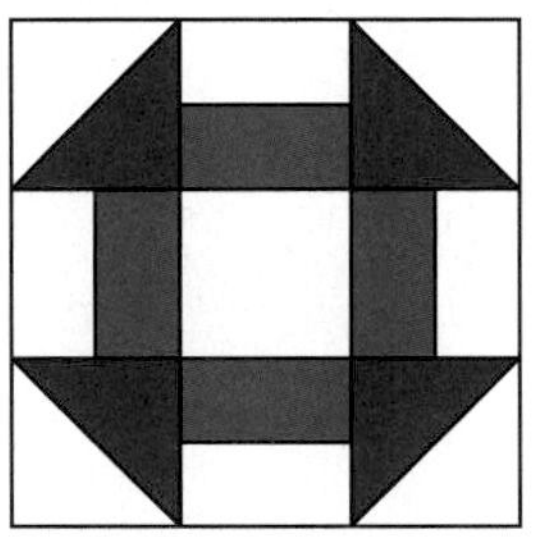

3. Name the two quadrilaterals that appear in the design of the spinning wheel. Classify the triangle according to the length of its sides.
4. How many lines of symmetry does this quilt block have?

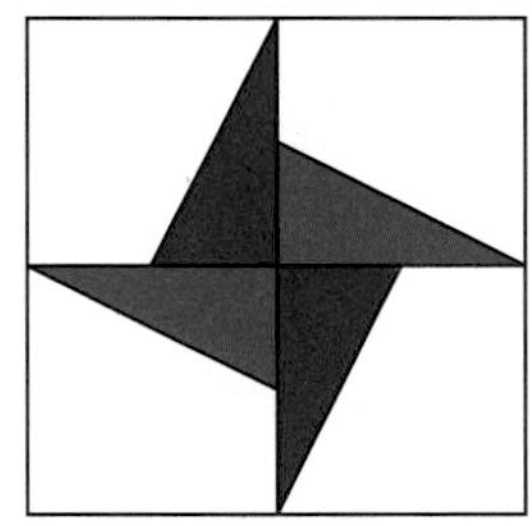

5. Look at the design of the weather vane. Name two polygons formed by the combination of triangles and rectangles.
6. When can a trapezoid have a line of symmetry?

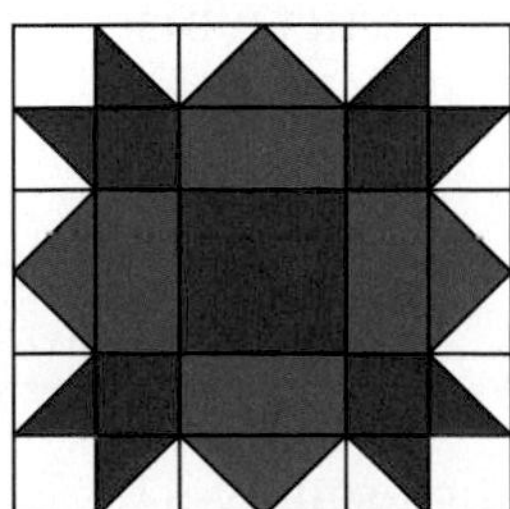

Showplace Convention Center

The Showplace Convention Center in High Point, North Carolina, has about 500,000 square feet of space and a central atrium that is 100 feet tall. The strength and beauty of the convention center lie in the use of plane geometry in its design. Architects and engineers used their knowledge of geometry to design a building that is both safe and beautiful.

1. The lines formed by the columns that support the convention center are in what relationship with each other?

2. Name a polygon that is part of the design for the supports of the roof.

3. Classify the angle formed by a column supporting the building and the surface of the floor on the second level of the convention center.

4. Suppose a visitor on the first level walks along a line from the front of the convention center to the back. Her sister, on the second level, walks along a line from one side of the convention center to the other. What is the relationship between the two lines?

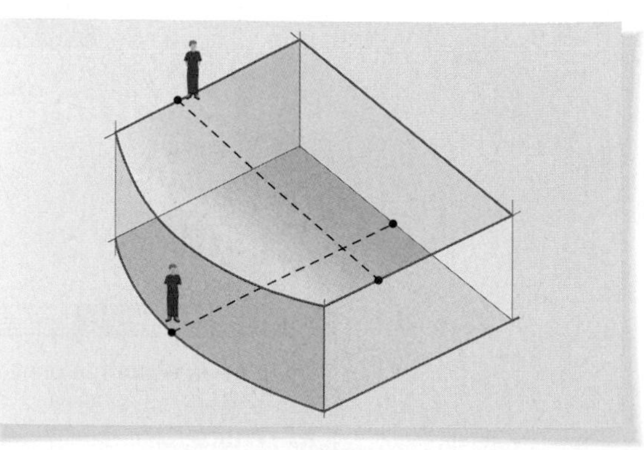

MATH-ABLES

Tangrams

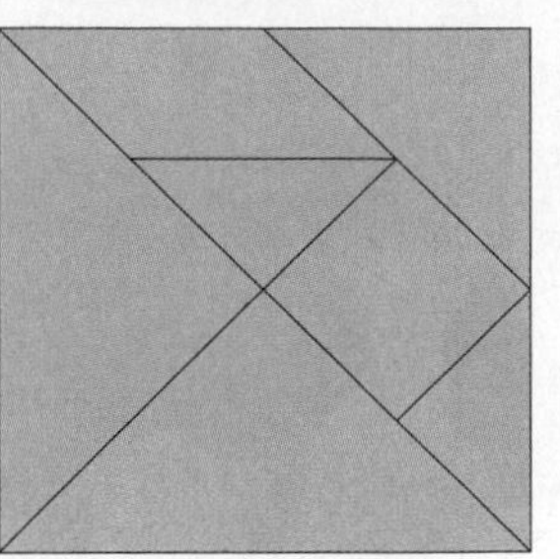

A tangram is an ancient Chinese puzzle. The seven shapes that make this square can be arranged to make many other figures. Copy the shapes that make this square, and then cut them apart. See if you can arrange the pieces to make the figures below.

internet connect

Go to ***go.hrw.com*** to print out Tangram puzzle pieces.
KEYWORD: MR4 Game7

Angles in Triangles

internet connect
Lab Resources Online
go.hrw.com
KEYWORD: MR4 TechLab7

The sum of the angle measures is the same for any triangle. You can use geometry software to find this sum and to check that the sum is the same for many different triangles.

Activity

1. Use the geometry software to make triangle *ABC*. Then use the angle measure tool to measure $\angle B$.

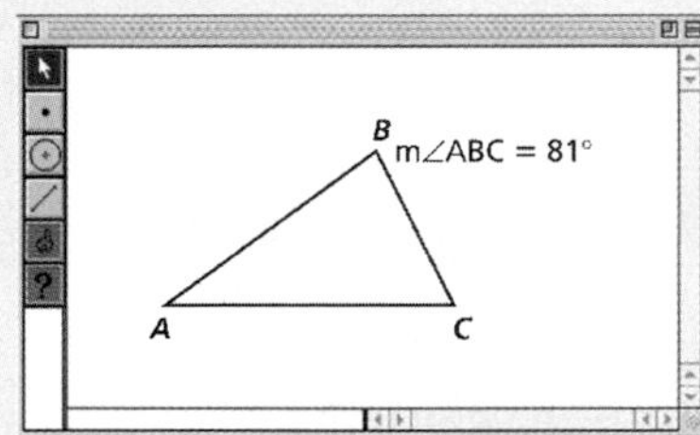

2. Use the angle measure tool to measure $\angle C$ and $\angle A$. Then use the calculator tool to add the measures of the three angles. Notice that the sum is 180°.

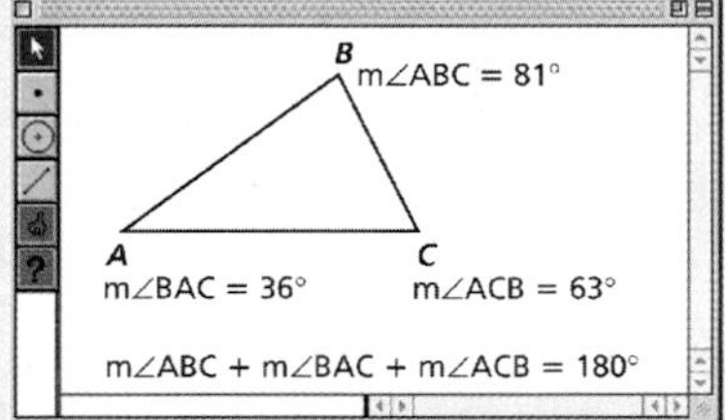

3. Select vertex *A* and drag it around to change the shape of triangle *ABC*. Watch the angle sum. Change the shape of the triangle again and then again. Be sure to make acute and obtuse triangles.

   Notice that the sum of the angle measures is always 180°, regardless of the triangle's shape.

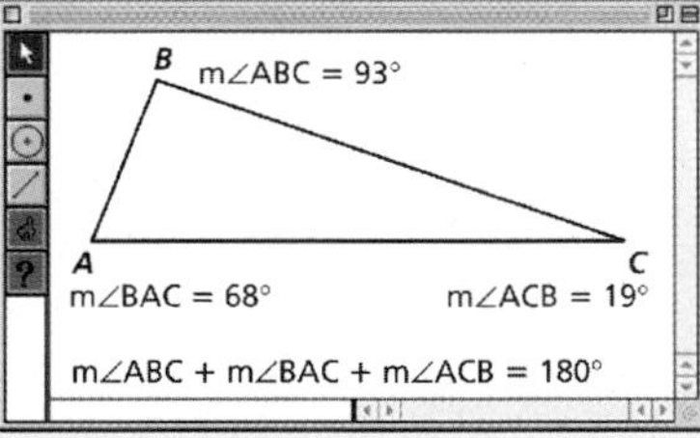

Think and Discuss

1. Can you use geometry software to draw a triangle with two obtuse angles? Explain.

Try This

Solve. Then use geometry software to check each answer.

1. In triangle *ABC*, $m\angle B = 49.15°$ and $m\angle A = 113.75°$. Find $m\angle C$.
2. Use geometry software to construct an acute triangle *XYZ*. Give the measures of its angles, and check that their sum is 180°.

Chapter 7

Study Guide and Review

Vocabulary

Choose the best term from the list above. Words may be used more than once.

1. A quadrilateral with exactly two parallel sides is called a(n) __?__.

2. A(n) __?__ is a closed plane figure formed by three or more line segments.

7-1 Points, Lines, and Planes (pp. 322–325)

EXAMPLE

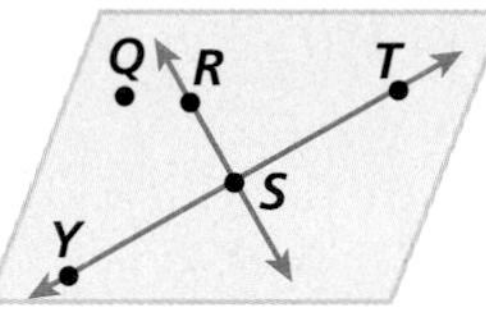

- **Use the diagram.**

  Name a line. $\overleftrightarrow{RS}$
  Name a line segment. $\overline{ST}$

EXERCISE

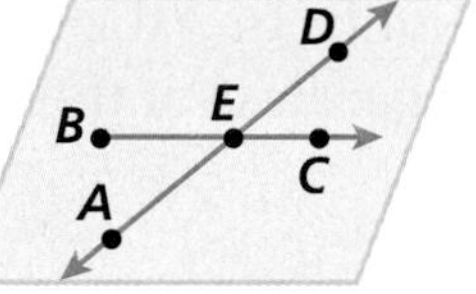

Use the diagram.

3. Name two lines.

7-2 Angles (pp. 326–329)

EXAMPLE

- **Classify each angle as acute, right, obtuse, or straight.**

  $m\angle A = 80°$
  $80° < 90°$, so $\angle A$ is acute.

EXERCISES

Classify each angle as acute, right, obtuse, or straight.

4. $m\angle x = 60°$

5. $m\angle x = 100°$

6. $m\angle x = 45°$

Study Guide and Review

7-3 Angle Relationships (pp. 332–335)

EXAMPLE

■ **Find the unknown angle measure.**

m∠a = 40° *Vertical angles are congruent.*

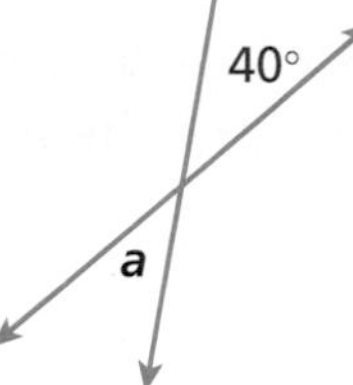

EXERCISES

Find each unknown angle measure.

7.

8.

7-4 Classifying Lines (pp. 336–339)

EXAMPLE

■ **Classify the lines.**

The red lines are parallel.
The blue lines are perpendicular.

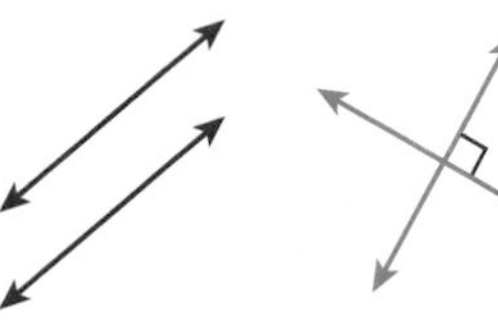

EXERCISES

Classify each pair of lines.

9.

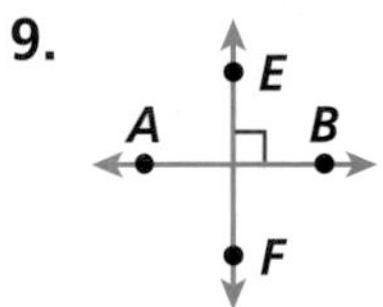

10.

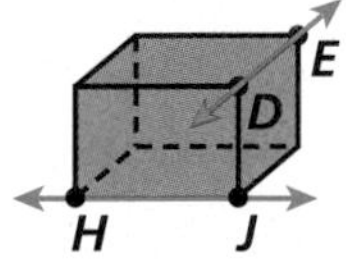

7-5 Triangles (pp. 344–347)

EXAMPLE

■ **Classify the triangle using the given information.**

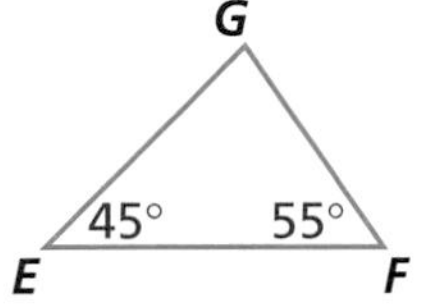

m∠G + 45° + 55° = 180°
m∠G = 80°, so △EFG is an acute triangle.

EXERCISES

Classify the triangle using the given information.

11.

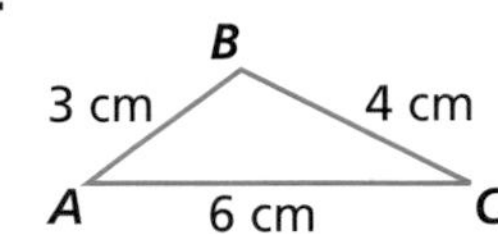

7-6 Quadrilaterals (pp. 348–351)

EXAMPLE

■ **Give the most exact name for the figure.**

The most exact name is rectangle.

EXERCISES

Give the most exact name for the figure.

12.

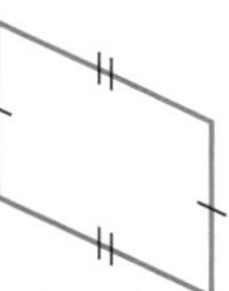

7-7 Polygons (pp. 352–355)

EXAMPLE

■ **Name the polygon and tell whether it appears to be regular or not regular.**

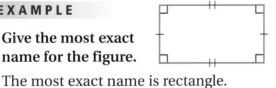

It is a regular octagon.

EXERCISES

Name each polygon and tell whether it appears to be regular or not regular.

13.

14. MAIN STREET

7-8 Geometric Patterns (pp. 356–359)

EXAMPLE

■ **Identify a possible pattern. Use the pattern to draw the missing figure.**

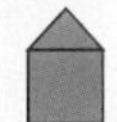 ? 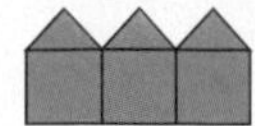

The missing figure might be 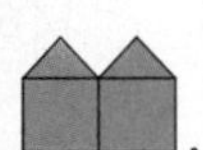.

EXERCISE

Identify a possible pattern. Use the pattern to draw the missing figure.

15.  ? 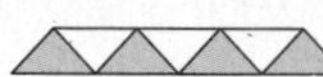

7-9 Congruence (pp. 362–364)

EXAMPLE

■ **Decide whether the figures are congruent. If not, explain.**

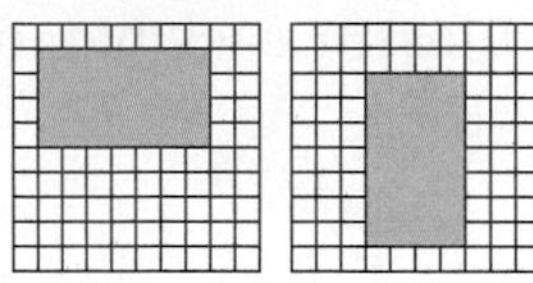

These figures are congruent.

EXERCISES

Decide whether the figures in each pair are congruent. If not, explain.

16. 

17. 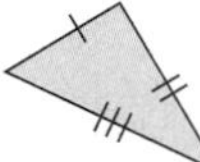

7-10 Transformations (pp. 365–368)

EXAMPLE

■ **Tell whether the transformation is a translation, rotation, or reflection.**

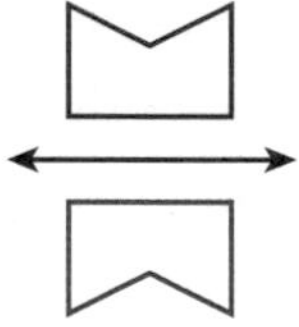

The transformation is a reflection.

EXERCISE

Tell whether the transformation is a translation, rotation, or reflection.

18. 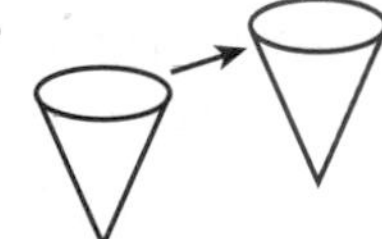

7-11 Symmetry (pp. 369–372)

EXAMPLE

■ **Determine whether the dashed line appears to be a line of symmetry.**

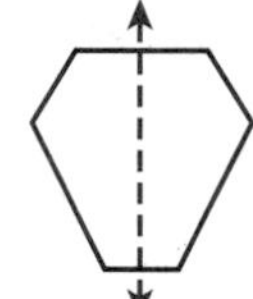

The line appears to be a line of symmetry.

EXERCISE

Determine whether the dashed line appears to be a line of symmetry.

19. 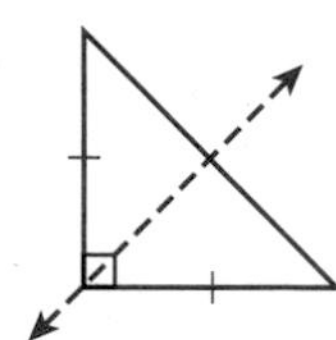

7-12 Tessellations (pp. 373–375)

EXAMPLE

■ **Identify whether the polygon can tessellate the plane.**

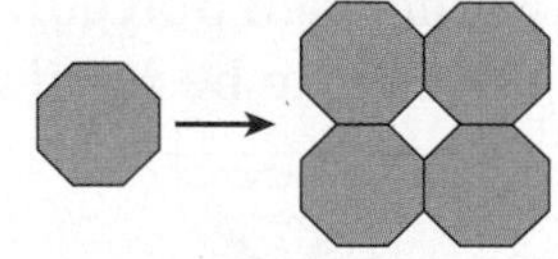

The octagon cannot tessellate the plane.

EXERCISES

Identify whether the polygon can tessellate the plane.

20. 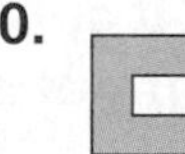

Chapter Test

Classify each pair of angles or lines.

1. 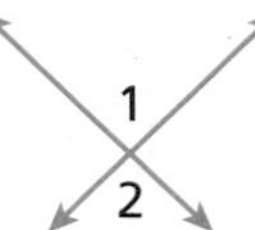

2. 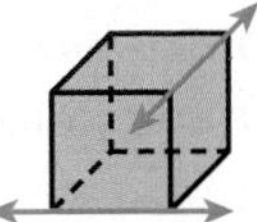

3. O S T P

Classify the triangles by angle and side measures.

4. 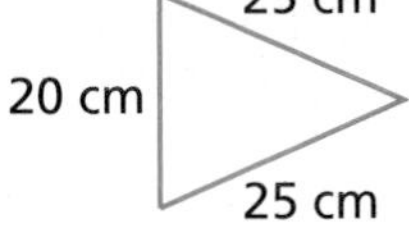

5. 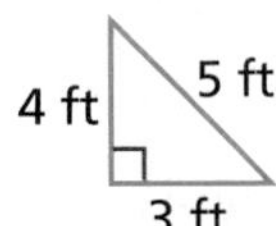

6. 16 km 16 km 16 km

Find the unknown angle measure.

7. 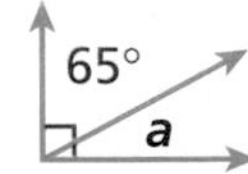

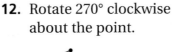

8. 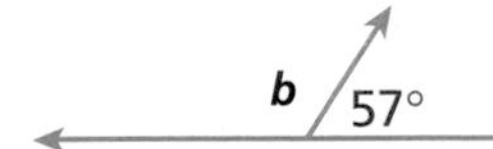

9. 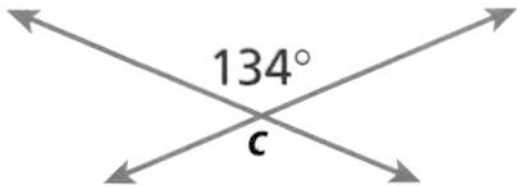

10. Triangle *ABC* has sides of equal length. The measure of $\angle A$ is 60°, and the measure of $\angle B$ is 60°. What is the measure of $\angle C$? Classify the triangle based on the measures of the angles and lengths of the sides.

Draw the indicated transformation.

11. Reflect across the line.

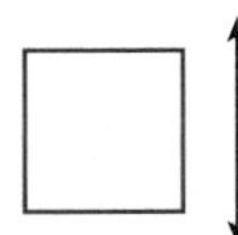

12. Rotate 270° clockwise about the point.

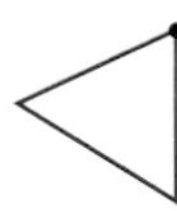

13. Translate $\frac{3}{4}$ in. right.

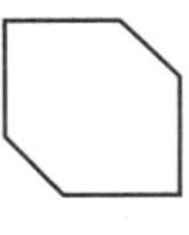

Make a drawing to show whether the figure can tessellate the plane.

14. 

15. 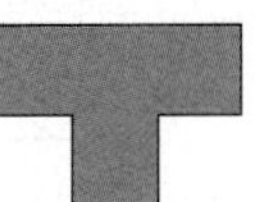

16. 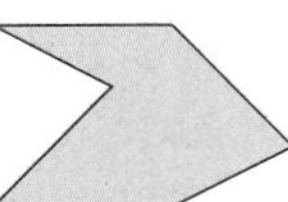

Decide whether the figures in each pair are congruent. If not, explain.

17. 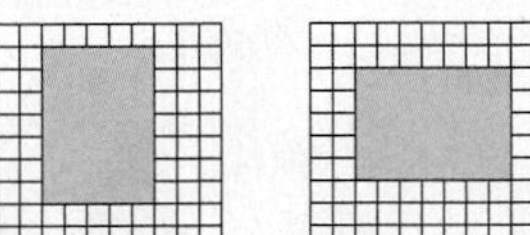

18. 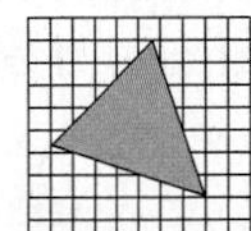

Identify a possible pattern. Use your pattern to draw the next figure.

19. 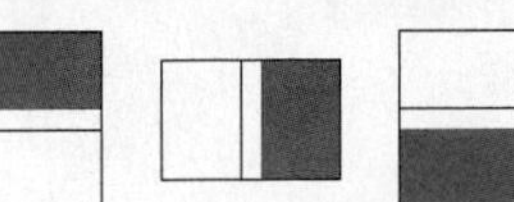

20. 

Chapter 7

Performance Assessment

Show What You Know

Create a portfolio of your work from this chapter. Complete this page and include it with your four best pieces of work from Chapter 7. Choose from your homework or lab assignments, mid-chapter quizzes, or any journal entries you have done. Put them together using any design you want. Make your portfolio represent what you consider your best work.

Short Response

1. Monica drew a quadrilateral. Then she drew a line segment connecting one pair of opposite corners. She saw that she had divided the quadrilateral into two congruent right isosceles triangles. Draw and classify all of the possible quadrilaterals she could have drawn.

2. José drew two intersecting lines. Of the four angles he formed, two adjacent angles are congruent. What can you conclude about the relationship between the two lines and the measures of each angle? Explain.

3. Nancy drew a quadrilateral with a right angle. The other three angles in her figure are congruent. Let x represent the measure of the other three angles. Write and solve an equation to find the value of x. Name the kind of quadrilateral Nancy drew.

Extended Problem Solving

4. Roger Penrose, a famous mathematician, used plane figures to create interesting patterns like the one below through tessellation.

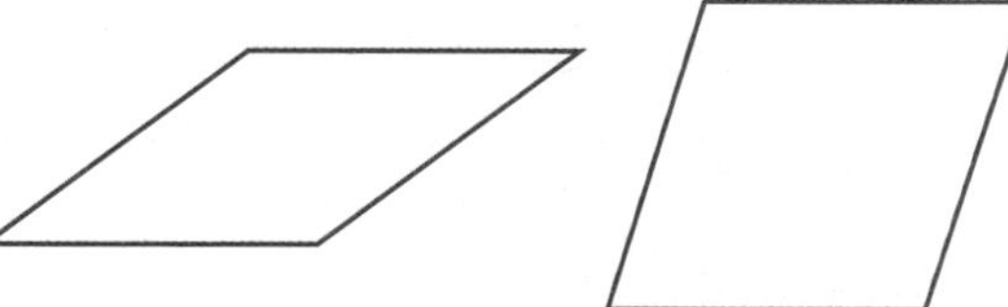

   a. Give the name that best describes each of the two plane figures at right.

   b. The obtuse angle in the first figure has a measure of 144°. Write an equation that could be used to determine the measure m of the acute angle in this figure. Solve for m. Show your work.

   c. Trace the two figures to create your own Penrose tiling.

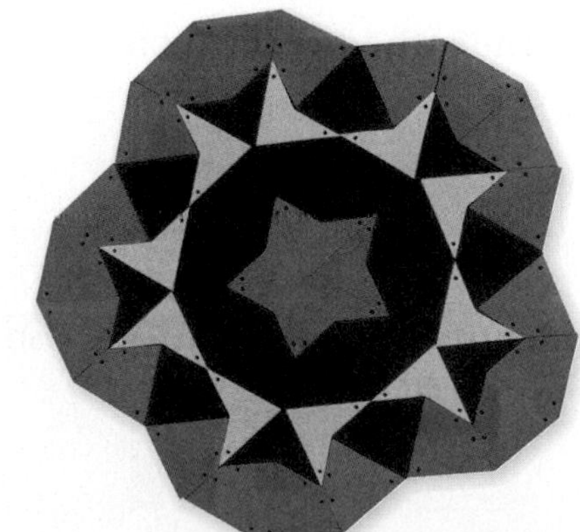

Performance Assessment

internet connect

EOG Test Practice Online
go.hrw.com Keyword: MR4 TestPrep

Getting Ready for EOG

Chapter 7

Cumulative Assessment, Chapters 1–7

1. Which of the following is an acute angle?

A 

B 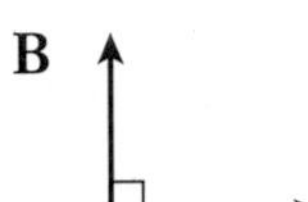

C

D 

2. Two angles whose measures have a sum of 180° are what type of angles?

A Congruent

B Vertical

C Complementary

D Supplementary

3. In which equation does $n = 3$?

A $n + 13 = 17$

B $4 = 12n$

C $4.8 - n = 1.2$

D $\frac{5}{6} \cdot n = 2\frac{1}{2}$

4. Which of the following numbers is divisible by 2, 3, and 6, but **not** 9?

A 312

B 306

C 288

D 256

5. Subtract. $6\frac{3}{4} - 3\frac{5}{8}$

A $3\frac{1}{2}$

B $3\frac{1}{8}$

C $2\frac{3}{4}$

D $2\frac{1}{4}$

6. Which expression has the greatest value?

A $2^3 + 20 \div 4$

B $3.85 \div 0.25$

C $15\frac{1}{3} \div 2$

D $4\frac{1}{4} \cdot 3\frac{2}{3}$

7. What term is used to describe quadrilaterals that have opposite sides that are parallel and congruent and opposite angles that are congruent?

A Rectangle

B Tessellation

C Parallelogram

D Trapezoid

8. What type of transformation is shown in the picture?

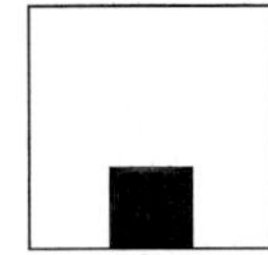 

A 90° counterclockwise rotation

B Translation to the right

C Horizontal reflection

D 180° counterclockwise rotation

TEST TAKING TIP!

The sum of the measures of the angles in a triangle is 180°.

9. ***SHORT RESPONSE*** For triangle ABC, $\angle A$ measures 35° and $\angle B$ measures 30°. Show the steps necessary to find the measure of $\angle C$, and classify the triangle by its angles.

10. ***SHORT RESPONSE*** Find the median and mode of the following data set. Explain how you found your answers.

13, 8, 9, 10, 12, 11, 13, 13, 10, 18

Chapter 8

Ratio, Proportion, and Percent

TABLE OF CONTENTS

Career *Fisheries Biologist*

A fisheries biologist interacts with nature and with people. Fisheries biologists complete surveys, improve habitats, monitor water conditions, and work with land developers.

Fisheries biologists often must determine the number of fish in a lake or pond. They use the tag, release, and recapture method to estimate this number.

$$\frac{\text{tagged number in recapture}}{\text{total number recaptured}} = \frac{\text{number originally tagged}}{\text{total number in lake}}$$

Lake	Tagged Number in Recapture	Total Number Recaptured	Number Originally Tagged
Duck	23	96	108
Los Dos Perros	32	40	56
Robyn	18	26	75

internet connect

Chapter Opener Online
go.hrw.com
KEYWORD: MR4 Ch8

ARE YOU READY?

Choose the best term from the list to complete each sentence.

1. A(n) $\underline{\quad ? \quad}$ is a three-sided polygon, and a(n) $\underline{\quad ? \quad}$ is a four-sided polygon.
2. A(n) $\underline{\quad ? \quad}$ is used to name a part of a whole.
3. When two numbers have the same value, they are said to be $\underline{\quad ? \quad}$.
4. When writing 0.25 as a fraction, 25 is the $\underline{\quad ? \quad}$ and 100 is the $\underline{\quad ? \quad}$.

fraction
numerator
denominator
equivalent
angle
triangle
quadrilateral
pentagon

Complete these exercises to review skills you will need for this chapter.

✔ Simplify Fractions

Write each fraction in simplest form.

5. $\frac{6}{10}$
6. $\frac{9}{12}$
7. $\frac{8}{6}$

✔ Write Equivalent Fractions

Write three equivalent fractions for each given fraction.

8. $\frac{4}{16}$
9. $\frac{5}{10}$
10. $\frac{5}{6}$

✔ Write Fractions as Decimals

Write each fraction as a decimal.

11. $\frac{3}{10}$
12. $\frac{3}{4}$
13. $\frac{5}{8}$
14. $\frac{11}{12}$

✔ Write Decimals as Fractions

Write each decimal as a fraction in simplest form.

15. 0.5
16. 0.35
17. 0.08
18. 0.12

✔ Multiply Decimals

Multiply.

19. $0.42 \cdot 10$
20. $0.3 \cdot 52$
21. $20.5 \cdot 0.25$
22. $6.75 \cdot 0.40$
23. $9.8 \cdot 0.2$
24. $0.8 \cdot 7.4$
25. $0.52 \cdot 0.64$
26. $0.75 \cdot 8.9$

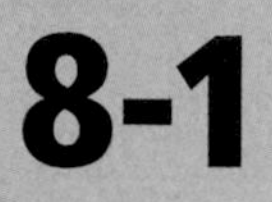

8-1 Ratios and Rates

Learn to write ratios and rates and to find unit rates.

Vocabulary

ratio

equivalent ratios

rate

unit rate

For a time, the Boston Symphony Orchestra was made up of 95 musicians.

Violins	29	Violas	12
Cellos	10	Basses	9
Flutes	5	Trumpets	3
Double reeds	8	Percussion	5
Clarinets	4	Harp	1
Horns	6	Trombones	3

You can compare the different groups by using ratios. A **ratio** is a comparison of two quantities using division.

Reading Math

Read the ratio $\frac{29}{12}$ as "twenty-nine to twelve."

For example, you can use a ratio to compare the number of violins with the number of violas. This ratio can be written in three ways.

Terms → $\frac{29}{12}$ 29 to 12 29:12

Notice that the ratio of **violins** to **violas**, $\frac{29}{12}$, is different from the ratio of **violas** to **violins**, $\frac{12}{29}$. The order of the terms is important.

Ratios can be written to compare a part to a part, a part to the whole, or the whole to a part.

EXAMPLE 1 Writing Ratios

Use the table above to write each ratio.

A flutes to clarinets

$\frac{5}{4}$ *or* 5 to 4 *or* 5:4 *Part to part*

B trumpets to total instruments

$\frac{3}{95}$ *or* 3 to 95 *or* 3:95 *Part to whole*

C total instruments to basses

$\frac{95}{9}$ *or* 95 to 9 *or* 95:9 *Whole to part*

Equivalent ratios are ratios that name the same comparison. You can find an equivalent ratio by multiplying or dividing both terms of a ratio by the same number.

EXAMPLE 2 Writing Equivalent Ratios

Write three equivalent ratios to compare the number of stars with the number of moons in the pattern.

$\frac{\text{number of stars}}{\text{number of moons}} = \frac{4}{6}$ *There are 4 stars and 6 moons.*

$\frac{4}{6} = \frac{4 \div 2}{6 \div 2} = \frac{2}{3}$ *There are 2 stars for every 3 moons.*

$\frac{4}{6} = \frac{4 \cdot 2}{6 \cdot 2} = \frac{8}{12}$ *If you double the pattern, there will be 8 stars and 12 moons.*

So $\frac{4}{6}$, $\frac{2}{3}$, and $\frac{8}{12}$ are equivalent ratios.

A **rate** compares two quantities that have different units of measure.

Suppose a 2-liter bottle of soda costs $1.98.

$\text{rate} = \frac{\text{price}}{\text{number of liters}} = \frac{\$1.98}{2 \text{ liters}}$ $1.98 for 2 liters

When the comparison is to one unit, the rate is called a **unit rate**.

Divide both terms by the second term to find the unit rate.

$\text{unit rate} = \frac{\$1.98}{2} = \frac{\$1.98 \div 2}{2 \div 2} = \frac{\$0.99}{1}$ $0.99 for 1 liter

When the prices of two or more items are compared, the item with the lowest unit rate is the best deal.

EXAMPLE 3 *Consumer Application*

A 2-liter bottle of soda costs $2.02. A 3-liter bottle of the same soda costs $2.79. Which is the better deal?

2-liter bottle		3-liter bottle	
$\frac{\$2.02}{2 \text{ liters}}$	*Write the rate.*	$\frac{\$2.79}{3 \text{ liters}}$	*Write the rate.*
$\frac{\$2.02 \div 2}{2 \text{ liters} \div 2}$	*Divide both terms by 2.*	$\frac{\$2.79 \div 3}{3 \text{ liters} \div 3}$	*Divide both terms by 3.*
$\frac{\$1.01}{1 \text{ liter}}$	*$1.01 for 1 liter*	$\frac{\$0.93}{1 \text{ liter}}$	*$0.93 for 1 liter*

The 3-liter bottle is the better deal.

Think and Discuss

1. **Explain** why the ratio 2 boys:5 girls is different from the ratio 5 girls:2 boys.
2. **Describe** how to determine what number to divide by when finding a unit rate.

8-1 Exercises

5.04

FOR EOG PRACTICE
see page 668

internet connect
Homework Help Online
go.hrw.com Keyword: MR4 8-1

GUIDED PRACTICE

See Example 1 **Use the table to write each ratio.**

1. music programs to art programs
2. arcade games to entire collection
3. entire collection to educational games

Jacqueline's Software Collection	
Educational games	16
Word processing	2
Art programs	10
Arcade games	10
Music programs	3

See Example 2 4. Write three equivalent ratios to compare the number of red hearts in the picture with the total number of hearts.

See Example 3 5. An 8-ounce bag of sunflower seeds costs $1.68. A 4-ounce bag of sunflower seeds costs $0.88. Which is the better deal?

INDEPENDENT PRACTICE

See Example 1 **Use the table to write each ratio.**

6. Redbirds to Blue Socks
7. right-handed Blue Socks to left-handed Blue Socks
8. left-handed Redbirds to total Redbirds

	Redbirds	Blue Socks
Left-Handed Batters	8	3
Right-Handed Batters	11	19

See Example 2 9. Write three equivalent ratios to compare the number of stars in the picture with the number of stripes.

See Example 3 10. Gina charges $28 for 3 hours of swimming lessons. Hector charges $18 for 2 hours of swimming lessons. Which instructor offers a better deal?

PRACTICE AND PROBLEM SOLVING

Write each ratio three different ways.

11. ten to seven
12. $\frac{24}{11}$
13. 4 to 30

Use the diagram of an oxygen atom and a boron atom for Exercises 14–17. Find each ratio. Then give two equivalent ratios.

Boron

Oxygen

Key: Proton, Neutron, Electron

14. oxygen protons to boron protons

15. boron neutrons to boron protons

16. boron electrons to oxygen electrons

17. oxygen electrons to oxygen protons

18. A lifeguard-training program includes 16 hours of instruction in basic first aid and 8 hours of instruction in cardiopulmonary resuscitation (CPR). Write the ratio of hours of CPR instruction to hours of first aid instruction.

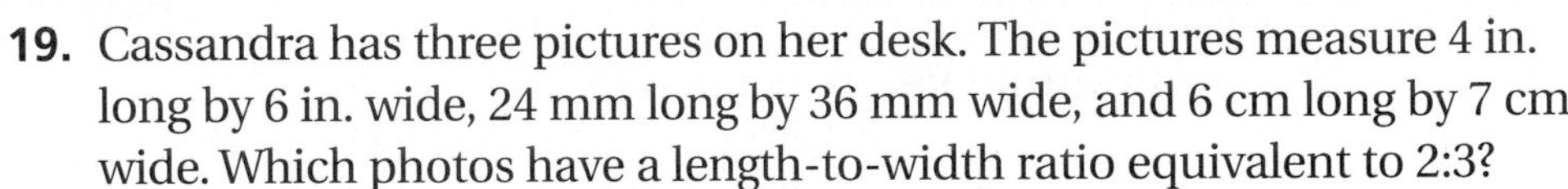

19. Cassandra has three pictures on her desk. The pictures measure 4 in. long by 6 in. wide, 24 mm long by 36 mm wide, and 6 cm long by 7 cm wide. Which photos have a length-to-width ratio equivalent to 2:3?

20. On which day did Alfonso run faster?

Alfonso's Runs		
Day	**Distance (m)**	**Time (min)**
Monday	1,020	6
Wednesday	1,554	9

21. ***EARTH SCIENCE*** Water rushes over Niagara Falls at the rate of 180 million cubic feet every 30 minutes. How much water goes over the falls in 1 minute?

22. ***WRITE ABOUT IT*** How are equivalent ratios like equivalent fractions?

23. ***WHAT'S THE QUESTION?*** The ratio of total students in Mr. Avalon's class to students in the class who have a blue backpack is 3 to 1. The answer is 1:2. What is the question?

24. ***CHALLENGE*** There are 36 performers in a dance recital. The ratio of men to women is 2:7. How many men are in the dance recital?

Spiral Review

Solve each equation. (Lesson 3-10)

25. $5.5 = 5c$ **26.** $d + 4.96 = 9$ **27.** $t - 12.5 = 39.04$

Write each improper fraction as a mixed number. (Lesson 4-7)

28. $\frac{13}{4}$ **29.** $\frac{70}{9}$ **30.** $\frac{41}{3}$ **31.** $\frac{75}{6}$

32. **EOG PREP** How many lines of symmetry are in a rectangle that is not a square? (Lesson 7-11)

A 1 B 2 C 4 D 6

Explore Proportions

Use with Lesson 8-2

internet connect
Lab Resources Online
go.hrw.com
KEYWORD: MR4 Lab8A

You can use counters to model equivalent ratios.

Activity 1

Find three ratios that are equivalent to $\frac{6}{12}$.

1. Show 6 red counters and 12 yellow counters.

2. Separate the red counters into two equal groups. Then separate the yellow counters into two equal groups.

3. Write the ratio of red counters in each group to yellow counters in each group.

$$\frac{3 \text{ red counters}}{6 \text{ yellow counters}} = \frac{3}{6}$$

4. Now separate the red counters into three equal groups. Then separate the yellow counters into three equal groups.

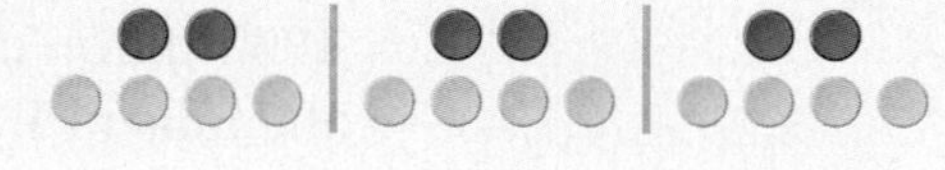

5. Write the ratio of red counters in each group to yellow counters in each group.

$$\frac{2 \text{ red counters}}{4 \text{ yellow counters}} = \frac{2}{4}$$

6. Now separate the red counters into six equal groups. Then separate the yellow counters into six equal groups.

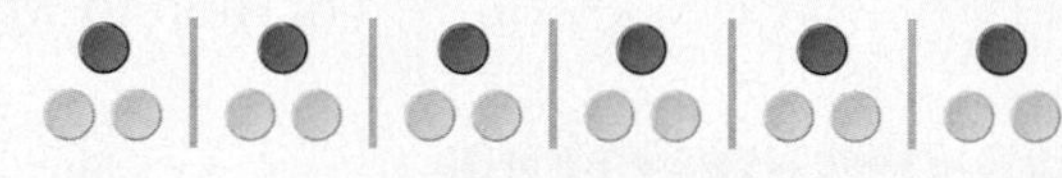

7. Write the ratio of red counters in each group to yellow counters in each group.

$$\frac{1 \text{ red counter}}{2 \text{ yellow counters}} = \frac{1}{2}$$

The three ratios you wrote are equivalent to $\frac{6}{12}$.

$$\frac{6}{12} = \frac{3}{6} = \frac{2}{4} = \frac{1}{2}$$

When you write an equation showing equivalent ratios, that equation is called a **proportion.**

Think and Discuss

1. How do the models show that the ratios are equivalent?

Try This

Use models to determine whether the ratios form a proportion.

1. $\frac{1}{3}$ and $\frac{4}{12}$
2. $\frac{3}{4}$ and $\frac{6}{9}$
3. $\frac{4}{10}$ and $\frac{2}{5}$

Activity 2

Write a proportion in which one of the ratios is $\frac{1}{3}$.

1. You must find a ratio that is equivalent to $\frac{1}{3}$. First show one red counter and three yellow counters.
2. Show one more group of one red counter and three yellow counters.

3. Write the ratio of red counters to yellow counters for the two groups.

$$\frac{\text{2 red counters}}{\text{6 yellow counters}} = \frac{2}{6}$$

4. The two ratios are equivalent. Write the proportion $\frac{1}{3} = \frac{2}{6}$.

You can find more equivalent ratios by adding more groups of one red counter and three yellow counters. Use your models to write proportions.

$$\frac{\text{3 red counters}}{\text{9 yellow counters}} = \frac{3}{9}$$

$$\frac{3}{9} = \frac{1}{3}$$

$$\frac{\text{4 red counters}}{\text{12 yellow counters}} = \frac{4}{12}$$

$$\frac{4}{12} = \frac{1}{3}$$

Think and Discuss

1. The models above show that $\frac{1}{3}$, $\frac{2}{6}$, $\frac{3}{9}$, and $\frac{4}{12}$ are equivalent ratios. Do you see a pattern in this list of ratios?
2. Use counters to find another ratio that is equivalent to $\frac{1}{3}$.

Try This

Use counters to write a proportion containing each given ratio.

1. $\frac{1}{4}$
2. $\frac{1}{5}$
3. $\frac{3}{7}$
4. $\frac{1}{6}$
5. $\frac{4}{9}$

8-2 Proportions

Learn to write and solve proportions.

Vocabulary

proportion

Have you ever heard water called H_2O? H_2O is the scientific formula for water. One molecule of water contains two hydrogen atoms (H_2) and one oxygen atom (O). No matter how many molecules of water you have, hydrogen and oxygen will always be in the ratio 2 to 1.

Water Molecules	1	2	3	4
$\frac{\text{Hydrogen}}{\text{Oxygen}}$	$\frac{2}{1}$	$\frac{4}{2}$	$\frac{6}{3}$	$\frac{8}{4}$

Notice that $\frac{2}{1}$, $\frac{4}{2}$, $\frac{6}{3}$, and $\frac{8}{4}$ are equivalent ratios.

A **proportion** is an equation that shows two equivalent ratios.

$$\frac{2}{1} = \frac{4}{2} \qquad \frac{4}{2} = \frac{8}{4} \qquad \frac{2}{1} = \frac{6}{3}$$

Read the proportion $\frac{2}{1} = \frac{4}{2}$ as "two is to one as four is to two."

EXAMPLE 1 Modeling Proportions

Write a proportion for the model.

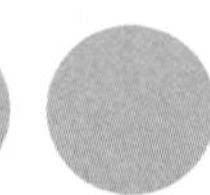

First write the ratio of triangles to circles.

$$\frac{\text{number of triangles}}{\text{number of circles}} = \frac{4}{2}$$

Next separate the triangles and the circles into two equal groups.

Now write the ratio of triangles to circles in each group.

$$\frac{\text{number of triangles in each group}}{\text{number of circles in each group}} = \frac{2}{1}$$

A proportion shown by the model is $\frac{4}{2} = \frac{2}{1}$.

CROSS PRODUCTS

Cross products in proportions are equal.

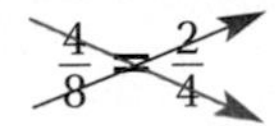 $\frac{4}{8} = \frac{2}{4}$	$\frac{3}{5} = \frac{9}{15}$	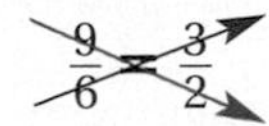 $\frac{9}{6} = \frac{3}{2}$	$\frac{14}{7} = \frac{2}{1}$
$8 \cdot 2 = 4 \cdot 4$	$5 \cdot 9 = 3 \cdot 15$	$6 \cdot 3 = 9 \cdot 2$	$7 \cdot 2 = 14 \cdot 1$
$16 = 16$	$45 = 45$	$18 = 18$	$14 = 14$

EXAMPLE 2 Using Cross Products to Complete Proportions

Find the missing value in the proportion $\frac{3}{4} = \frac{n}{16}$.

$\frac{3}{4} = \frac{n}{16}$ *Find the cross products.*

$4 \cdot n = 3 \cdot 16$ *The cross products are equal.*

$4n = 48$ *n is multiplied by 4.*

$\frac{4n}{4} = \frac{48}{4}$ *Divide both sides by 4 to undo the multiplication.*

$n = 12$

EXAMPLE  3 Measurement Application

Helpful Hint

In a proportion, the units must be in the same order in both ratios.

$\frac{tsp}{lb} = \frac{tsp}{lb}$

or $\frac{lb}{tsp} = \frac{lb}{tsp}$

The label from a bottle of pet vitamins shows recommended dosages. What dosage would you give an adult dog that weighs 15 lb?

Pet Vitamins

- **Adult dogs:** 1 tsp per 20 lb body weight
- **Puppies, pregnant dogs, or nursing dogs:** 1 tsp per 10 lb body weight
- **Cats:** 1 tsp per 12 lb body weight

$\frac{1 \text{ tsp}}{20 \text{ lb}} = \frac{v}{15 \text{ lb}}$ *Let v be the amount of vitamins for a 15 lb dog.*

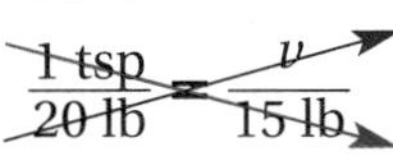

$\frac{1 \text{ tsp}}{20 \text{ lb}} = \frac{v}{15 \text{ lb}}$ *Write a proportion.*

$20 \cdot v = 1 \cdot 15$ *The cross products are equal.*

$20v = 15$ *v is multiplied by 20.*

$\frac{20v}{20} = \frac{15}{20}$ *Divide both sides by 20 to undo the multiplication.*

$v = \frac{3}{4}$ tsp *Write your answer in simplest form.*

You should give $\frac{3}{4}$ tsp of vitamins to a 15 lb dog.

Think and Discuss

1. **Tell** whether $\frac{7}{8} = \frac{4}{14}$ is a proportion. How do you know?
2. **Give an example** of a proportion. Tell how you know that it is a proportion.

8-2 Exercises

5.04

FOR EOG PRACTICE

see page 668

internet connect

Homework Help Online
go.hrw.com Keyword: MR4 8-2

GUIDED PRACTICE

See Example 1

1. Write a proportion for the model.

See Example 2

Find the missing value in each proportion.

2. $\frac{12}{9} = \frac{n}{3}$
3. $\frac{t}{5} = \frac{28}{20}$
4. $\frac{1}{c} = \frac{6}{12}$

See Example 3

5. Ursula is entering a bicycle race for charity. Her mother pledges \$0.75 for every 0.5 mile she bikes. If Ursula bikes 17.5 miles, how much will her mother donate?

INDEPENDENT PRACTICE

See Example 1

6. Write a proportion for the model.

See Example 2

Find the missing value in each proportion.

7. $\frac{3}{2} = \frac{24}{d}$
8. $\frac{p}{40} = \frac{3}{8}$
9. $\frac{6}{14} = \frac{x}{7}$

See Example 3

10. According to Ty's study guidelines, how many minutes of science reading should he do if his science class is 90 minutes long?

Ty's Study Guidelines

Class	Reading Time
Literature	35 minutes for every 50 minutes of class time
Science	20 minutes for every 60 minutes of class time
History	30 minutes for every 55 minutes of class time

PRACTICE AND PROBLEM SOLVING

Find the value of p in each proportion.

11. $\frac{18}{6} = \frac{6}{p}$
12. $\frac{4}{p} = \frac{48}{60}$
13. $\frac{p}{10} = \frac{15}{50}$
14. $\frac{21}{15} = \frac{p}{5}$
15. $\frac{3}{6} = \frac{p}{8}$
16. $\frac{15}{5} = \frac{9}{p}$
17. $\frac{150}{2} = \frac{p}{1}$
18. $\frac{1}{12} = \frac{0.8}{p}$
19. $\frac{8.1}{27} = \frac{p}{5}$

Social Studies LINK

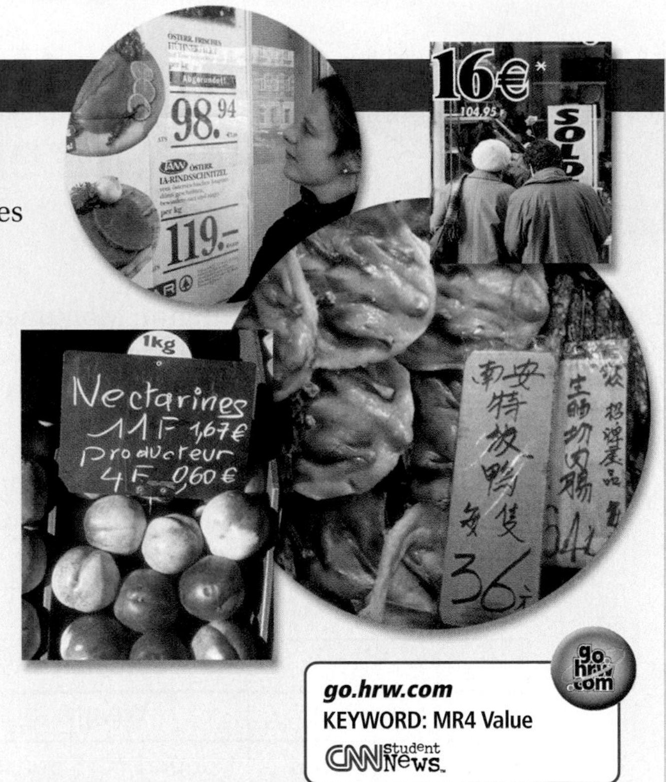

The value of the U.S. dollar as compared to the values of currencies from other countries changes every day. The graph shows the recent value of various currencies compared to the U.S. dollar. Use the graph for Exercises 20–25.

20. What is the value of 9.72 European euros in U.S. dollars?

21. You have $100 in U.S. dollars. Determine how much money this is in euros, Canadian dollars, renminbi, shekels, and Mexican pesos.

22. A watch in Israel costs 82 shekels. In the U.S., the watch costs $25. In which country does the watch cost less?

23. ***WHAT'S THE ERROR?*** A student set up the proportion $\frac{1}{8.28} = \frac{x}{30}$ to determine the value of 30 U.S. dollars in China. What is wrong with this proportion? Write the correct proportion, and find the missing value.

24. ***WRITE ABOUT IT*** Would you prefer to have five U.S. dollars or five Canadian dollars? Why?

25. ***CHALLENGE*** A dime is worth about how many Mexican pesos?

go.hrw.com
KEYWORD: MR4 Value
CNN Student News

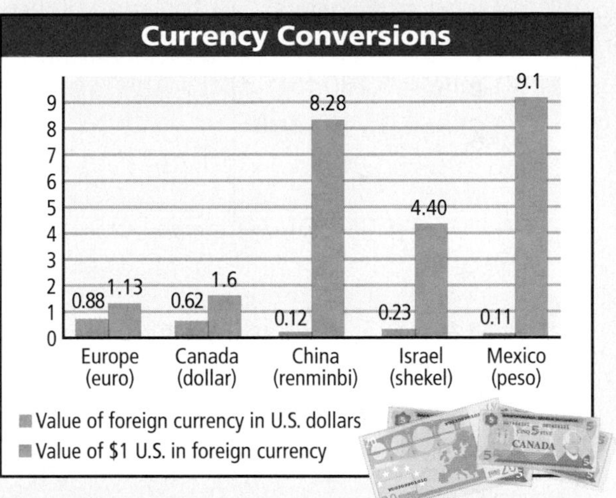

Spiral Review

26. The lengths of Ali's three jumps in the long jump were $9\frac{2}{3}$ feet, $9\frac{1}{2}$ feet, and $9\frac{5}{6}$ feet. Order these lengths from greatest to least. (Lesson 4-7)

Name the measure of central tendency described. (Lesson 6-2)

27. the number that appears the most often in a data set

28. the number in the middle of a data set that is in order from least to greatest

29. **EOG PREP** Which polygon has the greatest number of sides? (Lesson 7-7)

A Trapezoid **B** Quadrilateral **C** Hexagon **D** Triangle

8-3 Proportions and Customary Measurement

Learn to use proportions to make conversions within the customary system.

The Washington Monument is about 185 yards tall. This is almost equal to the length of two football fields. How many feet is this length?

You can use the information in the table below to write a proportion that will help you answer this question.

Common Customary Measurements			
Length	**Weight**	**Time**	**Capacity**
1 foot = 12 inches 1 yard = 36 inches 1 yard = 3 feet 1 mile = 5,280 feet 1 mile = 1,760 yards	1 pound = 16 ounces 1 ton = 2,000 pounds	1 minute = 60 seconds 1 hour = 60 minutes 1 day = 24 hours 1 week = 7 days 1 year = 12 months 1 year = 365 days 1 leap year = 366 days	1 cup = 8 fluid ounces 1 pint = 2 cups 1 quart = 2 pints 1 quart = 4 cups 1 gallon = 4 quarts 1 gallon = 16 cups

EXAMPLE 1 Using Proportions to Convert Measurements

A **Find the height in feet of the Washington Monument.**

$\frac{1 \text{ yd}}{3 \text{ ft}} = \frac{185 \text{ yd}}{x \text{ ft}}$ *1 yard is 3 feet. Write a proportion. Use a variable for the value you are trying to find.*

$3 \cdot 185 = 1 \cdot x$ *The cross products are equal.*

$555 = x$

The Washington Monument is 555 feet tall.

B **In March 1994, a rainbow was visible for 360 minutes over parts of the United Kingdom. How many hours was it visible?**

$\frac{1 \text{ hr}}{60 \text{ min}} = \frac{x \text{ hr}}{360 \text{ min}}$ *1 hour is 60 minutes. Write a proportion. Use a variable for the value you are trying to find.*

$60 \cdot x = 1 \cdot 360$ *The cross products are equal.*

$\frac{60x}{60} = \frac{360}{60}$ *x is multiplied by 60. Divide both sides by 60 to undo the multiplication.*

$x = 6$

The rainbow was visible for 6 hours.

C **The world's largest ice cream sundae weighed about 55,000 pounds. How many tons did it weigh?**

$\frac{1 \text{ ton}}{2{,}000 \text{ lb}} = \frac{x \text{ tons}}{55{,}000 \text{ lb}}$ — *1 ton is 2,000 pounds. Write a proportion. Use a variable for the value you are trying to find.*

$2{,}000 \cdot x = 1 \cdot 55{,}000$ — *The cross products are equal.*

$2{,}000x = 55{,}000$ — *x is multiplied by 2,000.*

$\frac{2{,}000x}{2{,}000} = \frac{55{,}000}{2{,}000}$ — *Divide both sides by 2,000 to undo the multiplication.*

$x = 27.5$

The ice cream sundae weighed about 27.5 tons.

Think and Discuss

1. **Describe** a situation in which you would need to convert measurements.
2. **Explain** how to set up a proportion to convert miles to yards.
3. **Tell** what is wrong with this proportion: $\frac{24 \text{ hr}}{1 \text{ day}} = \frac{x \text{ days}}{168 \text{ hr}}$. Then tell the correct way to write it.

8-3 Exercises

FOR EOG PRACTICE
see page 668

internet connect
Homework Help Online
go.hrw.com Keyword: MR4 8-3

GUIDED PRACTICE

See Example 

1. Linda cut off 1.5 feet of her hair to donate to an organization that makes wigs for children with cancer. How many inches of hair did she cut off?
2. An adult male of average size normally has about 6 quarts of blood in his body. Approximately how many cups of blood does the average adult male have in his body?

INDEPENDENT PRACTICE

See Example 

3. The steel used to make the Statue of Liberty weighs about 125 tons. How many pounds of steel were used to make the Statue of Liberty?
4. Ky will spend 105 days in Nepal as an exchange student. How many weeks will Ky spend in Nepal?
5. Lake Superior is about 1,302 feet deep at its deepest point. Find this depth in yards.

PRACTICE AND PROBLEM SOLVING

Fill in each missing value.

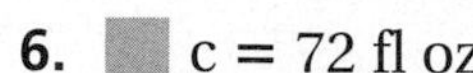

6. ■ c = 72 fl oz **7.** ■ in. = 4 yd **8.** 14,000 lb = ■ tons

9. ■ yd = 93 ft **10.** 20 min = ■ s **11.** 98 days = ■ weeks

Art LINK

Long-Term Parking is 65 feet tall and stands in front of a parking lot in Paris.

Compare. Write <, >, or =.

12. 18 ft ■ 220 in. **13.** 24 lb ■ 388 oz

14. 21 hr ■ $\frac{5}{6}$ day **15.** $\frac{1}{2}$ pt ■ 1 c

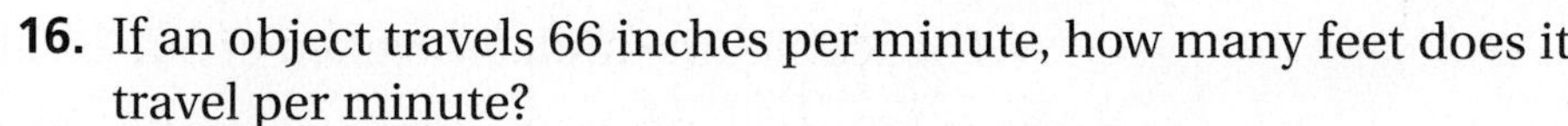

16. If an object travels 66 inches per minute, how many feet does it travel per minute?

17. If you drink 14 quarts of water per week, on average, how many pints do you drink per day?

18. ***ART*** In Paris, the sculpture *Long-Term Parking*, created by Armand Fernandez, contains 60 cars embedded in 3.5 million pounds of concrete. How many tons of concrete is this?

19. ***SPORTS*** The width of a tennis court is 27 feet.

a. How many yards wide is a tennis court?

b. How many inches wide is a tennis court?

c. An inch is equal to about 2.5 centimeters. Estimate the width in centimeters of a tennis court.

20. ***CHOOSE A STRATEGY*** A customer wanted a 25-foot piece of wire. The clerk incorrectly measured the wire with a yardstick that was 2 inches too short. How many inches were missing from the customer's piece of wire?

21. ***WRITE ABOUT IT*** Explain how to compare a weight given in ounces with a weight given in pounds.

22. ***CHALLENGE*** Human hair grows at an average rate of 12 cm per year. How many inches per month does the average human hair grow? (*Hint:* 2.5 cm is about 1 in.)

Spiral Review

Solve each equation. (Lessons 2-4, 2-5)

23. $d + 24 = 40$ **24.** $x - 15 = 5$ **25.** $9 + c = 44$

26. A new box of dominoes is opened, and 7 dominoes are added to it. The box now has 33 dominoes in it. Write an equation that can be used to find the number of dominoes in a new box. (Lesson 2-4)

27. **EOG PREP** Which length is the longest? (Lesson 3-4)

A 30 cm B 3,000 mm C 0.03 m D 3 cm

8-4 Similar Figures

Learn to use ratios to identify similar figures.

Vocabulary

similar

corresponding sides

corresponding angles

Two or more figures are **similar** if they have exactly the same shape. Similar figures may be different sizes.

Similar figures have corresponding sides and corresponding angles.

- **Corresponding sides** have lengths that are proportional.
- **Corresponding angles** are congruent.

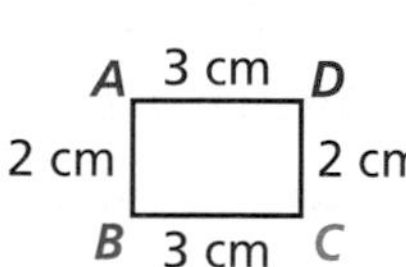

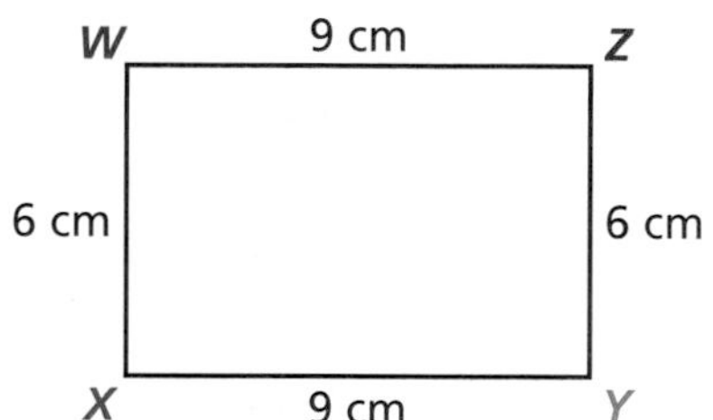

Corresponding sides:	**Corresponding angles:**
$\overline{AB}$ corresponds to $\overline{WX}$.	$\angle A$ corresponds to $\angle W$.
$\overline{BC}$ corresponds to $\overline{XY}$.	$\angle B$ corresponds to $\angle X$.
$\overline{CD}$ corresponds to $\overline{YZ}$.	$\angle C$ corresponds to $\angle Y$.
$\overline{AD}$ corresponds to $\overline{WZ}$.	$\angle D$ corresponds to $\angle Z$.

In the rectangles above, one proportion is $\frac{AB}{WX} = \frac{AD}{WZ}$, or $\frac{2}{6} = \frac{3}{9}$.

If you cannot use corresponding side lengths to write a proportion, or if corresponding angles are not congruent, then the figures are not similar.

EXAMPLE 1 Finding Missing Measures in Similar Figures

The two triangles are similar. Find the missing length x and the measure of $\angle A$.

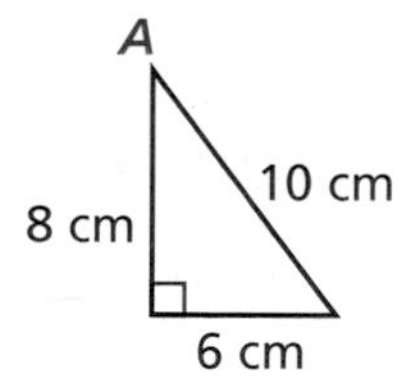

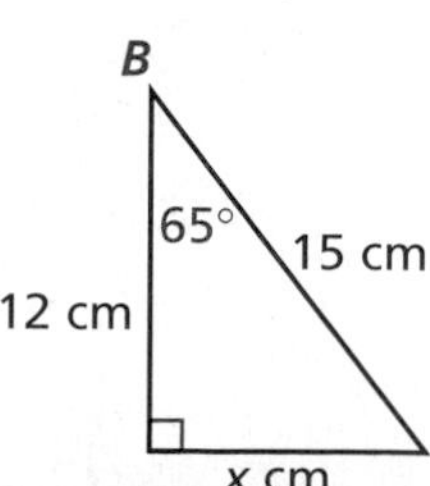

$\frac{8}{12} = \frac{6}{x}$ *Write a proportion using corresponding side lengths.*

$12 \cdot 6 = 8 \cdot x$ *The cross products are equal.*

$72 = 8x$ *x is multiplied by 8.*

$\frac{72}{8} = \frac{8x}{8}$ *Divide both sides by 8 to undo the multiplication.*

$9 \text{ cm} = x$

Angle A is congruent to angle B, and $m\angle B = 65°$.
$m\angle A = 65°$

EXAMPLE  2 PROBLEM SOLVING APPLICATION

The Boating Party **was painted by American artist Mary Cassatt. This reduction is similar to the actual painting. The height of the actual painting is 90.2 cm. To the nearest centimeter, what is the width of the actual painting?**

1 Understand the Problem

The **answer** will be the width of the actual painting.

List the **important information:**

- The actual painting and the reduction above are similar.
- The reduced painting is 4.6 cm tall and 6 cm wide.
- The actual painting is 90.2 cm tall.

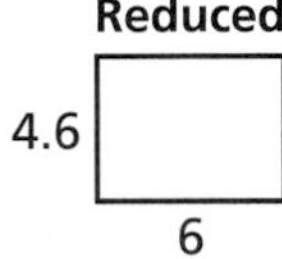

2 Make a Plan

Draw a diagram to represent the situation.
Use the corresponding sides to write a proportion.

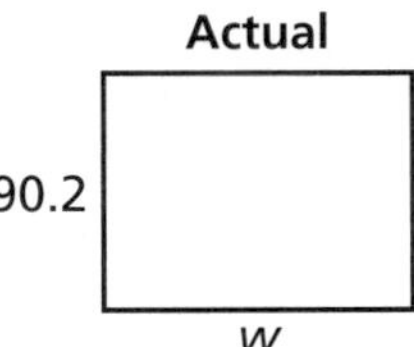

3 Solve

$\frac{4.6 \text{ cm}}{90.2 \text{ cm}} = \frac{6 \text{ cm}}{w \text{ cm}}$ *Write a proportion.*

$90.2 \cdot 6 = 4.6 \cdot w$ *The cross products are equal.*

$541.2 = 4.6w$ *w is multiplied by 4.6.*

$\frac{541.2}{4.6} = \frac{4.6w}{4.6}$ *Divide both sides by 4.6 to undo the multiplication.*

$118 \approx w$ *Round to the nearest centimeter.*

The width of the actual painting is about 118 cm.

4 Look Back

Estimate to check your answer. The ratio of the heights is about 5:90, or 1:18. The ratio of the widths is about 6:120, or 1:20. Since these ratios are close to each other, 118 cm is a reasonable answer.

Remember!

The symbol ≈ means "is approximately equal to."

Think and Discuss

1. **Name** two items in your classroom that appear to be similar figures.
2. **Describe** how similar figures are different from congruent figures.

8-4 Exercises

FOR EOG PRACTICE
see page 669

internet connect
Homework Help Online
go.hrw.com Keyword: MR4 8-4

5.04

GUIDED PRACTICE

See Example 1

1. The two triangles are similar. Find the missing length x and the measure of $\angle G$.

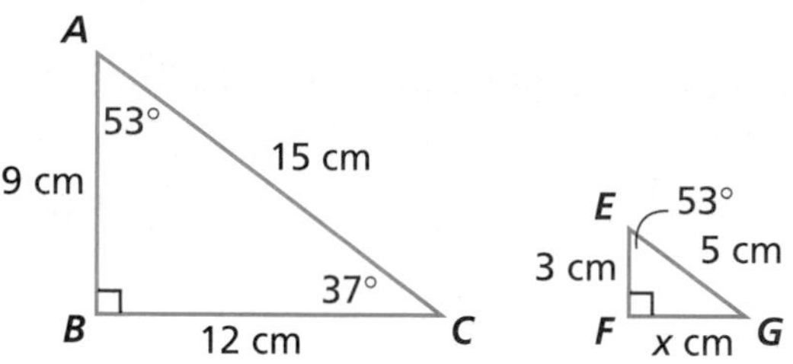

See Example 2

2. Pat's school photo package includes one large photo and several smaller photos. The large photo is similar to the photo at right. If the height of the large photo is 10 in., what is its width?

INDEPENDENT PRACTICE

See Example 1

3. The two triangles are similar. Find the missing length n and the measure of $\angle M$.

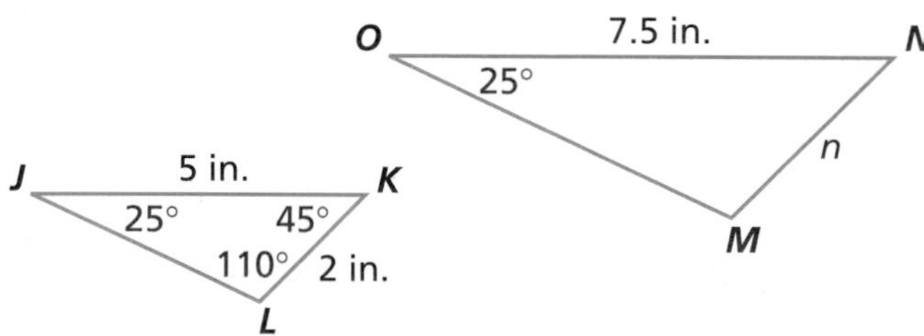

See Example 2

4. LeJuan swims in a pool that is similar to an Olympic-sized pool. LeJuan's pool is 30 m long by 8 m wide. The length of an Olympic-sized pool is 50 m. To the nearest meter, what is the width of an Olympic-sized pool?

PRACTICE AND PROBLEM SOLVING

Name the corresponding sides and angles for each pair of similar figures.

5. 6.

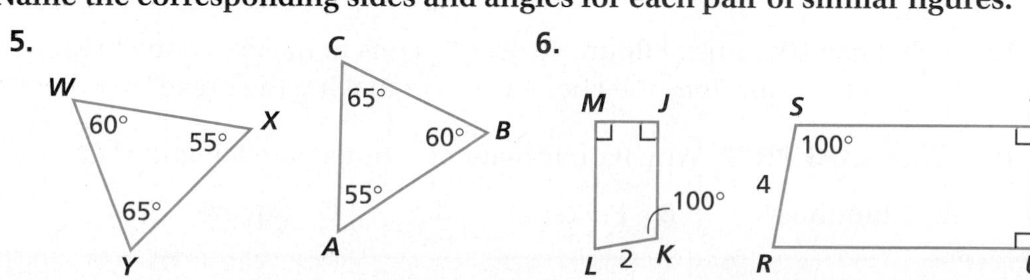

The figures in each pair are similar. Find the unknown measures.

7. E, 7 in., ?, 7 in., F, 10 in., D, 51°, 51°; L, ?, ?, 21 in., ?, 51°, M, 30 in., N

8. 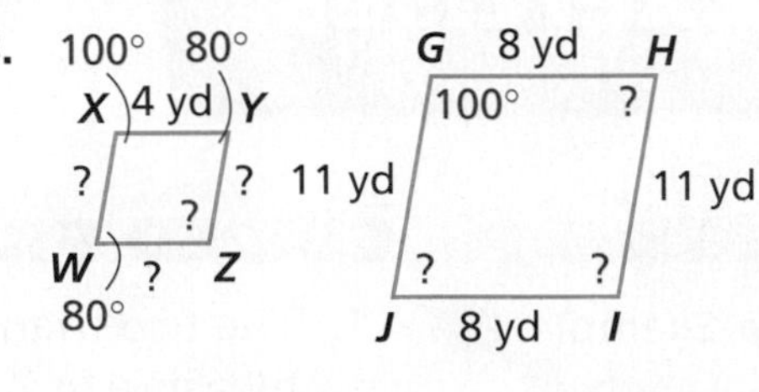

Tell whether the figures in each pair are similar. Explain your answers.

9. 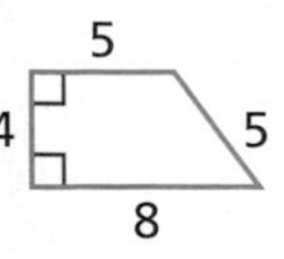

7.5, 6, 7.5, 12

10. 5, 53°, 3, 4

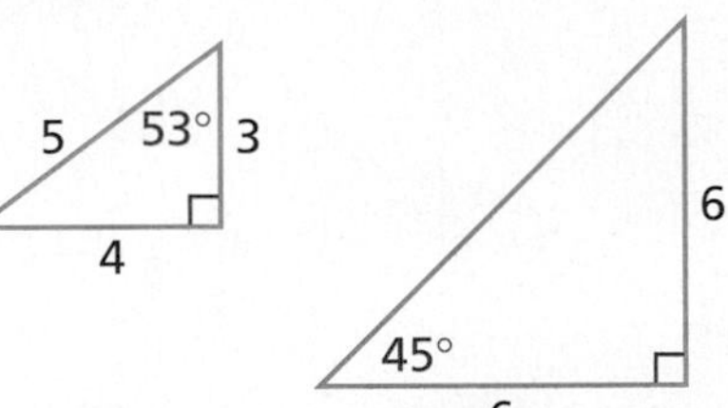

11. ***GRAPHIC ART*** Lenny designs billboards. He sketches his billboards before he paints them. The sketch and the billboard are similar. If the height of the billboard is 30 ft, what is the width to the nearest foot of the billboard?

12. ***WHAT'S THE ERROR?*** A student drew two rectangles. The dimensions of the rectangles are 10 in. by 9 in. and 5 in. by 3 in. The student said that the rectangles are similar. What's the error?

13. ***WRITE ABOUT IT*** Are all right triangles similar? Explain your answer.

14. ***CHALLENGE*** Draw two similar right triangles whose sides are in a ratio of 5:2.

Spiral Review

Write each decimal in standard form. (Lesson 3-1)

15. four tenths
16. thirty-one hundredths
17. ten and seven thousandths

18. A chef has $10\frac{1}{3}$ cups of flour. He uses $2\frac{1}{6}$ cups to make banana bread. How much flour does the chef have after making the bread? (Lesson 5-8)

19. **EOG PREP** Which quadrilateral is **not** a parallelogram? (Lesson 7-6)

A Rhombus B Rectangle C Square D Trapezoid

8-5 Indirect Measurement

Learn to use proportions and similar figures to find unknown measures.

Vocabulary

indirect measurement

Residents of Maine spent 14 days in 1999 building this enormous snowman. How could you measure the height of this snowman?

One way to find a height that you cannot measure directly is to use similar figures and proportions. This method is called **indirect measurement**.

Suppose that on a sunny day, the snowman cast a shadow that was 228 feet long. A 6-foot-tall person standing by the snowman cast a 12-foot-long shadow.

Both the person and the snowman form right angles with the ground, and their shadows are cast at the same angle. This means we can form two similar right triangles and use proportions to find the missing height.

EXAMPLE 1 Using Indirect Measurement

Use the similar triangles above to find the height of the snowman.

$\frac{6}{h} = \frac{12}{228}$ *Write a proportion using corresponding sides.*

$12 \cdot h = 6 \cdot 228$ *The cross products are equal.*

$12h = 1{,}368$ *h is multiplied by 12.*

$\frac{12h}{12} = \frac{1{,}368}{12}$ *Divide both sides by 12 to undo the multiplication.*

$h = 114$

The snowman was 114 feet tall.

EXAMPLE 2 *Measurement Application*

A lighthouse casts a shadow that is 36 m long when a meterstick casts a shadow that is 3 m long. How tall is the lighthouse?

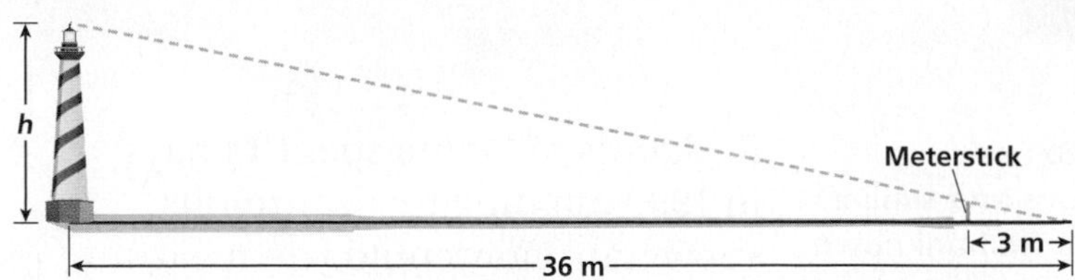

$\frac{h}{1} = \frac{36}{3}$ *Write a proportion using corresponding sides.*

$1 \cdot 36 = 3 \cdot h$ *The cross products are equal.*

$36 = 3h$ *h is multiplied by 3.*

$\frac{36}{3} = \frac{3h}{3}$ *Divide both sides by 3 to undo the multiplication.*

$12 = h$

The lighthouse is 12 m tall.

Think and Discuss

1. Name two items for which it would make sense to use indirect measurement to find their heights.

2. Name two items for which it would **not** make sense to use indirect measurement to find their heights.

8-5 Exercises

FOR EOG PRACTICE
see page 669

internet connect
Homework Help Online
go.hrw.com Keyword: MR4 8-5

5.04

GUIDED PRACTICE

See Example 1

1. Use the similar triangles to find the height of the flagpole.

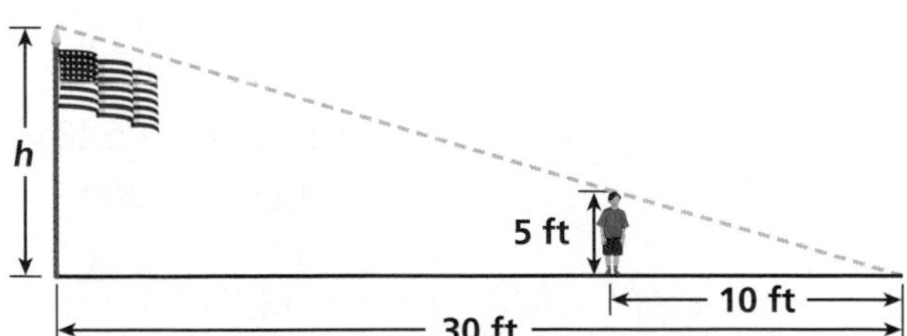

See Example 2

2. A tree casts a shadow that is 26 ft long. At the same time, a yardstick casts a shadow that is 4 ft long. How tall is the tree?

INDEPENDENT PRACTICE

See Example 1

3. Use the similar triangles to find the height of the lamppost.

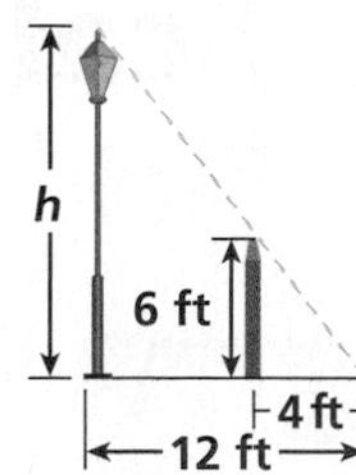

See Example 2

4. On a sunny day, the Eiffel Tower cast a shadow that was 328 feet long. A 6-foot-tall person standing by the tower cast a 2-foot-long shadow. How tall is the Eiffel Tower?

PRACTICE AND PROBLEM SOLVING

Find the unknown heights.

5. 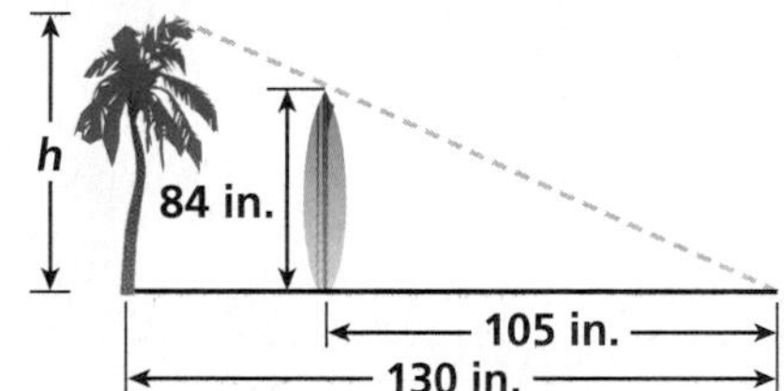

6. 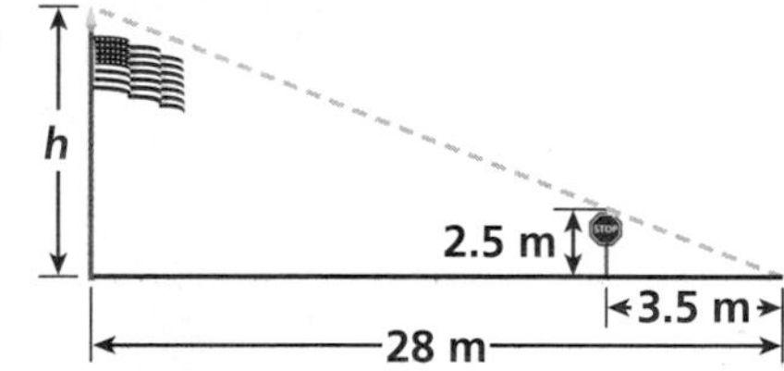

7. A statue casts a shadow that is 360 m long. At the same time, a person who is 2 m tall casts a shadow that is 6 m long. How tall is the statue?

8. ***WRITE ABOUT IT*** How are indirect measurements useful?

9. ***CHALLENGE*** A 5.5-foot-tall girl stands so that her shadow lines up with the shadow of a telephone pole. The tip of her shadow is even with the tip of the pole's shadow. If the length of the pole's shadow is 40 feet and the girl is standing 27.5 feet away from the pole, how tall is the telephone pole?

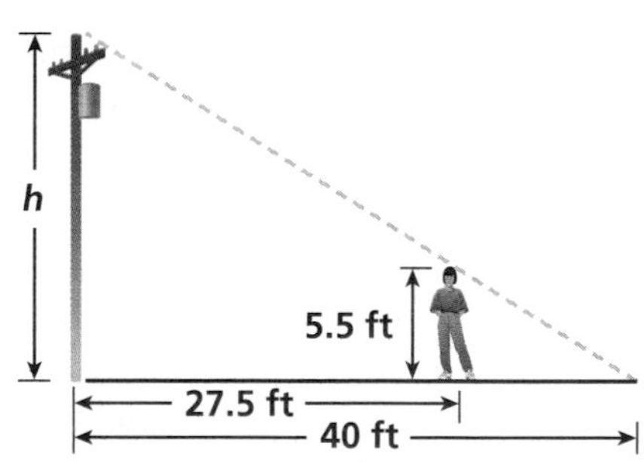

Spiral Review

Evaluate. (Lesson 1-4)

10. $(2 + 7 - 5) \div 2$ **11.** $10(6 - 3)$ **12.** $5 + 8 \cdot 7 - 1$ **13.** $5 + (8 + 2) - 3$

14. Make a stem-and-leaf plot of the following data: 85, 102, 89, 86, 104, 92, 103, 97, 91, 100. (Lesson 6-9)

15. **EOG PREP** A triangle has one right angle. What could be the measures of the other two angles? (Lesson 7-5)

A 30° and 15° B 70° and 20° C 60° and 120° D 100° and 90°

8-6 Scale Drawings and Maps

Learn to read and use map scales and scale drawings.

Vocabulary

scale drawing

scale

The map of Yosemite National Park shown above is a *scale drawing.* A **scale drawing** is a drawing of a real object that is proportionally smaller or larger than the real object. In other words, measurements on a scale drawing are in proportion to the measurements of the real object.

A **scale** is a ratio between two sets of measurements. In the map above, the scale is 1 in:2 mi. This ratio means that 1 inch on the map represents 2 miles in Yosemite National Park.

EXAMPLE 1 Finding Actual Distances

On the map, the distance between El Capitan and Panorama Cliff is 2 inches. What is the actual distance?

$\frac{1 \text{ in.}}{2 \text{ mi}} = \frac{2 \text{ in.}}{x \text{ mi}}$ *Write a proportion using the scale. Let x be the actual number of miles from El Capitan to Panorama Cliff.*

$2 \cdot 2 = 1 \cdot x$ *The cross products are equal.*

$4 = x$

The actual distance from El Capitan to Panorama Cliff is 4 miles.

Helpful Hint

In Example 1, think "1 inch is 2 miles, so 2 inches is how many miles?" This approach will help you set up proportions in similar problems.

EXAMPLE 2 *Astronomy Application*

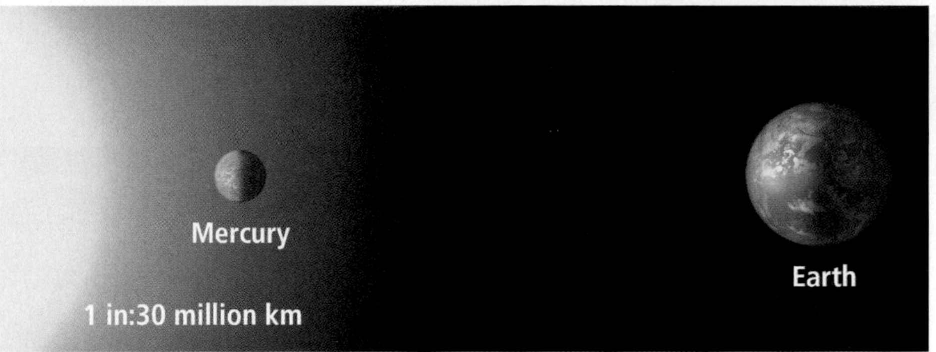

A **What is the actual distance from Mercury to Earth?**

Use your inch ruler to measure the distance from the center of Mercury to the center of Earth on the drawing. Mercury and Earth are about 3 inches apart.

$$\frac{1 \text{ in.}}{30 \text{ million km}} = \frac{3 \text{ in.}}{x \text{ million km}}$$ *Write a proportion. Let x be the actual distance from Mercury to Earth.*

$$30 \cdot 3 = 1 \cdot x$$ *The cross products are equal.*

$$90 = x$$

The actual distance from Mercury to Earth is about 90 million km.

B **The actual distance from Mercury to Venus is 50 million kilometers. How far apart should Mercury and Venus be drawn?**

$$\frac{1 \text{ in.}}{30 \text{ million km}} = \frac{x \text{ in.}}{50 \text{ million km}}$$ *Write a proportion. Let x be the distance from Mercury to Venus on the drawing.*

$$30 \cdot x = 1 \cdot 50$$ *The cross products are equal.*

$$30x = 50$$ *x is multiplied by 30.*

$$\frac{30x}{30} = \frac{50}{30}$$ *Divide both sides by 30 to undo the multiplication.*

$$x = 1\frac{2}{3}$$

Mercury and Venus should be drawn $1\frac{2}{3}$ inches apart.

Think and Discuss

1. **Give an example** of when you would use a scale drawing.
2. **Suppose** that you are going to make a scale drawing of your classroom with a scale of 1 inch:3 feet. Select a distance in your classroom and measure it. What will this distance be on your drawing?

8-6 Exercises

5.04

FOR EOG PRACTICE

see page 669

internet connect

Homework Help Online
go.hrw.com Keyword: MR4 8-6

GUIDED PRACTICE

See Example  1

1. On the map, the distance between the post office and the fountain is 6 cm. What is the actual distance?

See Example  2

2. What is the actual length of the car?
3. The actual height of the car is 1.6 meters. Is the car's height in the drawing correct?

Scale: 1 cm:0.8 m

INDEPENDENT PRACTICE

See Example  1

4. On the map of California, Los Angeles is 1.25 inches from Malibu. Find the actual distance from Los Angeles to Malibu.

See Example  2

5. Riverside, California, is 50 miles from Los Angeles. On the map, how far should Riverside be from Los Angeles?
6. A paramecium is a one-celled organism. The scale drawing at right is larger than an actual paramecium. Find the actual length of the paramecium.

Scale: 1 in:20 mi

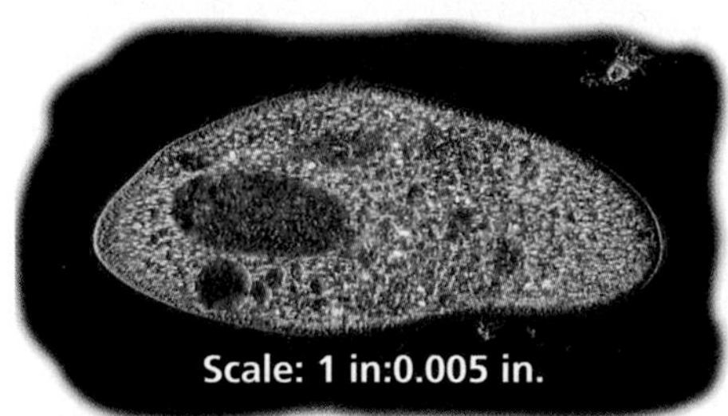

PRACTICE AND PROBLEM SOLVING

Suppose you are asked to make a scale drawing of a room. The scale is 1 in:4 ft. Use the actual lengths below to find the lengths in the drawing.

7. north wall: 12 ft
8. south wall: 8 ft
9. east wall: 5 ft
10. west wall: 10 ft
11. door width: 3.5 ft
12. window width: 2.5 ft

Social Studies LINK

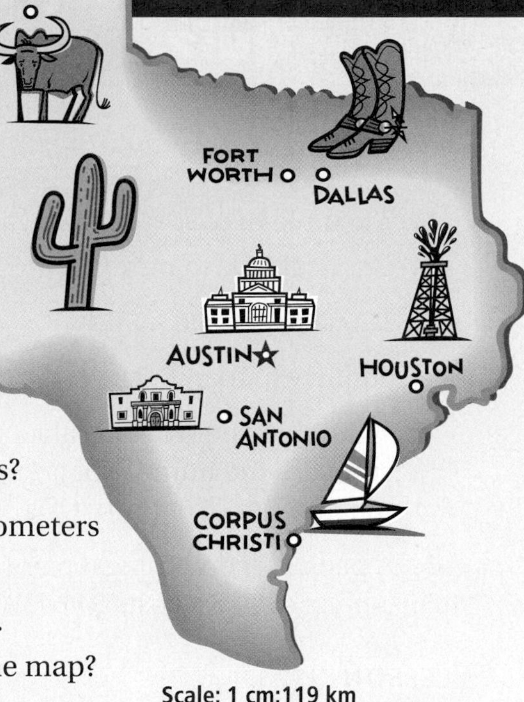

Texas is the second largest state in the country and is the largest state in the lower 48 states. It is more than 1,120 kilometers across. There is even a ranch in Texas that is larger than Rhode Island!

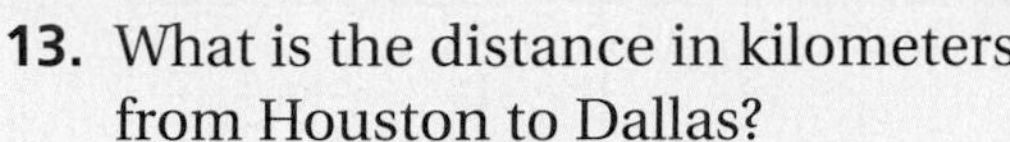

13. What is the distance in kilometers from Houston to Dallas?

14. What is the distance in kilometers from Corpus Christi to San Antonio?

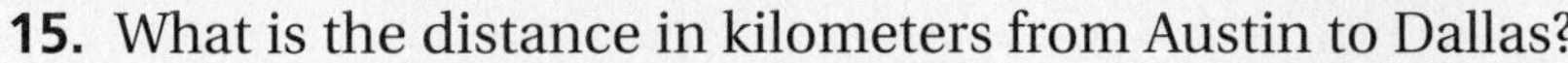

15. What is the distance in kilometers from Austin to Dallas?

16. Name two cities on the map that are more than 200 kilometers apart.

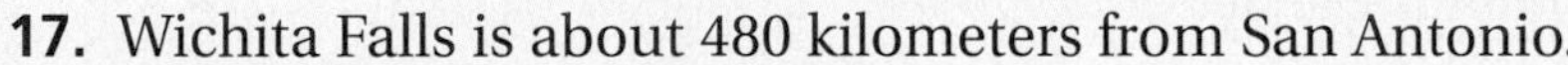

17. Wichita Falls is about 480 kilometers from San Antonio.
    a. About how far apart should these two cities be on the map?
    b. What else would you need to know to be able to place Wichita Falls on the map?

18. ***WRITE A PROBLEM*** Write a problem using the map and its scale.

19. ***WRITE ABOUT IT*** Explain how to find the actual distance between two cities if you know the distance on a map and the scale of the map.

20. ***CHALLENGE*** If you drive at a constant speed of 100 kilometers per hour, about how long will it take you to drive from Amarillo to San Antonio?

Spiral Review

21. Patrice had $5.89 in her pocket and $9.81 in her purse. She spent $12.99 on a new CD. How much money does Patrice have after her purchase? (Lesson 3-3)

Divide. Write each answer in simplest form. (Lesson 5-3)

22. $\frac{2}{3} \div \frac{1}{3}$
23. $\frac{9}{10} \div \frac{3}{4}$
24. $2\frac{3}{8} \div \frac{1}{4}$
25. $1\frac{1}{4} \div 2\frac{1}{3}$

26. **EOG PREP** A __?__ has two endpoints. (Lesson 7-1)

    A plane B line segment C line D ray

27. **EOG PREP** Which ratio is equivalent to 3:2? (Lesson 8-1)

    A 5:1 B 2:3 C 6:4 D 4:3

Mid-Chapter Quiz

Chapter 8 Mid-Chapter Quiz

LESSON 8-1 (pp. 392–395)

Use the table to write the following ratios.

Types of CDs in Mark's Music Collection			
Classical	4	Jazz	3
Country	9	Pop	14
Dance	8	Rock	10

1. classical CDs to rock CDs
2. dance CDs to jazz CDs
3. country to total CDs
4. Write three equivalent ratios to compare the number of jazz CDs with the number of country CDs.
5. A package containing 6 pairs of socks costs $6.89. A package containing 4 pairs of socks costs $4.64. Which is the better deal?

LESSON 8-2 (pp. 398–401)

Find the missing value in each proportion.

6. $\frac{1}{4} = \frac{n}{12}$ 7. $\frac{3}{n} = \frac{15}{25}$ 8. $\frac{n}{4} = \frac{18}{6}$ 9. $\frac{10}{4} = \frac{5}{n}$

LESSON 8-3 (pp. 402–404)

10. You have 26 feet of intestine in your body. How many inches of intestine do you have?
11. Death Valley is 282 feet below sea level. How many yards below sea level is it?

LESSON 8-4 (pp. 405–408)

12. The two triangles are similar. Find the missing length n and the measure of $\angle R$.

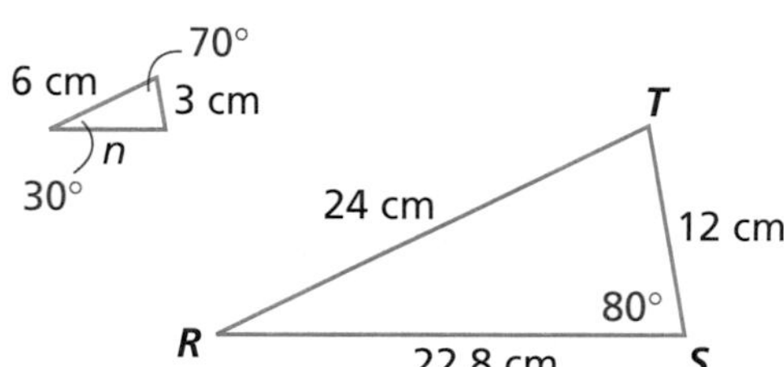

LESSON 8-5 (pp. 409–411)

13. A tree casts a shadow that is 18 feet long. At the same time, a 5-foot-tall person casts a shadow that is 3.6 feet long. How tall is the tree?

LESSON 8-6 (pp. 412–415)

Use the scale drawing to answer each question.

14. What is the actual length of the kitchen?
15. What are the actual length and width of bedroom 1?

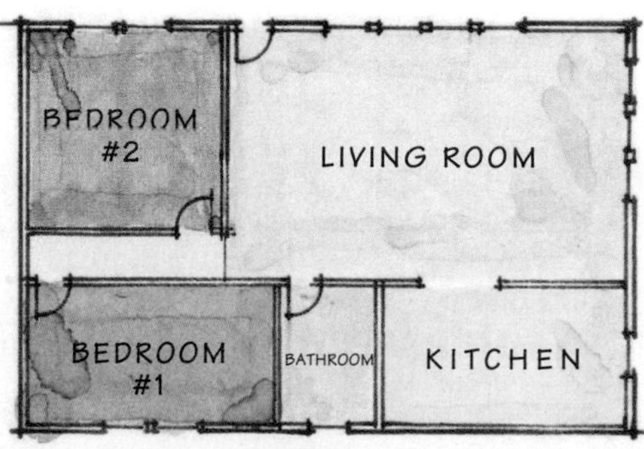

Focus on Problem Solving

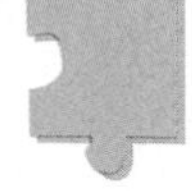

Make a Plan

- **Estimate or find an exact answer**

Sometimes an estimate is all you need to solve a problem, and sometimes you need to find an exact answer.

One way to decide whether you can estimate is to see if you can rewrite the problem using the words *at most, at least,* or *about.* For example, suppose Laura has \$30. Then she could spend *at most* \$30. She would not have to spend *exactly* \$30. Or, if you know it takes 15 minutes to get to school, you must leave your house *at least* (not exactly) 15 minutes before school starts.

Read the problems below. Decide whether you can estimate or whether you must find the exact answer. How do you know?

1. Alex is a radio station disc jockey. He is making a list of songs that should last no longer than 30 minutes total when played in a row. His list of songs and their playing times are given in the table. Does Alex have the right amount of music?

Song Title	Length (min)
Color Me Blue	4.5
Hittin' the Road	7.2
Stand Up, Shout	2.6
Top Dog	3.6
Kelso Blues	4.3
Smile on Me	5.7
A Long Time Ago	6.4

2. For every 10 minutes of music, Alex has to play 1.5 minutes of commercials. If Alex plays the songs on the list, how much time does he need to allow for commercials?

3. If Alex must play the songs on the list and the commercials in 30 minutes, how much music time does he need to cut to allow for commercials?

8-7 Percents

Learn to write percents as decimals and as fractions.

Vocabulary

percent

At a sales tax rate of 8%, the tax on this guitar and amplifier would be $36.56.

Most states charge sales tax on items you purchase. Sales tax is a percent of the item's price. A **percent** is a ratio of a number to 100.

You can remember that *percent* means "per hundred." For example, 8% means "8 per hundred," or "8 out of 100."

If a sales tax rate is 8%, the following statements are true:

- For every $1.00 you spend, you pay $0.08 in sales tax.
- For every $10.00 you spend, you pay $0.80 in sales tax.
- For every $100 you spend, you pay $8 in sales tax.

Because *percent* means "per hundred," 100% means "100 out of 100." This is why 100% is often used to mean "all" or "the whole thing."

EXAMPLE 1 Modeling Percents

Use a 10-by-10-square grid to model 8%.

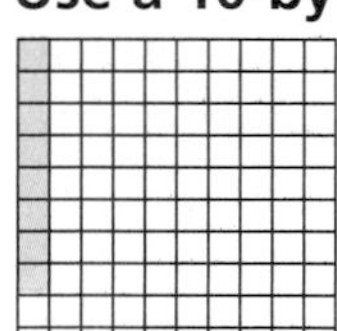

A 10-by-10-square grid has 100 squares.

8% means "8 out of 100," or $\frac{8}{100}$.

Shade 8 squares out of 100 squares.

EXAMPLE 2 Writing Percents as Fractions

Write 40% as a fraction in simplest form.

$40\% = \frac{40}{100}$ *Write the percent as a fraction with a denominator of 100.*

$\frac{40 \div 20}{100 \div 20} = \frac{2}{5}$ *Write the fraction in simplest form.*

Written as a fraction, 40% is $\frac{2}{5}$.

EXAMPLE 3 *Life Science Application*

Up to 55% of the heat lost by your body can be lost through your head. Write 55% as a fraction in simplest form.

$55\% = \frac{55}{100}$ — *Write the percent as a fraction with a denominator of 100.*

$\frac{55 \div 5}{100 \div 5} = \frac{11}{20}$ — *Write the fraction in simplest form.*

Written as a fraction, 55% is $\frac{11}{20}$.

EXAMPLE 4 Writing Percents as Decimals

Remember!

To divide by 100, move the decimal point two places to the left.

$24 \div 100 = 0.24$

Write 24% as a decimal.

$24\% = \frac{24}{100}$ — *Write the percent as a fraction with a denominator of 100.*

Write the fraction as a decimal.

$$\begin{array}{r} 0.24 \\ 100\overline{)24.00} \\ -\underline{200} \\ 400 \\ -\underline{400} \\ 0 \end{array}$$

Written as a decimal, 24% is 0.24.

EXAMPLE 5 *Earth Science Application*

The water frozen in glaciers makes up almost 75% of the world's fresh water supply. Write 75% as a decimal.

$75\% = \frac{75}{100}$ — *Write the percent as a fraction with a denominator of 100.*

$75 \div 100 = 0.75$ — *Write the fraction as a decimal.*

Written as a decimal, 75% is 0.75.

Think and Discuss

1. **Give an example** of a situation in which you have seen percents.
2. **Tell** how much sales tax you would have to pay on $1, $10, and $100 if your state had a 5% sales tax rate.
3. **Explain** how to write a percent as a fraction.
4. **Write** 100% as a decimal and as a fraction.

8-7 Exercises

FOR EOG PRACTICE
see page 670

internet connect
Homework Help Online
go.hrw.com Keyword: MR4 8-7

1.02a

GUIDED PRACTICE

See Example 1 **Use a 10-by-10-square grid to model each percent.**

1. 45% **2.** 3% **3.** 61%

See Example 2 **Write each percent as a fraction in simplest form.**

4. 25% **5.** 80% **6.** 54%

See Example 3 **7.** Belize is a country in Central America. Of the land in Belize, 92% is made up of forests and woodlands. Write 92% as a fraction in simplest form.

See Example 4 **Write each percent as a decimal.**

8. 72% **9.** 4% **10.** 90%

See Example 5 **11.** About 64% of the runways at airports in the United States are not paved. Write 64% as a decimal.

INDEPENDENT PRACTICE

See Example 1 **Use a 10-by-10-square grid to model each percent.**

12. 14% **13.** 98% **14.** 36%

See Example 2 **Write each percent as a fraction in simplest form.**

15. 20% **16.** 75% **17.** 11%

18. 5% **19.** 64% **20.** 31%

See Example 3 **21.** Nikki must answer 80% of the questions on her final exam correctly to pass her class. Write 80% as a fraction in simplest form.

See Example 4 **Write each percent as a decimal.**

22. 44% **23.** 13% **24.** 29%

25. 60% **26.** 92% **27.** 7%

See Example 5 **28.** Brett was absent 2% of the school year. Write 2% as a decimal.

PRACTICE AND PROBLEM SOLVING

Write each percent as a fraction in simplest form and as a decimal.

29. 23% **30.** 1% **31.** 49% **32.** 70%

33. 37% **34.** 85% **35.** 8% **36.** 63%

37. 94% **38.** 100% **39.** 0% **40.** 52%

Music LINK

The circle graph shows the percent of radio stations around the world that play each type of music listed. Use the graph for Exercises 41–48.

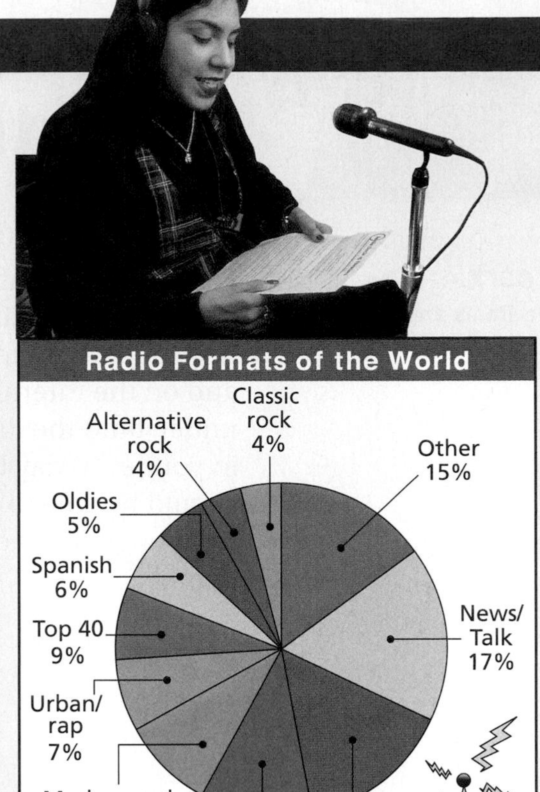

Source: Scholastic Kid's Almanac for the 21st Century

41. What fraction of the radio stations play easy listening music? Write this fraction in simplest form.

42. Use a 10-by-10-square grid to model the percent of radio stations that play country music. Then write this percent as a decimal.

43. Which type of music makes up $\frac{1}{20}$ of the graph?

44. Someone reading the graph said, "More than $\frac{1}{10}$ of the radio stations play top 40 music." Do you agree with this statement? Why or why not?

45. Suppose you converted all of the percents in the graph to decimals and added them. Without actually doing this, tell what the sum would be. Explain.

46. ***WRITE A PROBLEM*** Write a question about the circle graph that involves changing a percent to a fraction. Then answer your question.

47. ***WRITE ABOUT IT*** How does the percent of radio stations that play Spanish music compare with the fraction $\frac{1}{6}$? Explain.

48. ***CHALLENGE*** Name a fraction that is greater than the percent of radio stations that play Spanish music but less than the percent of radio stations that play urban/rap music.

Spiral Review

49. The heights of four plants are 139 cm, 208 cm, 144 cm, and 165 cm. Estimate the sum of the heights by rounding to the nearest ten. (Lesson 1-2)

Find the range, mean, median, and mode of each data set. (Lesson 6-2)

50. 22, 24, 22, 29, 33, 14 **51.** 87, 16, 19, 21, 23 **52.** 365, 180, 360, 720, 59

53. **EOG PREP** Which is the measure of an obtuse angle? (Lesson 7-2)

A 90° B 0° C 45° D 125°

54. **EOG PREP** Squares *ABCD* and *WXYZ* are congruent. The length of $\overline{AB}$ is 5 in. What is the perimeter of square *WXYZ*? (Lesson 7-9)

A 5 in. B 9 in. C 20 in. D 25 in.

8-8 Percents, Decimals, and Fractions

Learn to write decimals and fractions as percents.

Percents, decimals, and fractions appear in newspapers, on television, and on the Internet. To fully understand the data you see in your everyday life, you should be able to change from one number form to another.

Andrew Toos/Cartoon Resource.com

"Oh yes, a one-half of one percent allowance increase is quite a bit."

EXAMPLE 1 Writing Decimals as Percents

Write each decimal as a percent.

Method 1: Use place value.

A 0.3

$0.3 = \frac{3}{10}$ *Write the decimal as a fraction.*

$\frac{3 \cdot 10}{10 \cdot 10} = \frac{30}{100}$ *Write an equivalent fraction with 100 as the denominator.*

$\frac{30}{100} = 30\%$ *Write the numerator with a percent symbol.*

B 0.43

$0.43 = \frac{43}{100}$ *Write the decimal as a fraction.*

$\frac{43}{100} = 43\%$ *Write the numerator with a percent symbol.*

Method 2: Multiply by 100.

C 0.7431

$0.7431 \cdot 100$ *Multiply by 100.*

74.31% *Add the percent symbol.*

D 0.023

$0.023 \cdot 100$ *Multiply by 100.*

2.3% *Add the percent symbol.*

EXAMPLE 2 Writing Fractions As Percents

Write each fraction as a percent.

Method 1: Write an equivalent fraction with a denominator of 100.

Helpful Hint

When the denominator is a factor of 100, it is often easier to use method 1. When the denominator is not a factor of 100, it is usually easier to use method 2.

A $\frac{4}{5}$

$\frac{4 \cdot 20}{5 \cdot 20} = \frac{80}{100}$ — *Write an equivalent fraction with a denominator of 100.*

$\frac{80}{100} = 80\%$ — *Write the numerator with a percent symbol.*

Method 2: Use division to write the fraction as a decimal.

B $\frac{3}{8}$

$8\overline{)3.000}$ = 0.375 — *Divide the numerator by the denominator.*

$0.375 = 37.5\%$ — *Multiply by 100 by moving the decimal point right two places. Add the percent symbol.*

EXAMPLE 3 *Earth Science Application*

About $\frac{39}{50}$ of Earth's atmosphere is made up of nitrogen. About what percent of the atmosphere is nitrogen?

$\frac{39}{50}$

$\frac{39 \cdot 2}{50 \cdot 2} = \frac{78}{100}$ — *Write an equivalent fraction with a denominator of 100.*

$\frac{78}{100} = 78\%$ — *Write the numerator with a percent symbol.*

About 78% of Earth's atmosphere is made up of nitrogen.

Common Equivalent Fractions, Decimals, and Percents

Fraction	$\frac{1}{5}$	$\frac{1}{4}$	$\frac{1}{3}$	$\frac{2}{5}$	$\frac{1}{2}$	$\frac{3}{5}$	$\frac{2}{3}$	$\frac{3}{4}$	$\frac{4}{5}$
Decimal	0.2	0.25	$0.\overline{3}$	0.4	0.5	0.6	$0.\overline{6}$	0.75	0.8
Percent	20%	25%	$33.\overline{3}\%$	40%	50%	60%	$66.\overline{6}\%$	75%	80%

Think and Discuss

1. **Tell** which method you prefer for converting decimals to percents—using equivalent fractions or multiplying by 100. Why?
2. **Give** two different ways to write three-tenths.
3. **Explain** how to write fractions as percents using two different methods.

8-8 Exercises

FOR EOG PRACTICE
see page 670

internet connect
Homework Help Online
go.hrw.com Keyword: MR4 8-8

1.02a

GUIDED PRACTICE

See Example 1 **Write each decimal as a percent.**

1. 0.39 **2.** 0.125 **3.** 0.8 **4.** 0.112

See Example 2 **Write each fraction as a percent.**

5. $\frac{11}{25}$ **6.** $\frac{7}{8}$ **7.** $\frac{7}{10}$ **8.** $\frac{1}{2}$

See Example 3 **9.** Patti spent $\frac{3}{4}$ of her allowance on a new backpack. What percent of her allowance did she spend?

INDEPENDENT PRACTICE

See Example 1 **Write each decimal as a percent.**

10. 0.6 **11.** 0.55 **12.** 0.34 **13.** 0.308

14. 0.941 **15.** 0.01 **16.** 0.62 **17.** 0.02

See Example 2 **Write each fraction as a percent.**

18. $\frac{3}{5}$ **19.** $\frac{3}{10}$ **20.** $\frac{24}{25}$ **21.** $\frac{9}{20}$

22. $\frac{1}{8}$ **23.** $\frac{11}{16}$ **24.** $\frac{37}{50}$ **25.** $\frac{2}{5}$

See Example 3 **26.** About $\frac{1}{125}$ of the people in the United States have the last name *Johnson.* What percent of people in the United States have this last name?

PRACTICE AND PROBLEM SOLVING

Write each decimal as a percent and a fraction.

27. 0.04 **28.** 0.32 **29.** 0.45 **30.** 0.59

31. 0.81 **32.** 0.6 **33.** 0.39 **34.** 0.14

Write each fraction as a percent and as a decimal. Round to the nearest hundredth, if necessary.

35. $\frac{4}{5}$ **36.** $\frac{1}{3}$ **37.** $\frac{5}{6}$ **38.** $\frac{7}{12}$

39. $\frac{2}{30}$ **40.** $\frac{1}{25}$ **41.** $\frac{8}{11}$ **42.** $\frac{4}{15}$

Compare. Write <, >, or =.

43. 70% ■ $\frac{3}{4}$ **44.** $\frac{5}{8}$ ■ 6.25% **45.** 0.2 ■ $\frac{1}{5}$

46. 0.7 ■ 7% **47.** $\frac{9}{10}$ ■ 0.3 **48.** 37% ■ $\frac{3}{7}$

Order the numbers from least to greatest.

49. 45%, $\frac{21}{50}$, 0.43 **50.** $\frac{7}{8}$, 90%, 0.098 **51.** 0.7, 26%, $\frac{1}{4}$

52. 38%, $\frac{7}{25}$, 0.21 **53.** $\frac{9}{20}$, 14%, 0.125 **54.** 0.605, 17%, $\frac{5}{9}$

55. ***ENTERTAINMENT*** About 97 million households in the United States have at least one television. Use the table below to answer the questions that follow.

Entertainment LINK

This photo from 1953 shows one of the first color television cameras.

go.hrw.com KEYWORD: MR4 TV

CNN Student News

Television in the United States	
Fraction of households with at least one television	$\frac{49}{50}$
Percent of televisions that are color	99%
Fraction of households with three televisions	$\frac{19}{50}$
Percent of television owners with a VCR	82%
Fraction of television owners with basic cable	$\frac{2}{3}$

a. About what percent of television owners have basic cable?

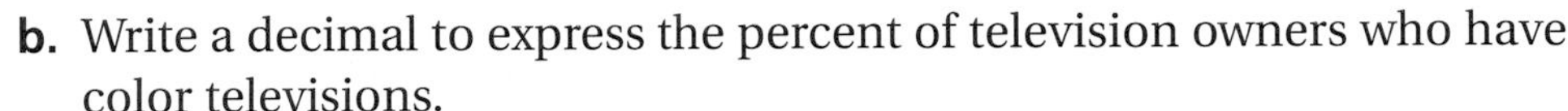

b. Write a decimal to express the percent of television owners who have color televisions.

c. What percent of television owners have three televisions?

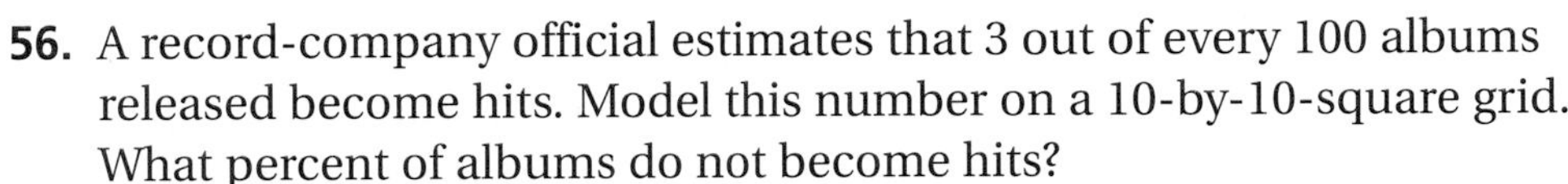

56. A record-company official estimates that 3 out of every 100 albums released become hits. Model this number on a 10-by-10-square grid. What percent of albums do not become hits?

57. ***WHAT'S THE QUESTION?*** Out of 25 students, 12 prefer to take their test on Monday, and 5 prefer to take their test on Tuesday. The answer is 32%. What is the question?

58. ***WRITE ABOUT IT*** Explain why 0.8 is equal to 80% and not 8%.

59. ***CHALLENGE*** The dimensions of a rectangle are 0.5 yard and 24% of a yard. What is the area of the rectangle? Write your answer as a fraction in simplest form.

Spiral Review

Solve each equation. (Lesson 2-6)

60. $7c = 77$ **61.** $12j = 228$ **62.** $22m = 176$ **63.** $41z = 205$

Draw each geometric figure. (Lesson 7-1)

64. $\overleftrightarrow{CD}$ **65.** $\overrightarrow{GM}$ **66.** $\overline{XY}$ **67.** point A

68. **EOG PREP** The sides of a number cube are numbered from 1 through 6. What is the ratio of multiples of 3 to multiples of 4? (Lesson 8-1)

A 1:1 B 2:1 C 1:2 D 2:2

8-9 Percent Problems

Learn to find the missing value in a percent problem.

The frozen-yogurt stand in the mall sells 420 frozen-yogurt cups per day, on average. Forty-five percent of the frozen-yogurt cups are sold to teenagers. On average, how many frozen-yogurt cups are sold to teenagers each day?

To answer this question, you will need to find 45% of 420.

To find the percent one number is of another, use this proportion:

$$\frac{\%}{100} = \frac{\text{is}}{\text{of}}$$

Because you are looking for **45% of 420**, 45 replaces the **percent sign** and 420 replaces "**of.**" The first denominator, 100, always stays the same. The "is" part is what you have been asked to find.

EXAMPLE 1 *Consumer Application*

How many frozen-yogurt cups are sold to teenagers each day?

First estimate your answer. Think: $45\% = \frac{45}{100}$, which is close to $\frac{1}{2}$.
So about $\frac{1}{2}$ of the 420 yogurt cups are sold to teenagers.

$\frac{1}{2} \cdot 420 = 210$ ← *This is the estimate.*

Now solve:

$\frac{45}{100} = \frac{y}{420}$ — *Let y represent the number of yogurt cups sold to teenagers.*

$100 \cdot y = 45 \cdot 420$ — *The cross products are equal.*

$100y = 18{,}900$ — *y is multiplied by 100.*

$\frac{100y}{100} = \frac{18{,}900}{100}$ — *Divide both sides of the equation by 100 to undo the multiplication.*

$y = 189$

Since 189 is close to your estimate of 210, 189 is a reasonable answer. About 189 yogurt cups per day are sold to teenagers.

Helpful Hint

Think: "45 out of 100 is how many out of 420?"

EXAMPLE 2 *Technology Application*

Heather is downloading a file from the Internet. So far, she has downloaded 75% of the file. If 30 minutes have passed since she started, how long will it take her to download the rest of the file?

$\frac{\%}{100} = \frac{\text{is}}{\text{of}}$

$\frac{75}{100} = \frac{30}{m}$ *75% of the file has downloaded, so 30 minutes **is** 75% **of** the total time needed.*

$100 \cdot 30 = 75 \cdot m$ *The cross products are equal.*

$3{,}000 = 75m$ *m is multiplied by 75.*

$\frac{3{,}000}{75} = \frac{75m}{75}$ *Divide both sides by 75 to undo the multiplication.*

$40 = m$

The time needed to download the entire file is 40 min. So far, the file has been downloading for 30 min. Because $40 - 30 = 10$, the remainder of the file will be downloaded in 10 min.

Instead of using proportions, you can also multiply to find a percent of a number.

EXAMPLE 3 Multiplying to Find a Percent of a Number

A **Find 20% of 150.**

20% = 0.20 *Write the percent as a decimal.*

$0.20 \cdot 150$ *Multiply using the decimal.*

30

So 30 is 20% of 150.

B **Find 5% of 90.**

5% = 0.05 *Write the percent as a decimal.*

$0.05 \cdot 90$ *Multiply using the decimal.*

4.5

So 4.5 is 5% of 90.

Think and Discuss

1. **Explain** why you must subtract 30 from 40 in Example 2.
2. **Give an example** of a time when you would need to find a percent of a number.

8-9 Exercises

FOR EOG PRACTICE

see page 671

internet connect
Homework Help Online
go.hrw.com Keyword: MR4 8-9

1.02a, b

GUIDED PRACTICE

See Example 1

1. Members of the drama club sold T-shirts for their upcoming musical. Of the 80 T-shirts sold, 55% were size medium. How many of the T-shirts sold were size medium?

See Example 2

2. Loni has read 25% of a book. If she has been reading for 5 hours, how many more hours will it take her to complete the book?

See Example 3

3. Find 12% of 56.
4. Find 65% of 240.
5. Find 85% of 115.
6. Find 70% of 54.

INDEPENDENT PRACTICE

See Example 1

7. Tamara collects porcelain dolls. Of the 24 dolls that she has, 25% have blond hair. How many of her dolls have blond hair?
8. Mr. Green has a garden. Of the 40 seeds he planted, 35% were vegetable seeds. How many vegetable seeds did he plant?

See Example 2

9. Kevin has mowed 40% of the lawn. If he has been mowing for 20 minutes, how long will it take him to mow the rest of the lawn?
10. Maggie ordered a painting. She paid 30% of the total cost when she ordered it, and she will pay the remaining amount when it is delivered. If she has paid $15, how much more does she owe?

See Example 3

11. Find 22% of 130.
12. Find 78% of 350.
13. Find 9% of 50.
14. Find 45% of 210.

PRACTICE AND PROBLEM SOLVING

Find the percent of each number.

15. 6% of 38
16. 20% of 182
17. 32% of 205
18. 14% of 88
19. 78% of 52
20. 31% of 345
21. 10% of 50
22. 1.5% of 800
23. ***GEOMETRY*** The width of a rectangular room is 75% of the length of the room. The room is 12 feet long.
    a. How wide is the room?
    b. The area of a rectangle is the product of the length and the width. What is the area of the room?

Someday you may do all of your schoolwork on a computer. You may even carry a computer to class with you, if your school becomes a digital school, like the one pictured above.

24. ***TECHNOLOGY*** Students were asked in a school survey about how they use their computers. The circle graph shows the results.

What Do You Do Most Often on Your Computer?

Other
Games 11%
Homework 28%
E-mail 34%
Other Internet 21%

a. If there are 850 students in the school, how many spend most of their computer time using e-mail?

b. Fifty-one students selected "other." What percent of the school population does this represent?

c. Which choices were selected by more than 200 students?

d. How many more students chose Internet than chose playing games?

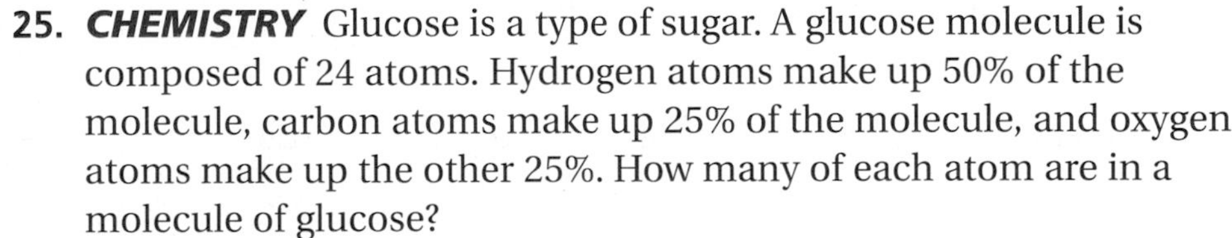

25. ***CHEMISTRY*** Glucose is a type of sugar. A glucose molecule is composed of 24 atoms. Hydrogen atoms make up 50% of the molecule, carbon atoms make up 25% of the molecule, and oxygen atoms make up the other 25%. How many of each atom are in a molecule of glucose?

26. ***WHAT'S THE ERROR?*** To find 80% of 130, a student set up the proportion $\frac{80}{100} = \frac{130}{x}$. Explain the error. Write the correct proportion, and find the missing value.

27. ***WRITE ABOUT IT*** Suppose you were asked to find 48% of 300 and your answer was 6.25. Would your answer be reasonable? How do you know? What is the correct answer?

28. ***CHALLENGE*** Mrs. Peterson makes ceramic figurines. She recently made 25 figurines. Of those figurines, 16 are animals. What percent of the figurines are **not** animals?

Spiral Review

Find each quotient. (Lesson 3-8)

29. 5.6 ÷ 0.8 **30.** 30.8 ÷ 1.4 **31.** 254.1 ÷ 0.35 **32.** 11.5 ÷ 0.05

Write the prime factorization of each number. (Lesson 4-2)

33. 38 **34.** 50 **35.** 120 **36.** 214

37. **EOG PREP** What is the least common denominator for $\frac{1}{2}$ and $\frac{5}{6}$? (Lesson 5-7)

A 12 B 6 C 2 D 4

38. **EOG PREP** Which is the movement of a figure around a point? (Lesson 7-10)

A Translation B Reflection C Rotation D Tessellation

Construct Circle Graphs

Use with Lesson 8-9

REMEMBER
The sum of the measures of the angles in any circle is 360°.

A circle graph shows parts of a whole. If you think of a complete circle as 100%, you can express sections of a circle graph as percents.

- Ms. Shipley's class earned \$400 at the school fair. What fraction of the \$400 did the class earn at the bake sale?
- What percent of the \$400 did the class earn at the bake sale?

Money Raised at School Fair

Activity

At Mazel Middle School, students were surveyed about their favorite types of TV programs. Make a circle graph to represent the results.

Students' Favorite Programs	
Type of Program	**Number of Students**
Science	25
Cooking	15
Sports	50
Sitcoms	150
Movies	60
Cartoons	200

1. Find the total number of students surveyed.
   $25 + 15 + 50 + 150 + 60 + 200 = 500$
2. Find the percent of the total represented by students who like science programs.
   $$\frac{25}{500} = 5\%$$
3. Since there are 360° in a circle, multiply 5% by 360°. This will give you an angle measure in degrees.
   $$0.05 \cdot 360° = 18°$$
4. Use a compass to draw a circle. Mark the center and use a straightedge to draw a line from the center to the edge of the circle.

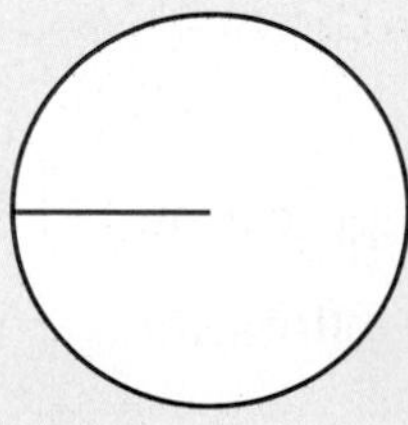

5. Use your protractor to draw an angle measuring 18°. The vertex of the angle will be the center of the circle, and one side will be the line that you drew. The section formed represents the percent of students who prefer science programs.

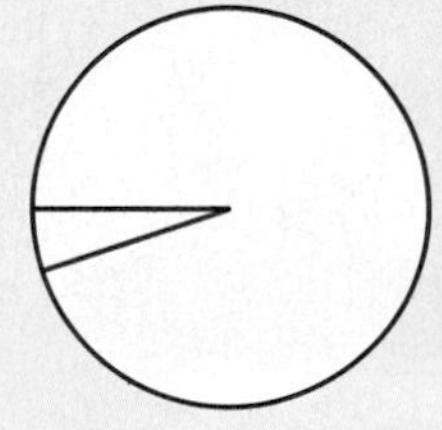

6. Repeat 2 through 5 for each type of program. Label each section, and give the graph a title.

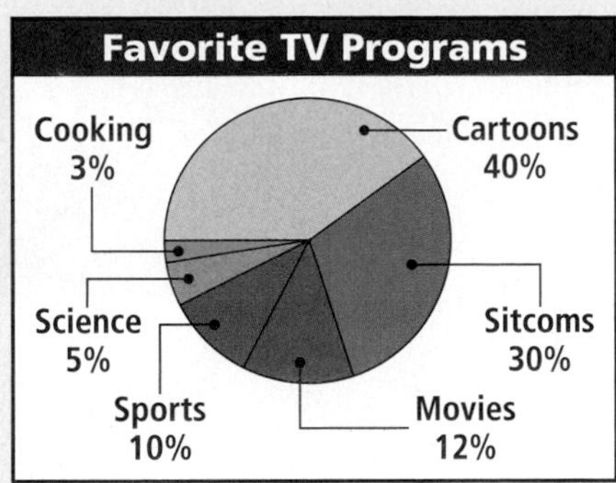

Think and Discuss

1. Looking at your circle graph, discuss five pieces of information you have learned about the TV habits of students at Mazel Middle School.
2. What does the whole circle represent?
3. Why do you need to know that there are 360° in a circle?
4. How does the size of each section of your circle graph relate to the percent that it represents?

Try This

1. People at a mall were surveyed about their favorite pets. Make a circle graph to display the results of the survey.

Favorite Pets	
Type of Pet	**Number of People**
Dog	225
Fish	150
Bird	112
Cat	198
Other	65

2. Collect data from your classmates about their favorite colors. Use the data to make a circle graph with no more than five sections.
3. The circle graph shows the results of a survey about what people in the United States like to eat for breakfast. If this survey included 1,500 people, how many people said they like to eat cereal for breakfast?

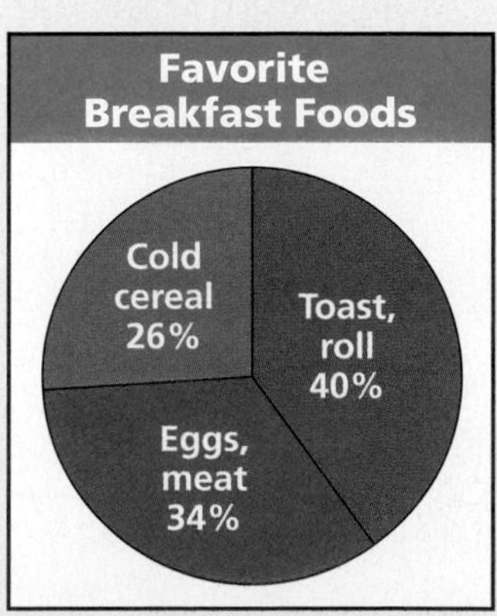

8-10 Using Percents

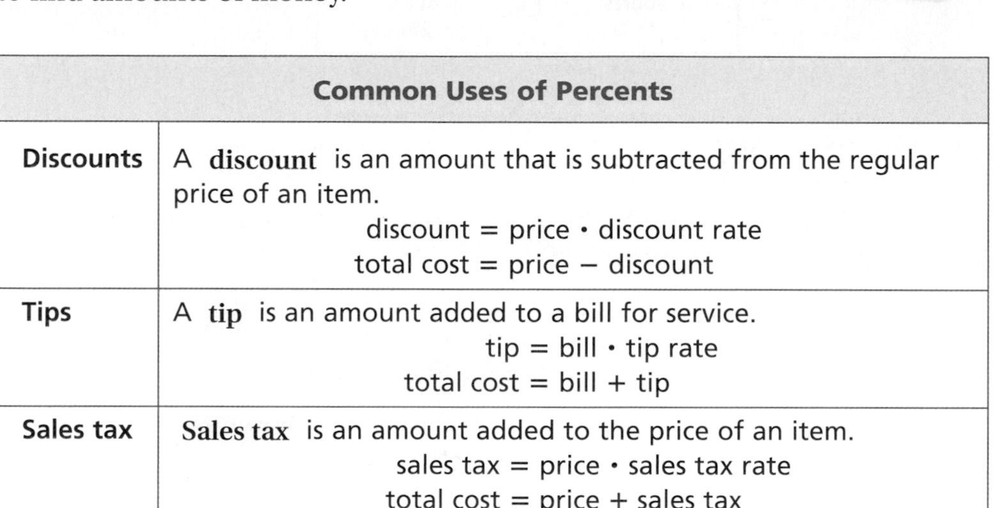

Learn to solve percent problems that involve discounts, tips, and sales tax.

Vocabulary

discount

tip

sales tax

Percents show up often in daily life. Think of examples that you have seen of percents—sales at stores, tips in restaurants, and sales tax on purchases. You can estimate percents such as these to find amounts of money.

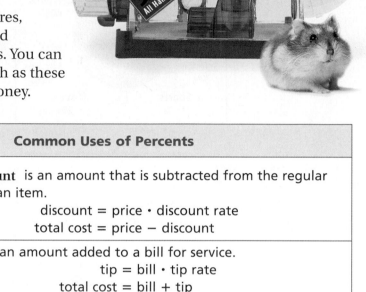

Common Uses of Percents	
Discounts	A **discount** is an amount that is subtracted from the regular price of an item. discount = price • discount rate total cost = price − discount
Tips	A **tip** is an amount added to a bill for service. tip = bill • tip rate total cost = bill + tip
Sales tax	**Sales tax** is an amount added to the price of an item. sales tax = price • sales tax rate total cost = price + sales tax

EXAMPLE 1 Finding Discounts

A music store sign reads "10% off the regular price." If Nichole wants to buy a CD whose regular price is $14.99, about how much will she pay for her CD after the discount?

Step 1: First round $14.99 to $15.

Step 2: Find 10% of $15 by multiplying 0.10 • $15. (**Hint:** Moving the decimal point one place left is a shortcut.)

10% of 15 = 0.10 • $15 = **$1.50**

The approximate discount is $1.50. Subtract this amount from $15.00 to estimate the cost of the CD.

$15.00 − **$1.50** = $13.50

Nichole will pay about $13.50 for the CD.

Remember!

To multiply by 0.10, move the decimal point one place left.

When estimating percents, use percents that you can calculate mentally.

- You can find 10% of a number by moving the decimal point one place to the left.
- You can find 1% of a number by moving the decimal point two places to the left.
- You can find 5% of a number by finding one-half of 10% of the number.

EXAMPLE 2 **Finding Tips**

Leslie's lunch bill is \$13.95. She wants to leave a tip that is 15% of the bill. About how much should her tip be?

Step 1: First round \$13.95 to \$14.

Step 2: Think: 15% = **10%** + 5%

10% of \$14 = 0.10 · \$14 = **\$1.40**

Step 3: 5% = **10%** ÷ 2

= **\$1.40** ÷ 2 = **\$0.70**

Step 4: 15% = **10%** + **5%**

= **\$1.40** + **\$0.70** = \$2.10

Leslie should leave about \$2.10 as a tip.

EXAMPLE 3 **Finding Sales Tax**

Marc is buying a scooter for \$79.65. The sales tax rate is 6%. About how much will the total cost of the scooter be?

Step 1: First round \$79.65 to \$80.

Step 2: Think: 6% = 6 · **1%**

1% of \$80 = 0.01 · \$80 = **\$0.80**

Step 3: 6% = 6 · **1%**

= 6 · **\$0.80** = \$4.80

The approximate sales tax is \$4.80. Add this amount to \$80 to estimate the total cost of the scooter.

\$80 + \$4.80 = \$84.80

Marc will pay about \$84.80 for the scooter.

Think and Discuss

1. **Tell** when it would be useful to estimate the percent of a number.
2. **Explain** how to estimate to find the sales tax of an item.

8-10 Exercises

FOR EOG PRACTICE
see page 671

internet connect
Homework Help Online
go.hrw.com Keyword: MR4 8-10

1.02a

GUIDED PRACTICE

See Example 1

1. Norine wants to buy a beaded necklace that is on sale for 10% off the marked price. If the marked price is $8.49, about how much will the necklace cost after the discount?

See Example 2

2. Alice and Wagner ordered a pizza to be delivered. The total bill was $12.15. They want to give the delivery person a tip that is 20% of the bill. About how much should the tip be?

See Example 3

3. A bicycle sells for $139.75. The sales tax rate is 8%. About how much will the total cost of the bicycle be?

INDEPENDENT PRACTICE

See Example 1

4. Peter has a coupon for 15% off the price of any item in a sporting goods store. He wants to buy a pair of sneakers that are priced at $36.99. About how much will the sneakers cost after the discount?
5. All DVDs are discounted 25% off the original price. The DVD that Marissa wants to buy was originally priced at $24.98. About how much will the DVD cost after the discount?

See Example 2

6. Michael's breakfast bill came to $7.65. He wants to leave a tip that is 15% of the bill. About how much should he leave for the tip?
7. Betty and her family went out for dinner. Their bill was $73.82. Betty's parents left a tip that was 15% of the bill. About how much was the tip that they left?

See Example 3

8. A computer game costs $36.85. The sales tax rate is 6%. About how much will the total cost be for this computer game?
9. Irene is buying party supplies. The cost of her supplies is $52.75. The sales tax rate is 5%. About how much will the total cost of her party supplies be?

PRACTICE AND PROBLEM SOLVING

10. An electronics store is going out of business. The sign on the door reads "All items on sale for 60% off the ticketed price." A computer has a ticketed price of $649, and a printer has a ticketed price of $199. What is the total cost of both items after the discount?
11. Jackie has $32.50 to buy a new pair of jeans. The pair she likes costs $38 but are marked "20% off ticketed price." The sales tax rate is 5%. Does Jackie have enough money to buy the jeans? Explain.

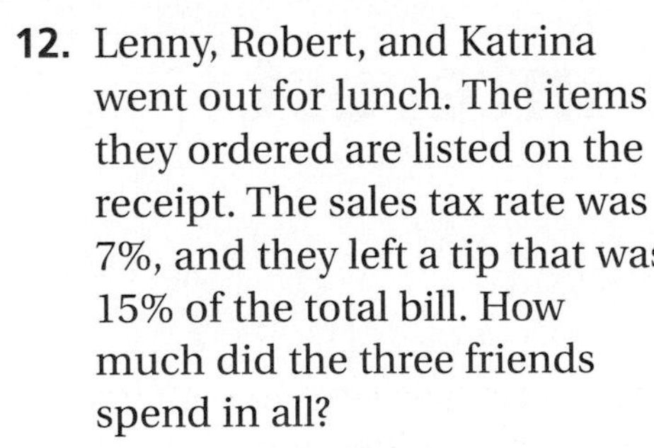

12. Lenny, Robert, and Katrina went out for lunch. The items they ordered are listed on the receipt. The sales tax rate was 7%, and they left a tip that was 15% of the total bill. How much did the three friends spend in all?

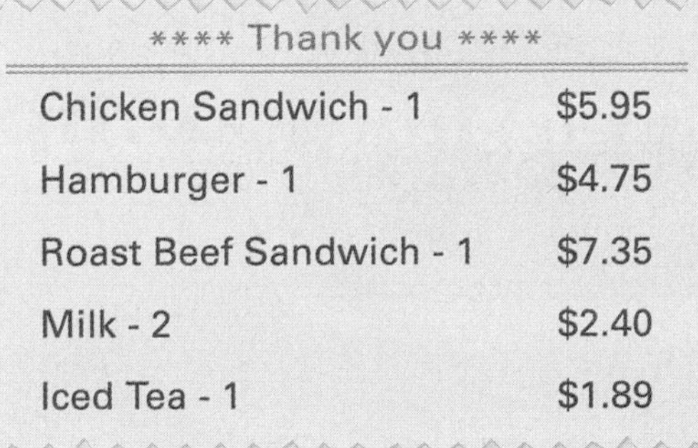

**** Thank you ****

Chicken Sandwich - 1	$5.95
Hamburger - 1	$4.75
Roast Beef Sandwich - 1	$7.35
Milk - 2	$2.40
Iced Tea - 1	$1.89

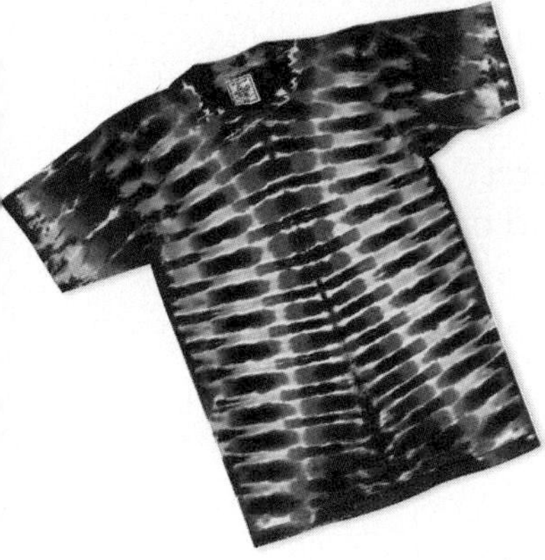

13. ***SOCIAL STUDIES*** The table shows the sales tax rate in some states.

State	Sales Tax Rate
Georgia	4%
Kentucky	6%
New York	4%
North Carolina	4.5%

   a. A shirt costs $18.95. Will the shirt cost more after sales tax in Georgia or in Kentucky? About how much more?

   b. A video game in North Carolina costs $59.75. The same video game in New York costs $60. After sales tax, in which state will the video game cost less? How much less?

14. ***WHAT'S THE ERROR?*** The original price of an item was $48.65. The item was discounted 40%. A customer calculated the price after the discount to be $19.46. What's the error? Give the correct price after the discount.

15. ***WRITE ABOUT IT*** Discuss the difference between a discount, sales tax, and a tip, in relation to the total cost. How does each affect the total cost? Give examples of situations in which each one is used.

16. ***CHALLENGE*** Suppose a jacket is discounted 50% off the original price and then discounted an additional 20%. Is this the same as discounting the jacket 70% off the original price? Explain, and give an example to support your answer.

Spiral Review

Give all the factors of each number. (Lesson 4-2)

17. 20 **18.** 40 **19.** 59 **20.** 85

Find the mean, median, and mode for each set of data. (Lesson 6-2)

21. 23, 24, 25, 22, 23, 28, 30, 21, 20 **22.** 70, 80, 78, 82, 90, 96, 74, 80

Use a protractor to draw an angle with each measure. (Lesson 7-2)

23. 120° **24.** 35° **25.** 90°

26. **EOG PREP** Which fraction is equivalent to 4%? (Lesson 8-7)

A $\frac{2}{5}$ B $\frac{1}{25}$ C $\frac{1}{4}$ D $\frac{4}{1}$

EXTENSION

Simple Interest

Learn to find simple interest.

Vocabulary

interest

principal

simple interest

When you save money in a savings account, you earn money that the bank adds to your account. The added money is called **interest**. The original amount you put into the account is the **principal**. Interest is a percentage of the principal.

One type of interest is called *simple interest.* **Simple interest** is a fixed percentage of the original principal and is often paid over a certain time period. For example, simple interest may be paid once per year or several times per year. In this section, we will assume that simple interest is paid once per year.

Simple Interest

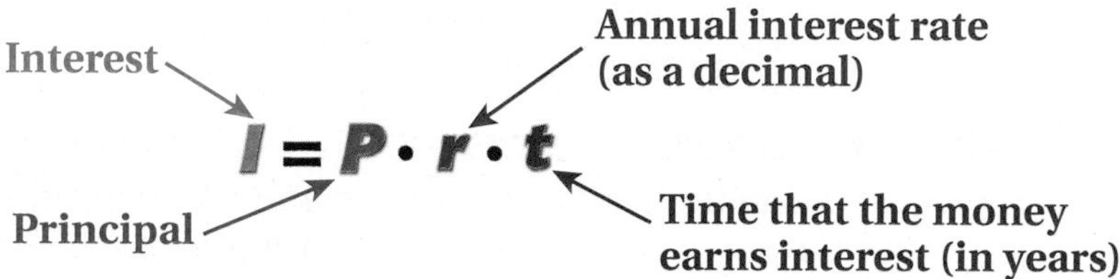

Note that interest rates are usually given as percents, but you must convert the rates to decimals when you use the simple interest formula above.

EXAMPLE 1 Finding Simple Interest

Alyssa put $250 in a savings account at a simple interest rate of 6% per year.

A **If she does not add money to or take money from her account, how much interest will she have earned at the end of 3 years?**

$I = P \cdot r \cdot t$ *P = $250, r = 0.06, t = 3 years*

$I = 250 \cdot 0.06 \cdot 3$ *Multiply.*

$I = \$45$

Alyssa will earn $45 in interest in 3 years.

B **How much money will be in her account after 3 years?**

To find the total amount in Alyssa's account after three years, add the interest to the principal.

$$\$250 + \$45 = \$295$$

Alyssa will have $295 in her account after 3 years.

EXTENSION

Exercises

1. Tamara put $425 in a savings account at a simple interest rate of 7% per year. How much interest will she have earned after 5 years?

2. Jerome put $75 in a savings account at a simple interest rate of 3% per year. How much interest will Jerome have earned after 1 year? How much money will he have in his account after 1 year?

Use the equation $I = P \cdot r \cdot t$ to find the missing amount.

3. principal = $320
   interest rate = 5% per year
   time = 2 years
   interest = ▢

4. principal = $150
   interest rate = 2% per year
   time = 7 years
   interest = ▢

5. principal = ▢
   interest rate = 4% per year
   time = 3 years
   interest = $30

6. principal = $456
   interest rate = 6% per year
   time = ▢
   interest = $109.44

7. Mr. Bruckner is saving to go on a vacation. He put $340 in a savings account at a simple interest rate of 4% per year. How much money will he have in the savings account after 2 years?

8. When you borrow money, the amount borrowed is the principal. Instead of receiving interest, you pay interest on the principal. Kendra borrowed $1,500 from the bank to buy a home computer. The bank is charging her a simple interest rate of 7% per year. How much interest will Kendra owe the bank after 1 year?

9. Mr. Pei paid $7,500 in interest over 20 years at 1% per year on a loan. How much money did he borrow?

10. Hunter put $165 in a savings account at a simple interest rate of 6% per year. Nicholas put $145 in a savings account at a simple interest rate of 7% per year. Who will have earned more interest after 3 years? How much more?

11. ***WRITE ABOUT IT*** Explain the difference between principal and interest.

12. ***WRITE ABOUT IT*** Would you prefer a high or low interest rate when you are borrowing money? When you are saving money? Explain.

13. ***CHALLENGE*** Madison put $200 in a savings account at an interest rate of 5%. Each year the interest is added to the principal, and then the new amount of interest is calculated. If Madison does not add money to or take money out of the account, how much will she have after 3 years?

Problem Solving on Location

NORTH CAROLINA

Sedalia

Charlotte Hawkins Brown

Charlotte Hawkins Brown began teaching in 1901, when she was only 18 years old. When she was 19, she established the Palmer Memorial Institute near Sedalia, North Carolina. The private school grew to include its own farm and 14 buildings on nearly 400 acres. During Dr. Brown's 50-year presidency, the school achieved worldwide recognition as students arrived from more than 40 states and several foreign countries. A museum bearing her name has been established on the site of the historic Palmer Memorial Institute.

Wedding portrait of Charlotte Hawkins Brown in 1911

1. If 3 out of every 5 of the nearly 400 acres were used by the institute to produce its own food, about how many acres were being farmed to feed the students and staff?

2. In planning a spring trip to the C. H. Brown Museum, the teachers want a chaperone-to-student ratio of 1 to 5. If there will be 90 students going, how many chaperones will be needed for the trip?

3. At the Palmer Memorial Institute, meals were served at Kimball Hall. If there were 100 students eating lunch one day and 45% were boys, how many girls were eating lunch that day?

4. The museum is on 40 acres of the approximately 400 acres that the school once occupied. What is the ratio of museum property to school property?

Vegetable Farming

Farmers in North Carolina plant thousands of acres of vegetables each year. After harvest, many bring their produce to farmers' markets and roadside markets throughout the state, where people can buy their produce fresh from the source.

In 2000, 141,300 acres of North Carolina land were planted with vegetables.

For 1–6, use the circle graph.

1. About what percent of the total acres were planted with sweet potatoes? Round your answer to the nearest whole percent.
2. What is the ratio of acres of snap peas to acres of sweet potatoes?
3. About how many acres were planted in 2000 with other vegetables?
4. How many more acres of cabbage were planted than acres of bell peppers?
5. About how many acres of cucumbers were planted in 2000?
6. About what percent of the total acres were planted with other vegetables? Round your answer to the nearest whole percent.

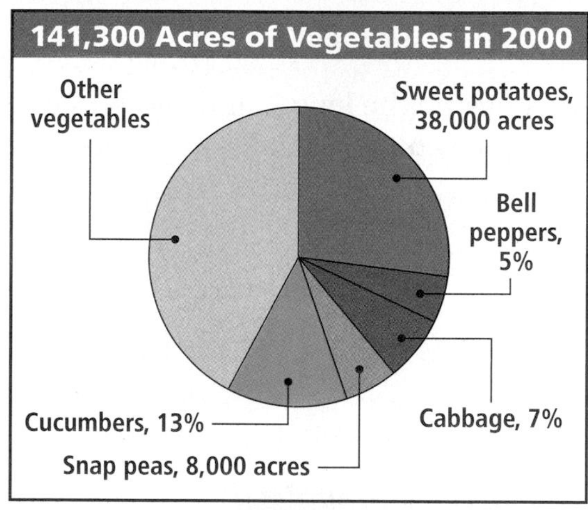

MATH-ABLES

The Golden Rectangle

Which rectangle do you find most visually pleasing?

Did you choose rectangle 3? If so, you agree with artists and architects throughout history. Rectangle 3 is a golden rectangle. Golden rectangles are said to be the most pleasing to the human eye.

In a golden rectangle, the ratio of the length of the longer side to the length of the shorter side is approximately equal to 1.6. In other words,

$$\frac{\text{length of longer side}}{\text{length of shorter side}} \approx \frac{1.6}{1}$$

Measure the length and width of each rectangle below. Which could be golden rectangles? Are they the most pleasing to your eye?

Triple Play

Number of players: 3–5

Deal five cards to each player. Place the remaining cards in a pile facedown. At any time, you may remove *triples* from your hand. A *triple* is a fraction card, a decimal card, and a percent card that are all equivalent.

On your turn, ask any other player for a specific card. For example, if you have the $\frac{3}{5}$ card, you might ask another player if he or she has the 60% card. If so, he or she must give it to you, and you repeat your turn. If not, take the top card from the deck, and your turn is over.

The first player to get rid of his or her cards is the winner.

internet connect

Go to ***go.hrw.com*** for a complete set of rules and game pieces.
KEYWORD: MR4 Game8

Fractions, Decimals, and Percents

internet connect
Lab Resources Online
go.hrw.com
KEYWORD: MR4 TechLab8

You can use your calculator to quickly change between fractions, decimals, and percents.

Activity

1. To write a decimal as a fraction on a graphing calculator, use the **FRAC** command from the **MATH** menu.

   Find the fraction equivalent of 0.225 by pressing 0.225 MATH 1 ENTER.

2. To write a percent as a fraction, first write the percent as a fraction whose denominator is 100. Then use the **FRAC** command to find the simplest form of the fraction.

   Find the fraction equivalent of 65% by pressing 65 ÷ 100 MATH 1 ENTER.

3. To write a fraction as a percent, multiply the fraction by 100.

   Find the percent equivalent of $\frac{11}{25}$ by pressing 11 ÷ 25 × 100 ENTER.

   $\frac{11}{25} = 44\%$

Think and Discuss

1. Use the **FRAC** command on a graphing calculator to find the fraction equivalent of 0.1428571429 by pressing 0.1428571429 MATH 1 ENTER. Describe what happens.

Try This

1. Write each percent as a fraction.

   a. 57.5% **b.** 32.5% **c.** 3.25% **d.** 1.65% **e.** 81.25%

2. Write each fraction as a percent.

   a. $\frac{7}{40}$ **b.** $\frac{3}{8}$ **c.** $\frac{19}{25}$ **d.** $\frac{3}{16}$ **e.** $\frac{17}{20}$

Chapter 8

Study Guide and Review

Vocabulary

Complete the sentences below with vocabulary words from the list above. Words may be used more than once.

1. A(n) __?__ is an amount subtracted from the regular price of an item.
2. The ratios 4:3 and 8:6 are __?__ because they name the same comparison.
3. A __?__ is a ratio of a number to 100.
4. In similar figures, __?__ are congruent.

8-1 Ratios and Rates (pp. 392–395)

EXAMPLE

■ **Write the ratio of hearts to diamonds.**

$\frac{\text{hearts}}{\text{diamonds}} = \frac{4}{8}$

EXERCISES

5. Write three equivalent ratios for 4:8.
6. Which is the better deal—an 8 oz package of pretzels for $1.92 or a 12 oz package of pretzels for $2.64?

8-2 Proportions (pp. 398–401)

EXAMPLE

■ **Find the value of n in $\frac{5}{6} = \frac{n}{12}$.**

$6 \cdot n = 5 \cdot 12$ *Cross products are equal.*

$\frac{6n}{6} = \frac{60}{6}$ *Divide both sides by 6.*

$n = 10$

EXERCISES

Find the value of n in each proportion.

7. $\frac{3}{5} = \frac{n}{15}$
8. $\frac{1}{n} = \frac{3}{9}$
9. $\frac{7}{8} = \frac{n}{16}$
10. $\frac{n}{4} = \frac{8}{16}$

Study Guide and Review

8-3 Proportions and Customary Measurement (pp. 402–404)

EXAMPLE

- **How many inches are in 5 feet?**

$\frac{12 \text{ in.}}{1 \text{ ft}} = \frac{x \text{ in.}}{5 \text{ ft}}$ *1 foot is 12 inches.*

$1 \cdot x = 12 \cdot 5$ *Cross products are equal.*

$x = 60$ There are 60 in. in 5 ft.

EXERCISES

11. Marc made 3 gallons of punch. How many cups of punch did he make?
12. Pam spent 150 minutes researching a project. How many hours is this?

8-4 Similar Figures (pp. 405–408)

EXAMPLE

- **The triangles are similar. Find b.**

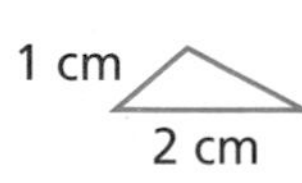

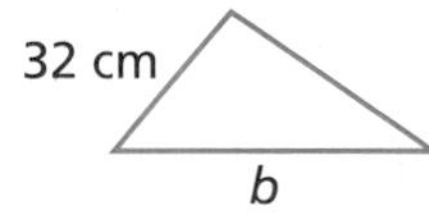

$\frac{1}{32} = \frac{2}{b}$ *Write a proportion.*

$32 \cdot 2 = 1 \cdot b$ *Cross products are equal.*

$64 \text{ cm} = b$

EXERCISES

13. The shapes are similar. Find n and m$\angle A$.

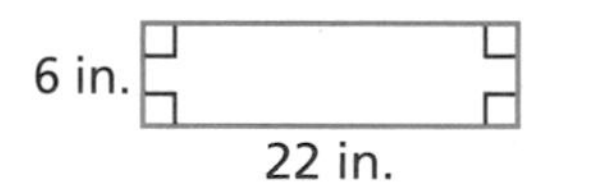

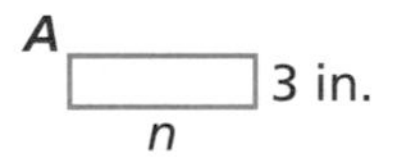

8-5 Indirect Measurement (pp. 409–411)

EXAMPLE

- **A tree casts a 12 ft shadow when a 6 ft man casts a 4 ft shadow. How tall is the tree?**

$\frac{h}{6} = \frac{12}{4}$ *Write a proportion.*

$6 \cdot 12 = 4 \cdot h$ *The cross products are equal.*

$\frac{72}{4} = \frac{4h}{4}$ *Divide both sides by 4.*

$18 = h$ The tree is 18 ft tall.

EXERCISES

14. Find the height of the building.

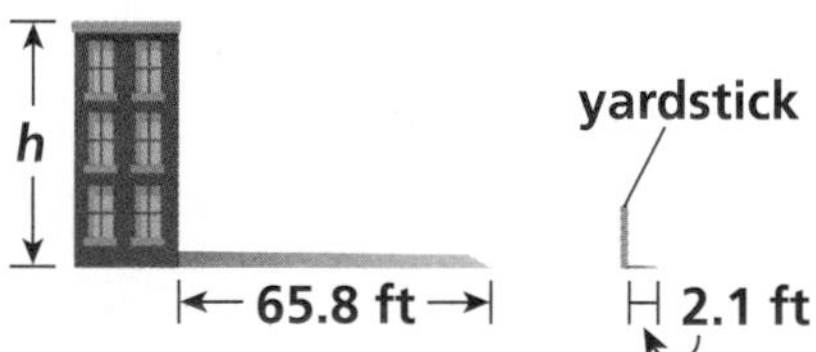

8-6 Scale Drawings and Maps (pp. 412–415)

EXAMPLE

- **Find the actual distance from A to B.**

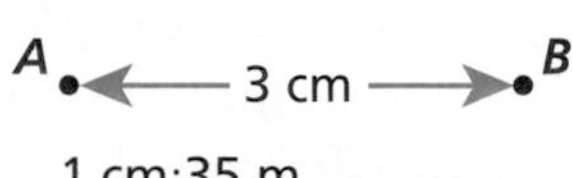

$\frac{1 \text{ cm}}{35 \text{ m}} = \frac{3 \text{ cm}}{x \text{ m}}$ *Write a proportion.*

$35 \cdot 3 = 1 \cdot x$ *Cross products are equal.*

$105 = x$

The actual distance is 105 m.

EXERCISES

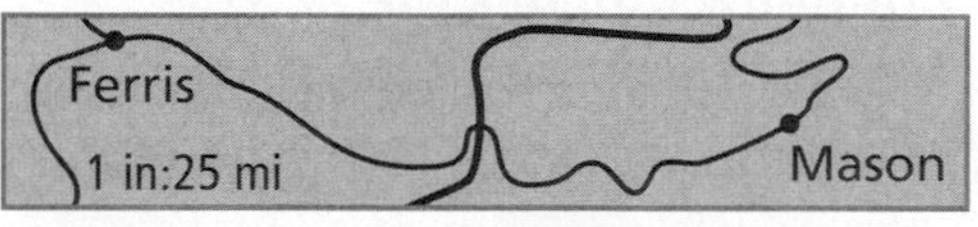

15. Find the actual distance from Ferris to Mason.
16. Renfield is 75 mi from Mason. About how far apart should Renfield and Mason be on the map?

Study Guide and Review

8-7 Percents (pp. 418–421)

EXAMPLE

- **Write 48% as a fraction in simplest form.**

$48\% = \frac{48}{100}$ $\frac{48 \div 4}{100 \div 4} = \frac{12}{25}$

EXERCISES

Write each as a fraction in simplest form.

17. 75% **18.** 6% **19.** 30%

Write each percent as a decimal.

20. 8% **21.** 65% **22.** 20%

8-8 Percents, Decimals, and Fractions (pp. 422–425)

EXAMPLES

- **Write 0.365 as a percent.**

0.365 = 36.5% *Multiply by 100.*

- **Write $\frac{3}{5}$ as a percent.**

$\frac{3 \cdot 20}{5 \cdot 20} = \frac{60}{100} = 60\%$

EXERCISES

Write each decimal or fraction as a percent.

23. 0.896 **24.** 0.70 **25.** 0.057

26. 0.12 **27.** $\frac{7}{10}$ **28.** $\frac{3}{12}$

29. $\frac{7}{8}$ **30.** $\frac{4}{5}$ **31.** $\frac{1}{16}$

8-9 Percent Problems (pp. 426–429)

EXAMPLE

- **Find 30% of 85.**

30% = 0.30 *Write 30% as a decimal.*

0.30 · 85 = 25.5 *Multiply.*

EXERCISES

32. Find 25% of 48. **33.** Find 33% of 18.

34. A total of 325 tickets were sold for the school concert, and 36% of these were sold to students. How many tickets were sold to students?

8-10 Using Percents (pp. 432–435)

EXAMPLE

- **A DVD costs $24.98. The sales tax is 5%. About how much is the tax?**

Step 1: Round $24.98 to $25.

Step 2: 5% = 5 · 1%
1% of $25 = 0.01 · $25 = **$0.25**

Step 3: 5% = 5 · 1%
= 5 · **$0.25** = $1.25

The tax is about $1.25.

EXERCISES

35. A sweater is marked 40% off the original price. The original price was $31.75. About how much is the sweater after the discount?

36. Barry and his friends went out for lunch. The bill was $28.68. About how much should they leave for a 15% tip?

37. Ana is purchasing a book for $17.89. The sales tax rate is 6%. About how much will she pay in sales tax?

Study Guide and Review

Chapter Test

Use the table to write the ratios.

Types of Videos in Richard's Collection			
Comedy	5	Cartoon	7
Drama	6	Exercise	3
Music	3	Documentary	2

1. three equivalent ratios to compare dramas to documentaries
2. documentaries to total videos
3. music videos to exercise videos
4. Which is a better deal—5 videos for \$29.50 or 3 videos for \$17.25?

Find the value of n in each proportion.

5. $\frac{5}{6} = \frac{n}{24}$
6. $\frac{8}{n} = \frac{12}{3}$
7. $\frac{n}{10} = \frac{3}{6}$
8. $\frac{3}{9} = \frac{4}{n}$
9. A cocoa recipe calls for 4 tbsp cocoa mix to make an 8 oz serving. How many tbsp of cocoa mix are needed to make a 15 oz serving?
10. Among states that have a shoreline, Pennsylvania has the shortest one. It is 89 miles long. How many yards long is it?
11. A 3-foot-tall mailbox casts a shadow that is 1.8 feet long. At the same time, a nearby street lamp casts a shadow that is 12 feet long. How tall is the street lamp?

Use the scale drawing for Exercises 12 and 13.

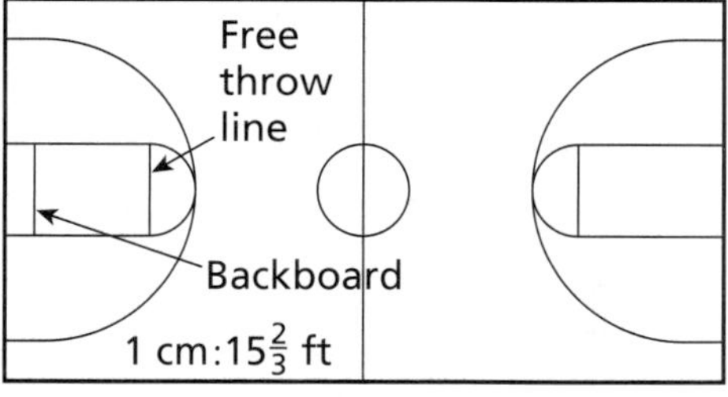

12. The length of the court in the drawing is 6 cm. How long is the actual court?
13. The free-throw line is always 15 feet from the backboard. Is the distance between the backboard and the free-throw line correct in the drawing? Explain.

Write each percent as a fraction in simplest form and as a decimal.

14. 66%
15. 90%
16. 5%
17. 18%

Write each decimal or fraction as a percent.

18. 0.546
19. 0.092
20. $\frac{14}{25}$
21. $\frac{1}{8}$

Find each percent.

22. 55% of 218
23. 30% of 310
24. 25% of 78
25. A bookstore sells paperback books at 20% off the listed price. If Brandy wants to buy a paperback book whose listed price is \$12.95, about how much will she pay for the book after the discount?

Chapter 8 Performance Assessment

Show What You Know

Create a portfolio of your work from this chapter. Complete this page and include it with your four best pieces of work from Chapter 8. Choose from your homework or lab assignments, mid-chapter quiz, or any journal entries you have done. Put them together using any design you want. Make your portfolio represent what you consider your best work.

Short Response

1. Find the unit rate for each of the following. Show your work. Which is the best deal?

3 for \$7.80 | 5 for \$13.25 | 10 for \$25.00 | 8 for \$19.84

2. Georgette purchased a sweater on sale for \$19.60. The original price was \$28.00. What percent of the original price was discounted? Show your work.

Extended Problem Solving

3. The small purple rectangle in the drawing is 8 millimeters wide and 18 millimeters tall. The larger purple rectangle is 18 millimeters wide and 25 millimeters tall.

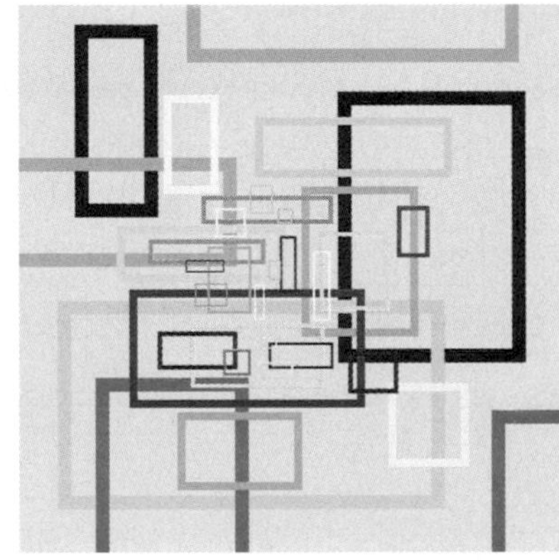

a. Are the two purple rectangles similar? Explain your answer.

b. A third rectangle is similar to the smaller purple rectangle. The width of the third rectangle is 14 millimeters. Let x represent the length of the third rectangle. Write an equation that could be used to find x.

c. Find the length of the third rectangle. Show your work.

Performance Assessment

internet connect

EOG Test Practice Online
go.hrw.com Keyword: MR4 TestPrep

Getting Ready for EOG

Chapter 8

Cumulative Assessment, Chapters 1–8

1. A quadrilateral whose opposite sides are parallel and that has four right angles is a __?__.
   A Rectangle
   B Rhombus
   C Parallelogram
   D Trapezoid

TEST TAKING TIP!

Estimate the correct answer before solving. You can often use your estimate to eliminate answer choices.

2. Find 8% of 215.
   A 1,720 C 17.2
   B 172 D 1.72

3. Which is an obtuse angle?

A 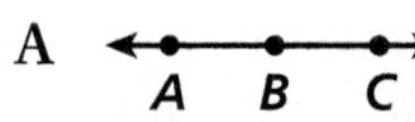

C 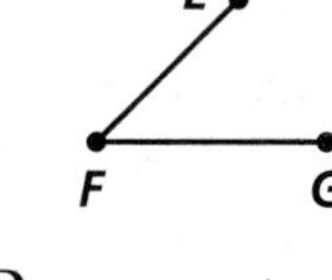

B 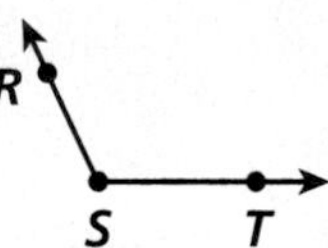

D 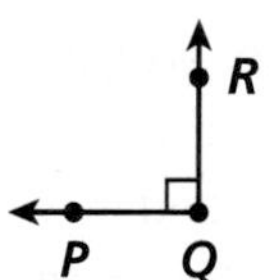

4. $3\frac{2}{7} + 2\frac{2}{3}$
   A $5\frac{2}{5}$ C $5\frac{4}{7}$
   B $5\frac{20}{21}$ D $5\frac{4}{21}$

5. Which is greatest?
   A $\frac{3}{8}$ C $\frac{2}{5}$
   B 5% D 0.5

6. Find the GCF of 6, 18, and 30.
   A 30 C 6
   B 18 D 3

7. Which ratio could be used to compare the number of triangles to the number of circles in the pattern?

▲▲▲●▲▲●●▲▲▲●

   A 3:9 C 2:1
   B 3:6 D 3:1

8. Use your ruler to help you find the actual distance between R and T.
   A $\frac{3}{5}$ mile C 5 miles
   B 3 miles D 15 miles

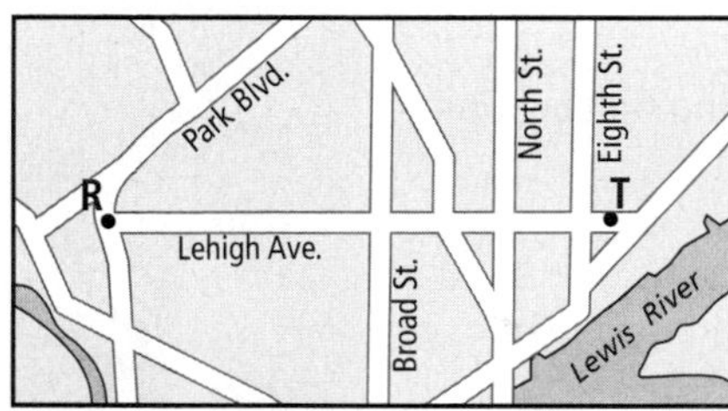

Scale 1 in:10 mi

9. ***SHORT RESPONSE*** Show how to write $\frac{3}{5}$ as a decimal and as a percent.

10. ***SHORT RESPONSE*** Marta purchased 12 gallons of gas for $13.08. Explain the term *unit rate* and give the unit rate for Marta's purchase.

Chapter 9

Integers

TABLE OF CONTENTS

internet connect

Chapter Opener Online
go.hrw.com
KEYWORD: MR4 Ch9

Continent	Highest Point (m)		Lowest Point (m)	
Africa	Mt. Kilimanjaro:	5,895	Lake Assal:	–156
Antarctica	Vinson Massif:	4,897	Bentley Subglacial Trench:	–2,538
Asia	Mt. Everest:	8,850	Dead Sea:	–411
Australia	Mt. Kosciusko:	2,228	Lake Eyre:	–12
Europe	Mt. Elbrus:	5,642	Caspian Sea:	–28
North America	Mt. McKinley:	6,194	Death Valley:	–86
South America	Mt. Aconcagua:	6,960	Valdes Peninsula:	–40

Career ***Geographer***

Geographers are interested in characteristics of our natural world, such as landforms, natural resources, and climate. Some geographers spend time in the field collecting information. Others create maps, charts, and graphs. Geographers use integers to express information such as high and low temperatures and elevations above and below sea level. The table lists the highest and lowest points on each continent.

ARE YOU READY?

Choose the best term from the list to complete each sentence.

1. When you __?__ a numerical expression, you find its value.
2. __?__ are the set of numbers 0, 1, 2, 3, 4, ...
3. A(n) __?__ is an exact location in space.
4. A(n) __?__ is a mathematical statement that two quantities are equal.

equation
evaluate
exponents
less than
point
whole numbers

Complete these exercises to review skills you will need for this chapter.

✔ Compare Whole Numbers

Write <, >, or = to compare the numbers.

5. 9 ■ 2	**6.** 4 ■ 5	**7.** 8 ■ 1	**8.** 3 ■ 3
9. 412 ■ 214	**10.** 1,076 ■ 1,074	**11.** 502 ■ 520	**12.** 9,123 ■ 9,001

✔ Whole Number Operations

Add or subtract.

13. $8 + 3$	**14.** $10 - 2$	**15.** $7 + 6$	**16.** $15 - 8$
17. $129 + 30$	**18.** $32 - 25$	**19.** $72 + 93$	**20.** $120 - 87$

Multiply or divide.

21. $3 \cdot 9$	**22.** $16 \div 4$	**23.** $6 \cdot 7$	**24.** $25 \div 5$
25. $119 \cdot 5$	**26.** $156 \div 6$	**27.** $249 \cdot 44$	**28.** $275 \div 25$

✔ Graph Ordered Pairs

Graph each ordered pair.

29. (1, 3)	**30.** (0, 5)	**31.** (3, 2)	**32.** (4, 0)
33. (6, 4)	**34.** (2, 5)	**35.** (0, 1)	**36.** (1, 0)

✔ Evaluate Expressions

Evaluate $n + 4$ for each value of n.

37. $n = 10$	**38.** $n = 5$	**39.** $n = 16$	**40.** $n = 27$
41. $n = 0$	**42.** $n = 4$	**43.** $n = 19$	**44.** $n = 33$

9-1 Understanding Integers

Learn to identify and graph integers, find opposites, and find the absolute value of an integer.

Vocabulary

positive number

negative number

opposites

integer

absolute value

The highest temperature recorded in the United States is 134°F, in Death Valley, California. The lowest recorded temperature is 80° below 0°F, in Prospect Creek, Alaska.

Positive numbers are greater than 0. They may be written with a positive sign (+), but they are usually written without it. So, the highest temperature can be written as +134°F or 134°F.

Negative numbers are less than 0. They are always written with a negative sign (−). So, the lowest temperature is written as −80°F.

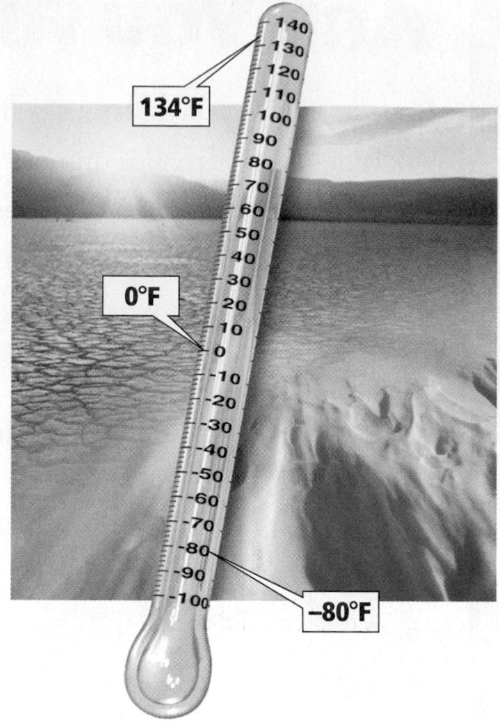

EXAMPLE 1 Identifying Positive and Negative Numbers in the Real World

Name a positive or negative number to represent each situation.

A **a gain of 20 yards in football**

Positive numbers can represent *gains* or *increases.*

+20

B **spending $75**

Negative numbers can represent *losses* or *decreases.*

−75

C **10 feet below sea level**

Negative numbers can represent values *below* or *less than* a certain value.

−10

You can graph positive and negative numbers on a number line.

On a number line, **opposites** are the same distance from 0 but on different sides of 0.

Integers are the set of all whole numbers and their opposites.

Remember!

The set of whole numbers includes zero and the counting numbers. {0, 1, 2, 3, 4, ...}

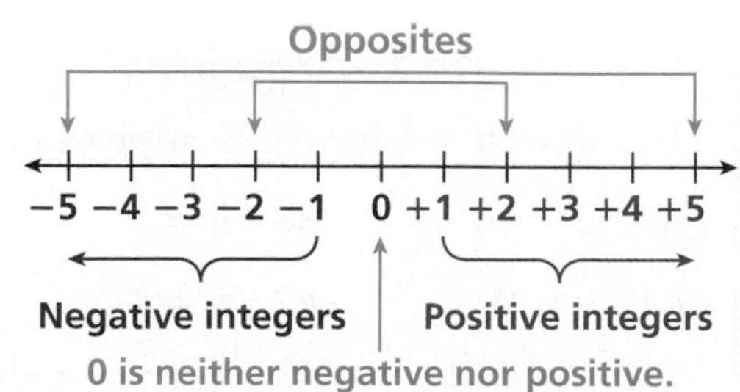

EXAMPLE 2 Graphing Integers

Graph each integer and its opposite on a number line.

A −4

−5 −4 −3 −2 −1 0 +1 +2 +3 +4 +5

+4 is the same distance from 0 as −4.

B +3

−4 −3 −2 −1 0 +1 +2 +3 +4

−3 is the same distance from 0 as +3.

C 0

−4 −3 −2 −1 0 +1 +2 +3 +4

Zero is its own opposite.

The **absolute value** of an integer is its distance from 0 on a number line. The symbol for absolute value is $| \ |$.

Reading Math

Read $|3|$ as "the absolute value of 3."

Read $|-3|$ as "the absolute value of negative 3."

$|-3| = 3 \quad |3| = 3$

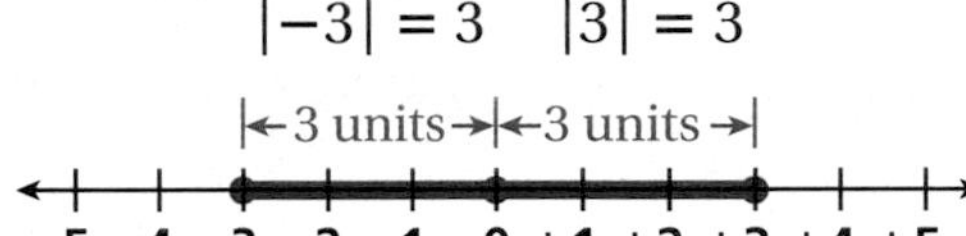

- Absolute values are never negative.
- Opposite integers have the same absolute value.
- $|0| = 0$

EXAMPLE 3 Finding Absolute Value

Use the number line to find the absolute value of each integer.

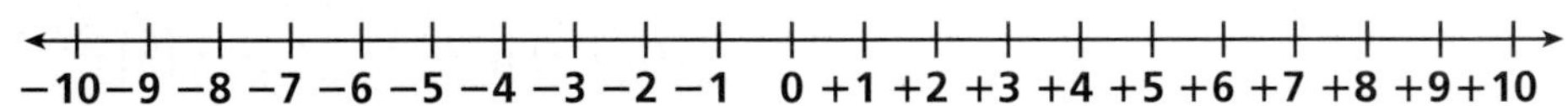

A $|5|$

5 — *5 is 5 units from 0, so $|5| = 5$.*

B $|-7|$

7 — *−7 is 7 units from 0, so $|-7| = 7$.*

Think and Discuss

1. Is –3.2 an integer? Why or why not?

2. Give the opposite of 14. What is the opposite of –11?

3. Name all the integers with an absolute value of 12.

9-1 Exercises

FOR EOG PRACTICE
see page 672

internet connect
Homework Help Online
go.hrw.com Keyword: MR4 9-1

1.01a

GUIDED PRACTICE

See Example 1 Name a positive or negative number to represent each situation.

1. an increase of 5 points
2. a loss of 15 yards

See Example 2 Graph each integer and its opposite on a number line.

3. −2
4. 1
5. −6
6. 9

See Example 3 Use the number line to find the absolute value of each integer.

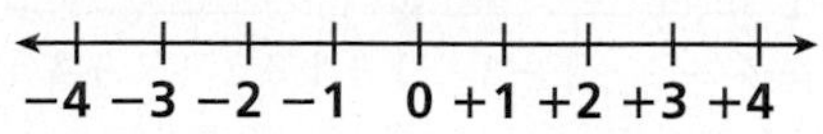

7. $|-3|$
8. $|4|$
9. $|1|$
10. $|2|$

INDEPENDENT PRACTICE

See Example 1 Name a positive or negative number to represent each situation.

11. earning $50
12. 20° below zero
13. 7 feet above sea level
14. a decrease of 39 points

See Example 2 Graph each integer and its opposite on a number line.

15. −5
16. 6
17. 2
18. −3

See Example 3 Use the number line to find the absolute value of each integer.

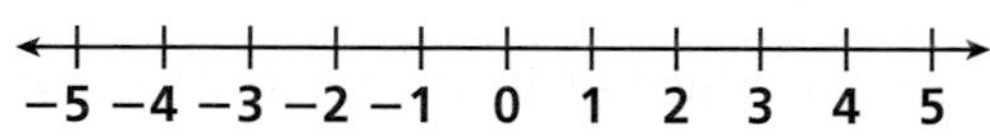

19. $|-4|$
20. $|-1|$
21. $|3|$
22. $|-2|$

PRACTICE AND PROBLEM SOLVING

Write a situation that each integer could represent.

23. +49
24. −83
25. −7
26. +15

Write the opposite of each integer.

27. −92
28. +75
29. −25
30. 0

Find the absolute value.

31. $|419|$
32. $|-189|$
33. $|723|$
34. $|-806|$
35. $|35|$
36. $|150|$
37. $|-295|$
38. $|-80|$

Underwater vehicles called submersibles are used to explore the deepest parts of the ocean. The *Alvin*, pictured above, was used to photograph and explore the *Titanic* in 1987.

39. ***EARTH SCIENCE*** The Mariana Trench is the deepest part of the Pacific Ocean, reaching a depth of 10,924 meters. Write the depth in meters of the Mariana Trench as an integer.

40. ***SPORTS*** When the Mountain Lions football team returned the kickoff, they gained 45 yards. Write an integer to represent this situation.

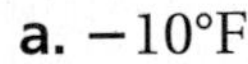

41. ***EARTH SCIENCE*** From June 21 to December 21, most of the United States loses 1 to 2 minutes of daylight each day. But on December 21, most of the country begins to gain 1 to 2 minutes of daylight each day. What integer could you write for a gain of 2 minutes? a loss of 2 minutes?

42. Match each temperature with the correct point on the thermometer.

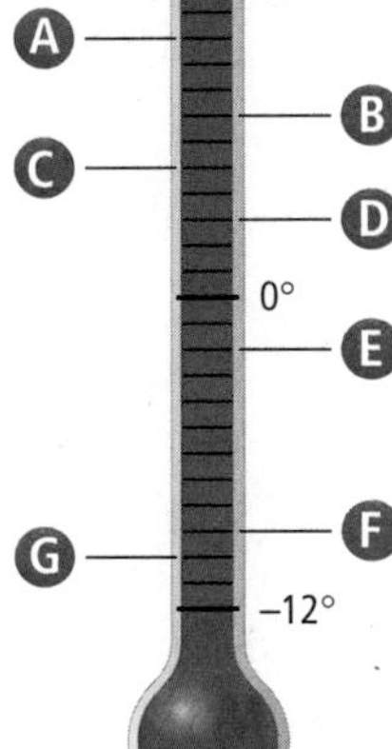

a. $-10°F$
b. 5°F
c. 10°F
d. $-2°F$
e. $-9°F$
f. 7°F
g. 3°F

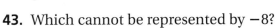

43. Which cannot be represented by -8?
A a temperature drop of 8°F
B a depth of 8 meters
C a growth of 8 centimeters
D a time 8 years ago

44. ***WRITE ABOUT IT*** Why do opposites have the same absolute value?

45. ***CHALLENGE*** What is the value of $|19 - 2|$? of $|2 - 11|$?

Spiral Review

Order the numbers in each set from least to greatest. (Lesson 1-1)

46. 1,945; 2,649; 1,495; 2,609

47. 17,465; 17,509; 17,395; 17,498

Solve each equation. (Lesson 2-4)

48. $n + 10 = 25$

49. $28 = 4 + x$

50. **EOG PREP** Which fraction is **not** in simplest form? (Lesson 4-5)

A $\frac{2}{3}$ B $\frac{17}{31}$ C $\frac{3}{9}$ D $\frac{1}{8}$

9-2 Comparing and Ordering Integers

Learn to compare and order integers.

The table shows three golfers' scores from a 2001 tournament.

Player	Score
David Berganio	+6
Sergio Garcia	−16
Tiger Woods	−4

In golf, the player with the lowest score wins the game. You can compare integers to find the winner of the tournament.

Sergio Garcia

EXAMPLE 1 Comparing Integers

Use the number line to compare each pair of integers. Write < or >.

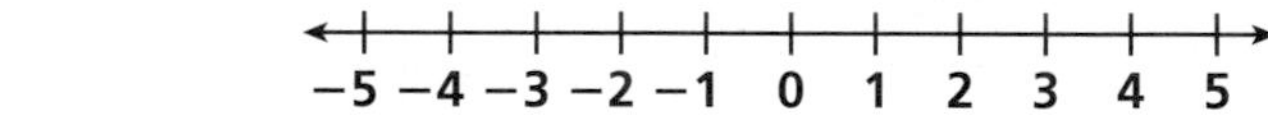

Remember!

Numbers on a number line increase in value as you move from left to right.

A $-4 \ \square \ 2$

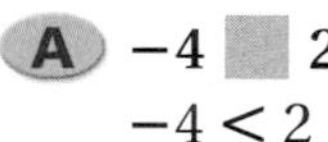

$-4 < 2$ *−4 is to the left of 2 on the number line.*

B $-3 \ \square \ -5$

$-3 > -5$ *−3 is to the right of −5 on the number line.*

C $0 \ \square \ -4$

$0 > -4$ *0 is to the right of −4 on the number line.*

EXAMPLE 2 Ordering Integers

Order the integers in each set from least to greatest.

A 4, −2, 1

Graph the integers on the same number line.

−5 −4 −3 −2 −1 0 1 2 3 4 5

Then read the numbers from left to right: −2, 1, 4.

Order the integers in each set from least to greatest.

B $-2, 0, 2, -5$

Graph the integers on the same number line.

Then read the numbers from left to right: $-5, -2, 0, 2$.

EXAMPLE 3 PROBLEM SOLVING APPLICATION

PROBLEM SOLVING

At a 2001 golf tournament, David Berganio scored +6, Sergio Garcia scored −16, and Tiger Woods scored −4. One of these three players was the winner of the tournament. Who won the tournament?

1 Understand the Problem

The **answer** will be the player with the *lowest* score.
List the **important information:**

- David Berganio scored +6.
- Sergio Garcia scored −16.
- Tiger Woods scored −4.

2 Make a Plan

You can draw a diagram to order the scores from least to greatest.

3 Solve

Draw a number line and graph each player's score on it.

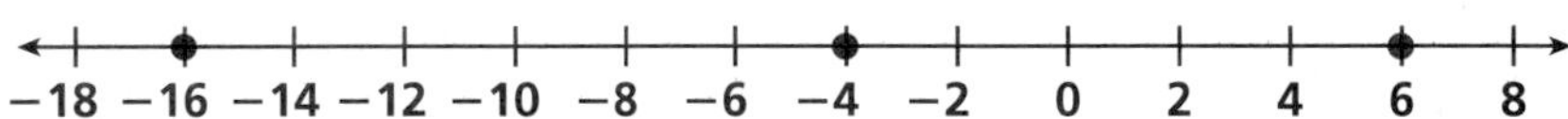

Sergio Garcia's score, –16, is farthest to the left, so it is the lowest score. Sergio Garcia won this tournament.

4 Look Back

Negative integers are always less than positive integers, so David Berganio cannot be the winner. Since Sergio Garcia's score of −16 is less than Tiger Woods's score of −4, Sergio Garcia won.

Sports LINK

In golf, *par* is the standard number of strokes needed to hit the ball into the hole. A score of −12 means "12 strokes under par." A score of +2 means "2 strokes over par."

go.hrw.com
KEYWORD: MR4 Golf

CNN Student News

Think and Discuss

1. **Tell** which is greater, a negative or a positive integer. Explain.
2. **Tell** which is greater, 0 or a negative integer. Explain.
3. **Explain** how to tell which of two negative integers is greater.

9-2 Exercises

FOR EOG PRACTICE

see page 672

internet connect

Homework Help Online
go.hrw.com Keyword: MR4 9-2

1.01b

GUIDED PRACTICE

See Example 1 **Use the number line to compare each pair of integers. Write < or >.**

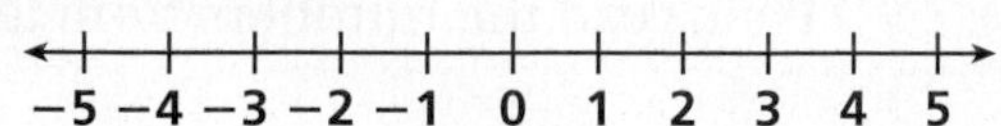

1. −4 ▒ −5
2. −2 ▒ 0
3. −1 ▒ 3

See Example 2 **Order the integers in each set from least to greatest.**

4. 9, 0, −2
5. 7, −4, 3, −5
6. 8, −6, −1, 10

See Example 3

7. At what time was the temperature the lowest?

Time	Temperature (°F)
10:00 P.M.	1
Midnight	−4
3:30 A.M.	−6
6:00 A.M.	1

INDEPENDENT PRACTICE

See Example 1 **Use the number line to compare each pair of integers. Write < or >.**

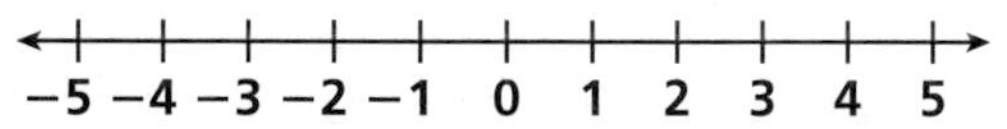

8. 0 ▒ 2
9. 4 ▒ −4
10. −3 ▒ −1

See Example 2 **Order the integers in each set from least to greatest.**

11. 11, −6, −3
12. 15, −8, 7
13. 5, −12, 0, 1
14. −9, 13, −1, −16
15. 24, −6, 7, −10, 4
16. 22, 0, −19, 8, −3

See Example 3

17. The table shows the depths of the world's three largest oceans. Which ocean is the deepest?

Ocean	Depth (ft)
Pacific	−36,200
Atlantic	−30,246
Indian	−24,442

PRACTICE AND PROBLEM SOLVING

Compare. Write < or >.

18. −30 ▒ 25
19. 0 ▒ −49
20. −16 ▒ −51
21. −64 ▒ −15
22. 77 ▒ 300
23. −28 ▒ 1

Order the integers in each set from least to greatest.

24. −39, 14, 21
25. −18, −9, −31
26. 0, −26, 43, −12
27. 15, −25, −4, 31
28. −67, 82, −73, −10, 20
29. 42, −27, 69, −50, 38

30. Which set of integers is written in order from greatest to least?

A $0, -4, -3, -1$

B $9, -9, -10, -15$

C $2, -4, 8, -16$

D $-8, -7, -6, -5$

31. ***EARTH SCIENCE*** The normal high temperature in January for Barrow, Alaska, is −7°F. The normal high temperature in January for Los Angeles is 68°F. Compare the two temperatures using $<$ or $>$.

32. ***GEOGRAPHY*** The table shows elevations for several natural features. Write the features in order from the least elevation to the greatest elevation.

Elevations of Natural Features	
Mt. Everest	29,022 ft
Mt. Rainier	14,410 ft
Kilimanjaro	19,000 ft
San Augustin Cave	–2,189 ft
Dead Sea	–1,296 ft

33. ***WHAT'S THE ERROR?*** Your classmate says that $0 < -91$. Explain why this is incorrect.

34. ***WRITE ABOUT IT*** Explain how you would order from least to greatest three numbers that include a positive number, a negative number, and zero.

35. ***CHALLENGE*** Write $<$ or $>$. $|-4|$ ▒ $|-3|$

Spiral Review

Evaluate each expression. (Lesson 1-4)

36. $2 \cdot (17 - 5) \div 4$

37. $(11 + 7) \div 3 - 5$

38. $3^2 \div 3 + (5 \cdot 3)$

39. $8 \cdot (14 \div 2) - 2^3$

40. $8^2 + (36 \div 4) - 5$

41. $4^2 \div (5 - 3) - 8$

Divide. (Lesson 3-7)

42. $1.40 \div 2$

43. $3.3 \div 3$

44. $0.85 \div 5$

45. $0.375 \div 3$

46. **EOG PREP** An isosceles triangle has at least ___?___ congruent sides. (Lesson 7-5)

A zero

B one

C two

D three

47. **EOG PREP** What is 35% written as a decimal? (Lesson 8-8)

A 35.0

B 3.5

C 0.035

D 0.35

9-3 The Coordinate Plane

Learn to locate and graph points on the coordinate plane.

Vocabulary

coordinate plane
axes
x-axis
y-axis
quadrants
origin
coordinates
x-coordinate
y-coordinate

A **coordinate plane** is formed by two number lines in a plane that intersect at right angles. The point of intersection is the zero on each number line.

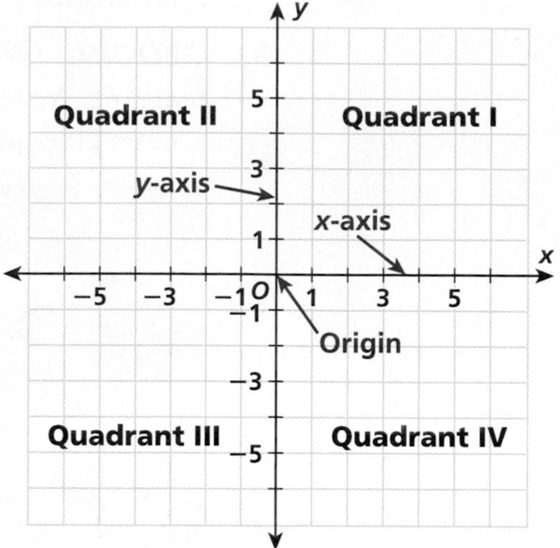

- The two number lines are called the **axes**.
- The horizontal axis is called the ***x*-axis**.
- The vertical axis is called the ***y*-axis**.
- The two axes divide the coordinate plane into four **quadrants**.
- The point where the axes intersect is called the **origin**.

EXAMPLE 1 Identifying Quadrants

Name the quadrant where each point is located.

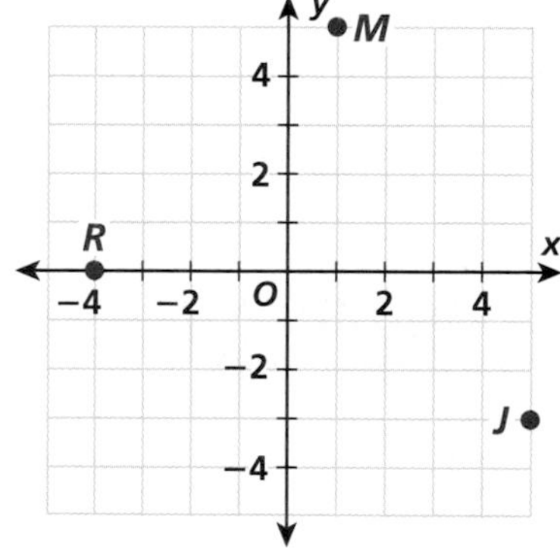

A *M*

Quadrant I

B *J*

Quadrant IV

C *R*

x-axis

no quadrant

Helpful Hint

Points on the axes are not in any quadrant.

An ordered pair gives the location of a point on a coordinate plane. The first number tells how far to move right (positive) or left (negative) from the origin. The second number tells how far to move up (positive) or down (negative).

The numbers in an ordered pair are called **coordinates**. The first number is called the ***x*-coordinate**. The second number is called the ***y*-coordinate**.

The ordered pair for the origin is (0, 0).

EXAMPLE 2 Locating Points on a Coordinate Plane

Give the coordinates of each point.

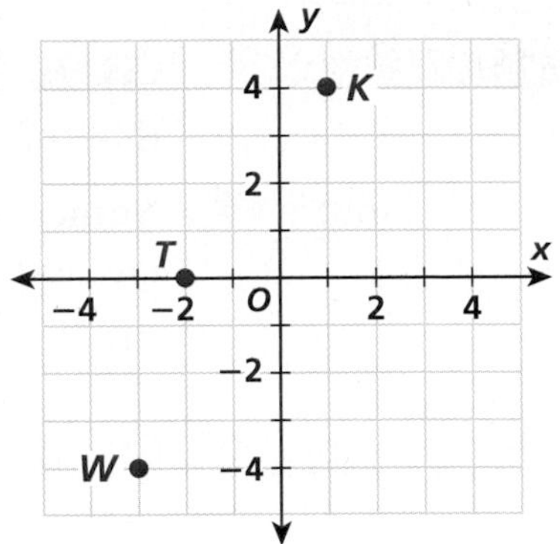

A K

From the origin, K is 1 unit right and 4 units up.

$(1, 4)$

B T

From the origin, T is 2 units left on the x-axis.

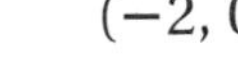

$(-2, 0)$

C W

From the origin, W is 3 units left and 4 units down.

$(-3, -4)$

EXAMPLE 3 Graphing Points on a Coordinate Plane

Graph each point on a coordinate plane.

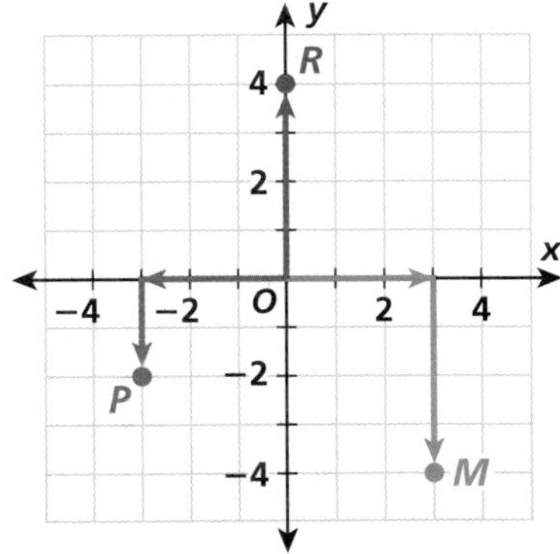

A $P(-3, -2)$

From the origin, move 3 units left and 2 units down.

B $R(0, 4)$

From the origin, move 4 units up.

C $M(3, -4)$

From the origin, move 3 units right and 4 units down.

Think and Discuss

1. Which number in an ordered pair tells you how far to move left or right from the origin? up or down?
2. **Describe** how graphing the point (5, 4) is similar to graphing the point (5, −4). How is it different?
3. **Tell** why it is important to start at the origin when you are graphing points.

9-3 Exercises

3.04

FOR EOG PRACTICE
see page 673

internet connect
Homework Help Online
go.hrw.com Keyword: MR4 9-3

GUIDED PRACTICE

Use the coordinate plane for Exercises 1–4.

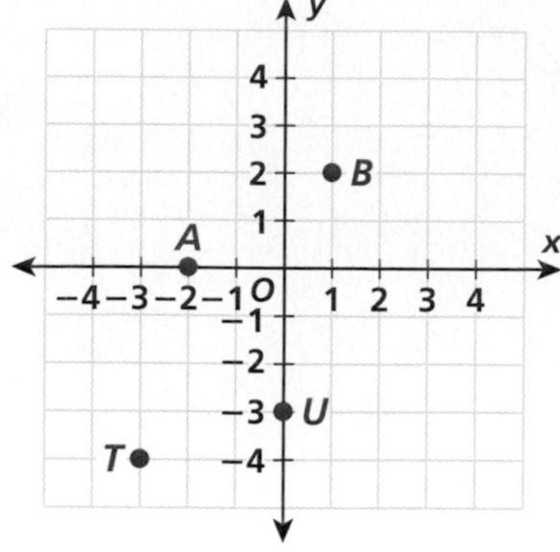

See Example 1 Name the quadrant where each point is located.

1. T
2. U

See Example 2 Give the coordinates of each point.

3. A
4. B

See Example 3 Graph each point on a coordinate plane.

5. $E(4, 2)$
6. $F(-1, -4)$

INDEPENDENT PRACTICE

Use the coordinate plane for Exercises 7–14.

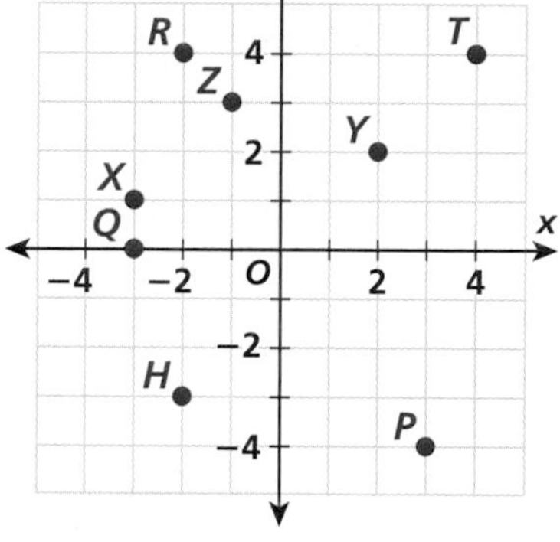

See Example 1 Name the quadrant where each point is located.

7. Q
8. X
9. Y
10. Z

See Example 2 Give the coordinates of each point.

11. P
12. R
13. T
14. H

See Example 3 Graph each point on a coordinate plane.

15. $L(0, 3)$
16. $M(3, -3)$
17. $V(-4, 3)$
18. $N(-2, -1)$

PRACTICE AND PROBLEM SOLVING

Name the quadrant where each ordered pair is located.

19. $(3, -1)$
20. $(2, 1)$
21. $(-2, 3)$

Graph each ordered pair.

22. $(0, -5)$
23. $(-4, -4)$
24. $(5, 0)$
25. $(-2, 2)$
26. $(0, -3)$
27. $(1, -4)$

Social Studies 

We use a coordinate system on Earth to find exact locations. The *equator* is like the x-axis, and the *prime meridian* is like the y-axis.

The lines that run east-west are *lines of latitude.* They are measured in degrees north and south of the equator.

The lines that run north-south are *lines of longitude.* They are measured in degrees east and west of the prime meridian.

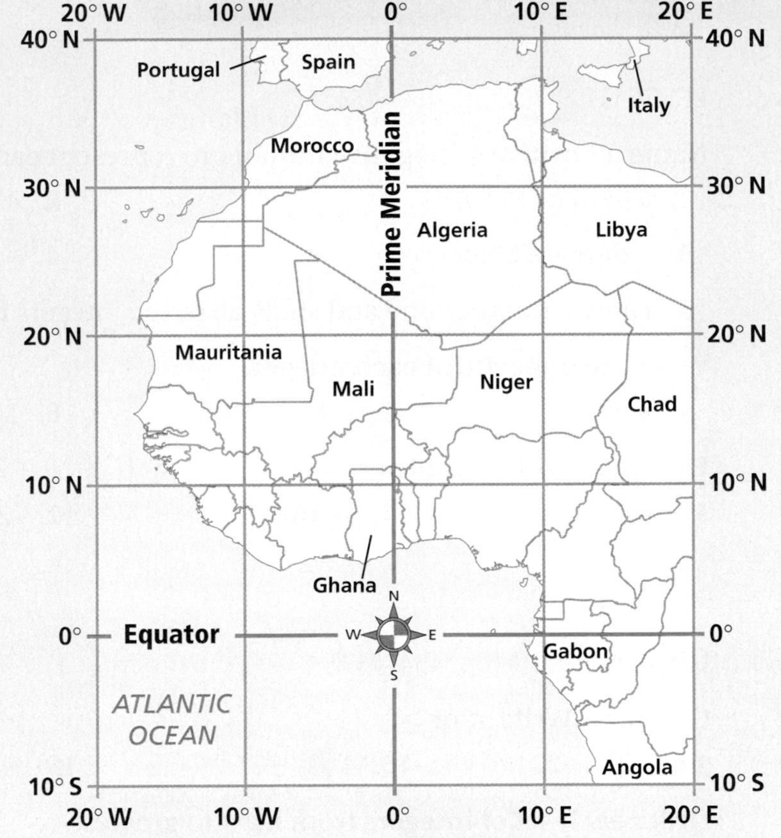

28. In what country is the location 0° latitude, 10° E longitude?

29. Give the coordinates of a location in Algeria.

30. Name two countries that lie along the 30° N line of latitude.

31. Where would you be if you were located at 10° S latitude, 10° W longitude?

32. ***WRITE ABOUT IT*** How is the coordinate system we use to locate places on Earth different from the coordinate plane? How is it similar?

33. ***CHALLENGE*** Begin at 10° S latitude, 20° E longitude. Travel 40° north and 20° west. What country would you be in now?

Spiral Review

Multiply. Write each answer in simplest form. (Lesson 5-1)

34. $\frac{2}{3} \cdot \frac{4}{7}$ **35.** $\frac{1}{5} \cdot \frac{3}{8}$ **36.** $\frac{3}{4} \cdot \frac{1}{2}$

Give the most exact name for each figure. (Lesson 7-6)

37. 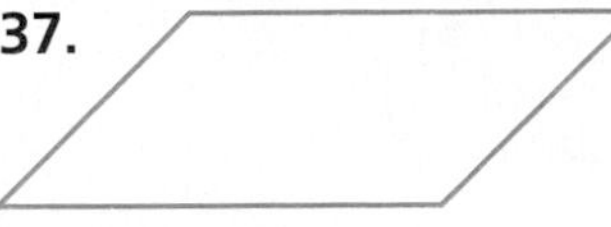

38. 

39. 

40. **EOG PREP** A __?__ compares a number to 100. (Lesson 8-7)

A rate **B** ratio **C** percent **D** proportion

Chapter 9

Mid-Chapter Quiz

LESSON 9-1 (pp. 450–453)

Name a positive or negative number to represent each situation.

1. a gain of 10 yards
2. 45 feet below sea level
3. 5 degrees below zero
4. earning \$50
5. Draw a number line and show all of the integers from −5 to 5.

Write the opposite of each integer.

6. 9
7. −17
8. 1
9. −20

Find the absolute value.

10. $|5|$
11. $|-16|$
12. $|2|$
13. $|-13|$

LESSON 9-2 (pp. 454–457)

Compare. Write < or >.

14. 9 ■ −22
15. 4 ■ −7
16. −8 ■ 2
17. −10 ■ −19

Order each set of integers from least to greatest.

18. 2, −7, 14
19. 25, −9, 4, −21
20. 10, 0, −23, −17, 8
21. During an archaeological dig, the farther down an object is found, the older it is. If pieces of jewelry are found at −7 ft, −17 ft, −4 ft, and −9 ft, which piece is oldest?

LESSON 9-3 (pp. 458–461)

Use the coordinate plane for problems 22–29.

Name the quadrant where each point is located.

22. *A*
23. *Y*
24. *J*
25. *C*

Give the coordinates of each point.

26. *H*
27. *I*
28. *W*
29. *B*

Graph each point on a coordinate plane.

30. $N(-5, -2)$
31. $M(2, 2)$
32. $P(-3, 0)$
33. $Q(4, -1)$
34. $R(-2, 6)$
35. $S(0, 4)$

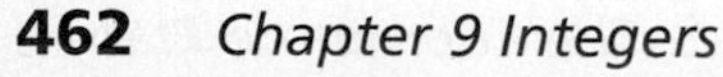

Focus on Problem Solving

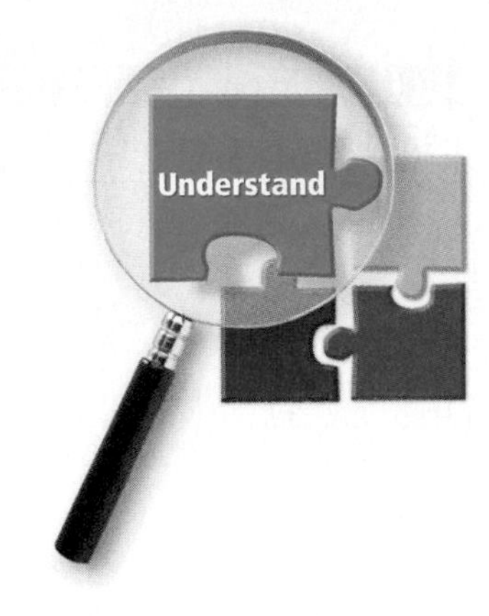

Understand the Problem

- **Restate the question**

After reading a real-world problem (perhaps several times), look at the question in the problem. Rewrite the question as a statement in your own words. For example, if the question is "How much money did the museum earn?" you could write, "Find the amount of money the museum earned."

Now you have a simple sentence telling you what you must do. This can help you understand and remember what the problem is about. This can also help you find the necessary information in the problem.

Read the problems below. Rewrite each question as a statement in your own words.

1. Israel is one of the hottest countries in Asia. A temperature of 129°F was once recorded there. This is the opposite of the coldest recorded temperature in Antarctica. How cold has it been in Antarctica?

2. The average recorded temperature in Fairbanks, Alaska, in January is about -10°F. In February, the average temperature is about -4°F. Is the average temperature lower in January or in February?

3. The south pole on Mars is made of frozen carbon dioxide, which has a temperature of -193°F. The coldest day recorded on Earth was -129°F, in Antarctica. Which temperature is lower?

4. The pirate Blackbeard's ship, the *Queen Anne's Revenge,* sank at Beauford Inlet, North Carolina, in 1718. In 1996, divers discovered a shipwreck believed to be the *Queen Anne's Revenge*. The ship's cannons were found 21 feet below the water's surface, and the ship's bell was found 20 feet below the surface. Were the cannons or the bell closer to the surface?

In this photo of Mars, different colors represent different temperature ranges. When the photo was taken, it was summer in the northern hemisphere and winter in the southern hemisphere.

-65°C — -120°C

Model Integer Addition

Use with Lesson 9-4

KEY

(yellow counter) = 1 (red counter) = −1

REMEMBER

Subtracting zero from a number does not change the number's value.

Two-color counters can be used to represent integers. Yellow counters represent positive numbers and red counters represent negative numbers.

Activity

Model with two-color counters.

1. $3 + 4$

$3 + 4 = 7$

2. $-5 + (-3)$

$-5 + (-3) = -8$

One red and one yellow counter together equal zero. Whenever you have a pair of red and yellow counters, you can remove them without changing the value of the model.

3. $3 + (-4)$

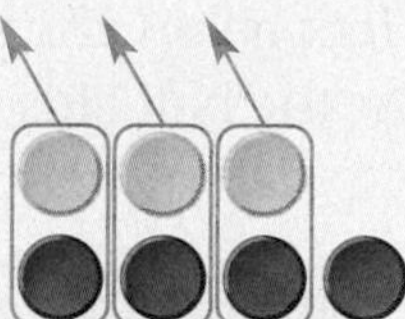

$3 + (-4) = -1$

Think and Discuss

1. When adding integers, would changing the order in which you add them affect the answer? Explain.
2. When can you remove counters from an addition model?

Try This

Model with two-color counters.

1. $-8 + (-4)$ 2. $-8 + 4$ 3. $8 + (-4)$ 4. $8 + 4$

9-4 Adding Integers

Learn to add integers.

One of the world's most active volcanoes is Kilauea, in Hawaii. Kilauea's base is 9 km below sea level. The top of Kilauea is 10 km above the base of the mountain.

You can add the integers −9 and 10 to find the height of Kilauea above sea level.

Adding Integers on a Number Line

Move **right** on a number line to add a **positive** integer.

Move **left** on a number line to add a **negative** integer.

EXAMPLE 1 Writing Integer Addition

Write the addition modeled on each number line.

Writing Math

Parentheses are used to separate addition, subtraction, multiplication, and division signs from negative integers.

$-2 + (-5) = -7$

A

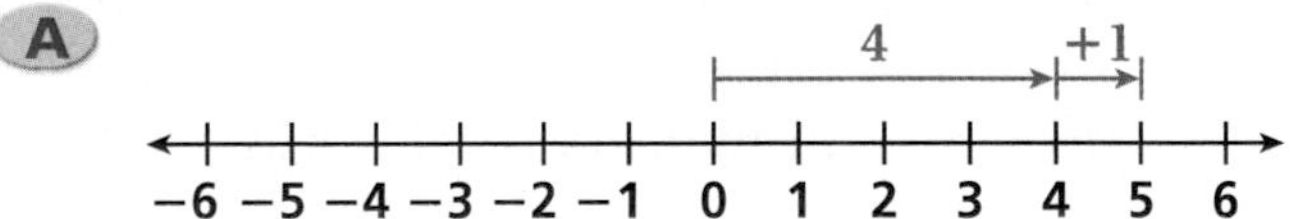

The addition modeled is $4 + 1 = 5$.

B

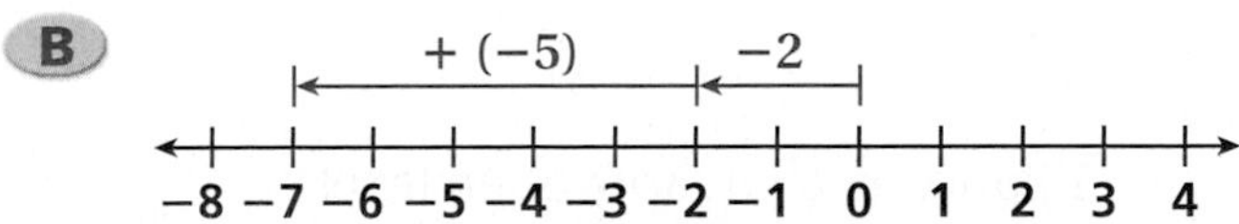

The addition modeled is $-2 + (-5) = -7$.

C

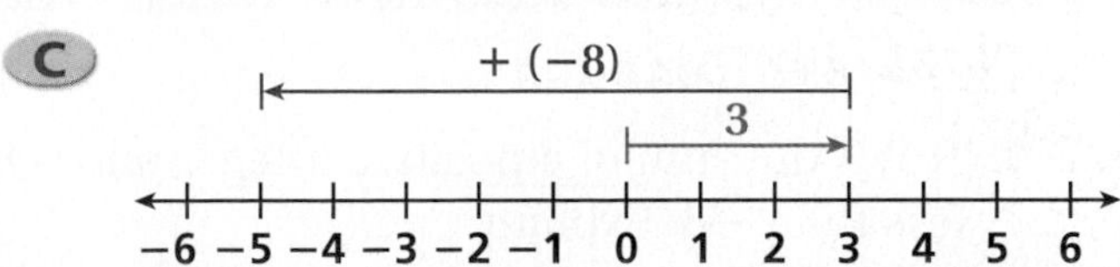

The addition modeled is $3 + (-8) = -5$.

EXAMPLE 2 Adding Integers

Find each sum.

A $6 + (-5)$

Think:

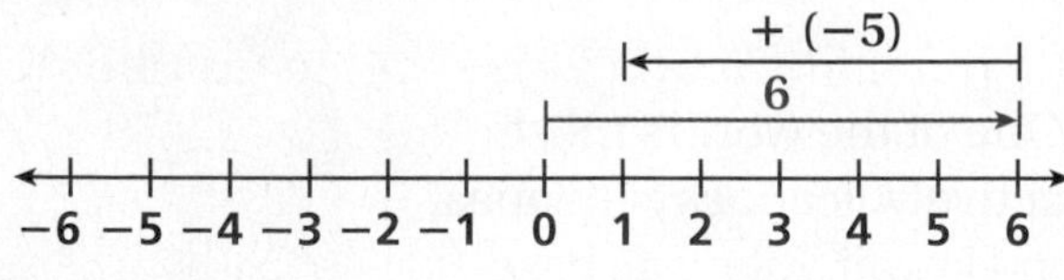

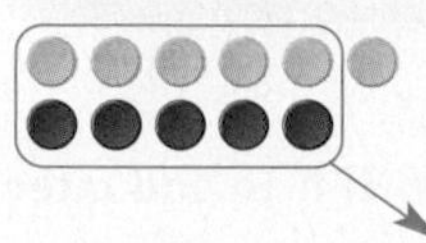

$6 + (-5) = 1$

B $-7 + 4$

Think:

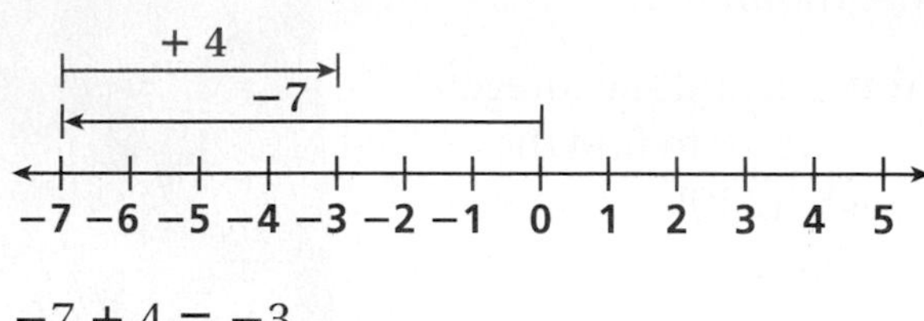

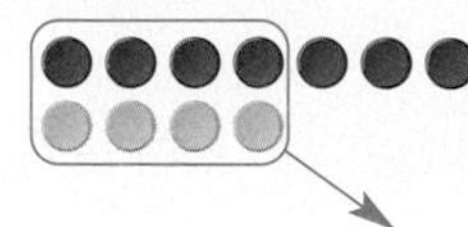

$-7 + 4 = -3$

EXAMPLE 3 Evaluating Integer Expressions

Evaluate $x + 3$ for each value of x.

A $x = 1$

$x + 3$	*Write the expression.*
$1 + 3$	*Substitute 1 for x.*
4	*Add.*

B $x = -9$

$x + 3$	*Write the expression.*
$-9 + 3$	*Substitute −9 for x.*
-6	*Add.*

EXAMPLE 4 *Earth Science Application*

The base of Kilauea is 9 km below sea level. The top is 10 km above the base. How high above sea level is Kilauea?

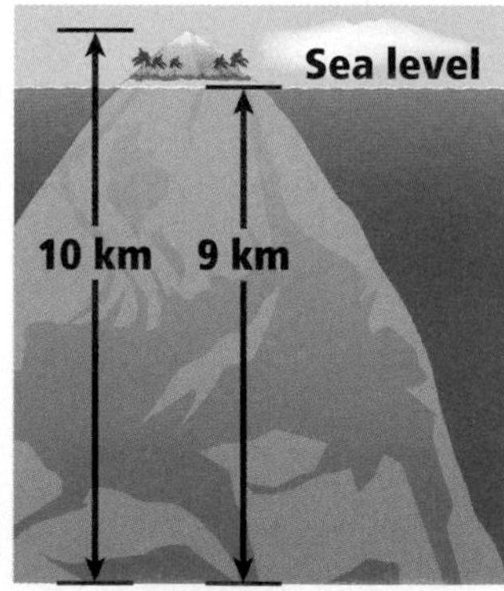

The base is **9 km below sea level** and the top is **10 km above the base.**

$-9 + 10$

1

Kilauea is 1 km above sea level.

Think and Discuss

1. **Tell** if the sum of a positive integer and −8 is greater than −8 or less than −8. Explain.
2. **Give** the sum of a number and its opposite.

9-4 Exercises

FOR EOG PRACTICE
see page 674

internet connect
Homework Help Online
go.hrw.com Keyword: MR4 9-4

1.01a, 5.02

GUIDED PRACTICE

See Example 1 **Write the addition modeled on the number line.**

1.

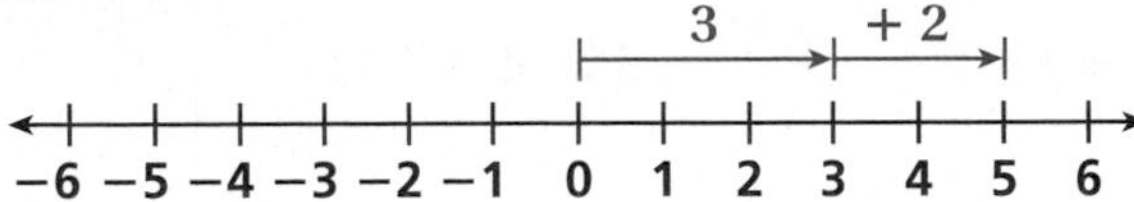

See Example 2 **Find each sum.**

2. $-5 + 9$ 3. $-3 + (-2)$ 4. $8 + (-7)$

See Example 3 **Evaluate $n + (-2)$ for each value of n.**

5. $n = -10$ 6. $n = 2$ 7. $n = -2$

See Example 4

8. A submarine at the water's surface dropped down 100 ft. After thirty minutes at that depth, it dove an additional 500 ft. What was its depth after the second dive?

INDEPENDENT PRACTICE

See Example 1 **Write the addition modeled on each number line.**

9.

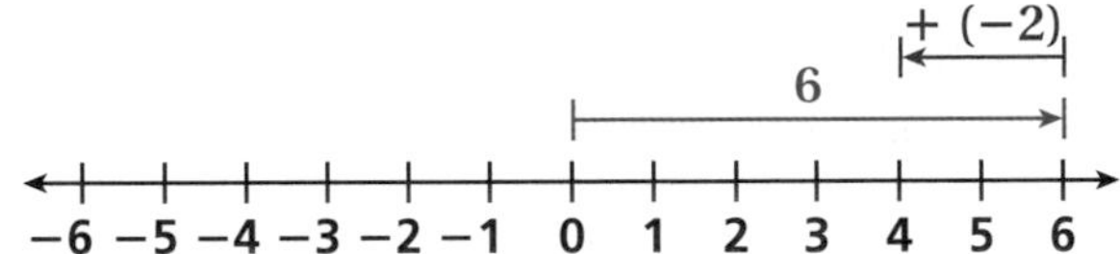

10.

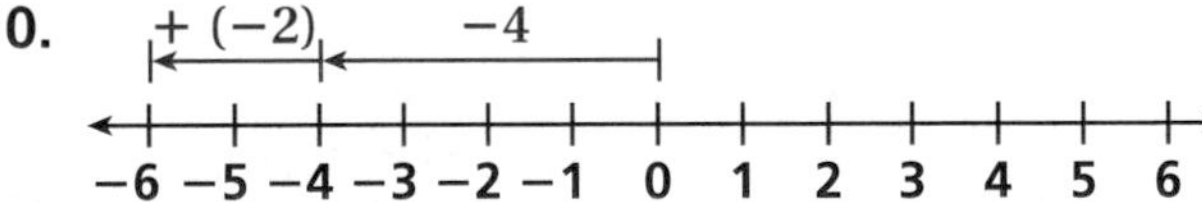

See Example 2 **Find each sum.**

11. $4 + 7$ 12. $2 + (-12)$ 13. $9 + (-9)$

14. $-8 + 2$ 15. $-2 + 8$ 16. $-1 + (-6)$

See Example 3 **Evaluate $-6 + a$ for each value of a.**

17. $a = -10$ 18. $a = 7$ 19. $a = -2$

20. $a = 4$ 21. $a = -9$ 22. $a = 8$

See Example 4

23. Jon works on a cruise ship and sleeps in a cabin that is 6 feet below sea level. The main deck is 35 feet above Jon's cabin. How far above sea level is the main deck?

PRACTICE AND PROBLEM SOLVING

Model each addition problem on a number line.

24. $3 + (-1)$ **25.** $-2 + (-4)$ **26.** $-6 + 5$

Find each sum.

27. $-18 + 25$ **28.** $8 + (-2)$ **29.** $-5 + -6$

30. $-6 + (-3)$ **31.** $4 + (-1)$ **32.** $20 + (-3)$

History LINK

Augustus was originally named Octavian, but the Roman senate gave him the title Augustus, meaning "revered one." He ruled the Roman Empire for more than 40 years.

Evaluate each expression for the given value of the variable.

33. $x + (-3); x = 7$ **34.** $-9 + n; n = 7$ **35.** $a + 5; a = -6$

36. $m + (-2); m = -4$ **37.** $-10 + x; x = -7$ **38.** $n + 19; n = -5$

39. ***EARTH SCIENCE*** The temperature at midnight was –2°F. During the next 4 hours, a decrease of 4°F was recorded. What was the temperature at 4 A.M.?

40. ***SPORTS*** In the 2001 U.S. Women's Open, Cristie Kerr had the following scores for the four rounds of golf: −1, +3, +1, and 0. What was her total score?

41. ***CHOOSE A STRATEGY*** The first Roman emperor, Augustus, was born in 63 B.C. and died in A.D. 14. How many years did he live? (*Hint*: Years B.C. are like negative numbers. Years A.D. are like positive numbers. There was no year 0.)

42. ***WRITE ABOUT IT*** When adding two integers, what will the sign of the answer be when

a. both integers are positive?

b. both integers are negative?

c. one integer is positive and the other is negative?

Explain your answers.

43. ***CHALLENGE*** Evaluate $-3 + (-2) + (-1) + 0 + 1 + 2 + 3 + 4$. Then use this pattern to find the sum of the integers from −10 to 11 and from −100 to 101.

Spiral Review

Find the GCF of each pair of numbers. (Lesson 4-3)

44. 12, 18 **45.** 12, 36 **46.** 25, 33 **47.** 45, 27

Find the LCM of each pair of numbers. (Lesson 5-5)

48. 9, 15 **49.** 4, 30 **50.** 17, 3 **51.** 10, 25

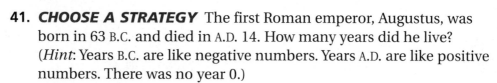

52. **EOG PREP** ___?___ are less than zero. (Lesson 9-1)

A Integers B Positive numbers C Absolute values D Negative numbers

Model Integer Subtraction

Use with Lesson 9-5

KEY

(light counter) = 1 (dark counter) = −1

REMEMBER

Adding zero to a number does not change the number's value.

(light counter) + (dark counter) = 0

internet connect
Lab Resources Online
go.hrw.com
KEYWORD: MR4 Lab9B

Activity

Model with two-color counters.

1 $3 - 2$

$3 - 2 = 1$

2 $-3 - (-2)$

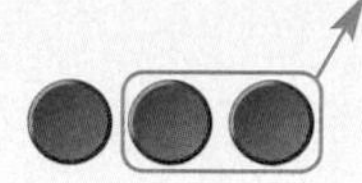

$-3 - (-2) = -1$

3 $3 - (-2)$

You do not have any red counters, so you cannot subtract −2. Add pairs of red and yellow counters until you have enough red counters to subtract.

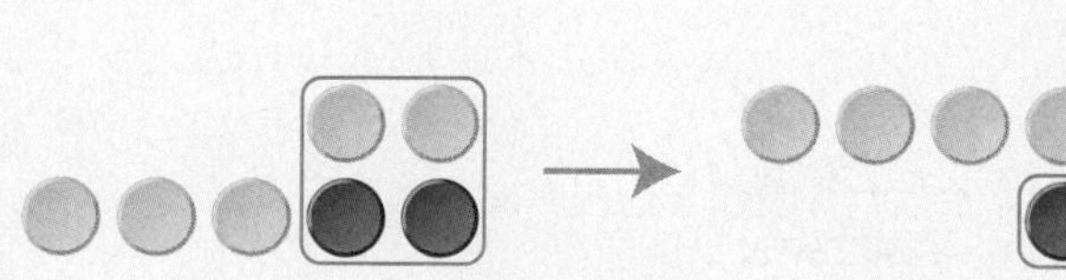

Add these.

Now you can subtract −2.

$3 - (-2) = 5$

Think and Discuss

1. How do you show subtraction with counters?

2. Why can you add pairs of red and yellow counters to a subtraction model?

Try This

Model with two-color counters.

1. $5 - 4$ **2.** $4 - (-5)$ **3.** $-4 - 5$ **4.** $-4 - (-5)$

9-5 Subtracting Integers

Learn to subtract integers.

On a number line, integer subtraction is the opposite of integer addition. Integer subtraction "undoes" integer addition.

Subtracting Integers on a Number Line
Move **left** on a number line to subtract a **positive** integer.   Move **right** on a number line to subtract a **negative** integer.

EXAMPLE 1 Writing Integer Subtraction

Write the subtraction modeled on each number line.

A

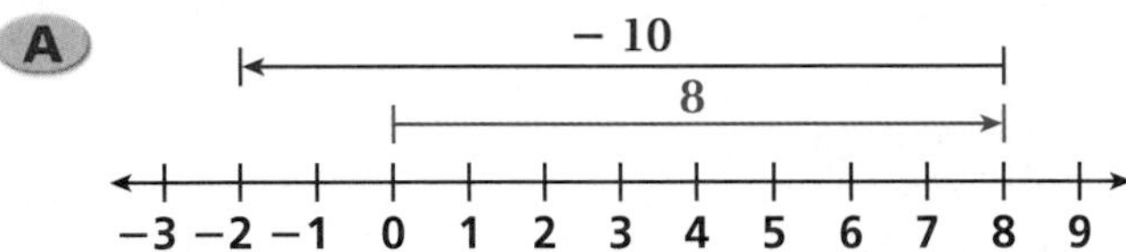

The subtraction modeled is $8 - 10 = -2$.

B

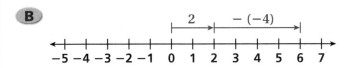

The subtraction modeled is $2 - (-4) = 6$.

EXAMPLE 2 Subtracting Integers

Find each difference.

A $7 - 4$

Think:

$7 - 4 = 3$

B $-8 - (-2)$

Think:

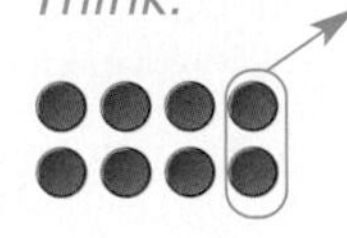

$-8 - (-2) = -6$

EXAMPLE 3 Evaluating Integer Expressions

Evaluate $x - (-4)$ for each value of x.

A $x = -4$

$x - (-4)$	*Write the expression.*
$-4 - (-4)$	*Substitute −4 for x.*
0	*Subtract.*

B $x = -5$

$x - (-4)$	*Write the expression.*
$-5 - (-4)$	*Substitute −5 for x.*
-1	*Subtract.*

Think and Discuss

1. In which direction do you move to add a positive integer? In which direction do you move to subtract a positive integer?

2. How do your answers to Example 1 help show that addition and subtraction are inverses?

9-5 Exercises

FOR EOG PRACTICE
see page 674

internet connect
Homework Help Online
go.hrw.com Keyword: MR4 9-5

1.01a, 5.02

GUIDED PRACTICE

See Example 1

1. Write the subtraction modeled on the number line.

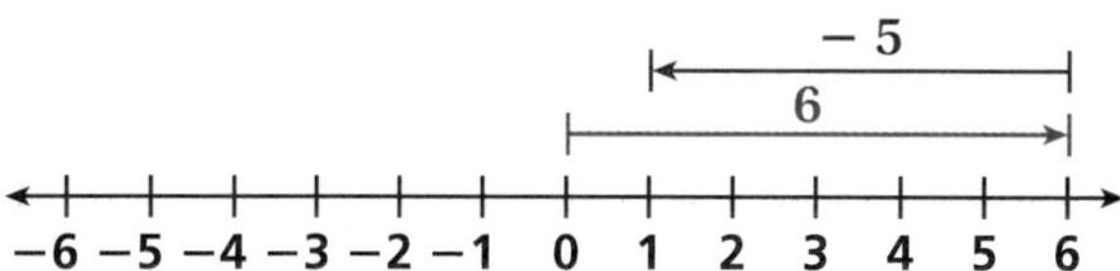

See Example 2

Find each difference.

2. $6 - 3$ **3.** $3 - 6$ **4.** $10 - (-4)$

See Example 3

Evaluate $n - (-6)$ for each value of n.

5. $n = -4$ **6.** $n = 2$ **7.** $n = -15$

INDEPENDENT PRACTICE

See Example 1

8. Write the subtraction modeled on the number line.

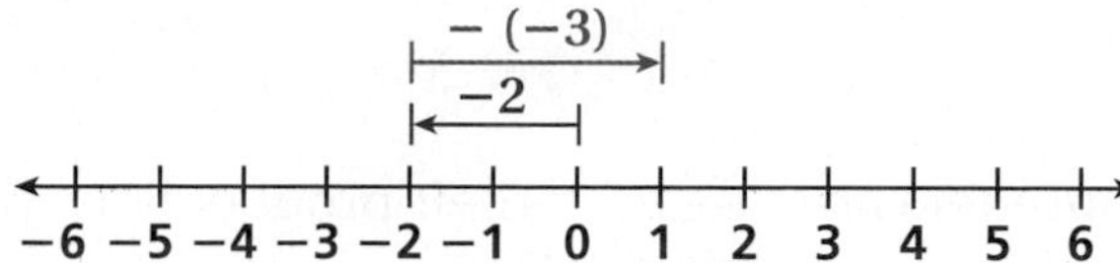

See Example 2

Find each difference.

9. $3 - 7$ **10.** $-4 - 9$ **11.** $2 - (-9)$

See Example 3 **Evaluate $m - (-3)$ for each value of m.**

12. $m = -1$
13. $m = 7$
14. $m = -8$
15. $m = 4$
16. $m = -9$
17. $m = -15$

PRACTICE AND PROBLEM SOLVING

Find each difference.

18. $-12 - (-6)$
19. $7 - (-3)$
20. $-4 - (-3)$
21. $13 - (-8)$
22. $4 - 12$
23. $2 - (-3)$

Evaluate each expression for the given value of the variable.

24. $n - (-10), n = 2$
25. $-6 - m, m = -9$
26. $x - 2, x = 6$

27. ***EARTH SCIENCE*** The surface of an underground water supply was 10 m below sea level. After one year, the depth of the water supply has decreased by 9 m. How far below sea level is the water's surface now?

28. ***CONSTRUCTION*** A 200-foot column holds an oil rig platform above the ocean's surface. The column rests on the ocean floor 175 feet below sea level. How high is the platform above sea level?

29. ***EARTH SCIENCE*** During summer 1997, NASA landed the *Pathfinder* on Mars. On July 9, *Pathfinder* reported a temperature of −1°F on the planet's surface. On July 10, it reported a temperature of 8°F. Find the difference between the temperature on July 10 and the temperature on July 9.

30. ***WHAT'S THE ERROR?*** Your friend says that $0 - (-4) = -4$. Explain why this is incorrect.

31. ***WRITE ABOUT IT*** Will the difference between two negative numbers ever be positive? Use examples to support your answer.

32. ***CHALLENGE*** This pyramid was built by subtracting integers. Two integers are subtracted from left to right, and their difference is centered above them. Find the missing numbers.

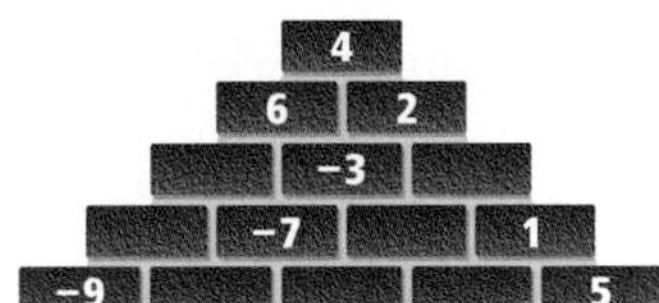

Spiral Review

Evaluate. (Lesson 1-3)

33. 3^3
34. 4^2
35. 5^3
36. 2^5

37. **EOG PREP** Two numbers are ___?___ if their product is 1. (Lesson 5-3)

A reciprocals
B opposites
C greatest common factors
D least common multiples

9-6 Multiplying Integers

Learn to multiply integers.

You have seen that you can multiply whole numbers to count items in equally sized groups.

There are three sets of twins in the sixth grade. How many sixth graders are twins?

A set of twins is 2 people.

$3 \cdot 2 = 6$ *3 sets of 2 is 6.*

So 6 students in the sixth grade are twins.

Multiplying with integers is similar.

Numbers	$3 \cdot 2$	$-3 \cdot 2$	$3 \cdot (-2)$	$-3 \cdot (-2)$
Words	3 groups of 2	**the opposite of** 3 groups of 2	3 groups of −2	**the opposite of** 3 groups of −2
Addition	$2 + 2 + 2$	$-(2 + 2 + 2)$	$(-2) + (-2) + (-2)$	$-[(-2) + (-2) + (-2)]$
Product	6	−6	−6	6

EXAMPLE 1 Multiplying Integers

Find each product.

A $4 \cdot 3$

$4 \cdot 3 = 12$ *Think: 4 groups of 3*

B $2 \cdot (-4)$

$2 \cdot (-4) = -8$ *Think: 2 groups of −4*

C $-5 \cdot 2$

$-5 \cdot 2 = -10$ *Think:* ***the opposite of*** *5 groups of 2*

D $-3 \cdot (-4)$

$-3 \cdot (-4) = 12$ *Think:* ***the opposite of*** *3 groups of −4*

Remember!

To find the opposite of a number, change the sign. The opposite of 6 is −6. The opposite of −4 is 4.

MULTIPLYING INTEGERS

If the signs are the same, the product is positive.

$4 \cdot 3 = 12$ $\quad -6 \cdot (-3) = 18$

If the signs are different, the product is negative.

$-2 \cdot 5 = -10$ $\quad 7 \cdot (-8) = -56$

The product of any number and 0 is 0.

$0 \cdot 9 = 0$ $\quad (-12) \cdot 0 = 0$

EXAMPLE 2 Evaluating Integer Expressions

Remember!

$5x$ means $5 \cdot x$.

Evaluate $5x$ for each value of x.

A $x = -4$

$5x$	*Write the expression.*
$5 \cdot (-4)$	*Substitute −4 for x.*
-20	*The signs are different, so the answer is negative.*

B $x = 0$

$5x$	*Write the expression.*
$5 \cdot 0$	*Substitute 0 for x.*
0	*Any number times 0 is 0.*

Think and Discuss

1. Explain how multiplying integers is similar to multiplying whole numbers. How is it different?

9-6 Exercises

1.01a, 5.02

FOR EOG PRACTICE see page 674

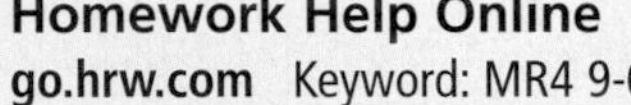

internet connect
Homework Help Online
go.hrw.com Keyword: MR4 9-6

GUIDED PRACTICE

See Example 1 **Find each product.**

1. $6 \cdot 4$ **2.** $5 \cdot (-2)$ **3.** $-3 \cdot 7$

4. $-9 \cdot (-1)$ **5.** $13 \cdot 0$ **6.** $-8 \cdot (-2)$

See Example 2 **Evaluate $3n$ for each value of n.**

7. $n = 3$ **8.** $n = -2$ **9.** $n = 11$

10. $n = -8$ **11.** $n = -12$ **12.** $n = 6$

INDEPENDENT PRACTICE

See Example 1 Find each product.

13. $5 \cdot 9$ 14. $-7 \cdot 6$ 15. $8 \cdot (-4)$

16. $-13 \cdot (-3)$ 17. $4 \cdot 12$ 18. $6 \cdot (-12)$

See Example 2 Evaluate $-4a$ for each value of a.

19. $a = 6$ 20. $a = 12$ 21. $a = 3$

22. $a = -10$ 23. $a = 7$ 24. $a = -15$

PRACTICE AND PROBLEM SOLVING

Multiply.

25. $-2 \cdot 3$ 26. $-4(9)$ 27. $-6 \cdot (-6)$ 28. $-5(-8)$

29. $-12 \cdot 2$ 30. $-9(9)$ 31. $-6 \cdot 7$ 32. $-6(25)$

Evaluate each expression for the given value of the variable.

33. $n \cdot (-7); n = -2$ 34. $-6 \cdot m; m = 4$ 35. $9x; x = 6$

36. $-5m; m = 5$ 37. $x \cdot 10; x = -9$ 38. $-8 \cdot n; n = -1$

39. $-15 \cdot x; x = 6$ 40. $-13n; n = -4$ 41. $m \cdot 14; m = -3$

42. ***EARTH SCIENCE*** When the moon, the sun, and Earth are in a straight line, spring tides occur on Earth. Spring tides may cause high and low tides to be two times as great as normal. If high tides at a certain location are usually 2 ft and low tides are usually −2 ft, what might the spring tides be?

43. ***WRITE ABOUT IT*** What is the sign of the product when you multiply three negative integers? four negative integers? Use examples to explain your answers.

44. ***CHALLENGE*** Name 2 integers whose product is −36 and whose sum is 0.

Spiral Review

Solve each equation. Check your answers. (Lesson 2-6)

45. $9y = 81$ 46. $70 = 10x$ 47. $6 \cdot 8 = n$ 48. $60 = 12m$

Write two equivalent ratios. (Lesson 8-1)

49. $\frac{1}{2}$ 50. $\frac{3}{12}$ 51. $\frac{2}{3}$ 52. $\frac{5}{15}$

53. **EOG PREP** On a coordinate plane, the point where the axes intersect is called the ___?___. (Lesson 9-3)

A quadrant B 0-axis C origin D coordinate point

9-7 Dividing Integers

Learn to divide integers.

Mona is a biologist studying an endangered species of wombat. Each year she records the change in the wombat population.

Year	Change in Population
1	−10
2	−5
3	−1
4	+4

Baby Australian wombat

One way to describe the change in the wombat population over time is to find the mean of the data in the table.

$$\frac{-10 + (-5) + (-1) + 4}{4} = \frac{-12}{4} = -12 \div 4 = \square$$

Remember!

To find the mean of a list of numbers:
1. Add all the numbers together.
2. Divide by how many numbers are in the list.

Multiplication and division are inverse operations. To solve a division problem, think of the related multiplication.

To solve $-12 \div 4$, think: What number times 4 equals -12?

$$-3 \cdot 4 = -12, \text{ so } -12 \div 4 = -3$$

The mean change in the wombat population is **−3**. So on average, the population **decreased by 3 wombats** per year.

EXAMPLE 1 Dividing Integers

Find each quotient.

A $12 \div (-3)$

Think: What number times −3 equals 12?

$-4 \cdot (-3) = 12$, so $12 \div (-3) = -4$.

B $-15 \div (-3)$

Think: What number times −3 equals −15?

$5 \cdot (-3) = -15$, so $-15 \div (-3) = 5$.

Because division is the inverse of multiplication, the rules for dividing integers are the same as the rules for multiplying integers.

DIVIDING INTEGERS

If the signs are the same, the quotient is positive.

$24 \div 3 = 8 \qquad -6 \div (-3) = 2$

If the signs are different, the quotient is negative.

$-20 \div 5 = -4 \qquad 72 \div (-8) = -9$

Zero divided by any integer equals 0.

$\frac{0}{14} = 0 \qquad \frac{0}{-11} = 0$

You cannot divide any integer by 0.

EXAMPLE 2 Evaluating Integer Expressions

Remember!

$\frac{x}{3}$ means $x \div 3$.

Evaluate $\frac{x}{3}$ for each value of x.

A $x = 6$

$\frac{x}{3}$ — *Write the expression.*

$\frac{6}{3} = 6 \div 3$ — *Substitute 6 for x.*

$= 2$ — *The signs are the same, so the answer is positive.*

B $x = -18$

$\frac{x}{3}$ — *Write the expression.*

$\frac{-18}{3} = -18 \div 3$ — *Substitute −18 for x.*

$= -6$ — *The signs are different, so the answer is negative.*

C $x = -12$

$\frac{x}{3}$ — *Write the expression.*

$\frac{-12}{3} = -12 \div 3$ — *Substitute −12 for x.*

$= -4$ — *The signs are different, so the answer is negative.*

Think and Discuss

Complete each sentence.

1. The quotient of two integers with like signs is __?__.

2. The quotient of two integers with unlike signs is __?__.

9-7 Exercises

FOR EOG PRACTICE

see page 675

internet connect

Homework Help Online
go.hrw.com Keyword: MR4 9-7

1.01a, 5.02

GUIDED PRACTICE

See Example 1 **Find each quotient.**

1. $64 \div 8$
2. $10 \div (-2)$
3. $-21 \div (-7)$

See Example 2 **Evaluate $\frac{m}{2}$ for each value of m.**

4. $m = -4$
5. $m = 20$
6. $m = -30$

INDEPENDENT PRACTICE

See Example 1 **Find each quotient.**

7. $45 \div 9$
8. $-42 \div 6$
9. $32 \div (-4)$
10. $-60 \div (-10)$
11. $-75 \div 15$
12. $22 \div 11$

See Example 2 **Evaluate $\frac{n}{4}$ for each value of n.**

13. $n = 4$
14. $n = -32$
15. $n = 12$
16. $n = 64$
17. $n = -92$
18. $n = 56$

PRACTICE AND PROBLEM SOLVING

Divide.

19. $-12 \div 2$
20. $\frac{16}{-4}$
21. $-6 \div (-6)$
22. $\frac{-30}{-3}$
23. $-45 \div 9$
24. $\frac{-35}{5}$

Evaluate each expression for the given value of the variable.

25. $n \div (-7)$; $n = -21$
26. $\frac{m}{3}$; $m = -15$
27. $\frac{x}{4}$; $x = 32$
28. $\frac{a}{3}$; $a = -9$
29. $w \div (-2)$; $w = -18$
30. $-48 \div n$; $n = -8$

31. The graph shows the low temperatures for 5 days in Fairbanks, Alaska.
    a. Find the mean low temperature for Monday, Tuesday, and Wednesday.
    b. Find the mean low temperature for all 5 days.

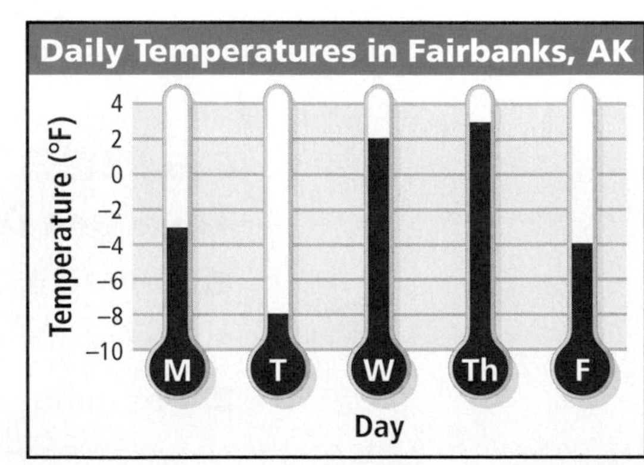

Life Science LINK

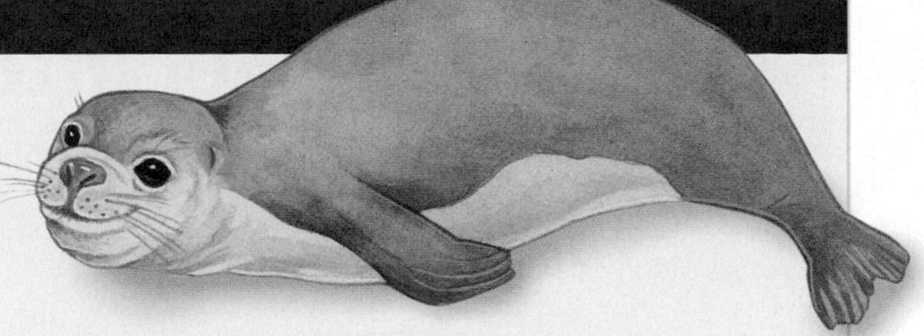

The Mediterranean monk seal is one of the world's rarest mammals. Monk seals have become endangered largely because divers hunt them for their skin and disturb their habitat.

Annette found this table in a science article about monk seals.

Changes in Population of Monk Seals							
Years	1966–1970	1971–1975	1976–1980	1981–1985	1986–1990	1991–1995	1996–2000
Change	−250	550	−300	−150	−50	100	200

32a. According to the table, what was the change in the monk seal population from 1966 to 1970?

b. What does this number mean?

33. Find the mean change per year from 1971 to 1975. (*Hint:* This is a range of 5 years, so divide by 5.) What does your answer mean?

34. Find the mean change per year from 1981 to 1990. What does your answer mean?

35. ***WRITE ABOUT IT*** Why is it important to use both positive and negative numbers when tracking the changes in a population?

36. ***CHALLENGE*** Suppose that there were 500 monk seals in 1966. How many were there in 2000?

Spiral Review

Solve each equation. Check your answers. (Lesson 3-10)

37. $4.2 + n = 6.7$ **38.** $x - 2.3 = 1.6$ **39.** $1.5w = 3.6$

Solve each equation. Check your answers. (Lesson 5-10)

40. $\frac{1}{2} + m = 2$ **41.** $3 - \frac{2}{3} = n$ **42.** $\frac{1}{5} + 6 = x$

43. **EOG PREP** ___?___ angles are always congruent. (Lesson 7-3)

A Complementary
B Adjacent
C Vertical
D Supplementary

44. **EOG PREP** Which of the following are equivalent ratios? (Lesson 8-1)

A $\frac{2}{3}, \frac{3}{2}$ B $\frac{1}{2}, \frac{4}{8}$ C $\frac{1}{5}, \frac{4}{5}$ D $\frac{3}{4}, \frac{5}{6}$

Model Integer Equations

Use with Lesson 9-8

KEY

(yellow tile) = 1

(red tile) = −1

(x-tile) = x

REMEMBER

You can add or subtract the same number on both sides of an equation.

Adding or subtracting zero does not change a number's value.

You can use algebra tiles to model equations. An equation mat represents the two sides of an equation. To find the value of the variable, get the x-tile by itself on one side of the mat. You may remove the same number of yellow tiles or the same number of red tiles from both sides.

Activity

Use algebra tiles to model and solve each equation.

1 $x + 2 = 6$

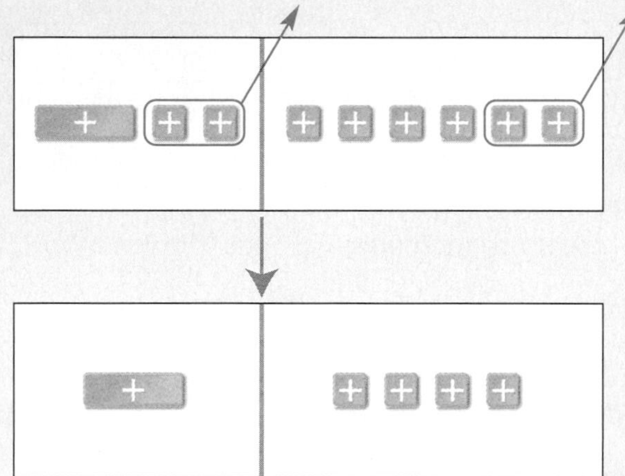

Remove 2 yellow tiles from both sides of the mat.

$x = 4$

2 $x - 3 = -5$

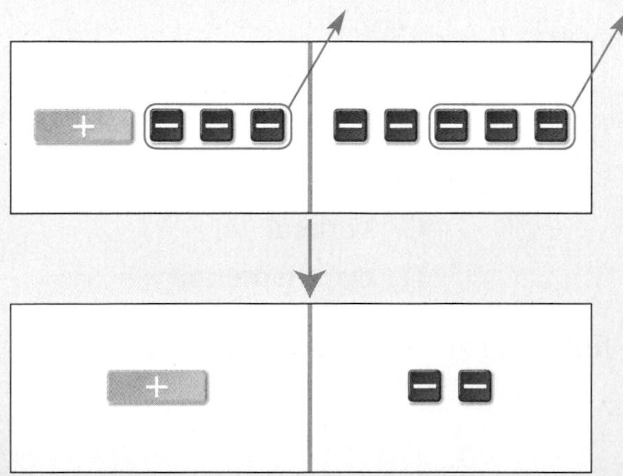

Use red tiles to model subtraction. Remove 3 red tiles from both sides of the mat.

$x = -2$

3 $x + 6 = 2$

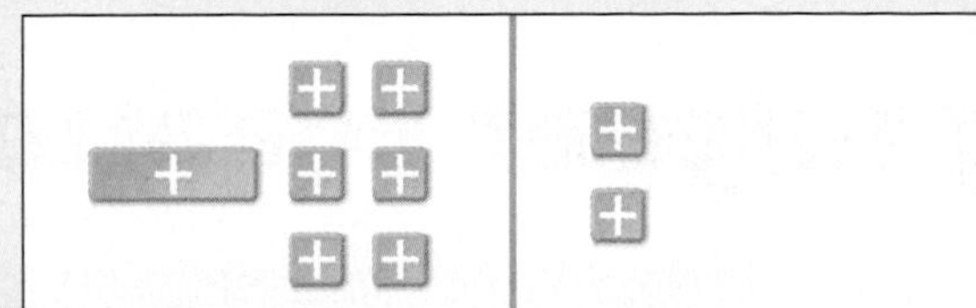

You do not have enough yellow tiles on the right side to remove 6 from both sides. Add pairs of red and yellow tiles to the right side until you have enough yellow tiles to subtract.

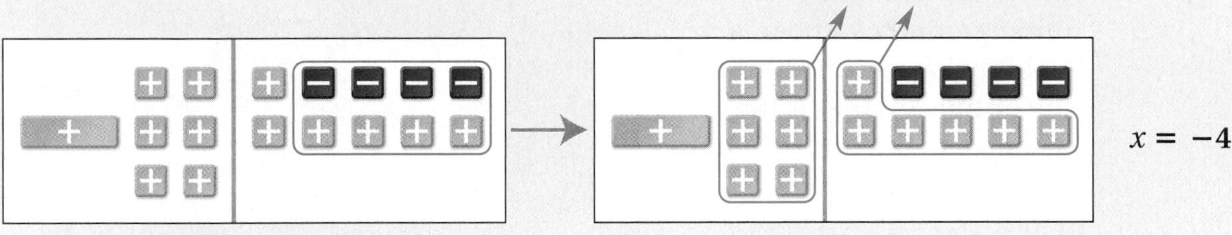
x = −4

Add these tiles.

Now you can remove 6 yellow tiles from both sides of the mat.

4 $3x = -9$

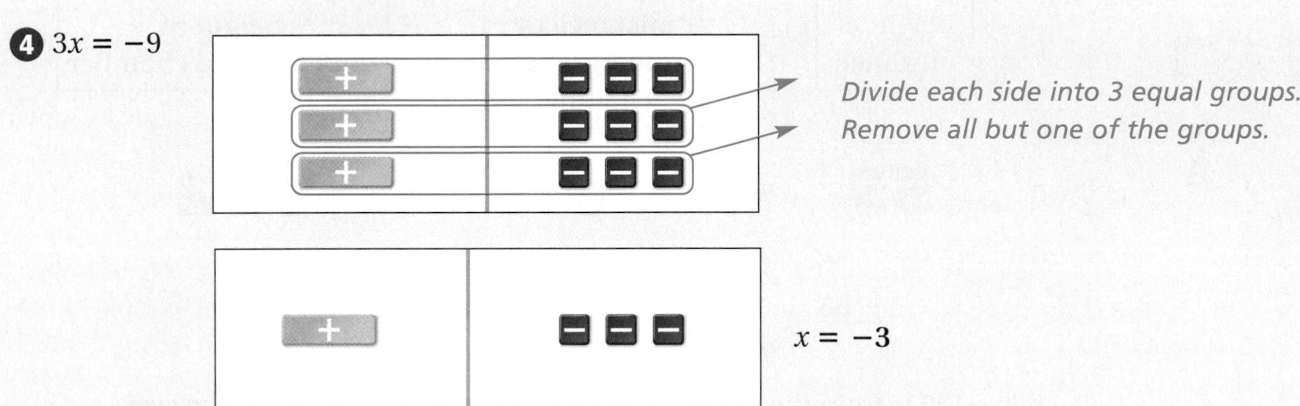
Divide each side into 3 equal groups.
Remove all but one of the groups.
x = −3

Think and Discuss

1. In **4**, why did you divide both sides into 3 groups?
2. Why can you add pairs of red and yellow tiles to an equation mat? Why is it not necessary to add them to both sides?
3. When you add zero to an equation, how do you know the number of red and yellow tiles to add?
4. How can you use algebra tiles to check your answers?

Try This

Use algebra tiles to model and solve each equation.

1. $x + 6 = 3$ **2.** $x - 1 = -8$ **3.** $2x = 14$ **4.** $4x = -8$

9-8 Solving Integer Equations

Learn to solve equations containing integers.

The entrance to the Great Pyramid of Khufu is 55 ft above ground. The underground chamber is 102 ft below ground. From the entrance, what is the distance to the underground chamber?

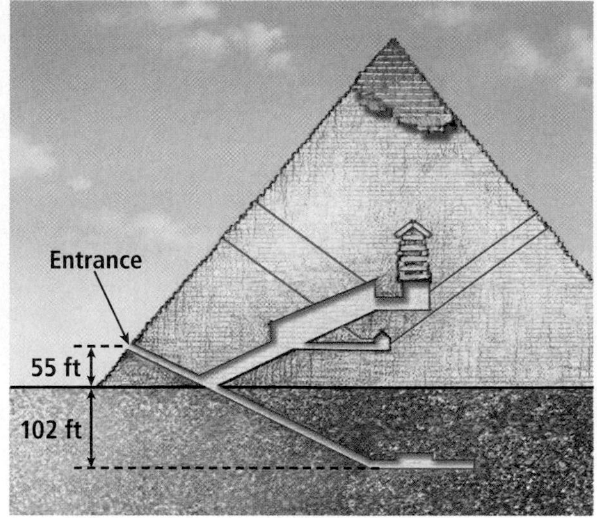

To solve this problem, you can use an equation containing integers.

height of entrance	+	distance to underground chamber	=	height of underground chamber
55	+	d	=	-102

$$55 + d = -102$$ *Write the equation.*

$$\underline{-55} \quad \underline{-55}$$ *Subtract 55 from both sides.*

$$d = -157$$

It is **−157 ft** from the entrance to the underground chamber. The sign is negative, which means you **go down 157 ft.**

EXAMPLE 1 Adding and Subtracting to Solve Equations

Helpful Hint

To solve this equation using algebra tiles, you can add four red tiles to both sides and then remove pairs of red and yellow tiles. This is because subtracting a number is the same as adding its opposite.

A **Solve $4 + x = -2$. Check your answer.**

$$4 + x = -2$$ *4 is added to x.*

$$\underline{-4} \quad \underline{-4}$$ *Subtract 4 from both sides to undo the addition.*

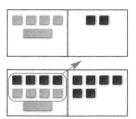

$$x = -6$$

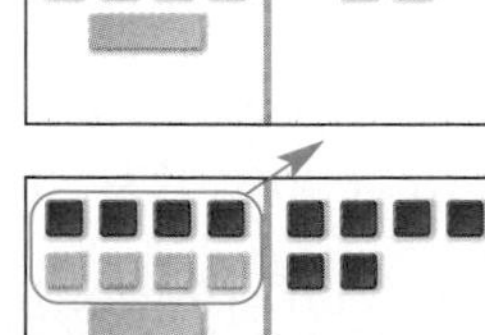

Check

$$4 + x = -2$$ *Write the equation.*

$$4 + (-6) \stackrel{?}{=} -2$$ *Substitute −6 for x.*

$$-2 \stackrel{?}{=} -2 \checkmark$$ *−6 is a solution.*

B Solve $y - 6 = -5$. Check your answer.

$$y - 6 = -5$$ 6 is subtracted from y.

$$\underline{+6} \quad \underline{+6}$$ Add 6 to both sides to undo the subtraction.

$$y = 1$$

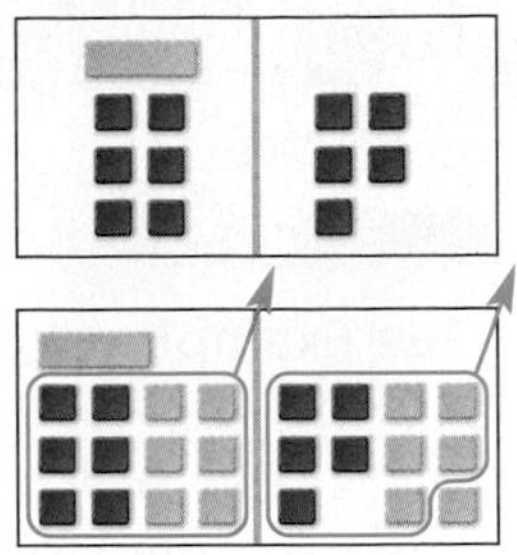

Check

$y - 6 = -5$ *Write the equation.*

$1 - 6 \stackrel{?}{=} -5$ *Substitute 1 for y.*

$-5 \stackrel{?}{=} -5$ ✔ *1 is a solution.*

EXAMPLE 2 Multiplying and Dividing to Solve Equations

Solve each equation. Check your answers.

A $-3a = 15$

$\frac{-3a}{-3} = \frac{15}{-3}$ *a is multiplied by −3. Divide both sides by −3 to undo the multiplication.*

$a = -5$

Check

$-3a = 15$ *Write the equation.*

$-3(-5) \stackrel{?}{=} 15$ *Substitute −5 for a.*

$15 \stackrel{?}{=} 15$ ✔ *−5 is a solution.*

B $\frac{b}{-4} = -2$

$-4 \cdot \frac{b}{-4} = -4 \cdot (-2)$ *b is divided by –4. Multiply both sides by –4 to undo the division.*

$b = 8$

Check

$\frac{b}{-4} = -2$ *Write the equation.*

$8 \div (-4) \stackrel{?}{=} -2$ *Substitute 8 for b.*

$-2 \stackrel{?}{=} -2$ ✔ *8 is a solution.*

Think and Discuss

1. **Tell** what operation you would use to solve $x + 12 = -32$.
2. **Tell** whether the solution to $-9t = -27$ will be positive or negative without actually solving the equation.
3. **Explain** how to check your answer to an integer equation.

9-8 Exercises

5.03

FOR EOG PRACTICE
see page 675

internet connect
Homework Help Online
go.hrw.com Keyword: MR4 9-8

GUIDED PRACTICE

See Example 1 Solve each equation. Check your answers.

1. $m - 3 = 9$
2. $a - 8 = -13$
3. $z - 12 = -3$

See Example 2

4. $-4b = 32$
5. $\frac{w}{3} = 18$
6. $5c = -35$

INDEPENDENT PRACTICE

See Example 1 Solve each equation. Check your answers.

7. $g - 9 = -5$
8. $v - 7 = 19$
9. $t - 13 = -27$
10. $x + 2 = -12$
11. $y + 9 = -10$
12. $20 + w = 10$

See Example 2

13. $6j = 48$
14. $7s = -49$
15. $\frac{a}{-2} = 26$
16. $\frac{m}{-12} = 4$
17. $\frac{k}{5} = -4$
18. $u \div 6 = -10$

PRACTICE AND PROBLEM SOLVING

Solve each equation. Check your answers.

19. $x - 12 = 5$
20. $w - 3 = -2$
21. $-7k = 28$
22. $\frac{m}{-3} = 5$
23. $a - 10 = 9$
24. $n - 19 = -22$
25. $13g = -39$
26. $s \div 6 = -3$
27. $24 + f = 16$
28. $d - 26 = 7$
29. $-6c = 54$
30. $h \div (-4) = 21$
31. $b - 17 = 15$
32. $u - 82 = -7$
33. $-8a = -64$
34. $\frac{t}{11} = -5$
35. $31 + j = -14$
36. $c + 23 = 10$
37. $15n = -60$
38. $z \div (-5) = -9$
39. $j - 20 = -23$

40. A submarine captain sets the following diving course: dive 200 ft, stop, and then dive another 200 ft. If this pattern is continued, how many dives will be necessary to reach a location 14,000 ft below sea level?

41. While exploring a cave, Lin noticed that the temperature dropped 4°F for every 30 ft that she descended. What is Lin's depth if the temperature is 8° lower than the temperature at the surface?

42. ***SPORTS*** After two rounds in the 2001 LPGA Champions Classic, Wendy Doolan had a score of –12. Her score in the second round was –8. What was her score in the first round?

Social Studies LINK

Bolivia actually has two capitals. La Paz is Bolivia's chief industrial city, and the congress meets there. However, the supreme court meets in Sucre, farther south.

Use the graph for Exercises 43 and 44.

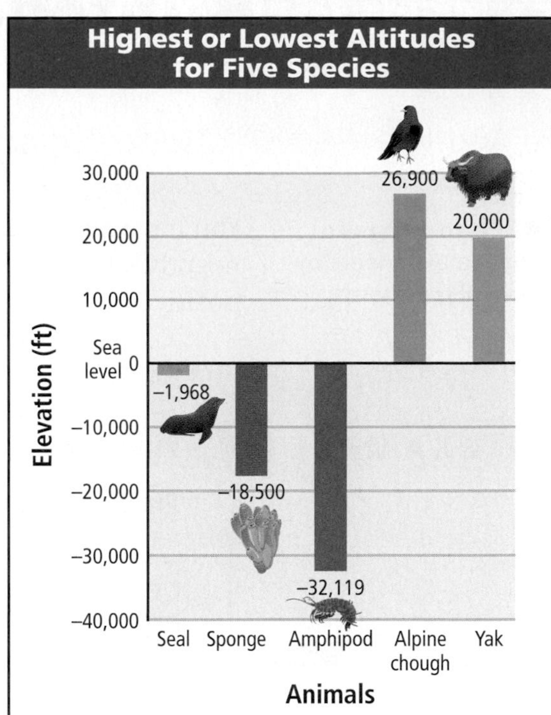

43. ***LIFE SCIENCE*** Scientists have found live bacteria at elevations of 135,000 ft. This is 153,500 ft above one of the animals in the graph. Which one? (*Hint:* Solve $x + 153{,}500 = 135{,}000$.)

44. The world's highest capital city is La Paz, Bolivia, with an elevation of 11,808 ft. The highest altitude that a yak has been found at is how much higher than La Paz? (*Hint:* Solve $11{,}808 + x = 20{,}000$.)

45. Carla is a diver. On Friday, she dove 5 times as deep as she dove on Monday. If she dove to −120 ft on Friday, how deep did she dive on Monday?

46. ***WRITE A PROBLEM*** Write a word problem that could be solved using the equation $x - 3 = -15$.

47. ***WRITE ABOUT IT*** Is the solution to $3n = -12$ positive or negative? How could you tell without solving the equation?

48. ***CHALLENGE*** Find each answer.

a. $12 \div (-3 \cdot 2) \div 2$ **b.** $12 \div (-3 \cdot 2 \div 2)$

Why are the answers different even though the numbers are the same?

Spiral Review

Write the prime factorization using exponents. (Lesson 4-2)

49. 76 **50.** 12 **51.** 16 **52.** 18

53. 21 **54.** 128 **55.** 156 **56.** 49

Add or subtract. Write each answer in simplest form. (Lesson 5-7)

57. $\frac{1}{2} + \frac{3}{4}$ **58.** $\frac{2}{3} - \frac{1}{5}$ **59.** $\frac{2}{5} + \frac{1}{2}$

60. $\frac{5}{6} - \frac{1}{3}$ **61.** $\frac{3}{7} + \frac{1}{4}$ **62.** $\frac{8}{9} - \frac{1}{6}$

63. **EOG PREP** A ___?___ is a rectangle with four congruent sides. (Lesson 7-6)

A triangle B square C rhombus D trapezoid

EXTENSION

Integer Exponents

Learn to recognize negative exponents by examining patterns.

You have already learned about positive exponents. Exponents can be negative, too. To determine the values of negative powers, write some positive powers and look for a pattern.

Remember!

Exponent

$10^3 = 10 \cdot 10 \cdot 10$

Base

EXAMPLE 1 Finding Patterns in Exponents

Find a pattern in the table.

Power	10^3	10^2	10^1	10^0	10^{-1}	10^{-2}
Value	1,000	100	10	1	$\frac{1}{10}$	$\frac{1}{100}$

÷ 10 ÷ 10 ÷ 10 ÷ 10 ÷ 10

One possible pattern is "divide by 10."

EXAMPLE 2 Using Patterns in Exponents

Find each value: 2^0, 2^{-1}, 2^{-2}, 2^{-3}.

Make a table like the one in Example 1. Write some powers of 2 that you know, and look for a pattern.

Power	2^3	2^2	2^1	2^0	2^{-1}	2^{-2}	2^{-3}
Value	8	4	2	■	■	■	■

÷ 2 ÷ 2 ÷ 2 ÷ 2 ÷ 2 ÷ 2

One possible pattern is "divide by 2."

$2^0 = 2 \div 2 = 1 \quad 2^{-1} = 1 \div 2 = \frac{1}{2} \quad 2^{-2} = \frac{1}{2} \div 2 = \frac{1}{4} \quad 2^{-3} = \frac{1}{4} \div 2 = \frac{1}{8}$

Look at the table in Example 2. There is another pattern.

$$2^{-1} = \frac{1}{2^1} = \frac{1}{2} \qquad 2^{-2} = \frac{1}{2^2} = \frac{1}{4} \qquad 2^{-3} = \frac{1}{2^3} = \frac{1}{8}$$

This pattern works for all negative exponents. A number raised to a negative exponent equals 1 divided by that number raised to the opposite (positive) exponent.

EXTENSION Exercises

Complete each table by extending the pattern.

1.

Power	3^3	3^2	3^1	3^0	3^{-1}	3^{-2}
Value	27	9	3	■	■	■

2.

Power	5^{-2}	5^{-1}	5^0	5^1	5^2	5^3
Value	■	■	■	5	25	125

3.

Power	6^3	6^2	6^1	6^0	6^{-1}	6^{-2}
Value	216	36	6	■	■	■

Find the missing exponent.

4. $81 = 9^{■}$ **5.** $\frac{1}{7} = 7^{■}$ **6.** $64 = 4^{■}$ **7.** $\frac{1}{64} = 8^{■}$

8. $49 = 7^{■}$ **9.** $\frac{1}{3} = 3^{■}$ **10.** $25 = 5^{■}$ **11.** $\frac{1}{49} = 7^{■}$

12. $64 = 2^{■}$ **13.** $\frac{1}{16} = 4^{■}$ **14.** $\frac{1}{64} = 4^{■}$ **15.** $\frac{1}{81} = 3^{■}$

Find each value.

16. 8^3 **17.** 3^{-3} **18.** 6^3 **19.** 9^{-3}

20. 7^{-3} **21.** 4^4 **22.** 1^{-8} **23.** 8^{-2}

24. 1^2 **25.** 5^{-3} **26.** 4^2 **27.** 1^{-3}

28. For each row of the table, find the number that is not equal to the other three.

a.	10	10^{-1}	$\frac{1}{10}$	0.1
b.	27	3^3	$\frac{1}{3}$	$3 \cdot 3 \cdot 3$
c.	$\frac{1}{25}$	5^{-2}	0.04	-25

29. What do you think is the value of any number raised to the 0 power?

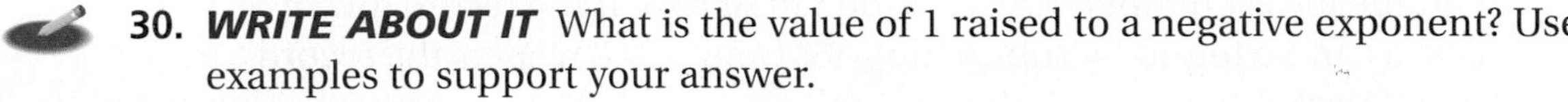

30. ***WRITE ABOUT IT*** What is the value of 1 raised to a negative exponent? Use examples to support your answer.

31. ***WRITE ABOUT IT*** You cannot raise 0 to a negative exponent. Why?

NORTH CAROLINA

Temperatures

North Carolina has tall mountains with deep forests, rolling hills with fertile farmland and ranch land, and sunny beaches on the Outer Banks. From these very different environments can come a wide range of temperatures.

North Carolina's record high temperature of 110°F was set on August 21, 1983, in Fayetteville. Its record low temperature of −34°F was set on January 25, 1985, on Mount Mitchell.

1. Find the difference between the record high and the record low temperatures for North Carolina.
2. If the high temperature on January 25, 1985, was reached after the temperature rose 39° from the record low of –34°F, what was the high temperature for that date?
3. Here is a useful rule for estimating temperature change: Temperatures change by about −4°F for every 1,000 feet in elevation that you climb. If professional climbers are 6,000 feet from the summit of Mount Mitchell and they record a temperature of 60°F, what can they expect the temperature to be at that time on the summit?
4. The low temperature of –34°F in North Carolina was much warmer than the lowest temperature recorded in Alaska. If the record low in Alaska can be found by subtracting 46° from −34°, what is the record low in Alaska?

Aquarius Underwater Lab

The National Undersea Research Center (NURC), at the University of North Carolina in Wilmington, began operating the underwater laboratory Aquarius in 1988. In 1992, Aquarius was relocated to the Florida Keys National Marine Sanctuary, although it is still operated by NURC.

Scientists live in Aquarius during ten-day research missions to study the coastal ocean. These aquanauts enjoy all the comforts of home, including a shower, refrigerator, and six sleeping bunks. Communication between the Aquarius habitat and shore takes place via computer, audio, and video.

Aquarius is typically located 50 feet below the ocean surface, but it can reach depths of up to 120 feet.

For 1–3, assume that Aquarius is at its typical depth.

1. An aquanaut stationed on Aquarius leaves the habitat to perform research on a coral reef that is 33 feet below Aquarius. How many feet below sea level is the coral reef? Write the answer as an integer.
2. An aquanaut from Aquarius dives 17 feet down from the lab. What is the aquanaut's final depth below sea level written as an integer?
3. If an aquanaut leaves Aquarius and swims toward the surface for 18 feet and then stops, what will be the depth of the scientist, written as an integer?
4. Find the difference between Aquarius's typical depth and its maximum depth.
5. Suppose the habitat is currently at -30 feet. NURC decides to move Aquarius to a new depth that is 3 times as deep. Write the new depth of the habitat as an integer.

MATH-ABLES

A Math Riddle

What coin doubles in value when half is subtracted?

To find the answer, graph each set of points. Connect each pair of points with a straight line.

1. (−8, 3) (−6, 3)	**2.** (−9, 1) (−7, 5)	**3.** (−7, 5) (−5, 1)	**4.** (−3, 1) (−3, 5)
5. (−1, 1) (−1, 5)	**6.** (−3, 3) (−1, 3)	**7.** (1, 1) (3, 5)	**8.** (3, 5) (5, 1)
9. (2, 3) (4, 3)	**10.** (6, 1) (6, 5)	**11.** (6, 1) (8, 1)	**12.** (9, 1) (9, 5)
13. (9, 5) (11, 5)	**14.** (9, 3) (11, 3)	**15.** (−9, −5) (−9, −1)	**16.** (−9, −1) (−7, −3)
17. (−7, −3) (−9, −5)	**18.** (−6, −1) (−6, −5)	**19.** (−6, −5) (−4, −5)	**20.** (−4, −5) (−4, −1)
21. (−4, −1) (−6, −1)	**22.** (−3, −1) (−3, −5)	**23.** (−3, −5) (−1, 5)	**24.** (1, −1) (1, −5)
25. (1, −5) (3, −5)	**26.** (4, −5) (6, −1)	**27.** (6, −1) (8, −5)	**28.** (5, −3) (7, −3)
29. (9, −5) (9, −1)	**30.** (9, −1) (11, −3)	**31.** (11, −3) (9, −3)	**32.** (9, −3) (11, −5)

Zero Sum

Each card contains either a positive number, a negative number, or 0. The dealer deals three cards to each player. On your turn, you may exchange one or two of your cards for new ones, or you may keep your three original cards. After everyone has had a turn, the player whose sum is closest to 0 wins the round and receives everyone's cards. The dealer deals a new round and the game continues until the dealer runs out of cards. The winner is the player with the most cards at the end of the game.

internet connect

go.hrw.com

Go to ***go.hrw.com*** for a complete set of rules and game pieces.
KEYWORD: MR4 Game9

Graph Points

internet connect
Lab Resources Online
go.hrw.com
KEYWORD: MR4 TechLab9

To graph on your calculator, you must first set the viewing window. To do this, press WINDOW.

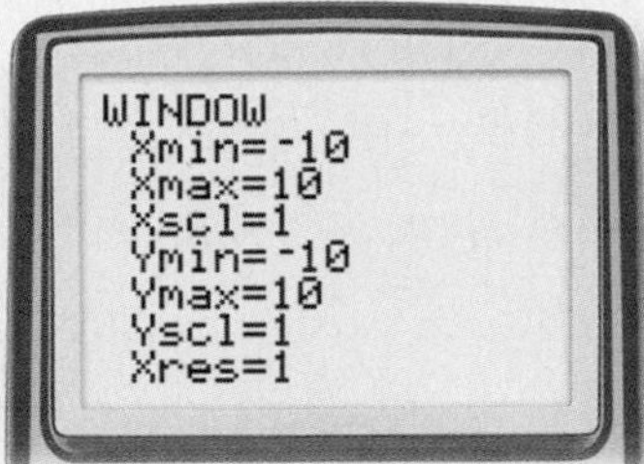

The calculator automatically uses the **standard window** shown at right unless you change the settings.

The x-values and y-values go from -10 to 10. These are set by **Xmin**, **Xmax**, **Ymin**, and **Ymax**.

Xscl and **Yscl** give the distance between tick marks. In the standard window, tick marks are 1 unit apart.

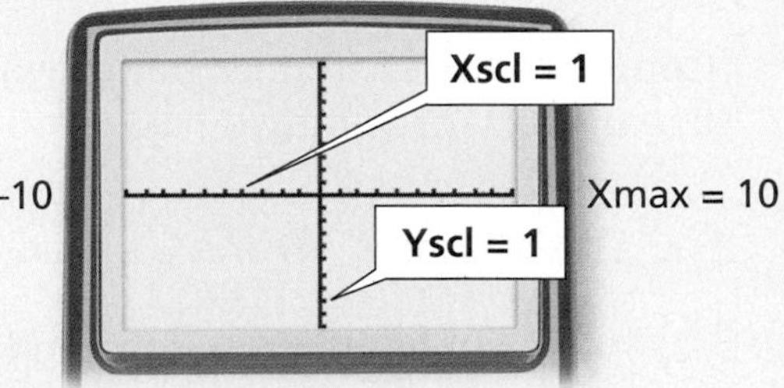

Activity

Graph the points (2, 5), (4, 3), $(-4, 6)$, and $(5, -7)$ on your calculator.

1. Access the **DRAW** menu by pressing 2nd PRGM (DRAW).

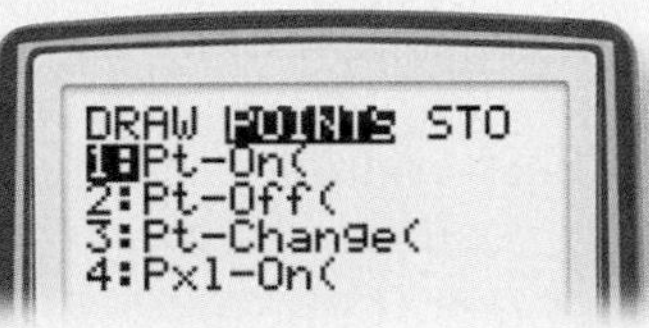

2. Press ▶ to highlight POINTS. Make sure 1: is highlighted to select **Pt-On**. Press ENTER.
3. The calculator gives you the opening parenthesis for the ordered pair. To plot (2, 5), press 2 , 5) ENTER.

4. To graph the next point, press 2nd MODE (QUIT) to quit the coordinate plane. Then repeat steps 1–3.

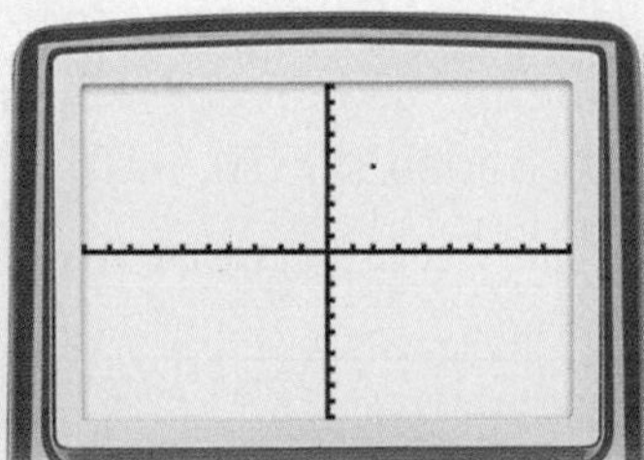

Think and Discuss

1. Suppose you want to graph (9, 16) on a graphing calculator. Should you use the standard window? Why or why not?

Try This

1. Graph (1, 3), (5, 3), $(-5, -2)$, $(3, -6)$, and $(-8, 1)$ on your calculator.

Chapter 9

Study Guide and Review

Vocabulary

Complete the sentences below with vocabulary words from the list above. Words may be used more than once.

1. The numbers −6 and 6 are called ___?___.
2. A coordinate plane is formed by the intersection of the ___?___ and the ___?___.
3. The axes separate the ___?___ into four ___?___.
4. Numbers greater than 0 are ___?___ and numbers less than 0 are ___?___.

9-1 Understanding Integers (pp. 450–453)

EXAMPLE

Name a positive or negative number to represent each situation.

- 15 feet below sea level −15
- a bank deposit of \$10 +10
- **Graph +4 on a number line.**

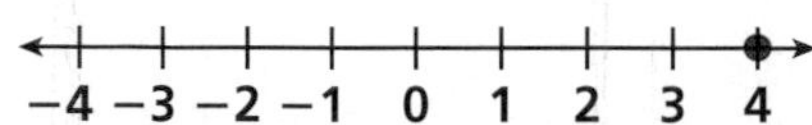

- **Use the number line to find |−3|.**

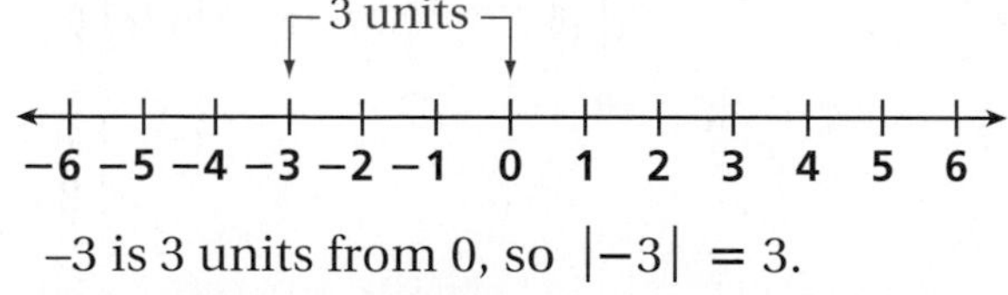

−3 is 3 units from 0, so |−3| = 3.

EXERCISES

Name a positive or negative number to represent each situation.

5. a raise of \$10
6. a loss of \$50

Graph each integer and its opposite on a number line.

7. −3
8. 1
9. 0

Use the number line to find the absolute value of each integer.

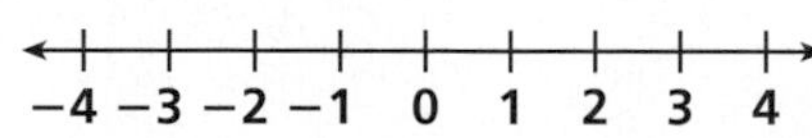

10. 2
11. −1
12. 0

9-2 Comparing and Ordering Integers (pp. 454–457)

EXAMPLE

- **Compare −2 and 3. Write < or >.**

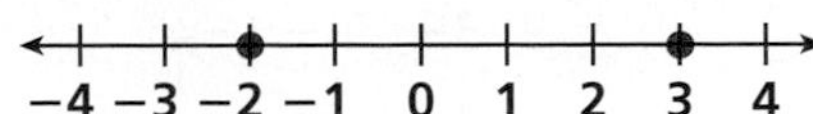

$-2 < 3$ *−2 is left of 3 on the number line.*

- **Order 3, −2, and 0 from least to greatest.**

−5 −4 −3 −2 −1 0 1 2 3 4 5

Read from left to right: −2, 0, 3.

EXERCISES

Compare. Write < or >.

13. 3 ▢ 4 **14.** −2 ▢ 5 **15.** 0 ▢ 6

16. −5 ▢ −7 **17.** 8 ▢ −11 **18.** −4 ▢ 0

Order each set of integers from least to greatest.

19. 2, −1, 4 **20.** −3, 0, 4

21. −3, 1, −2, 0 **22.** −6, −8, 0

23. 7, −4, −7 **24.** −1, 7, 3, −5

9-3 The Coordinate Plane (pp. 458–461)

EXAMPLE

- **Give the coordinates of *A* and identify the quadrant in which it lies.**

A is in the fourth quadrant. Its coordinates are (2, –3).

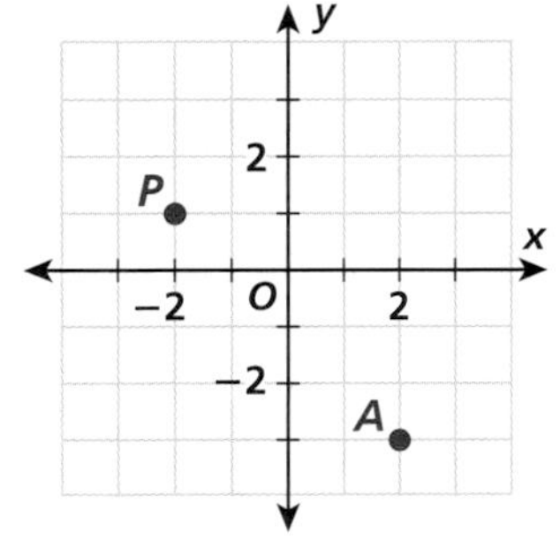

- **Graph *P*(−2, 1) on a coordinate plane.**

From (0, 0), move 2 units left and 1 unit up.

EXERCISES

Give the coordinates of each point.

25. *A* **26.** *C*

Give the quadrant in which each point lies.

27. *A* **28.** *B*

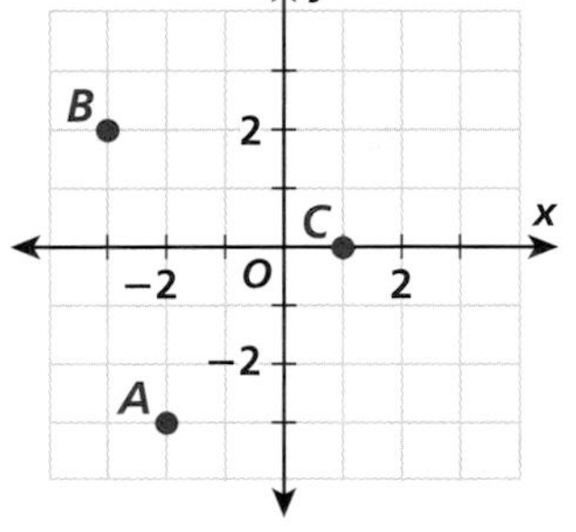

Graph each point on a coordinate plane.

29. *D*(−2, −1) **30.** *E*(0, 0) **31.** *F*(3, 4)

9-4 Adding Integers (pp. 465–468)

EXAMPLE

- **Find the sum: 3 + (−2).**

+ (−2)
3
−6 −5 −4 −3 −2 −1 0 1 2 3 4 5 6

3 + (–2) = 1

- **Evaluate *x* + (−2) when *x* = 9.**

9 + (−2) = 7

EXERCISES

Find each sum.

32. −4 + 2 **33.** 4 + (−4)

34. 3 + (−2) **35.** −3 + (−2)

Evaluate *x* + 3 for the following values.

36. *x* = −20 **37.** *x* = 5

9-5 Subtracting Integers (pp. 470–472)

EXAMPLE

- **Subtract the integers: $3 - (-2)$.**

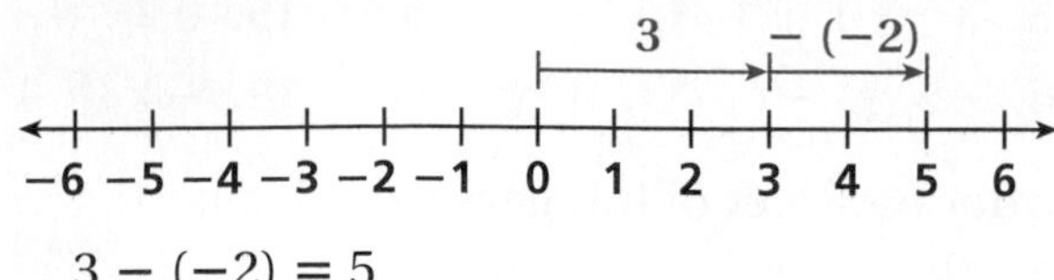

$3 - (-2) = 5$

- **Evaluate $n - 4$ for $n = -1$.**
$(-1) - 4 = -5$

EXERCISES

Find each difference.

38. $-6 - 2$
39. $5 - (-4)$
40. $5 - (-5)$
41. $-3 - (-3)$

Evaluate $x - (-1)$ for each value of x.

42. $x = 12$
43. $x = -7$

9-6 Multiplying Integers (pp. 473–475)

EXAMPLE

- **Find the product: $3 \cdot (-2)$.**
Think: 3 groups of −2
$3 \cdot (-2) = -6$

- **Evaluate $-2x$ for $x = -4$.**
$-2(-4) = 8$

EXERCISES

Find each product.

44. $5 \cdot (-2)$
45. $3 \cdot 2$
46. $-3 \cdot (-2)$
47. $-4 \cdot 2$

Evaluate each expression for the given value of the variable.

48. $n \cdot (-8)$; $n = 2$
49. $-9y$; $y = -5$

9-7 Dividing Integers (pp. 476–479)

EXAMPLE

- **$-24 \div 4$**
Think: $-6 \cdot 4 = -24$
$-24 \div 4 = -6$

- **Evaluate $\frac{x}{-2}$ for $x = 14$.**
$\frac{14}{-2} = -7$

EXERCISES

Find each quotient.

50. $6 \div (-2)$
51. $9 \div 3$
52. $-14 \div (-7)$
53. $-4 \div 2$

Evaluate each expression for the given value of the variable.

54. $n \div 2$; $n = -24$
55. $\frac{x}{-3}$; $x = 27$

9-8 Solving Integer Equations (pp. 482–485)

EXAMPLE

- **Solve the equation.**
$x + 4 = 18$

$$\begin{aligned} x + 4 &= 18 \\ \underline{-4} \quad &\underline{-4} \\ x \quad &= 14 \end{aligned}$$ *Subtract 4 from both sides.*

EXERCISES

Solve each equation. Check your answers.

56. $w - 5 = -1$
57. $\frac{a}{-4} = 3$
58. $2q = -14$
59. $x + 3 = -2$

Chapter Test

Chapter 9

Name a positive or negative number to represent each situation.

1. 30° below zero
2. a bank deposit of $75
3. an increase of 10 points
4. a loss of 5 yards

Write the opposite of each integer.

5. -3 **6.** 2 **7.** -19 **8.** 0

Find the absolute value.

9. $|-6|$ **10.** $|7|$ **11.** $|-20|$ **12.** $|11|$

Compare. Write $<$ or $>$.

13. -4 ▢ 4 **14.** 2 ▢ -9 **15.** -10 ▢ 8 **16.** -2 ▢ -12

Order each set of integers from least to greatest.

17. $21, -19, 34$
18. $-16, -2, 13, 46$
19. $-10, 0, 25, -7, 18$

Graph each point on a coordinate plane.

20. $A(2, 3)$
21. $C(-1, 3)$
22. $E(0, 1)$
23. $B(3, -2)$
24. $D(2, 0)$
25. $F(-1, -2)$

Add, subtract, multiply, or divide.

26. $-4 + 4$
27. $-2 - 9$
28. $-3 \cdot 8$
29. $12 \div (-3)$
30. $-48 \div (-4)$
31. $13 + (-9)$
32. $8 - (-11)$
33. $-7 \cdot (-6)$
34. $7 \cdot (-9)$
35. $-42 \div 2$
36. $-15 + (-10)$
37. $-31 - (-16)$

Evaluate each expression for the given value of the variable.

38. $n + 3, n = -10$
39. $9 - x, x = -9$
40. $m \cdot 4, m = -6$
41. $\frac{15}{a}, a = -3$
42. $(-11) + z, z = 28$
43. $w - (-8), w = 13$
44. $-7c, c = 13$
45. $n \div 4, n = -32$
46. $p + (-14), p = -22$

Solve each equation.

47. $\frac{b}{7} = -3$
48. $-9 \cdot f = -81$
49. $r - 14 = -32$
50. $y + 17 = -2$

Chapter Test

Chapter 9

Performance Assessment

Show What You Know

Create a portfolio of your work from this chapter. Complete this page and include it with your four best pieces of work from Chapter 9. Choose from your homework or lab assignments, mid-chapter quiz, or any journal entries you have done. Put them together using any design you want. Make your portfolio represent what you consider your best work.

Short Response

1. The high temperatures in Nome, Alaska, for one week were 5°F, 4°F, −2°F, −3°F, −1°F, 2°F, and 2°F. What was the average high temperature in Nome for that week? Show all your steps.

2. Lionel finished four rounds of a golf tournament with a score of −10. His scores on the first three rounds were −4, −2, and −6. What was his score on the last round? Show your work.

3. Mount McKinley is 20,320 feet above sea level. Death Valley is 20,602 feet lower than Mount McKinley. Write and solve an equation to find the elevation of Death Valley.

Extended Problem Solving

4. Asheka has a checking account and a savings account. There is no monthly service fee for her checking account, but she must pay $4.00 a month to maintain her savings account.
   a. Asheka's checking account had a balance of $50. She then deposited $50 and wrote checks for $40, $24, and $18. What is Asheka's new checking account balance?
   b. During June, Asheka deposited $1,250 in her savings account, withdrew $575, and paid the monthly fee. Write an equation that could be used to find the balance b at the beginning of June if the balance at the end of June was $968.
   c. Find Asheka's savings account balance at the beginning of June.

Performance Assessment

internet connect
EOG Test Practice Online
go.hrw.com Keyword: MR4 TestPrep

Getting Ready for EOG

Chapter 9

Cumulative Assessment, Chapters 1–9

1. Which is the correct value of 3^3?

A 6
B 9
C 12
D 27

TEST TAKING TIP!

You can solve the equation, or you can substitute each answer choice in the equation to check whether it is a solution.

2. What is the solution to $18 + g = 72$?

A $g = 4$
B $g = 54$
C $g = 90$
D $g = 1{,}296$

3. Which is the correct value of $18 - (2 + 4) \cdot 4$?

A −6
B 10
C 42
D 48

4. Which is the greatest common factor of 24, 32, and 40?

A 4
B 6
C 8
D 10

5. Which decimal is equivalent to 0.035?

A 0.35
B 0.0350
C 0.305
D 0.0305

6. Which is the value of $|-62|$?

A −62
B −4
C 0
D 62

7. Which set of numbers has a mean of 7?

A 4, 8, 9
B 3, 7, 8
C 7, 7, 10
D 7, 9, 11

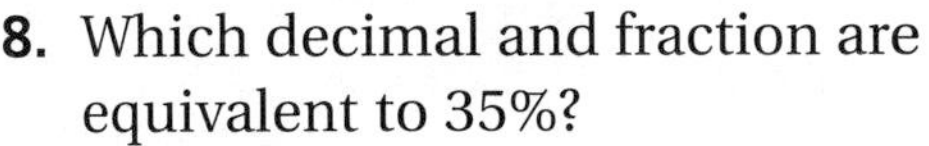

8. Which decimal and fraction are equivalent to 35%?

A $35.0, \frac{7}{20}$
B $0.35, \frac{7}{20}$
C $0.35, \frac{35}{10}$
D $3.5, \frac{35}{10}$

9. ***SHORT RESPONSE*** Darius drinks $1\frac{1}{2}$ cups of water three times a day. How many cups of water does Darius drink per week? Show your work.

10. ***SHORT RESPONSE*** The line graph below shows the temperature from 8 A.M. to noon. What was the mean change per hour in temperature from 8 A.M. to noon? Explain how you found your answer.

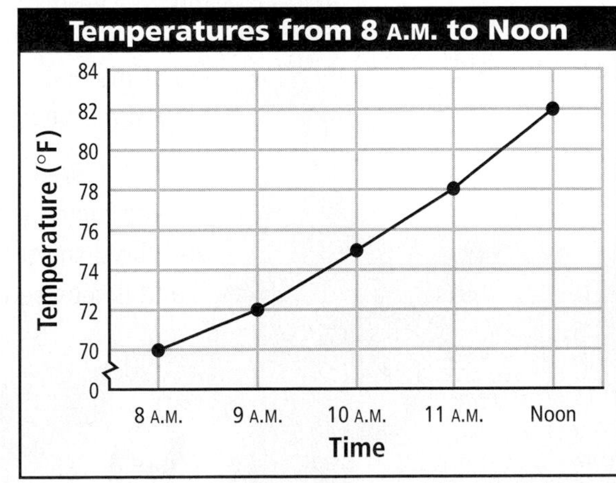

Getting Ready for EOG

Chapter

10

Perimeter, Area, and Volume

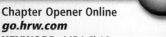

TABLE OF CONTENTS

Career ***Mathematician***

Some mathematicians apply their knowledge in areas such as airplane scheduling, medical safety, and automobile and industrial research. Other mathematicians prefer to study the concepts behind mathematics.

For hundreds of years, mathematicians have studied the relationship between the circumference and the diameter of a circle. This ratio is called *pi* and is represented by the Greek letter π.

internet connect

Chapter Opener Online
go.hrw.com
KEYWORD: MR4 Ch10

ARE YOU READY?

Choose the best term from the list to complete each sentence.

1. A(n) __?__ is a quadrilateral with opposite sides that are parallel and congruent.
2. Some customary units of length are __?__ and __?__. Some metric units of length are __?__ and __?__.
3. A(n) __?__ is a quadrilateral with side lengths that are all congruent and four right angles.
4. A(n) __?__ is a polygon with six sides.

square
feet
cube
meters
liters
hexagon
inches
parallelogram
trapezoid
centimeters
sphere

Complete these exercises to review skills you will need for this chapter.

✔ Add and Multiply Whole Numbers, Fractions, and Decimals

Find each sum or product.

5. $1.5 + 2.4 + 3.6 + 2.5$
6. $2 \cdot 3.5 \cdot 4$
7. $\frac{22}{7} \cdot 21$
8. $\frac{1}{2} \cdot 5 \cdot 4$
9. $3.2 \cdot 5.6$
10. $\frac{1}{2} \cdot 10 \cdot 3$
11. $(2 \cdot 5) + (6 \cdot 8)$
12. $2(3.5) + 2(1.5)$
13. $9(20 + 7)$

✔ Estimate Metric Lengths

Use a centimeter ruler to measure each line to the nearest centimeter.

14. ______________________________

15. __________________

✔ Identify Polygons

Name each polygon. Determine whether it appears to be regular or not regular.

16. 

17. 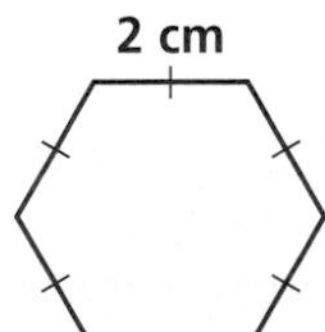

18. 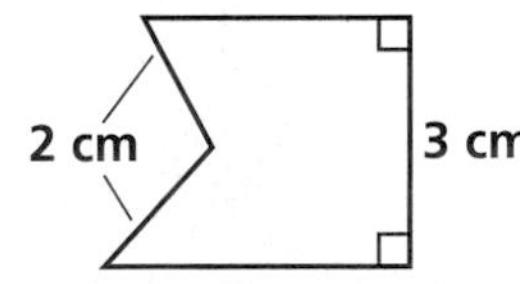

10-1 Finding Perimeter

Learn to find the perimeter and missing side lengths of a polygon.

Vocabulary

perimeter

One of the biggest finger paintings ever painted is *Ten Fingers, Ten Toes*. It is 8.53 meters wide and 10.66 meters long.

The **perimeter** of a figure is the distance around it. To find the perimeter of the painting you can add the lengths of the sides.

$8.53 + 10.66 + 8.53 + 10.66 = 38.38$

The perimeter of the painting is 38.38 meters.

EXAMPLE 1 Finding the Perimeter of a Polygon

Find the perimeter of the figure.

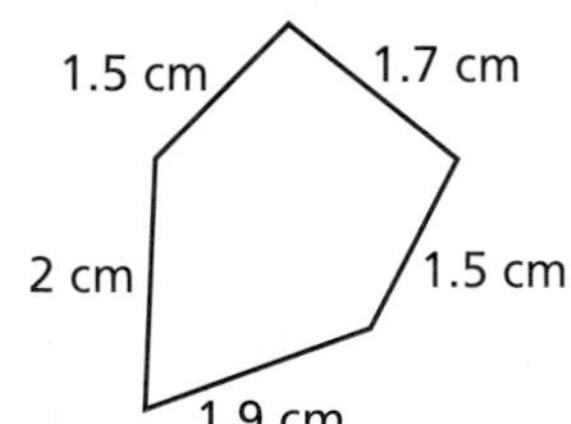

$1.5 + 1.7 + 1.5 + 1.9 + 2 = 8.6$

Add all the side lengths.

The perimeter is 8.6 cm.

PERIMETER OF A RECTANGLE

The opposite sides of a rectangle are equal in length. Find the perimeter of a rectangle by using the formula, in which ℓ is the length and w is the width.

$P = 2\ell + 2w$

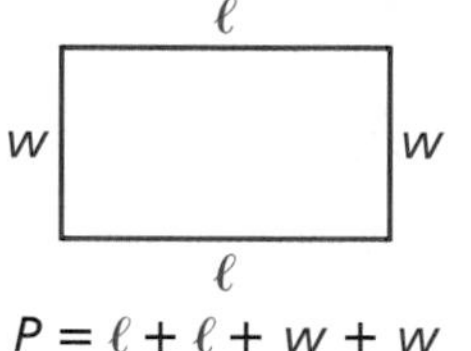

$P = \ell + \ell + w + w$

EXAMPLE 2 Using a Formula to Find Perimeter

Find the perimeter P of the rectangle.

$P = 2\ell + 2w$

$P = (2 \cdot 3) + (2 \cdot 2)$ *Substitute 3 for ℓ and 2 for w.*

$P = 6 + 4$ *Multiply.*

$P = 10$ *Add.*

The perimeter is 10 feet.

EXAMPLE 3 Finding Unknown Side Lengths and the Perimeter of a Polygon

Find each unknown measure.

A **What is the length of side a if the perimeter equals 105 m?**

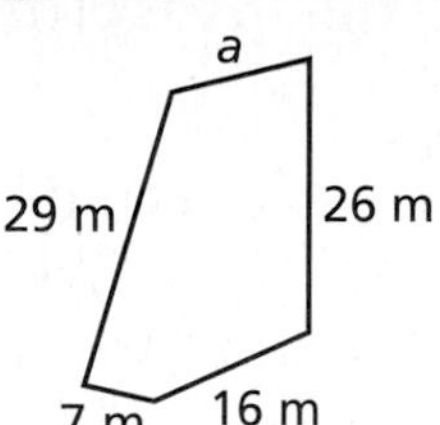

$P =$ sum of side lengths	
$105 = a + 26 + 16 + 7 + 29$	*Use the values you know.*
$105 = a + 78$	*Add the known lengths.*
$105 - 78 = a + 78 - 78$	*Subtract 78 from both sides.*
$27 = a$	

Side a is 27 m long.

B **What is the perimeter of the polygon?**

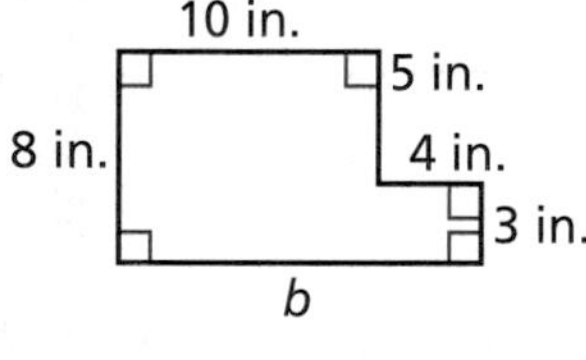

First find the unknown side length.
Find the sides opposite side b.
The length of side b = 10 + 4.
Side b is 14 in. long.

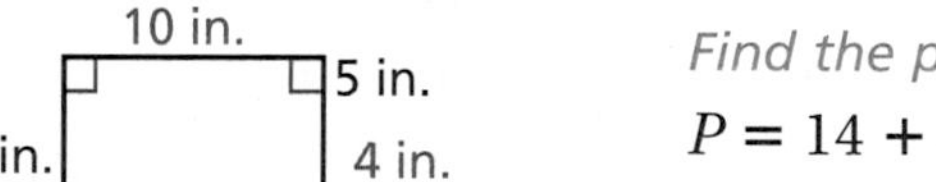

Find the perimeter.
$P = 14 + 8 + 10 + 5 + 4 + 3$
$P = 44$

The perimeter of the polygon is 44 in.

C **The width of a rectangle is 12 cm. What is the perimeter of the rectangle if the length is 3 times the width?**

3w
12 cm

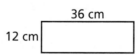

$\ell = 3w$	*Find the length.*
$\ell = (3 \cdot 12)$	*Substitute 12 for w.*
$\ell = 36$	*Multiply.*

36 cm
12 cm

$P = 2\ell + 2w$	*Use the formula for the perimeter of a rectangle.*
$P = 2(36) + 2(12)$	*Substitute 36 and 12.*
$P = 72 + 24$	*Multiply.*
$P = 96$	*Add.*

The perimeter of the rectangle is 96 cm.

Think and Discuss

1. Explain how to find the perimeter of a regular pentagon if you know the length of one side.

2. Tell what formula you can use to find the perimeter of a square.

10-1 Exercises

FOR EOG PRACTICE
see page 676

internet connect
Homework Help Online
go.hrw.com Keyword: MR4 10-1

2.02

GUIDED PRACTICE

See Example 1 Find the perimeter of each figure.

1. 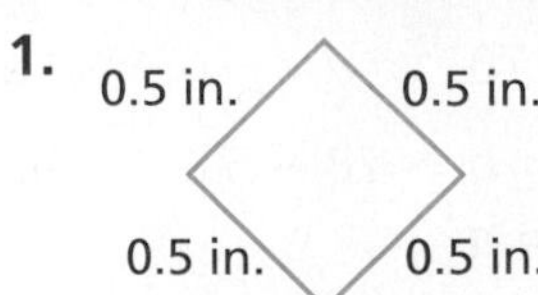

2. 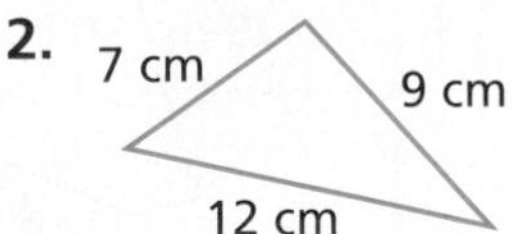

See Example 2 Find the perimeter P of each rectangle.

3. 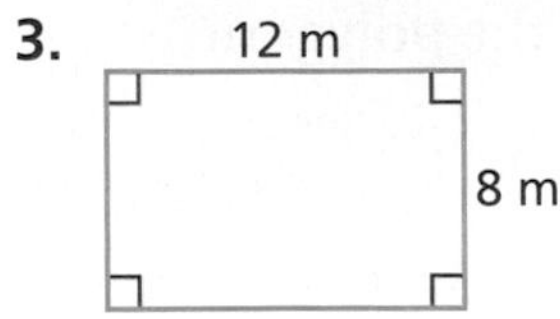

4. 

See Example 3 Find the unknown measure.

5. What is the length of side b if the perimeter equals 21 yd?

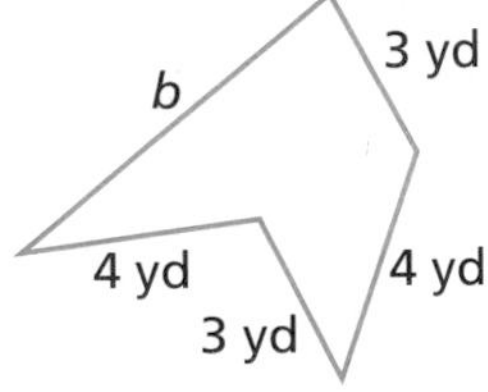

INDEPENDENT PRACTICE

See Example 1 Find the perimeter of each figure.

6. 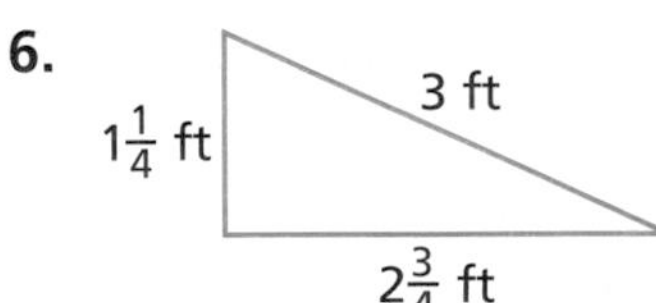

7. regular octagon

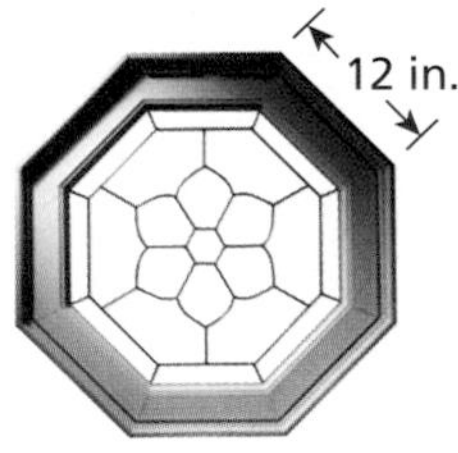

See Example 2 Find the perimeter P of each rectangle.

8. 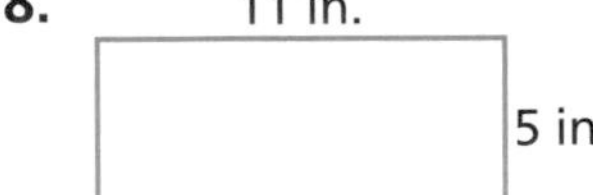

9. 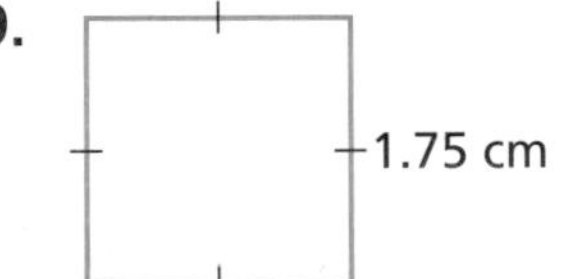

10. 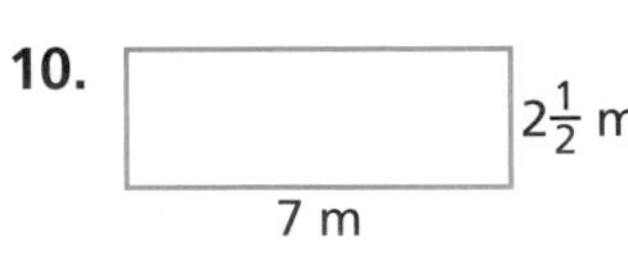

See Example 3 Find each unknown measure.

11. What is the perimeter of the polygon?

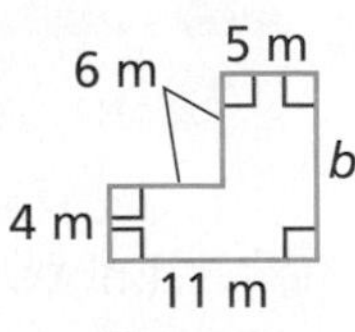

12. The width of a rectangle is 15 ft. What is the perimeter of the rectangle if the length is 5 ft longer than the width?

PRACTICE AND PROBLEM SOLVING

Use the figure *ACDEFG* for Exercises 13–15.

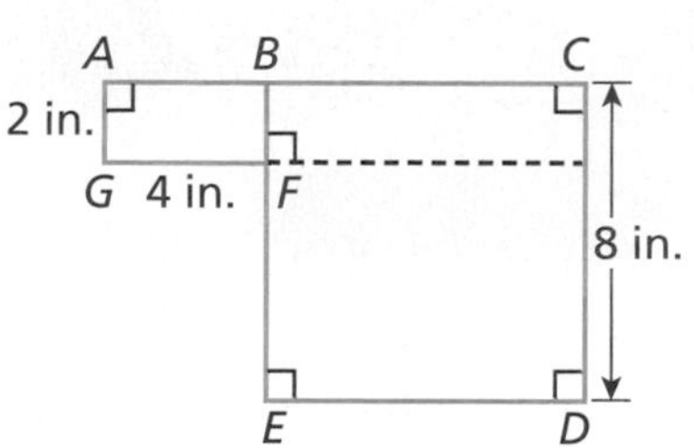

13. What is the length of side *FE*?
14. If the perimeter of rectangle *BCDE* is 34 in., what is the length of side *BC*?
15. Use your answer from Exercise 14 to find the perimeter of figure *ACDEFG*.

Find the perimeter of each figure.

16. a triangle with side lengths 6 in., 8 in., and 10 in.
17. a regular pentagon with side length $\frac{2}{5}$ km
18. ***SPORTS*** The diagram shows one-half of a badminton court.
    a. What are the dimensions of the whole court?
    b. What is the perimeter of the whole court?

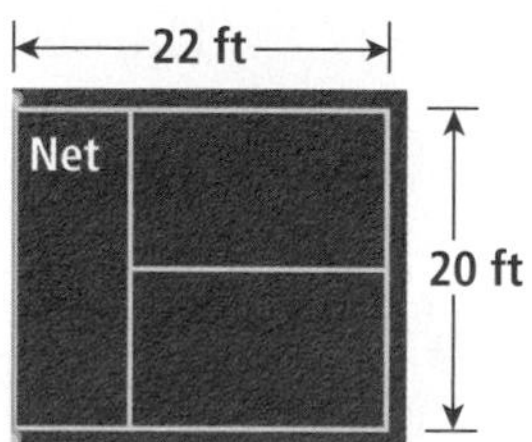

19. ***MEASUREMENT*** Use the map and a centimeter ruler to find the perimeter of the triangle formed by the three cities.

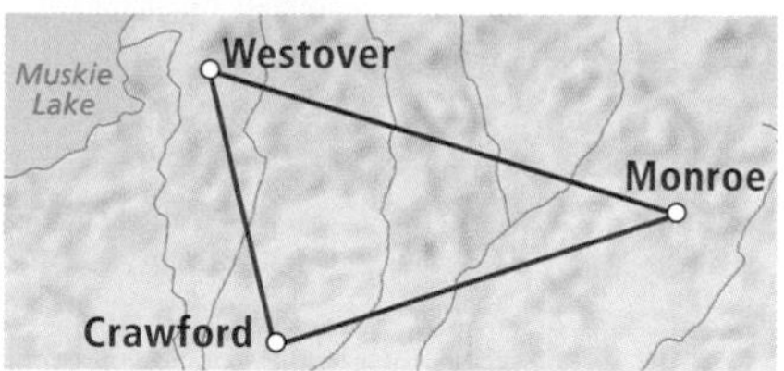

20. ***WHAT'S THE ERROR?*** A student found the perimeter of a 10-inch-by-13-inch rectangle to be 23 inches. Explain the student's error. Then find the correct perimeter.

21. ***WRITE ABOUT IT*** Explain how to find the unknown length of a side of a triangle that has a perimeter of 24 yd and two sides that measure 6 yd and 8 yd.

22. ***CHALLENGE*** The perimeter of a regular octagon is 20 m. What is the length of one side of the octagon?

Spiral Review

Find each sum or difference. (Lesson 3-3)

23. $30 - 5.32$
24. $80.37 + 15.125$
25. $100 - 25.65$
26. $200.6 + 62.78$

Solve each proportion. (Lesson 8-2)

27. $\frac{9}{15} = \frac{x}{5}$
28. $\frac{a}{20} = \frac{3}{15}$
29. $\frac{1}{7} = \frac{6}{k}$

30. **EOG PREP** Which decimal is equivalent to 85%? (Lesson 8-7)

A 85.0   B 8.5   C 0.85   D 0.085

10-2 Estimating and Finding Area

Learn to estimate the area of irregular figures and to find the area of rectangles, triangles, and parallelograms.

Vocabulary

area

When colonists settled the land that would become the United States, ownership boundaries were sometimes natural landmarks such as rivers, trees, and hills. Landowners who wanted to know the size of their property needed to find the areas of irregular shapes.

The **area** of a figure is the amount of surface it covers. We measure area in square units.

EXAMPLE 1 Estimating the Area of an Irregular Figure

Estimate the area of the figure.

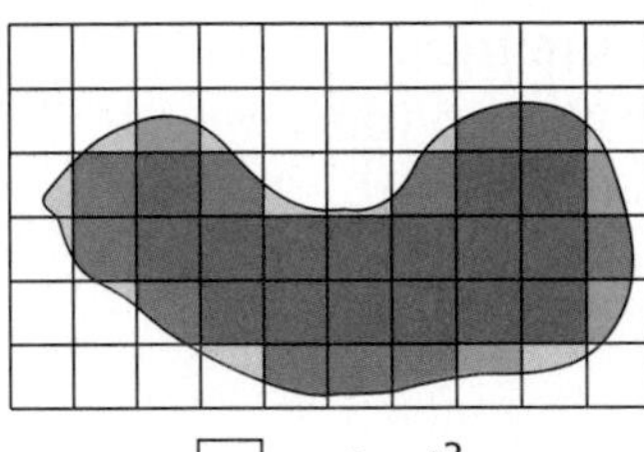

Count full squares: 16 red squares.

Count almost-full squares: 11 blue squares.

Count squares that are about half-full:
4 green squares ≈ 2 full squares.

Do not count almost empty yellow squares.

Add. 16 + 11 + 2 = 29

The area of the figure is about 29 mi^2.

AREA OF A RECTANGLE

To find the area of a rectangle, multiply the length by the width.

$$A = \ell w$$
$$A = 4 \cdot 3 = 12$$

The area of the rectangle is 12 square units.

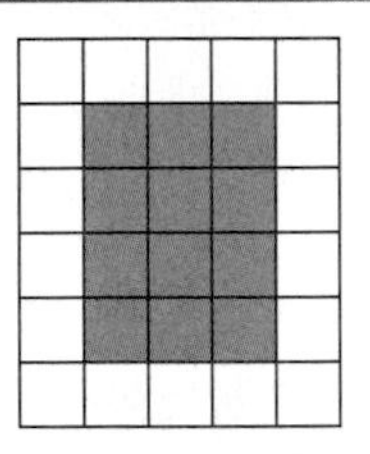

EXAMPLE 2 Finding the Area of a Rectangle

Find the area of the rectangle.

13 m

8 m

$A = \ell w$ *Write the formula.*

$A = 13 \cdot 8$ *Substitute **13** for ℓ.*

$A = 104$ *Substitute **8** for w.*

The area is 104 m^2.

You can use the formula for the area of a rectangle to write a formula for the area of a parallelogram. Imagine cutting off the triangle drawn in the parallelogram and sliding it to the right to form a rectangle.

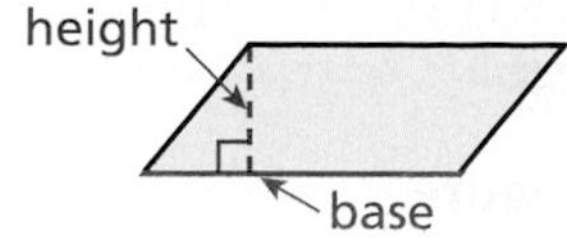

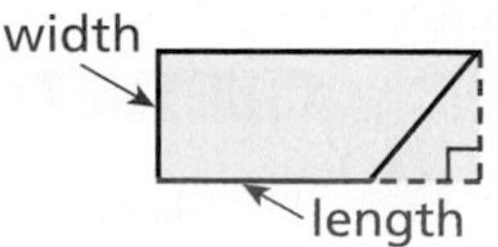

The area of a parallelogram = bh. The area of a rectangle = ℓw.

The **base** of the parallelogram is the **length** of the rectangle.
The **height** of the parallelogram is the **width** of the rectangle.

EXAMPLE 3 Finding the Area of a Parallelogram

Find the area of the parallelogram.

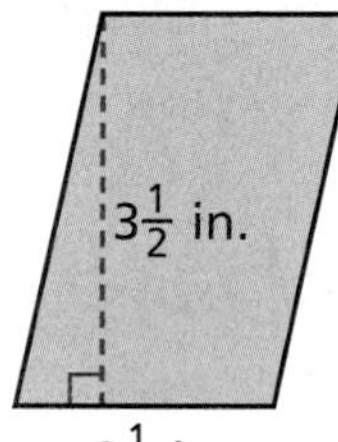

$A = bh$	*Write the formula.*
$A = 2\frac{1}{3} \cdot 3\frac{1}{2}$	*Substitute $2\frac{1}{3}$ for b and $3\frac{1}{2}$ for h.*
$A = \frac{7}{3} \cdot \frac{7}{2}$	*Multiply.*
$A = \frac{49}{6}$, or $8\frac{1}{6}$	The area is $8\frac{1}{6}$ in^2.

You can make a parallelogram out of two congruent triangles.

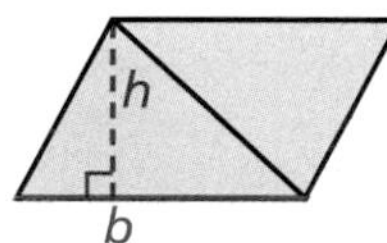

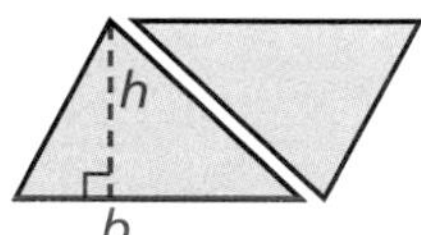

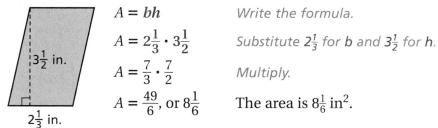

The area of each triangle is half the area of the parallelogram, so the formula for the area of a triangle is $A = \frac{1}{2}bh$.

EXAMPLE  4 Finding the Area of a Triangle

Find the area of the triangle.

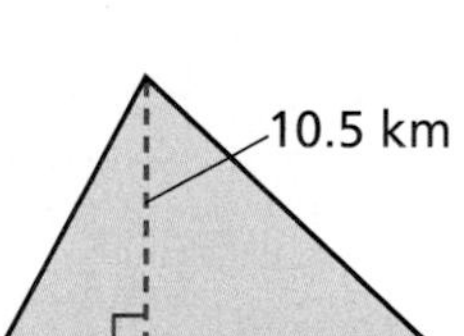

$A = \frac{1}{2}bh$	*Write the formula.*
$A = \frac{1}{2}(16.8 \cdot 10.5)$	*Substitute 16.8 for b.* *Substitute 10.5 for h.*
$A = \frac{1}{2}(176.4)$	*Multiply.*
$A = 88.2$	The area is 88.2 km^2.

Think and Discuss

1. **Explain** how the area of a triangle and the area of a rectangle that have the same base and the same height are related.
2. **Give** a formula for the area of a square.

10-2 Exercises

FOR EOG PRACTICE

see page 676

internet connect

Homework Help Online

go.hrw.com Keyword: MR4 10-2

2.02

GUIDED PRACTICE

See Example 1 Estimate the area of each figure.

1. 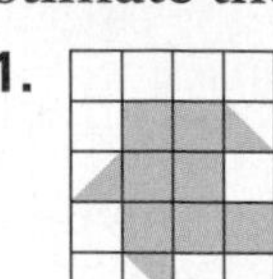

2. 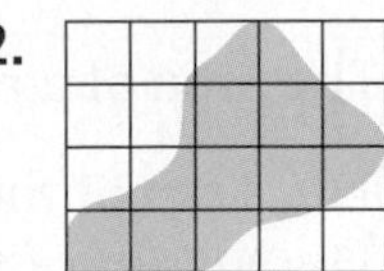

See Example 2 Find the area of each rectangle.

3. 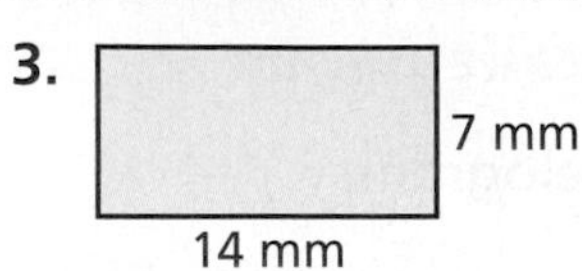

4. 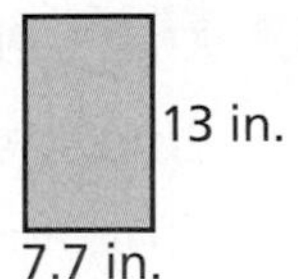

See Example 3 Find the area of each parallelogram.

5. 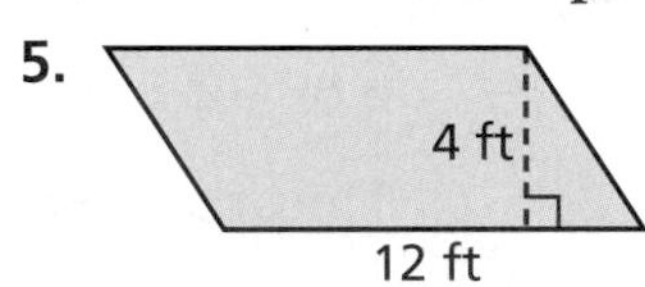

6. 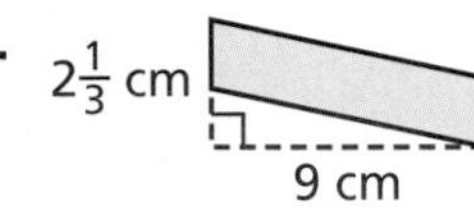

See Example 4 Find the area of each triangle.

7. 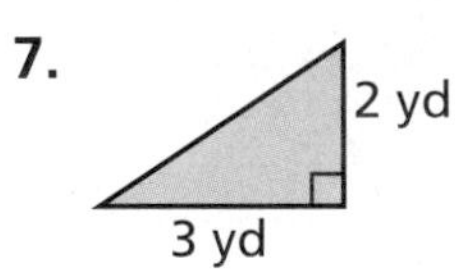

8. 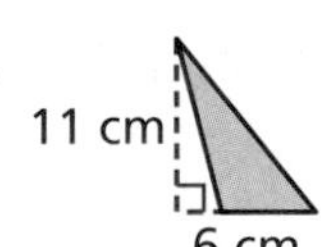

INDEPENDENT PRACTICE

See Example 1 Estimate the area of each figure.

9. 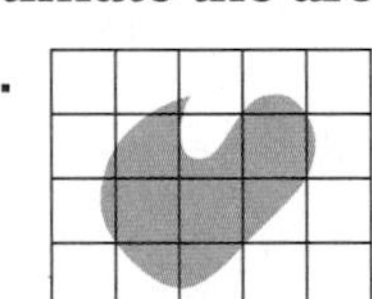

10. 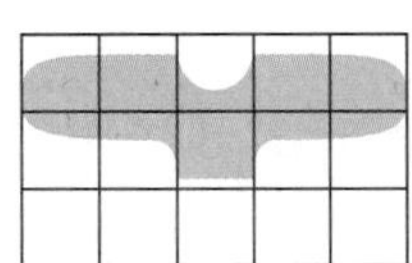

See Example 2 Find the area of each rectangle.

11. 5 mi, 25 mi

12. 1.5 m, 8.5 m

See Example 3 Find the area of each parallelogram.

13. 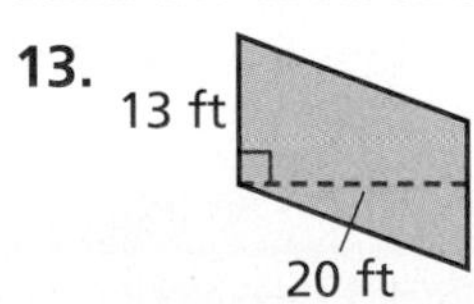

14. 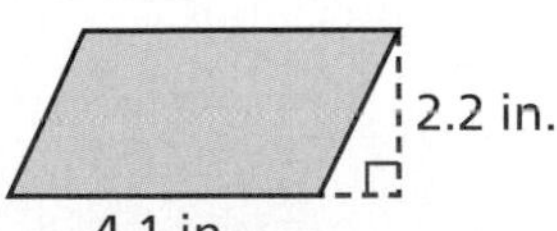

See Example 4 Find the area of each triangle.

15. 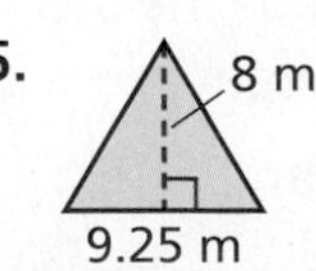

16. 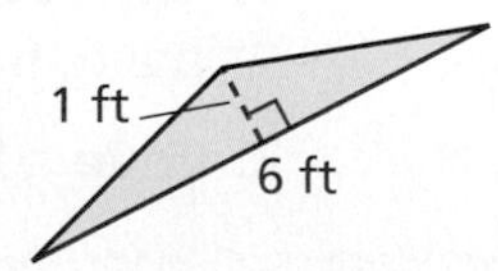

Social Studies 

Iceland has many active volcanoes and frequent earthquakes. There are more hot springs in Iceland than in any other country in the world. One spring, the Great Geysir, is capable of releasing about 250 liters of boiling water per second. The word *geyser* comes from the Great Geysir, which, though it rarely erupts anymore, can spray hot water up to 60 meters high.

Sightseers watch the eruption of Geyser Namafjall in the Myvatn Region of North Iceland.

Use the map for Exercises 17–18.

17. One square on the map represents 1,700 km^2. Which is a reasonable estimate for the area of Iceland?

A Less than 65,000 km^2

B Between 90,000 and 105,000 km^2

C Between 120,000 and 135,000 km^2

D Greater than 150,000 km^2

18. About 10% of the area of Iceland is covered with glaciers. Estimate the area covered by glaciers.

19. ***WRITE ABOUT IT*** The House is Iceland's oldest building. When it was built in 1765, the builders measured length in *ells*. The House is 14 ells wide and 20 ells long. Explain how to find the area in ells of the House.

20. ***CHALLENGE*** The length of one ell varied from country to country. In England, one ell was equal to $1\frac{1}{4}$ yd. Suppose the House were measured in English ells. Find the area in yards of the House.

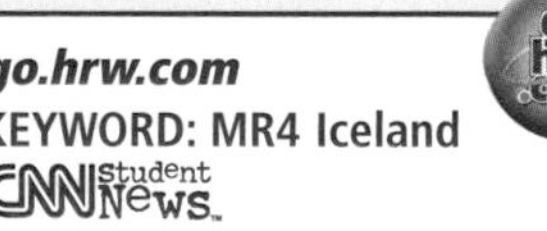

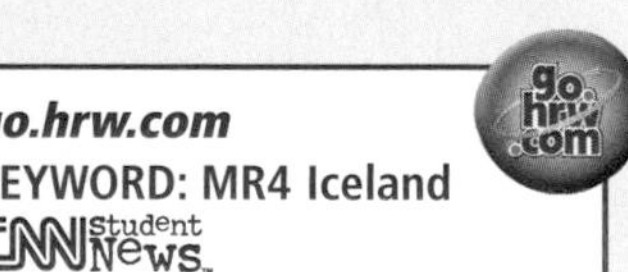

Spiral Review

21. Damien's favorite song is 4.2 minutes long. Jan's favorite song is 2.89 minutes long. Estimate the difference in the lengths of the songs. (Lesson 3-2)

Find each product. (Lesson 5-2)

22. $2\frac{2}{3} \times \frac{1}{8}$ **23.** $\frac{1}{4} \times 3\frac{1}{2}$ **24.** $1\frac{1}{4} \times 1\frac{2}{5}$ **25.** $2\frac{1}{5} \times 2\frac{2}{3}$

26. **EOG PREP** Which ratio is equivalent to $\frac{1}{20}$? (Lesson 8-1)

A 9:180 **B** 180 to 9 **C** 4 to 100 **D** 100:4

10-3 Break into Simpler Parts

Problem Solving Skill

Learn to break a polygon into simpler parts to find its area.

You know how to find the area of rectangles, parallelograms, and triangles. To help you find the areas of other polygons, first solve a simpler problem by breaking the polygons apart into rectangles, parallelograms, and triangles.

EXAMPLE 1 Finding Areas of Composite Figures

Find the area of each polygon.

A

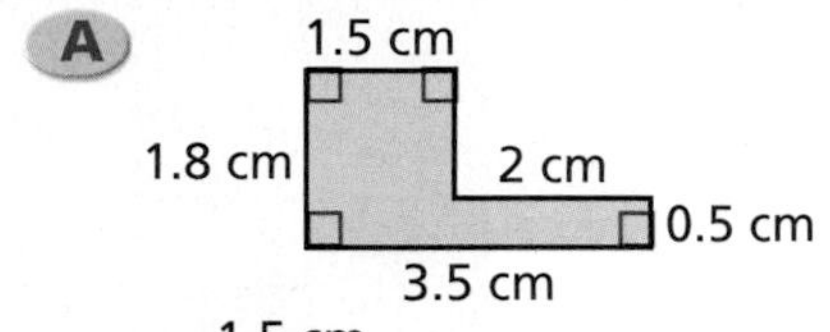

Think: Break the polygon apart into rectangles.

1.5 cm
1.8 cm
2 cm
0.5 cm

Find the area of each rectangle.

$A = \ell w$	$A = \ell w$	*Write the formula for the area of a rectangle.*
$A = \mathbf{1.8} \cdot \mathbf{1.5}$	$A = \mathbf{2} \cdot \mathbf{0.5}$	
$A = 2.7$	$A = 1$	

$2.7 + 1 = 3.7$ *Add to find the total area.*

The area of the polygon is 3.7 cm^2.

B

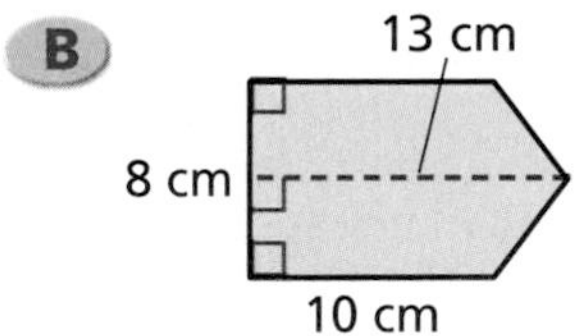

Think: Break the figure apart into a triangle and a rectangle.

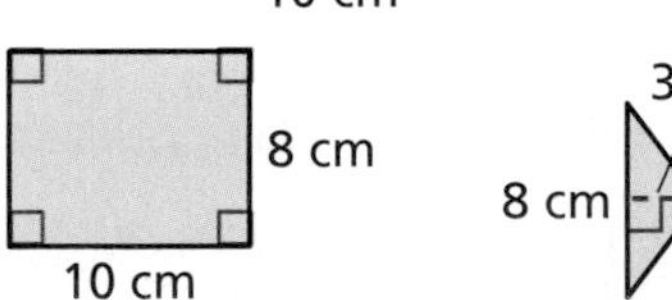

$A = \ell w$	$A = \frac{1}{2}bh$	*Find the area of each polygon.*
$A = \mathbf{8} \cdot \mathbf{10}$	$A = \frac{1}{2} \cdot \mathbf{8} \cdot \mathbf{3}$	
$A = 80$	$A = 12$	

$80 + 12 = 92$ *Add to find the total area of the figure.*

The area of the figure is 92 cm^2.

EXAMPLE 2 *ART APPLICATION*

Stan made a wall hanging. All the sides are 6 inches long, except for two longer sides that are each 12 inches. All the angles are right angles. What is the area of the wall hanging?

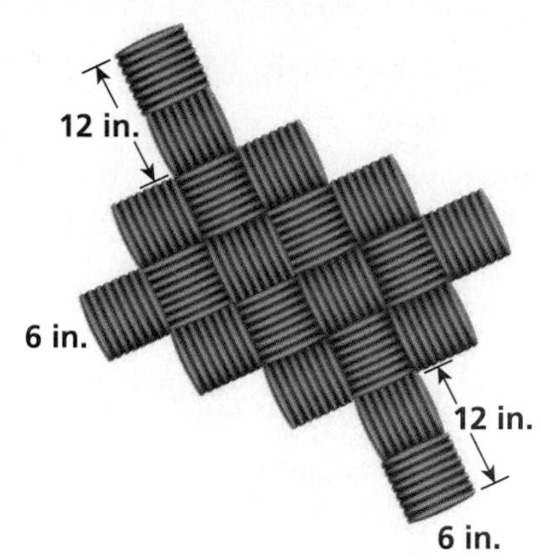

Think: Divide the wall hanging into 20 squares.

Find the area of one square that has a side length of 6 in.

$A = \ell w$ *Write the formula.*

$A = 6 \cdot 6 = 36$

$20 \cdot 36 = 720$ *Multiply to find the area of the 20 squares.*

The area of the wall hanging is 720 in^2.

Helpful Hint

You can also use the formula $A = s^2$, where s is the length of a side, to find the area of a square.

Think and Discuss

1. **Explain** how you can find the area of a regular octagon by breaking it apart into congruent triangles, if you know the area of one triangle.
2. **Explain** one way you can find the area of a trapezoid.

10-3 Exercises

2.02

FOR EOG PRACTICE
see page 677

internet connect
Homework Help Online
go.hrw.com Keyword: MR4 10-3

GUIDED PRACTICE

See Example 1 **Find the area of each polygon.**

1. 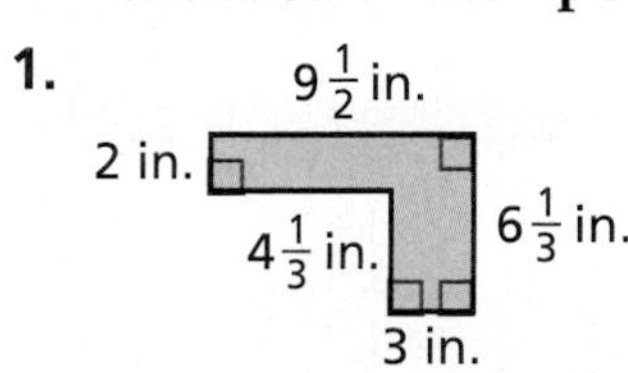

2. 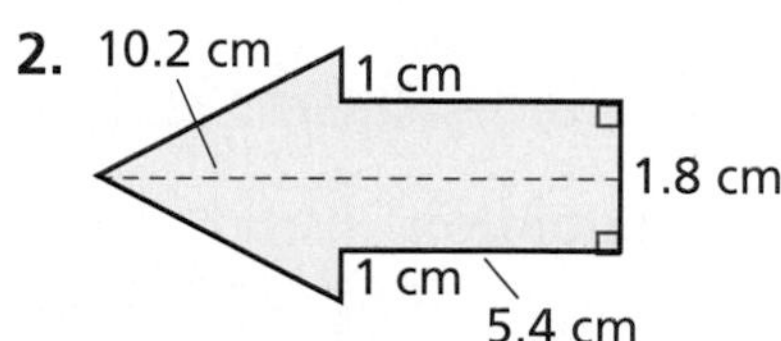

See Example 2

3. Gina used tiles to create a design. The lengths of all the sides are 3 cm, except for two longer sides that are 9 cm. What is the area of Gina's design?

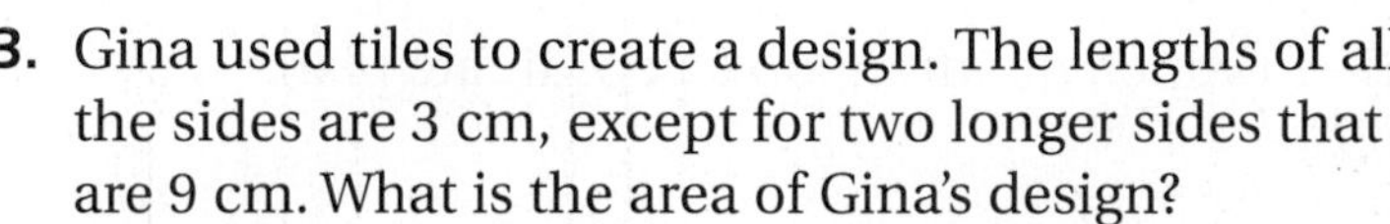

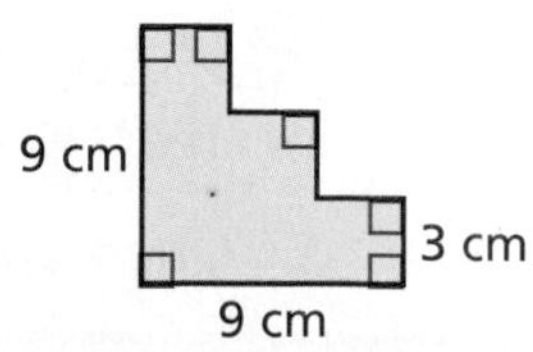

INDEPENDENT PRACTICE

See Example 1 **Find the area of each polygon.**

4. 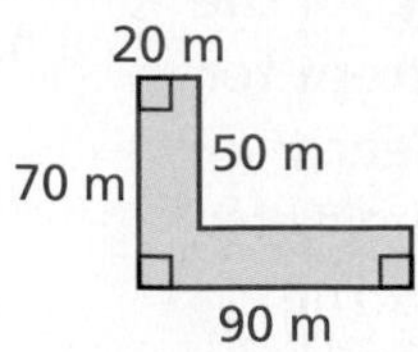

5. 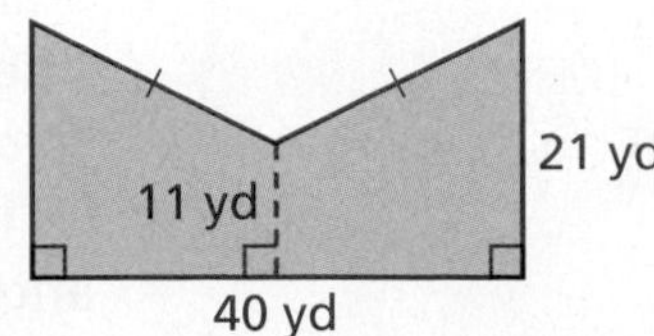

See Example 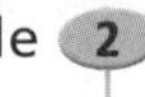

6. Edgar plants daffodils around a rectangular pond. The yellow part of the diagram shows where the daffodils are planted. What is the area of the yellow part of the diagram?

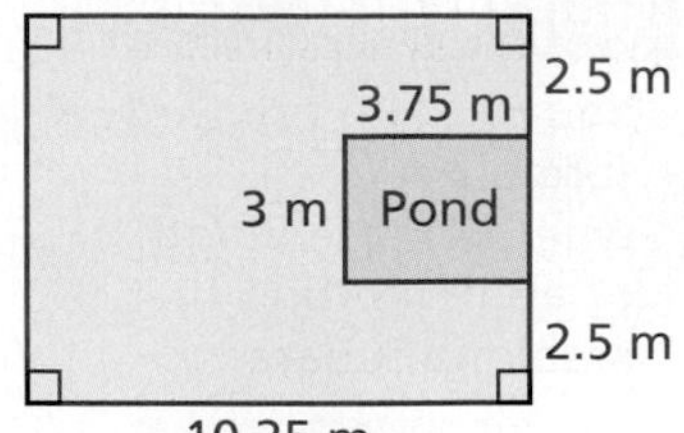

PRACTICE AND PROBLEM SOLVING

7. ***SOCIAL STUDIES*** The map shows the approximate dimensions of the state of South Australia.

   a. Look at the red figure outlining South Australia. Divide the figure into a right triangle and a rectangle.

   b. Find the total area of the rectangle and triangle.

   c. The total area of Australia is about 7.7 million km^2. About what fraction of the total area of Australia is the area of the state of South Australia?

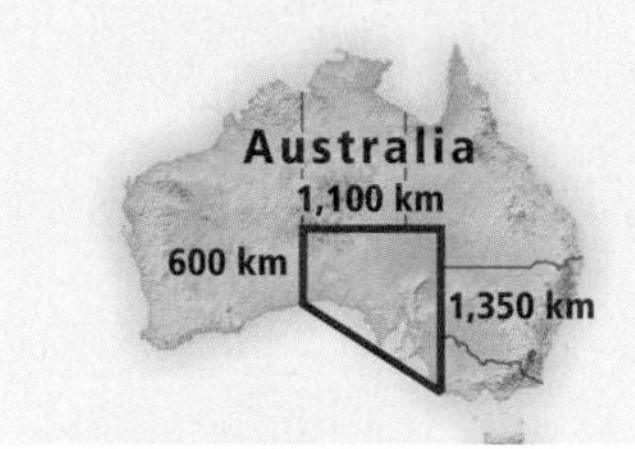

8. ***WRITE ABOUT IT*** Draw a figure that can be broken up into two rectangles. Label the lengths of each side. Explain how you can find the area of the figure. Then find the area.

9. ***CHALLENGE*** The perimeter of this figure is 42.5 cm. Find the area of this figure.

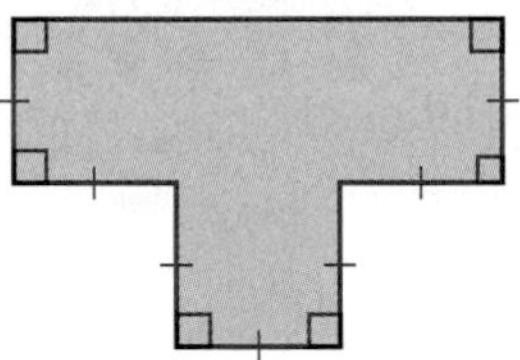

Spiral Review

Solve each equation. (Lesson 5-4)

10. $\frac{1}{4}x = 12$

11. $5\frac{1}{3}y = 9$

12. $\frac{2n}{5} = 3$

13. $3a = \frac{1}{7}$

14. **EOG PREP** In which quadrant on a coordinate plane is the point (–1, 2) located? (Lesson 9-3)

A Quadrant I B Quadrant II C Quadrant III D Quadrant IV

10-4 Comparing Perimeter and Area

Learn to make a model to explore how area and perimeter are affected by changes in the dimensions of a figure.

Ms. Cohn wants to enlarge this photo by doubling its length and width.

You can make a model on grid paper to see how the area and the perimeter of a figure change when its dimensions change.

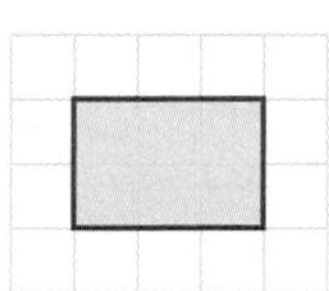

The original photo is a 3 in. × 2 in. rectangle.

perimeter = 10 in.
area = 6 in^2

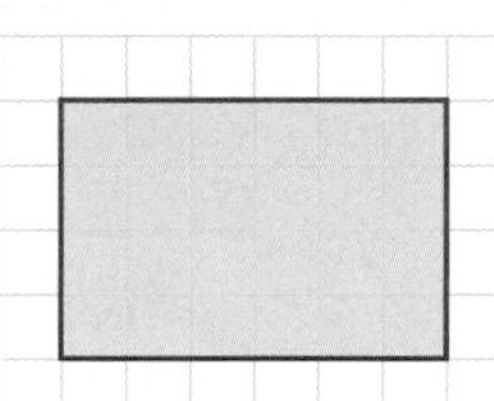

The enlarged photo will be a 6 in. × 4 in. rectangle.

perimeter = 20 in.
area = 24 in^2

If Ms. Cohn doubles the dimensions of the photo, the **perimeter** will also be doubled, and the **area** will be four times greater than the area of the original photo.

EXAMPLE 1 Changing Dimensions

Find how the perimeter and the area of the figure change when its dimensions change.

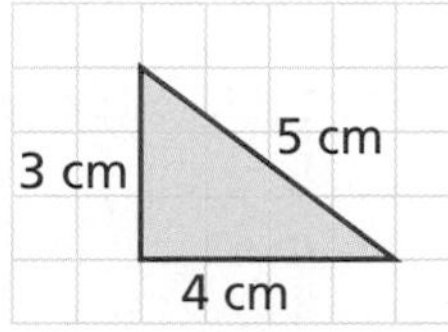

$P = 12$ cm
$A = 6$ cm^2

Divide each dimension by 2.

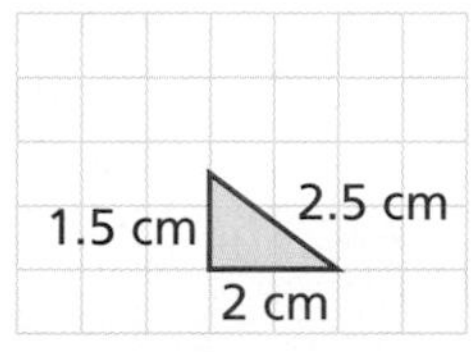

$P = 6$ cm
$A = 1.5$ cm^2

When the dimensions of the triangle are divided by 2, the **perimeter** is divided by 2, and the **area** is divided by 4, or 2^2.

EXAMPLE 2 Measurement Application

Use a centimeter ruler to measure the photo. Draw a rectangle whose sides are 3 times as long to enlarge the photo. How do the perimeter and the area change?

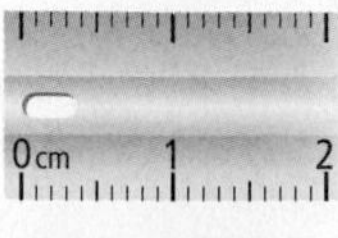

$P = 6$ cm
$A = 2$ cm^2

Multiply each dimension by 3.

$P = 18$ cm
$A = 18$ cm^2

When the dimensions of the rectangle are multiplied by 3, the **perimeter** is multiplied by 3, and the **area** is multiplied by 9, or 3^2.

Think and Discuss

1. **Explain** how the perimeter of a triangle changes when all the side lengths are doubled.
2. **Tell** how the area of a rectangle changes when all the side lengths are divided in half.

10-4 Exercises

FOR EOG PRACTICE
see page 677

internet connect
Homework Help Online
go.hrw.com Keyword: MR4 10-4

2.01, 2.02

GUIDED PRACTICE

See Example 

1. Find how the perimeter and the area of the figure change when its dimensions change.

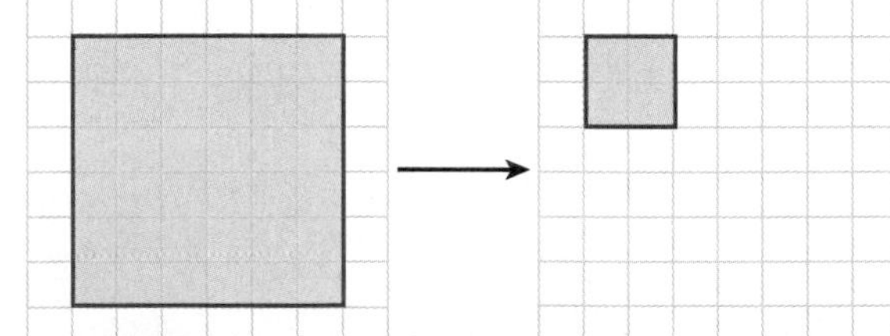

See Example 

2. Use a centimeter ruler to measure the rectangle. Then draw another rectangle with dimensions that are 2 times greater than the given rectangle. How do the perimeter and the area change when the dimensions change?

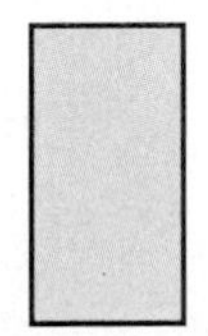

INDEPENDENT PRACTICE

See Example 

3. Find how the perimeter and the area of the figure change when its dimensions change.

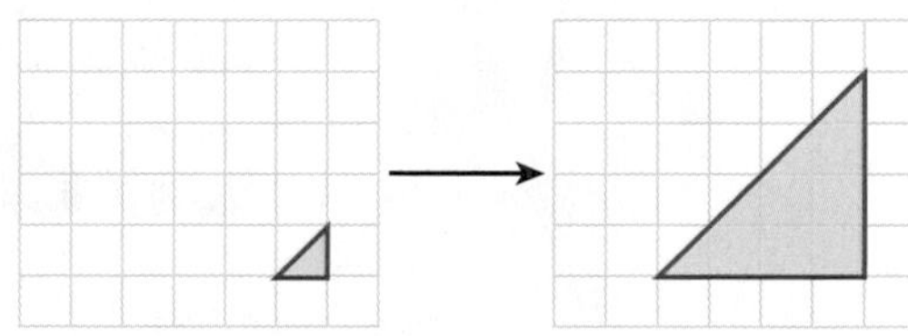

See Example 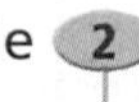

4. Use a centimeter ruler to measure the triangle. Then draw another triangle with dimensions that are half as great as the given triangle. How do the perimeter and the area change when the dimensions change?

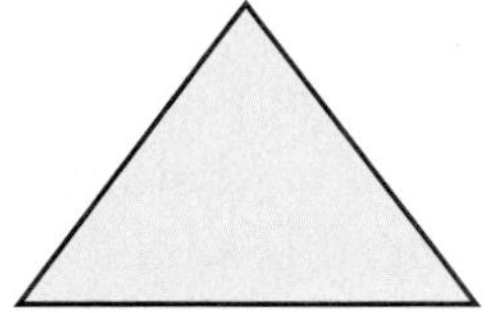

PRACTICE AND PROBLEM SOLVING

5. The schoolyard is a rectangle with a length of 120 ft and a width of 80 ft. The PE teacher plans to make a field in the schoolyard by dividing either the length or the width of the schoolyard in half.
   a. What will the area of the field be if she divides only one of the dimensions in half?
   b. What will the perimeter of the field be if she divides only the length in half? the width in half?

Use the table for Exercises 6–8.

6. Give an example of two frames, one with dimensions that are double the dimensions of the other.
7. George has a 3 in. × 4 in. photo. Which frame has dimensions that are 6 times greater than the dimensions of George's photo?
8. If George enlarges a 3 in. × 4 in. photo so that it is 12 in. × 16 in., how will its area change?

Photo Frames (in.)	
6 × 8	12 × 16
8 × 10	16 × 20
9 × 12	18 × 24

9. ***WRITE ABOUT IT*** What happens to the area and the perimeter of a rectangle when the length and width are multiplied by 4?

10. ***CHALLENGE*** A rectangle has a perimeter of 24 meters. If its length and width are whole numbers, what is its greatest possible area?

Spiral Review

Write each phrase as a numerical or algebraic expression. (Lesson 2-2)

11. 19 times 3
12. the quotient of g and 6
13. the sum of 5 and 9

14. **EOG PREP** Angle A and Angle B are supplementary. What is the measure of $\angle B$ if the measure of $\angle A$ is 75°? (Lesson 7-3)

A 15° B 25° C 105° D 150°

Discover Properties of Circles

Use with Lesson 10-5

Circles are not polygons because they are not made of line segments. The distance around a circle is not called the perimeter; it is called the *circumference.*

Activity 1

1. Use a compass to draw several different-sized circles, or trace around circular objects such as lids, cups, and plates.

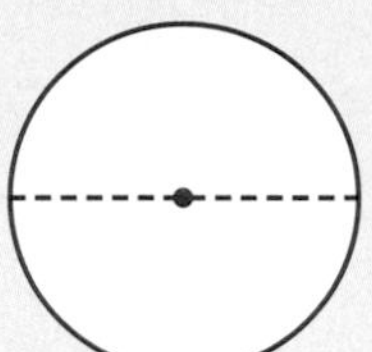

Diameter

2. For each circle, use a ruler to measure the distance across it through its center. Record this as the *diameter.*

3. Lay a piece of string around the circle and mark the string where it meets itself. Measure the length, and record it as the circumference.

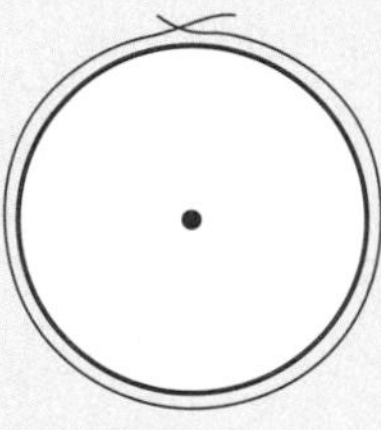

Circumference

4. Use a calculator to find the relationship between the circumference C, and the diameter d. Round this value to the nearest hundredth, and record it in the table.

	Circle 1	Circle 2	Circle 3	Circle 4
Circumference C				
Diameter d				
$\frac{C}{d}$				

The ratio of C to d is called *pi,* which is represented by the Greek letter π. You can write the equation for circumference as $C = \pi d$.

Think and Discuss

1. Give an approximation of π using the data in your table.

Try This

Use your approximation of π to find the circumference of each circle. Compare your answer to the given value of C.

1. 4 in. $C = 12.57$ in.

2. 3 cm $C = 9.42$ cm

You can use your approximation of π to learn about the area of a circle.

Activity 2

1 The *radius* of a circle is half of its diameter. Use a compass to draw a circle with a 2-inch radius. Cut your circle out and fold it three times as shown.

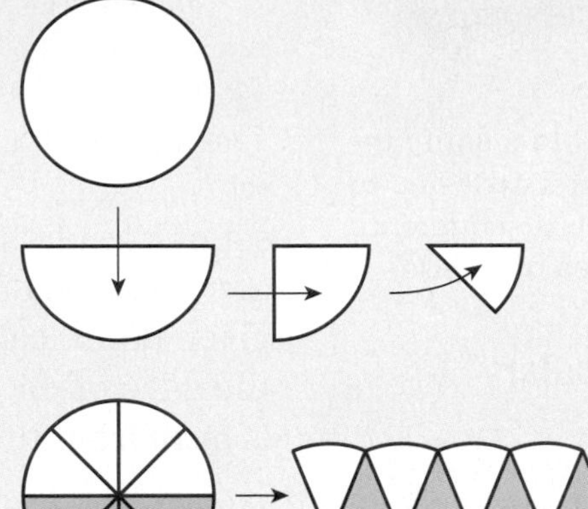

2 Unfold the circle, trace the folds, and shade one-half of the circle.

3 Cut along the folds, and fit the pieces together to make a figure that looks approximately like a parallelogram.

Think of this figure as a parallelogram. The base and height of the parallelogram relate to the parts of the circle.

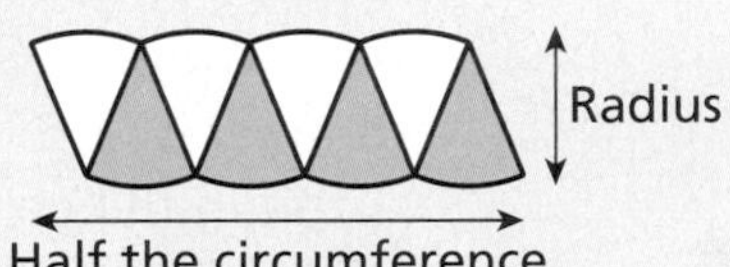

base (b) = $\frac{1}{2}$ the circumference of the circle, or πr

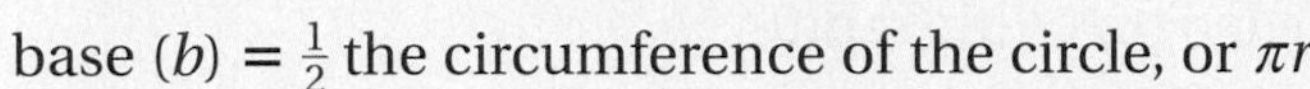

height (h) = the radius of the circle, or r

To find the area of a parallelogram, use the equation $A = bh$.

To find the area of a circle, use the equation $A = \pi r(r) = \pi r^2$.

Think and Discuss

1. Compare the lengths of all the diameters of a circle.
2. Compare the lengths of all the radii of a circle.

Try This

Find the area of each circle with the given measure.

1. $r = 4$ yd
2. $r = 3$ in.
3. $d = 10$ m

Find the area of each circle.

4.

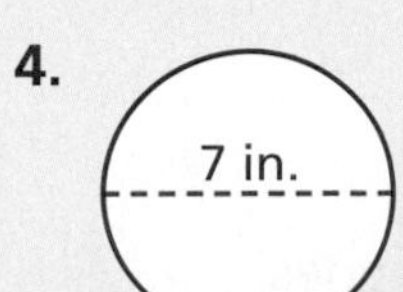

5.

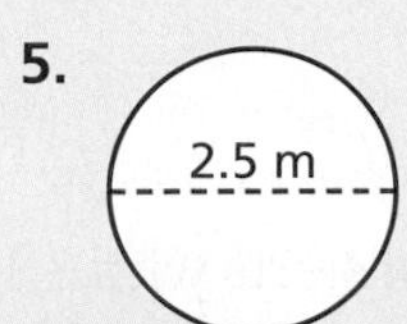

6.

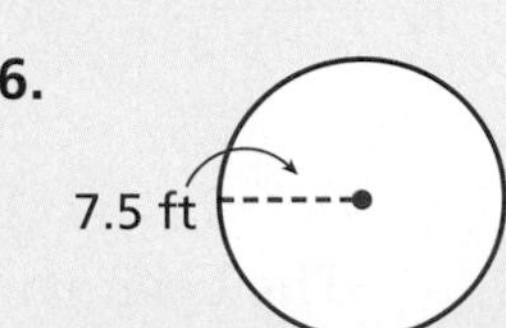

10-5 Circles

Learn to identify the parts of a circle and to find the circumference and area of a circle.

Vocabulary

circle

center

radius (radii)

diameter

circumference

pi

The shape of an inline skate wheel is a *circle*. A **circle** is the set of all points in a plane that are the same distance from a given point, called the **center**. At the edge of the wheel, every point is ***5 cm*** from the center.

Center

Radius

Diameter

Circumference

A line segment with one endpoint at the center of the circle and the other endpoint on the circle is a **radius** (plural: *radii*).

A *chord* is a line segment with both endpoints on a circle. A **diameter** is a chord that passes through the center of the circle. The length of the diameter is twice the length of the radius.

The radius of the wheel = ***5 cm.***

The diameter of the wheel = 2 · ***5 cm*** = ***10 cm.***

Like a polygon, a circle is a plane figure. But a circle is not a polygon because it is not made of line segments.

EXAMPLE 1 Naming Parts of a Circle

Name the circle, a diameter, and three radii.

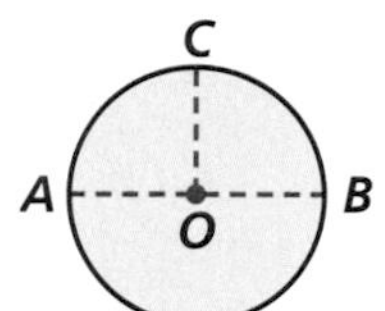

The circle is circle O.
$\overline{AB}$ is a diameter.
$\overline{OA}$, $\overline{OB}$, and $\overline{OC}$ are radii.

The distance around a circle is called the **circumference**.

The ratio of the circumference to the diameter, $\frac{C}{d}$, is the same for any circle. This ratio is represented by the Greek letter π, which is read "***pi***."

$$\frac{C}{d} = \pi$$

The decimal representation of *pi* starts with 3.14159265 . . . and goes on forever without repeating. We estimate *pi* using either 3.14 or $\frac{22}{7}$.

The formula for the circumference of a circle is $C = \pi d$, or $C = 2\pi r$.

EXAMPLE 2 Using the Formula for the Circumference of a Circle

Find each missing value to the nearest hundredth. Use 3.14 for *pi*.

A $d = 8$ ft; $C = ?$

$C = \pi d$ — *Write the formula.*

$C \approx 3.14 \cdot 8$ — *Replace π with **3.14** and **d** with **8**.*

$C \approx 25.12$ ft

B $r = 3$ cm; $C = ?$

$C = 2\pi r$ — *Write the formula.*

$C \approx 2 \cdot 3.14 \cdot 3$ — *Replace π with **3.14** and **r** with **3**.*

$C \approx 18.84$ cm

C $C = 37.68$ in.; $d = ?$

$C = \pi d$ — *Write the formula.*

$37.68 \approx 3.14d$ — *Replace **C** with **37.68**, and π with **3.14**.*

$\frac{37.68}{3.14} \approx \frac{3.14d}{3.14}$ — *Divide both sides by 3.14.*

12.00 in. $\approx d$

The formula for the area of a circle is $A = \pi r^2$.

EXAMPLE 3 Using the Formula for the Area of a Circle

Find the area of the circle. Use $\frac{22}{7}$ for *pi*.

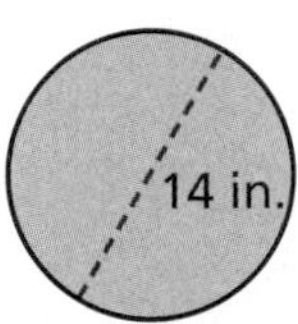

$d = 14$ in.; $A = ?$

$A = \pi r^2$ — *Write the formula to find the area.*

$r = d \div 2$ — *The length of the diameter is twice*

$r = 14 \div 2 = 7$ — *the length of the radius.*

$A \approx \frac{22}{7} \cdot 7^2$ — *Replace π with $\frac{22}{7}$ and **r** with **7**.*

$A \approx \frac{22}{\cancel{7}_1} \cdot \cancel{49}^7$ — *Use the GCF to simplify.*

$A \approx 154 \text{ in}^2$ — *Multiply.*

Think and Discuss

1. **Explain** how to find the radius in Example 2C.
2. **Tell** whether the approximation 3.14 is less than or greater than π.
3. **Express** $\frac{22}{7}$ as a decimal rounded to the nearest hundredth. Compare this number with 3.14.

10-5 Exercises

FOR EOG PRACTICE

see page 677

internet connect

Homework Help Online
go.hrw.com Keyword: MR4 10-5

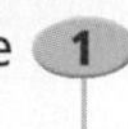

2.02, 3.02

GUIDED PRACTICE

See Example 1

1. Point *G* is the center of the circle. Name the circle, a diameter, and three radii.

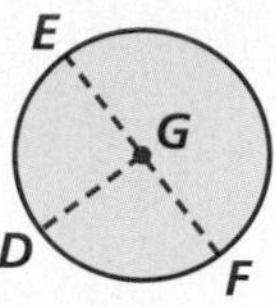

See Example 2

Find each missing value to the nearest hundredth. Use 3.14 for *pi*.

2. $C = \underline{?}$

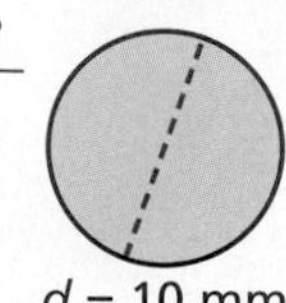

3. $C = \underline{?}$

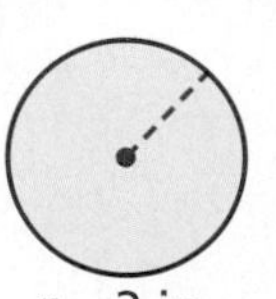

See Example 3

Find the area of each circle to the nearest hundredth. Use $\frac{22}{7}$ for *pi*.

4. $A = \underline{?}$

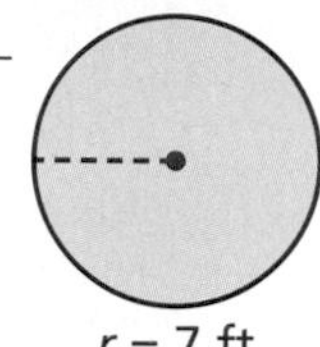

5. $A = \underline{?}$

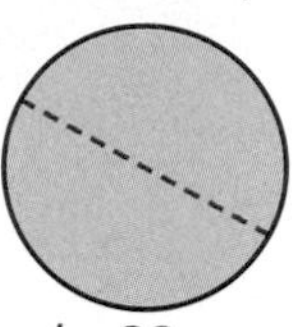

INDEPENDENT PRACTICE

See Example 1

6. Point *P* is the center of the circle. Name the circle, a diameter, and three radii.

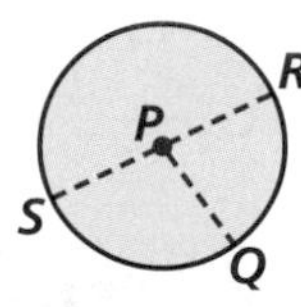

See Example 2

Find each missing value to the nearest hundredth. Use 3.14 for *pi*.

7. $C = \underline{?}$

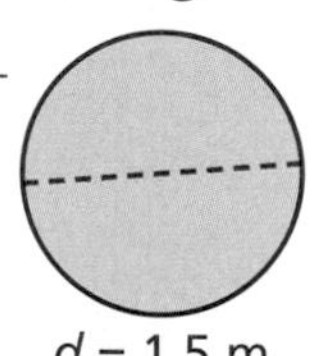

8. $C = \underline{?}$

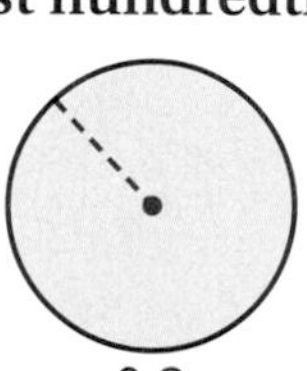

9. $d = \underline{?}$

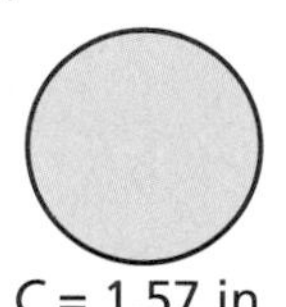

See Example 3

Find the area of each circle to the nearest hundredth. Use $\frac{22}{7}$ for *pi*.

10. $A = \underline{?}$

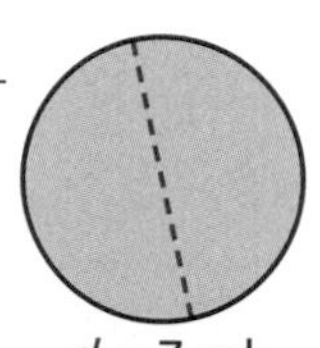

11. $A = \underline{?}$

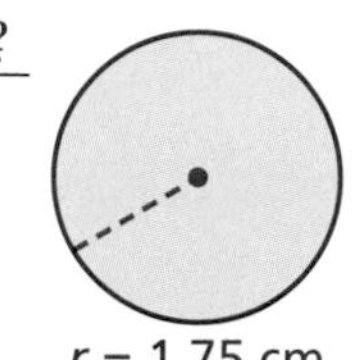

12. $A = \underline{?}$

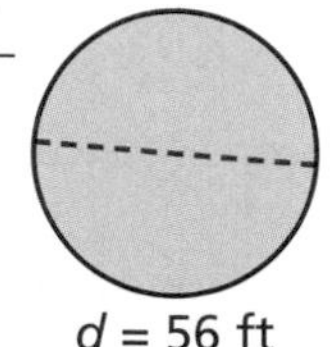

PRACTICE AND PROBLEM SOLVING

Fill in the blanks. Use 3.14 for *pi* and round to the nearest hundredth.

13. If $r = 7$ m, then $d = \underline{\quad ? \quad}$, $C = \underline{\quad ? \quad}$, and $A = \underline{\quad ? \quad}$.

14. If $d = 11.5$ ft, then $r = \underline{\quad ? \quad}$, $C = \underline{\quad ? \quad}$, and $A = \underline{\quad ? \quad}$.

15. If $C = 7.065$ cm, then $d = \underline{\quad ? \quad}$, $r = \underline{\quad ? \quad}$, and $A = \underline{\quad ? \quad}$.

Performers like Patsy Rosales have learned to twirl multiple Hula Hoops around their bodies. The world record for most Hula Hoops kept in revolution is held by Katja Lange, who kept 90 of them spinning.

16. ***HISTORY*** The first Hula Hoop® was introduced in 1958. It is one of the most popular toys in U.S. history. What is the circumference of a Hula Hoop with a 3 ft diameter? (Use 3.14 for π. Round the answer to the nearest hundredth.)

17. ***MEASUREMENT*** Draw a circle with center O and radius $\overline{AO}$ that is 2 inches long.

    a. Draw the diameter $\overline{AB}$ and give its length.

    b. What is the circumference of circle O? (Use 3.14 for π. Round to the nearest hundredth.)

    c. What is the area of circle O?

18. ***SPORTS*** The diameter of the circle that a shot-putter stands in is 7 ft. What is the area of the circle? (Use $\frac{22}{7}$ for π.)

19. a. Estimate the difference between the area of a pizza with a 6 in. diameter and a pizza with a 12 in. diameter. (Use 3.14 for π. Round to the nearest whole number.)

    b. Is the area of the 12 in. pizza about twice as large as the area of the 6 in. pizza? Explain.

20. ***WRITE A PROBLEM*** Write a problem that involves finding the circumference and the area of a circle when given the diameter. Give the answer to your problem.

21. ***WRITE ABOUT IT*** The circumference of a circle is 3.14 m. Explain how you can find the diameter and radius of the circle.

22. ***CHALLENGE*** What is the area of the shaded part of the figure? (Use 3.14 for π. Round the answer to the nearest hundredth.)

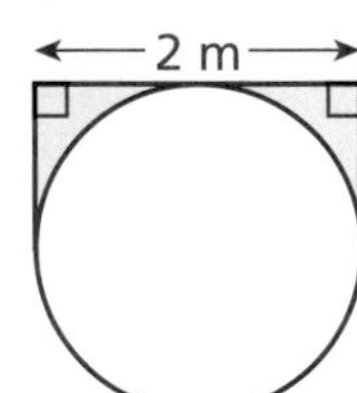

Spiral Review

Find each quotient. (Lesson 3-7)

23. 25.5 ÷ 5 **24.** 44.7 ÷ 3 **25.** 96.48 ÷ 6 **26.** 0.0378 ÷ 9

Order the fractions from greatest to least. (Lesson 4-6)

27. $\frac{1}{2}, \frac{3}{8}, \frac{5}{8}$ **28.** $\frac{3}{4}, \frac{10}{12}, \frac{1}{12}$ **29.** $\frac{3}{10}, \frac{3}{5}, \frac{7}{10}$

30. **EOG PREP** Which number is 20% of 360? (Lesson 8-9)

A 18 B 72 C 380 D 7,200

Chapter 10 Mid-Chapter Quiz

LESSON 10-1 (pp. 500–503)

Find the perimeter of each figure.

1.
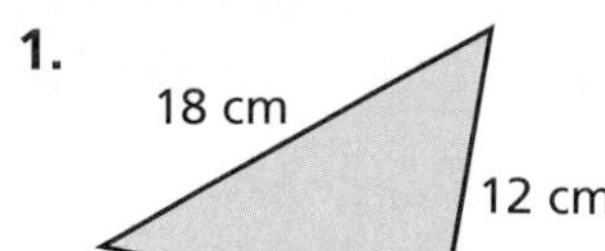

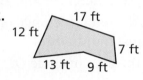

2. 17 ft, 12 ft, 7 ft, 13 ft, 9 ft

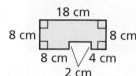

3. 18 cm, 8 cm, 8 cm, 8 cm, 4 cm, 2 cm

LESSON 10-2 (pp. 504–507)

Find the area of each figure.

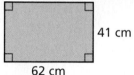

4. 41 cm, 62 cm

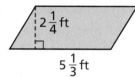

5. $2\frac{1}{4}$ ft, $5\frac{1}{3}$ ft

6.
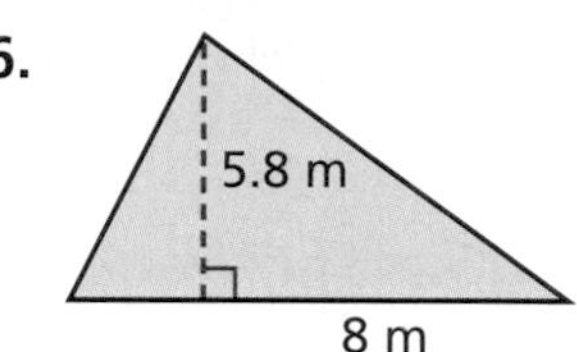

LESSON 10-3 (pp. 508–510)

Find the area of each polygon.

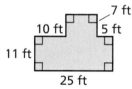

7. 7 ft, 10 ft, 5 ft, 11 ft, 25 ft

8.
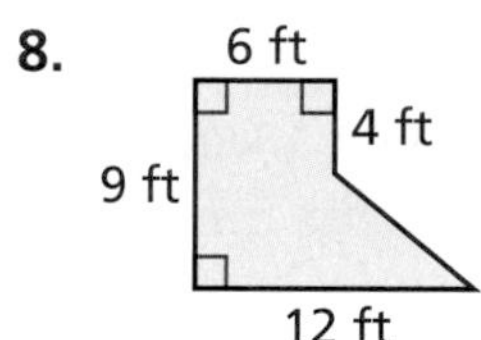

9.
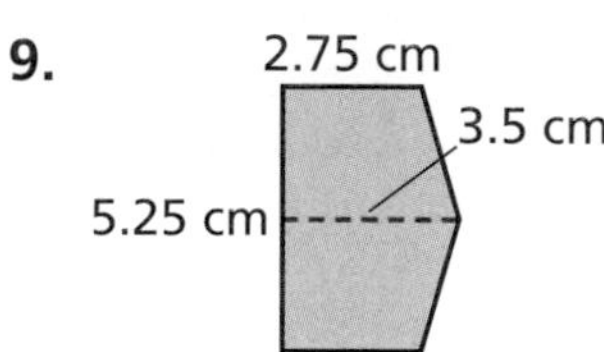

LESSON 10-4 (pp. 511–513)

10. The length and width of a rectangle are each multiplied by 4. Find how the perimeter and the area of the rectangle change.

LESSON 10-5 (pp. 516–519)

Name two radii and find the circumference and area for each circle. Use 3.14 for *pi* and round to the nearest hundredth.

11.
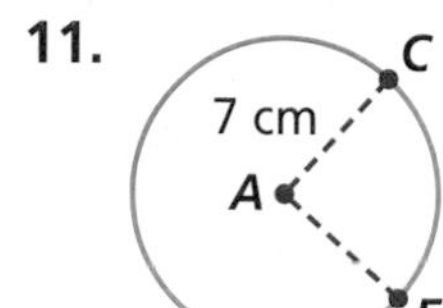

12.
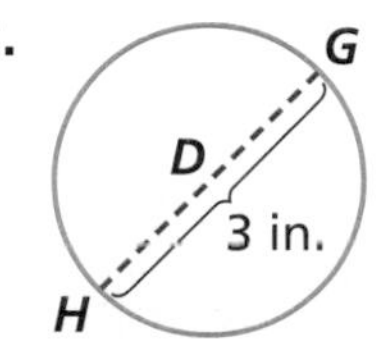

13.
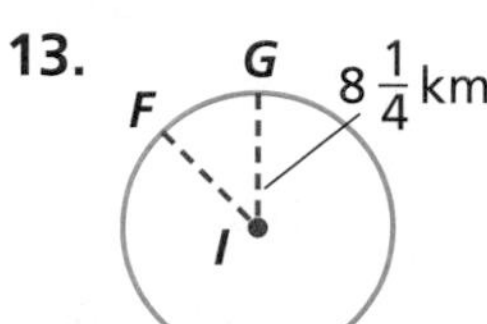

14.
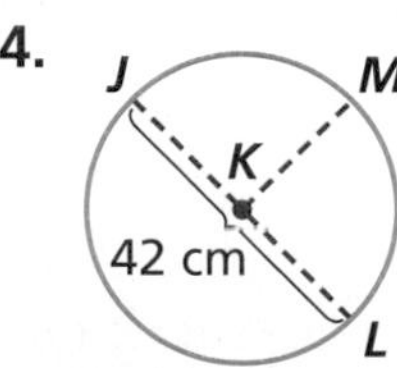

Find each missing value to the nearest hundredth. Use 3.14 for *pi*.

15. $r = 9$ in.; $C = \underline{?}$ **16.** $d = 20$ m; $C = \underline{?}$ **17.** $C = 37.68$ ft; $d = \underline{?}$

Find the area of each circle. Use $\frac{22}{7}$ for pi.

18. $d = 6$ cm; $A = \underline{?}$ **19.** $r = 14$ ft; $A = \underline{?}$ **20.** $r = 1\frac{3}{4}$ in.; $A = \underline{?}$

Focus on Problem Solving

Solve

- **Choose the operation**

Read the whole problem before you try to solve it. Determine what action is taking place in the problem. Then decide whether you need to add, subtract, multiply, or divide in order to solve the problem.

Action	Operation
Combining or putting together	Add
Removing or taking away Comparing or finding the difference	Subtract
Combining equal groups	Multiply
Sharing equally or separating into equal groups	Divide

Read each problem and determine the action taking place. Choose an operation, and then solve the problem.

1. There are 3 lily ponds in the botanical gardens. They are identical in size and shape. The total area of the ponds is 165 ft^2. What is the area of each lily pond?

2. The greenhouse is made up of 6 rectangular rooms with an area of 4,800 ft^2 each. What is the total area of the greenhouse?

3. A shady area with 17 different varieties of magnolia trees, which bloom from March to June, surrounds the plaza in Magnolia Park. In the center of the plaza, there is a circular bed of shrubs as shown in the chart. If the total area of the park is 625 ft^2, what is the area of the plaza?

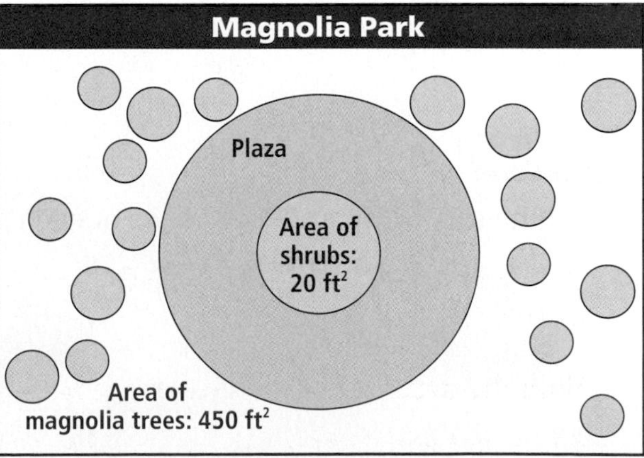

Draw Views of Solid Figures

Use with Lesson 10-6

internet connect
Lab Resources Online
go.hrw.com
KEYWORD: MR4 Lab10B

Activity 1

1. Draw a rectangular prism. Imagine that you are looking at the top of the prism, and draw what you would see. Draw the front and side views of the prism.

Top

Front

Side

All the faces of a rectangular prism are rectangles.

2. Stack centimeter cubes to make the solid figure shown. Draw the top, front, and side views.

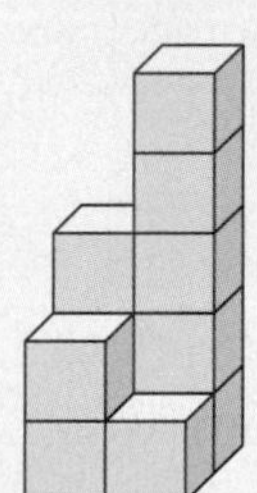

Top

Front

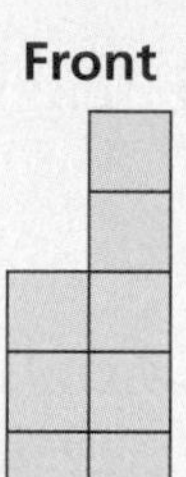

Side

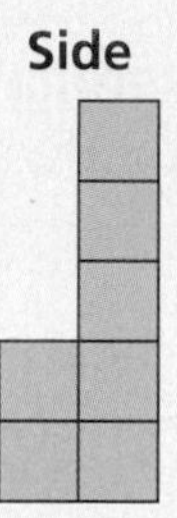

Each view shows a different configuration of squares representing the number of cubes you see.

Think and Discuss

1. Explain why a side view of a figure might change if you look at a different side.

Try This

Draw the top, front, and side views of each solid figure.

1.

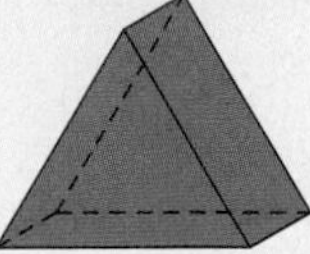

2.

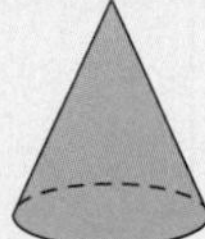

3.

You can use different views of a solid to identify the figure.

Activity 2

1 Name the solid figure that has the given views.

Top

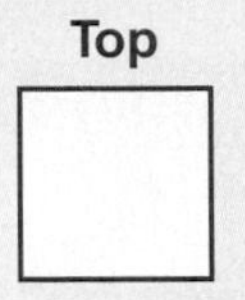

Front

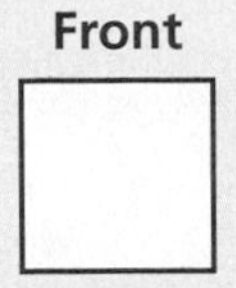

Side

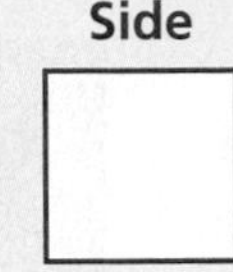

Each view of the solid shows a square.

The solid is a cube.

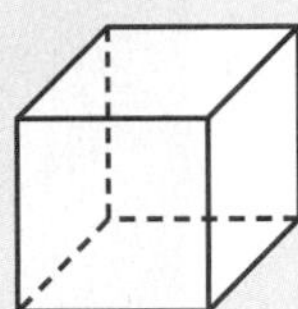

2 Name the solid figure that has the given views.

Top

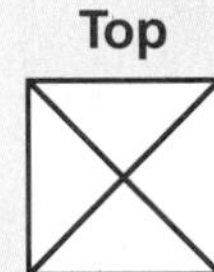

Front

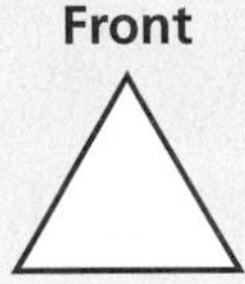

Side

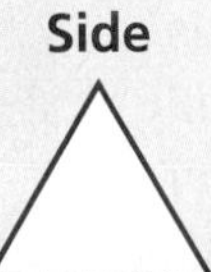

The top view shows that the base is a square. It also shows that the other faces come together at a point.

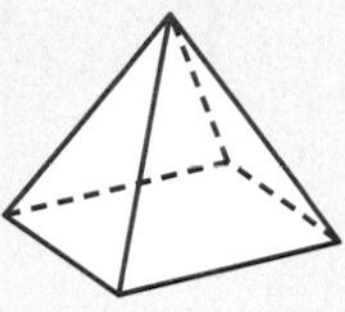

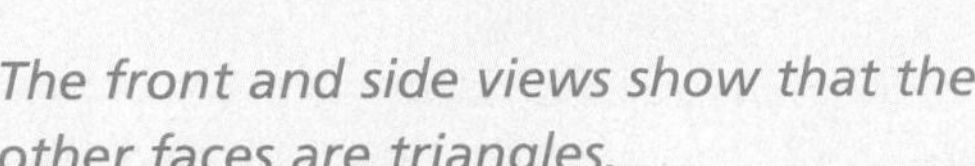

The front and side views show that the other faces are triangles.

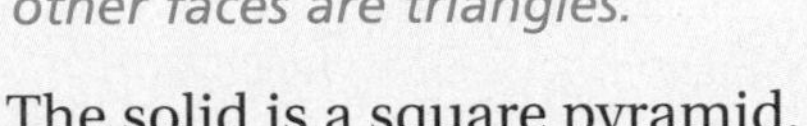

The solid is a square pyramid.

Think and Discuss

1. Explain which views show how tall a solid figure is.

Try This

Name each solid figure that has the given views.

1.

Top

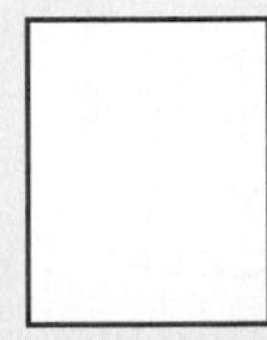

Front

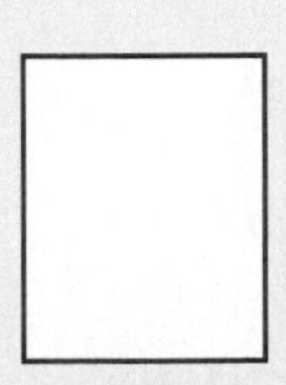

Side

2.

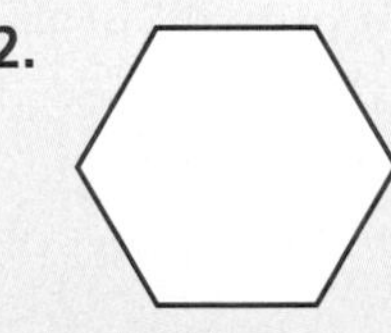

Top

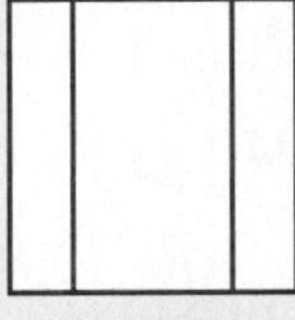

Front

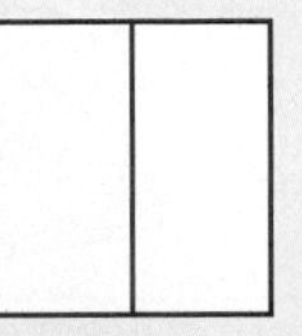

Side

10-6 Solid Figures

Learn to name solid figures.

Vocabulary
polyhedron
face
edge
vertex
prism
base
pyramid
cylinder
cone

A **polyhedron** is a three-dimensional object, or solid figure, with flat surfaces, called **faces**, that are polygons.

When two faces of a solid figure share a side, they form an **edge**. On a solid figure, a point at which three or more edges meet is a **vertex** (plural: *vertices*).

A cube is formed by 6 square faces. It has 8 vertices and 12 edges. The sculpture in front of this building is based on a cube. The artist's work is not a polyhedron because of the hole cut through the middle.

This sculpture, *Red Cube*, in front of the Marine Midland Bank in New York City was done by Isamu Noguchi.

EXAMPLE 1 **Identifying Faces, Edges, and Vertices**

Identify the number of faces, edges, and vertices on each solid figure.

A

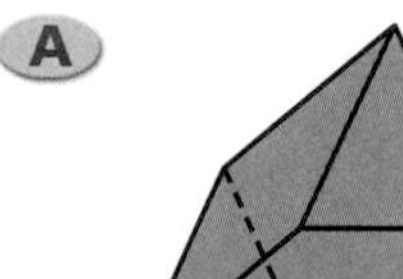

5 faces
9 edges
6 vertices

B

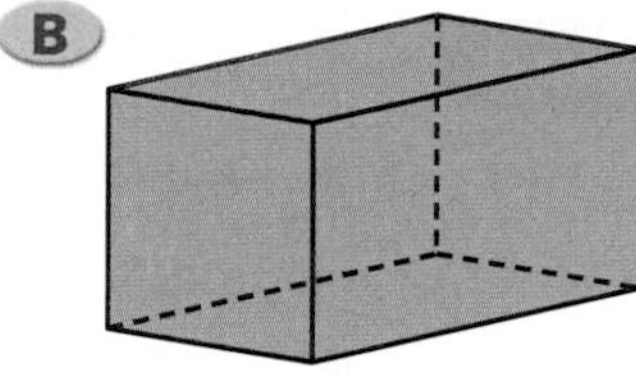

6 faces
12 edges
8 vertices

A **prism** is a polyhedron with two congruent, parallel **bases**, and other faces that are all parallelograms. A prism is named for the shape of its bases. A **cylinder** also has two congruent, parallel bases, but bases of a cylinder are circular. Cylinders are not polyhedra because not every surface is a polygon.

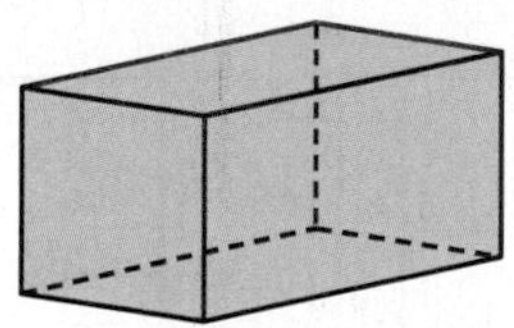

Rectangular prism

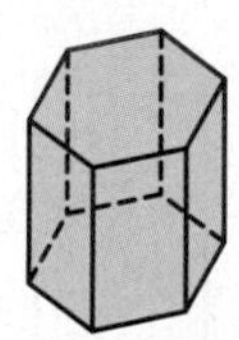

Hexagonal prism

Cylinder

Helpful Hint

The point of a cone is called its *vertex.*

A **pyramid** has one polygon shaped base, and the other faces are triangles that come to a point. A pyramid is named for the shape of its base. A **cone** has a circular base and a curved surface that comes to a point. Cones are not polyhedra because not every face is a polygon.

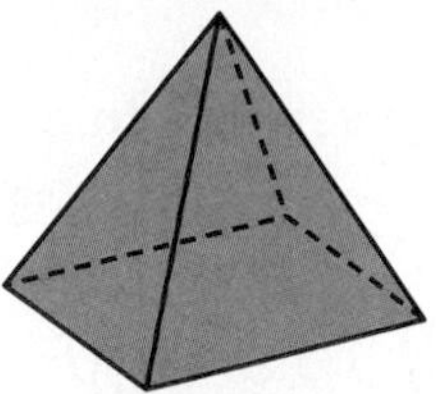

Square pyramid

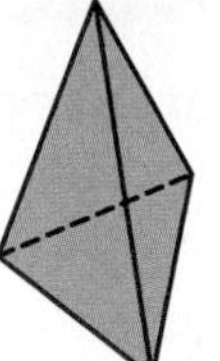

Triangular pyramid

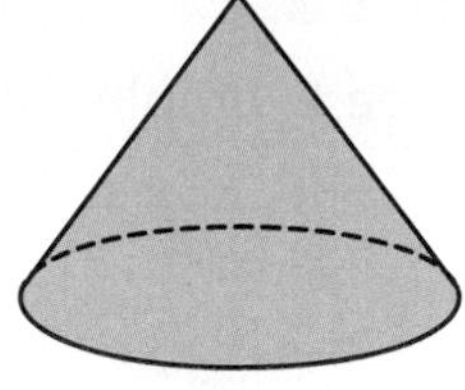

Cone

EXAMPLE 2 Naming Solid Figures

Name the solid figure represented by each object.

A

All the faces are flat and are polygons.

The figure is a polyhedron.

There are two congruent, parallel bases, so the figure is a prism.

The bases are triangles.

The figure is a triangular prism.

B

There is a curved surface.

The figure is not a polyhedron.

There is a flat, circular base.

The lateral surface comes to a point.

The figure represents a cone.

C

All the faces are flat and are polygons.

The figure is a polyhedron.

It has one base and the other faces are triangles that meet at a point, so the figure is a pyramid.

The base is a square.

The figure is a square pyramid.

Think and Discuss

1. **Tell** how a cylinder and a prism are alike and how they are different.
2. **Explain** how a cone and a pyramid are alike and how they are different.
3. **Name** a prism that has squares as its bases and as its faces.

10-6 Exercises

FOR EOG PRACTICE
see page 678

internet connect
Homework Help Online
go.hrw.com Keyword: MR4 10-6

GUIDED PRACTICE

See Example 1

Identify the number of faces, edges, and vertices in each solid figure.

1. 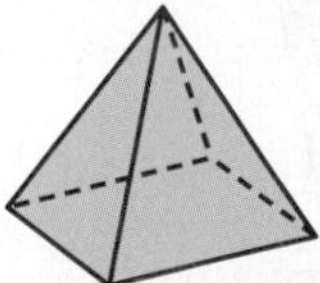

2. 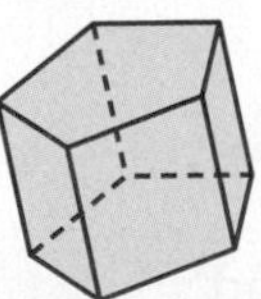

3. 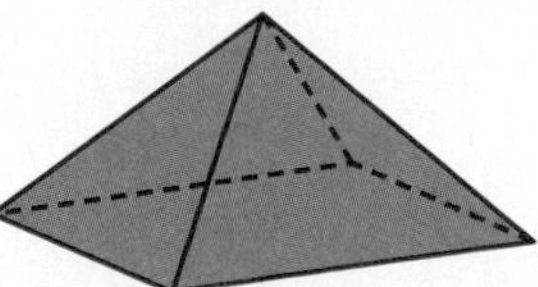

See Example 2

Name the solid figure represented by each object.

4. 

5. 

6. 

INDEPENDENT PRACTICE

See Example 1

Identify the number of faces, edges, and vertices in each solid figure.

7. 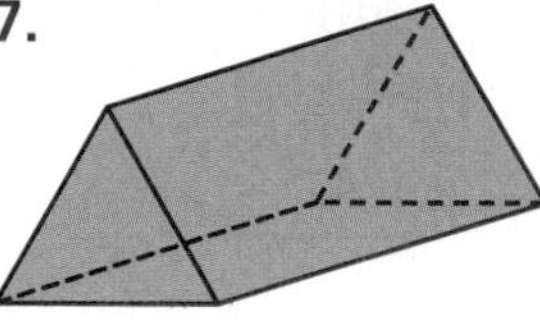

8. 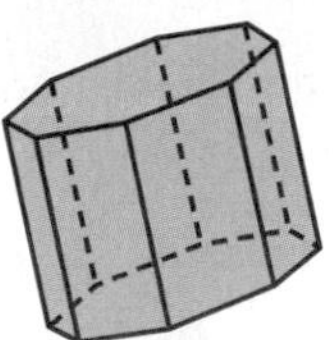

9. 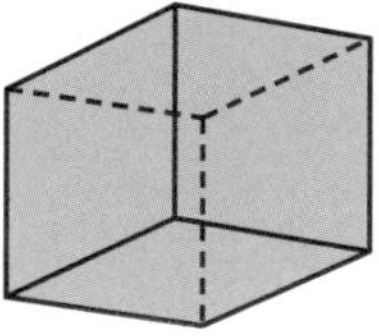

See Example 2

Name the solid figure represented by each object.

10. 

11. 

12. 

PRACTICE AND PROBLEM SOLVING

Name each figure and tell whether it is a polyhedron.

13. 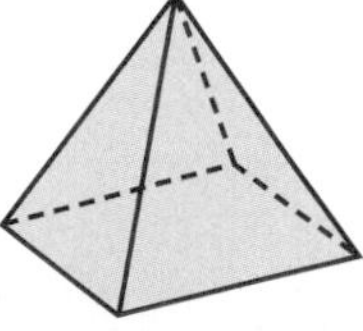

14. 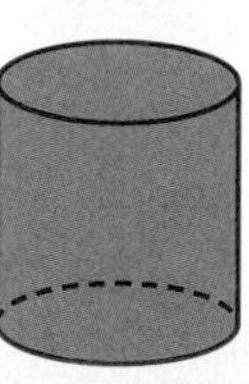

15. 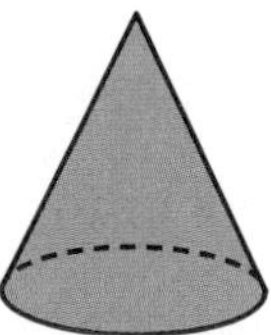

Write the letter of each figure described.

16. prism
17. has triangular faces
18. has 6 faces
19. has 5 vertices

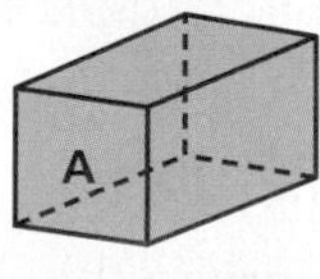

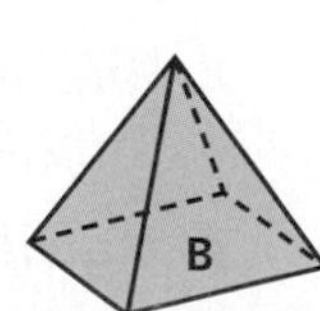

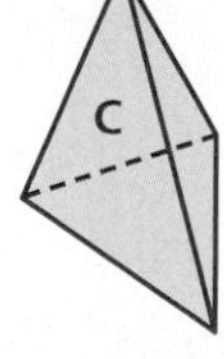

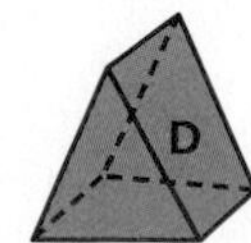

Write *true* or *false* for each statement.

20. A cone does not have a flat surface.

21. The bases of a cylinder are congruent.

22. All pyramids have five or more vertices.

23. All of the edges of a cube are congruent.

24. ***ARCHITECTURE*** Name the solid figure represented by each building.

a. 

b. 

c. 

25. ***HOBBIES*** Li makes candles with her mother. She made a candle in the shape of a pyramid that had 9 faces. How many sides did the base of the candle have? Name the polyhedron formed by the candle.

26. ***WHAT'S THE ERROR?*** A student says that any polyhedron can be named if the number of faces it has is known. What is the student's error?

27. ***WRITE ABOUT IT*** How are a cone and cylinder alike? How are they different?

28. ***CHALLENGE*** The top of a square pyramid is cut off, and the cut is made parallel to the base of the pyramid. What are the shapes of the faces of the new figure?

Spiral Review

Order the numbers from greatest to least. (Lesson 1-1)

29. 108, 24, 89, 75, 5, 91

30. 246, 235, 241, 36, 240

31. 19, 18, 15, 17, 13

Find each product. (Lesson 3-6)

32. 1.2×8

33. 0.05×0.6

34. 14×0.02

35. 22.1×22.1

36. **EOG PREP** Which of the following types of data should not be displayed using a line graph? (Lesson 6-7)

A The temperature each hour in one day

B A child's height on each of her birthdays

C The price of a computer from 1990 to 2000

D Students' favorite foods

Model Solid Figures

Use with Lesson 10-6

You can build a solid figure by cutting its faces from paper, taping them together, and then folding them to form the solid. A pattern of shapes that can be folded to form a solid figure is called a *net*.

Activity 1

1 To make a pattern for a rectangular prism follow the steps below.

a. Draw the following rectangles and cut them out:

Two 2 in. × 3 in. rectangles

Two 1 in. × 3 in. rectangles

Two 1 in. × 2 in. rectangles

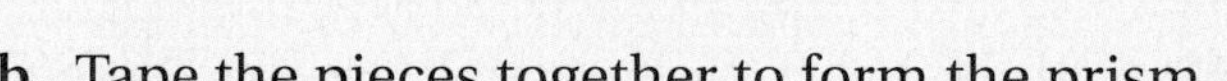

b. Tape the pieces together to form the prism.

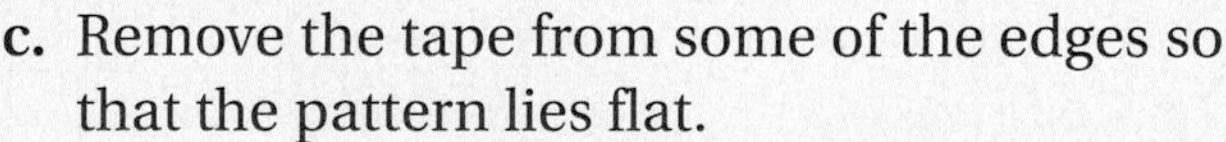

c. Remove the tape from some of the edges so that the pattern lies flat.

2 Name the solid figure that can be formed by the net shown.

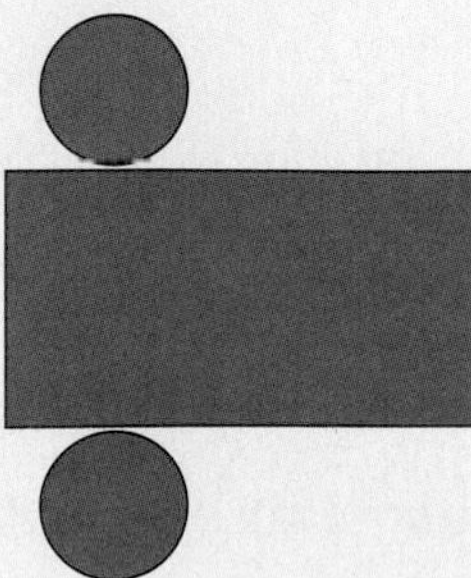

There are circular faces, so the solid is not a prism or a pyramid. Because there are two circular faces, the solid is not a cone. It could be a cylinder.

The net shows one rectangular face that can be curved to form a cylinder.

The net can form a cylinder.

3 Tell whether the net can form a prism.

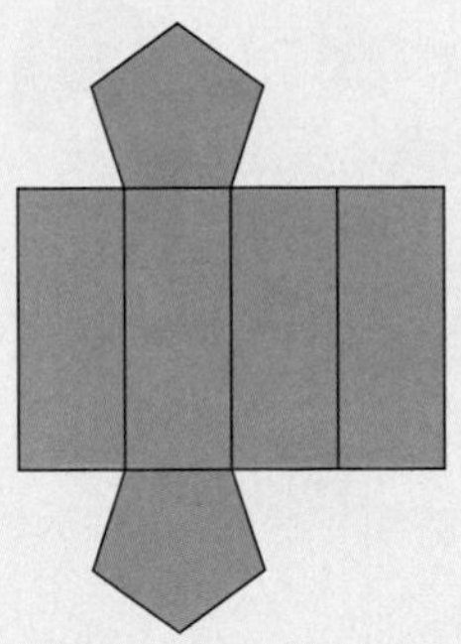

The net shows two pentagonal-shaped bases that appear to be congruent.

All other faces are rectangles. The net might form a pentagonal prism.

The base has five sides, but there are only four rectangular faces.

The net cannot form a prism.

Think and Discuss

1. Compare the nets for a rectangular prism and a cube.
2. Tell what shapes will always appear in a net for a pyramid.
3. Tell what shapes will always appear in a net for a prism.

Try This

Tell whether each net can be folded to form a cube. If not, explain.

1.

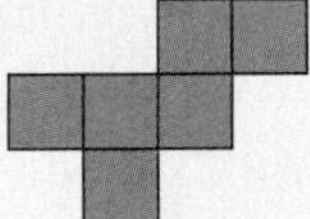

2.

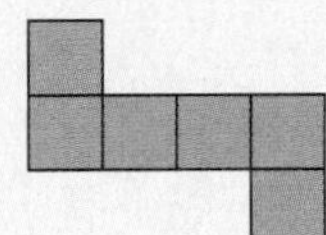

3.

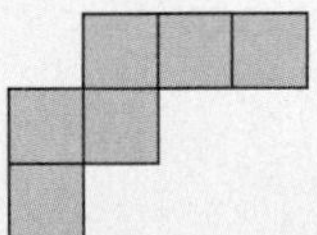

4.

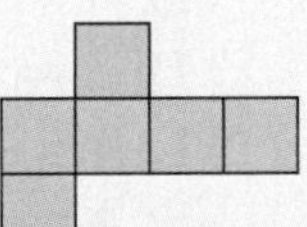

Name the solid that can be formed from each net.

5.

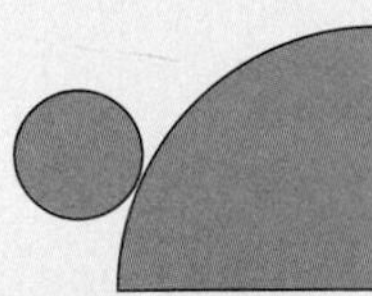

6.

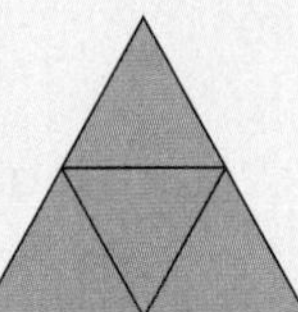

7.

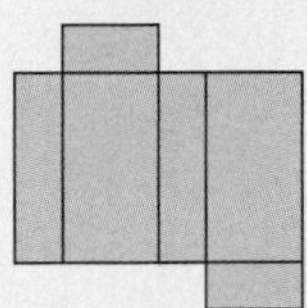

8.

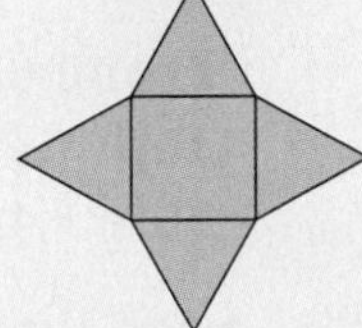

10-7 Surface Area

Learn to find the surface areas of prisms, pyramids, and cylinders.

Vocabulary

surface area

net

Katie made a toy for her cat to scratch by attaching carpet to the faces of a wooden box. The amount of carpet needed to cover the box is equal to the surface area of the box.

The **surface area** of a solid figure is the sum of the areas of its surfaces. To help you see all the surfaces of a solid figure, you can use a *net*. A **net** is the pattern made when the surface of a solid is layed out flat showing each face of the figure.

EXAMPLE 1 Finding the Surface Area of a Prism

Find the surface area S of each prism.

A Method 1: Use a net.

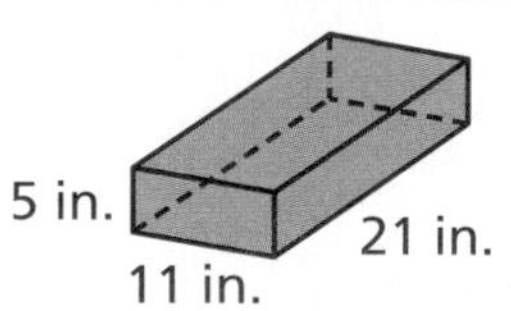

Draw a net to help you see each face of the prism.

Use the formula $A = \ell w$ to find the area of each face.

5 in.
5 in.
A
11 in.
11 in.
21 in.
B
C
D
E
F
5 in.

A: $A = 11 \times 5 = 55$
B: $A = 21 \times 11 = 231$
C: $A = 21 \times 5 = 105$
D: $A = 21 \times 11 = 231$
E: $A = 21 \times 5 = 105$
F: $A = 11 \times 5 = 55$

$S = 55 + 231 + 105 + 231 + 105 + 55 = 782$ *Add the areas of each face.*

The surface area is 782 in^2.

B Method 2: Use a three-dimensional drawing.

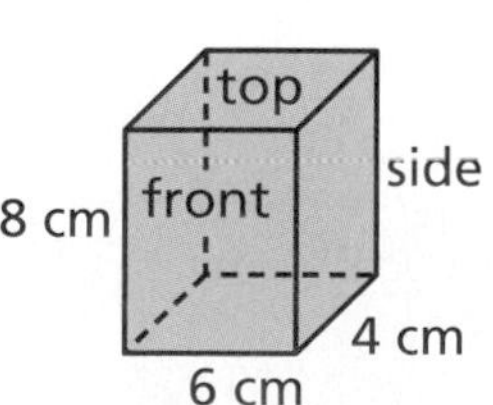

Find the area of the front, top, and side, and multiply each by 2 to include the opposite faces.

Front: $6 \times 8 = 48 \longrightarrow 48 \times 2 = 96$
Top: $6 \times 4 = 24 \longrightarrow 24 \times 2 = 48$
Side: $4 \times 8 = 32 \longrightarrow 32 \times 2 = 64$

$S = 96 + 48 + 64 = 208$ *Add the areas of the faces.*

The surface area is 208 cm^2.

The surface area of a pyramid equals the sum of the area of the base and the areas of the triangular faces. To find the surface area of a pyramid, think of its net.

EXAMPLE 2 Finding the Surface Area of a Pyramid

Find the surface area S of the pyramid.

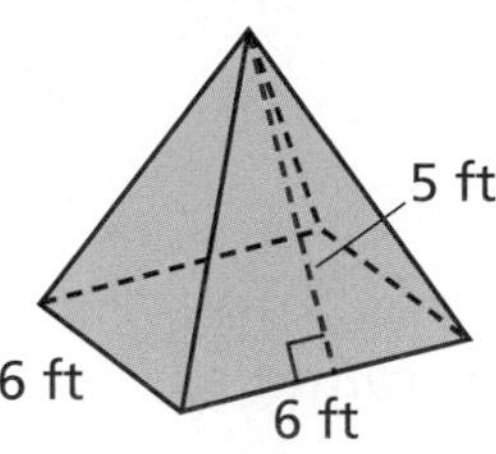

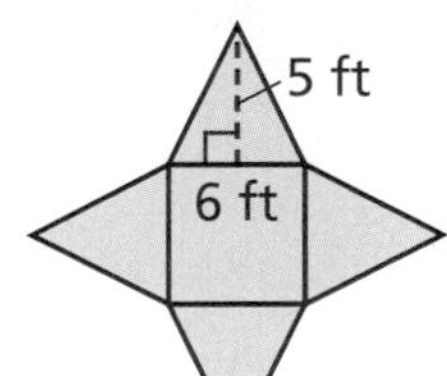

S = area of square + 4 × (area of triangular face)

$S = s^2 + 4 \times \left(\frac{1}{2}bh\right)$

$S = \mathbf{6}^2 + 4 \times \left(\frac{1}{2} \times \mathbf{6} \times \mathbf{5}\right)$ *Substitute.*

$S = 36 + 4 \times 15$

$S = 36 + 60$

$S = 96$

The surface area is 96 ft^2.

The surface area of a cylinder equals the sum of the area of its bases and the area of its curved surface.

EXAMPLE 3 Finding the Surface Area of a Cylinder

Find the surface area S of the cylinder. Use 3.14 for π, and round to the nearest hundredth.

Helpful Hint

To find the area of the curved surface of a cylinder, multiply its height by the circumference of the base.

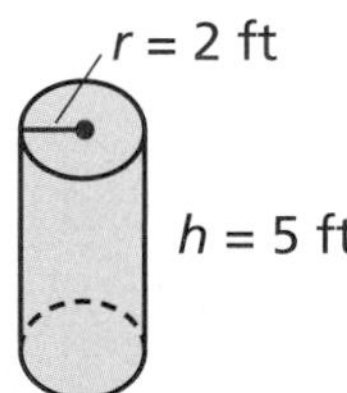

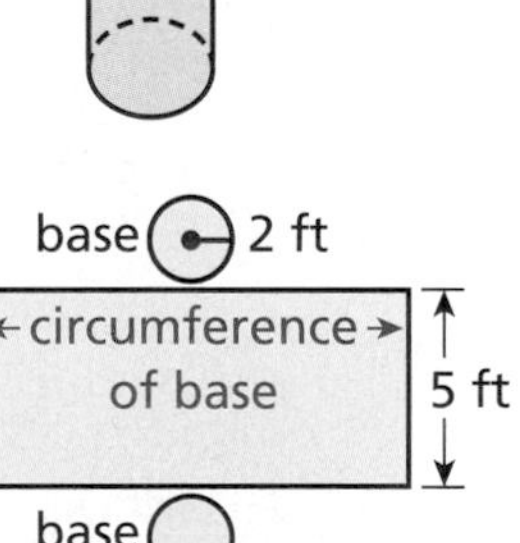

S = area of lateral surface + 2 × (area of each base)

$S = h \times (2\pi r) + 2 \times (\pi r^2)$

$S = \mathbf{5} \times (2 \times \pi \times \mathbf{2}) + 2 \times (\pi \times \mathbf{2}^2)$ *Substitute.*

$S = 5 \times 4\pi + 2 \times 4\pi$

$S \approx 5 \times 4(\mathbf{3.14}) + 2 \times 4(\mathbf{3.14})$ *Use 3.14 for π.*

$S \approx 5 \times 12.56 + 2 \times 12.56$

$S \approx 62.8 + 25.12$

$S \approx 87.92$

The surface area is about 87.92 ft^2.

Think and Discuss

1. **Describe** how to find the surface area of a pentagonal prism.
2. **Tell** how to find the surface area of a cube if you know the area of one face.

10-7 Exercises

FOR EOG PRACTICE
see page 678

internet connect
Homework Help Online
go.hrw.com Keyword: MR4 10-7

2.02

GUIDED PRACTICE

See Example 1 Find the surface area S of each prism.

1. 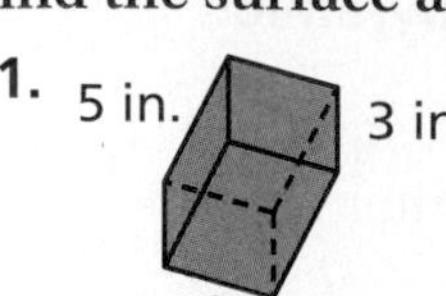

2. 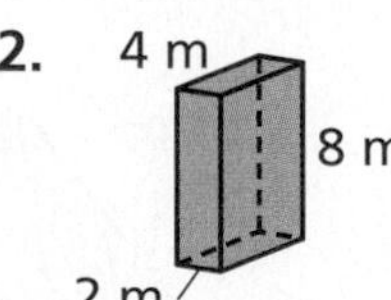

See Example 2 Find the surface area S of each pyramid.

3. 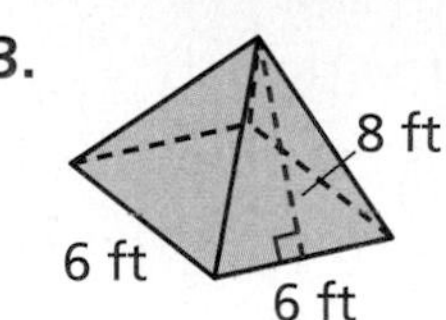

4. 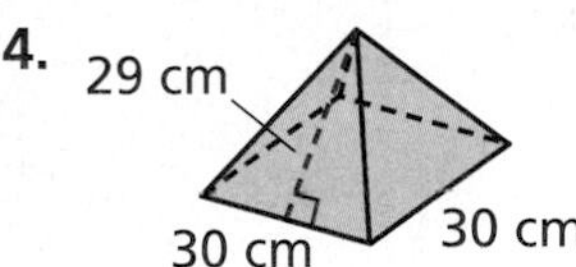

See Example 3 Find the surface area S of each cylinder. Use 3.14 for π, and round to the nearest hundredth.

5. 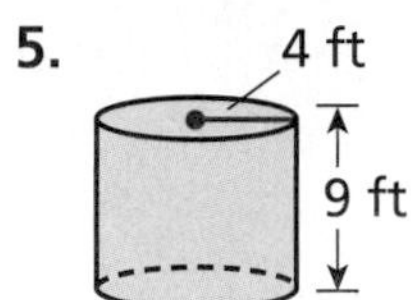

6. 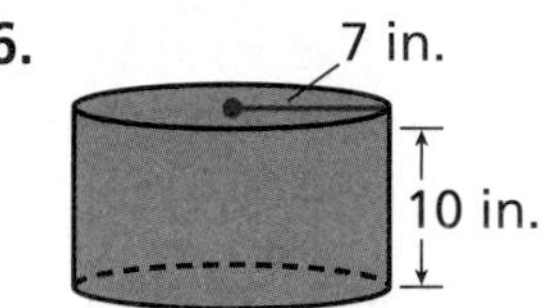

INDEPENDENT PRACTICE

See Example 1 Find the surface area S of each prism.

7. 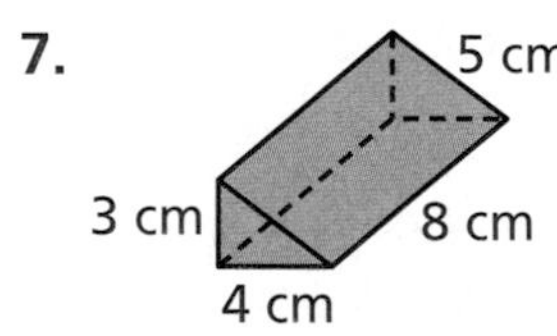

8. 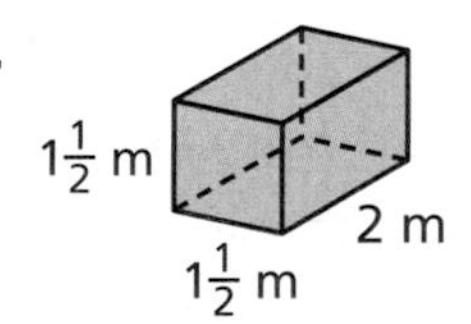

9. 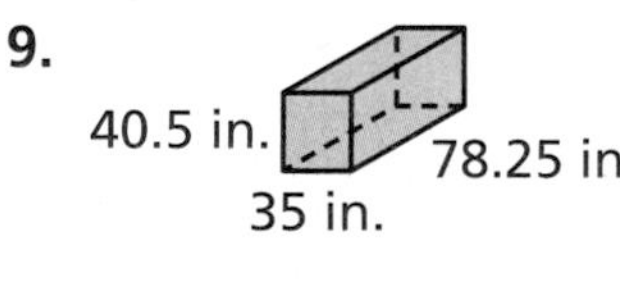

See Example 2 Find the surface area S of each pyramid.

10. 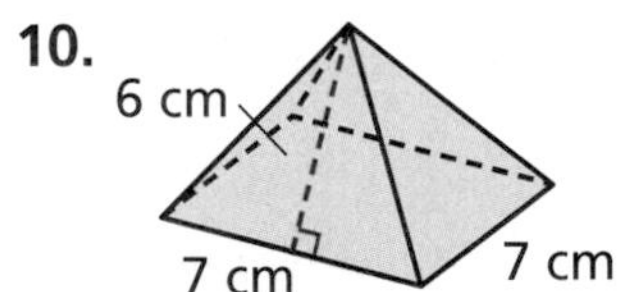

11. 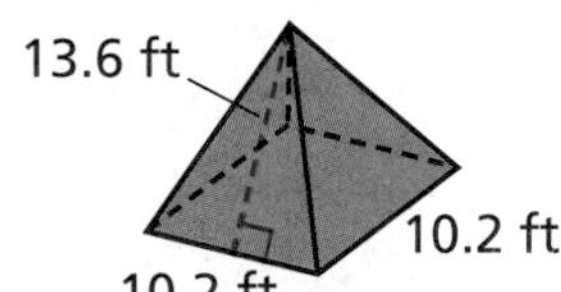

12. 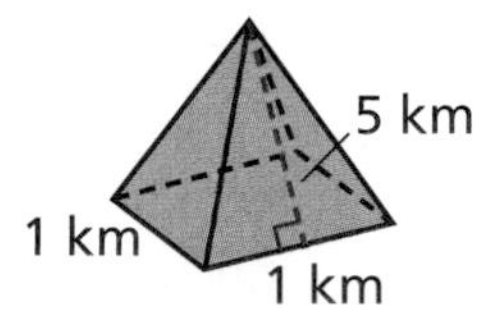

See Example 3 Find the surface area S of each cylinder. Use 3.14 for π, and round to the nearest hundredth.

13. 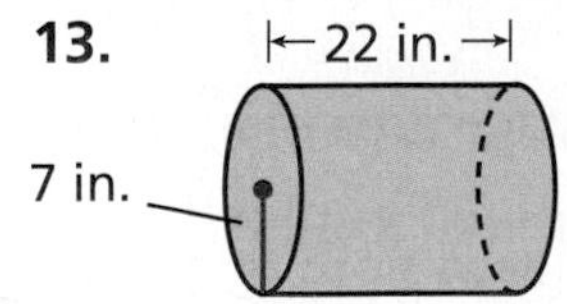

14. 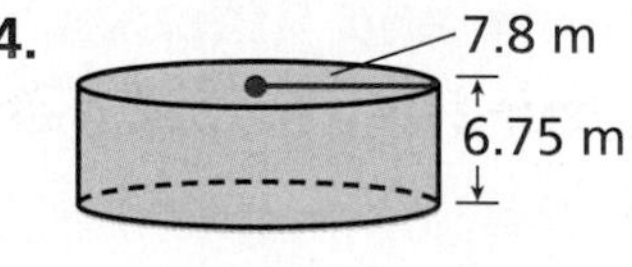

15. 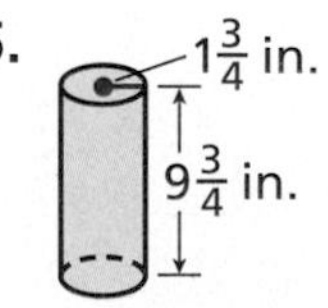

PRACTICE AND PROBLEM SOLVING

Find the surface area of each figure.

16.

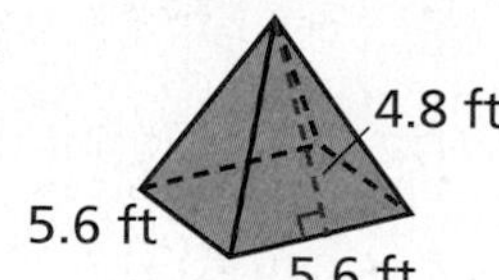

17.

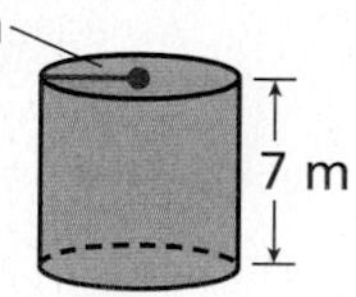

18.

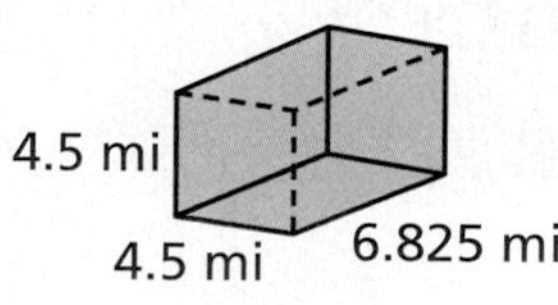

Architecture LINK

I. M. Pei is the architect of the pyramid-shaped addition to the Louvre in Paris, France.

go.hrw.com
KEYWORD: MR4 Pei

Find the surface area of each solid figure with the given measurements.

19. cube; $s = 1\frac{1}{2}$ km

20. square pyramid; base side = 12 m; triangular face height = 8 m

21. cylinder; $d = 10$ in.; $h = 6$ in.

22. ***ARCHITECTURE*** The entrance to the Louvre Museum is a glass-paned square pyramid. The width of the base is 34.2 m, and the height of the triangular sides is 27 m. What is the surface area of the glass?

23. Find the length, height, and surface area of each rectangular prism.

a. The length is half the width. The height is half the length. The width is 20 m.

b. The length is three times the height. The height is one-fourth the width. The width is 12 in.

24. ***WHAT'S THE QUESTION?*** The surface area of a cube is 150 cm^2. The answer is 5 cm. What is the question?

25. ***WRITE ABOUT IT*** How is finding the surface area of a rectangular pyramid different from finding the surface area of a triangular prism?

26. ***CHALLENGE*** This cube is made of 27 smaller cubes whose sides measure 1 in.

a. What is the surface area of the large cube?

b. Remove one small cube from each of the eight corners of the larger cube. What is the surface area of the solid formed?

Spiral Review

27. On a sunny day, a 4-foot-tall girl casts a shadow that is 7.2 feet long. She is standing near a tree that casts a shadow 25.56 feet long. How tall is the tree? (Lesson 8-5)

Compare. Write <, >, or =. (Lesson 9-2)

28. 0 ▢ −4 **29.** −345 ▢ 7 **30.** −12 ▢ −6 **31.** 14 ▢ 18

32. **EOG PREP** Which of the following is a solution to the equation $-7a = 42$? (Lesson 9-8)

A $a = 6$ B $a = 294$ C $a = 49$ D $a = -6$

10-8 Finding Volume

Learn to estimate and find the volumes of rectangular prisms and triangular prisms.

Vocabulary

volume

Volume is the number of cubic units needed to fill a space.

3 cm
2 cm
5 cm

It takes 10, or 5 · 2, centimeter cubes to cover the bottom layer of this rectangular prism.

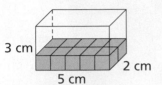

There are 3 layers of 10 cubes each. It takes 30, or 5 · 2 · 3, cubes to fill the prism.

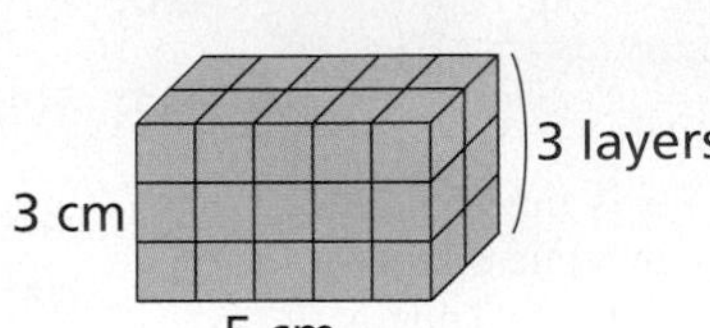

The volume of the prism is
$5 \text{ cm} \cdot 2 \text{ cm} \cdot 3 \text{ cm} = 30 \text{ cm}^3$.

EXAMPLE 1 Finding the Volume of a Rectangular Prism

Find the volume of the rectangular prism.

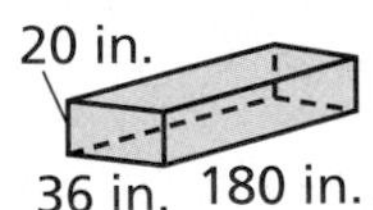

$V = \ell wh$	*Write the formula.*
$V = \mathbf{180} \cdot \mathbf{36} \cdot 20$	*ℓ = 180; w = 36; h = 20*
$V = 129{,}600 \text{ in}^3$	*Multiply.*

To find the volume of any prism, you can use the formula $V = Bh$, where B is the area of the base, and h is the prism's height. So, to find the volume of a triangular prism, B is the area of the triangular base and h is the height of the prism.

EXAMPLE 2 Finding the Volume of a Triangular Prism

Find the volume of each triangular prism.

A

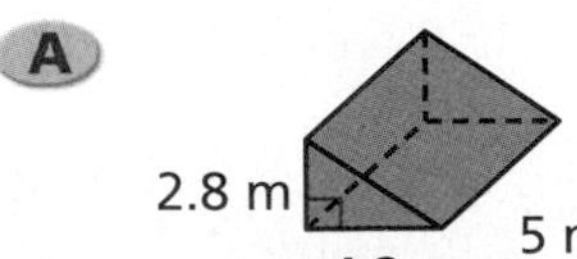

$V = Bh$	*Write the formula.*
$V = \left(\frac{1}{2} \cdot 2.8 \cdot 4.2\right) \cdot 5$	$B = \frac{1}{2} \cdot 2.8 \cdot 4.2;\ h = 5$
$V = 29.4 \text{ m}^3$	*Multiply.*

B

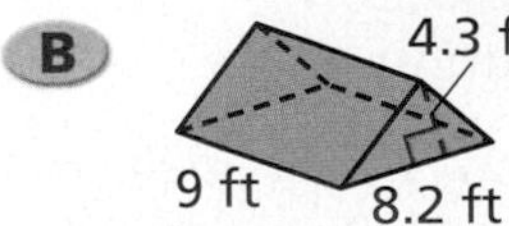

$V = Bh$	*Write the formula.*
$V = \left(\frac{1}{2} \cdot 8.2 \cdot 4.3\right) \cdot 9$	$B = \frac{1}{2} \cdot 8.2 \cdot 4.3;\ h = 9$
$V = 158.67 \text{ ft}^3$	*Multiply.*

EXAMPLE 3 PROBLEM SOLVING APPLICATION

A craft supplier ships 12 cubic trinket boxes in a case. What are the possible dimensions for a case of the trinket boxes?

1 Understand the Problem

The **answer** will be all possible dimensions for a case of 12 cubic boxes.

List the **important information:**

- There are 12 trinket boxes in a case.
- The boxes are cubic, or square prisms.

2 Make a Plan

You can make models using cubes to find the possible dimensions for a case of 12 trinket boxes.

3 Solve

Make different arrangements of 12 cubes.

The possible dimensions for a case of 12 cubic trinket boxes are the following: $1 \cdot 1 \cdot 12$, $1 \cdot 2 \cdot 6$, $1 \cdot 3 \cdot 4$, and $2 \cdot 2 \cdot 3$.

4 Look Back

Notice that each dimension is a factor of 12. Also, the product of the dimensions (length · width · height) is 12, showing that the volume of each case is 12 cubes.

Think and Discuss

1. **Explain** how to find the height of a rectangular prism if you know its length, width, and volume.
2. **Tell** how to use mental math strategies to find the volume of a rectangular prism with dimensions 2 cm, 15 cm, and 5 cm.

10-8 Exercises

FOR EOG PRACTICE
see page 679

GUIDED PRACTICE

See Example 1

Find the volume of each rectangular prism.

1.

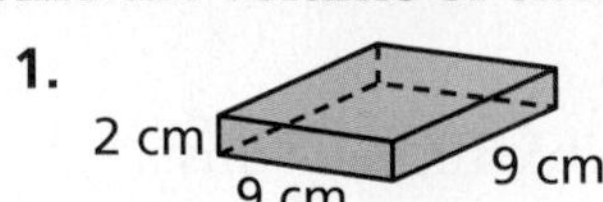

2.

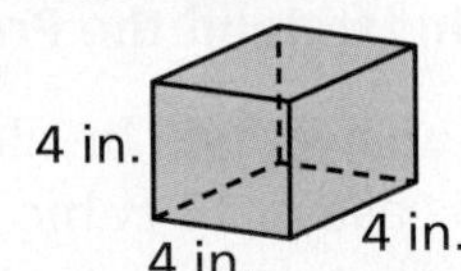

3.

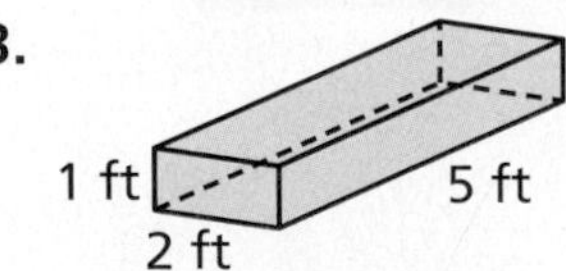

See Example 2

Find the volume of each triangular prism.

4.

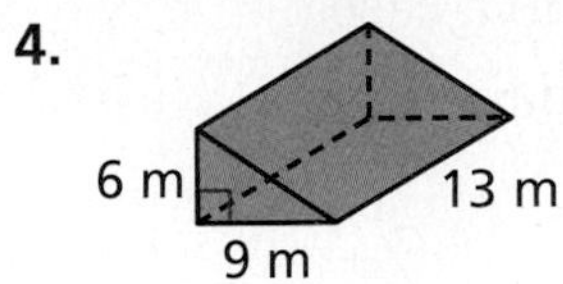

5.

4 ft
20 ft
8 ft

6.

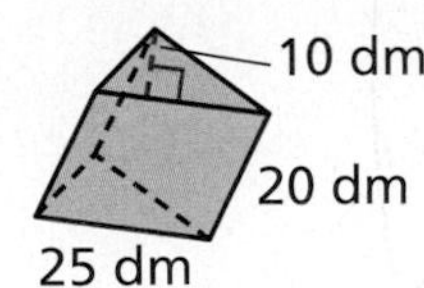

See Example 3

7. A toy company packs 10 cubic boxes of toys in a case. What are the possible dimensions for a case of toys?

INDEPENDENT PRACTICE

See Example 1

Find the volume of each rectangular prism.

8.

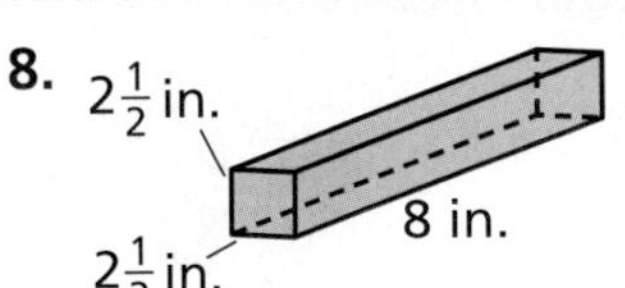

9.

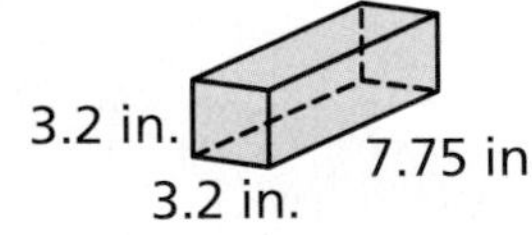

10.

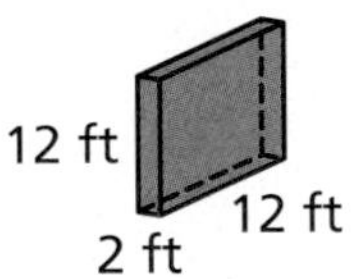

See Example 2

Find the volume of each triangular prism.

11.

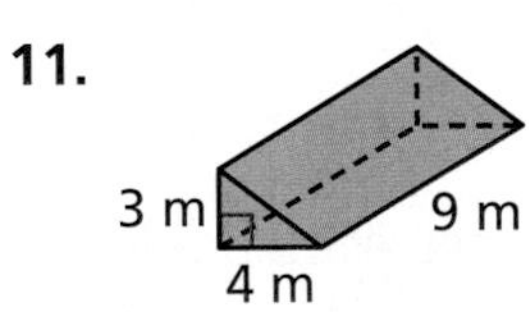

12.

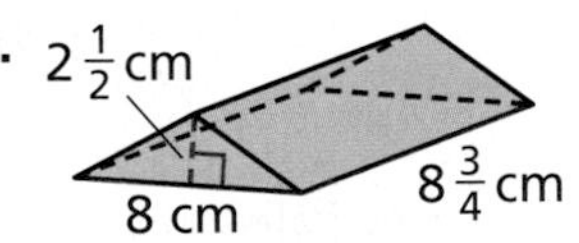

13.

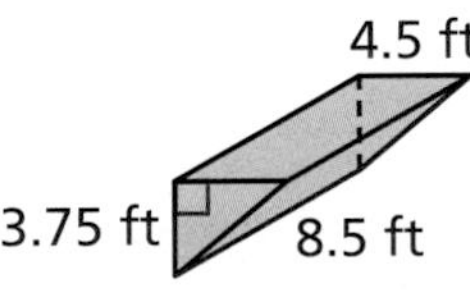

See Example 3

14. A printing company packs 15 cubic boxes of business cards in a larger shipping box. What are the possible dimensions for the shipping box?

PRACTICE AND PROBLEM SOLVING

Find the volume of each figure.

15.

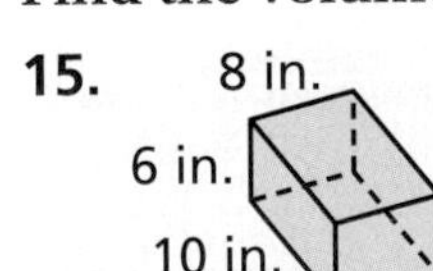

16.

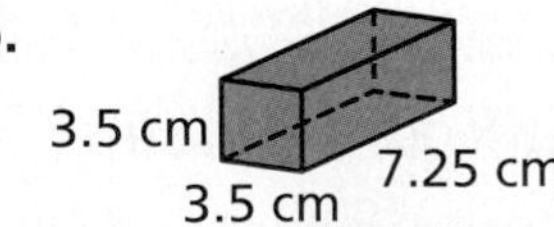

17.

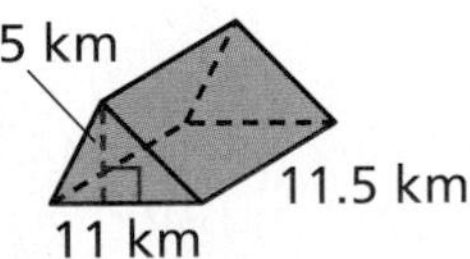

Find the missing measurement for each rectangular prism.

18. ℓ = __?__; $w = 25$ m; $h = 4$ m; $V = 300\text{ m}^3$

19. $\ell = 9$ ft; w = __?__; $h = 5$ ft; $V = 900\text{ ft}^3$

Physical Science 

The density of a substance is a measure of its mass per unit of volume. The density of a particular substance is always the same. The formula for density is the mass of a substance divided by its volume, or $D = \frac{m}{V}$.

Iron filings are attracted by a magnet.

Copper is used in color-coded telephone wires.

20. Find the volume of each substance in the table.

21. Calculate the density of each substance in the table.

22. Water has a density of 1 g/cm^3. A substance whose density is less than that of water will float. Which of the substances in the table will float in water?

23. A fresh egg has a density of approximately 1.2 g/cm^3. A spoiled egg has a density of about 0.9 g/cm^3. How can you tell whether an egg is fresh without cracking it open?

24. Alicia has a solid rectangular prism of a substance she believes is gold. The dimensions of the prism are 2 cm by 1 cm by 2 cm, and the mass is 20.08 g.

a. Find the volume of Alicia's substance.

b. Is the substance that Alicia has gold? Explain.

Rectangular Prisms

Substance	Length (cm)	Width (cm)	Height (cm)	Mass (g)
Copper	2	1	5	89.6
Gold	$\frac{2}{3}$	$\frac{3}{4}$	2	19.32
Iron pyrite	0.25	2	7	17.57
Pine	10	10	3	120
Silver	2.5	4	2	210

25. ***WRITE ABOUT IT*** In a science lab, you are given a prism of copper. You determine that its dimensions are 4 cm, 2 cm, and 6 cm. Without weighing the prism, how can you determine its mass? Explain your answer.

26. ***CHALLENGE*** A solid rectangular prism of silver has a mass of 84 g. What are some possible dimensions of the prism?

Gold is used to make many pieces of jewelry.

Spiral Review

Find the mean of each set. (Lesson 6-2)

27. 0, 5, 2, −3, 7, 1

28. 6, 6, 6, 6, 6, 6, 6, 6, 6, 6, 6, 6, 6

Find 20% of each number. (Lesson 8-9)

29. 200

30. 50

31. 15

32. 3,000

33. **EOG PREP** Which sum is negative? (Lesson 9-4)

A −9 + 10

B 17 + (−4)

C 0 + 5

D −3 + (−5)

10-9 Volume of Cylinders

Learn to find volumes of cylinders.

Thomas Edison invented the first phonograph in 1877. The main part of this phonograph was a cylinder with a 4-inch diameter and a height of $3\frac{3}{8}$ inches.

To find the volume of a cylinder, you can use the same method as you did for prisms: Multiply the area of the base by the height.

volume of a cylinder = area of base × height

The area of the circular base is πr^2, so the formula is $V = Bh = \pi r^2 h$.

EXAMPLE 1 Finding the Volume of a Cylinder

Find the volume *V* of each cylinder to the nearest cubic unit.

A

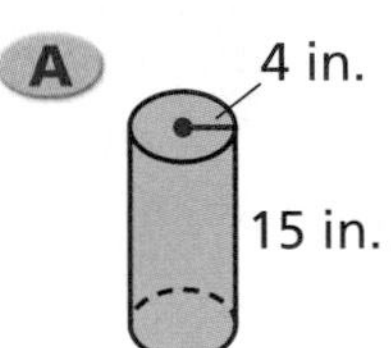

$V = \pi r^2 h$	*Write the formula.*
$V \approx \mathbf{3.14} \times \mathbf{4}^2 \times 15$	*Replace π with **3.14**, r with **4**, and h with 15.*
$V \approx 753.6$	*Multiply.*

The volume is about 754 in^3.

B

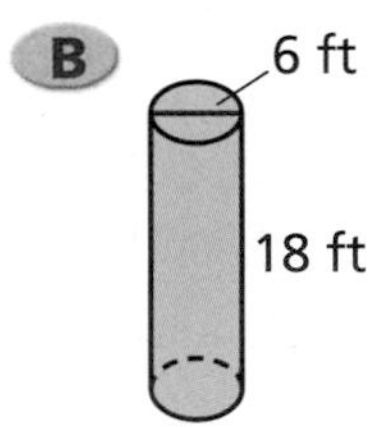

6 ft ÷ 2 = 3 ft	*Find the radius.*
$V = \pi r^2 h$	*Write the formula.*
$V \approx \mathbf{3.14} \times \mathbf{3}^2 \times 18$	*Replace π with **3.14**, r with **3**, and h with 18.*
$V \approx 508.68$	*Multiply.*

The volume is about 509 ft^3.

C

$r = \frac{h}{6} + 1$

$h = 24$ cm

$r = \frac{h}{6} + 1$	*Find the radius.*
$r = \frac{24}{6} + 1 = 5$	*Substitute 24 for h.*
$V = \pi r^2 h$	*Write the formula.*
$V \approx \mathbf{3.14} \times \mathbf{5}^2 \times 24$	*Replace π with **3.14**, r with **5**, and h with 24.*
$V \approx 1{,}884$	*Multiply.*

The volume is about 1,884 cm^3.

EXAMPLE 2 *Music Application*

The cylinder in Edison's first phonograph had a 4 in. diameter and a height of about 3 in. The standard phonograph manufactured 21 years later had a 2 in. diameter and a height of 4 in. Estimate the volume of each cylinder to the nearest cubic inch.

A **Edison's first phonograph**

4 in. ÷ 2 = 2 in.	*Find the radius.*
$V = \pi r^2 h$	*Write the formula.*
$V \approx 3.14 \times 2^2 \times 3$	*Replace π with 3.14, r with 2, and h with 3.*
$V \approx 37.68$	*Multiply.*

The volume of Edison's first phonograph was about 38 in^3.

B **Edison's standard phonograph**

2 in. ÷ 2 = 1 in.	*Find the radius.*
$V = \pi r^2 h$	*Write the formula.*
$V \approx \frac{22}{7} \times 1^2 \times 4$	*Replace π with $\frac{22}{7}$, r with 1, and h with 4.*
$V \approx \frac{88}{7} = 12\frac{4}{7}$	*Multiply.*

The volume of the standard phonograph was about 13 in^3.

EXAMPLE 3 Comparing Volumes of Cylinders

Find which cylinder has the greater volume.

Cylinder 1: $V = \pi r^2 h$

$V \approx 3.14 \times 6^2 \times 12$

$V \approx 1{,}356.48 \text{ cm}^3$

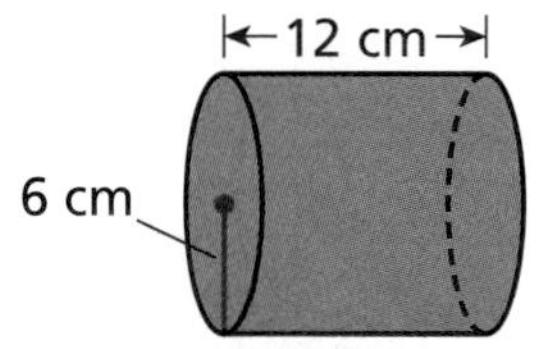

Cylinder 2: $V = \pi r^2 h$

$V \approx 3.14 \times 4^2 \times 16$

$V \approx 803.84 \text{ cm}^3$

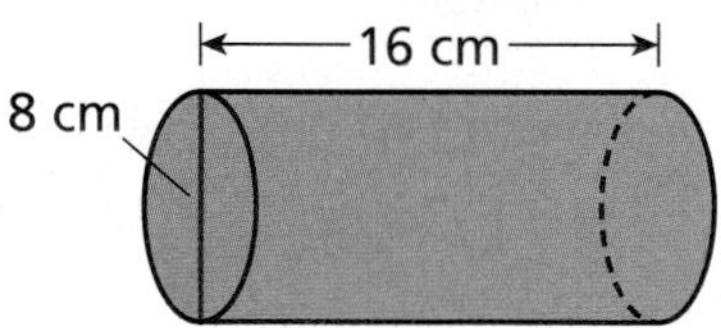

Cylinder 1 has the greater volume because $1{,}356.48 \text{ cm}^3 > 803.84 \text{ cm}^3$.

Think and Discuss

1. **Explain** how the formula for the volume of a cylinder is similar to the formula for the volume of a rectangular prism.
2. **Explain** which parts of a cylinder are represented by πr^2 and h in the formula $V = \pi r^2 h$.

10-9 Exercises

FOR EOG PRACTICE
see page 679

internet connect
Homework Help Online
go.hrw.com Keyword: MR4 10-9

GUIDED PRACTICE

See Example 1

Find the volume V of each cylinder to the nearest cubic unit.

1.

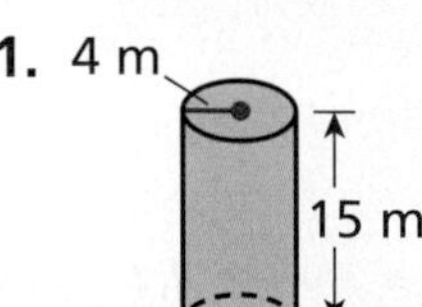

2.

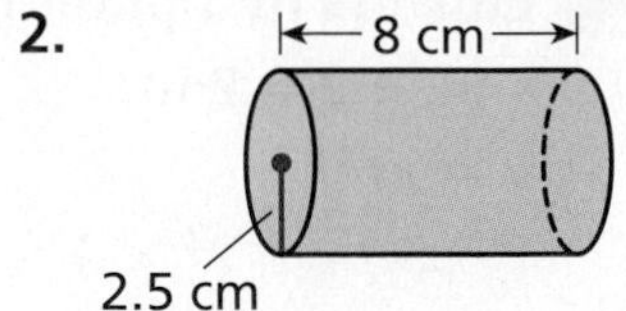

3.

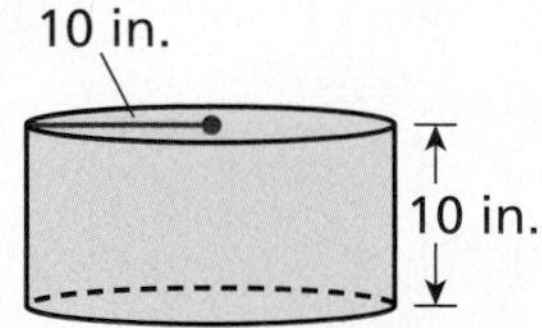

See Example 2

4. A cylindrical bucket with a diameter of 4 inches is filled with rainwater to a height of 2.5 inches. Estimate the volume of the rainwater to the nearest cubic inch.

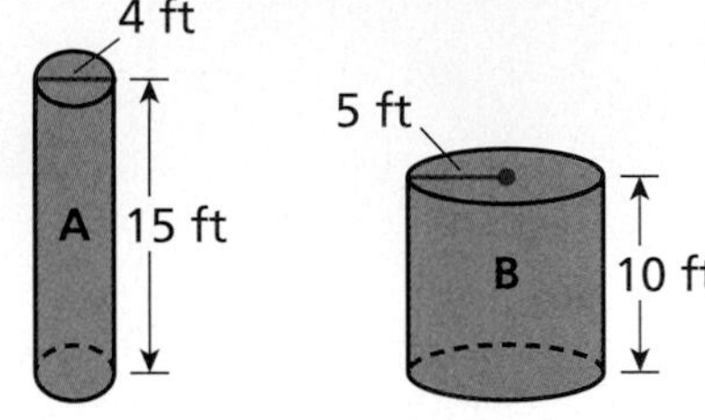

See Example 3

5. Find which cylinder, A or B, in the diagram has the greater volume.

INDEPENDENT PRACTICE

See Example 1

Find the volume V of each cylinder to the nearest cubic unit.

6.

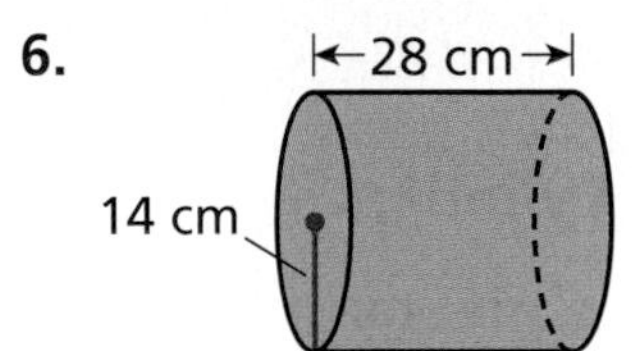

7.

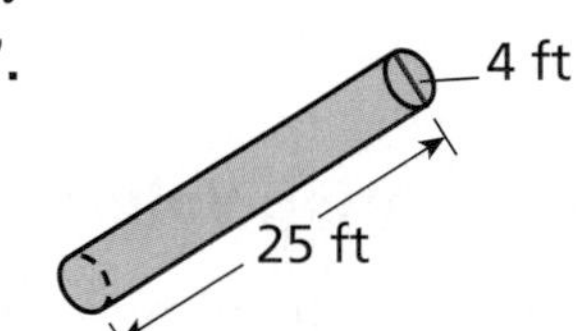

8.

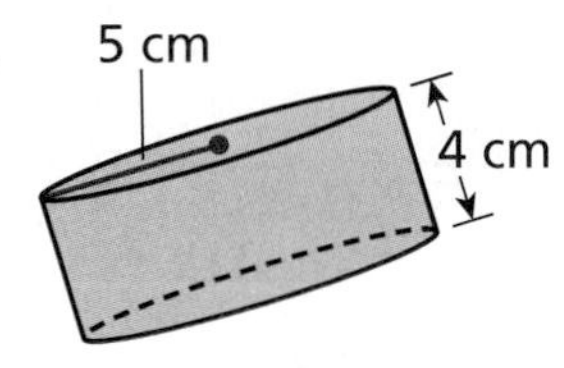

See Example 2

9. Wooden dowels are solid cylinders of wood. One dowel has a radius of 1 cm, and another dowel has a radius of 3 cm. Both dowels have a height of 10 cm. Estimate the volume of each dowel to the nearest cubic cm.

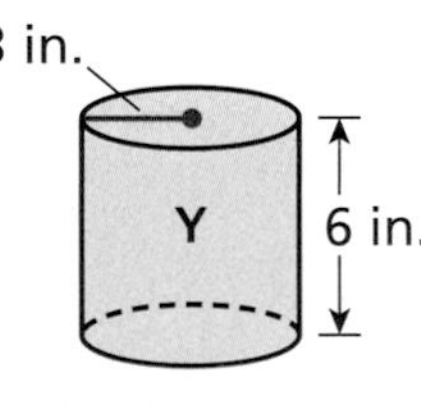

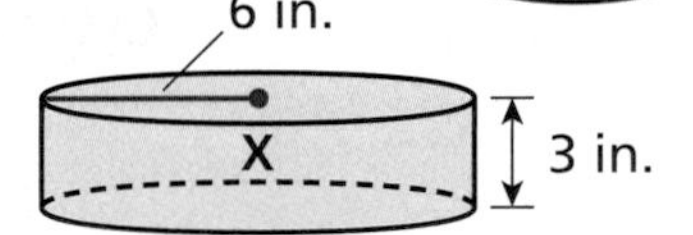

See Example 3

10. Find which cylinder, X or Y, in the diagram has the greater volume.

PRACTICE AND PROBLEM SOLVING

Find the volume of each cylinder to the nearest cubic unit.

11.

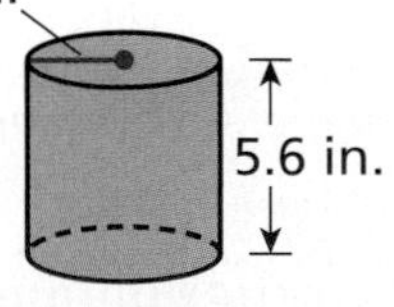

12.

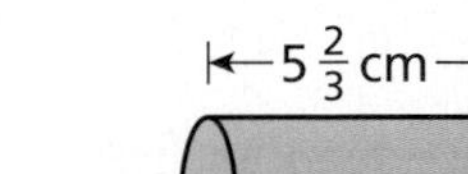

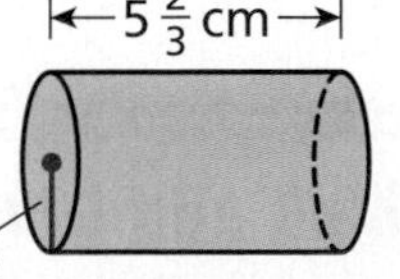

13.

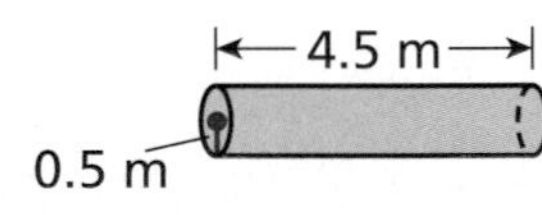

Find the volume of each cylinder using the information given.

14. $r = 6$ cm; $h = 6$ cm

15. $d = 4$ in.; $h = 8$ in.

16. $r = 7.5$ ft; $h = 11.25$ ft

17. $d = 12\frac{1}{4}$ yd; $h = 5\frac{3}{5}$ yd

Find the volume of each interior cylinder to the nearest cubic unit.

18. 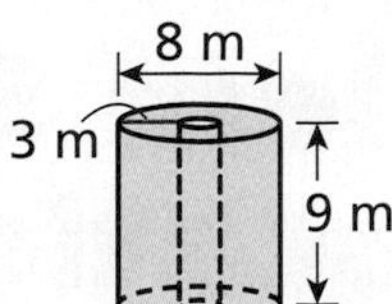

19. 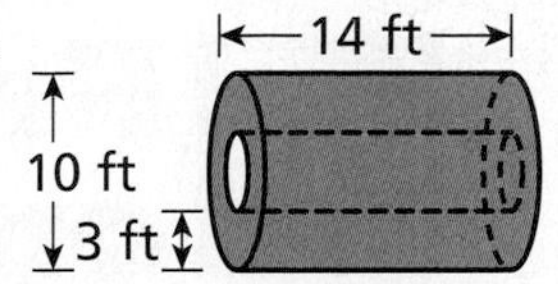

20. 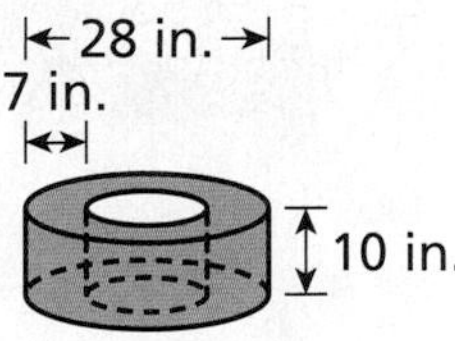

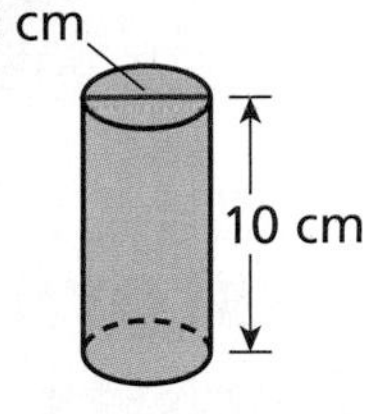

21. ***MEASUREMENT*** Could this blue can hold 200 cm^3 of juice? How do you know?

22. ***GARDENING*** Kyle wants to fill a cylindrical planter with soil that costs $0.01 per cubic inch. The diameter of the planter is 8 inches and the height is 6 inches. About how much will Kyle spend on soil?

23. ***SCIENCE*** A scientist filled a cylindrical beaker with 942 mm^3 of a chemical solution. The area of the base of the cylinder is 78.5 mm^2. What is the height of the solution?

24. ***CHOOSE A STRATEGY*** Fran, Gene, Helen, and Ira have cylinders with different volumes. Gene's cylinder holds more than Fran's. Ira's cylinder holds more than Helen's, but less than Fran's. Whose cylinder has the largest volume? What color cylinder does each person have?

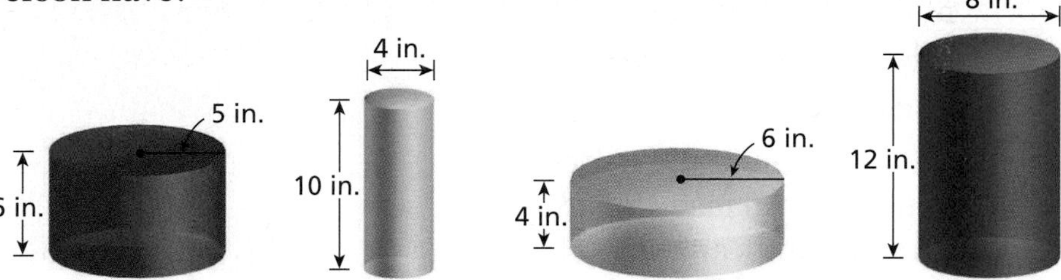

25. ***WRITE ABOUT IT*** Explain why volume is expressed in cubic units of measurement.

26. ***CHALLENGE*** Find the volume of the shaded area.

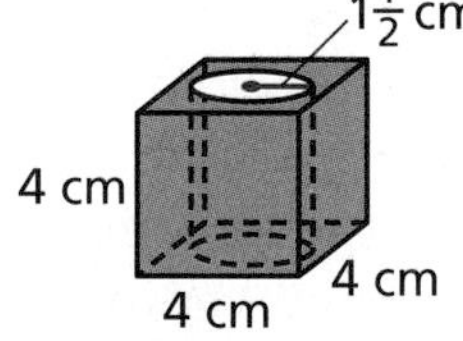

Spiral Review

Determine whether the given value is a solution to each equation. (Lesson 2-3)

27. $2x + 3 = 10; x = 4$ **28.** $5(b - 3) = 25; b = 8$

Write a fraction whose value is between the given fractions. (Lesson 4-6)

29. $\frac{1}{2}$ and $\frac{5}{6}$ **30.** $\frac{2}{3}$ and $\frac{5}{12}$ **31.** $\frac{3}{5}$ and $\frac{9}{10}$

32. **EOG PREP** Which of the following is the best deal? (Lesson 8-1)

A 2 lb for $8.40 **B** 3 lb for $12.50 **C** 4 lb for $17.00 **D** 5 lb for $21.00

Problem Solving on Location

NORTH CAROLINA

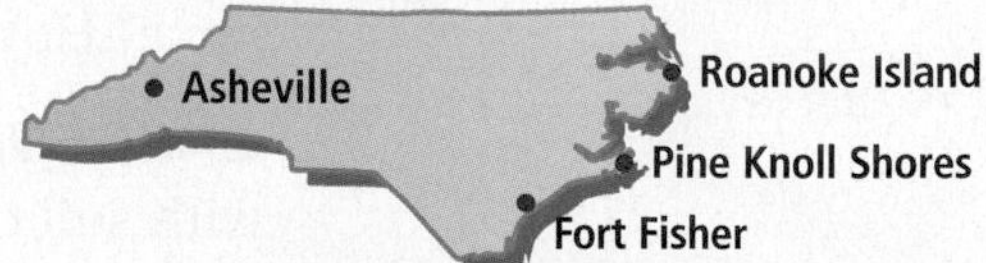

Historical Significant Events Imagery

The Historical Significant Events Imagery (HSEI) database, in Asheville, North Carolina, has hundreds of selected satellite images capturing some of the most important weather and environmental events of the last 30 years. Dust storms, floods, forest fires, and hurricanes have been photographed by satellite, and anyone with a computer can download the images.

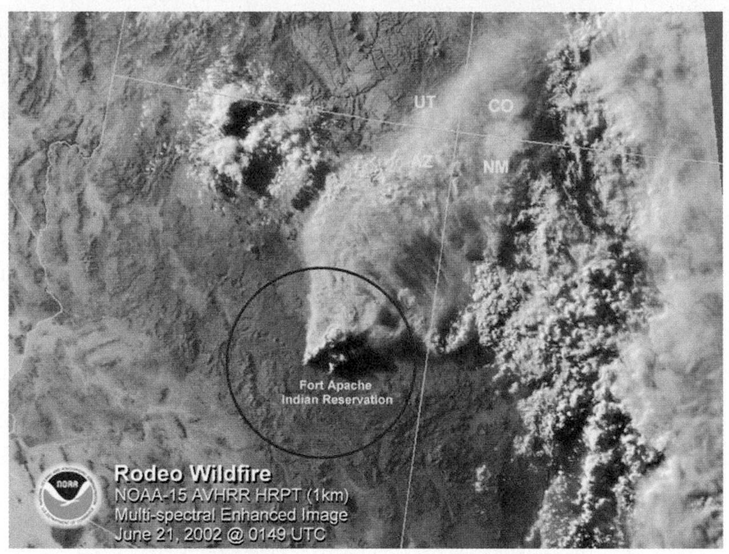

For 1–3, use the drawings on the satellite images.

1. The circle drawn around the Rodeo wildfire on the Fort Apache Indian Reservation represents an area 180 mi in diameter. What is the circumference and area represented by the circle? Use 3.14 for π.
2. The rectangle drawn on the image of hurricane Dennis defines an area that was under a severe thunderstorm warning. The dimensions represented by the rectangle are 150 mi by 55 mi. Find the perimeter and area represented by the rectangle.
3. There was danger of high seas in a region shaped like a right triangle. The triangle that defined the region had a base of 200 mi and a height of 75 mi. Find the area of the triangle.
4. A geostationary satellite that captures images of Earth is shaped like a rectangular prism measuring approximately 6.6 ft by 6.9 ft by 7.5 ft. Find the volume of the satellite.

North Carolina Aquariums

North Carolina Aquariums use hundreds of thousands of gallons of water to educate and entertain visitors. The aquariums at Roanoke Island, Pine Knoll Shores, and Fort Fisher display how important aquatic and marine life are to the state of North Carolina. The three aquariums, based in their own regions along the coast, focus on specific themes unique to their area.

1. The aquarium at Pine Knoll Shores has a building that measures 35,000 ft^2. If the base of the building were a rectangle 125 ft wide, how long would the 35,000 ft^2 building be?
2. If the curator at Pine Knoll Shores designed an aquarium in the shape of a cylinder that was 4 ft in diameter and 4 ft tall, could the aquarium hold 350 gal of water? (One cubic foot contains about 7.48 gal.)

For 3–5, use the information that 1 acre is 43,560 ft^2.

3. If a section of land is 198 ft wide by 220 ft long, what is the perimeter of the land? Is the section of land 1 acre?
4. Find the number of acres contained in the 85,000 ft^2 building at Fort Fisher. Round the answer to the nearest hundredth.
5. The aquarium at Roanoke Island sits on 14 acres of land. How many square feet are in 14 acres?

MATH-ABLES

Polygon Hide-and-Seek

Use the figure to name each polygon described.

1. an obtuse scalene triangle
2. a right isosceles triangle
3. a parallelogram with no right angles
4. a trapezoid with two congruent sides
5. a pentagon with three congruent sides

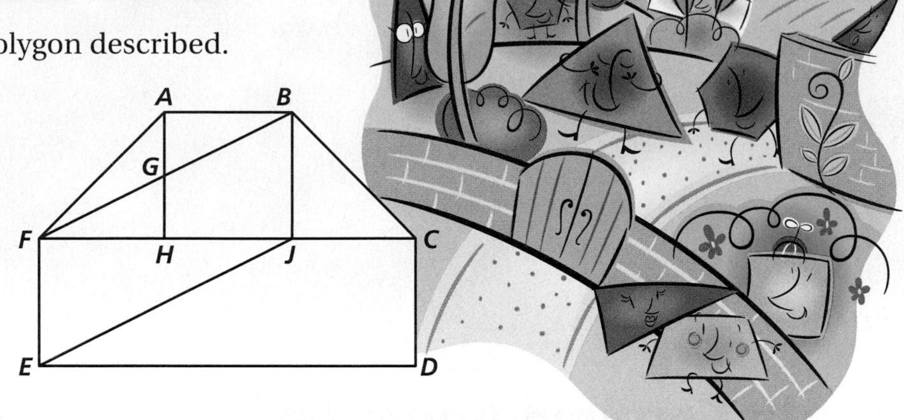

Poly-Cross Puzzle

You will use the names of the figures below to complete a crossword puzzle.

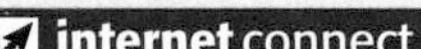

internet connect

Go to ***go.hrw.com*** for a blank crossword puzzle.
KEYWORD: MR4 Game10

ACROSS

1.
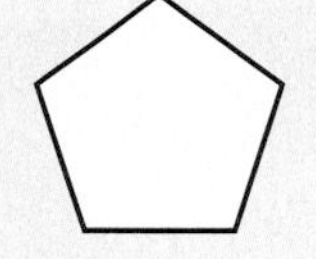

2.
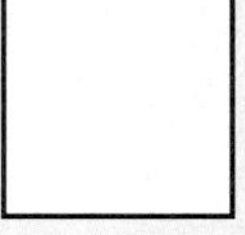

3.
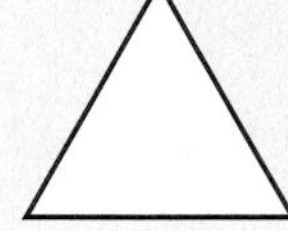

4.
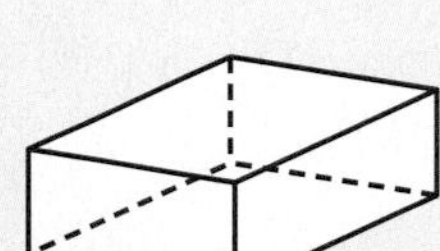

5.
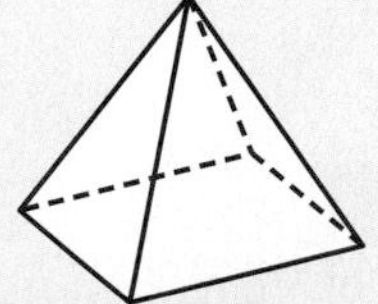

6.
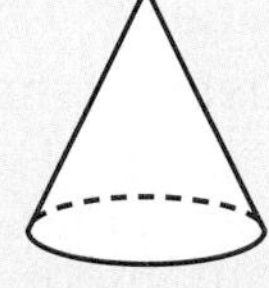

DOWN

1.

7.
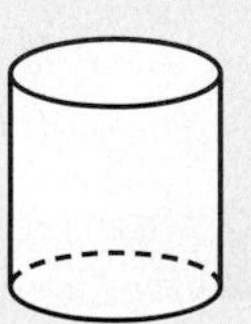

8.
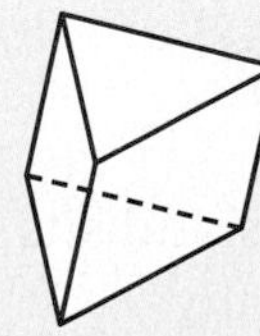

Area and Perimeter

internet connect
Lab Resources Online
go.hrw.com
KEYWORD: MR4 TechLab10

Geometry software can be used to explore geometric formulas.

Activity

1 Use your geometry software to explore the formula for the area of a rectangle, $A = \ell \cdot w$.

a. Construct a rectangle *ABCD*. Choose four points and connect them with line segments, making sure the opposite sides are parallel.

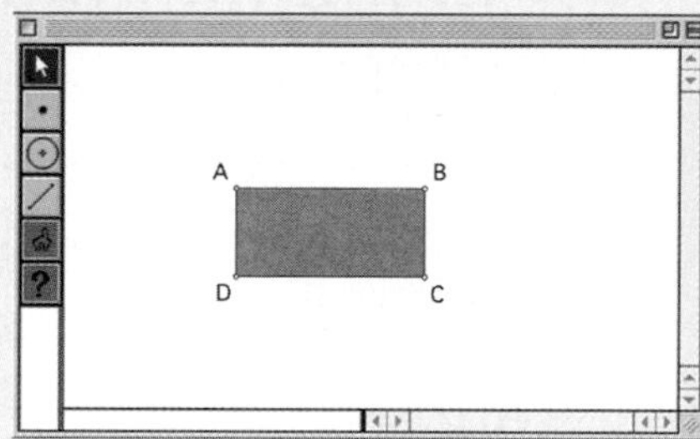

b. Use the distance tool to measure the length of sides $\overline{AB}$ and $\overline{CB}$.

Select the interior of the rectangle, and then use the area tool to measure the area.

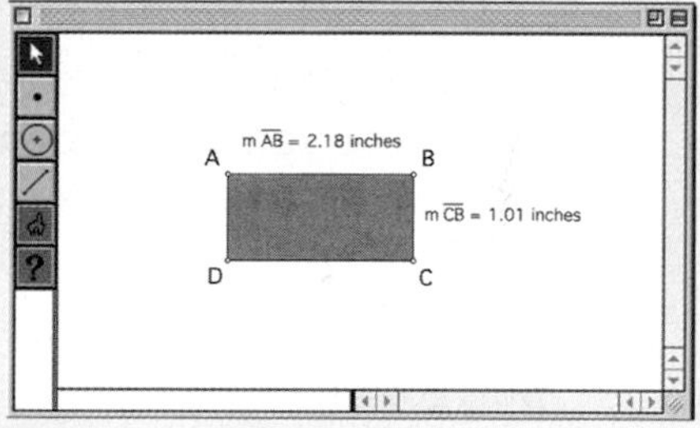

c. Use a calculator or paper and pencil to find the product of the side lengths. Round to the hundredths place.
$2.18 \cdot 1.01 \approx 2.20$

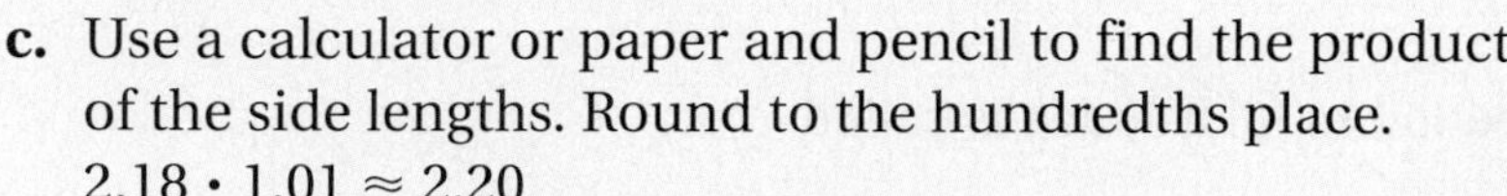

Notice that the geometry software rounds the product to 2.21, which is close to 2.20.
So $Area = AB \cdot CB = \ell \cdot w$.

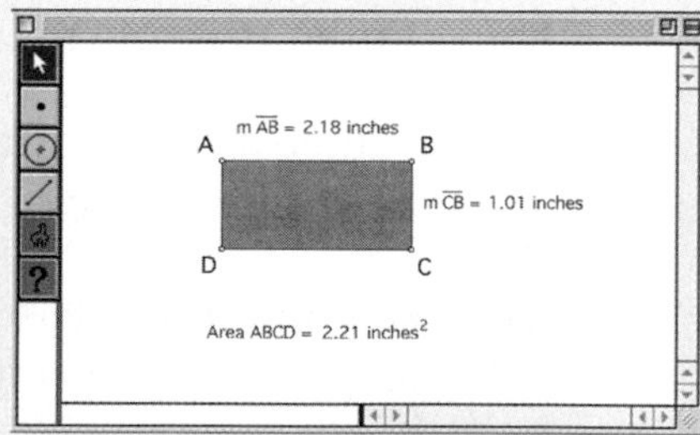

Think and Discuss

1. Tell whether the perimeter *P* of rectangle *ABCD* is equal to $2 \cdot (AB + CB)$.
2. Determine whether the area of rectangle *ABCD* divided by 2 is equal to the perimeter.

Try This

1. Use geometry software to construct a triangle *ABC* where $m\angle B = 90°$.
   - **a.** Measure the area of the triangle and the lengths of sides $\overline{AB}$ and $\overline{CB}$. Find $\frac{1}{2} \cdot AB \cdot CB$.
   - **b.** Drag angle *A*, making sure $m\angle B = 90°$. Do this three more times to construct triangles with different areas and side lengths. For each triangle, find $\frac{1}{2} \cdot AB \cdot CB$. What do you conclude?

Chapter 10

Study Guide and Review

Vocabulary

area	504	net	530
base	524	perimeter	500
center	516	*pi*	516
circle	516	polyhedron	524
circumference	516	prism	524
cone	525	pyramid	525
cylinder	524	radius (radii)	516
diameter	516	surface area	530
edge	524	vertex	524
face	524	volume	534

Complete the sentences below with vocabulary words from the list above. Words may be used more than once.

1. A __?__ is a three-dimensional object with flat faces that are polygons.
2. The number of cubic units needed to fill a space is called __?__.
3. The distance around a figure is called the __?__, and the distance around a circle is called the __?__.
4. A line segment that passes through the center of a circle and has both endpoints on the circle is a __?__.

10-1 Finding Perimeter (pp. 500–503)

EXAMPLE

- **Find the perimeter of the figure.**

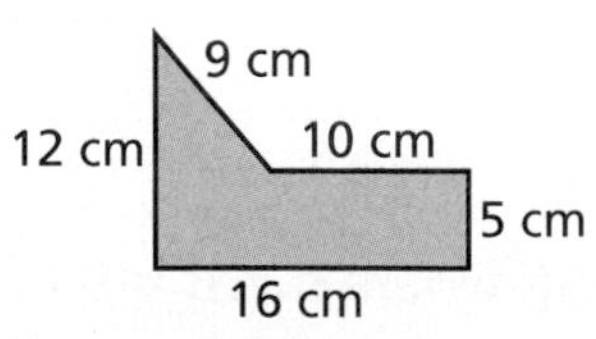

$P = 9 + 10 + 5 + 16 + 12$ *Add all the side lengths.*
$P = 52$
The perimeter is 52 cm.

EXERCISES

5. Find the perimeter.

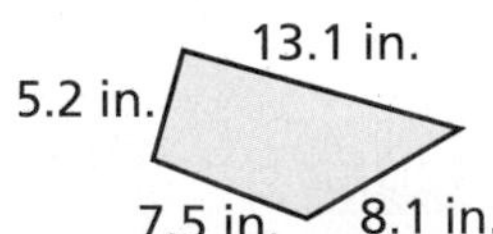

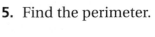

6. What is the length of n if the perimeter is 20 ft?

4 ft 1 ft 1 ft 1 ft 3 ft 3 ft n

10-2 Estimating and Finding Area (pp. 504–507)

EXAMPLE

- **Find the area of the rectangle.**

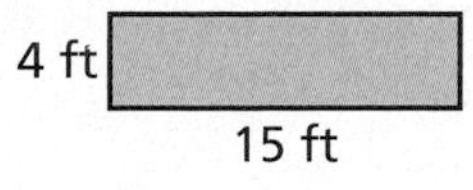

$A = \ell w$

$A = 15 \cdot 4 = 60$

The area is 60 ft^2.

EXERCISES

Find the area of each figure.

7. 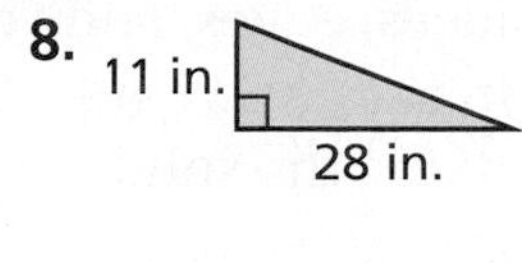

8.

10-3 Break into Simpler Parts (pp. 508–510)

EXAMPLE

- **Find the area of the polygon.**

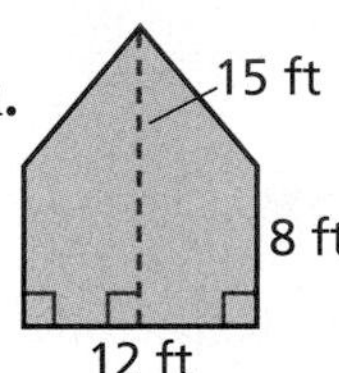

$A = 8 \cdot 12 = 96$

$A = \frac{1}{2} \cdot 12 \cdot 7 = 42$

The area of the figure is $42 \text{ ft}^2 + 96 \text{ ft}^2 = 138 \text{ ft}^2$.

EXERCISES

Find the area of each polygon.

9. 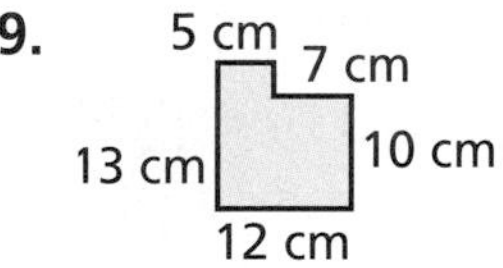

10. 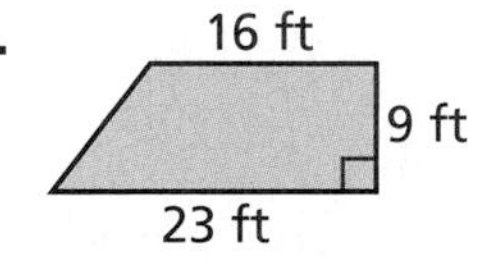

10-4 Comparing Perimeter and Area (pp. 511–513)

EXAMPLE

- **Find how the perimeter and area of a rectangle change when the dimensions change.**

When the dimensions of the rectangle are multiplied by x, the perimeter is multiplied by x, and the area is multiplied by x^2.

EXERCISES

Find how the perimeter and area change when the dimensions change.

11.

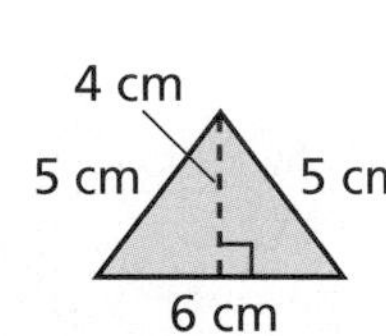

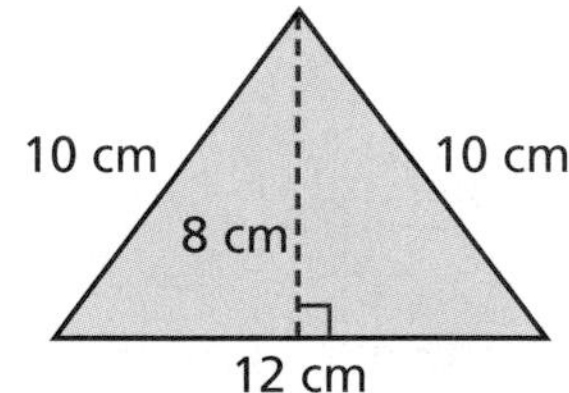

10-5 Circles (pp. 516–519)

EXAMPLE

- **Find the circumference and the area of the circle. Use 3.14 for *pi*. Round to the nearest hundredth.**

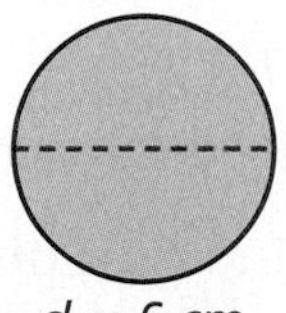

$C = \pi d$

$C \approx 3.14 \cdot 6$

$C \approx 18.84$ cm

$A = \pi r^2$

$r = d \div 2 = 6 \div 2 = 3$

$A \approx 3.14 \cdot 3^2$

$A \approx 3.14 \cdot 9 \approx 28.26 \text{ cm}^2$

EXERCISES

Find each missing value to the nearest hundredth. Use 3.14 for *pi*.

12. $d = 10$ ft; $C =$? **13.** $r = 8$ cm; $C =$?

14. $C = 28.26$ m; $d =$? **15.** $C = 69.08$ ft; $r =$?

16. Find the area of a circle with radius 7 cm. Use $\frac{22}{7}$ for *pi*.

17. Find the area of a circle with diameter 28 cm. Use $\frac{22}{7}$ for *pi*.

10-6 Solid Figures (pp. 524–527)

EXAMPLE

- **Identify the number of faces, edges, and vertices in the solid figure. Then name the solid.**

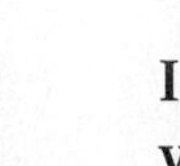

5 faces; 9 edges; 6 vertices

There are two congruent parallel bases, so the figure is a prism. The bases are triangles.

The solid is a triangular prism.

EXERCISES

Identify the number of faces, edges, and vertices in each solid figure. Then name the solid.

18.

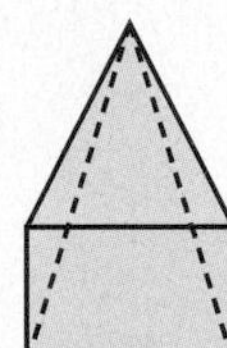

19.

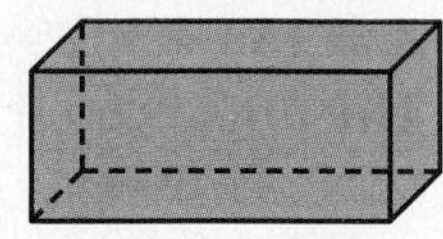

10-7 Surface Area (pp. 530–533)

EXAMPLE

- **Find the surface area S of the cylinder.**

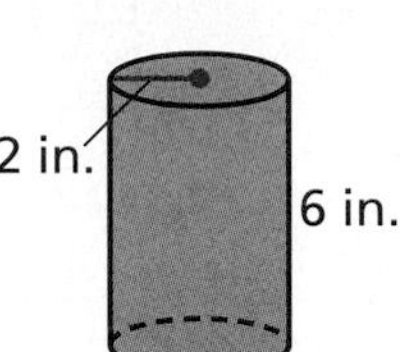

$S = h \cdot (2\pi r) + 2 \cdot (\pi r^2)$

$S \approx 6 \cdot (2 \cdot 3.14 \cdot 2) + 2 \cdot (3.14 \cdot 2^2)$

$S \approx 100.48 \text{ in}^2$

EXERCISES

Find the surface area S of each solid.

20.

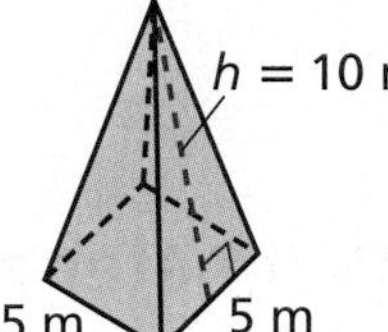

21.

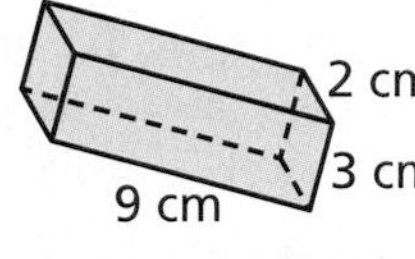

10-8 Finding Volume (pp. 534–537)

EXAMPLE

- **Find the volume of the rectangular prism.**

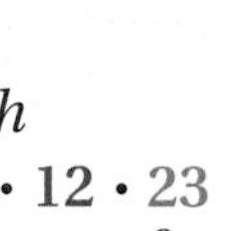

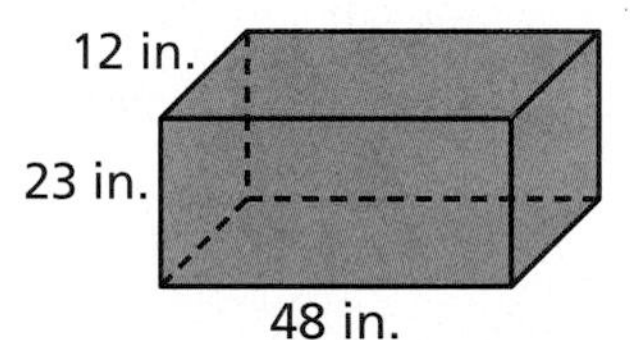

$V = \ell wh$

$V = 48 \cdot 12 \cdot 23$

$V = 13{,}248 \text{ in}^3$

EXERCISES

Find the volume of each prism.

22.

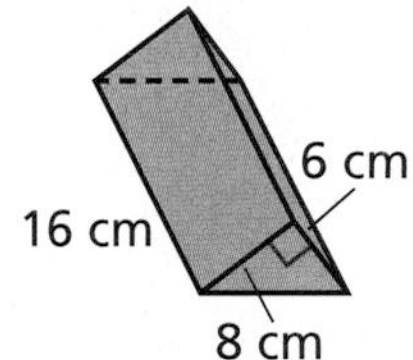

23.

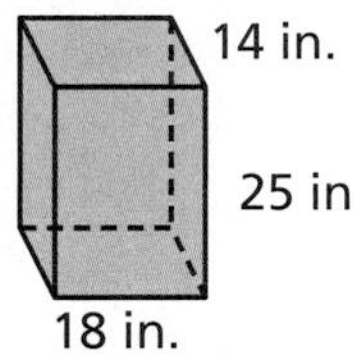

10-9 Volume of Cylinders (pp. 538–541)

EXAMPLE

- **Find the volume of the cylinder to the nearest cubic unit.**

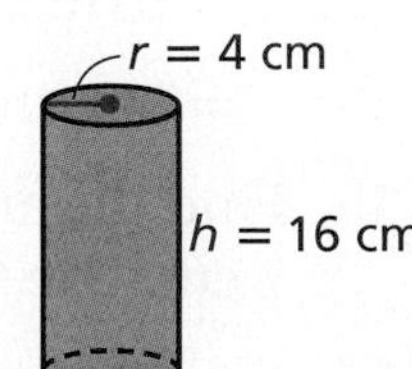

$V \approx 3.14 \cdot 4^2 \cdot 16$

$V \approx 803.84 \text{ cm}^3$

The volume is about 804 cm^3.

EXERCISES

Find the volume of each cylinder to the nearest cubic unit.

24. h = 12.5 m, r = 3 m

25.

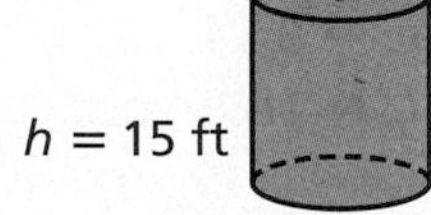

Chapter Test

Chapter 10

Find the perimeter and area of each figure.

1. 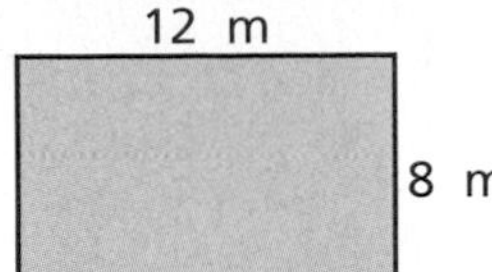

2. 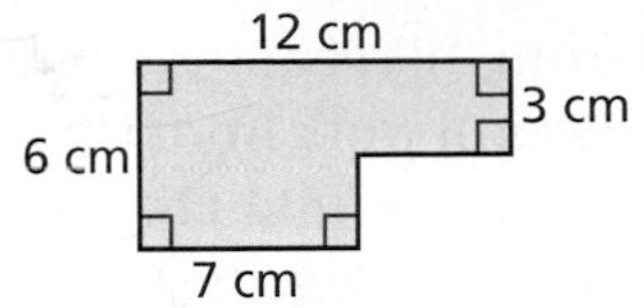

3. 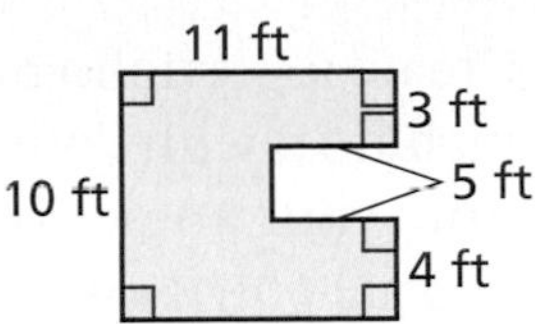

4. Find how the perimeter and the area of a rectangle change when the length and width are doubled.

Name the circle and two radii for each circle. Then find the area and the circumference of each circle.

5. 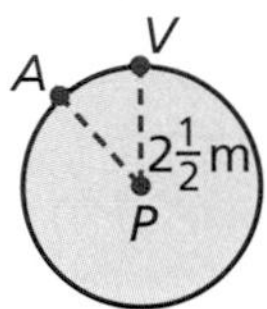

6. 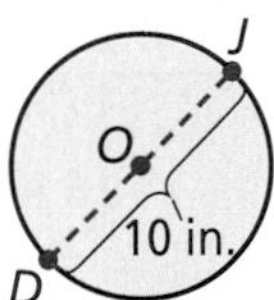

7. 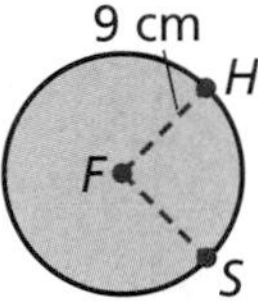

Identify the number of faces, edges, and vertices in each solid figure. Then tell whether each figure is a polyhedron and name the solid.

8. 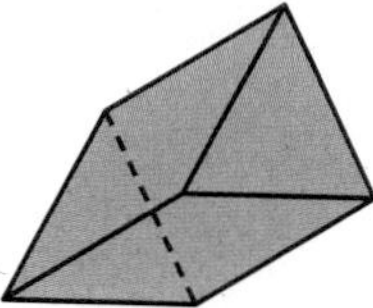

9. 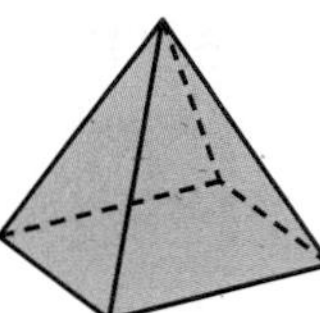

10. 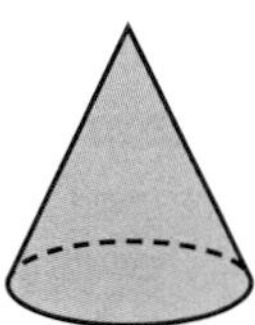

Find the surface area *S* of each solid.

11. 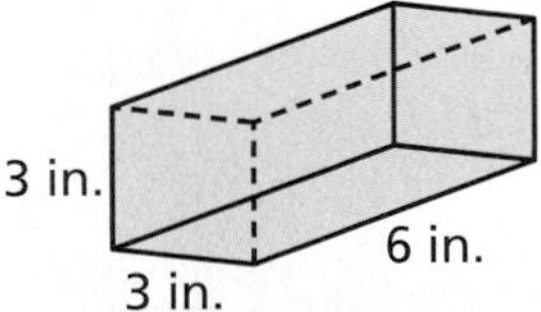

12. 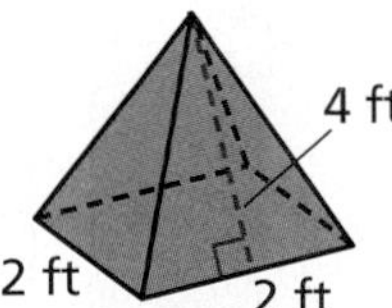

13. 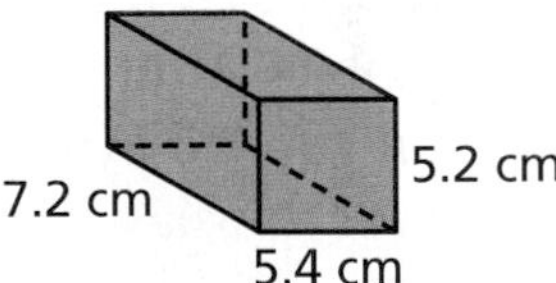

Find the volume *V* of each solid.

14. 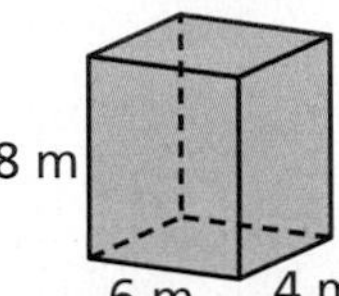

15. 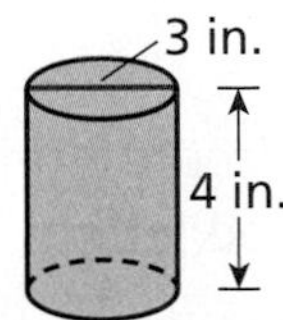

16. 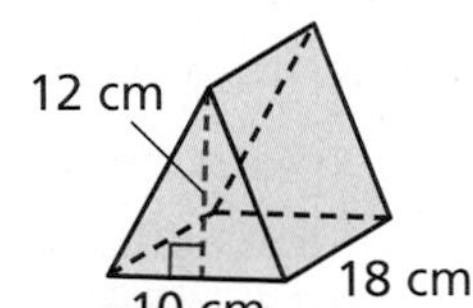

17. Patricia has two cylinder-shaped jars. Jar A has a radius of 6 cm and a height of 9 cm. Jar B has a diameter of 8 cm and a height of 17 cm. Which jar has the greater volume? How much greater?

Chapter Test

Chapter 10

Performance Assessment

Show What You Know

Create a portfolio of your work from this chapter. Complete this page and include it with your four best pieces of work from Chapter 10. Choose from your homework or lab assignments, mid-chapter quiz, or any journal entries you have done. Put them together using any design you want. Make your portfolio represent what you consider your best work.

Short Response

1. Find the perimeter and area of each figure. Show your work. Which figure has the greatest perimeter? Which has the greatest area?

a. 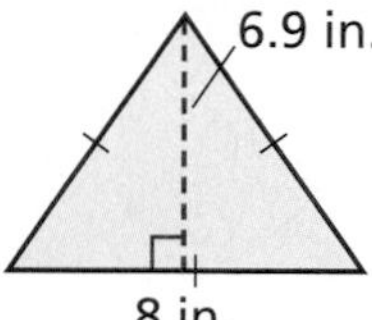

b. 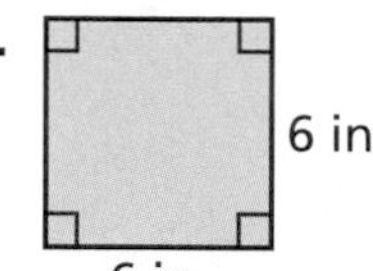

c. 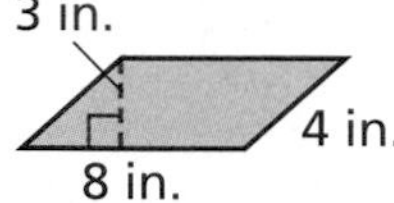

2. George will use 32 feet of fencing to enclose a rectangular garden. George wants the greatest possible area for his garden. What dimensions, in whole feet, will his garden be? Explain how you found your answer.

Extended Problem Solving

3. The ancient Sumerians kept records by etching onto clay tablets. After etching the tablet, they rolled cylindrical stones over the clay. The stones were carved with unique images whose impressions were then transferred onto the tablets. These impressions proved that the writings were originals.

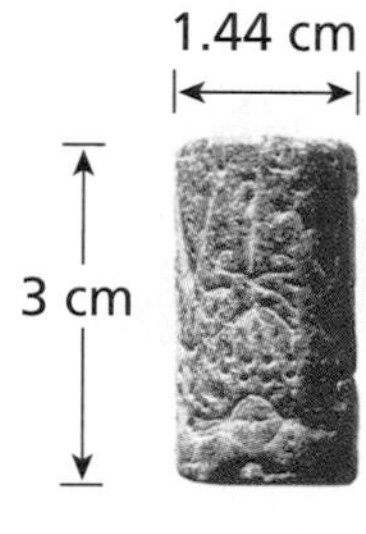

a. Explain the relationship between the circumference of the base of the cylindrical stone and the indicated length of the clay impression.

b. Find the indicated length of the clay impression. Use 3.14 for π, and round your answer to the nearest hundredth. Show your work.

c. Find the volume of the cylindrical stone. Use 3.14 for π, round your answer to the nearest hundredth, and show your work.

internet connect
EOG Test Practice Online
go.hrw.com Keyword: MR4 TestPrep

Getting Ready for EOG

Chapter 10

Cumulative Assessment, Chapters 1–10

TEST TAKING TIP!
Perimeter is the distance around a figure, and area is the amount of surface a figure covers.

1. What is the perimeter and the area of a rectangle with length 8 cm and width 6 cm?
   A $P = 14$ cm, $A = 48$ cm^2
   B $P = 28$ cm, $A = 48$ cm^2
   C $P = 48$ cm, $A = 14$ cm^2
   D $P = 48$ cm, $A = 28$ cm^2

2. Which expression has the greatest value?
   A 20% · 150
   C $4\frac{1}{2} \cdot 5\frac{2}{3}$
   B $2^2 \cdot 5 + 8$
   D $4 \div 0.12$

3. Which value is equivalent to 521 cm?
   A 5,210 m
   C 5.21 m
   B 0.521 m
   D 0.0521 km

4. Which is a polyhedron?
   A 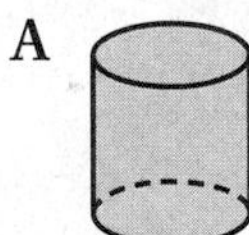
   C 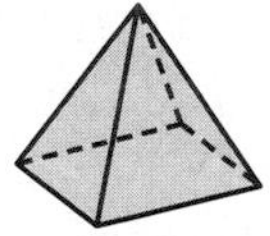
   B 
   D 

5. Evaluate $3\frac{1}{3} \cdot 2\frac{4}{5} + 1\frac{2}{3}$.
   A $7\frac{14}{15}$
   C 11
   B $4\frac{1}{3}$
   D $5\frac{1}{2}$

6. What is the measure of angle a?

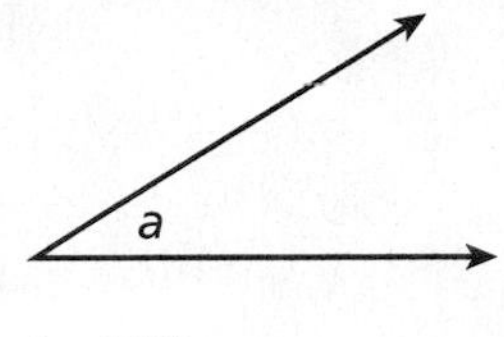

   A 32°
   C 68°
   B 58°
   D 148°

7. What is the prime factorization of 126?
   A $2 \cdot 63$
   C $1 \cdot 2 \cdot 3 \cdot 21$
   B $2 \cdot 7 \cdot 9$
   D $2 \cdot 3^2 \cdot 7$

8. Which figure below has the greatest volume?
   A F
   C H
   B G
   D J

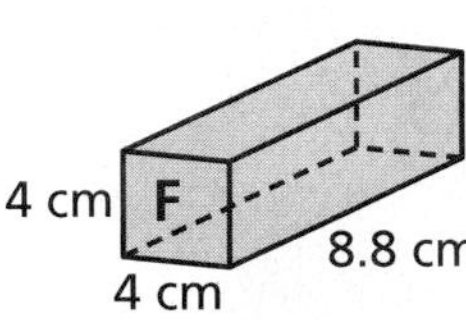

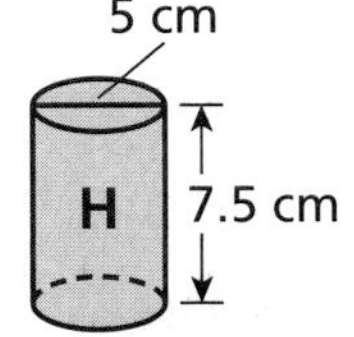

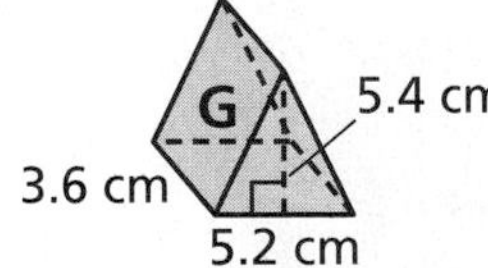

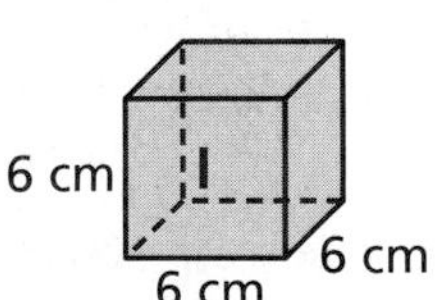

9. ***SHORT RESPONSE*** Aaron plans to buy a shirt that is on sale for 40% off the ticketed price. The ticketed price is $24.00, and sales tax is 7%. How much will Aaron pay for the shirt? Show your work.

10. ***SHORT RESPONSE*** Use prime factorization to find the LCM of 10, 12, and 15. Show your work.

Chapter

11

Probability

TABLE OF CONTENTS

internet connect

Chapter Opener Online
go.hrw.com
KEYWORD: MR4 Ch11

Interest Earned on $100 Investment

Years Invested	Interest (compounded annually)			
	7%	8%	9%	10%
1	$7	$8	$9	$10
2	$14	$17	$19	$21
5	$40	$47	$54	$61
10	$97	$116	$137	$159

Career ***Financial Advisor***

We all must decide how much money to spend and how much to invest and save for the future. Financial advisors help people make these decisions.

Financial advisors must understand the relationship between risk and earnings. An investment with a high probability of returning a profit is less risky than an investment with a lower probability of returning a profit. However, riskier investments may return larger profits. The table lists returns for different investments with different interest rates. Which investment is the most risky? Which do you think is the safest?

ARE YOU READY?

Choose the best term from the list to complete each sentence.

1. The denominator of a fraction represents the ___?___, and the numerator represents the ___?___.
2. A ___?___ can be used to show all the possible combinations.
3. A ___?___ is a comparison of two quantities by division.
4. Tally marks in a table show the ___?___, or total, for each result.
5. A ratio of a number to 100 is called a ___?___.

ratio
fraction
percent
frequency
tree diagram
table
part
whole

Complete these exercises to review skills you will need for this chapter.

✔ Model Fractions

Write the fraction in simplest form that represents the shaded portion.

6. 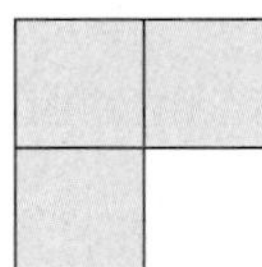

7. 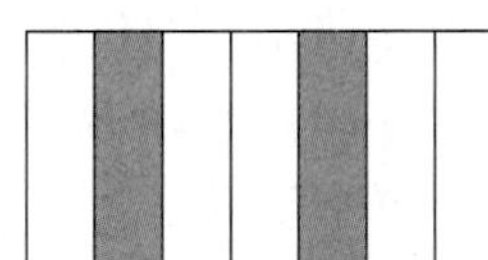

8. 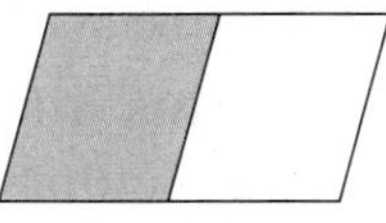

✔ Write Fractions as Decimals

Write each fraction as a decimal.

9. $\frac{9}{10}$ **10.** $\frac{1}{2}$ **11.** $\frac{12}{25}$ **12.** $\frac{11}{20}$

✔ Compare Fractions, Decimals, and Percents

Compare. Write <, >, or =.

13. 0.35 ■ 0.4 **14.** 0.25 ■ 25% **15.** $\frac{3}{5}$ ■ 0.7 **16.** 0.5 ■ $\frac{23}{50}$

✔ Write Ratios

Write each ratio.

17. blue circles to total circles

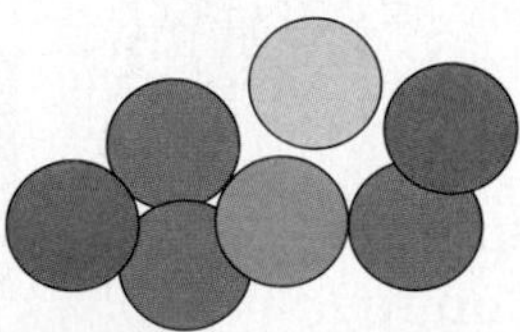

18. squares to triangles

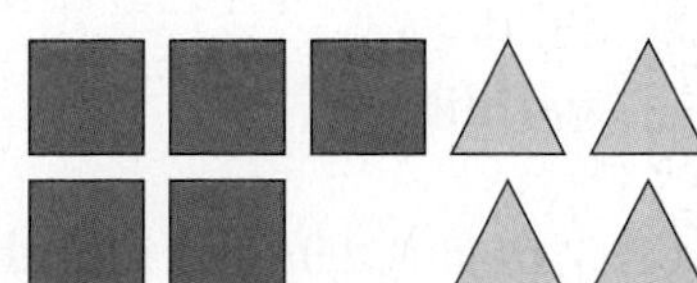

11-1 Introduction to Probability

Learn to estimate the likelihood of an event and to write and compare probabilities.

Vocabulary
probability

The weather report gives a 5% chance of rain today. Will you wear your raincoat? What if the report gives a 95% chance of rain?

In this situation, you are using probability to help make a decision. **Probability** is the measure of how likely an event is to occur. In this case, both 5% and 95% are probabilities of rain.

Probabilities are written as fractions or decimals from 0 to 1 or as percents from 0% to 100%. The higher an event's probability, the more likely that event is to happen.

- Events with a probability of 0, or 0%, never happen.
- Events with a probability of 1, or 100%, always happen.
- Events with a probability of 0.5, or 50%, have the same chance of happening as of not happening.

Impossible	Unlikely	As likely as not	Likely	Certain
0 0%		0.5 $\frac{1}{2}$ 50%		1 100%

A 95% chance of rain means rain is highly likely. A 5% chance of rain means rain is highly unlikely.

EXAMPLE 1 Estimating the Likelihood of an Event

Helpful Hint

A standard number cube is numbered from 1 to 6.

Write *impossible, unlikely, as likely as not, likely,* or *certain* to describe each event.

A **The month of June has 30 days.**
certain

B **A coin toss comes up heads.**
as likely as not

C **You roll a 9 on a standard number cube.**
impossible

D **This spinner lands on red.**
likely

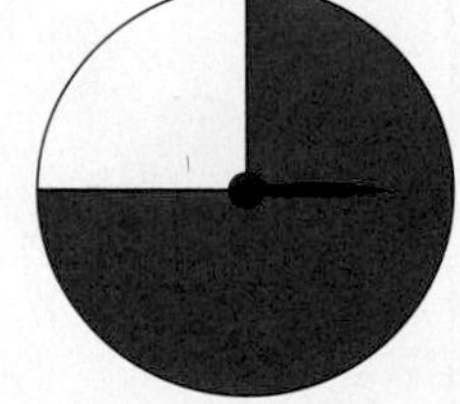

EXAMPLE 2 Writing Probabilities

A **The weather report gives a 35% chance of rain for tomorrow. Write this probability as a decimal and as a fraction.**

35% = 0.35 *Write as a decimal.*

$35\% = \frac{35}{100} = \frac{7}{20}$ *Write as a fraction in simplest form.*

B **The chance that Ethan is chosen to represent his class in the student council is 0.6. Write this probability as a fraction and as a percent.**

$0.6 = \frac{6}{10} = \frac{3}{5}$ *Write as a fraction in simplest form.*

0.6 = 60% *Write as a percent.*

C **There is a $\frac{9}{25}$ chance of getting a green gumball out of a certain machine. Write this probability as a decimal and as a percent.**

$\frac{9}{25} = 9 \div 25 = 0.36$ *Write as a decimal.*

$\frac{9}{25} = \frac{9 \cdot 4}{25 \cdot 4} = \frac{36}{100} = 36\%$ *Write as a percent.*

Helpful Hint

In Example 2C, after you find the decimal form of $\frac{9}{25}$, you can use it to find the percent.

0.36 = 36%

EXAMPLE 3 Comparing Probabilities

A **On a flowering plant called the four o'clock, there is a 50% chance the flowers will be pink, a 25% chance the flowers will be white, and a 25% chance the flowers will be red. Is it more likely that the flowers will be pink or white?**

Compare: 50% > 25%

The flowers are more likely to be pink than white.

B **When you spin this spinner, there is a 25% chance that it will land on red, a 50% chance that it will land on yellow, and a 25% chance that it will land on blue. Is it more likely to land on red or on blue?**

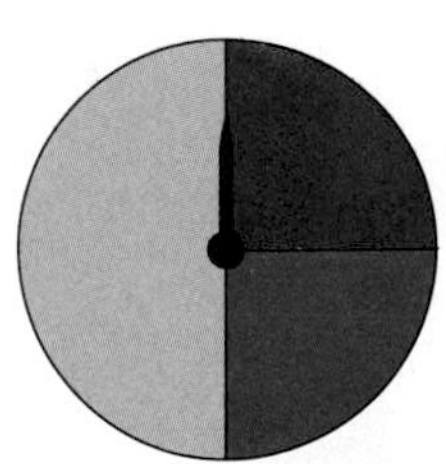

Compare: 25% = 25%

It is as likely to land on red as on blue.

Think and Discuss

1. **Give an example** of a situation that involves probability.
2. **Name** events that can be described by each of the following terms: *impossible, likely, as likely as not, unlikely,* and *certain.*

11-1 Exercises

FOR EOG PRACTICE see page 680

internet connect — Homework Help Online go.hrw.com Keyword: MR4 11-1

1.02a, 1.03

GUIDED PRACTICE

See Example 1

Write *impossible, unlikely, as likely as not, likely,* or *certain* to describe each event.

1. This year has 12 months.

2. You win the lottery.

See Example 2

3. Suppose that the chance of reaching into a bag of coins and selecting a quarter is 40%. Write this probability as a decimal and as a fraction.

See Example 3

4. If there are two children in a family, there is a 25% chance that both children are boys, a 25% chance that both children are girls, and a 50% chance that one child is a boy and the other is a girl. Which is more likely, that both children are boys or that one child is a boy and the other is a girl?

INDEPENDENT PRACTICE

See Example 1

Write *impossible, unlikely, equally likely, likely,* or *certain* to describe each event.

5. The spinner at right lands on green.

6. The spinner at right lands on blue.

7. You guess one winning number between 1 and 500.

8. You correctly guess one of eight winning numbers between 1 and 10.

See Example 2

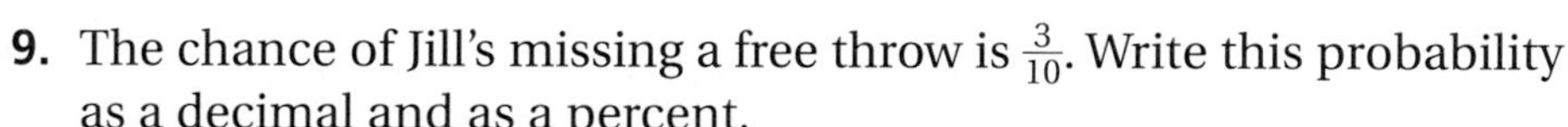

9. The chance of Jill's missing a free throw is $\frac{3}{10}$. Write this probability as a decimal and as a percent.

See Example 3

10. If you choose from a bag of mixed nuts, there is a 45% chance of choosing a peanut, a 20% chance of choosing a pecan, a 15% chance of choosing a cashew, and a 20% chance of choosing a walnut. Is it less likely that you will choose a pecan or a cashew from the bag?

PRACTICE AND PROBLEM SOLVING

Describe the events as *impossible, unlikely, as likely as not, likely,* or *certain*.

11. The probability of winning a game is $\frac{2}{3}$.

12. The probability of being chosen for a team is 0.09.

13. There is a 50% chance of snow today.

14. Your chances of being struck by lightning are $\frac{1}{2{,}000{,}000}$.

Life Science 

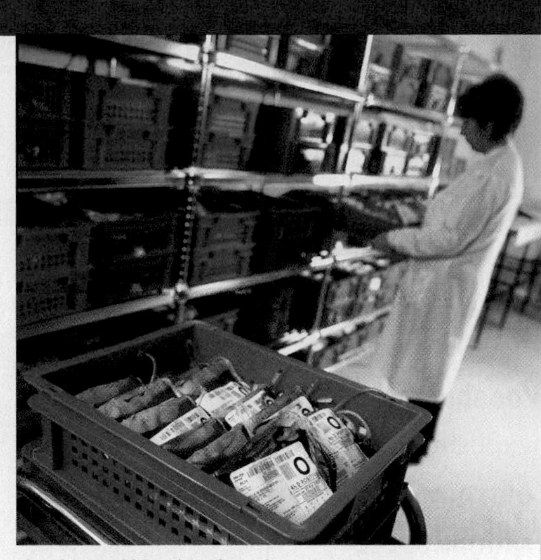

The donated blood in these bags is Type O.

Each year, millions of people donate blood. This blood is given to patients with certain bleeding disorders, people who have been in serious accidents, or patients who lose a lot of their own blood during surgery.

There are eight different human blood types, which are shown in the chart, along with the percent of people who have each type. It is very important that people receive the right type of blood. If they do not, their bodies will not recognize the foreign blood cells and will attack the cells.

15. How would you describe the probability of a person having AB positive blood: impossible, unlikely, as likely as not, likely, or certain? Explain.

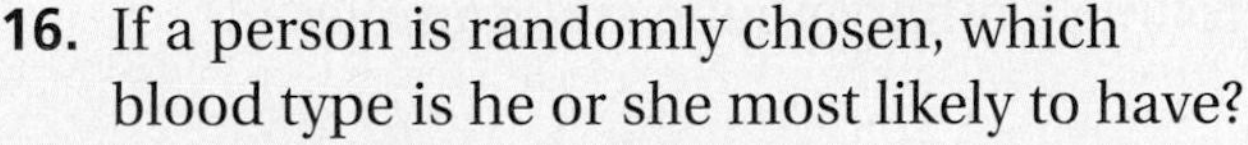

16. If a person is randomly chosen, which blood type is he or she most likely to have?

17. If a person is randomly chosen, which blood type is he or she least likely to have?

18. Write the probability that a randomly chosen person will have A negative blood as a decimal and as a fraction in simplest form.

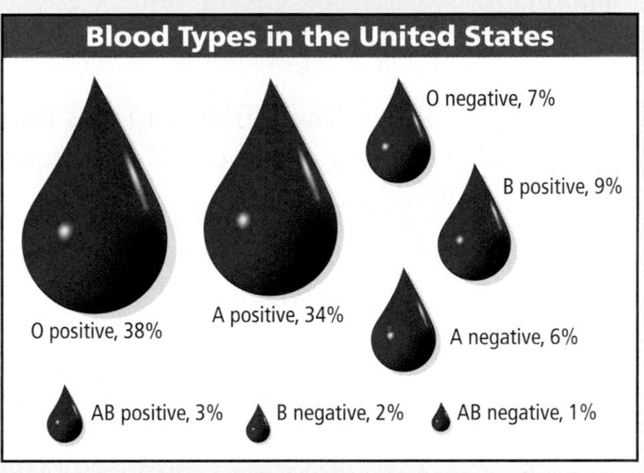

19. ***WRITE ABOUT IT*** Blood banks especially encourage people with certain types of blood to donate. Which blood types do you think these are? Explain.

20. ***CHALLENGE*** A person with AB positive blood can safely receive O, A, B, or AB blood. What is the probability that a randomly chosen person could donate blood to a person with AB positive blood?

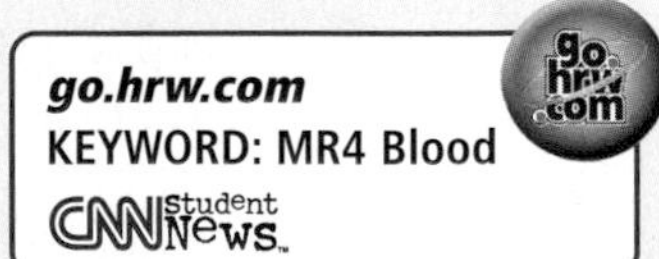

Spiral Review

21. Juanita needs $\frac{3}{4}$ cup of pineapple juice to make punch. She has only half that amount. How much pineapple juice does Juanita have? (Lesson 5-1)

Name the quadrant on a coordinate plane where each point is located. (Lesson 9-3)

22. $(4, -6)$ **23.** $(-1, 5)$ **24.** $(2, 3)$ **25.** $(-2, -4)$

26. **EOG PREP** The perimeter of a square is 24 m. What is the length of one side of the square? (Lesson 10-1)

A 6 m B 8 m C 12 m D 16 m

11-2 Experimental Probability

Learn to find the experimental probability of an event.

Vocabulary
experiment
outcome
sample space
experimental probability

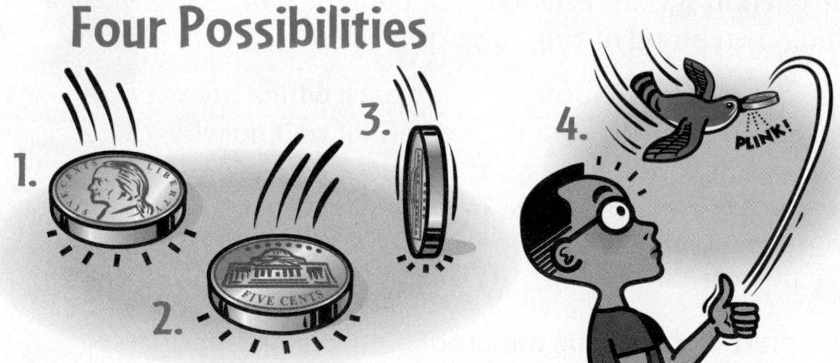

An **experiment** is an activity involving chance that can have different results. Flipping a coin and rolling a number cube are examples of experiments.

The different results that can occur are called **outcomes** of the experiment. If you are flipping a coin, heads is one possible outcome.

The **sample space** of an experiment is the set of all possible outcomes. You can use { } to show sample spaces. When a coin is being flipped, {heads, tails} is the sample space.

EXAMPLE 1 Identifying Outcomes and Sample Spaces

For each experiment, identify the outcome shown and the sample space.

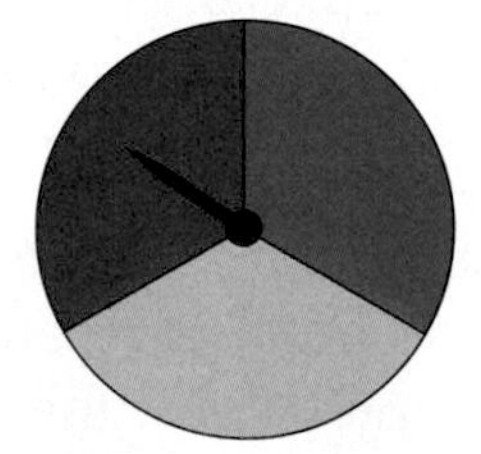

A **spinning a spinner**
outcome shown: red
sample space: {red, blue, yellow}

B **tossing two coins**
outcomes shown: heads, tails (H, T)
sample space: {HH, HT, TH, TT}

Performing an experiment is one way to estimate the probability of an event. If an experiment is repeated many times, the **experimental probability** of an event is the ratio of the number of times the event occurs to the total number of times the experiment is performed.

EXPERIMENTAL PROBABILITY

$$\text{probability} \approx \frac{\text{number of times the event occurs}}{\text{total number of trials}}$$

EXAMPLE 2 Finding Experimental Probability

For one month, Tosha recorded the time at which her school bus arrived. She organized her results in a frequency table.

Time	7:00–7:04	7:05–7:09	7:10–7:15
Frequency	8	9	3

Writing Math

The probability of an event can be written as *P*(event). *P*(blue) means "the probability that blue will be the outcome."

A **Find the experimental probability that the bus will arrive between 7:00 and 7:04.**

$$P(\text{between 7:00 and 7:04}) \approx \frac{\text{number of times the event occurs}}{\text{total number of trials}}$$

$$= \frac{8}{20} = \frac{2}{5}$$

B **Find the experimental probability that the bus will arrive before 7:10.**

$$P(\text{before 7:10}) \approx \frac{\text{number of times the event occurs}}{\text{total number of trials}}$$

$$= \frac{8+9}{20}$$ *Before 7:10 includes 7:00–7:04 and 7:05–7:09.*

$$= \frac{17}{20}$$

EXAMPLE 3 Comparing Experimental Probabilities

Ian tossed a cone 30 times and recorded whether it landed on its base or on its side. Based on Ian's experiment, which way is the cone more likely to land?

On its side

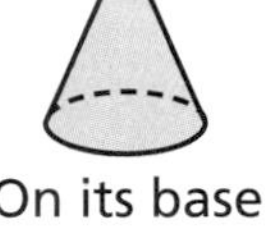
On its base

Outcome	On its base	On its side
Frequency	~~IIII~~ II	~~IIII~~ ~~IIII~~ ~~IIII~~ ~~IIII~~ III

$$P(\text{base}) \approx \frac{\text{number of times the event occurs}}{\text{total number of trials}} = \frac{7}{30}$$

$$P(\text{side}) \approx \frac{\text{number of times the event occurs}}{\text{total number of trials}} = \frac{23}{30}$$

Find the experimental probability of each outcome.

$$\frac{7}{30} < \frac{23}{30}$$ *Compare the probabilities.*

It is more likely that the cone will land on its side.

Think and Discuss

1. **Explain** whether you and a friend will get the same experimental probability for an event if you perform the same experiment.
2. **Tell** why it is important to repeat an experiment many times.

11-2 Exercises

FOR EOG PRACTICE
see page 680

internet connect
Homework Help Online
go.hrw.com Keyword: MR4 11-2

4.01, 4.02, 4.03, 4.04

GUIDED PRACTICE

See Example 1

1. Identify the outcome and the sample space shown on the spinner.

Josh recorded the number of hits his favorite baseball player made in each of 15 games. He organized his results in a frequency table.

Number of Hits	0	1	2	3
Frequency	4	8	2	1

See Example 

2. Find the experimental probability that this player will get one hit in a game.

See Example 

3. Based on Josh's results, is this player more likely to get two hits in a game or no hits in a game? How many hits will this player most likely get in a game?

INDEPENDENT PRACTICE

See Example 

For each experiment, identify the outcome shown and the sample space.

4.

5.

Jennifer has a bag of marbles. She removed one marble, recorded the color, and placed it back in the bag. She repeated this process several times and recorded her results in the table.

See Example 

6. Find the experimental probability that a marble selected from the bag will be red.

7. Find the experimental probability that a marble selected from the bag will not be black.

Color	Frequency
White	卌
Red	///
Yellow	卌
Black	卌 卌 //

See Example 

8. Based on Jennifer's experiment, which color marble is she most likely to select from the bag?

PRACTICE AND PROBLEM SOLVING

Identify the sample space for each situation.

9. Fe has three clean T-shirts—a yellow one, a red one, and a green one. Without looking, she pulls one from her laundry basket.

10. You roll a number cube whose sides are all numbered 7.

11. The principal will choose two of the three finalists in the science fair to attend a banquet. The three finalists are Anna, Joel, and Roseann.

12.

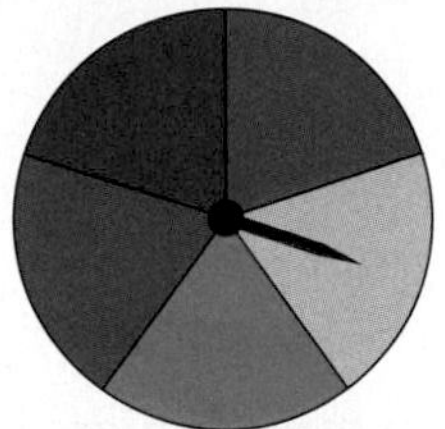

13.

14. ***WEATHER*** Janet recorded the high temperature every day in January. She recorded her results in a frequency table.

Temperature (°F)	26–35	36–45	46–55	56–65
Number of Days	10	9	11	1

a. According to Janet's results, what is the probability that a day in January will be warmer than 55°F?

b. Describe this probability as certain, likely, as likely as not, unlikely, or impossible.

15. ***WRITE ABOUT IT*** Conduct an experiment in which you toss a coin 100 times. Keep a tally of the number of times the coin shows heads. According to your results, what is the experimental probability that it will show heads? Compare your results with a classmate. Did you both get the same experimental probability? Why or why not?

16. ***CHALLENGE*** Suppose you roll two number cubes and add the two numbers that come up. What do you think the most likely sum would be? (*Hint:* Perform an experiment.)

Spiral Review

Find the next three numbers in each sequence. (Lesson 1-7)

17. 1, 3, 5, 7, …　　**18.** 2, 4, 6, 8, 10, …　　**19.** 1, 4, 9, 16, …

20. **EOG PREP** Which expression is **not** equal to half of n? (Lesson 8-8)

A $0.5n$　　B 5% of n　　C $\frac{n}{2}$　　D $n \div 2$

21. **EOG PREP** Choose the greatest amount. (Lesson 8-9)

A 45% of 200　　B 60% of 190　　C 50% of 150　　D 100% of 110

Simulations

Use with Lesson 11-2

internet connect
Lab Resources Online
go.hrw.com
KEYWORD: MR4 Lab11A

A **simulation** is a model of an experiment that would be difficult or inconvenient to actually perform. In this lab, you will conduct simulations.

Activity 1

A cereal company is having a contest. To win a prize, you must collect six different cards that spell *YOU WIN*. One of the six letters is put into each cereal box. The letters are divided equally among the boxes. How many boxes do you think you will have to buy to collect all six cards?

1. Since there are six different cards that are evenly distributed, you can use a number cube to simulate collecting the letters. Each of the numbers from 1 to 6 will represent a letter. A roll of the number cube will simulate buying one box of cereal, and the number rolled will represent the letter inside that box.

1	2	3	4	5	6
Y	O	U	W	I	N
/	卌 /	////	//	//	/

2. Roll the number cube, and keep track of the numbers you roll. Continue to roll the number cube until you have rolled every number at least once.

Think and Discuss

1. Look at the results in the table above. What was the last number rolled? How do you know?
2. How many rolls did it take to get all six numbers in your simulation?
3. How many boxes of cereal do you think you would have to buy to get all six letters? If you bought this many boxes, would you be sure to win? Explain.

Try This

1. Repeat the simulation three more times. Record your results.
2. Combine your data with the data of 5 of your classmates. Find the mean number of rolls from all 6 sets of data.
3. How many boxes of cereal do you think you would have to buy to get all six letters? Is this number different from what you thought after the first simulation? Explain.

Activity 2

Amy is a basketball player who usually makes $\frac{1}{2}$ of the baskets that she attempts. Suppose she makes 20 shots in each game. If she plays ten games, in how many games do you think she will make at least four baskets in a row?

1. There are two possible outcomes every time Amy shoots the ball—either she will make the basket or she will miss. Since Amy makes $\frac{1}{2}$ of her shots, you can toss a coin to simulate one shot. Let heads represent making the basket, and let tails represent missing.
2. Toss the coin 20 times to simulate one game. Keep track of your results.
3. Repeat 2 nine more times to simulate ten games.

Trial	Results
1	THTHHTTHTTHTHTTHHHTT
2	HHTTTHTHTHHHHHTTHTHT
3	HTTTHTTTHTHTTTHTTHTT
4	HTHTHTHTHTTHTHTTTTTT
5	THTTTTHHTHTHHTHTTHTT
6	HTTHTHHHHHTHHHHHHHHH
7	TTHHTTHHHTHTHHTTHTTT
8	HTTHTTHTTTHHTTHTTHTT
9	HHHTTTTTHHHHHTHHTHHT
10	HTTHHTTHHHTHHTHTHHHH

Think and Discuss

1. Why does tossing a coin 20 times represent only one trial?
2. Do any of your sequences contain four or more heads in a row? How many?
3. In how many games do you think Amy will make at least four baskets in a row? Out of *every* ten games, will Amy always make at least four baskets in a row this number of times?
4. You can use your simulation to find the experimental probability that Amy will make at least four or more baskets in a row. Divide the number of trials in which the coin came up heads at least four times in a row by the total number of trials. What is the experimental probability that Amy will make at least four baskets in a row?
5. Suppose Amy made only $\frac{1}{3}$ of her shots. Would you still be able to use a coin as a simulation? Why or why not?

Try This

1. In a group of ten families that each have four children, how many families do you think will have two girls and two boys? Make a prediction, and then design and carry out a simulation to answer this question. (Assume that having a boy and having a girl are equally likely events.) Was your prediction close?
2. Use your results from the previous problem to give the experimental probability that a family with four children will have two girls and two boys.
3. Think of an experiment, and design your own simulation to model it.

11-3 Theoretical Probability

Learn to find the theoretical probability of an event.

Vocabulary

theoretical probability

equally likely

fair

Another way to estimate probability of an event is to use **theoretical probability**. One situation in which you can use theoretical probability is when all outcomes have the same chance of occurring. In other words, the outcomes are **equally likely**.

Equally likely outcomes

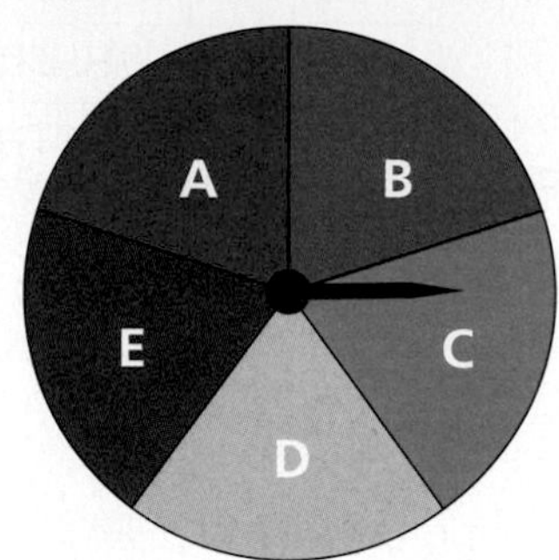

There is the same chance that the spinner will land on any of these letters.

Not equally likely outcomes

There is a greater chance that the spinner will land on 1 than on any other number.

An experiment with equally likely outcomes is said to be **fair**. You can usually assume that experiments involving items such as coins and number cubes are fair.

THEORETICAL PROBABILITY

$$\text{probability} \approx \frac{\text{number of ways event can occur}}{\text{total number of possible outcomes}}$$

EXAMPLE 1 **Finding Theoretical Probability**

A **What is the probability that a fair coin will land heads up?**

There are two possible outcomes when flipping a coin, heads or tails. Both are equally likely because the coin is fair.

$$P(\text{heads}) = \frac{\blacksquare}{\text{2 possible outcomes}}$$

There is only one way for the coin to land heads up.

$$P(\text{heads}) = \frac{\text{1 way event can occur}}{\text{2 possible outcomes}}$$

$$P(\text{heads}) = \frac{\text{1 way event can occur}}{\text{2 possible outcomes}} = \frac{1}{2}$$

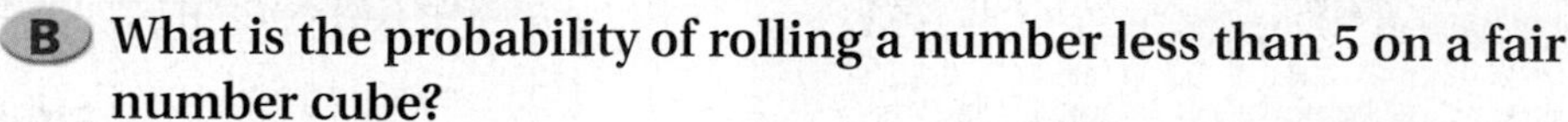

B **What is the probability of rolling a number less than 5 on a fair number cube?**

There are six possible outcomes when a fair number cube is rolled: 1, 2, 3, 4, 5, or 6. All are equally likely.

$$P(\text{less than } 5) = \frac{\blacksquare}{\text{6 possible outcomes}}$$

There are 4 ways to roll a number less than 5: 1, 2, 3, or 4.

$$P(\text{less than } 5) = \frac{\text{4 ways event can occur}}{\text{6 possible outcomes}}$$

$$P(\text{less than } 5) = \frac{\text{4 ways event can occur}}{\text{6 possible outcomes}} = \frac{4}{6} = \frac{2}{3}$$

Think about a single experiment, such as tossing a coin. There are two possible outcomes, heads or tails. What is $P(\text{heads}) + P(\text{tails})$?

Experimental Probability (coin tossed 10 times)			Theoretical Probability
H	T		
𝍸 I	IIII	$P(\text{heads}) = \frac{6}{10}$ $P(\text{tails}) = \frac{4}{10}$ $\frac{6}{10} + \frac{4}{10} = \frac{10}{10} = 1$	$P(\text{heads}) = \frac{1}{2}$ $P(\text{tails}) = \frac{1}{2}$ $\frac{1}{2} + \frac{1}{2} = \frac{2}{2} = 1$

No matter how you determine the probabilities, their sum is 1.

This is true for any experiment—the probabilities of the individual outcomes add to 1 (or 100%, if the probabilities are given as percents).

EXAMPLE 2 Finding Probabilities of Events Not Happening

Suppose there is a 10% chance of rain today. What is the probability that it will not rain?

In this situation, there are two possible outcomes, either it will rain or it will not rain.

$$P(\text{rain}) + P(\text{not rain}) = 100\%$$

$$10\% + P(\text{not rain}) = 100\%$$

$$\underline{-10\%} \qquad \underline{-10\%} \qquad \textit{Subtract 10\% from each side.}$$

$$P(\text{not rain}) = 90\%$$

Think and Discuss

1. Give an example of a fair experiment. Give an example of an unfair experiment.

2. Suppose there is a 50% chance of snow and a 10% chance of sleet. What is the probability that neither one will occur?

11-3 Exercises

FOR EOG PRACTICE

see page 681

internet connect

Homework Help Online
go.hrw.com Keyword: MR4 11-3

4.03, 4.04

GUIDED PRACTICE

See Example 1

1. What is the probability that a fair coin will land tails up?
2. What is the probability of randomly choosing a vowel from the letters *A, B, C, D,* and *E*?

See Example 2

3. The probability that a spinner will land on blue is 26%. What is the probability that it will not land on blue?
4. Suppose you have an unfair number cube and the probability of rolling a 2 is 0.7. What is the probability that you will not roll a 2?

INDEPENDENT PRACTICE

See Example 1

5. What is the probability of rolling the number 3 on a fair number cube?
6. What is the probability of rolling a number that is a multiple of 3 on a fair number cube?
7. Find the probability that a yellow marble will be chosen from a bag that contains 3 green marbles, 2 red marbles, and 4 yellow marbles.

See Example 2

8. Suppose there is an 81% chance of snow today. What is the probability that it will not snow?
9. On a game show, the chance that the spinner will land on the winning color is 0.04. Find the probability that it will not land on the winning color.

PRACTICE AND PROBLEM SOLVING

A standard number cube is rolled. Find each probability.

10. *P*(4)
11. *P*(not 3)
12. *P*(1, 2, or 3)
13. *P*(number greater than 0)
14. *P*(odd number)
15. *P*(number divisible by 5)
16. *P*(prime number)
17. *P*(negative number)

Nine pieces of paper with the numbers 1, 2, 2, 3, 4, 4, 5, 6, and 6 printed on them are placed in a bag. A student chooses one without looking. Compare the probabilities. Write <, >, or =.

18. *P*(1) ■ *P*(5)
19. *P*(3) ■ *P*(2)
20. *P*(4) ■ *P*(5 or 6)
21. *P*(3 or 5) ■ *P*(6)
22. *P*(even number) ■ *P*(odd number)
23. *P*(multiple of 3) ■ *P*(number less than 4)

Games LINK

Mah Jong was developed in China. Although the game's exact history is unknown, Mah Jong may be over 2,000 years old. Today the tiles are often made of plastic, but traditional tiles are still made of bamboo, bone, and ivory.

go.hrw.com
KEYWORD: MR4 Game

For Exercises 24–29, A represents an event. The probability that A will happen is given. Find the probability that A will not happen.

24. $P(A) = 47\%$ **25.** $P(A) = 0.9$ **26.** $P(A) = \frac{7}{12}$

27. $P(A) = \frac{5}{8}$ **28.** $P(A) = 0.23$ **29.** $P(A) = 100\%$

30. ***GAMES*** Mah Jong is a traditional Chinese game played with 144 decorated tiles—36 Bamboo tiles, 36 Circle tiles, 36 Character tiles, 16 Wind tiles, 12 Dragon tiles, and 8 bonus tiles. The tiles are the same shape and size, and are all blank on the back. Suppose the tiles are all placed face down and you choose one. What is the probability that you will choose a Wind tile? Write your answer as a fraction in simplest form.

31. ***GEOMETRY*** This net can be folded to make a solid figure. The solid figure can then be rolled like a number cube. Give the probability of rolling each number with the solid figure.

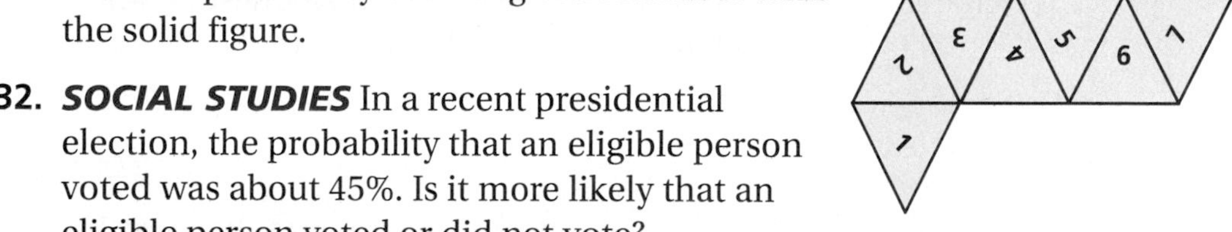

32. ***SOCIAL STUDIES*** In a recent presidential election, the probability that an eligible person voted was about 45%. Is it more likely that an eligible person voted or did not vote?

33. Molly has a bag with eight marbles in it: four red, two green, and two yellow. Without looking, she draws a green marble and does not put it back into the bag. What is the probability of her drawing a yellow marble after the green marble has been removed?

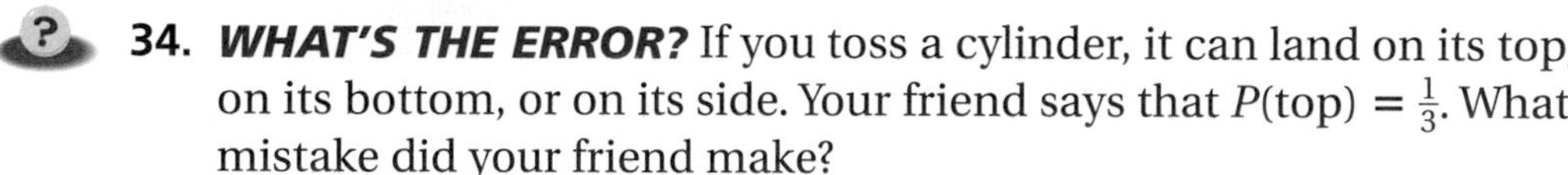

34. ***WHAT'S THE ERROR?*** If you toss a cylinder, it can land on its top, on its bottom, or on its side. Your friend says that $P(\text{top}) = \frac{1}{3}$. What mistake did your friend make?

35. ***WRITE ABOUT IT*** Toss a coin 20 times and record your results. According to your experiment, what is the probability that the coin shows tails? What is the theoretical probability that it shows tails? How do the two probabilities compare? Repeat the experiment, but this time toss the coin 50 times. Now how do the probabilities compare?

36. ***CHALLENGE*** Suppose you perform an experiment in which you toss a fair coin and roll a fair number cube. Find the theoretical probability that heads *and* 3 will be the outcomes.

Spiral Review

Find each length in meters. (Lesson 3-4)

37. 20 cm **38.** 4 mm **39.** 9,000 km **40.** 100 km

41. Make a table that shows the number of days in each month of a non-leap year. (Lesson 6-1)

42. **EOG PREP** Which quotient is greatest? (Lesson 9-7)

A $-10 \div 5$ B $-8 \div (-2)$ C $15 \div (-5)$ D $-10 \div (-5)$

Chapter 11 Mid-Chapter Quiz

LESSON 11-1 (pp. 554–557)

For problems 1 and 2, write *impossible, unlikely, as likely as not, likely,* or *certain* to describe the event.

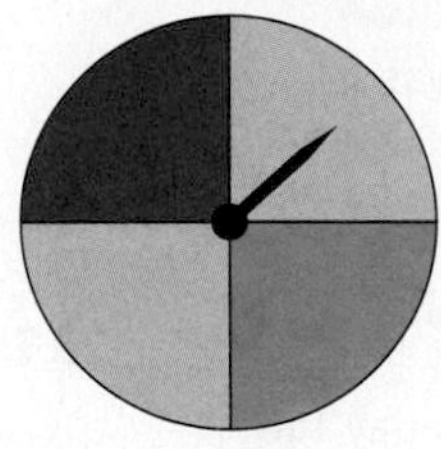

1. This spinner lands on blue.
2. You roll an even number on a fair number cube.
3. Mitch entered a contest to win concert tickets. The chance that Mitch will win is 0.15. Write this probability as a fraction and as a percent.
4. The chance of rain is 33% on Tuesday, 45% on Wednesday, and 35% on Thursday. On which day is it most likely to rain?

LESSON 11-2 (pp. 558–561)

For each experiment, identify the outcome shown and the sample space.

5. 

6. 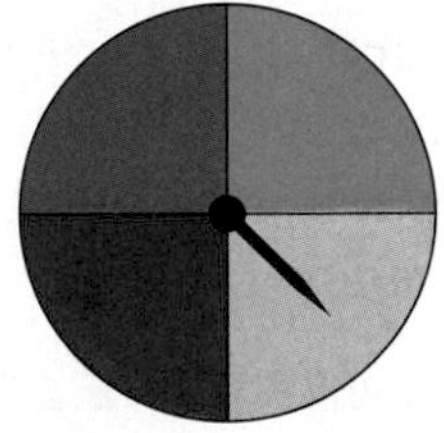

7. Gregory surveyed students to determine their favorite colors. His results are recorded in the table. Find the experimental probability that a student's favorite color is blue.

Color	Red	Yellow	Blue	Purple
Frequency	13	11	16	22

8. Find the experimental probability that a student's favorite color is red or yellow.
9. Jeremy recorded the number of times a spinner landed on each number. Based on Jeremy's experiment, on which number is the spinner most likely to land?

Outcome	1	2	3
Frequency	𝍸 II	𝍸 𝍸 II	𝍸 I

LESSON 11-3 (pp. 564–567)

10. What is the probability that this spinner will land on 2?
11. What is the probability of rolling a number less than 3 on a fair number cube?
12. Kirk has a 33% chance of scoring in the basketball game. What is the probability that Kirk will **not** score in the game?

Focus on Problem Solving

Look Back

- **Estimate to check that your answer is reasonable**

When you have finished solving a problem, take a minute to reread the problem. See if your answer makes sense. Make sure that your answer is reasonable given the situation in the problem.

One way to do this is to estimate the answer before you begin solving the problem. Then when you get your final answer, compare it with your original estimate. If your answer is not close to your estimate, check your work again.

Each problem below has an answer given, but it is not right. How do you know that the answer is not reasonable? Give your own estimate of the correct answer.

1. A rental car agency has 55 blue cars, 32 red cars, and 70 white cars. A customer is given a car at random. How many color outcomes are possible?

   Answer: 2,100

2. A box has 120 marbles. If the probability of drawing a blue marble is $\frac{3}{8}$ and the probability of drawing a red marble is $\frac{5}{8}$, how many of each color are in the box?

   Answer: 100 blue marbles and 20 red marbles

3. A store manager decides to survey one out of every ten shoppers. How many would be surveyed out of 350 shoppers?

   Answer: 3 shoppers

4. Sue has just started to collect old dimes. She has six dimes from 1941, five dimes from 1932, and one dime from 1930. If she chooses one dime at random, what is the probability that it is from before 1932?

   Answer: 50%

11-4 Make an Organized List

Problem Solving Strategy

Learn to make an organized list to find all possible outcomes.

DNA is a substance found in cells. It is part of what makes up chromosomes, where genetic information is stored.

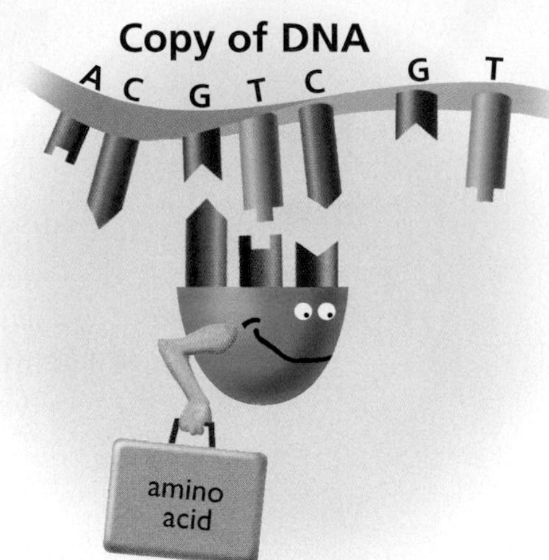

DNA contains four chemical bases, which are abbreviated A, T, G, and C. Sequences of three bases code for specific amino acids. How could you figure out how many different sequences are possible?

When you have to find many possibilities, one way to find them all is to make an organized list. A tree diagram is one way to organize information.

EXAMPLE 1 Using a Tree Diagram

At a circus, the clowns have two choices of clown suits—polka dots or stripes. They have three choices of wigs—pigtails, rainbow curly hair, or blue hair. How many different costumes can the clowns wear?

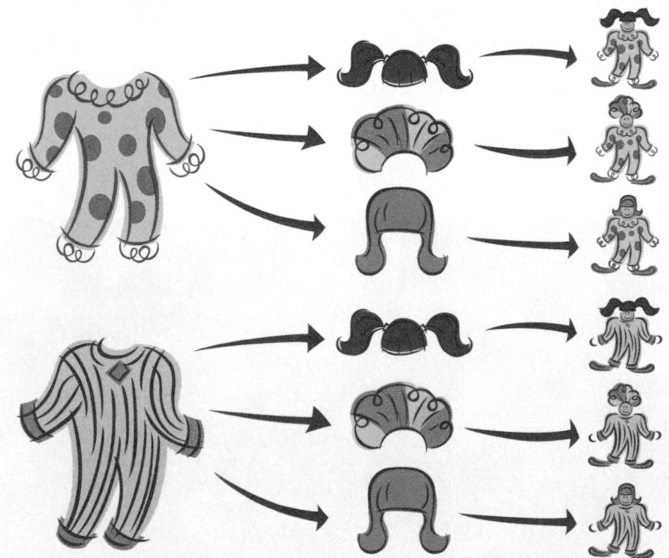

Follow each branch on the tree diagram to find all of the possible outcomes. There are 6 different costume combinations.

EXAMPLE 2 PROBLEM SOLVING APPLICATION

DNA contains four bases, A, T, G, and C. A sequence of three bases codes for an amino acid. A base may appear more than once in a sequence. For example, CCC codes for the amino acid proline. The order of the bases is also important. CAT codes for the amino acid histidine, but TAC codes for tyrosine, and CTA codes for leucine. How many different sequences of three can be formed from the four bases?

1 Understand the Problem

List the **important information:**

- There are four bases, A, T, G, and C.
- A base may repeat in a sequence.
- The order of the bases is important.

2 Make a Plan

You can make an organized list to keep track of the sequences.

3 Solve

First, find all the sequences that begin with A.

- List all that begin with AA. *AAA, AAT, AAG, AAC*
- List all that begin with AT. *ATA, ATT, ATG, ATC*
- List all that begin with AG. *AGA, AGT, AGG, AGC*
- Finally, list all that begin with AC. *ACA, ACT, ACG, ACC*

There are 16 sequences that begin with A. To find the sequences that begin with T, replace the beginning A in each sequence above with T. There are 16 sequences that begin with T. The same is true for sequences that begin with C and G.

$$16 + 16 + 16 + 16 = 64$$

There are 64 sequences of three bases that can be made.

4 Look Back

You could have continued to list all 64 possibilities. But looking for a pattern in the list shortened the amount of work needed.

Science LINK

There are 64 sequences of three bases, called *codons,* but cells use only 20 amino acids. Most amino acids have more than one codon. For example, TAT and TAC both code for the amino acid tyrosine.

Think and Discuss

1. **Explain** the advantages of an organized list over a random list.
2. **Describe** how you can check whether your list is accurate.

11-4 Exercises

FOR EOG PRACTICE

see page 682

internet connect

Homework Help Online
go.hrw.com Keyword: MR4 11-4

4.01

GUIDED PRACTICE

See Example 1

1. Carl can choose turkey, tacos, or pasta for his main dish at lunch. His choices for a side dish are fruit and salad. How many different lunch combinations are available to him?

See Example 2

2. Patrice, Jason, Kenya, Leon, and Brice are auditioning for the school play. The director has two roles available, a doctor and a teacher. Each can be played by either a boy or a girl. How many different ways can the two roles be assigned?

INDEPENDENT PRACTICE

See Example 1

3. Mr. Li is offering a make-up science test. He can give the test on Monday, Tuesday, or Thursday, before school, during lunch, or after school. How many different times can Mr. Li give his make-up test?

4. The Outdoor Club is planning its annual Spring Festival. The members must vote to choose the day of the event and the main activity. The event can take place on Saturday or Sunday. The main activity can be a foot race, a bicycle race, a hike, a swim, or a scavenger hunt. How many different combinations for the day and activity are there?

See Example 2

5. Greta's apartment building is protected by a security system that requires a pass code to let in residents. The code is made up of numbers from 1 to 3. The code is three digits long, and a digit can repeat. How many different pass codes are possible?

PRACTICE AND PROBLEM SOLVING

6. Keisha will choose a shirt and a skirt or a pair of pants from her closet to wear to school. Find the number of different outfits she can make if she has

   a. 3 shirts, 3 pants, and 3 skirts.
   b. 7 shirts, 5 pants, and 3 skirts.
   c. 4 shirts, 2 pants, and 2 skirts.
   d. 5 shirts, 2 pants, and 6 skirts.
   e. 2 shirts, 4 pants, and 3 skirts.
   f. 6 shirts, 7 pants, and 4 skirts.

7. A middle school is purchasing new basketball jerseys. Each jersey will have a two-digit number on it. The possible digits are 0, 1, 2, 3, 4, and 5, and a digit may appear twice on a jersey. How many different two-digit numbers are possible?

8. One burger restaurant offers a single, double, or triple burger on either a plain or sesame seed bun. Another restaurant offers single and double burgers and four different choices of buns. Does the first or second restaurant offer more possible burger combinations?

9. Omar is redecorating his bedroom. He can choose one paint color, one border, and one type of brush.

   a. How many different combinations of paint, border, and brush are possible?

   b. If Omar found another brush that he could use, how many different combinations would be possible?

10. Japanese children play a game called *Jan-Ken-Pon*. You may know it as Rock, Paper, Scissors. Two players shout at the same time, "*jan-ken-pon!*" On "*pon!*" each player shows one of three hand positions—closed fist (*gu*), open hand palm down (*pa*), or index and middle finger extended to form a V (*choki*). How many different outcomes are possible in this game?

11. ***CHOOSE A STRATEGY*** At a meeting, each person shook hands with every other person exactly one time. There were a total of 28 handshakes. How many people were at the meeting?

12. ***WRITE A PROBLEM*** Write a question that involves making an organized list. Then answer your question.

13. ***WRITE ABOUT IT*** Suppose you are going to choose one boy and one girl from your class for a group project. How can you find the number of possible combinations? Explain.

14. ***CHALLENGE*** A sailor has five flags: blue, green, red, orange, and yellow. Suppose she wants to fly three flags, but their order is not important; red, orange, yellow is the same as yellow, orange, red. How many different combinations of flags are possible?

Spiral Review

Evaluate. (Lesson 1-3)

15. 13^2 **16.** 4^3 **17.** 2^5 **18.** 3^4

Find each sum or difference. (Lesson 5-7)

19. $\frac{1}{9} + \frac{1}{3}$ **20.** $\frac{11}{12} - \frac{5}{6}$ **21.** $\frac{1}{5} + \frac{3}{10} - \frac{1}{15}$

22. **EOG PREP** Choose the circle with the greatest circumference. (Lesson 10-5)

A 5 in.

B 

C 

D 3 in.

11-5 Compound Events

Learn to list all the outcomes and find the theoretical probability of a compound event.

Vocabulary

compound event

If a family is going to have four children, there are 16 possibilities for the birth order of the children based on gender (boy, B, or girl, G).

BBBB, BBBG, BBGB, BBGG,
BGBB, BGBG, BGGB, BGGG,
GBBB, GBBG, GBGB, GBGG,
GGBB, GGBG, GGGB, GGGG

A **compound event** consists of two or more single events. For example, the birth of one child is a single event. The births of four children make up a compound event.

EXAMPLE 1 Finding Probabilities of Compound Events

Theresa rolls a fair number cube and then flips a fair coin.

A **Find the probability that the number cube will show an odd number and that the coin will show tails.**

First find all of the possible outcomes.

Number Cube

Coin		1	2	3	4	5	6
	H	1, H	2, H	3, H	4, H	5, H	6, H
	T	1, T	2, T	3, T	4, T	5, T	6, T

There are 12 possible outcomes, and all are equally likely.

Three of the outcomes have an odd number and tails:

1, T; 3, T; and 5, T.

$$P(\text{odd, tails}) = \frac{\text{3 ways event can occur}}{\text{12 possible outcomes}}$$

$$= \frac{3}{12}$$

$$= \frac{1}{4} \quad \textit{Write your answer in simplest form.}$$

B **Find the probability that the number cube will show a 2 and that the coin will show heads.**

Only one outcome is 2, H.

$$P(2, \text{H}) = \frac{\text{1 way event can occur}}{\text{12 possible outcomes}}$$

$$= \frac{1}{12}$$

C **The following experiment is going to be performed.**

Step 1: Toss a fair coin. **Step 2:** Spin the spinner. **Step 3:** Choose a marble.

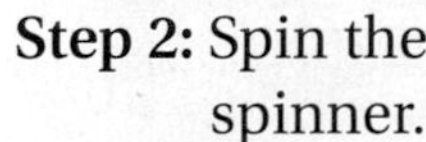

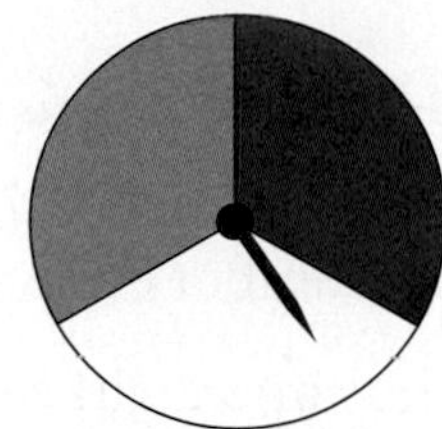

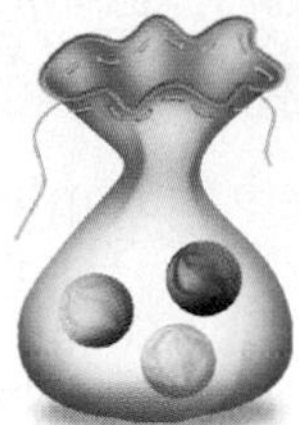

What is the probability that the coin will show heads, the spinner will land on orange, and a red marble will be chosen?

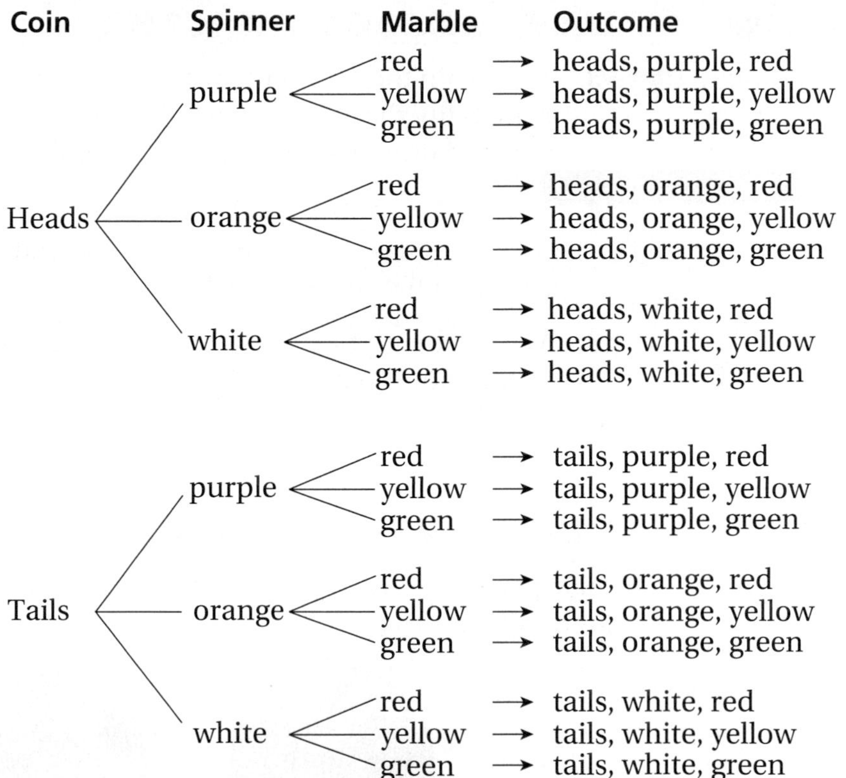

There are 18 equally likely outcomes.

$$P(\text{heads, orange, red}) = \frac{1 \text{ way event can occur}}{18 \text{ possible outcomes}}$$

$$= \frac{1}{18}$$

Think and Discuss

1. **Give an example** of a compound event.
2. **Explain** any pattern you noticed while finding the number of possible outcomes in a compound event.

11-5 Exercises

FOR EOG PRACTICE
see page 682

internet connect
Homework Help Online
go.hrw.com Keyword: MR4 11-5

4.01, 4.02, 4.03, 4.04

GUIDED PRACTICE

See Example 1

1. Patrick rolled a fair number cube twice. Find the probability that the number cube will show an even number both times.
2. A boy and a girl each flip a coin. What is the probability that the boy's coin will show heads and the girl's coin will show tails?

INDEPENDENT PRACTICE

See Example 1

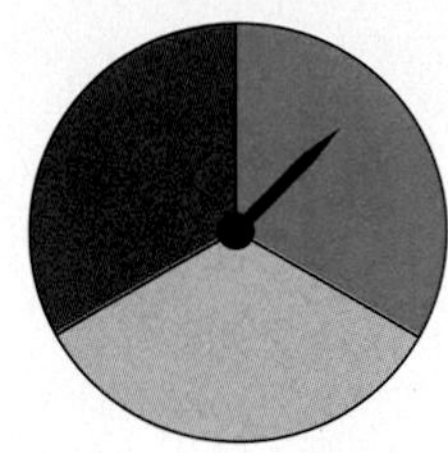

3. If you spin the spinner twice, what is the probability that it will land on green on the first spin and on purple on the second spin?
4. What is the probability that the spinner will land on either green or purple on the first spin and yellow on the second spin?
5. What is the probability that the spinner will land on the same color twice in a row?

PRACTICE AND PROBLEM SOLVING

An experiment involves spinning each spinner once. Find each probability.

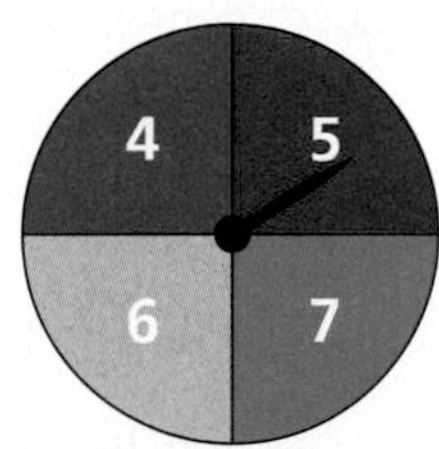

6. P(2 on spinner 1 and 5 on spinner 2)
7. P(not 1 on spinner 1 and not 7 on spinner 2)
8. P(even number on both spinners)
9. P(odd number on spinner 1 and even number on spinner 2)
10. P(number on spinner 2 is greater than number on spinner 1)
11. P(same number on both spinners)
12. P(a multiple of 3 on both spinners)
13. P(different number on each spinner)

A fair number cube is rolled, and a fair coin is tossed. Compare the probabilities. Write <, >, or =.

14. P(3 and tails) ▒ P(5 and heads)

15. P(even number and tails) ▒ P(odd number and heads)

16. P(number less than 3 and tails) ▒ P(odd number and tails)

17. P(number greater than 5 and heads) ▒ P(prime number and tails)

18. ***LIFE SCIENCE*** If a cat has 5 kittens, what is the probability that they are all female? What is the probability that they are all male? (Assume that having a male and having a female are equally likely events.)

19. A jar contains tiles that are numbered 1, 2, 3, 4, and 5. Danny removes a tile from the jar, replaces the tile, and draws a second tile. What is the probability that Danny will draw the same number both times?

20. The students in Jared's class have ID numbers made up of two digits from 1 through 6. The same digit can be used twice. In fact, the same digit is used twice in Jared's ID number. If Jared rolls 2 number cubes, what is the probability that he does **not** roll his ID number?

21. ***WHAT'S THE ERROR?*** One of your classmates said, "If you flip a coin and roll a number cube, the probability of getting heads and a 3 is $\frac{1}{2} + \frac{1}{6} = \frac{2}{3}$." What mistake did your classmate make? Explain how to find the correct answer.

22. ***WRITE ABOUT IT*** Describe a situation that involves a compound event.

23. ***CHALLENGE*** You roll a number cube six times. What is the probability of rolling the numbers 1 through 6 in order?

Spiral Review

Solve each equation. (Lesson 5-10)

24. $g + \frac{3}{10} = \frac{2}{5}$ **25.** $7m = \frac{1}{2}$ **26.** $\frac{2}{3}p = \frac{1}{6}$

27. Angles A and B are supplementary. The measure of angle A is 38 degrees. What is the measure of angle B? (Lesson 7-3)

28. **EOG PREP** Which of the following is always negative? (Lesson 9-1)

A The opposite of a number

B An integer

C The absolute value of a number

D A number less than zero

Explore Permutations and Combinations

Use with Lesson 11-5

For a compound event, you often must count the arrangements of individual outcomes. To do this, you must know whether the order of the outcomes in these arrangements matters. With three outcomes *A, B,* and *C,* when is *A-B-C* different from *C-B-A,* and when is it considered to be the same?

Activity 1

In how many different arrangements can Ellen, Susan, and Jeffrey sit in a row?

1 Write each name on 6 index cards. You will have a total of 18 cards. Show all the different ways the cards can be arranged in a row.

Arrangement	1	2	3	4	5	6
First Seat	Ellen	Ellen	Susan	Susan	Jeffrey	Jeffrey
Second Seat	Susan	Jeffrey	Jeffrey	Ellen	Susan	Ellen
Third Seat	Jeffrey	Susan	Ellen	Jeffrey	Ellen	Susan

There are 6 different ways that these three people can sit in a row.

Notice that the order of the students in the different arrangements is important. "Ellen, Susan, Jeffrey" is different from "Ellen, Jeffrey, Susan." An arrangement in which order is important is called a **permutation.**

Think and Discuss

1. Think of another situation in which the order in an arrangement is important. Can you think of a situation in which the order would **not** be important? Explain.

Try This

1. Cindy, Laurie, Marty, and Joel are running for president of their class. The person who gets the second greatest amount of votes will be the vice president. How many different ways can the election turn out?

Activity 2

1 Abe, Babe, Cora, and Dora are going to work on a project in groups of 2. How many different ways can they pair off?

Write each name on 3 index cards. You will have a total of 12 cards. Show all pairings.

Abe	Babe		Babe	Cora		Cora	Dora
Abe	Cora		Babe	Dora			
Abe	Dora						

There are 6 different possible pairs.

Notice that in this situation, the order in the pairs is not important. "Abe, Cora" is the same as "Cora, Abe." When order is not important, the arrangements are called **combinations.**

Think and Discuss

Tell whether each of the following is a permutation or a combination. Explain.

1. There are 20 horses in a race. Ribbons are given for first, second, and third place. How many possible ways can the ribbons be awarded?
2. There are 20 violin players trying out for the school band and 6 players will be chosen. How many different ways could students be selected for the band?
3. Connie has 10 different barrettes. She wears 2 each day. How many ways can she choose 2 barrettes each morning?
4. Yoko belongs to a book club, and she has just received 25 new books. How many possible ways are there for them to be placed on the shelf?

Try This

1. The video club is sponsoring a double feature. How many ways can club members choose 2 movies from a list of 6 possibilities?
2. Ms. Baker must pick a team of 3 students to send to the state mathematics competition. She has decided to choose 3 students from the 5 with the highest grades in her class. Ms. Baker can either send 3 equal representatives, or she can send a captain, an assistant captain, and a secretary. Which choice results in more possible teams? Explain. Find the number of teams possible for each choice.

11-6 Making Predictions

Learn to use probability to predict events.

Vocabulary

prediction

The Old Farmer's Almanac, first published in 1792, predicts weather, sunrise and sunset times, and tides.

A **prediction** is a guess about something in the future. The predictions in *The Old Farmer's Almanac* are based on several factors, such as the cycles of the Sun and Moon. Another way to make a prediction is to use probability.

EXAMPLE 1 Using Probability to Make Predictions

A An airline claims that its flights have a 92% probability of being on time. Out of 1,000 flights, how many would you predict will be on time?

You can write a proportion. Remember that *percent* means "per hundred."

$\frac{92}{100} = \frac{x}{1{,}000}$ *Think: 92 out of 100 is how many out of 1,000?*

$100 \cdot x = 92 \cdot 1{,}000$ *The cross products are equal.*

$100x = 92{,}000$ *x is multiplied by 100.*

$\frac{100x}{100} = \frac{92{,}000}{100}$ *Divide both sides by 100 to undo the multiplication.*

$x = 920$

You can predict that about 920 of 1,000 flights will be on time.

B If you roll a number cube 24 times, how many times do you expect to roll a 5?

$P(\text{rolling a 5}) = \frac{1}{6}$

$\frac{1}{6} = \frac{x}{24}$ *Think: 1 out of 6 is how many out of 24?*

$6 \cdot x = 1 \cdot 24$ *The cross products are equal.*

$6x = 24$ *x is multiplied by 6.*

$\frac{6x}{6} = \frac{24}{6}$ *Divide both sides by 6 to undo the multiplication.*

$x = 4$

You can expect to roll a 5 about 4 times.

EXAMPLE  2 PROBLEM SOLVING APPLICATION

A stadium sells yearly parking passes. If you have a parking pass, you can park at the stadium for any event during that year.

The managers of the stadium estimate that the probability that a person with a pass will attend any one event is 80%. The parking lot has 300 spaces. If the managers want the lot to be full at every event, how many passes should they sell?

1 Understand the Problem

The **answer** will be the number of parking passes they should sell.
List the **important information:**

- *P*(person with pass attends event) = 80%
- There are 300 parking spaces.

2 Make a Plan

The managers want to fill all 300 spaces. But, on average, only 80% of parking pass holders will attend. So 80% of pass holders must equal 300. You can write an equation to find this number.

3 Solve

$$\frac{80}{100} = \frac{300}{x}$$ *Think: 80 out of 100 is 300 out of how many?*

$$100 \cdot 300 = 80 \cdot x$$ *The cross products are equal.*

$$30{,}000 = 80x$$ *x is multiplied by 80.*

$$\frac{30{,}000}{80} = \frac{80x}{80}$$ *Divide both sides by 80 to undo the multiplication.*

$$375 = x$$

The managers should sell 375 parking passes.

4 Look Back

If the managers sold only 300 passes, the parking lot would not usually be full because only about 80% of the people with passes will attend any one event. The managers should sell more than 300 passes, so 375 is a reasonable answer.

Think and Discuss

1. Tell whether you expect to be exactly right if you make a prediction based on probability. Explain your answer.

11-6 Exercises

FOR EOG PRACTICE
see page 683

internet connect
Homework Help Online
go.hrw.com Keyword: MR4 11-6

1.04a, d, 4.06

GUIDED PRACTICE

See Example 1

1. A local newspaper states that 12% of the city's residents have volunteered at an animal shelter. Out of 5,000 residents, how many would you predict have volunteered at the animal shelter?

2. If you roll a fair number cube 30 times, how many times would you expect to roll a number that is a multiple of 3?

See Example 2

3. Airlines routinely overbook flights, which means that they sell more tickets than there are seats on the planes. They do this because ticketed customers sometimes do not show up for flights, and the airlines want to fill the planes. Suppose an airline estimates that 93% of customers will show up for a particular flight. If the plane seats 186 people, how many tickets should the airline sell?

INDEPENDENT PRACTICE

See Example 1

4. The U.S. Bureau of Engraving and Printing prints paper money. The bureau estimates that 45% of the bills printed on any given day are \$1 bills. Out of 500 bills printed in one day, how many would you predict are \$1 bills?

5. If you flip a coin 64 times, how many times do you expect the coin to show tails?

6. A bag contains 2 black chips, 5 red chips, and 4 white chips. You pick a chip from the bag, record its color, and put the chip back in the bag. If you repeat this process 99 times, how many times do you expect to remove a red chip from the bag?

See Example 2

7. The director of a blood bank is eager to increase his supply of O negative blood, because O negative blood can be given to people with any blood type. The probability that a person has O negative blood is 7%. The director would like to have 9 O negative donors each day. How many total donors does the director need to find each day to reach his goal of O negative donors?

PRACTICE AND PROBLEM SOLVING

8. A random survey of 50 people in Harrisburg indicates that 10 of them know the name of the mayor of their neighboring city.
   a. Out of 5,500 Harrisburg residents, how many would you expect to know the name of the mayor of the neighboring city?
   b. Out of 600 Harrisburg residents, how many would you predict do not know the name of the mayor of the neighboring city?

Social Studies LINK

Canada's population consists of several diverse groups. The Native Canadians lived in Canada before the Europeans arrived. The French were the first Europeans to settle successfully in Canada. There were also British settlers, and after the American Revolution, many Americans who had remained loyal to the king of England fled to Canada.

The graph shows the results of a survey of 400 Canadian citizens.

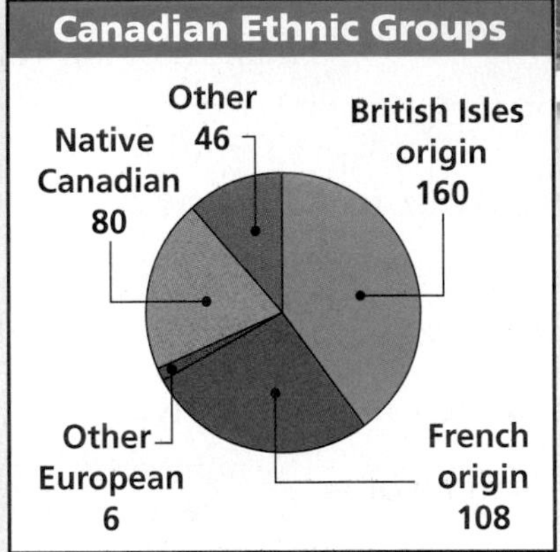

9. Out of 75 Canadians, how many would you predict are of French origin?

10. A random group of Canadians includes 18 Native Canadians. How many total Canadians would you predict are in the group?

11. Predict the number of people of British origin in a group of 50 Canadians.

12. **WHAT'S THE ERROR?** A student said that in any group of Canadians, 20 of them will be Native Canadians. What mistake did this student make?

13. **WRITE ABOUT IT** How could you predict the number of people of French *or* Native Canadian origin in a group of 150 Canadians?

14. **CHALLENGE** In a group of Canadians, 15 are in the Other European origin category. Predict how many Canadians in the same group are **not** in that category.

Spiral Review

Solve each equation. (Lesson 9-8)

15. $\frac{y}{-10} = 12$ 16. $\frac{p}{25} = -4$ 17. $\frac{j}{-3} = -15$

Find the GCF of each set of numbers. (Lesson 4-3)

18. 8 and 12 19. 18 and 42 20. 5 and 80 21. 2, 9, and 13

22. **EOG PREP** Which solid figure has the greatest number of faces? (Lesson 10-6)

A Cube B Triangular prism C Cone D Octagonal prism

EXTENSION

Odds

Learn to find the odds for and against a specified outcome.

Vocabulary

odds

Odds in favor of an event are written as a ratio of the number of ways the event can happen to the number of ways the event can fail to happen.

ODDS IN FAVOR OF AN EVENT

$$\text{odds in favor} = \frac{\text{number of ways event can happen}}{\text{number of ways event can fail to happen}}$$

EXAMPLE 1 Finding Odds in Favor of an Event

The band is selling raffle tickets for $2.00 each. Suppose you bought 3 tickets, and a total of 500 raffle tickets were sold.

A **What are the odds in favor of your winning?**

$$\text{odds in favor} = \frac{\text{number of ways event can happen}}{\text{number of ways event can fail to happen}}$$

$$= \frac{3}{497}$$ *You have 3 tickets that could be drawn. There are 497 other tickets.*

The odds in favor of your winning the raffle are 3 to 497.

B **What is the probability that you will win the raffle?**

$$P(\text{win}) = \frac{\text{number of ways event can happen}}{\text{number of possible outcomes}} = \frac{3}{500}$$

You can also calculate the odds against an event happening.

ODDS AGAINST AN EVENT

$$\text{odds against} = \frac{\text{number of ways event can fail to happen}}{\text{number of ways event can happen}}$$

EXAMPLE 2 Finding Odds Against an Event

There are 615 students in Carl's class, and Mr. Rosenweig will draw one name at random to attend the governor's lunch.

Helpful Hint

The two numbers given as the odds will add up to the total number of possible outcomes.

A **What are the odds against Carl's being chosen?**

$$\text{odds against} = \frac{\text{number of ways event can fail to happen}}{\text{number of ways event can happen}}$$

$$= \frac{614}{1}$$ *There are 614 other students who can be chosen. There is 1 way Carl can be chosen.*

The odds against Carl's being chosen are 614 to 1.

B **What is the probability that Carl will not be chosen?**

$$P(\text{not chosen}) = \frac{\text{number of ways event can happen}}{\text{number of possible outcomes}} = \frac{614}{615}$$

EXTENSION Exercises

There are 30 students in Darian's class. His teacher put each student's name in a hat and chose one name.

1. What are the odds in favor of Darian's name being chosen?
2. What is the probability that Darian's name will be chosen?
3. What are the odds against Darian's name being chosen?
4. What is the probability that Darian's name will not be chosen?

A fair number cube is rolled. Find the odds in favor of and against each of the following outcomes.

5. rolling a 4
6. rolling a number greater than 2
7. rolling a number divisible by 3
8. rolling a factor of 12
9. rolling an odd number
10. rolling a 1, a 2, a 3, a 4, or a 6

11. The surface of Earth is about 70% water and 30% land. If you stopped a spinning globe with your finger, what are the odds in favor of your finger landing on a body of water?

The circle graph shows the origins of bananas in the United States. Use the graph for Exercises 12 and 13.

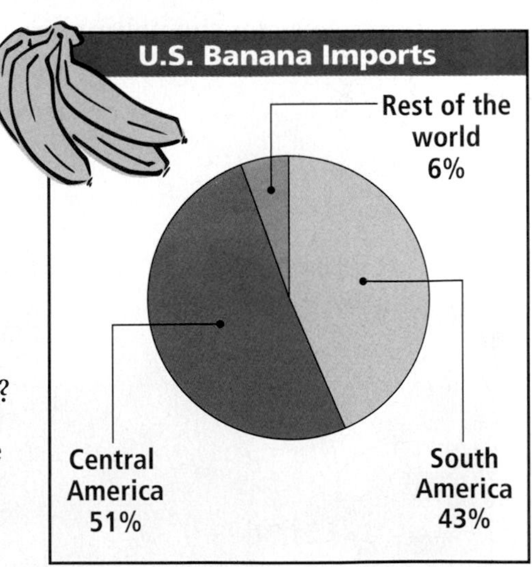

12. If you are eating a banana, what are the odds in favor of its being from Central America?
13. What are the odds against the banana's being from somewhere other than Central or South America?
14. The odds against winning a prize are 2 to 9. Are you more likely to win a prize or not to win a prize? Explain your answer.

15. ***CHALLENGE*** The probability of winning a game is 0.25. What are the odds against winning the game?

Problem Solving on Location

NORTH CAROLINA

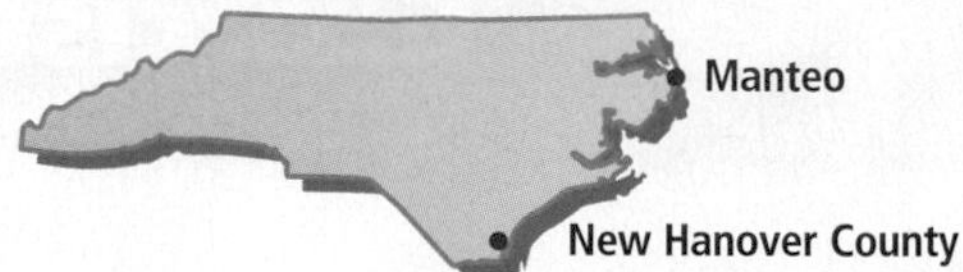

Airlie Gardens

It has been said that in the 1920s, Airlie Gardens, in New Hanover County, North Carolina, had half a million azaleas, 5,000 camellias, and numerous other exotic plants from all over the world. Today, the 67-acre garden is still known for its flowers, statuary, and 10 acres of freshwater lakes. The salt marshes and lakes attract swans, geese, egrets, herons, ibis, ducks, and kingfishers.

1. If you are equally likely to see any one of the seven birds attracted to Airlie Gardens, what is the probability that the first bird you see will be a swan?
2. The staff of Airlie Gardens is planning for 500 visitors in the garden this weekend. If they plan to give away a free gift to 5 lucky visitors, what is one visitor's probability of winning a gift?
3. Ticket sales on one day included 100 adult, 75 senior, and 25 student tickets. If the pattern of ticket sales continues, what is the probability that the next ticket sold will be a senior ticket?
4. Three volunteers are scheduled to work on Saturday, with each volunteer coming in at a different time. How many different ways can the volunteers be scheduled to arrive for work?

Lt. Commander Tommy Blackburn landing the first F4U Corsair on the deck of the USS *Midway*

Dare County Regional Airport

The Dare County Regional Airport, in Manteo, North Carolina, is a small airport with a big history. In 1943, a squadron of Navy pilots commanded by Lieutenant Commander Tommy Blackburn trained for combat at the Dare airport. The squadron, named the Jolly Rogers, went on to become the most famous Naval squadron in WWII history. During the same time that Naval pilots were training, the Civil Air Patrol flew antisubmarine missions out of Dare. Today, the role played by Dare County Regional Airport in WWII can be seen in the museum housed in the west end of the terminal building.

1. One day, the Jolly Rogers were using 36 planes for training and the Civil Air Patrol was using 12 planes for antisubmarine missions. What is the theoretical probability that on that day a plane seen flying over the Dare Airport was associated with the Jolly Rogers?
2. Four planes flying in formation are called a division. If the planes are flying in a straight line, how many ways might the planes be arranged?
3. If pilots had to use a three-letter code made up of *A*, *B*, and *C* to enter the hangar area each day and if each letter could be repeated, how many different codes would be possible?

MATH-ABLES

Probability Brain Teasers

Can you solve these riddles that involve probability? Watch out—some of them are tricky!

1. In Wade City, 5% of the residents have unlisted phone numbers. If you selected 100 people at random from the town's phone directory, how many of them would you predict have unlisted numbers?

2. Amanda has a drawer that contains 24 black socks and 18 white socks. If she reaches into the drawer without looking, how many socks does she have to pull out in order to be *certain* that she will have two socks of the same color?

3. Dale, Melvin, Carter, and Ken went out to eat. Each person ordered something different. When the food came, the waiter could not remember who had ordered what, so he set the plates down at random in front of the four friends. What is the probability that exactly three of the boys got what they ordered?

Round and Round and Round

This is a game for two players.

The object of this game is to determine which of the three spinners is the winning spinner (lands on the greater number most often).

Both players choose a spinner and spin at the same time. Record which spinner lands on the greater number. Repeat this 19 times, keeping track of which spinner wins each time. Repeat this process until you have played spinner A against spinner B, spinner B against spinner C, and spinner A against spinner C. Spin each pair of spinners 20 times and record the results.

Which spinner wins more often, A or B?
Which spinner wins more often, B or C?
Which spinner wins more often, A or C?
Is there anything surprising about your results?

internet connect

Go to ***go.hrw.com*** for a complete set of rules and game pieces.
KEYWORD: MR4 Game11

Random Numbers

internet connect
Lab Resources Online
go.hrw.com
KEYWORD: MR4 TechLab11

Your calculator can randomly generate numbers using the **randInt** function. This is helpful when you cannot actually perform an experiment. Instead, you can use random numbers to simulate it.

Activity

A dodecahedron is a 12-sided solid figure in which all faces are congruent. Imagine a dodecahedron whose faces are numbered from 1 to 12. Roll this dodecahedron 25 times and give your experimental probability of rolling a 6.

You may not have a dodecahedron handy to perform this experiment, but your calculator can simulate it.

MATH NUM CPX PRB
1:rand
2:nPr
3:nCr
4:!
5:randInt(
6:randNorm(
7:randBin(

1. Press MATH and use the right arrow key to highlight PRB.
   Use the down arrow key to highlight 5:. Press ENTER.
2. To have the calculator give you a number from 1 to 12, press 1 , 12) ENTER. The number that appears represents the number you have "rolled" on the dodecahedron.
3. To generate another random number, press ENTER. Continue to press ENTER as many times as you wish to represent rolling the dodecahedron.

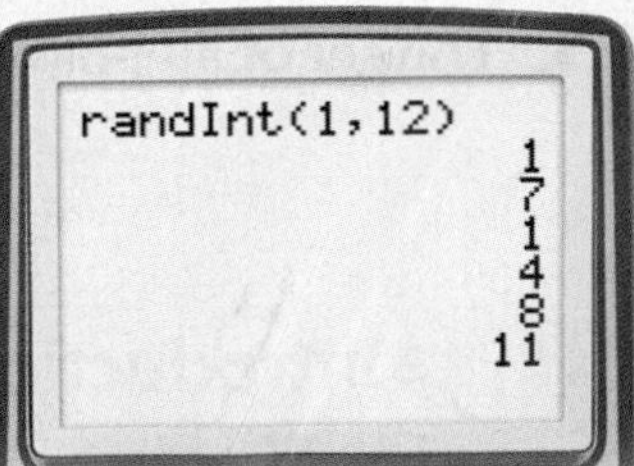

Think and Discuss

1. How could you use your calculator to simulate rolling a standard number cube? flipping a coin?
2. What kinds of experiments do you think random numbers would be most useful to simulate?

Try This

1. Simulate rolling a dodecahedron 50 times, and give the experimental probability of rolling a 9.
2. Design your own experiment, and describe how you would use your calculator to simulate it. Then perform the simulation several times, and give your experimental probabilities of each outcome.

Chapter 11

Study Guide and Review

Vocabulary

Complete the sentences below with vocabulary words from the list above. Words may be used more than once.

1. When all outcomes have the same probability of occurring, the outcomes are ___?___.
2. A(n) ___?___ is an activity involving chance that can have different results. Each possible result is called a(n) ___?___.
3. The measure of how likely an event is to occur is the event's ___?___.
4. ___?___ is the ratio of the number of ways an event can occur to the total number of possible outcomes.
5. The set of all possible outcomes for an experiment is the ___?___.

11-1 Introduction to Probability (pp. 554–557)

EXAMPLE

- **Is it impossible, unlikely, as likely as not, likely, or certain that the spinner will land on yellow?**

Half of the spinner is yellow, so it is as likely to land on yellow as not.

EXERCISES

6. Is it impossible, unlikely, as likely as not, likely, or certain that next week will have 7 days?
7. There is a 75% chance that George will win a race. Write this probability as a decimal and as a fraction.
8. Barry has a 30% chance of picking a black sock and a 50% chance of picking a white sock from his drawer. Which color sock is he more likely to pick?

Study Guide and Review

11-2 Experimental Probability (pp. 558–561)

EXAMPLE

■ **Margie recorded the number of times a spinner landed on each color. Based on Margie's experiment, on which color is the spinner most likely to land?**

Outcomes	Red	Blue	Green
Frequency	~~IIII~~ ~~IIII~~ IIII	IIII	~~IIII~~ II

$P(\text{red}) \approx \frac{14}{25}$ $P(\text{blue}) \approx \frac{4}{25}$ $P(\text{green}) \approx \frac{7}{25}$

The spinner will most likely land on red.

EXERCISES

9. One day, the cafeteria supervisor recorded the number of students who chose each type of beverage. She organized her results in a table. Find the experimental probability that a student will choose juice.

Beverage	Juice	Milk	Water
Frequency	20	37	18

11-3 Theoretical Probability (pp. 564–567)

EXAMPLE

■ **What is the probability of rolling a 4 on a fair number cube?**

There are six possible outcomes when a number cube is rolled: 1, 2, 3, 4, 5, or 6. All are equally likely because the number cube is fair.

$$P = \frac{\text{number of ways event can occur}}{\text{total number of possible outcomes}}$$

$$P(4) = \frac{\text{1 way event can occur}}{\text{6 possible outcomes}} = \frac{1}{6}$$

EXERCISES

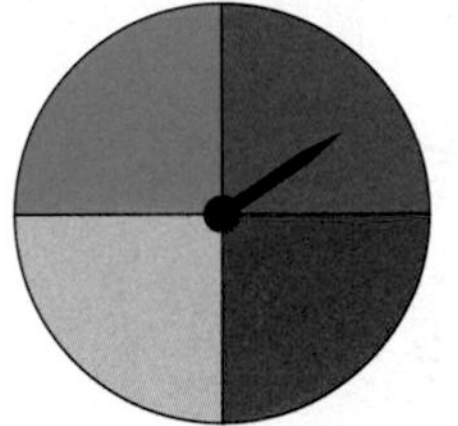

10. What is the probability that the spinner will land on yellow?

11. What is the probability of rolling a number greater than 3 on a fair number cube?

12. There is a 25% chance of choosing a purple marble from a bag. Find the probability of choosing a marble that is **not** purple.

11-4 Make an Organized List (pp. 570–573)

EXAMPLE

■ **Liz is wrapping a gift. She can use gold or silver paper and either a red or white ribbon. From how many different combinations can Liz choose?**

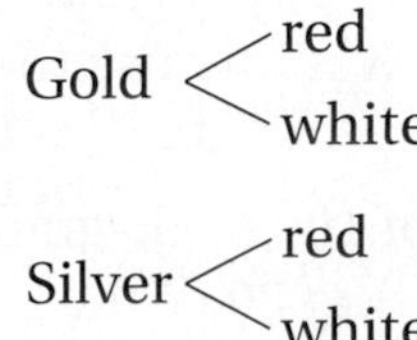

Follow each branch to find all outcomes.
There are 4 different combinations.

EXERCISES

13. The local restaurant has a lunch special in which you can pick an appetizer, a sandwich, and a drink. How many different lunch-special combinations are there if you have the following choices?

appetizers: soup or salad
sandwiches: turkey, roast beef, or ham
drinks: juice, milk, or iced tea

Study Guide and Review

11-5 Compound Events (pp. 574–577)

EXAMPLE

■ **What is the probability of spinning red or blue and having the coin land heads up?**

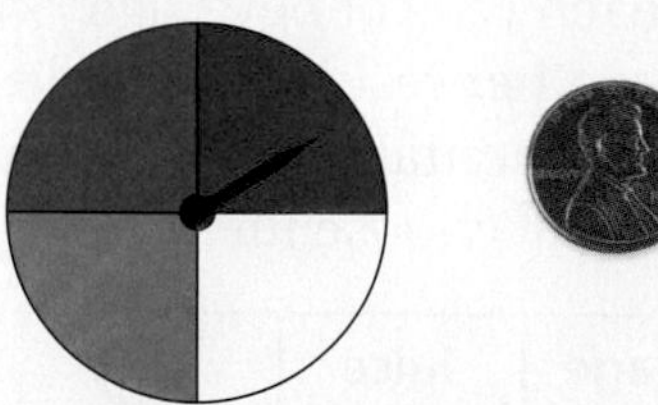

	Red	Blue	Green	White
Heads	red, H	blue, H	green, H	white, H
Tails	red, T	blue, T	green, T	white, T

There are 8 possible outcomes, and all are equally likely.

$$P(\text{red or blue, H}) = \frac{2 \text{ ways event can occur}}{8 \text{ possible outcomes}}$$
$$= \frac{2}{8} = \frac{1}{4}$$

EXERCISES

14. Find the probability that a blue marble will be chosen, the first coin will show heads, and the second coin will show tails.

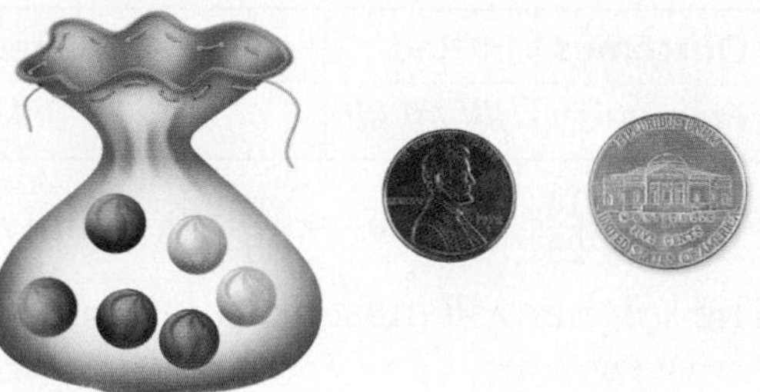

15. Jacob rolled a fair number cube, flipped a fair penny, and then flipped a fair quarter. Find the probability that the number cube will show an even number and both coins will show heads.

11-6 Making Predictions (pp. 580–583)

EXAMPLE

■ **If you spin the spinner 30 times, how many times do you expect it to land on red?**

$P(\text{red}) = \frac{1}{3}$

$\frac{1}{3} = \frac{x}{30}$

$3 \cdot x = 1 \cdot 30$ *The cross products are equal.*

$3x = 30$ *x is multiplied by 3.*

$\frac{3x}{3} = \frac{30}{3}$ *Divide both sides by 3 to undo the multiplication.*

$x = 10$

You can expect it to land on red about 10 times.

EXERCISES

16. About 2% of the items produced by a company are defective. Out of 5,000 items, how many can you predict will be defective?

17. If you roll a fair number cube 50 times, how many times can you expect to roll an even number?

18. A survey of 500 teenagers indicated that 175 of them use their computers regularly. Out of 4,500 teenagers, predict how many use their computers regularly.

19. A survey of 100 sixth-grade students indicated that 20 of them take music lessons. Out of 500 sixth-grade students, predict how many take music lessons.

Chapter 11

Chapter Test

For Exercises 1 and 2, write *impossible, unlikely, as likely as not, likely,* or *certain* to describe each event.

1. You roll a 3 on a fair number cube.
2. You pick a blue marble from a bag of 5 white marbles and 20 blue marbles.
3. There is a 12% chance of rain tomorrow. Write this probability as a decimal and as a fraction.
4. The probability that Mark will be selected for a scholarship is 0.8. Write this probability as a percent and as a fraction.
5. Iris asked 60 students what time they go to bed. Her results are in the table. Estimate the probability that a student chosen at random goes to bed at 8:30 P.M.

Time (P.M.)	8:00	8:30	9:00	9:30
Frequency	12	24	18	6

6. Estimate the probability that a student chosen at random goes to bed before 8:30 P.M.
7. Josh threw darts at a dartboard 10 times. Assume that he threw the darts randomly and did not aim. Based on his results, what is the probability that a dart will land in the center circle?

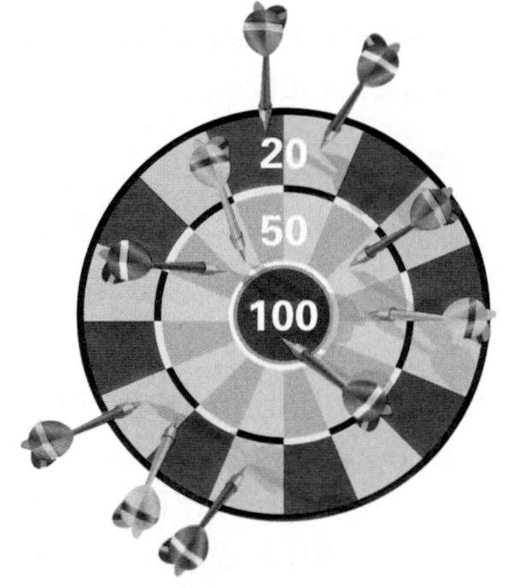

8. What is the probability of rolling an even number greater than 2 on a fair number cube?
9. The baseball game has a 64% chance of being rained out. What is the probability that it will **not** be rained out?
10. Peter has four photos to arrange in a frame. How many different ways can he arrange the photos?

11. Marsha can wear jeans or black pants with a red, blue, or white shirt. How many different outfits can she choose from?
12. If you roll a number cube 36 times, how many times do you expect to roll an even number?
13. Find the probability that you will pick a blue marble from both bags and that the spinner will land on blue.

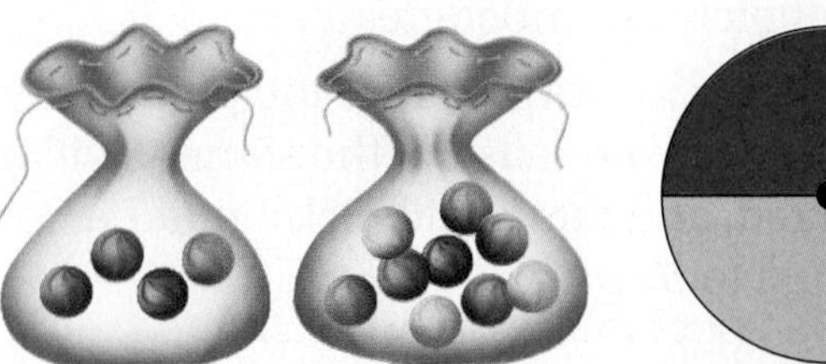

Chapter Test

Chapter 11

Performance Assessment

Show What You Know

Create a portfolio of your work from this chapter. Complete this page and include it with your four best pieces of work from Chapter 11. Choose from your homework or lab assignments, mid-chapter quiz, or any journal entries you have done. Put them together using any design you want. Make your portfolio represent what you consider your best work.

Short Response

1. A restaurant offers a choice of roast beef, chicken, or fish, with broccoli, carrots, or corn, and a soup or salad. Make an organized list to determine how many different meal combinations the restaurant offers.

2. Find the theoretical probability that the spinner will land on each color. Which of the spinners is fair? Explain. Which spinner is most likely to land on green?

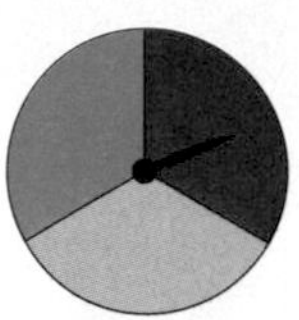

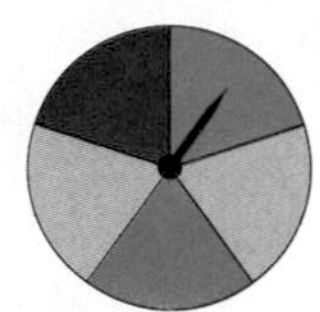

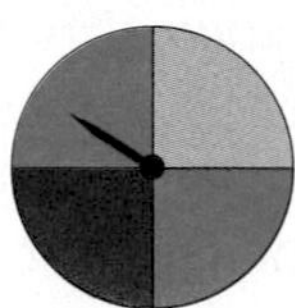

3. What is the probability that a green marble will not be chosen from a bag that contains 2 yellow marbles, 2 green marbles, and 3 red marbles? Explain two different methods that you can use to determine the answer.

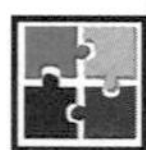

Extended Problem Solving

4. Jamie is performing an experiment in which she tosses the coin, spins the spinner, and then chooses a card.
   a. How many possible outcomes are there?
   b. What is the probability that the coin will land heads up, the spinner will land on red, and Jamie will pick a number greater than 5?
   c. Explain two ways to find the probability that the coin will land tails up, the spinner will not land on red, and Jamie will pick a number less than or equal to 5.

internet connect
EOG Test Practice Online
go.hrw.com Keyword: MR4 TestPrep

Getting Ready for EOG

Chapter 11

Cumulative Assessment, Chapters 1–11

TEST TAKING TIP!
Do not waste a lot of time on a question that is giving you trouble. Skip it and come back to it later if you have time.

1. What is the probability that the spinner lands on 1 or 2?

A $\frac{1}{4}$ C $\frac{1}{8}$

B $\frac{1}{2}$ D $\frac{3}{8}$

2. Which set is ordered from least to greatest?

A −4, −8, 0, 2 C 3, 1, −4, −5

B −6, −2, 1, 5 D −2, −5, 6, 8

3. Which data set has a mean of 15?

A 5, 17, 16, 16, 15, 7, 8

B 11, 15, 15, 17, 14, 12

C 21, 10, 17, 16, 14, 16, 11

D 25, 10, 10, 24, 18, 12

4. What is the GCF of 8, 16, and 20?

A 2 C 160

B 4 D 20

5. Which expression has a value of 12?

A $6 \cdot (-2)$ C $-3 \cdot (-4)$

B $-15 + (-3)$ D $22 - (-10)$

6. Mr. Rodriguez is remodeling his den. He can choose hardwood, carpet, or tiles for the floor and wallpaper, paint, or paneling for the walls. From how many different combinations can he choose?

A 5 C 8

B 6 D 9

7. A computer is on sale for 25% off the ticketed price of $986.00. How much will the computer cost after the discount?

A $246.50 C $961.00

B $739.50 D $961.35

8. What is the name for the transformation shown?

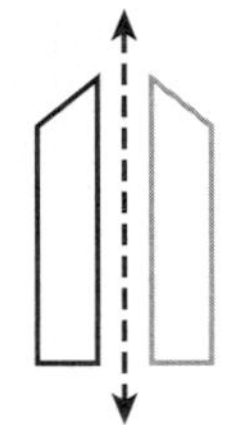

A translation C reflection

B rotation D tessellation

9. ***SHORT RESPONSE*** Lily is purchasing shelves for her new bookcase. She has 45 books to place in the bookcase. If 12 books fit on a shelf, how many shelves should she purchase? Explain your answer.

10. ***SHORT RESPONSE*** Sam flipped a fair coin and rolled a fair number cube. What is the probability that the coin will show heads and the cube will show a number greater than 2? Show how you found your answer.

Chapter

12

Functions and Coordinate Geometry

TABLE OF CONTENTS

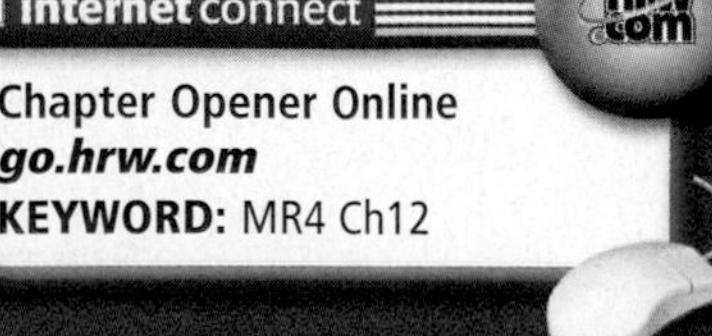

internet connect

Chapter Opener Online

go.hrw.com

KEYWORD: MR4 Ch12

Speed (m/s)	Energy Use (calories/hr)		
	Cycling	Walking	Running
1	105	120	—
2	114	210	—
3	135	420	420
4	147	—	540
5	195	—	720

Career ***Sports Physiologist***

Sports physiologists study people's oxygen use and muscle fatigue from exercise by evaluating their breathing rates and energy use. Sports physiologists can create special training and conditioning programs for athletes. The table shows the amount of energy used by a person weighing 110 pounds during different activities performed at different speeds.

ARE YOU READY?

Choose the best term from the list to complete each sentence.

1. The x-axis on a coordinate grid is the __?__ axis, and the y-axis is the __?__ axis.
2. The point (1, 2) is written as a(n) __?__.
3. A mathematical statement that says two quantities are equal is a(n) __?__.
4. A(n) __?__ is a mathematical phrase that includes only numbers and operation symbols.
5. A(n) __?__ is formed by two number lines in a plane that intersect at right angles.

coordinate plane
equation
horizontal
numerical expression
ordered pair
vertical
origin

Complete these exercises to review skills you will need for this chapter.

✔ Equations

Solve each equation.

6. $4t = 32$
7. $2b + 4 = 12$
8. $15 = 6r - 3$
9. $3x = 72$
10. $23 = 4a - 5$
11. $12m + 3 = 63$

✔ Ordered Pairs

Give the coordinates of each point.

12. A
13. Z
14. C
15. G
16. M
17. H

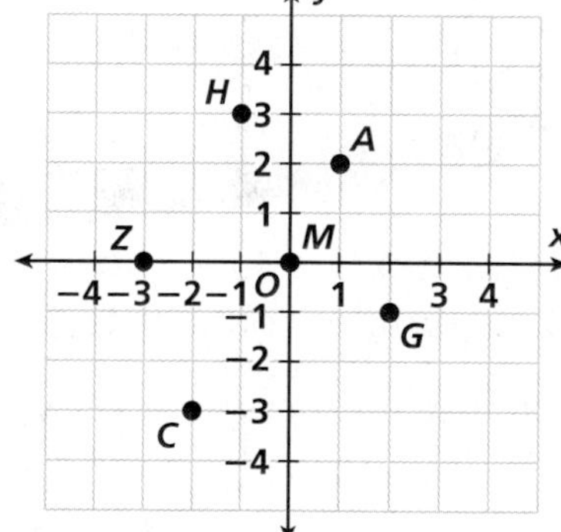

✔ Transformations

Tell whether each transformation is a translation, a rotation, or a reflection.

18.
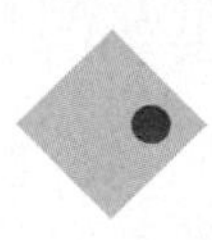

19.
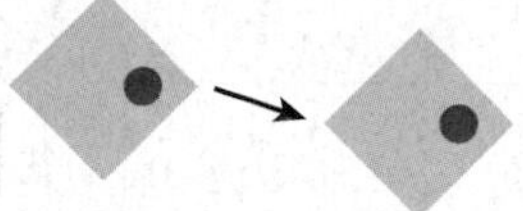

20.
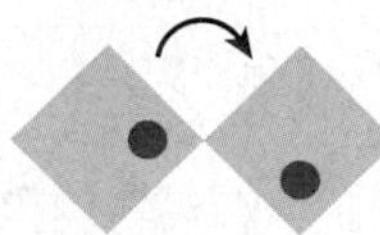

12-1 Tables and Functions

Learn to use data in a table to write an equation for a function and to use the equation to find a missing value.

Vocabulary

function

input

output

A baseball pitch thrown too high, low, or wide is considered outside the strike zone. A pitcher threw a ball 4 inches too low. How far in centimeters was the ball outside the strike zone? Make a table to show how the number of centimeters increases as the number of inches increases.

Andrew Toos/Cartoon Resource.com

"Come on, ump, that pitch was at least four centimeters outside!"

	Inches	Centimeters	
	1	2.54	
+1	2	5.08	+2.54
+1	3	7.62	+2.54
+1	4	10.16	+2.54

The number of centimeters is 2.54 times the number of inches. Let x represent the number of inches and y represent the number of centimeters. Then the equation $y = 2.54x$ relates centimeters to inches.

A **function** is a rule that relates two quantities so that each **input** value corresponds exactly to one **output** value.

When the input is 4 in., the output is 10.16 cm. So the ball was 10.16 centimeters outside the strike zone.

You can use a function table to show some of the values for a function.

EXAMPLE 1 Writing Equations from Function Tables

Write an equation for a function that gives the values in the table. Use the equation to find the value of y for the indicated value of x.

x	3	4	5	6	7	10
y	7	9	11	13	15	■

y is 2 times $x + 1$. *Compare x and y to find a pattern.*

$y = 2x + 1$ *Use the pattern to write an equation.*

$y = 2(10) + 1$ *Substitute 10 for x.*

$y = 20 + 1 = 21$ *Use your function rule to find y when x = 10.*

Helpful Hint

When all the y-values are greater than the corresponding x-values, use addition and/or multiplication in your equation.

You can write equations for functions that are described in words.

EXAMPLE 2 Translating Words into Math

Write an equation for the function. Tell what each variable you use represents.

The length of a rectangle is 5 times its width.

ℓ = length of rectangle *Choose variables for the equation.*

w = width of rectangle

$\ell = 5w$ *Write an equation.*

EXAMPLE 3 PROBLEM SOLVING APPLICATION

Car washers tracked the number of cars they washed and the total amount of money they earned. They charged the same price for each car they washed. They earned \$60 for 20 cars, \$66 for 22 cars, and \$81 for 27 cars. Write an equation for the function.

1 Understand the Problem

The **answer** will be an equation that describes the relationship between the number of cars washed and the money earned.

2 Make a Plan

You can make a table to display the data.

3 Solve

Let c be the number of cars. Let m be the amount of money earned.

c	20	22	27
m	60	66	81

m is equal to 3 times c. *Compare c and m.*

$m = 3c$ *Write an equation.*

4 Look Back

Substitute the c and m values in the table to check that they are solutions of the equation $m = 3c$.

$m = 3c$ (20, 60)
$60 \stackrel{?}{=} 3 \cdot 20$
$60 \stackrel{?}{=} 60$ ✔

$m = 3c$ (22, 66)
$66 \stackrel{?}{=} 3 \cdot 22$
$66 \stackrel{?}{=} 66$ ✔

$m = 3c$ (27, 81)
$81 \stackrel{?}{=} 3 \cdot 27$
$81 \stackrel{?}{=} 81$ ✔

Think and Discuss

1. Explain how you find the y-value when the x-value is 20 for the function $y = 5x$.

12-1 Exercises

5.03, 5.04

FOR EOG PRACTICE
see page 684

internet connect
Homework Help Online
go.hrw.com Keyword: MR4 12-1

GUIDED PRACTICE

See Example 1

Write an equation for a function that gives the values in each table. Use the equation to find the value of y for the indicated value of x.

1.

x	1	2	3	6	9
y	−5	−4	−3	0	

2.

x	3	4	5	6	10
y	16	21	26	31	

See Example 2

Write an equation for the function. Tell what each variable you use represents.

3. Jen is 6 years younger than her brother.

See Example 3

4. Brenda sells balloon bouquets. She charges the same price for each balloon in a bouquet. The cost of a bouquet with 6 balloons is $3, with 9 balloons is $4.50, and with 12 balloons is $6. Write an equation for the function.

INDEPENDENT PRACTICE

See Example 1

Write an equation for a function that gives the values in each table. Use the equation to find the value of y for the indicated value of x.

5.

x	0	1	2	5	7
y	0	4	8	20	

6.

x	4	5	6	7	12
y	−2	0	2	4	

See Example 2

Write an equation for the function. Tell what each variable you use represents.

7. The cost of a case of bottled juices is $2 less than the cost of twelve individual bottles.

8. The population of New York is twice as large as the population of Michigan.

See Example 3

9. Oliver is playing a video game. He earns the same number of points for each prize he captures. He earned 1,050 points for 7 prizes, 1,500 points for 10 prizes, and 2,850 points for 19 prizes. Write an equation for the function.

PRACTICE AND PROBLEM SOLVING

Write an equation for a function that gives the values in each table, and then find the missing terms.

10.

x	2	3	5	9	11	14
y	−6	−10	−18	−34	−42	

11.

x	−1	0	1	2	5	7
y		3.4	4.4	5.4		10.4

Write an equation for each function. Define the variables that you use.

12.

The Denominators
$125.00
plus $55 per hour

13.

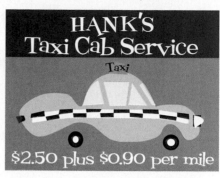

HANK'S
Taxi Cab Service
Taxi
$2.50 plus $0.90 per mile

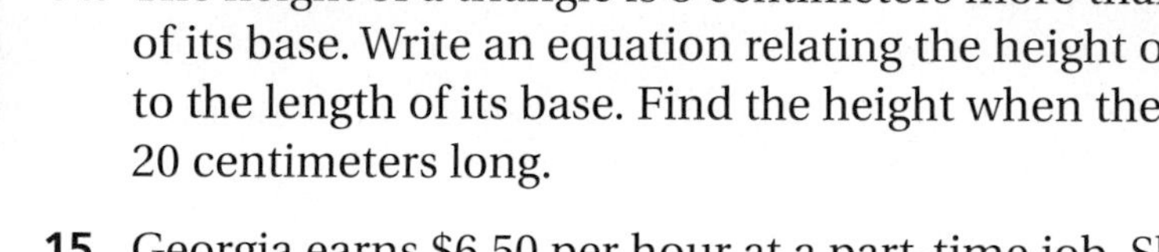

14. The height of a triangle is 5 centimeters more than twice the length of its base. Write an equation relating the height of the triangle to the length of its base. Find the height when the base is 20 centimeters long.

15. Georgia earns $6.50 per hour at a part-time job. She wants to buy a sweater that costs $58.50. Write an equation relating the number of hours she works to the amount of money she earns. Find how many hours Georgia needs to work to buy the sweater.

Career LINK

Graphic designers work in media such as magazines, brochures, product packaging, and the Internet to visually communicate the appeal of a company or its products.

go.hrw.com
KEYWORD: MR4 Designer
CNN Student News

Use the table for Exercises 16–18.

16. ***GRAPHIC DESIGN*** Margo is designing a Web page displaying similar rectangles. She uses a linear pattern to determine the dimensions of each rectangle. Use the table to write an equation relating the width of a rectangle to the length of a rectangle. Find the length of a rectangle that has a width of 250 pixels.

Width (pixels)	Length (pixels)
30	95
40	125
50	155
60	185

17. ***WHAT'S THE ERROR?*** Margo predicted that the length of a rectangle with a width of 100 pixels would be 310 pixels. Explain the error she made. Then find the correct length.

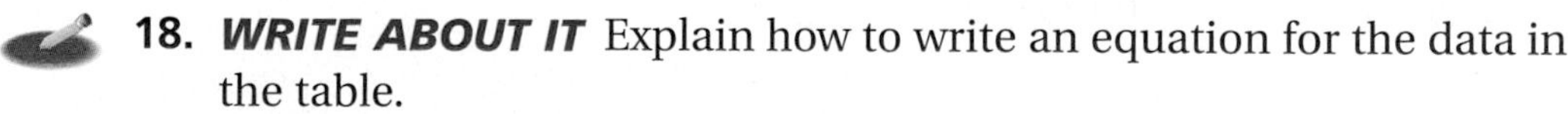

18. ***WRITE ABOUT IT*** Explain how to write an equation for the data in the table.

19. ***CHALLENGE*** Write an equation that would give the same y-values as $y = 2x + 1$ for $x = 0, 1, 2, 3$.

Spiral Review

Evaluate each expression for $x = 5$. (Lesson 2-1)

20. $x + 7$ **21.** $-4x$ **22.** $3x + 6$ **23.** $x^2 + 5$

24. **EOG PREP** Jazmine played a game of chance 100 times and won 24 times. Which of the following is the best estimate of the experimental probability of winning the game? (Lesson 11-6)

A 5% B 25% C 50% D 75%

Explore Linear and Nonlinear Relationships

Use with Lesson 12-2

You can learn about linear and nonlinear relationships by looking at patterns.

Activity

1 This model shows stage 1 to stage 3 of a pattern.

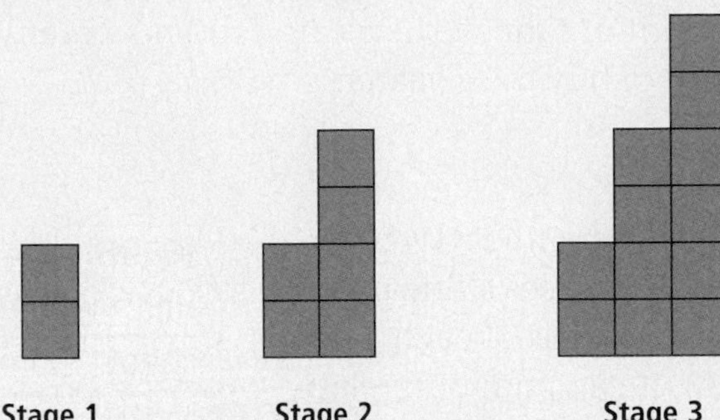

a. Use square tiles or graph paper to model stages 4, 5, and 6.

b. Record each stage and the perimeter of each figure in a table.

c. Graph the ordered pairs (x, y) from the table on a coordinate plane.

Stage (x)	Perimeter (y)
1	6
2	12
3	18
4	24
5	30
6	36

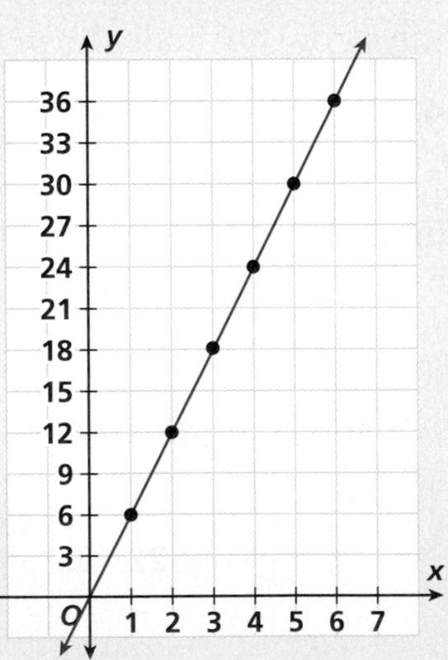

If you connected the points you graphed, you would draw a straight line. This shows that the relationship between the stage and the perimeter of the figure is linear. The equation for this line is $y = 3x$.

2 This table shows the ordered pairs for stages 1, 4, and 9 of a pattern.

Stage (x)	Square Root (y)
1	1
4	2
9	3
16	4
25	5
36	6

$1 = 1 \cdot 1$ or $1 = 1^2$

$4 = 2 \cdot 2$ or $4 = 2^2$

$9 = 3 \cdot 3$ or $9 = 3^2$

The square root of 4 is 2, which is written like this: $\sqrt{4} = 2$.

a. You can model this by arranging squares in 1-by-1, 2-by-2, and 3-by-3 blocks.

b. Record each stage number and the number's square root in a table. Graph the ordered pairs (x, y) from the table on a coordinate plane.

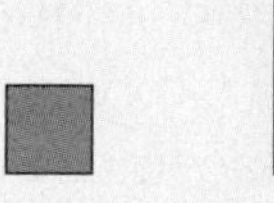
Stage 1

Stage 4

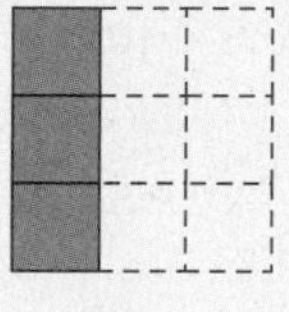
Stage 9

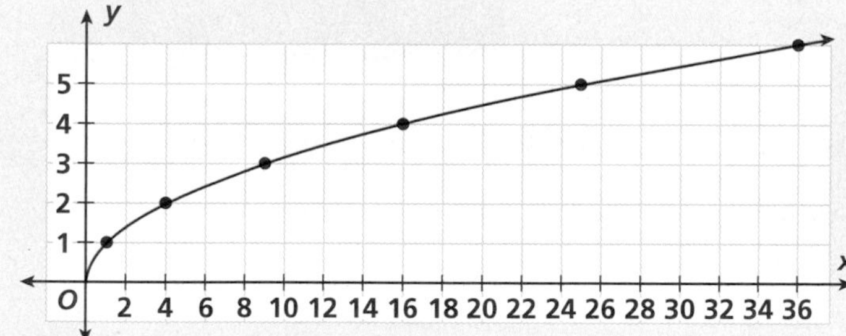

If you connect the graphed points, you draw a curved line. This shows that the relationship between the stage number and that number's square root is nonlinear. The equation for this curve is $y = \sqrt{x}$.

Think and Discuss

1. Explain what pattern you see in the y-values of the ordered pairs from the graph above.

Try This

Use the x-values 1, 2, 3, and 4 to find ordered pairs for each equation. Then graph the equation. Tell whether the relationship between x and y is linear or nonlinear.

1. $y = 2 + x$

2. $y = 4x$

3. $y = x^3$

4. $y = x + 4$

5. $y = x(2 + x)$

6. $y = x + x$

12-2 Graphing Functions

Learn to represent linear functions using ordered pairs and graphs.

Vocabulary

linear equation

Christa is ordering CDs online. Each CD costs $16, and the shipping and handling charge is $6 for the whole order.

The total cost, y, depends on the number of CDs, x. This function is described by the equation $y = 16x + 6$.

To find solutions of an equation with two variables, first choose a replacement value for one variable and then find the value of the other variable.

EXAMPLE 1 Finding Solutions of Equations with Two Variables

Use the given x-values to write solutions of the equation $y = 16x + 6$ as ordered pairs.

Make a function table by using the given values for x to find values for y.

Write these solutions as ordered pairs.

x	$16x + 6$	y
1	$16(1) + 6$	22
2	$16(2) + 6$	38
3	$16(3) + 6$	54
4	$16(4) + 6$	70

(x, y)
(1, 22)
(2, 38)
(3, 54)
(4, 70)

Check if an ordered pair is a solution of an equation by putting the x and y values into the equation to see if they make it a true statement.

EXAMPLE 2 Checking Solutions of Equations with Two Variables

Determine whether the ordered pair is a solution to the given equation.

(8, 16); $y = 2x$

$y = 2x$ *Write the equation.*

$16 \stackrel{?}{=} 2(8)$ *Substitute 8 for x and 16 for y.*

$16 \stackrel{?}{=} 16$ ✔

So (8, 16) is a solution of $y = 2x$.

You can also graph the solutions of an equation on a coordinate plane. When you graph the ordered pairs of some functions, they form a straight line. The equations that express these functions are called **linear equations**.

EXAMPLE 3 Reading Solutions on Graphs

Use the graph of the linear function to find the value of y for the given value of x.

$x = 1$

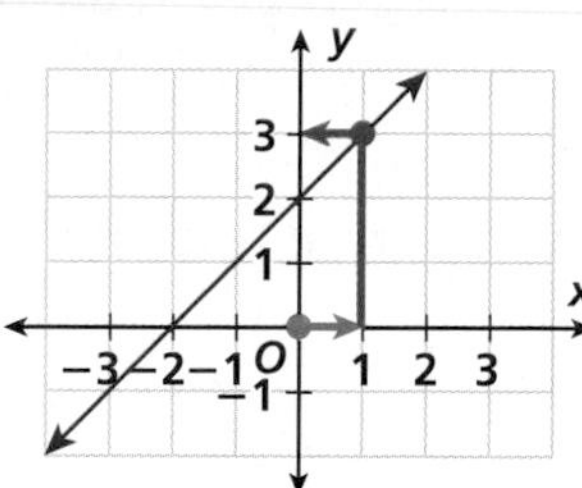

Start at the origin and move 1 unit right. Move up until you reach the graph. Move left to find the y-value on the y-axis.

When $x = 1$, $y = 3$. The ordered pair is (1, 3).

EXAMPLE 4 Graphing Linear Functions

Graph the function described by the equation.

$y = 2x + 1$

Make a function table.

x	$2x + 1$	y
−1	2(−1) + 1	−1
0	2(0) + 1	1
1	2(1) + 1	3

Write the solutions as ordered pairs.

(x, y)

(−1, −1)

(0, 1)

(1, 3)

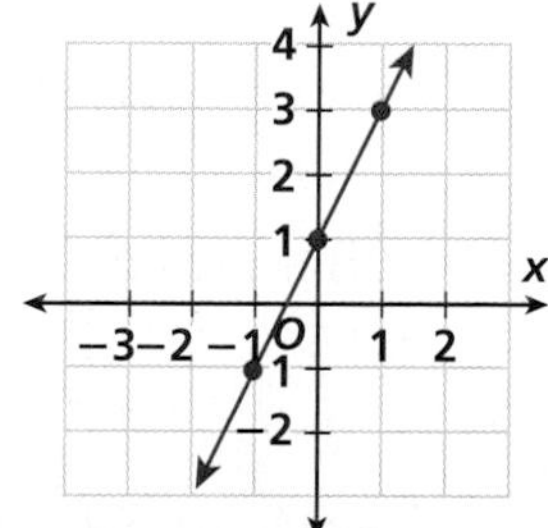

Graph the ordered pairs on a coordinate plane. Draw a line through the points to represent all the values of x you could have chosen and the corresponding values of y.

Think and Discuss

1. **Explain** why the points in Example 4 are not the only points on the graph. Name two points that you did not plot.
2. **Tell** whether the equation $y = 10x - 5$ describes a linear function.

12-2 Exercises

FOR EOG PRACTICE
see page 685

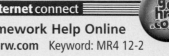

internet connect
Homework Help Online
go.hrw.com Keyword: MR4 12-2

5.02, 5.04

GUIDED PRACTICE

See Example 1 — **Use the given x-values to write solutions of each equation as ordered pairs.**

1. $y = 6x + 2$ for $x = 1, 2, 3, 4$ **2.** $y = -2x$ for $x = 1, 2, 3, 4$

See Example 2 — **Determine whether each ordered pair is a solution of the given equation.**

3. $(2, 12); y = 4x$ **4.** $(5, 9); y = 2x - 1$

See Example 3 — **Use the graph of the linear function to find the value of y for each given value of x.**

5. $x = 1$ **6.** $x = 0$

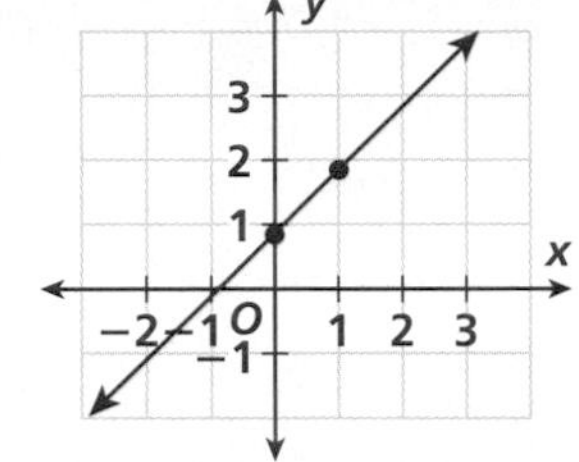

See Example 4 — **Graph the function described by each equation.**

7. $y = x + 3$ **8.** $y = 3x - 1$

INDEPENDENT PRACTICE

See Example 1 — **Use the given x-values to write solutions of each equation as ordered pairs.**

9. $y = -4x + 1$ for $x = 1, 2, 3, 4$ **10.** $y = 5x - 5$ for $x = 1, 2, 3, 4$

See Example 2 — **Determine whether each ordered pair is a solution of the given equation.**

11. $(3, -10); y = -6x + 8$ **12.** $(-8, 1); y = 7x - 15$

See Example 3 — **Use the graph of the linear function to find the value of y for each given value of x.**

13. $x = -2$ **14.** $x = 1$

15. $x = 0$ **16.** $x = -1$

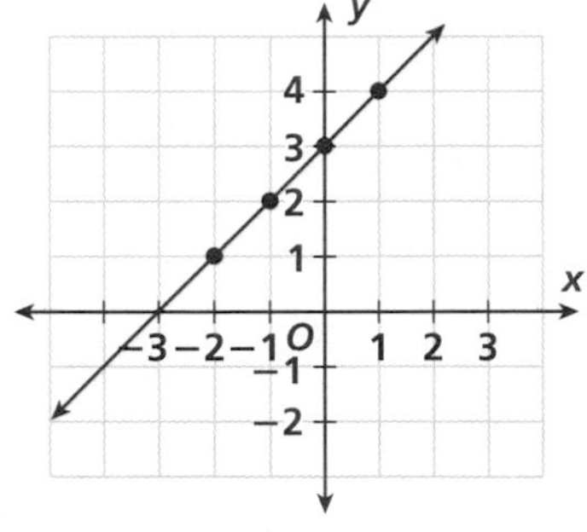

See Example 4 — **Graph the function described by each equation.**

17. $y = 4x + 1$ **18.** $y = -x - 2$

PRACTICE AND PROBLEM SOLVING

Complete each table, and then use the table to graph the function.

19. $y = x - 2$

x	−1	0	1	2
y				

20. $y = 2x - 4$

x	−1	0	1	2
y				

21. Which of the ordered pairs below is not a solution of $y = 4x + 9$?
(1, 14), (0, 9), (−1, 5), (−2, 1), (2, 17)

Physical Science 

Temperature can be expressed according to different scales. The Kelvin scale is divided into units called kelvins, and the Celsius scale is divided into degrees Celsius.

The table shows several temperatures recorded in degrees Celsius and their equivalent measures in kelvins.

Equivalent Temperatures	
Celsius (°C)	**Kelvin (K)**
−100	173
−50	223
0	273
50	323
100	373

A technician preserves brain cells in this tank of liquid nitrogen, which is at −196°C, for later research. Scientists hope to understand more about how the brain works and how some diseases start.

22. Write an equation for a function that gives the values in the table. Define the variables that you use.

23. Graph the function described by your equation.

24. Use your equation to find the equivalent Kelvin temperature for –54°C.

25. Use your equation to find the equivalent Celsius temperature for 77 kelvins.

26. ***WHAT'S THE QUESTION?*** The answer is −273°C. What is the question?

27. ***WRITE ABOUT IT*** Explain how to use your equation to determine whether 75°C is equivalent to 345 kelvins. Then determine whether the temperatures are equivalent.

28. ***CHALLENGE*** How many ordered-pair solutions exist for the equation you wrote in Exercise 22?

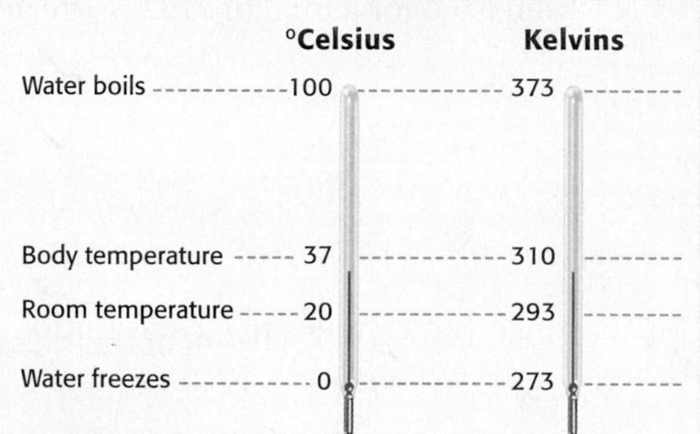

Spiral Review

Write each percent as a fraction in simplest form. (Lesson 8-7)

29. 25% **30.** 78% **31.** 40% **32.** 99%

In which quadrant would you find each point? (Lesson 9-3)

33. (5, −1) **34.** (2, 14) **35.** (−6, 3) **36.** (−9, −15)

37. **EOG PREP** The dimensions of four rectangular prisms are given below. Which prism has the greatest volume? (Lesson 10-8)

A 9 cm × 1.5 cm × 1.5 cm

B 3 cm × 3 cm × 2 cm

C 2 cm × 9 cm × 1.1 cm

D 7 cm × 2 cm × 1 cm

Chapter 12

Mid-Chapter Quiz

Mid-Chapter Quiz

LESSON 12-1 (pp. 598–601)

Write an equation for a function that gives the values in each table. Use the equation to find the value of *y* for each indicated value of *x*.

1.

x	2	3	4	5	8
y	7	9	11	13	■

2.

x	1	4	5	6	8
y	■	18	23	28	38

Write an equation for the function. Tell what each variable you use represents.

3. The number of plates is 5 less than 3 times the number of cups.
4. The time Rodney spends running is 10 minutes more than twice the time he spends stretching.
5. The height of a triangle is twice the length of its base.
6. A store manager tracked T-shirt sales. The store charges the same price for each T-shirt. On Monday, 5 shirts were sold for a total of \$60. On Tuesday, 8 shirts were sold for a total of \$96. On Wednesday, 11 shirts were sold for a total of \$132. Write an equation for the function.

LESSON 12-2 (pp. 604–607)

Use the given *x*-values to write solutions of each equation as ordered pairs.

7. $y = 4x + 6$ for $x = 1, 2, 3, 4$
8. $y = 10x - 7$ for $x = 2, 3, 4, 5$

Determine whether each ordered pair is a solution of the given equation.

9. $(3, 7)$; $y = 2x + 1$
10. $(5, 1)$; $y = x - 5$
11. $(4, 8)$; $y = 5x - 12$
12. $(9, 6)$; $y = \frac{1}{3}x + 4$

Use the graph of the linear function at right to find the value of *y* for each given value of *x*.

13. $x = 3$
14. $x = 0$
15. $x = -1$
16. $x = -2$

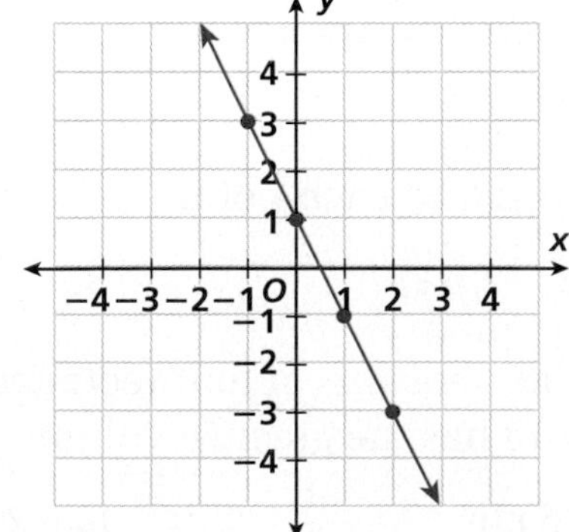

Graph the function described by each equation.

17. $y = x + 5$
18. $y = 3x + 2$
19. $y = x - 3$
20. $y = -2x$

Focus on Problem Solving

Look Back

- **Check that the question is answered**

Sometimes a problem asks you to go through a series of steps and then give the answer. When you read a question, ask yourself what you need to find to answer it. After you have solved the problem, read the question again and make sure you have completely answered it.

Read each problem. Follow all the directions in the problem and perform each required step. Then check that you have completely answered the question.

1. If all the x- and y-coordinates of A, B, C, and D were multiplied by 3, graphed, and connected, what would the area of the new figure be?

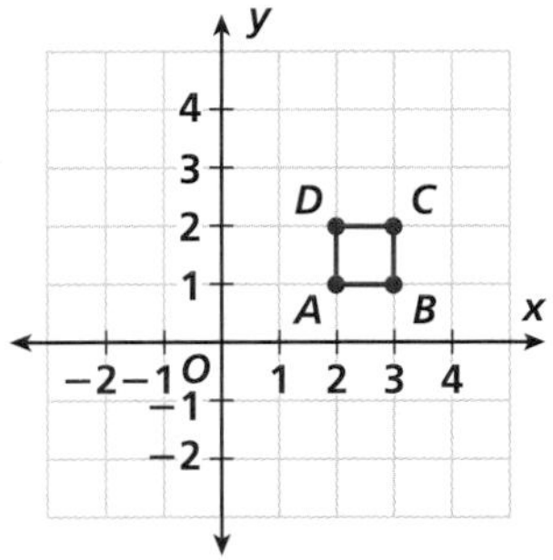

2. If all the x- and y-coodinates of P, Q, and R were multiplied by 4, graphed, and connected, how many units long would side $\overline{PQ}$ be in the new figure?

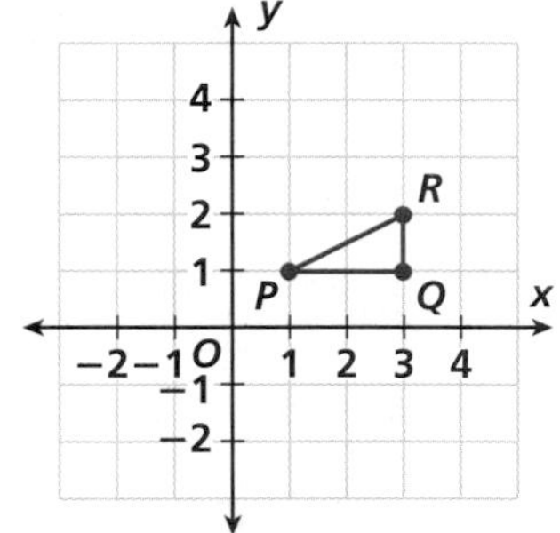

3. Vicky drew a cat's face on the coordinate grid at right. Then she stretched the cat's face vertically by a factor of 2 to create a new face. Give the new coordinates of A, B, C, D, and E in the new cat's face.

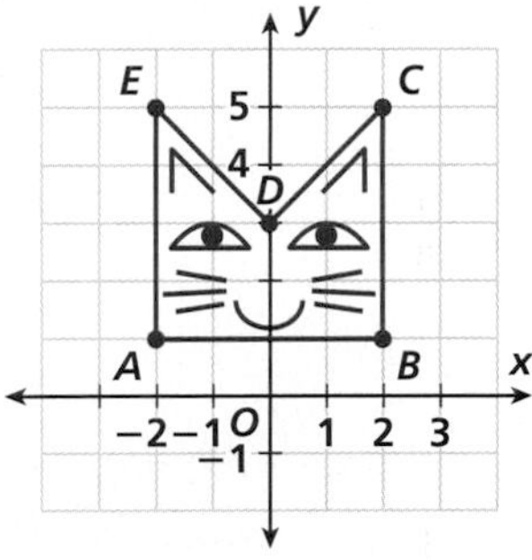

4. Ralph transformed the cat face below. The new coordinates are $A'(-2, -1)$, $B'(2, -1)$, $C'(2, -5)$, $D'(0, -3)$, and $E'(-2, -5)$. Graph $A'B'C'D'E'$ to create the new cat's face. What kind of transformations did Ralph use to create the new cat's face?

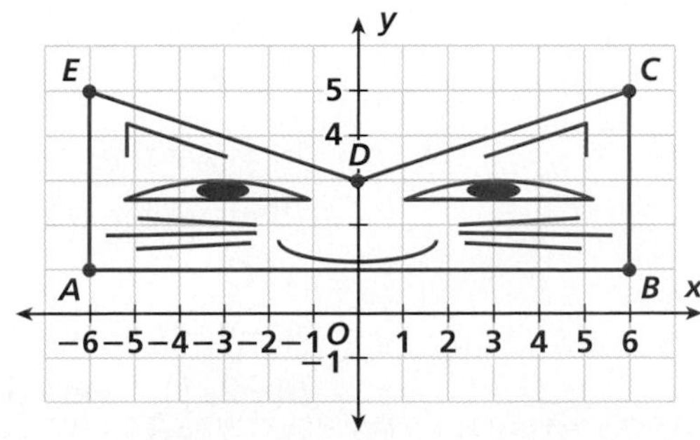

12-3 Graphing Translations

Learn to use translations to change the positions of figures on a coordinate plane.

Have you ever seen a marching band that stays in formation while moving forward, backward, or sideways? The moves may have been planned using a coordinate plane.

A translation is a movement of a figure along a straight line. You can translate a figure on a coordinate plane by sliding it horizontally, vertically, or diagonally.

EXAMPLE 1 Translating Figures on a Coordinate Plane

Reading Math

After a transformation is performed on a figure, A' labels the image, or new location, of A. Read A' as "A prime."

Give the coordinates of the vertices of the figure after the given translation.

Translate triangle ABC 6 units right and 5 units down.

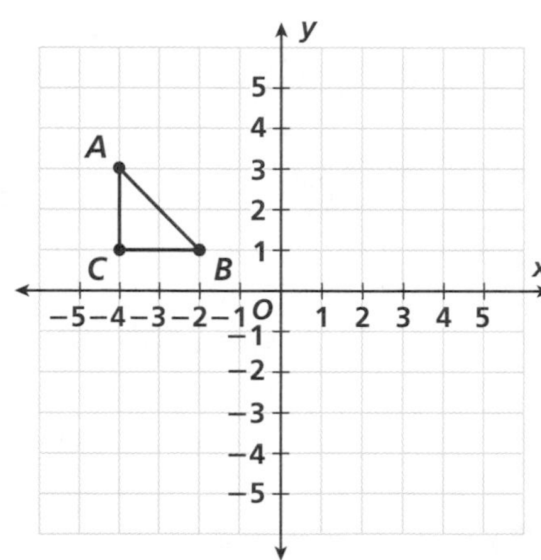

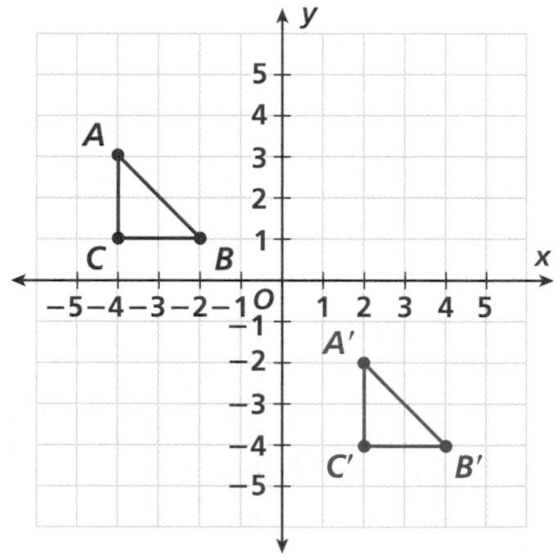

To move the triangle 6 units right, add 6 to each of the x-coordinates.

To move the triangle 5 units down, subtract 5 from each of the y-coordinates.

ABC		$A'B'C'$
$A(-4, 3)$ →	$A'(-4 + \mathbf{6}, 3 - \mathbf{5})$ →	$A'(2, -2)$
$B(-2, 1)$ →	$B'(-2 + \mathbf{6}, 1 - \mathbf{5})$ →	$B'(4, -4)$
$C(-4, 1)$ →	$C'(-4 + \mathbf{6}, 1 - \mathbf{5})$ →	$C'(2, -4)$

EXAMPLE 2 *Music Application*

Members of a marching band begin in a square formation, represented by the square *FGHJ*. Then they move 4 steps left and 3 steps forward. Give the coordinates of the vertices of the square after such a translation.

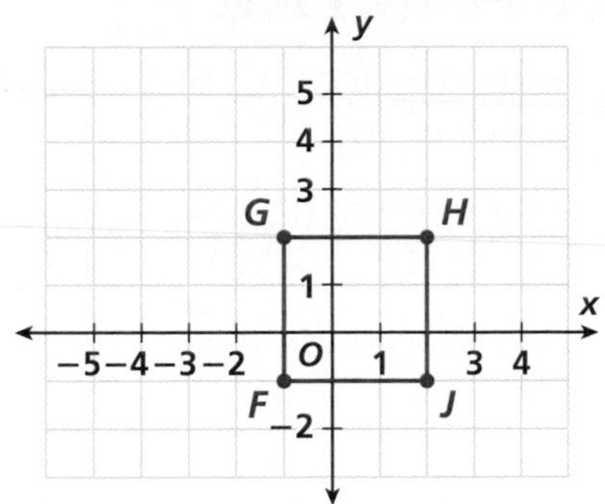

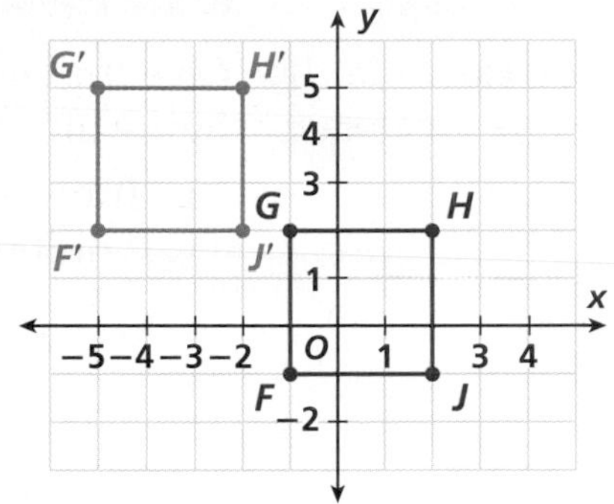

To move 4 steps left, subtract 4 from the x-coordinates.

To move 3 steps forward, add 3 to the y-coordinates.

FGHJ				*F'G'H'J'*
$F(-1, -1)$	→	$F'(-1 - 4, -1 + 3)$	→	$F'(-5, 2)$
$G(-1, 2)$	→	$G'(-1 - 4, 2 + 3)$	→	$G'(-5, 5)$
$H(2, 2)$	→	$H'(2 - 4, 2 + 3)$	→	$H'(-2, 5)$
$J(2, -1)$	→	$J'(2 - 4, -1 + 3)$	→	$J'(-2, 2)$

Think and Discuss

1. Tell how the location of a point changes when the x-coordinate increases.

2. Tell how the location of a point changes when the y-coordinate decreases.

12-3 Exercises

3.03, 3.04

FOR EOG PRACTICE
see page 686

internet connect
Homework Help Online
go.hrw.com Keyword: MR4 12-3

GUIDED PRACTICE

See Example 1

Give the coordinates of the vertices of each figure after the given translation.

1. Translate triangle *RST* 3 units up.

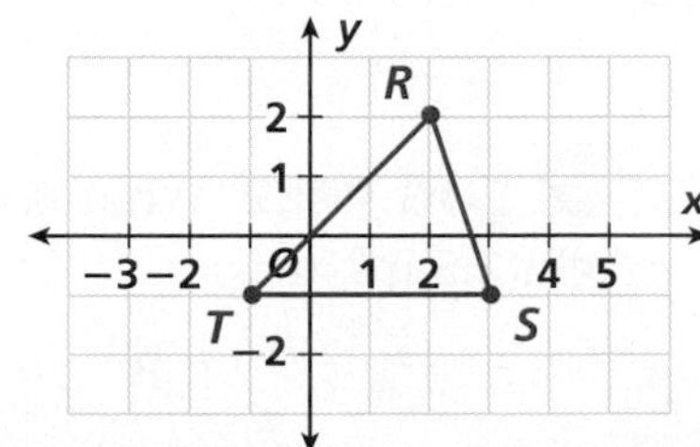

See Example 2

2. The graph shows the position of a fountain on the plans for a new park. The architect wants to move it 2 units right and 3 units down. Give the coordinates of the vertices of the fountain after this translation.

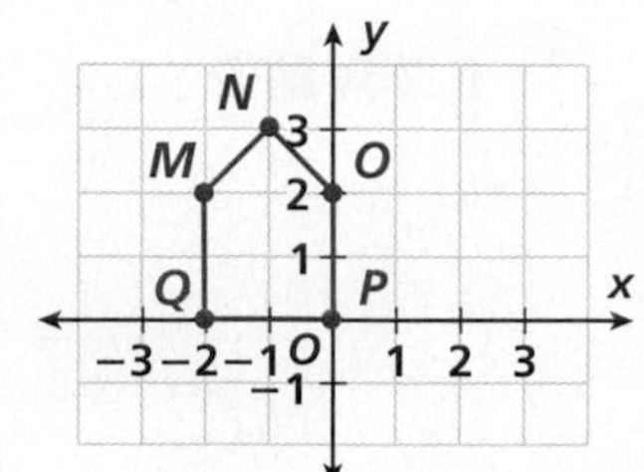

INDEPENDENT PRACTICE

See Example 1

Give the coordinates of the vertices of each figure after the given translation.

3. Translate triangle *RST* 5 units up and 2 units left.

4. Translate triangle *RST* 6 units right and 1 unit down.

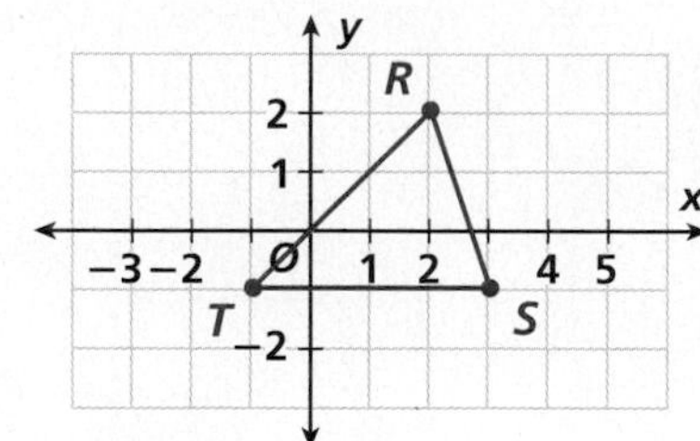

See Example 2

5. The placement of a rug in a room is represented on the graph. The rug is moved 7 units left and 4 units down. Give the coordinates of the vertices of the rug after this translation.

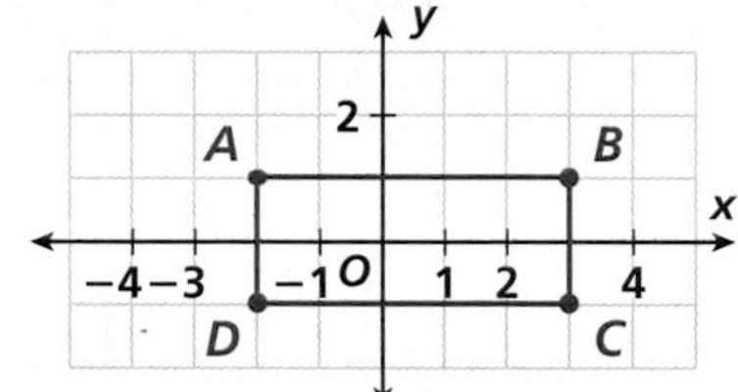

PRACTICE AND PROBLEM SOLVING

Give the coordinates of the trapezoid after each given translation.

6. Translate the trapezoid 1 unit up and 8 units left.

7. Translate the trapezoid 2 units down and 2 units right.

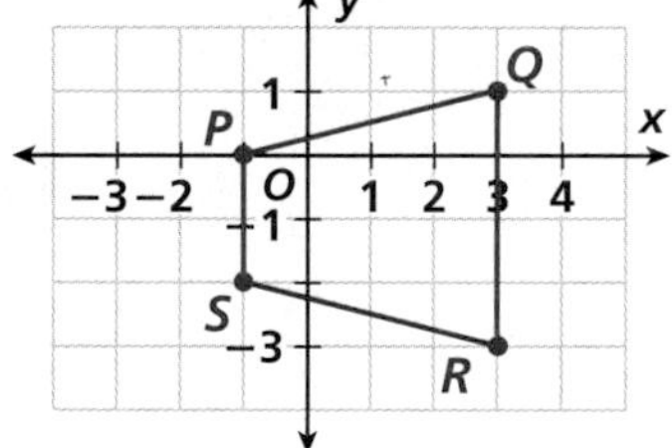

8. Rectangle *ABCD* was translated 2 units down and 5 units left. The new coordinates for the rectangle are $A'(1, 2)$, $B'(4, 2)$, $C'(4, 1)$, and $D'(1, 1)$. What were the coordinates of *A*, *B*, *C*, and *D*?

9. ***WRITE ABOUT IT*** Explain how to find the new coordinates of a point that is translated 3 units right and 2 units down.

10. ***CHALLENGE*** Give examples of how you could translate rectangle $A'B'C'D'$ from Exercise 8 so that the entire figure is in quadrant IV.

Spiral Review

Solve for *x*. (Lesson 9-8)

11. $x + 10 = -2$

12. $x - 20 = -5$

13. $-9x = 45$

14. **EOG PREP** What is the area of a triangle with base 8 in. and height 2.5 in? (Lesson 10-2)

A 10.5 in^2 B 10 in^2 C 13 in^2 D 20 in^2

12-4 Graphing Reflections

Learn to use reflections to change the positions of figures on a coordinate plane.

Textile designers create patterns for printed, woven, or knitted fabrics. The design might include a shape and the image of that shape's reflection.

To design a fabric pattern, a designer might use a coordinate plane. A figure can be reflected on a coordinate plane across the *x*-axis or the *y*-axis.

EXAMPLE 1 **Reflecting Figures on a Coordinate Plane**

Remember!

When a figure is flipped over a line, the new image is a reflection of the original.

Give the coordinates of the figure after the given reflection.

Reflect parallelogram *ABCD* across the *y*-axis.

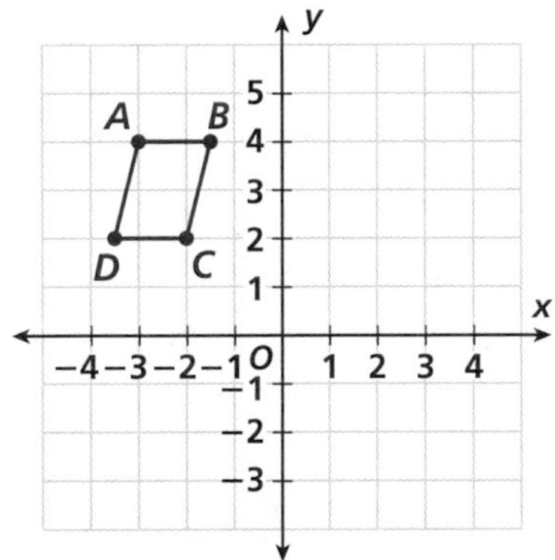

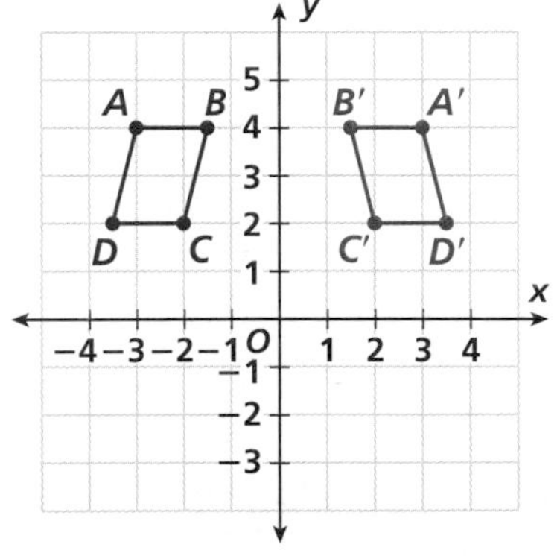

To reflect the parallelogram across the y-axis, write the opposites of the x-coordinates.

The y-coordinates do not change.

ABCD		*A′B′C′D′*
$A(-3, 4)$	→	$A'(3, 4)$
$B\left(-1\frac{1}{2}, 4\right)$	→	$B'\left(1\frac{1}{2}, 4\right)$
$C(-2, 2)$	→	$C'(2, 2)$
$D\left(-3\frac{1}{2}, 2\right)$	→	$D'\left(3\frac{1}{2}, 2\right)$

EXAMPLE 2 Design Application

A designer is using a stencil that is shaped like the figure below. A pattern is made by reflecting the figure across the *x*-axis. Give the coordinates of the vertices of the figure after the given reflection.

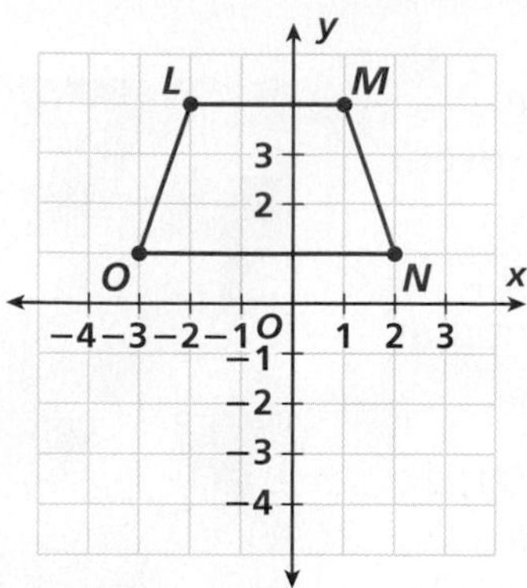

To reflect the trapezoid across the x-axis, write the opposites of the y-coordinates. The x-coordinates do not change.

$LMNO$		$L'M'N'O'$
$L(-2, 4)$	→	$L'(-2, -4)$
$M(1, 4)$	→	$M'(1, -4)$
$N(2, 1)$	→	$N'(2, -1)$
$O(-3, 1)$	→	$O'(-3, -1)$

Think and Discuss

1. **Tell** why the x-coordinates of a figure do not change when the figure is reflected across the x-axis.
2. **Decribe** how the x-coordinates of a figure change when the figure is reflected across the y-axis.

12-4 Exercises

3.03, 3.04

FOR EOG PRACTICE
see page 686

internet connect
Homework Help Online
go.hrw.com Keyword: MR4 12-4

GUIDED PRACTICE

See Example 1

Give the coordinates of the vertices of each figure after the given reflection.

1. Reflect triangle PQR across the y-axis.

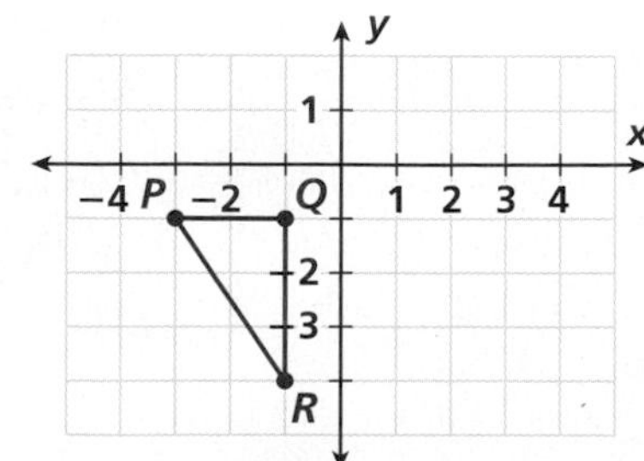

See Example 

2. Some cheerleaders stood in a formation similar to trapezoid *ABCD*. Then, they moved to form a second figure represented by reflecting *ABCD* across the *x*-axis. Give the coordinates of the vertices of the new trapezoid.

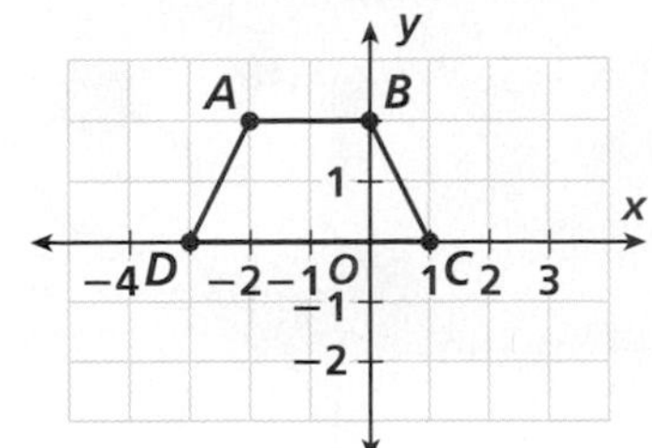

INDEPENDENT PRACTICE

See Example 1

Give the coordinates of the vertices of each figure after the given reflection.

3. Reflect triangle *NQP* across the *x*-axis.

See Example 2

4. Patricia created a plan for her new garden by graphing the parallelogram *JKLM*. She changed her mind and decided to reflect the figure across the *y*-axis. Give the new coordinates of the vertices of the figure.

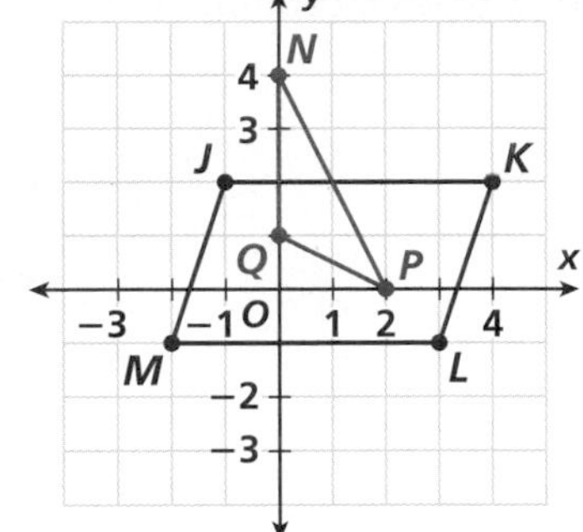

PRACTICE AND PROBLEM SOLVING

Give the coordinates of the square after each given reflection.

5. Reflect square *ABCD* across the *y*-axis.

6. Reflect square *ABCD* across the *x*-axis.

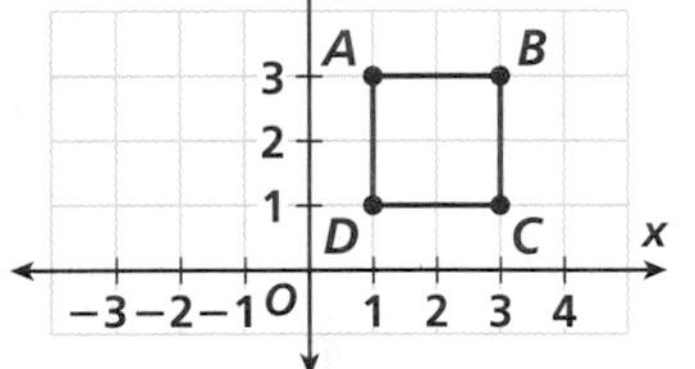

7. Tell what word is spelled when the cross-stitch piece at right is reflected across the *x*-axis.

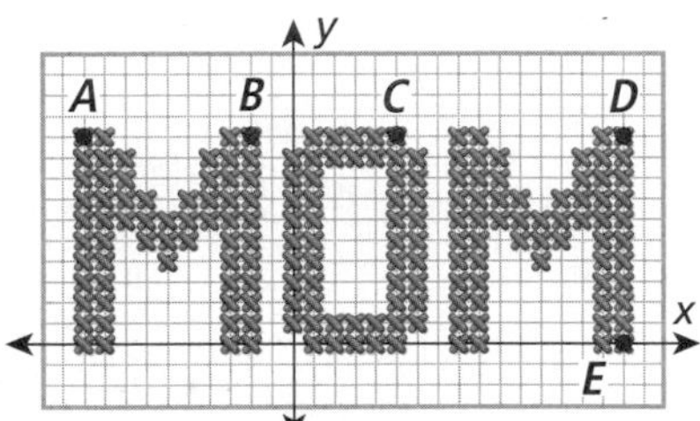

8. ***WRITE ABOUT IT*** Explain how the coordinates of an image change when the image is reflected across the *y*-axis.

9. ***CHALLENGE*** Reflect square *ABCD* from Exercises 5 and 6 across $\overline{BD}$. Give the new coordinates of the vertices.

Spiral Review

Write each number in scientific notation. (Lesson 3-5)

10. 2,345

11. 100

12. 56,700

13. **EOG PREP** The probability of a spinner with three sections landing on green is 56%, and the probability of it landing on yellow is 0.24. What is the probability of it landing on blue if blue is the other color? (Lesson 11-3)

A 20% B 24% C 30% D 44%

12-5 Graphing Rotations

Learn to use rotations to change positions of figures on a coordinate plane.

Swimmers on synchronized swimming teams perform routines that involve making designs in the water with their bodies. They can rotate a design by changing their positions in the water.

You can rotate a figure about the origin or another point on a coordinate plane.

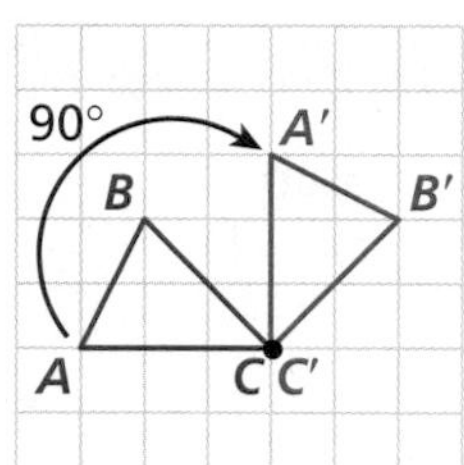

90° rotation

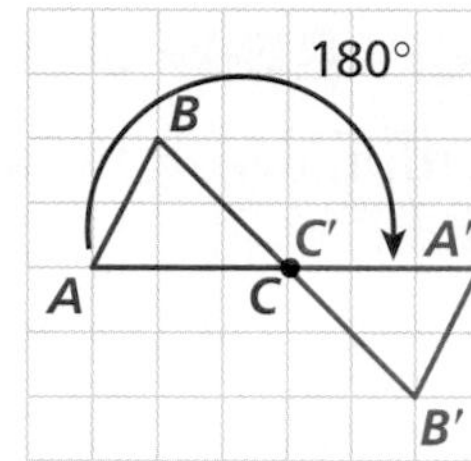

180° rotation

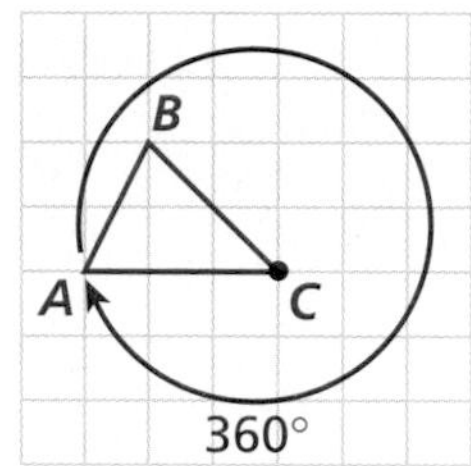

360° rotation

EXAMPLE 1 Rotating Figures on a Coordinate Plane

Give the coordinates of the vertices of the figure after the given rotation.

Remember!

A rotation is the movement of a figure about a point. Rotating a figure "about the origin" means that the origin is the center of rotation.

Rotate trapezoid *ABCD* clockwise 90° about the origin.

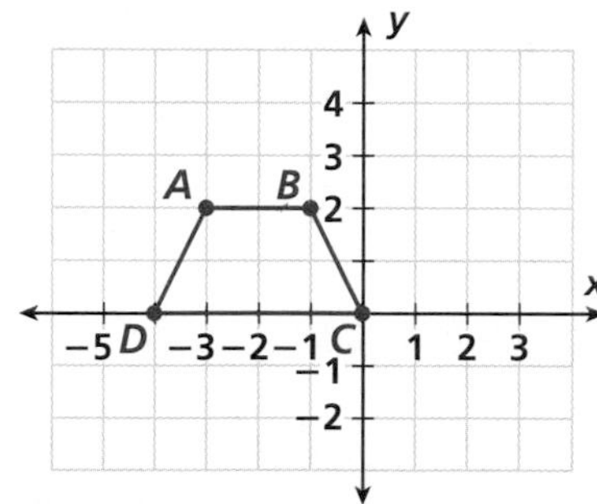

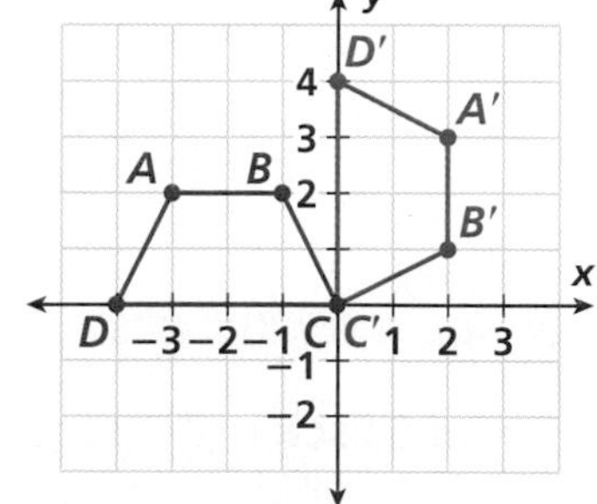

The new x-coordinates are the old y-coordinates.

The new y-coordinates are the opposites of the old x-coordinates.

ABCD		*A′B′C′D′*
A(–3, 2)	→	*A′*(2, 3)
B(–1, 2)	→	*B′*(2, 1)
C(0, 0)	→	*C′*(0, 0)
D(–4, 0)	→	*D′*(0, 4)

EXAMPLE 2 *Sports Application*

A synchronized swimming team forms this figure. The swimmers rotate the figure without changing its size or shape.

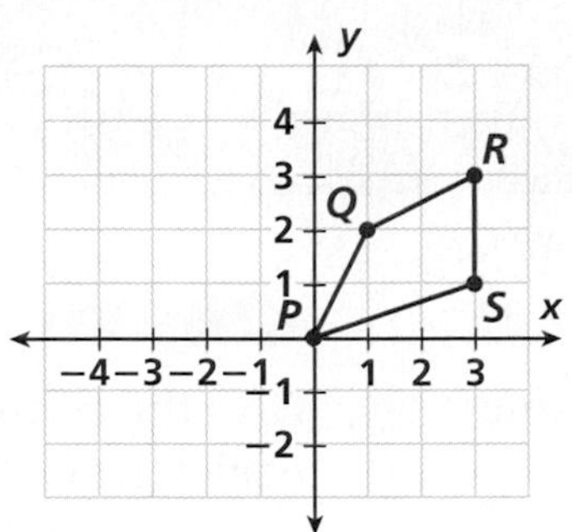

A **Give the coordinates of the vertices of the figure after a clockwise rotation of 90° about the origin.**

The new x-coordinates are the old y-coordinates.

The new y-coordinates are the opposites of the old x-coordinates.

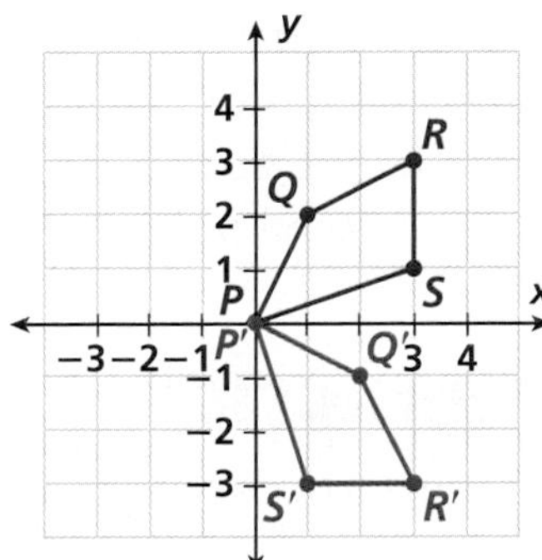

PQRS		*P'Q'R'S'*
P(0, 0)	→	*P'*(0, 0)
Q(1, 2)	→	*Q'*(2, −1)
R(3, 3)	→	*R'*(3, −3)
S(3, 1)	→	*S'*(1, −3)

B **Give the coordinates of the vertices of the figure after a counterclockwise rotation of 180° about the origin.**

The new x-coordinates are the opposites of the old x-coordinates.

The new y-coordinates are the opposites of the old y-coordinates.

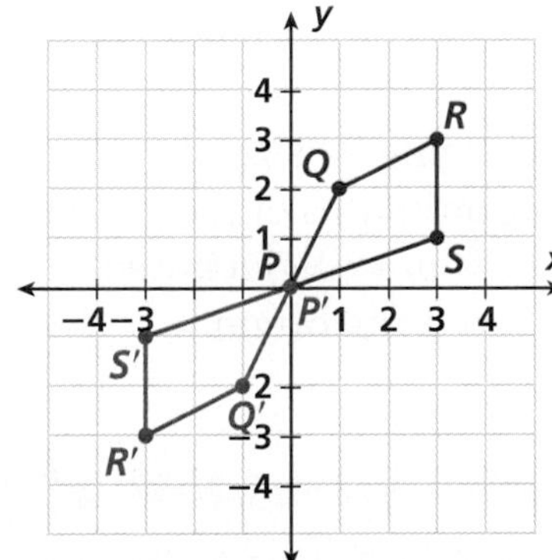

PQRS		*P'Q'R'S'*
P(0, 0)	→	*P'*(0, 0)
Q(1, 2)	→	*Q'*(−1, −2)
R(3, 3)	→	*R'*(−3, −3)
S(3, 1)	→	*S'*(−3, −1)

Think and Discuss

1. **Tell** how the *x*- and *y*-coordinates would change if you rotated figure *ABCD* from Example 1 counterclockwise 90° about the origin.
2. **Tell** what the coordinates would be if you rotated figure *P'Q'R'S'* from Example 2A clockwise 90° about the origin.

12-5 Exercises

FOR EOG PRACTICE

see page 687

internet connect

Homework Help Online
go.hrw.com Keyword: MR4 12-5

3.03, 3.04

GUIDED PRACTICE

See Example 1

Give the coordinates of the vertices of each figure after the given rotation.

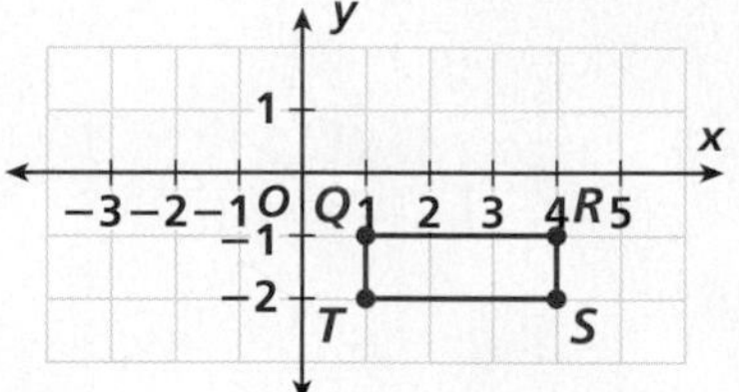

1. Rotate rectangle *QRST* clockwise 180° about the origin.
2. Rotate rectangle *QRST* counterclockwise 90° about the origin.

See Example 2

3. Sean is using triangular tiles that look like the triangle on the graph. If he rotates the tiles, he can create a pattern. Give the coordinates of the vertices of triangle *ABC* after a clockwise rotation of 90° about point *A*.

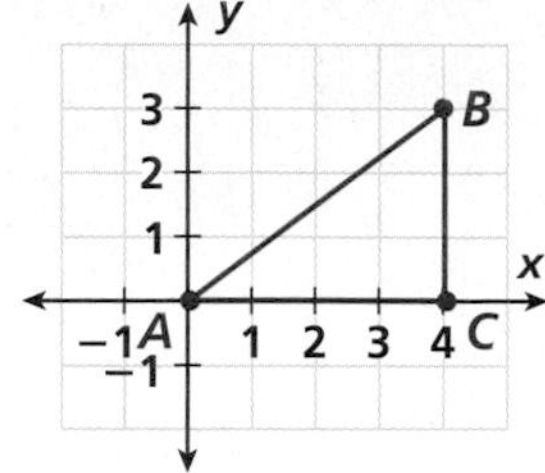

INDEPENDENT PRACTICE

See Example 1

Give the coordinates of the vertices of each figure after the given rotation.

4. Rotate rectangle *LMNP* counterclockwise 180° about the origin.
5. Rotate rectangle *LMNP* clockwise 90° about the origin.

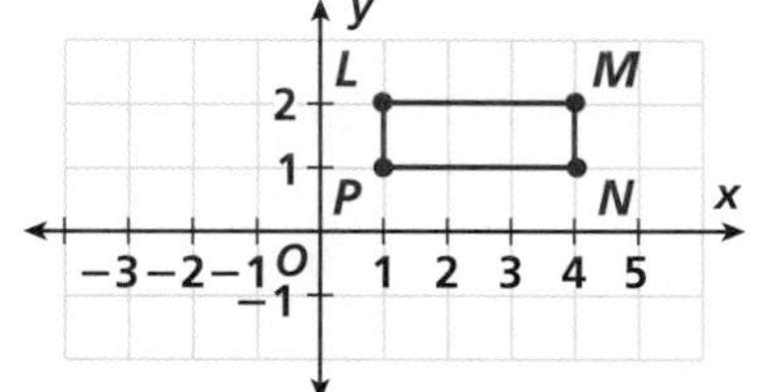

See Example 2

6. A group of sky divers forms a figure like this in the air. The divers rotate the figure without changing its size or shape. Give the coordinates of the vertices of the figure after a clockwise rotation of 180° about the origin.

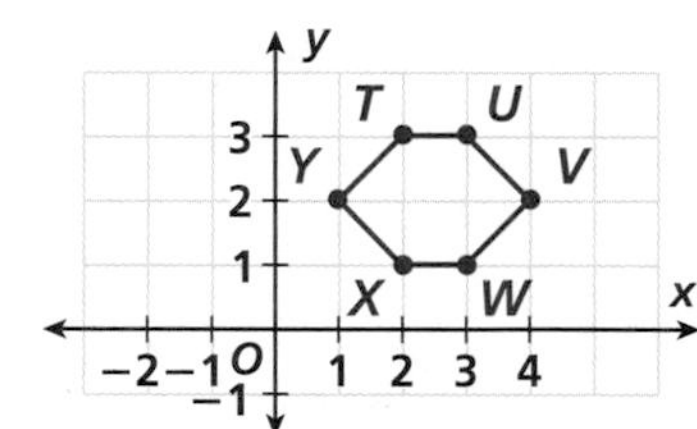

PRACTICE AND PROBLEM SOLVING

Use the graph for Exercises 7–10. Give the coordinates of the vertices of each figure after the given transformation.

7. Rotate square *ABCD* clockwise 90° about the origin.
8. Rotate square *ABCD* counterclockwise 270° about the origin.
9. Reflect triangle *BCE* across the *x*-axis.
10. Translate trapezoid *ABED* up 2 units and left 5 units.

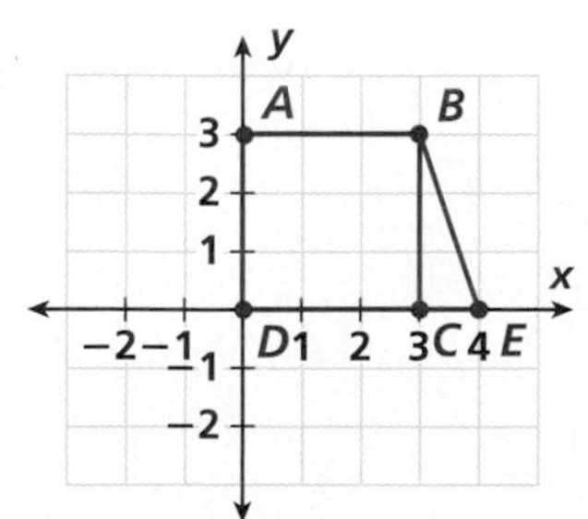

Art LINK

Some works of art, like *A Fish Story,* by Gustave Verbeek, are optical illusions. When this picture is rotated 180°, a different image appears.

11. ***ART*** Use the points labeled on the graph at right to rotate the design 180° clockwise about the origin. Tell what word is spelled after this rotation.

12. Graph the points $A(2, 6)$, $B(-3, 6)$, $C(-3, 1)$, and $D(2, 1)$. Then join the points to form rectangle $ABCD$.
    a. Translate the rectangle 3 units up. What are the coordinates of the vertices of $A'B'C'D'$?
    b. Rotate $A'B'C'D'$ 360° about the origin. Now what are the coordinates of its vertices?

Use the graph of the rhombus for Exercises 13–16.

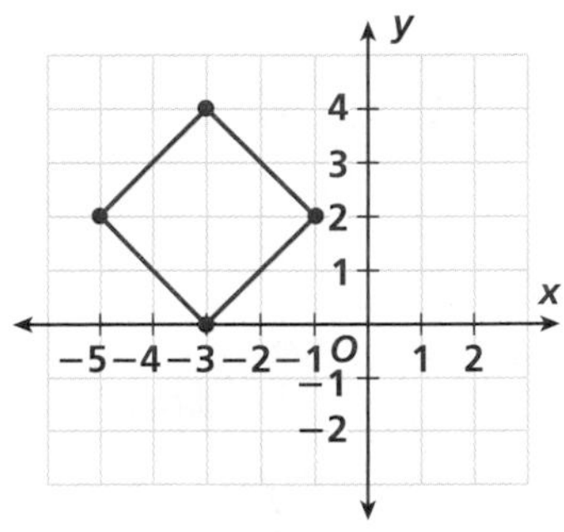

13. Lara translated the rhombus 2 units right and then rotated it clockwise 180° about the origin. Mike rotated the original rhombus clockwise 180° about the origin and then translated it 2 units right. Are the new coordinates of Lara's rhombus the same as the new coordinates of Mike's rhombus? Explain.

14. ***WRITE A PROBLEM*** Write a problem about rotation using the rhombus. Solve your problem.

15. ***WRITE ABOUT IT*** Describe a translation, a reflection, and a rotation that can be performed on the rhombus.

16. ***CHALLENGE*** Describe two different transformations that would not change the coordinates of the rhombus.

Spiral Review

Order each set of numbers from least to greatest. (Lesson 3-1)

17. 1.2, 0.445, 1.06, 0.9 **18.** 2.45, 2.678, 2.007, 2.02 **19.** 7.99, 7.999, 7.9, 7.09

20. **EOG PREP** Choose the coordinates that are the farthest to the right of the origin on a coordinate plane. (Lesson 9-3)

A (0, 12) B (−19, 7) C (7, 0) D (4, 15)

21. **EOG PREP** How many different 4-digit numbers can be made using the digits 5, 3, 2, and 7? (Lesson 11-4)

A 4 B 12 C 16 D 24

12-6 Stretching and Shrinking

Learn to visualize and show the results of stretching or shrinking a figure.

By increasing or decreasing the size of one dimension of a figure, the look of a design can change.

A funhouse mirror distorts your reflection because of its curved surfaces. The parts that curve inward stretch your image, and the parts that curve outward shrink your image.

EXAMPLE 1 Stretching Figures

Write the dimensions of each part of the figure. Stretch the figure as stated, and give the new dimensions of each part.

Remember!

Vertical means "straight up or down." *Horizontal* means "straight across," such as from left to right.

A **Increase the horizontal dimensions of the face, eyes, and mouth by a factor of 3.**

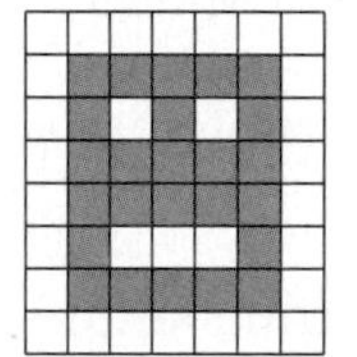

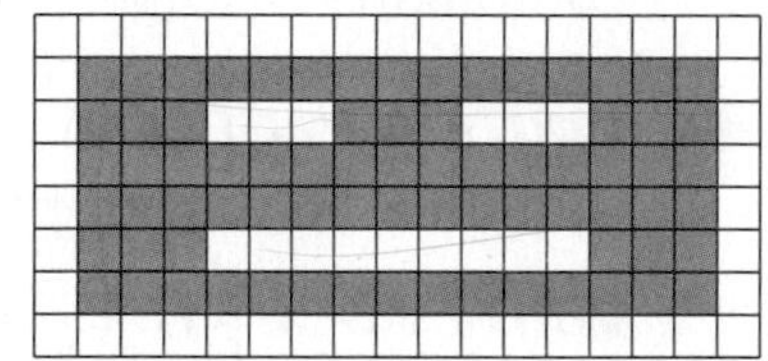

	Original Dimensions	New Dimensions
Face	Vertical 6 Horizontal 5	Vertical 6 Horizontal 15
Eyes	Vertical 1 Horizontal 1	Vertical 1 Horizontal 3
Mouth	Vertical 1 Horizontal 3	Vertical 1 Horizontal 9

B **Increase the vertical dimensions of the face, eyes, and mouth by a factor of 2.**

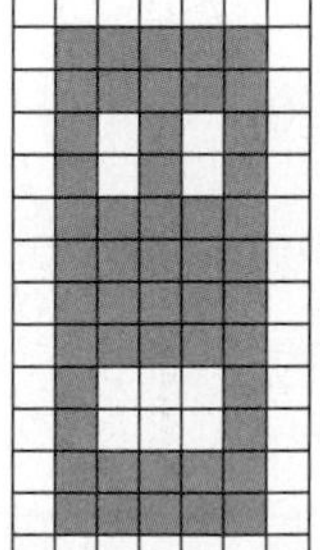

	Original Dimensions	New Dimensions
Face	Vertical 6 Horizontal 5	Vertical 12 Horizontal 5
Eyes	Vertical 1 Horizontal 1	Vertical 2 Horizontal 1
Mouth	Vertical 1 Horizontal 3	Vertical 2 Horizontal 3

EXAMPLE 2 Shrinking Figures

Write the dimensions of each figure. Shrink the figure as stated and give the new dimensions.

A Decrease the vertical dimensions by multiplying by $\frac{1}{3}$.

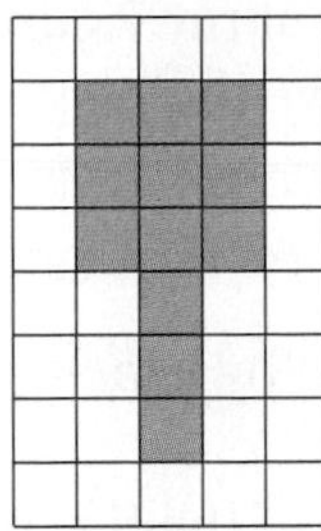

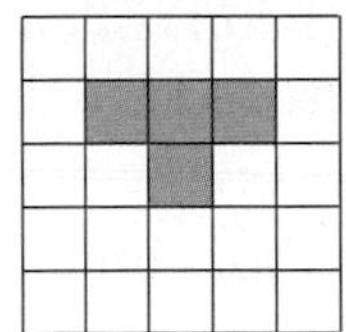

Original Dimensions		New Dimensions	
Vertical	6	Vertical	2
Horizontal	3	Horizontal	3

B Decrease the horizontal dimensions by multiplying by $\frac{1}{2}$.

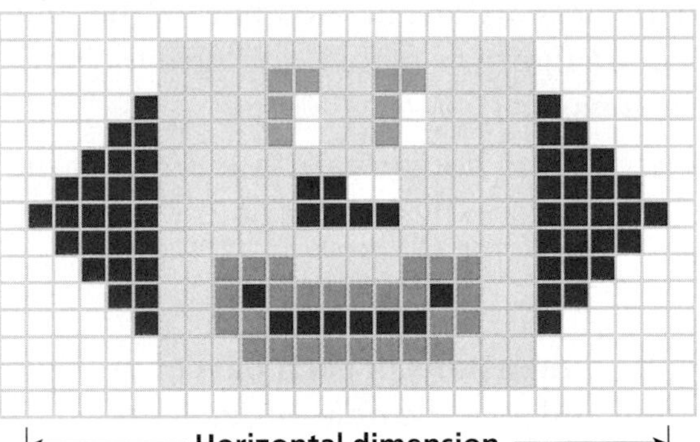

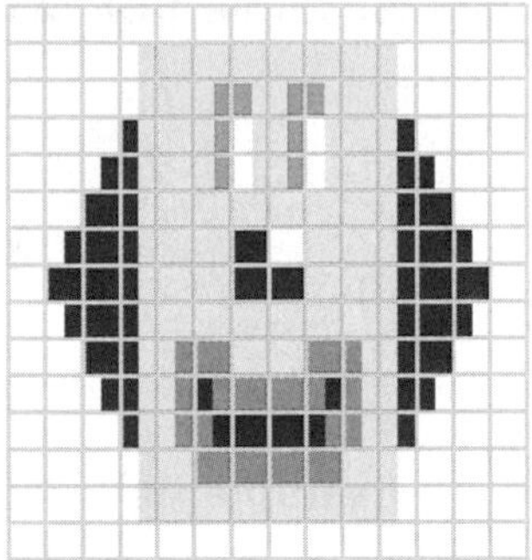

Original Dimensions		New Dimensions	
Vertical	13	Vertical	13
Horizontal	24	Horizontal	12

Think and Discuss

1. **Describe** what happens to a figure when the horizontal dimension is stretched.
2. **Tell** whether a figure whose vertical dimension has been shrunk by being multiplied by $\frac{1}{3}$ is similar to the original figure. Explain.

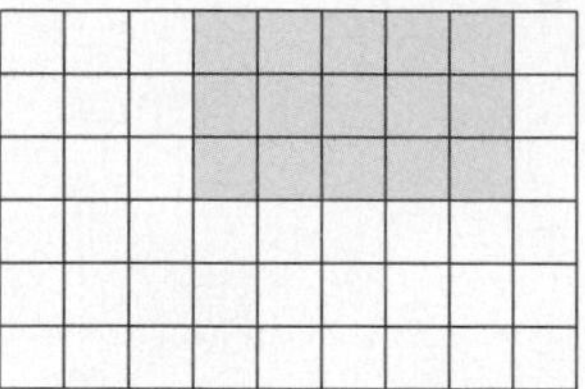

FOR EOG PRACTICE

see page 687

internet connect
Homework Help Online
go.hrw.com Keyword: MR4 12-6

1.04b, 3.03, 3.04

GUIDED PRACTICE

See Example 1 **Write the dimensions of each part of the figure. Stretch the figure as stated, and give the new dimensions.**

1. Increase the horizontal dimension by a factor of 3.
2. Increase the vertical dimension by a factor of 10.

See Example 2 **Write the dimensions of each part of the figure above. Shrink the figure as stated, and give the new dimensions.**

3. Decrease the horizontal dimensions by multiplying by $\frac{1}{5}$.
4. Decrease the vertical dimensions by multiplying by $\frac{1}{3}$.

INDEPENDENT PRACTICE

See Example 1 **Write the dimensions of each part of the figure. Stretch the figure as stated, and give the new dimensions.**

5. Increase the horizontal dimensions of the shaded region by a factor of 5.

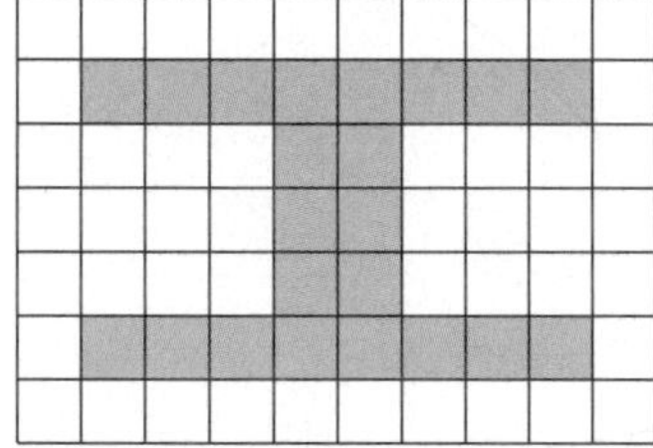

See Example 2 **Write the dimensions of each part of the figure above. Shrink the figure as stated, and give the new dimensions.**

6. Decrease the vertical dimensions of the shaded region by multiplying by $\frac{1}{3}$.
7. Decrease the horizontal dimensions of the shaded region by multiplying by $\frac{1}{2}$.

PRACTICE AND PROBLEM SOLVING

8. When Craig put his basketball jersey in the dryer, it looked like the picture.
   a. What was the length of the jersey before he put it in the dryer?
   b. When Craig took the jersey out of the dryer, he noticed that it had shrunk in length only and was now only $\frac{5}{6}$ as long as it was. Find how many inches the jersey shrank.

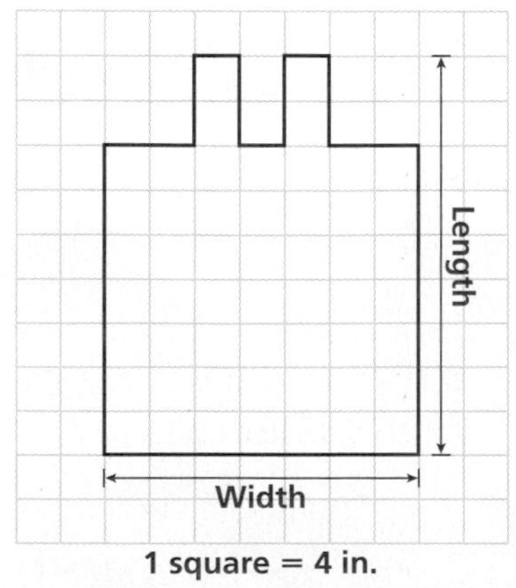

1 square = 4 in.

Social Studies LINK

The United Nations, or U.N., is an international organization established in 1945. The U.N. now includes more than 180 nations. Samoa and Tonga are Pacific island nations that are members of the U.N. Samoa joined on December 15, 1976, and Tonga became a member on September 14, 1999.

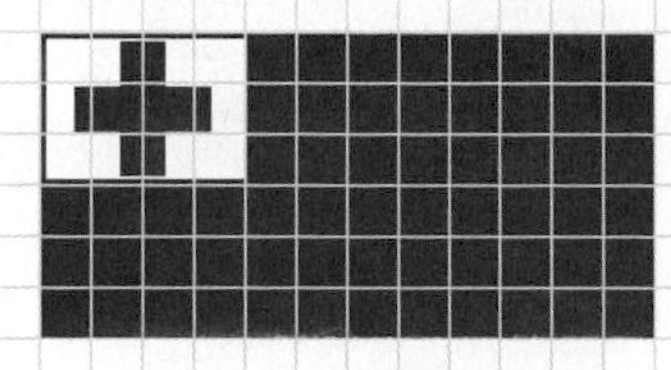

Tonga

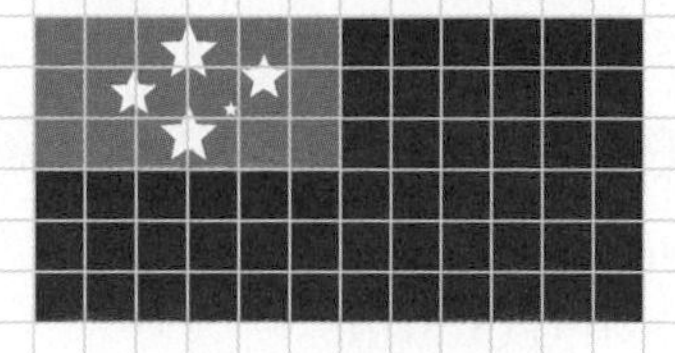

Samoa

9. **a.** Draw the flag of Tonga, and increase both the horizontal and vertical dimensions by a factor of 4.
   b. What is the perimeter of the cross on the Tongan flag before and after the change?

10. The horizontal dimension of the Tongan flag was increased by a factor of 4, and the vertical dimension was increased by a factor of 2. Is the cross on the new flag similar to the cross in the original flag? Explain.

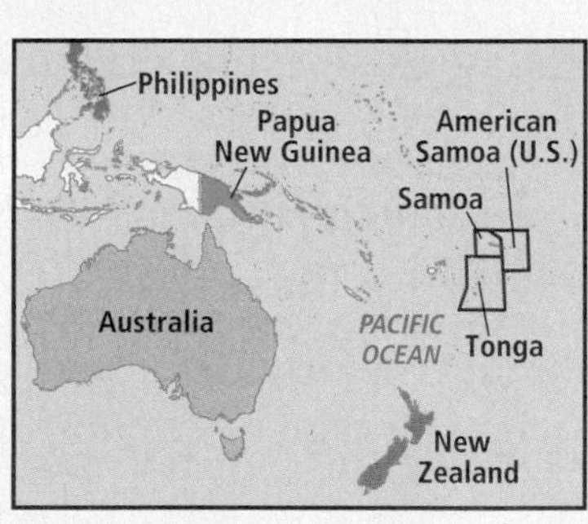

11. ***CHOOSE A STRATEGY*** Which of the following would give the Samoan flag an area of 36 square units?
   A vertical increase by a factor of 2
   B horizontal decrease by a factor of 3
   C vertical decrease by a factor of 2

12. ***WRITE ABOUT IT*** Explain which dimensions of the Tongan flag you could change so that the new flag from Exercise 10 is similar to the original flag.

13. ***CHALLENGE*** A Samoan flag was made with a perimeter of 180 units. By what whole-number factor were the horizontal and vertical dimensions increased?

Spiral Review

Determine whether each number is divisible by 2, 3, or 5. (Lesson 4-1)

14. 155 **15.** 14 **16.** 99 **17.** 2,345

18. **EOG PREP** A circle has a diameter of 12 in. What is the circumference of the circle rounded to the nearest hundredth? (Use 3.14 for π.) (Lesson 10-5)

   A 15.14 in **B** 37.68 in **C** 113.04 in **D** 150.72 in

Problem Solving on Location

NORTH CAROLINA

Cherokee Living Museum

The Oconaluftee Indian Village, near Cherokee, North Carolina, offers a glimpse of Cherokee life and culture more than 300 years ago. On a walking tour, visitors can view demonstrations by skilled weavers, sculptors, and potters. Guides also give detailed information about important aspects of the village, including the ceremonial grounds and the Council House.

1. Part of a design used to weave a basket has been drawn on the coordinate grid. Reflect the basket design *ABCD* across the *y*-axis. Give the coordinates of the new points.
2. Rotate the original design counterclockwise 90° about the origin, and give the coordinates of the new points.
3. Translate the original design 3 units up and 4 units right. Give the coordinates of the new points.
4. Rotate the original design clockwise 90° about the origin, and give the coordinates of the new points.

North Carolina Barbecue

North Carolina barbecue is legendary. Like most legends, it involves a dispute. In this case, the dispute involves the eastern style and the western style of barbecuing. There are strong opinions within the state about whether the whole hog gets barbecued, or just the shoulder. There are also disagreements about whether the sauce should include tomatoes or not. There seems to be no end to the dispute or to the legend involving barbecue, North Carolina style.

1. One recipe for barbecue sauce calls for 5 times as much red pepper as cayenne pepper. Write an equation for this function, and tell what each variable represents.

For 2–3, use the table.

The table shows the relationship between the number of barbecue plates sold p and the amount of money paid for the plates m.

p	1	2	3	4
m	\$3.75	\$7.50	\$11.25	\$15.00

2. Write an equation for the function that gives the values in this table. Use the equation to find the amount of money paid for 15 plates of barbecue.

3. How many plates will have to be sold for \$225.00 to be made?

4. This year, Rick sold 5 less than 8 times the number of barbecue plates he sold last year. Write an equation relating the sales for this year to the sales for last year. Use y for this year's sales and x for last year's.

MATH-ABLES

Logic Puzzle

Each day from Monday through Friday, Mayuri, Naomi, Brett, Thomas, and Angela took turns picking a restaurant for lunch. They ate at restaurants that serve either Chinese food, hamburgers, pizza, seafood, or tacos. Use the clues below to determine which student picked the restaurant on each day and which restaurant the student picked.

1. Angela skipped Friday's lunch to play in a basketball game.
2. Brett picked the restaurant on Wednesday.
3. The students ate tacos on Friday.
4. Naomi is allergic to seafood and volunteered to pick the first restaurant.
5. Thomas picked a hamburger restaurant on the day before another student chose a pizza restaurant.

You can use a chart like the one below to help you solve this puzzle. Place an *O* in a square for something that is true and an *X* in a square for something that cannot be true. Remember that when you place an *O* in a square, you can put *X*'s in the rest of the squares in that row and column. The information from the first two clues has been entered for you.

		Student					Restaurant				
		Mayuri	Naomi	Brett	Thomas	Angela	Seafood	Pizza	Hamburger	Chinese	Tacos
Day	Monday			X							
	Tuesday			X							
	Wednesday	X	X	O	X	X					
	Thursday			X							
	Friday			X		X					
Restaurant	Seafood										
	Pizza										
	Hamburgers										
	Chinese										
	Tacos										

internet connect

Go to *go.hrw.com* for a blank logic puzzle chart.
KEYWORD: MR4 Game12

Use Graphs to Estimate Solutions

internet connect
Lab Resources Online
go.hrw.com
KEYWORD: MR4 TechLab12

You can use graphs to estimate solutions to equations.

Activity

1 Use a graphing calculator to estimate the solution to the equation $x - 3 = 4$.

a. Press Y= and enter $x - 3$ for **Y1** and 4 for **Y2.** These are the left and right sides of the equation $x - 3 = 4$.
Press ZOOM 6 to select **ZStandard.** This sets the view of the x-axis and y-axis from –10 to 10.

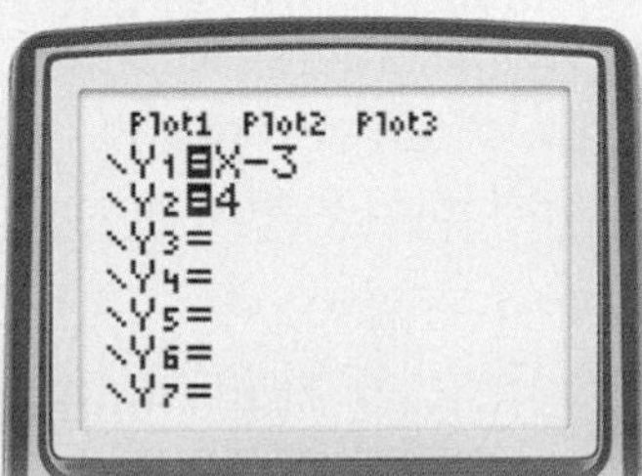

b. There is one graph of a line representing $y = x - 3$ and a second graph of a line representing $y = 4$.

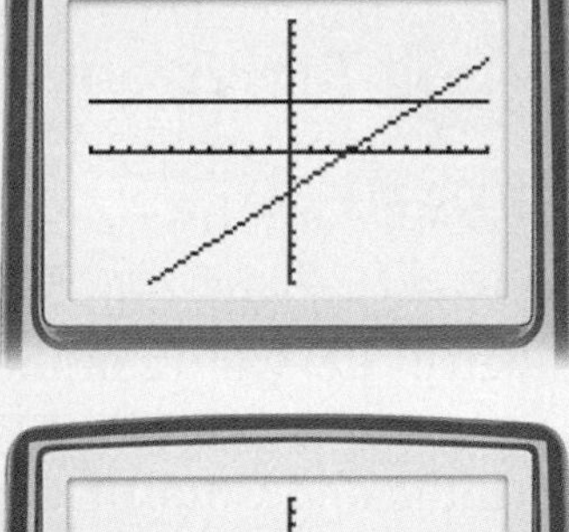

c. The expression $x - 3$ and the number 4 have equal values at the point where their graphs intersect.

To find the coordinates of the point of intersection, press 2nd CALC TRACE 5. A flashing cursor appears. Use the arrow keys to move the cursor near the intersection and press ENTER. Do this for both graphs.

At the bottom of the window, a guess is shown for the value. Press ENTER again to see the coordinates of the point.

The point of intersection is (7, 4).

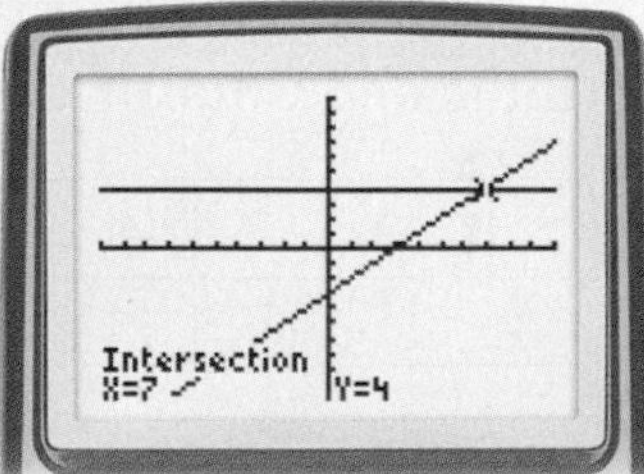

Think and Discuss

1. Tell how you would solve $2x + 5 = 6x - 3$ by using a graphing calculator.

Try This

Use a graphing calculator to estimate the solution to each equation.

1. $x - 5 = 2$

2. $x + 3 = -3$

3. $3\frac{1}{2}x = 7\frac{1}{4}$

4. $x - 1.75 = 6.35$

Chapter 12

Study Guide and Review

Vocabulary

Complete the sentences below with vocabulary words from the list above. Words may be used more than once.

1. For the equation $y = 3x$, the ___?___ is 12 when the ___?___ is 4.
2. When you graph the ordered pairs of a function and a straight line is formed, the equation of the function is called a ___?___.
3. A rule that relates two quantities so that each value of x corresponds with exactly one value of y is called a ___?___.

12-1 Tables and Functions (pp. 598–601)

EXAMPLE

■ **Write an equation for a function that gives the values in the table. Use the equation to find the value of y for the indicated value of x.**

x	2	3	4	5	6	12
y	5	8	11	14	17	

y is 3 times x minus 1. *Compare x and y to find a pattern.*

$y = 3x - 1$ *Use the pattern to write an equation.*

$y = 3(12) - 1$ *Substitute 12 for x.*

$y = 36 - 1$

$y = 35$ *Use your function rule to find y when x = 12.*

When x is 12, y is 35.

EXERCISES

Write an equation for a function that gives the values in each table. Use the equation to find the value of y for each indicated value of x.

4.

x	2	3	4	5	6	8
y	6	8	10	12	14	

5.

x	20	18	16	14	12	6
y	11	10	9	8	7	

6.

x	1	3	5	7	9	11
y	3	13	23	33	43	

Write an equation to describe the function. Tell what each variable you use represents.

7. The length of a rectangle is 4 times its width.

12-2 Graphing Functions (pp. 604–607)

EXAMPLE

- **Graph the function described by the equation $y = 3x + 4$.**

Make a function table.

x	$3x + 4$	y
−2	3(−2) + 4	−2
−1	3(−1) + 4	1
0	3(0) + 4	4

Write the solutions as ordered pairs.

(x, y)

$(-2, -2)$

$(-1, 1)$

$(0, 4)$

Graph the ordered pairs on a coordinate plane.

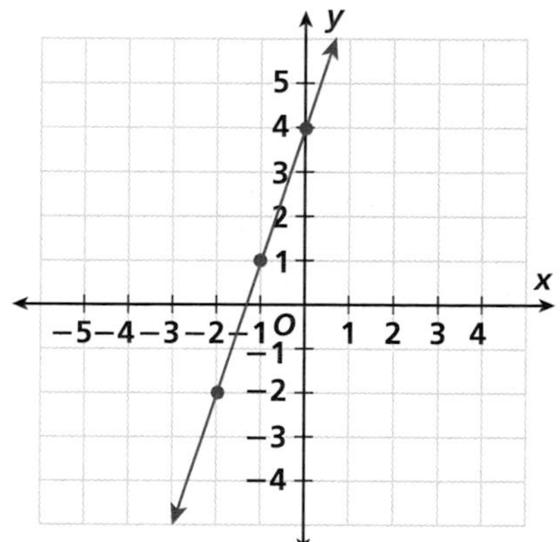

EXERCISES

Use the given x-values to write solutions of each equation as ordered pairs.

8. $y = 2x - 5$ for $x = 1, 2, 3, 4$
9. $y = x + 7$ for $x = 1, 2, 3, 4$

Determine whether each ordered pair is a solution to the given equation.

10. $(3, 12)$; $y = 5x - 3$
11. $(6, 14)$; $y = 2x + 3$
12. Use the graph of the linear function to find the value of y when x is 2.

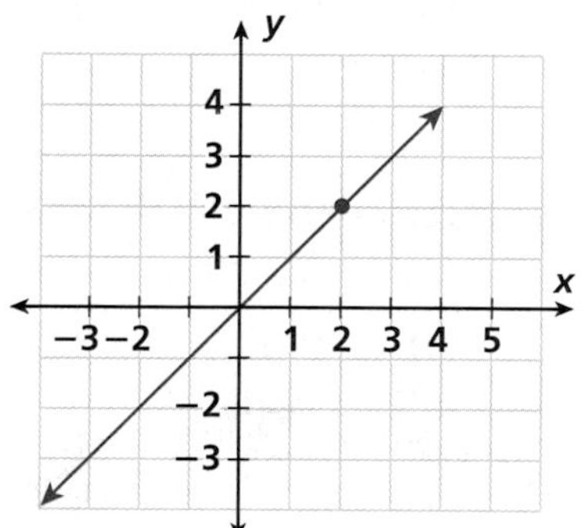

13. Graph the function described by the equation $y = 2x + 1$.

12-3 Graphing Translations (pp. 610–612)

EXAMPLE

- **Give the coordinates of the figure after the given translation.**

Translate triangle RST 3 units left and 1 unit up.

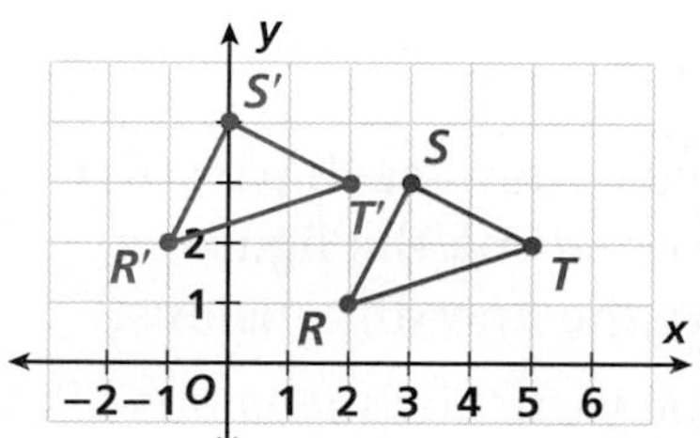

$R(2, 1) \rightarrow R'(2 - 3, 1 + 1) \rightarrow R'(-1, 2)$

$S(3, 3) \rightarrow S'(3 - 3, 3 + 1) \rightarrow S'(0, 4)$

$T(5, 2) \rightarrow T'(5 - 3, 2 + 1) \rightarrow T'(2, 3)$

EXERCISES

Give the coordinates of the figure after the given translation.

14. Translate square $ABCD$ 3 units right and 2 units down.
15. Translate triangle FGH 1 unit left and 4 units up.
16. Translate triangle FGH 2 units right and 1 unit down.

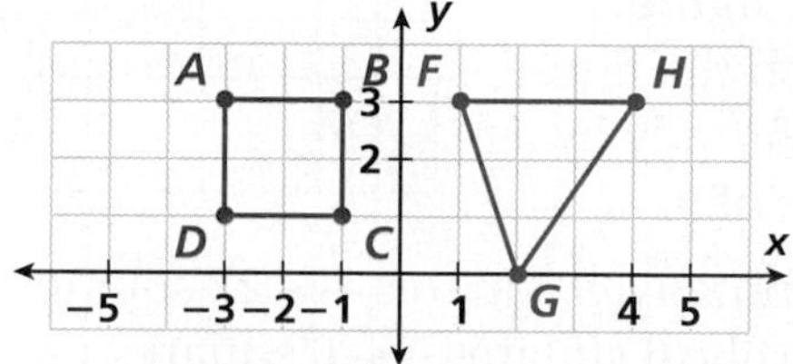

Study Guide and Review

12-4 Graphing Reflections (pp. 613–615)

EXAMPLE

- **Give the coordinates of the vertices of the figure after the given reflection.**

  Reflect figure *LMNO* across the *y*-axis.

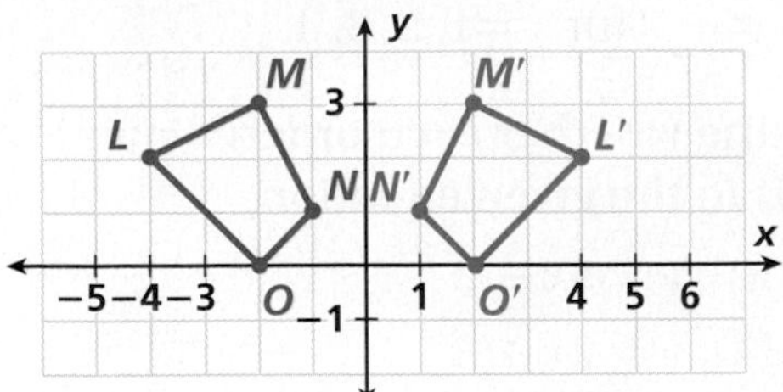

$L(-4, 2) \rightarrow L'(4, 2)$ $N(-1, 1) \rightarrow N'(1, 1)$
$M(-2, 3) \rightarrow M'(2, 3)$ $O(-2, 0) \rightarrow O'(2, 0)$

EXERCISES

Give the coordinates of the vertices of the figure after the given reflection.

17. Reflect parallelogram *ABCD* across the *y*-axis.

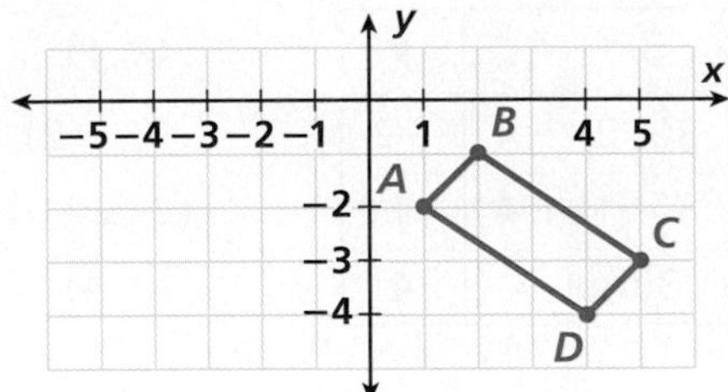

12-5 Graphing Rotations (pp. 616–619)

EXAMPLE

- **Give the coordinates of the figure after the given rotation.**

  Rotate triangle *ABC* clockwise 90° about the origin.

  $A(2, 1) \rightarrow A'(1, -2)$
  $B(2, 3) \rightarrow B'(3, -2)$
  $C(4, 1) \rightarrow C'(1, -4)$

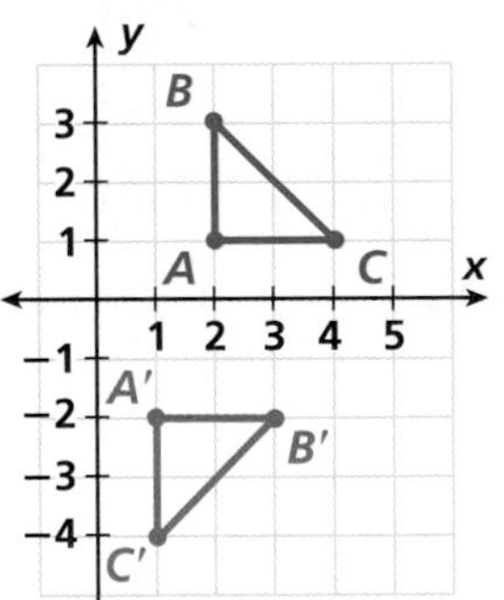

EXERCISES

Give the coordinates of the figure after the given rotation.

18. Rotate parallelogram *ABCD* clockwise 90° about the origin.

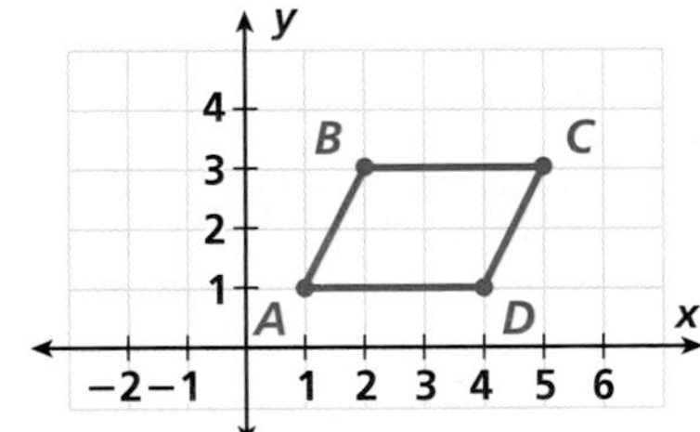

12-6 Stretching and Shrinking (pp. 620–623)

EXAMPLE

- **Write the dimensions of the figure. Stretch the figure horizontally by a factor of 3.**

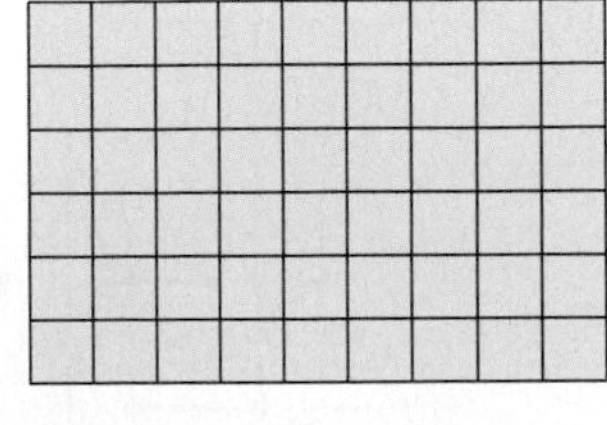

horizontally: 9 squares → 27 squares
vertically: 6 squares → 6 squares

EXERCISES

Write the dimensions of each part of the figure. Stretch or shrink the figure as stated, and give the new dimensions.

19. Increase the vertical dimensions of the figure in the example by a factor of 2.

Study Guide and Review

Chapter Test

Chapter 12

Write an equation for a function that gives the values in each table. Use the equation to find the value of y for each indicated value of x.

1.

x	2	3	4	5	6	7
y	■	8	11	14	17	20

2.

x	1	2	3	4	5	9
y	8	10	12	14	16	■

Write an equation to describe the function. Tell what each variable you use represents.

3. The number of buttons on the jacket is 4 more than the number of zippers.
4. The length of a parallelogram is 2 in. more than twice the height.
5. The number of cards is 6 less than the number of envelopes.
6. The width of the rectangle is 4 cm less than the length.

Use the given x-values to write solutions of each equation as ordered pairs. Then graph the equation.

7. $y = 5x - 3$ for $x = 1, 2, 3, 4$

8. $y = 2x - 3$ for $x = 0, 1, 2, 3$

Determine whether each ordered pair is a solution of the given equation.

9. $(2, 5)$; $y = 3x - 1$

10. $(0, 6)$; $y = 6x$

11. $(-3, -5)$; $y = 2x + 1$

Use the graph of the linear function to find the value of y for each indicated value of x.

12. $x = 1$

13. $x = -1$

14. $x = 3$

15. $x = -3$

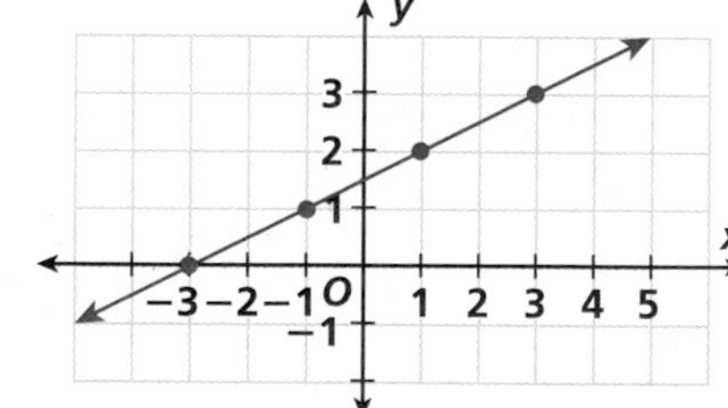

Give the coordinates of the vertices of each figure after the given transformation.

16. Translate triangle ABC 3 units right and 2 units down.
17. Reflect triangle ABC across the y-axis.
18. Rotate triangle ABC clockwise 90° about the origin.

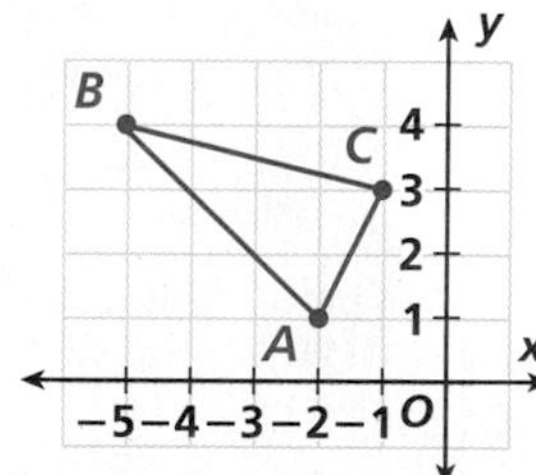

Write the dimensions of each part of the figure. Stretch or shrink the figure as stated, and give the new dimensions.

19. Increase the horizontal dimensions by a factor of 2.
20. Decrease the vertical dimensions by multiplying by $\frac{1}{3}$.

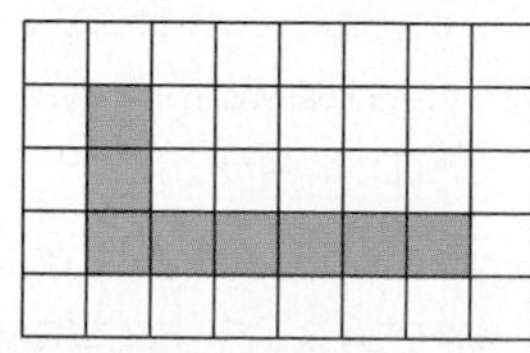

Chapter Test

Chapter 12

Performance Assessment

Show What You Know

Create a portfolio of your work from this chapter. Complete this page and include it with your four best pieces of work from Chapter 12. Choose from your homework or lab assignments, mid-chapter quiz, or any journal entries you have done. Put them together using any design you want. Make your portfolio represent what you consider your best work.

Short Response

1. Draw quadrilateral *ABCD* on a coordinate plane with *A*(1,2), *B*(1,5), *C*(5,5), and *D*(5,2). What is the area of the quadrilateral? Describe two different methods you could use to find the answer.

2. On a coordinate plane, draw quadrilateral *EFGH* with *E*(1,2), *F*(2,5), *G*(4,4), and *H*(3,1). Reflect the figure across the x-axis. Explain how to find the coordinates of the vertices of the figure after the reflection.

Extended Problem Solving

3. A store sold 52 folk-art masks in September for a total price of $624. In October, sales totaled $492 for 41 masks. In November, sales totaled $456 for 38 masks. All of the masks cost the same.
   a. Make a table to display the data, and then graph the data. Is the function linear?
   b. Write an equation to represent the function. Indicate what each variable represents.
   c. In December, the store sold 67 masks. What was the total sales amount in December?

Performance Assessment

internet connect
EOG Test Practice Online
go.hrw.com Keyword: MR4 TestPrep

Getting Ready for EOG

Chapter 12

Cumulative Assessment, Chapters 1–12

1. Which ordered pair is a solution of the equation $y = 3x + 2$?

A (1, 6) C (2, 8)
B (5, 1) D (−2, 4)

2. What is 15% of 130?

A 19.5 C 1.35
B 13 D 1.95

TEST TAKING TIP!
To reflect a figure across the *y*-axis, write the opposites of the *x*-coordinates. The *y*-coordinates do not change.

3. A figure has the following vertices: $A(1, -2)$, $B(2, -5)$, $C(6, -3)$. What are the coordinates after the figure is reflected across the *y*-axis?

A $A'(1, 2)$, $B'(2, 5)$, $C'(6, 3)$
B $A'(-2, -1)$, $B'(-5, -2)$, $C'(-3, -6)$
C $A'(-1, 2)$, $B'(-2, 5)$, $C'(-6, 3)$
D $A'(-1, -2)$, $B'(-2, -5)$, $C'(-6, -3)$

4. What kind of transformation is shown?

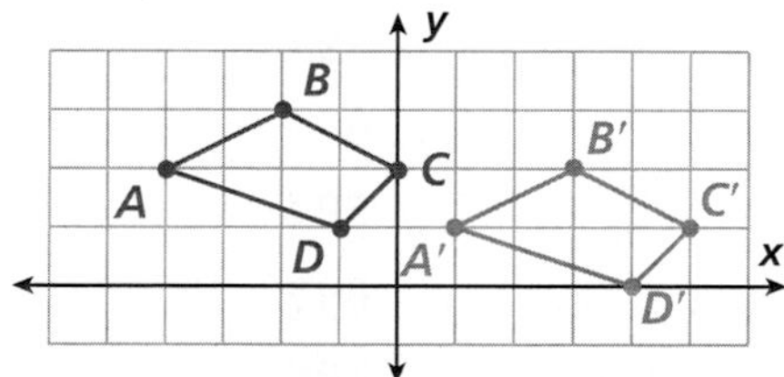

A Translation C Reflection
B Rotation D Tessellation

5. What is the probability of rolling a number less than 3 on a fair number cube?

A $\frac{1}{2}$ C $\frac{2}{3}$
B $\frac{1}{3}$ D $\frac{1}{6}$

6. Multiply $3\frac{5}{6} \cdot 1\frac{2}{3}$.

A $3\frac{5}{9}$ C $6\frac{7}{18}$
B $5\frac{1}{2}$ D $4\frac{7}{9}$

7. Which set is in order from least to greatest?

A 4, 2, −1, 0 C −3, −1, 2, 4
B 0, −1, 3, 5 D −3, −5, 1, 4

8. Which of the following numbers is divisible by 2, 3, 4, and 6, but **not** by 9?

A 216 C 822
B 414 D 912

9. ***SHORT RESPONSE*** A cylinder has a diameter of 6 in. and a height of 8 in. What is the cylinder's volume? Use 3.14 for π, and round to the nearest hundredth. Show your work.

10. ***SHORT RESPONSE*** Explain how to find the coordinates of triangle *ABC* after it is rotated clockwise 90° about the origin.

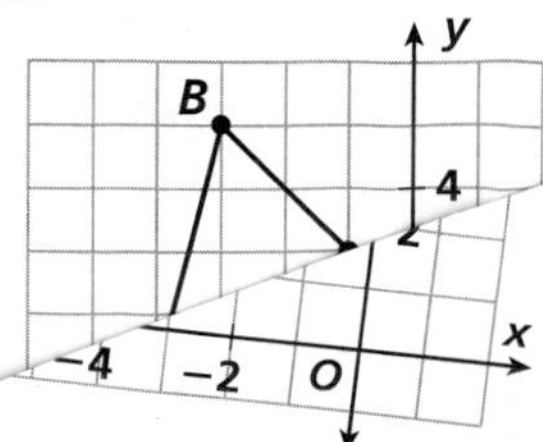

Getting Ready for EOG

Student Handbook

Exponent

Base 2^4

Student Handbook

EOG Practice ▪ Chapter 1

1A Whole Numbers and Exponents

LESSON 1-1

1. Which of the following countries has the *greatest* area?

   A Brazil: 3,286,500 square miles

   B Canada: 3,851,788 square miles

   C Nigeria: 356,669 square miles

   D United States: 3,717,792 square miles

2. Which of the following sets of numbers is ordered from *least* to *greatest*?

   A 783; 772; 1,702

   B 1,308; 10,318; 10,301

   C 1,780; 1,356; 1,345

   D 24,560; 24,615; 24,829

3. Which of the following sets of numbers is ordered from *greatest* to *least*?

   A 5,667; 5,676; 5,700

   B 29,992; 22,922; 22,929

   C 34,903; 32,788; 32,679

   D 86,127; 82,305; 82,497

4. What is eighty billion, seven hundred fourteen million, three hundred twenty-nine thousand, sixteen in standard form?

   A 8,071,432,916

   B 8,714,329,016

   C 80,714,329,016

   D 80,700,143,002,916

5. ***SHORT RESPONSE*** Write the number 9,018,355,109 in expanded form and in word form.

6. ***SHORT RESPONSE*** Explain how to use place value to compare 8,306,351 with 8,315,986. Which number is *less*?

7. ***SHORT RESPONSE*** Plot the numbers 3,102; 438; 507; and 1,908 on a number line. Then write them in order from *greatest* to *least*.

LESSON 1-2

8. What is the *estimate* of 292,801 − 156,127? Round to the hundred thousands place.

   A 100,000 C 137,000

   B 130,000 D 200,000

9. What is the *estimate* of 14,325 + 25,629? Round to the hundreds place.

   A 39,000 C 39,960

   B 39,900 D 40,000

10. What is the *estimate* of 9,210 − 396? Round to the hundreds place.

    A 8,600 C 8,810

    B 8,800 D 9,600

11. What is the approximate value of 23,618 + 37,518? Round to the ten thousands place.

    A 50,000 C 61,000

    B 60,000 D 62,000

12. What is the approximate value of 52,087 − 35,210? Round to the ten thousands place.

    A 10,000 C 17,000

    B 16,880 D 20,000

13. ***SHORT RESPONSE*** Mr. Peterson needs topsoil for his garden. The rectangular garden is 78 inches long and 48 inches wide. A bag of topsoil covers an area of 500 square inches. About how many bags of topsoil should Mr. Peterson buy? Explain how you determined your answer.

14. ***SHORT RESPONSE*** Natalie's family is having a picnic at an amusement park. The park is 153 miles from Natalie's house. If the family drives 54 miles per hour, about how long will it take them to get to the park? Explain how you found your answer.

EOG Practice

EOG Practice ■ Chapter 1

1A Whole Numbers and Exponents

LESSON 1-3

15. What is $2 \times 2 \times 2 \times 2$ written in exponential form?
    A 2^4 C 16
    B 4^2 D $10 + 6$

16. What is $3 \times 3 \times 3 \times 3$ written in exponential form?
    A 3^4 C 81
    B 4^3 D $80 + 1$

17. What is $5 \times 5 \times 5 \times 5 \times 5 \times 5$ written in exponential form?
    A 5^6
    B 6^5
    C 15,625
    D $10{,}000 + 5{,}000 + 600 + 20 + 5$

18. What is $7 \times 7 \times 7 \times 7$ written in exponential form?
    A 7^4
    B 4^7
    C 2,401
    D $2{,}000 + 400 + 1$

19. Which is equal to 729?
    A 3^9 C 9^4
    B $9 \times 9 \times 9$ D 81^9

20. Which is *not* equal to 64?
    A 2^8 C 8^2
    B 4^3 D 64^1

21. What is the value of 4^3?
    A 7 C 64
    B 12 D 81

22. What is the value of 10^6?
    A 60
    B 1,000,000
    C 10,000,000
    D 60,466,176

23. What is the value of 3^6?
    A 9 C 216
    B 18 D 729

24. What is the value of 5^5?
    A 10 C 125
    B 25 D 3,125

25. Which of the following choices is *not* the same as 7^4?
    A $7 \times 7 \times 7 \times 7$
    B 2,401
    C $4 \times 4 \times 4 \times 4 \times 4 \times 4 \times 4$
    D $7^2 \times 7^2$

26. ***SHORT RESPONSE*** Write the expression 16^3 as repeated multiplication. What is the value of this expression? Show your work.

27. ***SHORT RESPONSE*** Write the expression 2^9 as repeated multiplication. What is the value of this expression? Show your work.

28. ***SHORT RESPONSE*** Jerry has 3 children, each of whom has 3 children. Each of Jerry's grandchildren also has 3 children. Draw a diagram to help you write an exponential expression that represents the number of great-grandchildren Jerry has. What is the value of this expression?

29. ***SHORT RESPONSE*** Cassie claims that the value of 2^3 is the same as the value of 3^2 because the product of 2×3 is the same as the product of 3×2. Use repeated multiplication to explain why her reasoning is *not* correct.

30. ***SHORT RESPONSE*** If 2^2 has a value of 4, explain why 3^3 does *not* have a value of 9.

31. ***SHORT RESPONSE*** A student made the mistake of writing $8^3 = 24$. What mistake could the student be making and what could help correct the error?

EOG Practice ■ Chapter 1

1B Using Whole Numbers

LESSON 1-4

1. In the expression $15 + 7 \times 3$, which operation would be performed first?

   A addition
   B subtraction
   C multiplication
   D division

2. In the expression $3 \times 3 + (13 - 5)$, which operation would be performed first?

   A addition
   B subtraction
   C multiplication
   D division

3. In the expression $10 \div 5 \times 2 - 8$, which operation would be performed first?

   A addition
   B subtraction
   C multiplication
   D division

4. What is the value of $4^2 - 12 \div 3 + (7 - 6)$?

   A 1   C 12
   B 5   D 13

5. What is the value of $8 + 3 \times 4^2$?

   A 56   C 176
   B 152   D 400

6. What is the value of $12 \div 4 + 2 \times 3$?

   A 6   C 9
   B 8   D 15

7. ***SHORT RESPONSE*** Evaluate the expression $10 \times (28 - 14) \div 7 + 3$. Show your work, and use the order of operations to explain each step.

8. ***SHORT RESPONSE*** Evaluate the expression $(3 + 6) \times 18 \div 2 + 7$. Show your work, and use the order of operations to explain each step.

LESSON 1-5

9. What is the value of $15 + 7 + 23 + 5$?

   A 40   C 225
   B 50   D 600

10. What is the value of $4 \times 13 \times 5$?

    A 22   C 60
    B 33   D 260

11. Which expression is equal to 9×73?

    A $9 \times (70 + 3)$ C $70 + (9 \times 3)$
    B $(9 \times 70) + 3$ D $9 + (70 \times 3)$

12. Which expression is equal to 5×54?

    A $50 + (5 \times 4)$
    B $5 \times 50 + 4$
    C $5 + (50 \times 4)$
    D $(5 \times 50) + (5 \times 4)$

13. The equation $(3 + 6) + 17 = 3 + (6 + 17)$ is an example of which property?

    A Associative Property of Addition
    B Commutative Property of Addition
    C Distributive Property
    D Identity Property of Zero

14. The equation $43 + 28 + 17 = 17 + 43 + 28$ is an example of which property?

    A Associative Property of Addition
    B Commutative Property of Addition
    C Distributive Property
    D Identity Property of Zero

15. ***SHORT RESPONSE*** Explain how to use the Distributive Property to find the product of 7×26. What is the value of this expression?

16. ***SHORT RESPONSE*** Explain how to use mental math to find the sum of $22 + 16 + 18 + 34$. What is the value of this expression?

EOG Practice ■ Chapter 1

1B Using Whole Numbers

LESSON 1-6

Month	Rainy Days	Month	Rainy Days
Jan	6	Jul	15
Feb	5	Aug	9
Mar	7	Sep	17
Apr	14	Oct	14
May	12	Nov	8
Jun	10	Dec	5

17. The table shows the number of days it rained each month. How many total days did it rain during the year?

A 108 C 121

B 117 D 122

18. The choir director assigns choir members to quartets for a performance. If the director forms 8 quartets, how many members are in the choir?

A 8 C 24

B 12 D 32

19. One year, the lowest temperature in a city was 11°F. The highest temperature that same year was 89°F. What is the difference between the highest and lowest temperatures?

A 77°F C 80°F

B 78°F D 88°F

20. Margaret has 6 nieces. She gives each niece 3 presents. How many total presents does she give to her nieces?

A 2 C 9

B 6 D 18

21. On Saturday, Rebecca earned $24 baby-sitting. On Sunday, she earned $18 baby-sitting. How much more did Rebecca earn Saturday than Sunday?

A $6 C $16

B $8 D $42

22. ***SHORT RESPONSE*** Heather is a member of a dance company. She practices 14 hours a week. Write an expression that could be used to determine the number of hours Heather practices each year. (*Hint:* There are 52 weeks in a year.) Evaluate your expression.

23. ***SHORT RESPONSE*** Mr. Fowler's art class has signed up to bake cookies for the school bake sale. Each of the 31 students in the class bakes a dozen cookies. Write an expression that could be used to determine the number of cookies the class contributes to the sale. (*Hint:* There are 12 cookies in a dozen.) Evaluate your expression.

LESSON 1-7

24. What are the next three terms in the sequence 8, 16, 32, ■, ■, ■, . . .?

A 40, 48, 56 C 56, 96, 160

B 48, 64, 80 D 64, 128, 256

25. What are the next three terms in the sequence 1, 3, 6, 10, 15, ■, ■, ■, . . .?

A 17, 19, 21 C 21, 27, 33

B 17, 20, 22 D 21, 28, 36

26. What are the missing terms in the sequence 496, 248, 260, ■, 142, 71, ■, . . .?

A 130, 35.5 C 272, 35.5

B 130, 83 D 272, 83

27. What are the missing terms in the sequence 1, 8, 4, 32, 16, ■, 64, 512, ■, . . .?

A 8, 256 C 128, 256

B 8, 4,096 D 128, 4,096

28. ***SHORT RESPONSE*** Describe in words a pattern in the sequence 6, 11, 16, ■, ■, ■, ... What are the next three terms?

29. ***SHORT RESPONSE*** Describe in words a pattern in the sequence 7, 21, 63, ■, ■, ■, ... What are the next three terms?

EOG Practice

EOG Practice ■ Chapter 2

2A Understanding Variables and Expressions

LESSON 2-1

1. What is the value of $5x + 9$ when $x = 3$?
   A 17
   B 24
   C 34
   D 60

2. What is the value of $59 - 7x$ when $x = 8$?
   A 3
   B 17
   C 56
   D 416

3. What is the value of $23 + y$ when $y = 37$?
   A 50
   B 60
   C 74
   D 107

y	$23 + y$
17	40
27	50
37	■

4. What is the value of $w \times 3 + 10$ when $w = 6$?
   A 28
   B 34
   C 78
   D 79

w	$w \times 3 + 10$
4	22
5	25
6	■

5. What are the missing values in the table?
   A 6; 7
   B 7; 8
   C 40; 48
   D 56; 64

x	$x \div 8$
40	5
48	■
56	■

6. What are the missing values in the table?
   A 3; 1
   B 12; 16
   C 12; 22
   D 18; 22

c	$12 + c \div 2$
4	14
12	■
20	■

7. What are the missing values in the table?
   A 8; 18; 28
   B 9; 18; 29
   C 9; 19; 29
   D 10; 20; 30

a	$2 \times a - 1$
5	■
10	■
15	■

8. What are the missing values in the table?
   A 18; 8; 6
   B 18; 9; 6
   C 51; 48; 45
   D 57; 60; 63

b	$54 \div b$
3	■
6	■
9	■

9. ***SHORT RESPONSE*** Find an expression for the table. Explain how you found the expression, and verify that the expression fits each pair of numbers in the table.

t	■
7	35
8	40
9	45

10. ***SHORT RESPONSE*** Find an expression for the table. Explain how you found the expression, and verify that the expression fits each pair of numbers in the table.

s	■
66	11
54	9
36	6

11. ***SHORT RESPONSE*** Find an expression for the table. Explain how you found the expression, and verify that the expression fits each pair of numbers in the table.

g	■
52	26
60	30
68	34

12. ***SHORT RESPONSE*** Find an expression for the table. Explain how you found the expression, and verify that the expression fits each pair of numbers in the table.

m	■
2	6
4	12
6	18

EOG Practice ■ Chapter 2

2A Understanding Variables and Expressions

LESSON 2-2

13. Which expression represents the phrase "the product of 7 and 12"?

A $7 + 12$

B 7×12

C $7 - 12$

D $7 \div 12$

14. Which algebraic expression represents the phrase "14 more than x"?

A $x + 14$

B $14 - x$

C $x - 14$

D $14 \times x$

15. Which algebraic expression represents the phrase "the quotient of n and 8"?

A $n - 8$

B $n \times 8$

C $8 \div n$

D $\frac{n}{8}$

16. Which expression represents the phrase "the sum of 322 and 18"?

A 322×18

B $322 + 18$

C $322 - 18$

D $\frac{322}{18}$

17. Which of the following is a phrase for $y \div 4$?

A 4 less than y

B y less than 4

C the quotient of y and 4

D the quotient of 4 and y

18. Which of the following is a phrase for $52 - p$?

A p less than 52

B 52 less than p

C the quotient of 52 and p

D the sum of p and 52

19. Which of the following is a phrase for (23)(6)?

A the sum of 23 and 6

B 23 divided by 6

C the product of 23 and 6

D the difference of 23 and 6

20. Which of the following is a phrase for $h + 96$?

A the difference of h and 96

B the product of h and 96

C 96 more than h

D 96 minus h

21. ***SHORT RESPONSE*** Jodie's class is dividing into groups of 3 students for a project. Let x represent the total number of students in the class. Write an expression that represents the number of groups in the class. What action in the problem tells you which operation to use?

22. ***SHORT RESPONSE*** Joseph cut 18 inches from a piece of kite string. Let y represent the remaining length of the string. Write an expression that represents the original length of the kite string. What action in the problem tells you which operation to use?

23. ***SHORT RESPONSE*** Let p represent the number of players on a team. Write an expression that represents how many players will be on 65 teams. What action in the problem tells you which operation to use?

24. ***SHORT RESPONSE*** Earth has a diameter of 7,926 miles. Let d represent the diameter of the Moon, which is smaller than the diameter of Earth. Write an expression that represents how much larger the diameter of Earth is than the diameter of the Moon. What action in the problem tells you which operation to use?

25. ***SHORT RESPONSE*** Marion scored 82 more points than Jody in a contest. Let j represent the number of points that Jody scored. Write an expression that represents the number of points Marion scored. What action in the problem tells you which operation to use?

EOG Practice ■ Chapter 2

2B Equations

LESSON 2-3

1. Which of the following has a solution of 22?

   A $14 + t = 35$

   B $t - 9 = 13$

   C $2t = 54$

   D $\frac{t}{3} = 11$

2. Which of the following has a solution of 19?

   A $a + 15 = 34$

   B $a - 7 = 15$

   C $5a = 105$

   D $\frac{a}{3} = 6$

3. Which of the following is a solution to the equation $3x - 5 = 7$?

   A 3 C 5

   B 4 D 6

4. Which of the following is a solution to the equation $x + 12 = 21$?

   A 7 C 9

   B 8 D 10

5. ***SHORT RESPONSE*** Rachel says she is 5 feet tall. Her friend measures her height as 60 inches. Are the two measurements equivalent? Explain how you determined your answer.

6. ***SHORT RESPONSE*** A recipe calls for 2 cups of flour. Tim measured 24 tablespoons of flour. Is this the correct amount of flour? Explain how you determined your answer. (*Hint:* There are 16 tablespoons in a cup.)

LESSON 2-4

7. What is the solution to the equation $r + 13 = 36$?

   A 21 C 33

   B 23 D 49

8. What is the solution to the equation $52 = 24 + n$?

   A 28 C 76

   B 32 D 78

9. If $y + 9 = 52$, what is the value of $2y - 12$?

   A 43 C 74

   B 61 D 110

10. If $19 = m + 2$, what is the value of $m - 9$?

    A 8 C 17

    B 12 D 21

11. ***SHORT RESPONSE*** Let n represent the number of players who have signed up for a soccer league as of Saturday. On Sunday, 19 new players sign up, bringing the total number of players to 63. Write an equation that can be used to find the value of n. Solve your equation.

12. ***SHORT RESPONSE*** Molly had 17 collectible dolls before her birthday. After her birthday, she had 25 total dolls. Let d represent the number of dolls Molly received for her birthday. Write an equation that can be used to find the value of d. Solve your equation.

13. ***SHORT RESPONSE*** Towns A, B, and C are located along Main Road, as shown on the map. Town A is 34 miles from town C Town B is 12 miles from town C Write an equation that can be used to find the distance d between town A and town B. Solve your equation.

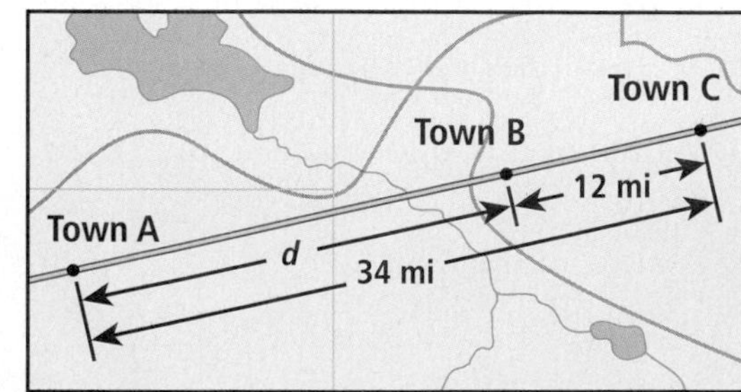

EOG Practice ■ Chapter 2

2B Equations

LESSON 2-5

14. What is the solution to the equation $z - 5 = 9$?

A 4 C 13
B 6 D 14

15. What is the solution to the equation $17 = v - 14$?

A 3 C 21
B 7 D 31

16. If $w - 6 = 24$, what is the value of $\frac{w}{3}$?

A 6 C 10
B 9 D 30

17. If $14 = y - 8$, what is the value of $3 \cdot (y + 5)$?

A 6 C 33
B 22 D 81

18. ***SHORT RESPONSE*** Reggie withdrew $175 from his bank account to go shopping. After his withdrawal, there was $234 left in his account. Write an equation that could be used to find how much money Reggie had in his account before his withdrawal. Solve your equation.

19. ***SHORT RESPONSE*** Cameron ate 13 pieces of candy. After he ate the candy, there were 47 pieces left in the bag. Write an equation that could be used to find how many pieces of candy were in the bag before Cameron ate any. Solve your equation.

LESSON 2-6

20. What is the solution to $4y = 20$?

A 4 C 16
B 5 D 80

21. What is the solution to the equation $72 = 9g$?

A 8 C 63
B 9 D 648

22. If $3t = 21$, what is the value of $t + 9$?

A 7 C 63
B 16 D 72

23. If $90 = 6h$, what is the value of $3h - 12$?

A 15 C 540
B 33 D 1,608

24. ***SHORT RESPONSE*** The area of a rectangle is 54 square inches. Its width is 6 inches. Write and solve an equation that could be used to find the length of the rectangle.

25. ***SHORT RESPONSE*** A squirrel can run 36 miles in 3 hours. Write an equation that could be used to find the number of miles a squirrel can run in 1 hour. Solve your equation.

LESSON 2-7

26. What is the solution to the equation $\frac{a}{3} = 12$?

A 4 C 24
B 9 D 36

27. What is the solution to the equation $6 = \frac{n}{4}$?

A 2 C 24
B 10 D 36

28. If $5 = t \div 5$, what is the value of t^2?

A 0 C 25
B 1 D 625

29. If $x \div 2 = 10$, what is the value of $6x - 3$?

A 5 C 27
B 20 D 117

30. ***SHORT RESPONSE*** Irene likes to run and ride a bike for exercise. Each day, she runs for $\frac{1}{3}$ the time that she rides her bike. Yesterday, Irene ran for 15 minutes. Write an equation that could be used to find how many minutes she rode her bike. Solve your equation.

EOG Practice ■ Chapter 3

3A Understanding Decimals

LESSON 3-1

1. What is the value of the 8 in the number 0.189?

   A eight hundreds C eight tenths

   B eight tens D eight hundredths

2. What is the value of the 7 in the number 1.4397?

   A seven thousandths

   B seven ten-thousandths

   C seven hundredths

   D seven thousands

3. Which of the following sets of decimals is ordered from *least* to *greatest*?

   A 3.8; 3.89; 3.08

   B 3.89; 3.8; 3.08

   C 3.08; 3.89; 3.8

   D 3.08; 3.8; 3.89

4. Which of the following is *greater* than 0.081?

   A 0.09 C 0.0724

   B 0.08 D 0.009

5. Which of the following is *less* than 0.504?

   A 0.703 C 0.6

   B 0.510 D 0.060

6. ***SHORT RESPONSE*** Joshua ran 1.45 miles, and Jasmine ran 1.5 miles. Who ran farther? Use place value to explain your answer.

7. ***SHORT RESPONSE*** Write five and three thousandths in standard form and in expanded form.

8. ***SHORT RESPONSE*** Write 0.6 + 0.003 + 0.0008 in standard form and in words.

9. ***SHORT RESPONSE*** Write 1.327 in expanded form and in words.

LESSON 3-2

10. What is the estimate of 4.7609 + 7.2471? Round to the tenths place.

    A 11 C 12

    B 11.9 D 12.1

11. What is the estimate of 5.856 − 1.3497? Round to the hundredths place.

    A 4.5 C 4.52

    B 4.51 D 4.6

12. Which is the best estimate for the product of 31.22 and 4.91?

    A 124 C 150

    B 128 D 160

13. Which is the best estimate for the quotient of 20.84 and 3.201?

    A 4 C 7

    B 5 D 9

14. What is the approximate value of the expression 9.518 × 11.1102?

    A 85 C 120

    B 100 D 190

15. What is the approximate value of the expression 47.36 ÷ 7.66?

    A 4 C 8

    B 6 D 9

16. ***SHORT RESPONSE*** *Estimate* a range for the sum of 8.38 + 24.92 + 4.8. Explain in words how you found your answer.

17. ***SHORT RESPONSE*** *Estimate* a range for the sum of 38.27 + 2.99 + 15.32. Explain in words how you found your answer.

18. ***SHORT RESPONSE*** Bernice and 7 of her friends went to dinner. The total bill was $71.56. They agreed to split the bill evenly. Explain in words how to estimate the amount of money each person owed for dinner. About how much money did Bernice owe?

EOG Practice ▪ Chapter 3

3A Understanding Decimals

LESSON 3-3

19. What is the sum of $1.65 + 4.53 + 3.2$?

A 9.2 C 9.38
B 9.23 D 9.4

20. What is the difference between 7 and 0.6?

A 1.0 C 6.4
B 6.3 D 7.4

21. What is the value of the expression $9.08 + 10.12$?

A 19.10 C 19.92
B 19.2 D 20.20

22. What is the value of the expression $5.91 - 4.003$?

A 1.088 C 1.907
B 1.88 D 1.913

23. What is the value of $6.35 - s$ when $s = 3.2$?

A 3.15 C 3.33
B 3.2 D 6.03

24. What is the value of $2.2 + t$ when $t = 6.82$?

A 8.02 C 9.02
B 8.84 D 9.04

25. ***SHORT RESPONSE*** Amelia spent $5.14, $5.83, and $6.98 for lunches during the week. Write an expression that could be used to determine the total amount Amelia spent. Evaluate your expression.

26. ***SHORT RESPONSE*** Richard had $64.15 in his wallet. He gave $34.97 to his friend. Write an expression that could be used to determine how much money was left in Richard's wallet. Evaluate your expression.

LESSON 3-4

27. What is the product of 2,318 and 1,000?

A 23,180 C 2,318,000
B 231,800 D 23,180,000

28. What is the value of the expression 34.5×10^4?

A 3,450 C 345,000
B 34,500 D 3,450,000

29. What is the value of the product 0.079×100?

A 0.00079 C 7.9
B 0.79 D 790

30. What is the value of the expression $6{,}210 \div 100$?

A 0.621 C 621
B 62.1 D 621,000

31. How many kilograms are in 2,640 grams?

A 2.64 kilograms
B 264 kilograms
C 264,000 kilograms
D 2,640,000 kilograms

32. What is 230 millimeters converted to meters?

A 0.023 meter C 2,300 meters
B 0.23 meter D 23,000 meters

33. ***SHORT RESPONSE*** Describe two methods to find the quotient of 59,840 and 1,000. Then find the value.

34. ***SHORT RESPONSE*** Explain in words how to find the number of milliliters that are contained in 0.6 liter. Then find the number of milliliters.

35. ***SHORT RESPONSE*** Terry wants to cut a 0.45-meter piece of ribbon from a spool, but she has a centimeter ruler. Show the steps necessary to find how many centimeters she should measure.

36. ***SHORT RESPONSE*** Manuel's backpack weighs about 5,750 grams. How many kilograms does his backpack weigh? Show your work.

EOG Practice ■ Chapter 3

3B Multiplying and Dividing Decimals

LESSON 3-5

1. What is 60,000 written in scientific notation?

   A 0.6×10^5 C 6×10^5

   B 6×10^4 D 60×10^3

2. What is 423,800 written in scientific notation?

   A 0.4238×10^6 C 4.238×10^2

   B 4.238×10^5 D 42.38×10^4

3. What is 1.425×10^4 written in standard form?

   A 1.425 C 14,250

   B 1,425 D 14,250,000

4. What is 2.1×10^5 written in standard form?

   A 2.1 C 210,000

   B 21,000 D 2,100,000

5. ***SHORT RESPONSE*** Explain how to write 5.632×10^8 in standard form.

6. ***SHORT RESPONSE*** Explain how to write 8,500,000 in scientific notation.

LESSON 3-6

7. What is the value of the expression $0.5 \cdot 0.7$?

   A 0.035 C 3.5

   B 0.35 D 35

8. What is the value of $0.3 \cdot 0.06$?

   A 0.018 C 1.8

   B 0.18 D 18

9. What is the product of $3.8 \cdot 0.4$?

   A 0.0152 C 1.52

   B 0.152 D 15.2

10. What is the value of $1.6 \cdot 2.3$?

    A 0.0368 C 3.68

    B 0.368 D 36.8

11. What is the product of 0.9 and 5.4?

    A 0.486 C 48.6

    B 4.86 D 486

12. What is the product of 1.2 and 7.0?

    A 0.84 C 8.4

    B 8.2 D 82

LESSON 3-7

13. Find the value of the expression $0.564 \div 12$.

    A 0.0417 C 0.417

    B 0.047 D 0.47

14. Find the value of the expression $22.32 \div 18$.

    A 0.124 C 1.24

    B 0.179 D 1.79

15. What is the value of the expression $18.5 \div 5$?

    A 0.217 C 2.17

    B 0.37 D 3.7

16. What is the value of the expression $48.78 \div 9$?

    A 5.312 C 53.12

    B 5.42 D 542

17. ***SHORT RESPONSE*** Amanda made \$44.59 pet sitting. She worked for 7 days and earned the same amount of money each day. Write an expression that could be used to find the amount of money Amanda earned each day. Evaluate your expression.

18. ***SHORT RESPONSE*** Bart and his two brothers earned \$55.02 mowing a lawn. Write an expression that could be used to find the amount each boy gets if they split the money evenly. Evaluate your expression.

EOG Practice ■ Chapter 3

3B Multiplying and Dividing Decimals

LESSON 3-8

19. Which number is equal to $4.5 \div 0.9$?

A 0.05 C 5.0
B 0.5 D 50

20. What is 59.7 divided by 0.4?

A 1.4925 C 149.25
B 14.925 D 1,492.5

21. What is 1.665 divided by 0.09?

A 0.185 C 18.5
B 1.85 D 185

22. What is the value of $28.98 \div 1.4$?

A 2.07 C 20.7
B 2.7 D 27

23. ***SHORT RESPONSE*** Jackie won $83.20 in a contest. She decides to give an equal portion of the money to each of her 8 friends. Write an expression that could be used to find the amount of money each friend will receive. Evaluate your expression.

24. ***SHORT RESPONSE*** Mariah's car used 14.5 gallons of gas to travel 442.25 miles. Write an expression that could be used to find the number of miles per gallon Mariah's car uses on average. Evaluate your expression.

LESSON 3-9

25. Sue has 317 flyers to mail. Each flyer needs one stamp. She buys books of stamps that contain 20 stamps each. How many books will she need to mail the flyers?

A 15 C 17
B 16 D 18

26. John wants to send holiday cards to 44 of his relatives. The cards are sold in packs of 8. How many packs does John need to buy?

A 4 C 6
B 5 D 7

27. Karla is making bracelets for her friends. She uses 7.5 inches of string for each bracelet. If she has 36.8 inches of string, how many bracelets can she make?

A 4 C 29
B 5 D 30

28. Paolo's goal is to read a 315-page book in 3 weeks. If he reads the same amount 7 days a week, how many pages would he have to read each day?

A 15 C 45
B 16 D 105

29. ***SHORT RESPONSE*** Jocelyn has 3.5 yards of ribbon. She needs 0.6 yard of ribbon to make one bow. How many bows can Jocelyn make? Explain how you determined your answer.

30. ***SHORT RESPONSE*** Louie cuts a piece of wood that is 46.8 cm long into 4 equal pieces. Write and evaluate an expression that could be used to determine the length of each section.

LESSON 3-10

31. What is the solution to $\frac{p}{3} = 1.8$?

A 0.6 C 4.8
B 1.2 D 5.4

32. What is the solution to $5t = 24.5$?

A 4.9 C 29.5
B 19.5 D 122.5

33. ***SHORT RESPONSE*** A rectangle with an area of 41 cm^2 has a length of 8.2 cm. Write and solve an equation that could be used to determine the width of the rectangle.

34. ***SHORT RESPONSE*** The area of Henry's kitchen is 168 ft^2. The cost of tile is $4.62 per square foot. Write and evaluate an expression that represents the total cost to tile the kitchen.

EOG Practice ▪ Chapter 4

4A Number Theory

LESSON 4-1

1. Which of the following is prime?
   A 9 C 101
   B 51 D 256

2. Which of the following is composite?
   A 37 C 111
   B 53 D 373

3. By which of the following is 174 *not* divisible?
   A 2 C 4
   B 3 D 6

4. By which of the following is 852 *not* divisible?
   A 3 C 6
   B 4 D 9

5. Which of the following is divisible by 3?
   A 68 C 312
   B 152 D 406

6. Which of the following is divisible by 4?
   A 510 C 3,918
   B 734 D 6,100

7. Which of the following is divisible by 6?
   A 56 C 87
   B 78 D 94

8. ***SHORT RESPONSE*** Explain how to determine whether a number is prime or composite. Is 97 prime or composite? Is 153 prime or composite? Use 97 and 153 as examples in your explanation.

9. ***SHORT RESPONSE*** Explain how to determine whether a number is divisible by 6. Is 1,638 divisible by 6? Is 12,680 divisible by 6?

LESSON 4-2

10. What is the prime factorization of 48?
    A $2 \cdot 3$ C $2^4 \cdot 3$
    B $2^3 \cdot 3$ D $2 \cdot 3^4$

11. What is the prime factorization of 150?
    A $2 \cdot 3 \cdot 5$ C $2^2 \cdot 3 \cdot 5$
    B $2 \cdot 3 \cdot 5^2$ D $2^3 \cdot 5^5$

12. Which number is a factor of 28?
    A 3 C 7
    B 5 D 9

13. Which number is a factor of 38?
    A 3 C 14
    B 7 D 19

14. Which number is a factor of 24?
    A 5 C 8
    B 7 D 9

15. Of which number is 6 a factor?
    A 193 C 503
    B 312 D 974

16. Which is *not* a factor of 32?
    A 1 C 8
    B 3 D 16

17. Which is *not* a factor of 45?
    A 3 C 9
    B 5 D 12

18. ***SHORT RESPONSE*** What is the prime factorization of 99? Draw a factor tree to support your answer.

19. ***SHORT RESPONSE*** What is the prime factorization of 72? Draw a ladder diagram to support your answer.

4A Number Theory

LESSON 4-3

20. What is the GCF of the number of squares and the number of circles?

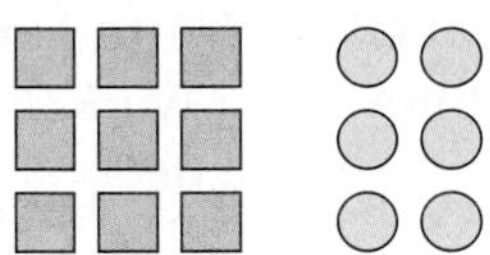

A 1 C 8
B 3 D 16

21. What is the GCF of 16 and 40?

A 2 C 5
B 4 D 8

22. What is the GCF of 8, 20, and 32?

A 2 C 10
B 4 D 20

23. What is the GCF of 26, 65, and 78?

A 2 C 5
B 3 D 13

24. What is the GCF of 27, 36, and 54?

A 2 C 6
B 3 D 9

25. What is the GCF of 22 and 68?

A 2 C 11
B 3 D 17

26. Of which set of numbers is 6 the GCF?

A 12, 28 C 36, 64
B 22, 68 D 42, 72

27. Of which set of numbers is 3 the GCF?

A 12, 42, 72
B 24, 39, 162
C 24, 60, 96
D 27, 36, 54

28. Of which set of numbers is 13 the GCF?

A 14, 28, 63
B 16, 56, 72
C 26, 65, 78
D 35, 60, 85

29. ***SHORT RESPONSE*** Alice has 42 red beads and 24 white beads. Draw a factor tree for both numbers. What is the greatest number of bracelets Alice can make if each bracelet has the same number of red beads and the same number of white beads and every bead is used? Explain in words how you found your answer.

30. ***SHORT RESPONSE*** The flower shop has 16 red roses and 28 white carnations. Draw a factor tree for both numbers. What is the greatest number of bouquets that can be made if each bouquet has the same number of roses and the same number of carnations and every flower is used? Explain in words how you found your answer.

31. ***SHORT RESPONSE*** Lisa has 15 bars of strawberry-scented soap, 30 bars of almond-scented soap, and 25 bars of orange-scented soap. She wants to make gift baskets. Each basket must have the same number of each scented soap. Draw a factor tree for each number. What is the greatest number of gift baskets Lisa can make if every bar of soap is used? Explain in words how you found your answer.

32. ***SHORT RESPONSE*** Jeremy is making balloon arrangements for a party. He has 12 red balloons, 14 blue balloons, and 16 yellow balloons. Each arrangement must have the same number of each color balloon. Jeremy uses every balloon to make the greatest number of arrangements possible. Draw a factor tree for each number. How many red balloons will be in each arrangement? Explain in words how you found your answer.

EOG Practice ■ Chapter 4

4B Understanding Fractions

LESSON 4-4

1. Which of the following sets of numbers is ordered from *least* to *greatest*?

   A $\frac{3}{5}$, 0.3, 0.53 C 0.3, 0.53, $\frac{3}{5}$

   B 0.3, $\frac{3}{5}$, 0.53 D 0.53, $\frac{3}{5}$, 0.3

2. Which of the following sets of numbers is ordered from *greatest* to *least*?

   A $\frac{3}{4}$, $\frac{2}{3}$, 0.68 C 0.68, $\frac{2}{3}$, $\frac{3}{4}$

   B $\frac{2}{3}$, 0.68, $\frac{3}{4}$ D $\frac{3}{4}$, 0.68, $\frac{2}{3}$

3. What fraction is represented on the decimal grid?

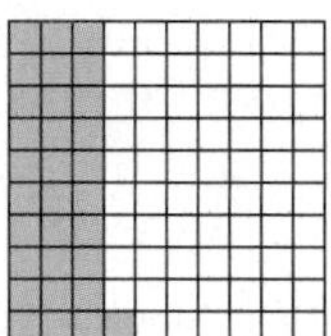

   A $\frac{13}{100}$ C $\frac{31}{100}$

   B $\frac{3}{10}$ D $\frac{41}{100}$

4. Which of the following is equivalent to 2.52?

   A $2\frac{52}{100}$ C $\frac{52}{200}$

   B $2\frac{52}{10}$ D $\frac{2}{52}$

5. Which of the following is equivalent to 1.9?

   A $\frac{9}{10}$ C $\frac{1}{9}$

   B $\frac{19}{100}$ D $1\frac{9}{10}$

6. Which decimal is equivalent to $1\frac{7}{8}$?

   A 0.875 C 1.78

   B 1.375 D 1.875

7. Which decimal is equivalent to $6\frac{3}{5}$?

   A 6.3 C 6.6

   B 6.35 D 6.67

8. ***SHORT RESPONSE*** Order the numbers 0.8, 0.67, and $\frac{7}{8}$ from *least* to *greatest*, and plot the numbers on a number line.

9. ***SHORT RESPONSE*** Write $\frac{5}{6}$ and $\frac{5}{9}$ as decimals. Show your work. Explain how you know whether a decimal repeats.

LESSON 4-5

10. Which fraction is *not* equivalent to $\frac{3}{12}$?

    A $\frac{1}{4}$ C $\frac{4}{16}$

    B $\frac{2}{8}$ D $\frac{5}{15}$

11. Which fraction is *not* equivalent to $\frac{8}{20}$?

    A $\frac{1}{4}$ C $\frac{4}{10}$

    B $\frac{2}{5}$ D $\frac{16}{40}$

12. Which fraction is equivalent to $\frac{6}{16}$?

    A $\frac{1}{10}$ C $\frac{3}{8}$

    B $\frac{3}{4}$ D $\frac{12}{36}$

13. What is $\frac{2}{6}$ written in simplest form?

    A 0.333… C $\frac{1}{3}$

    B $\frac{3}{9}$ D $\frac{1}{4}$

14. What value of x will make the fractions $\frac{6}{7}$ and $\frac{x}{28}$ equivalent?

    A 10 C 24

    B 18 D 27

15. What value of x will make the fractions $\frac{8}{12}$ and $\frac{4}{x}$ equivalent?

    A 3 C 8

    B 6 D 10

16. ***SHORT RESPONSE*** What is the GCF of 51 and 85? How can you use it to write $\frac{51}{85}$ in simplest form?

EOG Practice

EOG Practice ■ Chapter 4

4B Understanding Fractions

LESSON 4-6

17. Which fraction makes the statement ■ $< \frac{3}{5}$ true?

A $\frac{2}{1}$ C $\frac{2}{3}$

B $\frac{3}{4}$ D $\frac{2}{5}$

18. Which fraction makes the statement $\frac{5}{6} <$ ■ true?

A $\frac{3}{4}$ C $\frac{7}{8}$

B $\frac{1}{2}$ D $\frac{2}{3}$

19. Which fraction is *greater* than $\frac{9}{15}$?

A $\frac{3}{4}$ C $\frac{3}{5}$

B $\frac{2}{5}$ D $\frac{6}{12}$

20. Which fraction is *greater* than $\frac{5}{7}$?

A $\frac{1}{2}$ C $\frac{3}{4}$

B $\frac{2}{5}$ D $\frac{6}{12}$

21. Which set of fractions is ordered from *least* to *greatest*?

A $\frac{1}{6}, \frac{3}{7}, \frac{1}{3}$ C $\frac{1}{6}, \frac{1}{3}, \frac{3}{7}$

B $\frac{3}{7}, \frac{1}{6}, \frac{1}{3}$ D $\frac{1}{3}, \frac{1}{6}, \frac{3}{7}$

22. Which set of fractions is ordered from *greatest* to *least*?

A $\frac{1}{2}, \frac{5}{8}, \frac{7}{12}$ C $\frac{5}{8}, \frac{1}{2}, \frac{7}{12}$

B $\frac{7}{12}, \frac{5}{8}, \frac{1}{2}$ D $\frac{5}{8}, \frac{7}{12}, \frac{1}{2}$

23. ***SHORT RESPONSE*** Mary sprinted $\frac{1}{3}$ mile. Natasha sprinted $\frac{9}{27}$ mile. Who ran farther? Explain in words how you determined your answer.

24. ***SHORT RESPONSE*** Natalie lives $\frac{1}{6}$ mile from school. Peter lives $\frac{3}{10}$ mile from school. Who lives closer to the school? Explain in words how you determined your answer.

LESSON 4-7

25. What is $3\frac{3}{4}$ written as an improper fraction?

A $\frac{10}{4}$ C $\frac{15}{4}$

B $\frac{12}{4}$ D $\frac{21}{4}$

26. What is $6\frac{5}{7}$ written as an improper fraction?

A $\frac{18}{7}$ C $\frac{47}{7}$

B $\frac{40}{7}$ D $\frac{65}{7}$

27. Which improper fraction is equivalent to $3\frac{2}{9}$?

A $\frac{14}{9}$ C $\frac{29}{9}$

B $\frac{21}{9}$ D $\frac{38}{9}$

28. Which mixed number is equivalent to $\frac{27}{10}$?

A 2.7 C $2\frac{7}{10}$

B $2\frac{17}{10}$ D $7\frac{2}{10}$

29. Which mixed number is equivalent to $\frac{13}{5}$?

A $\frac{3}{5}$ C $2\frac{3}{5}$

B $1\frac{3}{5}$ D $3\frac{1}{5}$

30. Which mixed number is equivalent to $\frac{41}{3}$?

A $2\frac{1}{3}$ C $13\frac{2}{3}$

B $4\frac{1}{3}$ D $14\frac{1}{3}$

31. ***SHORT RESPONSE*** Brett's favorite soup recipe calls for $\frac{14}{4}$ cups of chicken broth. Explain how to write this improper fraction as a mixed number.

32. ***SHORT RESPONSE*** An experiment calls for $\frac{7}{2}$ teaspoons of baking soda. Paul measures $5\frac{1}{2}$ teaspoons of baking soda. Explain how to determine if Paul measured the correct amount.

EOG Practice ▪ Chapter 4

4C Introduction to Fraction Operations

LESSON 4-8

1. What is the value of the expression modeled below?

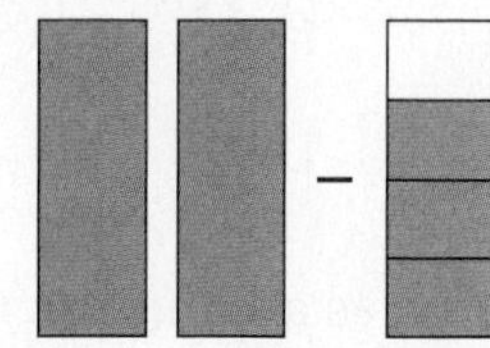

A $1\frac{1}{3}$ C 1

B $1\frac{1}{4}$ D $\frac{3}{4}$

2. What is the value of $10\frac{4}{5} - 2\frac{2}{5}$?

A $\frac{2}{5}$ C $8\frac{2}{5}$

B $7\frac{2}{5}$ D $8\frac{6}{5}$

3. What is the value of the expression $2\frac{2}{8} + 1\frac{1}{8}$?

A $\frac{3}{8}$ C $3\frac{3}{8}$

B $1\frac{1}{8}$ D 4

4. What is the value of $6\frac{2}{3} - x$ when $x = 5\frac{1}{3}$?

A $1\frac{1}{3}$ C $11\frac{1}{3}$

B 2 D 12

5. What is the value of $3\frac{6}{12} + y$ when $y = 2\frac{5}{12}$?

A $\frac{11}{12}$ C $5\frac{11}{12}$

B $1\frac{1}{12}$ D 6

6. What is the value of $2 - \frac{3}{4}$?

A $\frac{1}{4}$ C $2\frac{1}{4}$

B $1\frac{1}{4}$ D $2\frac{3}{4}$

7. What is the value of $7\frac{3}{4} + 4\frac{3}{4}$?

A 3 C 11

B $4\frac{1}{2}$ D $12\frac{1}{2}$

8. Which expression has a value of $2\frac{2}{3}$?

A $1\frac{1}{3} + 1\frac{2}{3}$ C $4\frac{1}{3} - 2\frac{1}{3}$

B $2\frac{5}{6} - \frac{2}{6}$ D $3\frac{5}{6} - 1\frac{1}{6}$

9. Which expression has a value of 10?

A $2\frac{3}{4} + 8\frac{1}{4}$ C $7\frac{4}{6} + 2\frac{1}{6}$

B $5\frac{7}{10} + 4\frac{3}{10}$ D $13\frac{5}{6} - 3\frac{1}{6}$

10. Which expression has a value of $14\frac{1}{4}$?

A $6\frac{3}{4} + 8\frac{3}{4}$ C $8\frac{5}{8} + 5\frac{5}{8}$

B $8\frac{1}{4} + 7$ D $19\frac{3}{4} - 5\frac{1}{4}$

11. Find the value of the expression $8\frac{5}{8} + 5\frac{5}{8}$.

A $13\frac{1}{4}$ C $14\frac{1}{4}$

B 14 D $15\frac{1}{4}$

12. ***SHORT RESPONSE*** Katelyn used $\frac{7}{9}$ of a carton of cotton balls while working on an art project. Write an expression that could be used to determine what portion of the carton remains. Evaluate your expression.

13. ***SHORT RESPONSE*** Gerry walked $\frac{5}{8}$ mile and then ran $\frac{3}{8}$ mile. Draw a diagram to represent the fractional distances Gerry walked and ran. What is the total distance Gerry covered?

14. ***SHORT RESPONSE*** Mike is $5\frac{3}{4}$ feet tall. He stands on a step stool that is $\frac{3}{4}$ foot tall. Write an expression that could be used to determine the combined height of Mike and the step stool. Evaluate your expression.

15. ***SHORT RESPONSE*** A pie was cut into 16 pieces. Then 5 pieces were eaten on Friday, and 3 pieces were eaten on Saturday. Draw a diagram of the pie, and shade the pieces that were eaten. What fraction of the pie is left on Sunday?

EOG Practice ■ Chapter 4

4C Introduction to Fraction Operations

LESSON 4-9

16. Which expression shows the correct first step to take when multiplying 5 and $\frac{7}{8}$?

A $\frac{1}{5} \cdot \frac{7}{8}$ C $\frac{5}{1} \cdot \frac{7}{8}$

B $\frac{1}{5} \cdot \frac{8}{7}$ D $\frac{5}{1} \cdot \frac{8}{7}$

17. What is the product of 6 and $\frac{2}{7}$?

A $\frac{12}{42}$ C $1\frac{6}{7}$

B $1\frac{5}{7}$ D $\frac{44}{7}$

18. Find the value of the expression $3x$ for $x = \frac{11}{12}$.

A $\frac{33}{36}$ C $2\frac{1}{4}$

B $1\frac{1}{6}$ D $2\frac{3}{4}$

19. Find the value of the expression $4y$ for $y = \frac{2}{10}$.

A $\frac{8}{40}$ C $\frac{3}{5}$

B $\frac{2}{5}$ D $\frac{4}{5}$

20. Find the value of $3 \cdot \frac{3}{7}$ in simplest form.

A $\frac{9}{21}$ C $1\frac{2}{7}$

B $\frac{6}{7}$ D $\frac{24}{7}$

21. Find the value of the expression $2 \cdot \frac{3}{6}$.

A $\frac{6}{12}$ C 1

B $\frac{5}{6}$ D $\frac{15}{6}$

22. What is the value of the expression $8 \cdot \frac{3}{4}$?

A $\frac{24}{32}$ C 6

B $1\frac{1}{2}$ D $\frac{35}{4}$

23. Which expression has a value of $2\frac{4}{5}$?

A $4 \cdot \frac{7}{10}$ C $2 \cdot \frac{4}{5}$

B $\frac{4}{5} \cdot 5$ D $9 \cdot \frac{3}{20}$

24. Which expression has a value of $1\frac{7}{8}$?

A $3 \cdot \frac{2}{4}$ C $8 \cdot \frac{7}{16}$

B $7 \cdot \frac{3}{12}$ D $3 \cdot \frac{5}{8}$

25. Which expression does *not* have a value of $\frac{1}{2}$?

A $3 \cdot \frac{1}{6}$ C $1 \cdot \frac{4}{8}$

B $\frac{1}{4} \cdot 2$ D $4 \cdot \frac{1}{12}$

26. Which multiplication expression is represented in the model below?

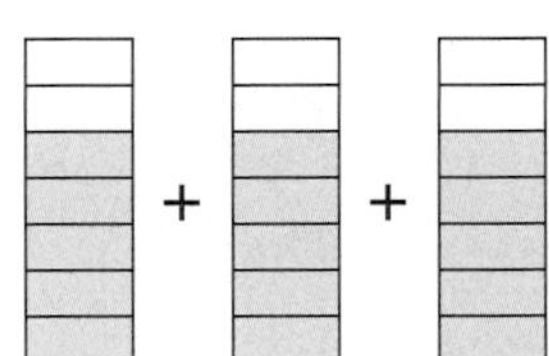

A $3 \cdot \frac{7}{5}$ C $5 \cdot \frac{3}{7}$

B $7 \cdot \frac{3}{5}$ D $3 \cdot \frac{5}{7}$

27. Find the value of the expression $7 \cdot \frac{6}{11}$.

A $\frac{13}{18}$ C $3\frac{9}{11}$

B $1\frac{2}{11}$ D $\frac{83}{11}$

28. ***SHORT RESPONSE*** There are 16 players on the baseball team. Of these players, $\frac{1}{4}$ are girls. Write an expression that could be used to determine how many girls play on the baseball team. Evaluate your expression.

29. ***SHORT RESPONSE*** Connie is using 16 flowers to make an arrangement. Of the 16 flowers, $\frac{3}{8}$ are roses. Write an expression that could be used to determine how many roses are in her arrangement. Evaluate your expression.

30. ***SHORT RESPONSE*** Sue made a beaded necklace. Of the 54 beads she used, $\frac{1}{6}$ were white. Write an expression that could be used to determine how many white beads Sue used. Evaluate your expression.

EOG Practice

EOG Practice ▪ Chapter 5

5A Multiplying and Dividing Fractions

LESSON 5-1

1. What is the value of $\frac{1}{10} \cdot \frac{5}{6}$ in simplest form?

   A $\frac{1}{12}$ C $\frac{5}{16}$

   B $\frac{1}{8}$ D $\frac{5}{60}$

2. What is the value of $\frac{8}{9} \cdot \frac{3}{4}$ in simplest form?

   A $\frac{1}{6}$ C $\frac{6}{9}$

   B $\frac{2}{3}$ D $\frac{3}{4}$

3. What is the product of $\frac{7}{10}$ times $\frac{3}{7}$ in simplest form?

   A $\frac{1}{7}$ C $\frac{21}{70}$

   B $\frac{3}{10}$ D $\frac{10}{17}$

4. Which expression has a value of $\frac{1}{20}$?

   A $\frac{1}{10} \cdot \frac{1}{10}$ C $\frac{3}{10} \cdot \frac{1}{3}$

   B $\frac{1}{2} \cdot \frac{2}{5}$ D $\frac{1}{4} \cdot \frac{1}{5}$

5. Which expression has a value of $\frac{5}{12}$?

   A $\frac{6}{8} \cdot \frac{4}{5}$ C $\frac{3}{10} \cdot \frac{5}{7}$

   B $\frac{7}{12} \cdot \frac{3}{5}$ D $\frac{3}{4} \cdot \frac{5}{9}$

6. Which expression does *not* have a value of $\frac{1}{4}$?

   A $\frac{3}{11} \cdot \frac{11}{12}$ C $\frac{1}{2} \cdot \frac{7}{14}$

   B $\frac{2}{3} \cdot \frac{3}{4}$ D $\frac{2}{5} \cdot \frac{5}{8}$

7. ***SHORT RESPONSE*** Joey spent $\frac{5}{12}$ of a day playing sports. He spent $\frac{6}{7}$ of that time playing football. Write an expression that could be used to determine what fraction of the day Joey spent playing football. Evaluate your expression.

8. ***SHORT RESPONSE*** Ty picked $\frac{7}{8}$ of the fruit on his tree. He canned $\frac{3}{4}$ of the fruit he picked. Write and evaluate an expression that could be used to determine what fraction of the fruit on his tree was canned.

LESSON 5-2

9. What is the value of $\frac{1}{4} \cdot 1\frac{2}{3}$?

   A $\frac{5}{12}$ C $\frac{11}{12}$

   B $\frac{6}{7}$ D $1\frac{1}{6}$

10. What is the product of $2\frac{3}{5}$ and $\frac{1}{3}$?

    A $\frac{13}{15}$ C $1\frac{3}{4}$

    B $\frac{14}{15}$ D $2\frac{1}{5}$

11. What is $\frac{2}{3}$ of $1\frac{2}{3}$?

    A $\frac{4}{9}$ C $2\frac{1}{3}$

    B $1\frac{1}{9}$ D $3\frac{1}{3}$

12. Which expression has a value of $1\frac{1}{11}$?

    A $6 \cdot 2\frac{1}{7}$ C $3\frac{1}{6} \cdot \frac{6}{12}$

    B $\frac{1}{3} \cdot 3\frac{1}{2}$ D $\frac{8}{11} \cdot 1\frac{1}{2}$

13. Which expression has a value that is *not* a whole number?

    A $1\frac{5}{6} \cdot 1\frac{1}{11}$ C $\frac{8}{9} \cdot 6\frac{1}{3}$

    B $8\frac{1}{3} \cdot \frac{3}{5}$ D $1\frac{1}{3} \cdot 2\frac{1}{4}$

14. Which expression has a value *greater* than 10?

    A $1\frac{3}{4} \cdot 5\frac{1}{5}$ C $\frac{2}{5} \cdot 4\frac{1}{2}$

    B $4 \cdot 2\frac{6}{7}$ D $\frac{2}{9} \cdot 2\frac{1}{4}$

15. ***SHORT RESPONSE*** Mary's tomato weighed $2\frac{1}{2}$ times as much as Jay's tomato. Jay's tomato weighed $2\frac{1}{4}$ lb. Write and evaluate an expression that could be used to determine the weight of Mary's tomato.

EOG Practice ■ Chapter 5

5A Multiplying and Dividing Fractions

LESSON 5-3

16. Which is *not* equivalent to the reciprocal of $\frac{2}{13}$?

A $\frac{13}{2}$ C $6\frac{1}{13}$

B $6\frac{1}{2}$ D $\frac{26}{4}$

17. What is the reciprocal of $\frac{8}{5}$?

A $\frac{1}{40}$ C $1\frac{3}{5}$

B $\frac{5}{8}$ D $1\frac{4}{5}$

18. What is the value of the expression $\frac{1}{6} \div 3$?

A $\frac{1}{18}$ C $\frac{2}{9}$

B $\frac{1}{9}$ D $\frac{1}{2}$

19. What is the value of the expression $1\frac{4}{5} \div 1\frac{1}{4}$?

A $\frac{25}{36}$ C 2

B $1\frac{11}{25}$ D $2\frac{1}{5}$

20. ***SHORT RESPONSE*** Kim cut $6\frac{1}{3}$ yards of ribbon into pieces that are each $\frac{1}{3}$ yard long. Write an expression that could be used to determine how many pieces of ribbon Kim has. Evaluate your expression.

21. ***SHORT RESPONSE*** Kelly has $3\frac{3}{5}$ quarts of sherbet. She serves each of her 8 friends an equal portion, and no sherbet is left over. Write an expression that could be used to determine how much sherbet each friend gets. Evaluate your expression.

LESSON 5-4

22. What is the solution to the equation $\frac{3}{5}a = 12$?

A $7\frac{1}{5}$ C 15

B $11\frac{2}{5}$ D 20

23. What is the solution to the equation $\frac{3}{7} = 6b$?

A $\frac{1}{14}$ C $5\frac{4}{7}$

B $2\frac{4}{7}$ D 14

24. What is the solution to the equation $\frac{5}{12}x = 3$?

A $\frac{4}{5}$ C $2\frac{7}{12}$

B $1\frac{1}{4}$ D $7\frac{1}{5}$

25. What is the solution to the equation $3y = \frac{7}{9}$?

A $\frac{7}{27}$ C $2\frac{2}{9}$

B $\frac{7}{18}$ D $2\frac{1}{3}$

26. If $3y = \frac{6}{7}$, what is $2y + 3$?

A 3 C $5\frac{2}{5}$

B $3\frac{4}{7}$ D $6\frac{2}{3}$

27. If $\frac{5}{6}r = \frac{15}{16}$, what is $2r + \frac{1}{2}$?

A $\frac{25}{32}$ C $2\frac{1}{16}$

B $\frac{9}{8}$ D $2\frac{3}{4}$

28. ***SHORT RESPONSE*** Joanie used $\frac{2}{3}$ of a box of invitations to invite friends to her birthday party. She sent out 12 invitations. Write and solve an equation that could be used to find how many total invitations were in the box.

29. ***SHORT RESPONSE*** The coach gave $\frac{3}{4}$ of the football team extra laps to run, so 33 players had to run extra laps. Write and solve an equation that could be used to find how many players are on the team.

30. ***SHORT RESPONSE*** Mary won $\frac{3}{8}$ of the games she played with her sister. Mary won 6 games. Write and solve an equation that could be used to find the total number of games Mary and her sister played.

EOG Practice ■ Chapter 5

5B Adding and Subtracting Fractions

LESSON 5-5

1. What is the LCM of 9 and 15?

   A 15 C 45
   B 18 D 90

2. What is the LCM of 12 and 16?

   A 6 C 48
   B 24 D 96

3. What is the LCM of 10 and 12?

   A 30 C 90
   B 60 D 120

4. Of which 2 numbers is 60 the LCM?

   A 2, 6 C 6, 8
   B 5, 6 D 10, 12

5. ***SHORT RESPONSE*** Barrettes are sold in packs of 6. Ponytail holders are sold in packs of 2. Use prime factorization to find the LCM of 2 and 6. There are 18 girls on the dance team. What is the *least* number of packs they could buy so that each girl has a barrette and a ponytail holder and none are left over? Explain in words how you determined your answer.

6. ***SHORT RESPONSE*** Buttons are sold in packs of 12, and snaps are sold in packs of 18. Use prime factorization to find the LCM of 12 and 18. What is the *least* number of packs of each Margo could buy so that she has the same number of buttons and snaps? How many buttons will she have? Explain in words how you determined your answer.

LESSON 5-6

7. ***SHORT RESPONSE*** When estimating with fractions, would you round $\frac{7}{8}$ to 0, $\frac{1}{2}$, or 1? Draw fraction bars to justify your answer. *Estimate* the sum of $\frac{7}{8}$ and $\frac{7}{15}$ by rounding each fraction to 0, $\frac{1}{2}$, or 1.

Use the table for problems 8 and 9.

Michael's Work Schedule	
Day	**Hours Worked**
Monday	$4\frac{5}{6}$
Tuesday	$5\frac{1}{4}$
Thursday	$6\frac{1}{10}$
Friday	$4\frac{5}{12}$

8. Which is the best estimate for the hours Michael worked on Monday and Tuesday?

   A 9 C 10
   B $9\frac{1}{2}$ D $10\frac{1}{2}$

9. About how many more hours did Michael work on Thursday than on Friday?

   A $\frac{1}{2}$ C $1\frac{1}{2}$
   B 1 D $2\frac{1}{2}$

LESSON 5-7

10. Find the value of $\frac{3}{5} + \frac{2}{3}$.

    A $\frac{1}{3}$ C $1\frac{4}{15}$
    B $\frac{5}{8}$ D $1\frac{9}{15}$

11. What is the difference of $\frac{7}{8}$ and $\frac{1}{6}$?

    A $\frac{3}{7}$ C $\frac{3}{4}$
    B $\frac{17}{24}$ D 3

12. ***SHORT RESPONSE*** Thomas has $\frac{3}{7}$ yard of wood trim. He purchases an additional $\frac{1}{2}$ yard. Write an expression that could be used to determine the total length of trim Thomas has. Evaluate your expression.

EOG Practice ■ Chapter 5

5B Adding and Subtracting Fractions

LESSON 5-8

13. What is the sum of $18\frac{1}{3}$ and $16\frac{1}{6}$?

A $24\frac{1}{2}$ C $34\frac{1}{3}$

B $34\frac{1}{9}$ D $34\frac{1}{2}$

14. What is the difference of $12\frac{1}{2}$ and $8\frac{2}{5}$?

A $3\frac{9}{10}$ C $4\frac{3}{10}$

B $4\frac{1}{10}$ D $4\frac{1}{3}$

15. ***SHORT RESPONSE*** Karl drives $7\frac{1}{4}$ miles to work. Meredith drives $10\frac{1}{3}$ miles to work. Draw a number line and plot these distances. Write an expression that could be used to determine how much farther Meredith drives than Karl does. Evaluate your expression.

16. ***SHORT RESPONSE*** Sam has $4\frac{3}{4}$ feet of red string, $1\frac{1}{8}$ feet of blue string, and $3\frac{2}{3}$ feet of green string. Write an expression that could be used to determine the total length of string that Sam has. Evaluate your expression.

LESSON 5-9

17. What is the value of $4\frac{2}{5} - 2\frac{9}{10}$?

A $\frac{1}{2}$ C $1\frac{1}{2}$

B $1\frac{2}{5}$ D $2\frac{1}{2}$

18. What is the value of $9\frac{1}{6} - 5\frac{5}{6}$?

A $3\frac{1}{6}$ C $4\frac{1}{3}$

B $3\frac{1}{3}$ D $4\frac{2}{3}$

19. Which expression has a value of $2\frac{1}{2}$?

A $11\frac{1}{6} - 9\frac{1}{2}$ C $4 - 1\frac{3}{4}$

B $8\frac{1}{3} - 5\frac{5}{6}$ D $3\frac{1}{5} - \frac{2}{5}$

20. ***SHORT RESPONSE*** The scarf that Monica knit was 5 feet long. To correct a mistake, she had to unravel $1\frac{7}{12}$ feet. Write an expression that could be used to determine how long the scarf is now. Evaluate your expression.

21. ***SHORT RESPONSE*** Mary's cat weighs $8\frac{1}{4}$ pounds. Nancy's cat weighs $7\frac{4}{5}$ pounds. Write an expression that could be used to determine the difference between the two cats' weights. Evaluate your expression.

LESSON 5-10

22. What is the solution to the equation $a + 5\frac{3}{10} = 9$?

A $3\frac{1}{7}$ C $4\frac{7}{10}$

B $3\frac{7}{10}$ D $14\frac{3}{10}$

23. What is the solution to the equation $1\frac{3}{8} = x - 2\frac{1}{4}$?

A $1\frac{1}{2}$ C $3\frac{1}{3}$

B $1\frac{7}{8}$ D $3\frac{5}{8}$

24. If $6\frac{5}{6} = t + 1\frac{2}{3}$, what is $3t - 2$?

A $5\frac{1}{6}$ C $13\frac{1}{2}$

B $5\frac{1}{2}$ D $15\frac{1}{2}$

25. ***SHORT RESPONSE*** Lauren had $1\frac{7}{10}$ yards of fabric. After buying another piece of fabric, she now has a total of $4\frac{1}{4}$ yards. Write an equation that could be used to determine how much fabric she bought. Solve your equation.

26. ***SHORT RESPONSE*** Lindsey is building a doghouse. She cut $3\frac{5}{8}$ feet of wood from a single board, leaving a piece $2\frac{3}{8}$ feet long. Write an equation that could be used to find the length of the original board. Solve your equation.

EOG Practice ■ Chapter 6

6A Organizing Data

LESSON 6-1

1. ***SHORT RESPONSE*** Each year, a community holds a 5 km race. In 1998, 1,345 people participated in the race. In 1999, 1,415 people participated. In 2000, 1,532 people participated. In 2001, 1,607 people participated, and in 2002, 1,781 people participated. Use the data to make a table.

2. ***SHORT RESPONSE*** Five quizzes were given this grading period. The class average on the first quiz was 85, the average on the second was 91, the average on the third was 87, the average on the fourth was 90, and the average on the fifth was 89. Use the data to make a table.

3. ***SHORT RESPONSE*** Use the career data of the three professional basketball players below to make a table. Then use your table to tell which player had the most points, rebounds, and assists. In 1,560 games, Kareem Abdul-Jabbar scored 38,387 points, grabbed 17,440 rebounds, and made 5,660 assists. In 897 games, Larry Bird scored 21,791 points, grabbed 8,974 rebounds, and made 5,695 assists. In 963 games, Bill Russell scored 14,522 points, grabbed 21,620 rebounds, and made 4,100 assists.

4. ***SHORT RESPONSE*** List five different ways to organize the table of basketball facts presented in problem 3.

LESSON  **6-2**

Points Scored				
16	18	23	13	15

5. What is the range of the points scored in the table?

   A 9   C 10.5

   B 10   D 11

Use the table for problems 6–9.

Number of Hours Spent on Homework								
Week	1	2	3	4	5	6	7	8
Hours	8	9	9	9	7	8	0	2

6. What is the range of the hours spent on homework?

   A 9   C 11

   B 10   D 12

7. What is the mean number of hours spent on homework?

   A 6   C 7

   B 6.5   D 8.5

8. What is the median of the hours spent on homework?

   A 2   C 8

   B 7   D 9

9. What is the mode for the hours spent on homework?

   A 8   C 10

   B 9   D 11

LESSON 6-3

Use the table for problems 10–11.

Lisa's Test Scores			
82	78	95	87

10. What is the median of Lisa's test scores?

    A 84.5   C 85

    B 84.8   D 85.7

11. On the next test, Lisa scored a 92. What is the mean with the new test score?

    A 85.5   C 86.8

    B 86   D 86.9

EOG Practice ■ Chapter 6

6A Organizing Data

Use the table for problems 12–15.

Number of Hours Spent Practicing								
Week	1	2	3	4	5	6	7	8
Hours	8	9	9	9	11	8	0	2

12. Which numbers in the table are outliers?

A 0, 2 C 8, 2
B 7, 8 D 9, 0

13. What is the mean number of hours spent practicing without the outliers?

A 6.6 C 9
B 7.9 D 9.4

14. What is the median of the data without the outliers?

A 8 C 9
B 8.5 D 9.5

15. What is the mode of the data without the outliers?

A 7 C 8.5
B 8 D 9

The daily temperatures for the first eight days of April were 52°F, 63°F, 61°F, 54°F, 52°F, 55°F, 68°F, and 75°F. Use the data for problems 16–21.

16. What is the range of the data set?

A 21 C 23
B 22 D 24

17. What is the mean of the temperatures?

A 60 C 75
B 65 D 78

18. What is the median of the data set?

A 43 C 50
B 45 D 58

19. What is the mode of the set of temperatures?

A 50 C 53
B 52 D 54

20. If a temperature of 28°F were recorded on the ninth day of April, what would be the mean temperature for the nine days?

A $56.\overline{4}$ C 68
B 65 D 69

21. ***SHORT RESPONSE*** Why is the mode of the data not affected by the addition of the ninth day's temperature of 28°F?

Use the table for problems 22–24.

Hours Spent Riding to School Each Week				
5	4.5	4	3	2.5
3	4	3	5	2

22. What is the mean number of hours spent riding to school?

A 3.5 C 4
B 3.6 D 4.8

23. What is the median of the hours spent riding?

A 2.9 C 3.5
B 3 D 4.5

24. What is the mode of the data?

A 2.5 C 3.5
B 3 D 4

25. ***SHORT RESPONSE*** The mean of a data set is 47.5. One more number is added to the set, and the mean does not change. Describe a number that could be added to a set and cause no change in the mean.

EOG Practice ■ Chapter 6

6B Displaying and Interpreting Data

LESSON 6-4

Use the bar graph for Exercises 1–5.

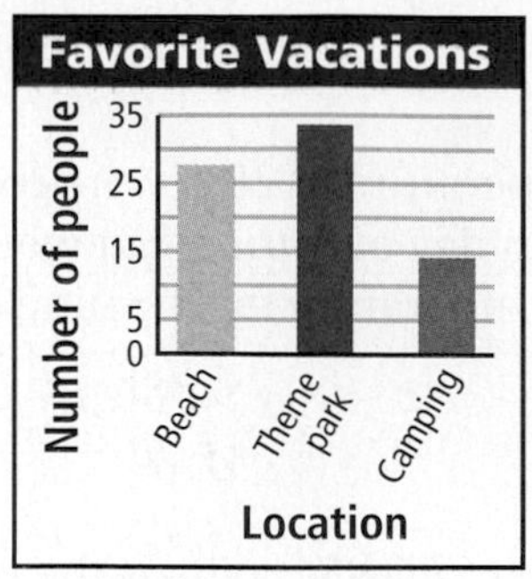

1. Which type of vacation received the most votes?
   A beach
   B beach and camping
   C camping
   D theme park

2. Which type of vacation received about 15 votes?
   A beach
   B camping
   C theme park
   D none

3. According to the bar graph, about how many more votes did theme parks receive than camping?
   A 15 C 25
   B 20 D 55

4. According to the bar graph, about how many votes were cast for theme parks and camping?
   A 40 C 60
   B 50 D 90

5. According to the bar graph, about how many votes were cast for all vacation choices?
   A 60 C 85
   B 75 D 90

LESSON 6-5

Use the table for Exercises 6–8.

Heights of Students (in.)							
63	58	48	60	60	65	56	57
56	62	61	58	59	55	64	50

6. If the data in the table was used to create a frequency table with intervals, which interval could be first?
   A 1–5
   B 20–40
   C 45–49
   D 65–69

7. ***SHORT RESPONSE*** Use the data from the table to make a frequency table with intervals, and draw a histogram.

8. ***SHORT RESPONSE*** Two students used the data to make histograms with different intervals. Explain one advantage the histogram with more intervals has over the other.

Use the data for problems 9–10.

baseball	basketball	soccer
tennis	baseball	football
baseball	football	soccer
basketball	basketball	baseball
baseball	basketball	baseball

9. ***SHORT RESPONSE*** Each student in Mr. Ander's class recorded the sport he or she plays. Make a tally table to organize the data. Which sport do most students play?

10. ***SHORT RESPONSE*** Use the tally table from problem 9 to make a cumulative frequency table.

EOG Practice ■ Chapter 6

6B Displaying and Interpreting Data

LESSON 6-6

Use the coordinate grid for problems 11–16.

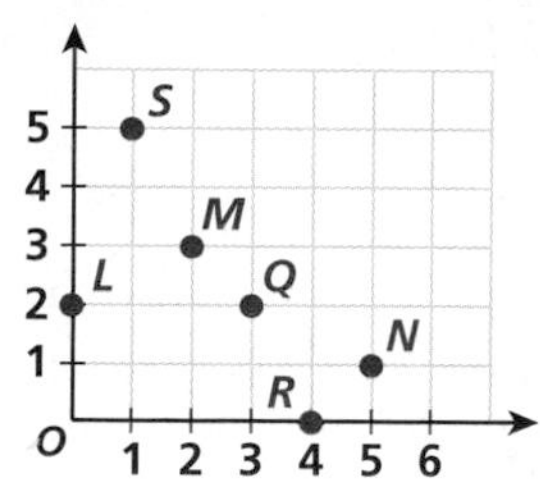

11. Which ordered pair represents the location of *L*?

A (0, 0) C (2, 2)
B (0, 2) D (2, 0)

12. Which ordered pair represents the location of *M*?

A (2, 2) C (3, 2)
B (2, 3) D (3, 3)

13. Which ordered pair represents the location of *S*?

A (0, 6) C (2, 6)
B (1, 5) D (5, 1)

14. Which point is represented by the ordered pair (4, 0)?

A *L* C *R*
B *M* D *S*

15. Which point is represented by the ordered pair (3, 2)?

A *L* C *Q*
B *M* D *R*

16. Which point is represented by the ordered pair (5, 1)?

A *M* C *R*
B *N* D *S*

LESSON 6-7

17. ***SHORT RESPONSE*** Use the data in the table to make a double-line graph.

Toy Sales				
	January	March	May	July
Store A	$460	$580	$950	$1,200
Store B	$520	$450	$880	$1,250

LESSON 6-8

18. ***SHORT RESPONSE*** Explain why this graph is misleading. Use the data to draw a bar graph that is *not* misleading.

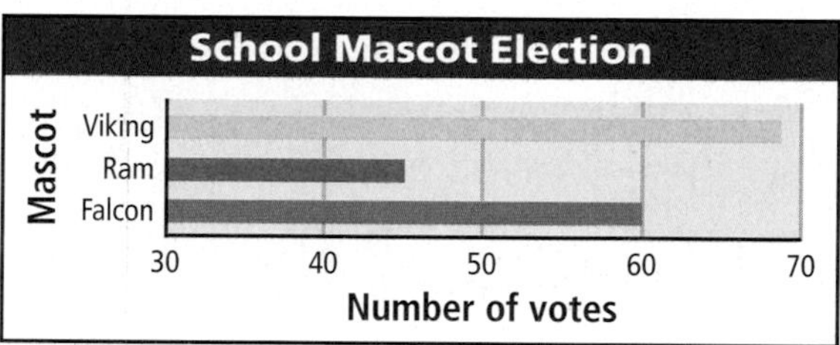

LESSON 6-9

Use the data for problem 19.

48	27	35	24
35	34	36	49

19. ***SHORT RESPONSE*** Make a stem-and-leaf plot of the data. Then use your stem-and-leaf plot to find the mean, median, and mode of the data.

Use the table for problem 20.

Time Spent Doing Homework (min)				
15	35	60	65	15
10	35	60	20	35

20. What is the median time spent doing homework?

A 60 min C 15 min
B 35 min D 20 min

EOG Practice ■ Chapter 7

7A Lines and Angles

LESSON 7-1

Use the diagram for Exercises 1–4.

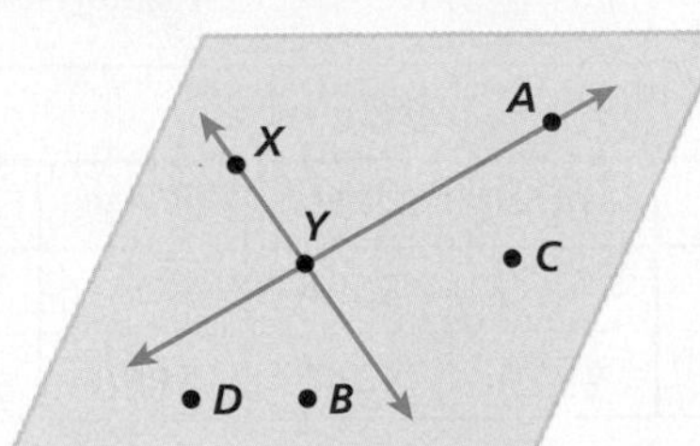

1. What is the correct notation for the line segment between points A and Y?

   A $\overline{AY}$　　C AY
   B $\overleftrightarrow{AY}$　　D $\overrightarrow{AY}$

2. What is the correct notation for the line segment between points X and Y?

   A $\overline{XY}$　　C $\overleftrightarrow{XY}$
   B XY　　D $\overrightarrow{XY}$

3. What is the correct notation for the ray starting at point A and passing through point Y?

   A AY　　C $\overline{YA}$
   B $\overrightarrow{YA}$　　D $\overrightarrow{AY}$

4. Which is the correct notation for the point sharing two lines?

   A X　　C Y
   B A　　D B

5. ***SHORT RESPONSE*** Give an example of something that could be considered a line segment. Give an example of something that could be considered a line.

6. ***SHORT RESPONSE*** Describe the figure that has the notation $\overrightarrow{XY}$. Make a drawing to illustrate your answer.

7. ***SHORT RESPONSE*** Explain the difference between $\overrightarrow{YX}$ and $\overrightarrow{XY}$. Make a drawing to illustrate your answer.

LESSON 7-2

8. The measure of $\angle ABC$ is 82°. What type of angle is $\angle ABC$?

   A acute
   B obtuse
   C right
   D straight

9. The measure of $\angle XYZ$ is 180°. What type of angle is $\angle XYZ$?

   A acute
   B obtuse
   C right
   D straight

10. The measure of $\angle A$ is 101°. What type of angle is $\angle A$?

    A acute
    B obtuse
    C right
    D straight

11. Which of the following is the measure of an acute angle?

    A 90°　　C 85°
    B 91°　　D 95°

12. Which of the following is the measure of an obtuse angle?

    A 91°　　C 45°
    B 32°　　D 6°

13. Which of the following is the measure of a right angle?

    A 45°　　C 100°
    B 90°　　D 180°

14. ***SHORT RESPONSE*** Explain how a straight angle and a right angle are different.

7A Lines and Angles

LESSON 7-3

15. What is the measure of ∠a?

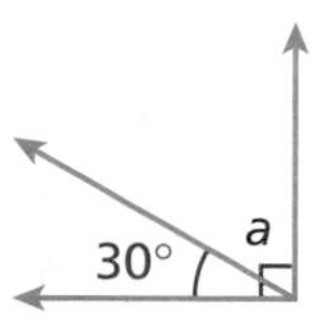

A 20° C 60°
B 30° D 70°

16. What is the measure of ∠b?

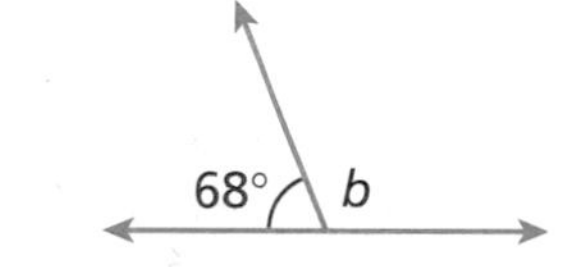

A 22° C 132°
B 112° D 168°

17. What is the measure of ∠c?

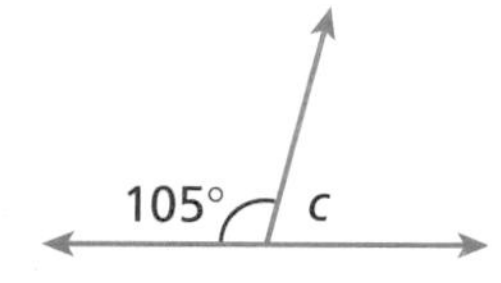

A 45° C 75°
B 70° D 95°

18. What is the measure of ∠d?

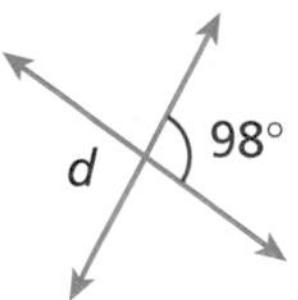

A 82° C 99°
B 98° D 102°

19. ***SHORT RESPONSE*** Frank is building a right triangle from wooden strips. He has joined two strips to form the right angle, and he has cut one end of the third piece to form an angle of 37° in the triangle. What angle must be formed at the other end of the third piece? Show your work.

LESSON  7-4

20. Which of the following describes the relationship between $\overleftrightarrow{AB}$ and $\overleftrightarrow{CD}$?

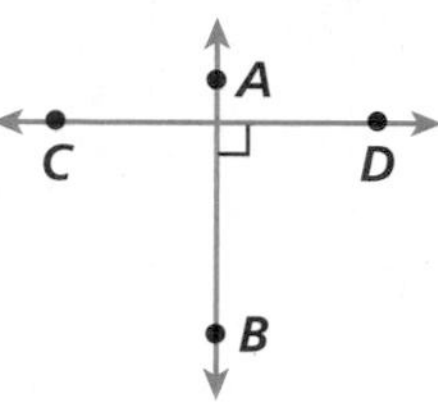

A congruent C perpendicular
B parallel D skew

21. Which of the following describes the relationship between $\overleftrightarrow{LO}$ and $\overleftrightarrow{MN}$?

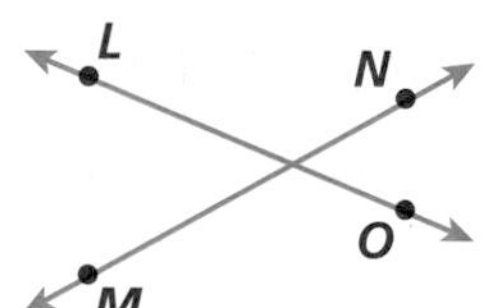

A intersecting C perpendicular
B parallel D skew

22. Which of the following describes the relationship between $\overleftrightarrow{CD}$ and $\overleftrightarrow{XY}$?

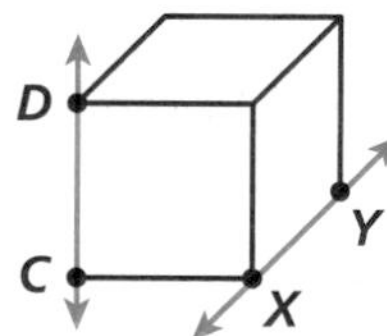

A intersecting C perpendicular
B parallel D skew

23. Which of the following describes the relationship between $\overleftrightarrow{PQ}$ and $\overleftrightarrow{UV}$?

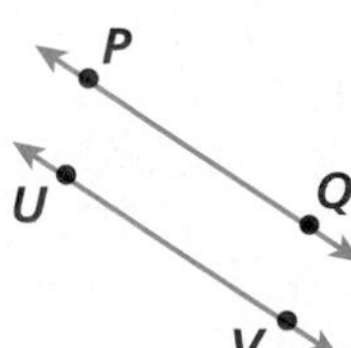

A intersecting C perpendicular
B parallel D skew

EOG Practice ■ Chapter 7

7B Polygons

LESSON 7-5

Use the diagram for problems 1–3.

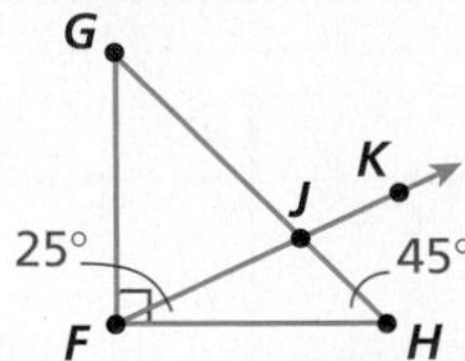

1. What is the measure of $\angle FJH$?

   A 60° C 100°
   B 110° D 90°

2. What is the measure of $\angle FJG$?

   A 45° C 70°
   B 55° D 65°

3. What is the measure of $\angle FGH$?

   A 30° C 40°
   B 35° D 45°

4. The sum of the lengths of the sides is 14 in. What is the classification of the triangle?

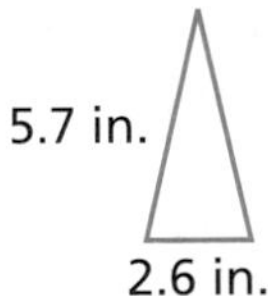

   A isosceles
   B scalene
   C obtuse
   D right

5. The sum of the lengths of the sides is 45 ft. What is the classification of the triangle?

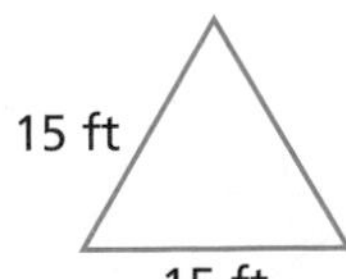

   A isosceles
   B scalene
   C obtuse
   D equilateral

LESSON 7-6

6. What is the most descriptive name for the figure?

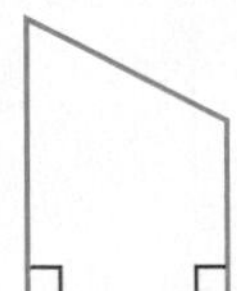

   A cube C rhombus
   B rectangle D trapezoid

7. What is the most descriptive name for the figure?

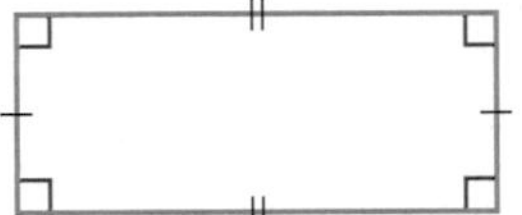

   A parallelogram C rectangle
   B rhombus D square

8. What is the most descriptive name for the figure?

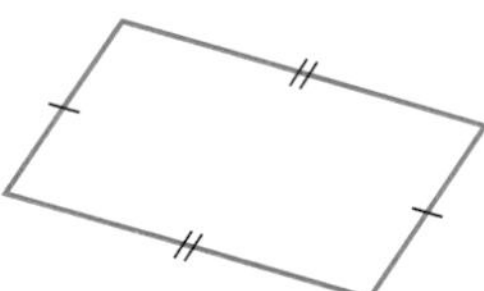

   A cube C rectangle
   B parallelogram D trapezoid

9. Which of the following choices is *not* a name for the figure?

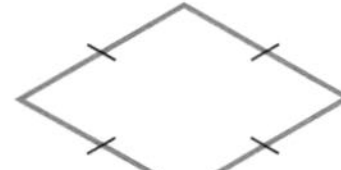

   A parallelogram C rhombus
   B quadrilateral D square

10. ***SHORT RESPONSE*** Kristi drew a quadrilateral. She classified it as a rectangle. Sara classified it as a rhombus. Explain how they can both be correct.

EOG Practice

EOG Practice ■ Chapter 7

7B Polygons

LESSON 7-7

11. What is the correct name for the polygon?

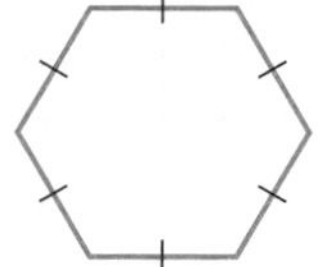

A hexagon
B octagon
C pentagon
D trapezoid

12. What is the correct name for the polygon?

A hexagon
B octagon
C pentagon
D quadrilateral

13. What is the correct name for the polygon?

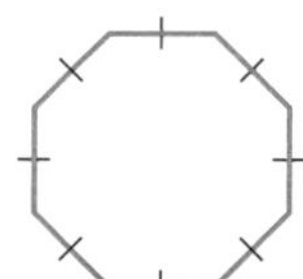

A hexagon
B octagon
C pentagon
D trapezoid

14. What is the correct name for the polygon?

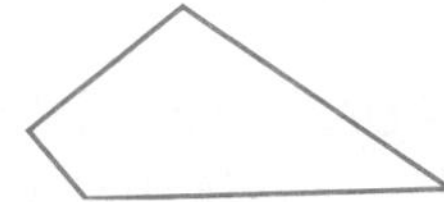

A hexagon
B octagon
C pentagon
D quadrilateral

15. ***SHORT RESPONSE*** The sum of the lengths of the sides of a regular pentagon is 35 centimeters. Write an equation that could be used to determine the length of the sides of the polygon.

LESSON 7-8

16. ***SHORT RESPONSE*** Describe a possible pattern for the series of figures. Draw the next possible figure in the series.

17. ***SHORT RESPONSE*** Describe a possible pattern for the series of figures. Draw the next possible figure in the series.

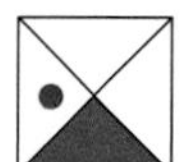 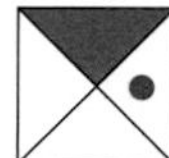 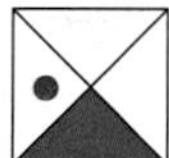 ? 

18. ***SHORT RESPONSE*** Describe a possible pattern for the series of figures. Draw the next possible figure in the series.

19. ***SHORT RESPONSE*** Describe a possible pattern for the series of figures. Draw the next possible figure in the series.

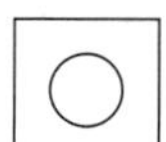    ?

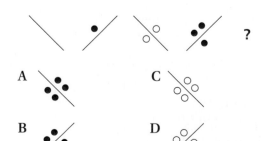

20. Find a possible pattern for the series of figures. Which figure could be next in the series?

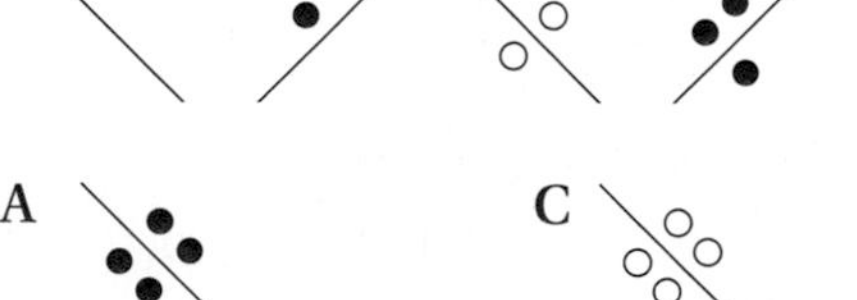

A 

C 

B

D

EOG Practice ■ Chapter 7

7C Polygon Relationships

LESSON 7-9

1. ***SHORT RESPONSE*** What does it mean for two figures to be congruent? Are the figures below congruent? Explain your answer.

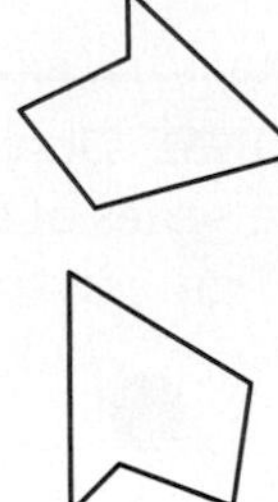

2. ***SHORT RESPONSE*** Explain why the two figures below are *not* congruent.

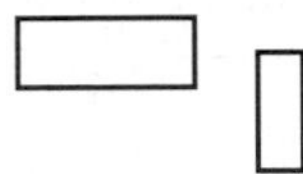

LESSON 7-10

3. What transformation is represented in the drawing?

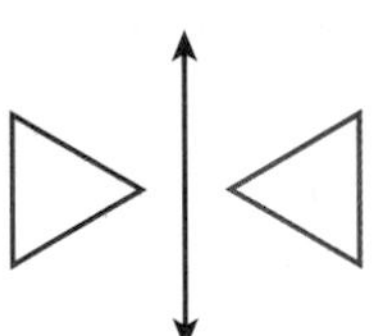

A indication
B reflection
C rotation
D translation

4. What transformation is represented in the drawing?

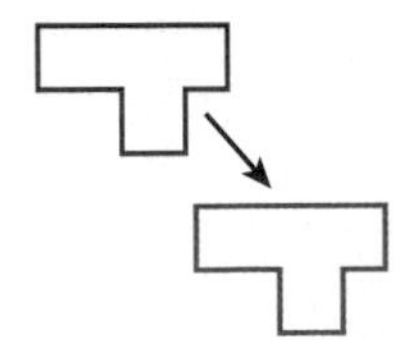

A indication
B reflection
C rotation
D translation

5. What transformation is represented in the drawing?

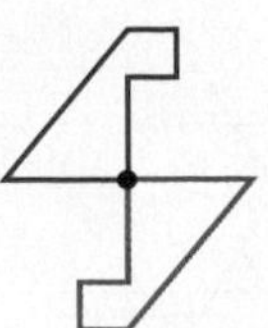

A indication
B reflection
C rotation
D translation

6. ***SHORT RESPONSE*** Copy the figure below and translate the figure 2 cm to the right.

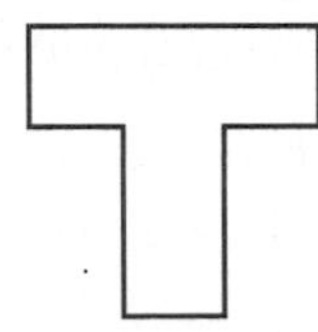

7. ***SHORT RESPONSE*** Copy the figure below and rotate the figure 90° clockwise about the point.

8. ***SHORT RESPONSE*** Use a separate sheet of paper, and draw a vertical reflection of the figure across the line

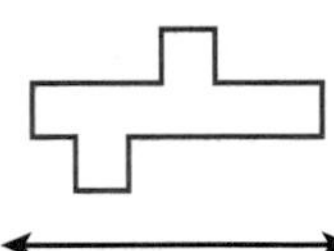

9. ***SHORT RESPONSE*** Explain how to translate a figure.

10. ***SHORT RESPONSE*** James drew a figure and rotated it 45°. He then translated the new figure 4 centimeters to the left and reflected it horizontally across a line. Are the original figure and the new figure congruent? Explain your answer.

EOG Practice ■ Chapter 7

7C Polygon Relationships

LESSON 7-11

11. ***SHORT RESPONSE*** Give the number of lines of symmetry in the polygon.

12. ***SHORT RESPONSE*** Give the number of lines of symmetry in the polygon.

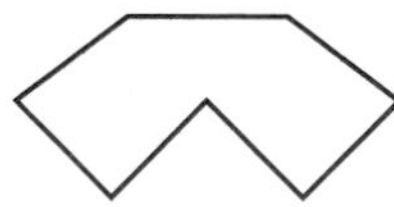

13. ***SHORT RESPONSE*** Explain how to tell if the line drawn through the figure is a line of symmetry.

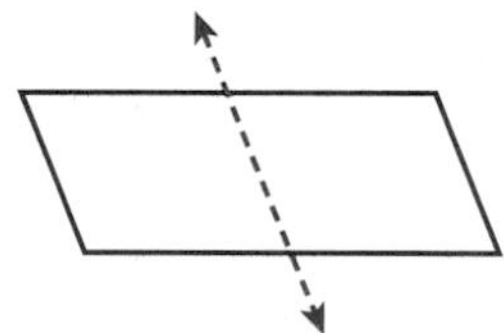

14. ***SHORT RESPONSE*** Give the number of lines of symmetry in the polygon.

15. ***SHORT RESPONSE*** Describe how to check whether a line drawn through a figure is a line of symmetry.

16. How many lines of symmetry can be drawn through a square?

    A 2

    B 4

    C 6

    D 8

LESSON 7-12

17. ***SHORT RESPONSE*** Make a drawing to show whether the shape can tessellate the plane.

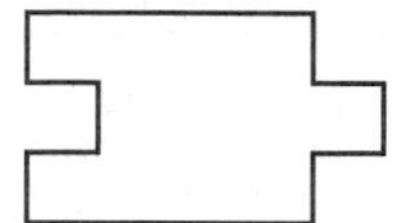

18. ***SHORT RESPONSE*** Make a drawing to show whether the shape can tessellate the plane.

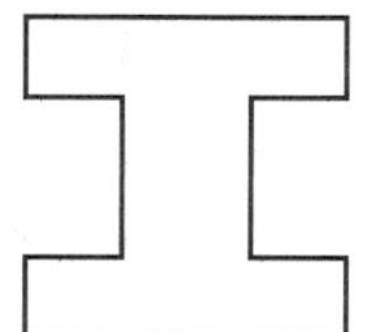

19. ***SHORT RESPONSE*** Make a drawing to show whether the shape can tessellate the plane.

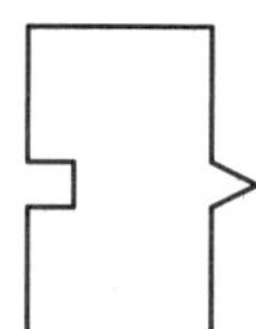

20. ***SHORT RESPONSE*** On a separate sheet of paper, draw a hexagon that is tessellating a plane.

21. ***SHORT RESPONSE*** Can all regular polygons tessellate a plane? Choose two regular polygons and make a drawing to support your answer.

22. Which of the following regular polygons *cannot* tessellate a plane?

    A hexagon

    B pentagon

    C square

    D triangle

EOG Practice ▪ Chapter 8

8A Ratios and Proportions

LESSON 8-1

Use the table for problems 1–3.

Types of Books in Doug's Collection			
Reference	10	Comic	7
Mystery	8	Poetry	5
Biography	3	Cooking	4

1. What is the ratio of cooking books to poetry books?

   A 4:5 C 5:4
   B 4:9 D 5:9

2. Which is *not* a ratio that compares the number of biography books with the number of cooking books?

   A 3:4 C 12 to 9
   B $\frac{6}{8}$ D 15:20

3. What is the ratio of biography books to total books?

   A 3:36 C 36:3
   B 3:37 D 37:3

4. ***SHORT RESPONSE*** A pack of 12 pens costs $5.52. A pack of 8 pens costs $3.92. Which is the better deal? Explain your answer.

5. ***SHORT RESPONSE*** Lucille's car uses 13.5 gallons of gas to travel 386.1 miles. Write an expression that could be used to determine how many miles the car can travel using 1 gallon of gas. Evaluate your expression.

LESSON 8-2

6. What is the value of n in the proportion $\frac{5}{4} = \frac{n}{12}$?

   A 11 C 14
   B 13 D 15

7. What is the value of n in the proportion $\frac{2}{9} = \frac{4}{n}$?

   A 6 C 15
   B 11 D 18

8. What is the value of n in the proportion $\frac{6}{10} = \frac{n}{5}$?

   A 1 C 3
   B 2 D 4

9. In which proportion is 7 *not* the correct value of n?

   A $\frac{3}{n} = \frac{12}{28}$ C $\frac{3}{n} = \frac{6}{28}$
   B $\frac{n}{9} = \frac{63}{81}$ D $\frac{n}{4} = \frac{49}{28}$

10. ***SHORT RESPONSE*** To make 2 quarts of punch, Jenny adds 16 grams of juice mix to 2 quarts of water. Write a proportion that could be used to determine the amount of mix Jenny needs to make 3 quarts of punch. Solve your proportion.

11. ***SHORT RESPONSE*** A cookie recipe that yields 3 dozen cookies requires 2 eggs. Write a proportion that could be used to determine the number of eggs needed for 12 dozen cookies. Solve your proportion.

LESSON 8-3

12. The length of a marathon is 26.2 miles. How long is a marathon in yards?

    A 0.0149 yard
    B 2,829.6 yards
    C 46,112 yards
    D 138,336 yards

13. Jeff spent $6\frac{1}{2}$ hours playing basketball. For how many seconds did Jeff play basketball?

    A 0.1083 second
    B 390 seconds
    C 23,400 seconds
    D 1,404,000 seconds

EOG Practice

EOG Practice ■ Chapter 8

8A Ratios and Proportions

14. ***SHORT RESPONSE*** The highest island peak is Puncak Jaya, in Indonesia, at 16,503 feet tall. Write a proportion that could be used to determine the height of Puncak Jaya in yards. Solve your proportion.

15. ***SHORT RESPONSE*** Emily's baby sister weighs 18.3 pounds. Write a proportion that could be used to determine Emily's sister's weight in ounces. Solve your proportion.

LESSON 8-4

The two triangles below are similar. Use the triangles for problems 16 and 17.

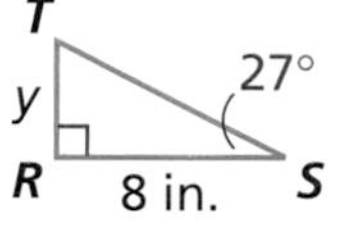

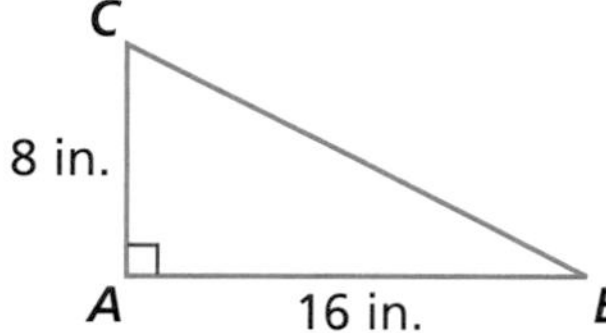

16. What is the missing length y?

    A 1 inch
    B 4 inches
    C 8 inches
    D 16 inches

17. What is the measure of $\angle B$?

    A 27°
    B 63°
    C 90°
    D 243°

18. ***SHORT RESPONSE*** For an art project, Tara drew two similar right triangles. One of the angles in the first triangle measures 35°. Draw the two triangles. What are the measurements of the three angles in the second triangle? Explain how you found your answer.

19. ***SHORT RESPONSE*** The two rectangular flower beds in Peggy's front yard are similar. The larger flower bed is 16 feet long by 5 feet wide. The length of the smaller flower bed is 8 feet. Draw a diagram of the two flower beds, and label the dimensions. Write a proportion that could be used to determine the width of the smaller flower bed. Solve your proportion.

LESSON 8-5

20. ***SHORT RESPONSE*** A telephone pole casts a shadow that is 32 yards long. At the same time, a yardstick casts a shadow that is 4 yards long. Write a proportion that could be used to determine the height of the telephone pole. Solve your proportion.

21. ***SHORT RESPONSE*** A 120-foot-tall lighthouse casts a shadow that is 300 feet long. At the same time, Jim's shadow is 12.5 feet long. Draw a diagram to show this scenario. Write a proportion that could be used to determine Jim's height. Solve your proportion.

LESSON 8-6

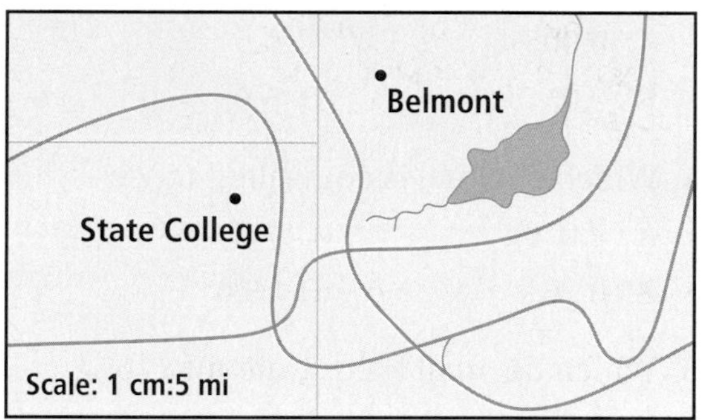

Use the map for problems 22 and 23.

22. On the map, the distance from State College to Belmont is 2 cm. What is the actual distance between the two cities?

    A 2 miles
    B 5 miles
    C 6 miles
    D 10 miles

23. Henderson City is 83 miles from State College. How many centimeters from State College should Henderson City be placed on the map?

    A 16.6 cm
    B 17.5 cm
    C 41.5 cm
    D 415 cm

8B Percents

LESSON 8-7

1. Which fraction is equivalent to 34%?

A $\frac{17}{500}$ C $\frac{17}{50}$

B $\frac{8}{25}$ D $\frac{3}{4}$

2. Which fraction is equivalent to 12%?

A $\frac{3}{250}$ C $\frac{3}{20}$

B $\frac{3}{25}$ D $\frac{1}{2}$

3. Which fraction is equivalent to 8%?

A $\frac{2}{25}$ C $\frac{1}{8}$

B $\frac{1}{10}$ D $\frac{4}{5}$

4. Which decimal is equivalent to 5%?

A 0.005 C 0.5

B 0.05 D 5.0

5. Which decimal is equivalent to 76%?

A 0.076 C 7.6

B 0.76 D 76.0

6. Which decimal is equivalent to 70%?

A 0.07 C 7

B 0.7 D 70

7. ***SHORT RESPONSE*** Michael's baseball team won 85% of its games. Explain in words how to write 85% as a fraction in simplest form. Write the equivalent fraction.

8. ***SHORT RESPONSE*** At the toy store, sales increased by 26%. Explain in words how to write 26% as a decimal. Write the equivalent decimal.

LESSON 8-8

9. Which percent is equivalent to 0.092?

A 0.092% C 9.2%

B 0.92% D 92%

10. Which percent is equivalent to 0.4?

A 0.4% C 40%

B 4% D 400%

11. Which percent is equivalent to 0.735?

A 0.735% C 73.5%

B 7.35% D 735%

12. Which percent is equivalent to $\frac{2}{5}$?

A 2.5% C 40%

B 4% D 97%

13. Which percent is equivalent to $\frac{7}{8}$?

A 7.8% C 87.5%

B 8.75% D 99%

14. Which percent is equivalent to $\frac{7}{16}$?

A 4.375% C 43.75%

B 7.16% D 91%

15. What percent is modeled by the grid below?

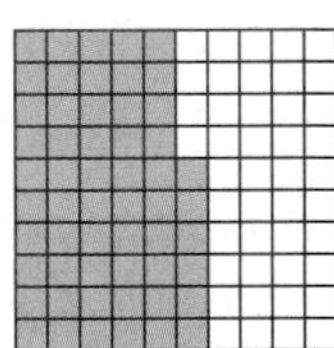

A 0.56% C 56%

B 0.65% D 65%

16. ***SHORT RESPONSE*** In Mrs. Piper's class, $\frac{19}{20}$ of the students have a pet. Explain in words how to write this fraction as a percent. What percent of the students in the class have pets?

17. ***SHORT RESPONSE*** In Andrew's poster collection, 0.305 of the posters are framed. Explain in words how to write this decimal as a percent. What percent of the posters in Andrew's collection are framed?

EOG Practice

EOG Practice ■ Chapter 8

8B Percents

LESSON 8-9

18. What is 30% of 98?

A 2.94 C 294
B 29.4 D 2,940

19. What is 15% of 220?

A 3.3 C 330
B 33 D 3,300

20. Which expression is equal to 2.16?

A 5% of 72 C 8% of 60
B 6% of 35 D 12% of 18

21. Which expression is equal to a whole number?

A 1% of 360 C 9% of 50
B 8% of 64 D 25% of 12

22. Jane has read 70% of the books on her bookshelf. If she has read 28 books, how many books are on her bookshelf?

A 19 C 30
B 20 D 40

23. Mr. Hernandez estimates that 65% of the flowers in his garden are roses. If there are 36 flowers in his garden, about how many are *not* roses?

A 8 C 23
B 13 D 55

24. ***SHORT RESPONSE*** A theater sold a total of 570 tickets for a new movie. Of those tickets, 30% were children's tickets. Write an expression that could be used to determine how many children's tickets were sold. Evaluate your expression.

25. ***SHORT RESPONSE*** Kathy has listened to 80% of the music on a CD Write an expression that could be used to determine how many more minutes of music are left on the CD if 26 minutes have passed. Evaluate your expression.

LESSON 8-10

26. Patricia is buying new roller skates that cost \$59.99. The sales tax rate is 7%. About how much will the total cost of the roller skates be?

A \$4.20 C \$64.20
B \$42.00 D \$106.00

27. Margo and her three friends went to dinner. The bill was \$34.62. They left a tip that was 15% of the bill. About how much was the tip?

A \$3.50 C \$5.25
B \$5.00 D \$7.00

28. Ashley wants to buy a sweater regularly priced at \$19.95. It is on sale for 25% off the regular price. About how much will she pay for the sweater after the discount?

A \$5 C \$16
B \$15 D \$25

29. Pete has \$50.00. He plans to purchase three books, which are priced at a total of \$39.97. If the sales tax rate is 8%, how much change should Pete receive from the cashier?

A \$3.20 C \$6.83
B \$4.00 D \$43.17

30. ***SHORT RESPONSE*** Aaron wants to buy a new CD player that is regularly priced at \$75.95. It is on sale for 10% off the regular price. The sales tax rate is 7%. Write expressions that could be used to determine the total cost of the CD player. Evaluate your expressions.

31. ***SHORT RESPONSE*** Sara spent \$102.38 on school clothes at the mall. The total of the ticket prices on the items she bought was \$94.58. How much money did Sara pay in sales tax? Write an equation that could be used to determine the sales tax rate. Solve your equation.

EOG Practice ■ Chapter 9

9A Integers

LESSON 9-1

SHORT RESPONSE **Write a positive or negative number to represent each situation.**

1. 120 feet below sea level
2. saving $22
3. a decrease of 5°F
4. a loss of 8 yd

SHORT RESPONSE **Graph each integer and its opposite on a number line.**

5. +1
6. −5
7. −3
8. +2

SHORT RESPONSE **Write the absolute value of each integer.**

9. −15
10. 11
11. −2
12. 25

13. Which integer is closest to zero?

A −5	C 3
B −4	D 4

14. Which integer is farthest from zero?

A −6	C 5
B −7	D 8

15. ***SHORT RESPONSE*** Death Valley, California, the country's lowest point, has an elevation of −282 feet. The city of Long Beach, California, has an elevation of 170 feet. Which location is farther from sea level? Use absolute value to explain your answer.

16. ***SHORT RESPONSE*** Describe the meaning of absolute value.

LESSON 9-2

17. Which symbol makes the expression 15 ■ −19 true?

A <	C =
B >	D ≅

18. Which symbol makes the expression −7 ■ −10 true?

A <	C =
B >	D ≅

19. Which symbol makes the expression −8 ■ 8 true?

A <	C =
B >	D ≅

20. Which of the following is *greater* than 7?

A 9	C −7
B 3	D −10

21. Which of the following is *less* than −8?

A −9	C 0
B −2	D 5

22. Which integer is *greater* than −9?

A −25	C −10
B −14	D 4

23. Which integer is *less* than −12?

A −16	C 5
B −10	D 8

SHORT RESPONSE **Order the integers in each set from least to greatest.**

24. −6, 5, −2
25. 12, −25, 10
26. −1, −3, 4, 0

27. ***SHORT RESPONSE*** On Monday the temperature was 3° F. On Tuesday the temperature was −4° F. On Wednesday the temperature was −1° F. On Thursday the temperature was 2° F. Which day had the coldest temperature?

EOG Practice

EOG Practice ▪ Chapter 9

9A Integers

LESSON 9-3

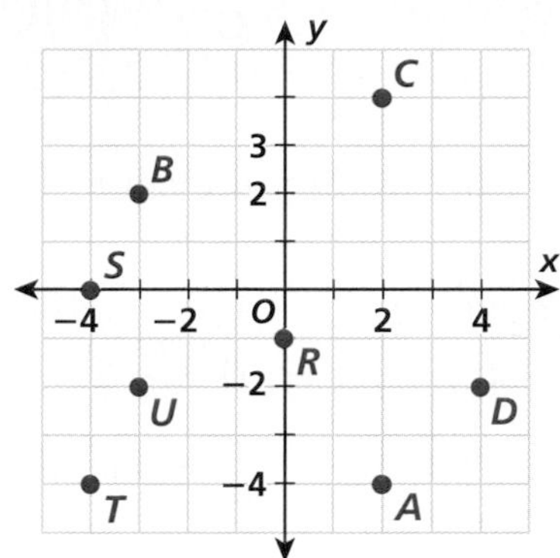

Use the coordinate plane for problems 28–33.

28. Which quadrant contains point *A*?

A I　　**C** III

B II　　**D** IV

29. Which quadrant contains point *B*?

A I　　**C** III

B II　　**D** IV

30. Which quadrant contains point *C*?

A I　　**C** III

B II　　**D** IV

31. Which quadrant contains point *T*?

A I　　**C** III

B II　　**D** IV

32. Which point is located in Quadrant II?

A *D*　　**C** *B*

B *C*　　**D** *A*

33. Which point is located in Quadrant IV?

A *R*　　**C** *B*

B *C*　　**D** *A*

34. ***SHORT RESPONSE*** Greg has located five points on a coordinate plane that has all four quadrants. He said that none of the points are in any of the quadrants. Explain how this could be possible.

Use the coordinate plane for problems 35–40.

35. What ordered pair is the location of point *S*?

A (0, −4)　　**C** (−4, 0)

B (4, 0)　　**D** (−4, 1)

36. What ordered pair is the location of point *B*?

A (−3, 2)　　**C** (−3, −2)

B (3, 2)　　**D** (3, −2)

37. What ordered pair is the location of point *R*?

A (−1, 0)　　**C** (0, −1)

B (1, 0)　　**D** (0, 1)

38. Which point has the coordinates (2, 4)?

A *T*　　**C** *R*

B *S*　　**D** *C*

39. Which point has the coordinates (2, −4)?

A *A*　　**C** *C*

B *B*　　**D** *T*

40. Which point has coordinates (−4, −4)?

A *D*　　**C** *S*

B *R*　　**D** *T*

SHORT RESPONSE **Graph each point on a coordinate plane.**

41. *M*(2, −1)

42. *W*(−4, −2)

43. *A*(2, 3)

44. *H*(−1, 3)

45. ***SHORT RESPONSE*** Graph and connect the points (5, 5), (5, −3), and (−2, −3) on a coordinate plane. What kind of figure do you have?

46. ***SHORT RESPONSE*** Graph the points (2, 3), (2, −3), and (−4, 3). What fourth ordered pair will make the figure a square?

EOG Practice ▪ Chapter 9

9B Integer Operations

LESSON 9-4

SHORT RESPONSE **Find each sum. Draw a number line and graph your answers.**

1. $3 + (-4)$
2. $-5 + 8$
3. $6 + (-2)$
4. $-9 + 4$

5. What is the value of $y + 2$ when $y = -5$?

   A 7 C 3
   B −7 D −3

6. What is the value of $-4 + x$ when $x = -2$?

   A 6 C −6
   B −2 D 2

7. Which expression has a value of 2 when $a = -7$.

   A $a + 9$ C $-5 + a$
   B $4 + a$ D $a + 15$

8. Which expression has a value of 5 when $n = 15$?

   A $n + 5$ C $-5 + n$
   B $-10 + n$ D $-15 + n$

LESSON 9-5

SHORT RESPONSE **Find each difference. Draw a number line and graph your answers.**

9. $10 - 6$
10. $-5 - (-3)$
11. $12 - (-4)$
12. $-3 - 2$

13. What is the value of $a - (-5)$ when $a = -6$?

   A 11 C −1
   B −11 D 1

14. What is the value of $x - 8$ when $x = -4$?

   A −12 C 4
   B 12 D −4

15. What is the value of $y - 2$ when $y = -9$?

   A −11 C 7
   B −7 D 11

16. Which expression has a value of 4 when $n = 2$?

   A $-2 - n$ C $n - (-2)$
   B $n - 4$ D $n - 2$

17. Which expression has a value of −13 when $x = -4$?

   A $x - (-9)$ C $13 - x$
   B $x - 9$ D $17 - x$

LESSON 9-6

SHORT RESPONSE **Find each product.**

18. $5 \cdot (-2)$
19. $-3 \cdot (-7)$
20. $-4 \cdot 4$
21. $8 \cdot (-9)$

22. What is the value of $3x$ when $x = -9$?

   A −27 C −12
   B 3 D −3

23. What is the value of $-6m$ when $m = -5$?

   A −30 C −11
   B 30 D 11

24. What is the value of $-5n$ when $n = 7$?

   A 35 C −35
   B −30 D 30

25. Which expression has a value of −12 when $a = -3$?

   A $-6a$ C $2a$
   B $-4a$ D $4a$

26. Which expression has a value of 8 when $c = -2$?

   A $8c$ C $-4c$
   B $4c$ D $-2c$

EOG Practice ■ Chapter 9

9B Integer Operations

LESSON 9-7

SHORT RESPONSE **Find each quotient.**

26. $20 \div (-4)$

27. $-48 \div (-6)$

28. $-24 \div 8$

29. $-18 \div (-2)$

30. What is $\frac{w}{3}$ when $w = 12$?

A -4 C 4
B 36 D -36

31. What is $\frac{n}{4}$ when $n = -60$?

A -15 C 15
B -240 D 240

32. What is $\frac{m}{5}$ when $m = -75$?

A -10 C 5
B 25 D -15

33. Which expression has a value of 7 when $y = -42$?

A $\frac{y}{-35}$ C $\frac{y}{6}$
B $\frac{y}{-6}$ D $\frac{y}{35}$

34. Which expression has a value of -4 when $r = -12$?

A $\frac{r}{-9}$ C $\frac{r}{3}$
B $\frac{r}{-3}$ D $\frac{r}{9}$

35. Which expression does *not* have a value of -2?

A $\frac{16}{-8}$ C $\frac{8}{-4}$
B $\frac{10}{5}$ D $\frac{-6}{3}$

LESSON 9-8

36. What value of m is the solution to $6 + m = -6$?

A $m = 0$ C $m = -1$
B $m = 12$ D $m = -12$

37. What value of y is the solution to $5 + y = 1$?

A $y = -6$ C $y = -10$
B $y = -4$ D $y = -3$

38. What value of b is the solution to $b - 8 = -6$?

A $b = 2$ C $b = -2$
B $b = 12$ D $b = -14$

39. If $-6 + m = -2$, what is $m - 7$?

A -15 C -3
B -8 D 4

40. If $x - 2 = -9$, what is $2x + 5$?

A -17 C -9
B -11 D -7

41. If $\frac{n}{6} = 4$, what is $-3n + 2$?

A -70 C 10
B -28 D 24

SHORT RESPONSE **Solve each equation. Show your work, or explain in words how you found each answer.**

42. $5 + y = 1$

43. $b - 8 = -6$

44. $7r = -42$

45. $\frac{x}{-2} = -6$

46. ***SHORT RESPONSE*** Without solving the equation, explain how you can tell if the solution to $3x = -12$ is positive or negative.

47. ***SHORT RESPONSE*** Robert is a diver. On Sunday, he dove 16 feet deeper than he dove on Saturday. He dove to -87 feet on Saturday. Write an equation that could be used to find the depth he dove to on Sunday. Solve your equation.

EOG Practice ▪ Chapter 10

10A Perimeter, Area, and Circumference

LESSON 10-1

1. What is the perimeter of the rectangle?

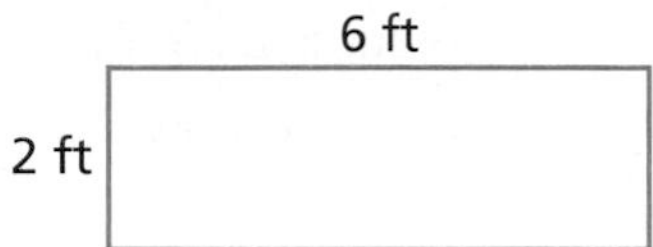

A 24 yd
B 18 yd
C 22 yd
D 20 yd

2. Find the perimeter of the rectangle.

A 20 ft
B 16 ft
C 18 ft
D 14 ft

3. What is the perimeter of the rectangle?

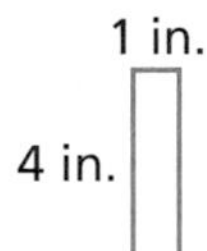

A 9 in.
B 8 in.
C 12 in.
D 10 in.

4. ***SHORT RESPONSE*** What is the value of *b* if the perimeter is 82 cm? Show your work, or explain in words how you determined your answer.

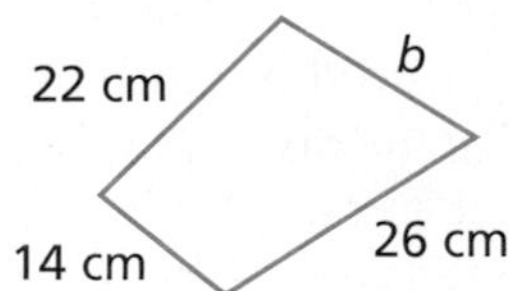

5. What is the perimeter of the polygon?

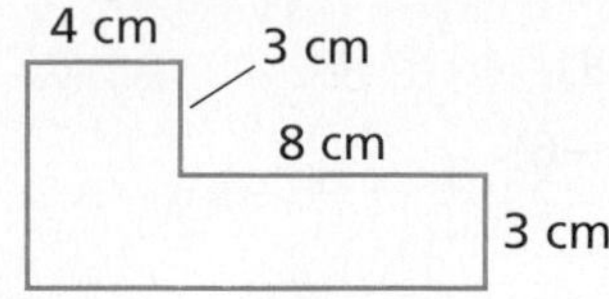

A 36 cm
B 38 cm
C 32 cm
D 30 cm

LESSON 10-2

6. What is the area of the rectangle?

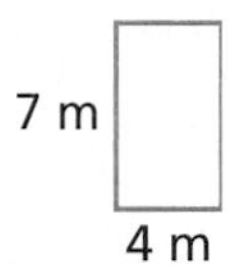

A 22 square meters
B 28 square meters
C 18 square meters
D 24 square meters

7. What is the area of the parallelogram?

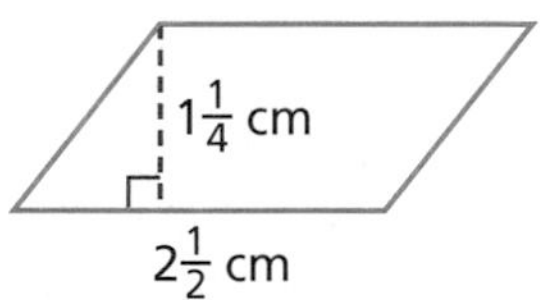

A $3\frac{1}{4}$ square centimeters

B $3\frac{1}{2}$ square centimeters

C $3\frac{1}{8}$ square centimeters

D 3 square centimeters

8. What is the area of the triangle?

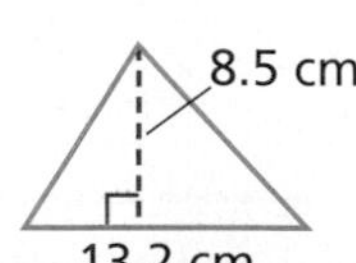

A 112.2 square centimeters
B 51.1 square centimeters
C 56.1 square centimeters
D 54 square centimeters

EOG Practice ■ Chapter 10

10A Perimeter, Area, and Circumference

LESSON 10-3

9. What is the area of the polygon?

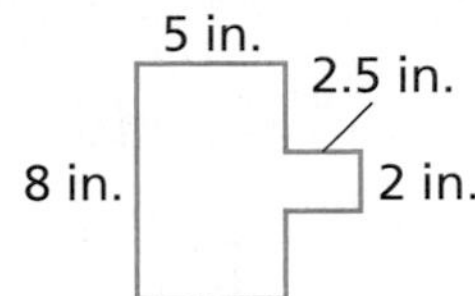

A 36 in^2

B 38 in^2

C 40 in^2

D 45 in^2

10. What is the area of the polygon?

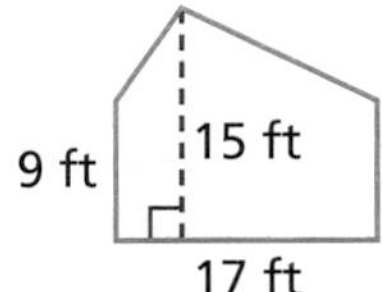

A 204 ft^2

B 300 ft^2

C 200 ft^2

D 220 ft^2

LESSON 10-4

11. ***SHORT RESPONSE*** Divide each measurement by 3. State how the perimeter and area change as the measurements change.

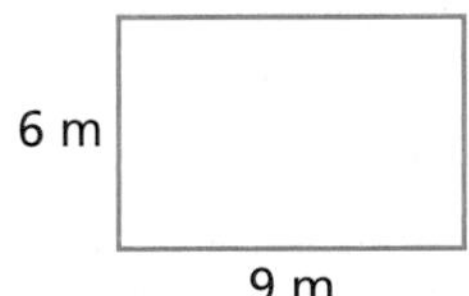

12. ***SHORT RESPONSE*** Multiply each measurement by 2. State how the perimeter and area change as the measurements change.

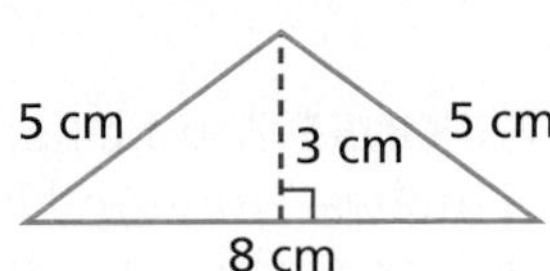

LESSON 10-5

Use 3.14 for π.

13. The diameter of a circle is 6 ft. What is the circle's circumference?

A 11.23 ft

B 18.84 ft

C 21.3 ft

D 19.45 ft

14. The radius of a circle is 12 cm. What is the circle's circumference?

A 71.41 cm

B 76.1 cm

C 80.5 cm

D 75.36 cm

15. The circumference of a circle is 17.27 in. What is the circle's diameter?

A 5.5 in.

B 4.9 in.

C 10.2 in.

D 9.45 in.

16. The radius of a circle is 10 cm. What is the circle's circumference?

A 69.4 cm

B 60.5 cm

C 62.8 cm

D 73.3 cm

17. The circumference of a circle is 29.83 in. What is the circle's diameter?

A 10.5 in.

B 9.5 in.

C 12.2 in.

D 9.35 in.

18. ***SHORT RESPONSE*** Find the circumference of a circle with a diameter of 25 miles. Use 3.14 for π. Show your work, or explain in words how you found your answer.

10B Volume and Surface Area

LESSON 10-6

1. ***SHORT RESPONSE*** Identify the number of faces, edges, and vertices on the rectangular prism.

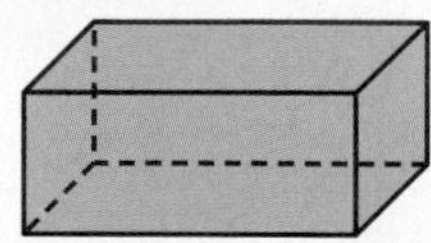

2. ***SHORT RESPONSE*** Identify the number of faces, edges, and vertices on the square pyramid.

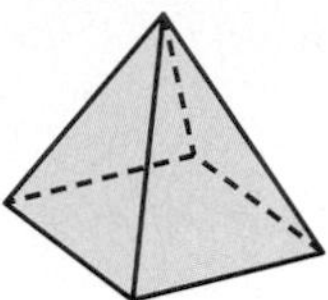

3. ***SHORT RESPONSE*** Identify the number of faces, edges, and vertices on the cylinder.

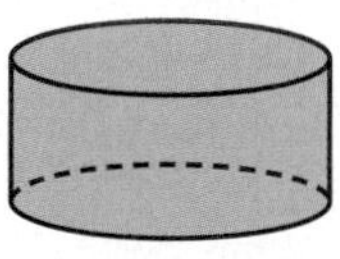

4. What is the name of a polyhedron that has 6 surfaces that are all squares?
   - A square pyramid
   - B cylinder
   - C cube
   - D cone

5. A prism is a polyhedron with two parts that are congruent and parallel. What is the name of the two parts?
   - A edges
   - B vertices
   - C bases
   - D angles

6. What is the correct name of a solid figure with one circular base and one vertex?
   - A pyramid
   - B cylinder
   - C cube
   - D cone

LESSON 10-7

7. Find the surface area S of the rectangular prism.

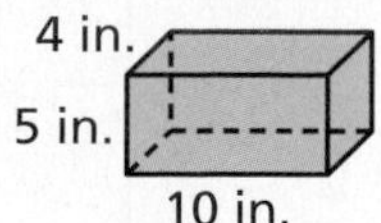

   - A $S = 200 \text{ in}^2$
   - B $S = 100 \text{ in}^2$
   - C $S = 240 \text{ in}^2$
   - D $S = 220 \text{ in}^2$

8. Find the surface area S of the square pyramid.

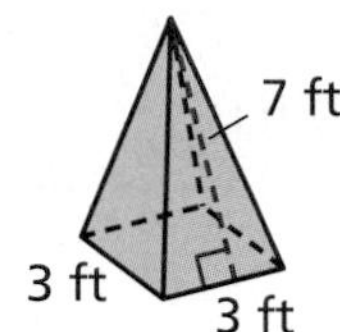

   - A $S = 75 \text{ ft}^2$
   - B $S = 51 \text{ ft}^2$
   - C $S = 48.2 \text{ ft}^2$
   - D $S = 55.1 \text{ ft}^2$

9. Find the surface area S of the cylinder.

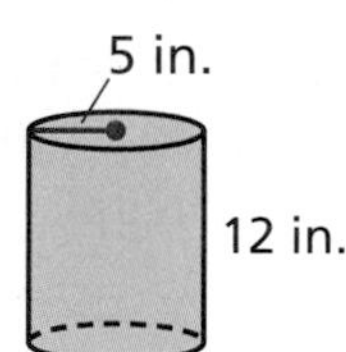

   - A $S = 545 \text{ in}^2$
   - B $S = 498 \text{ in}^2$
   - C $S = 608.1 \text{ in}^2$
   - D $S = 533.8 \text{ in}^2$

10. ***SHORT RESPONSE*** Explain how to find the area of the curved surface of a cylinder.

11. ***SHORT RESPONSE*** A cube has a side length of 5 inches. Draw and label the cube. What is the area of the base of the cube? What is the surface area of the cube? Show your work.

EOG Practice

EOG Practice ■ Chapter 10

10B Volume and Surface Area

LESSON 10-8

12. What is the volume of the rectangular prism?

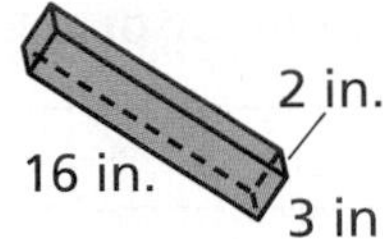

A 108 in^3
B 99 in^3
C 96 in^3
D 90 in^3

13. What is the volume of the triangular prism?

A 11.08 cm^3
B 16.4 cm^3
C 14.64 cm^3
D 10.8 cm^3

14. What is the volume of the triangular prism?

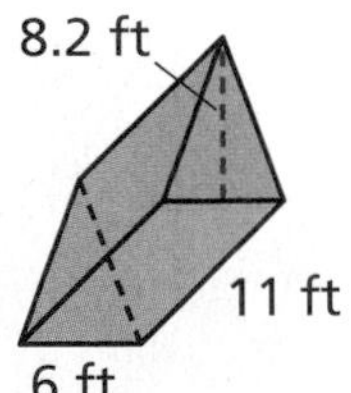

A 311.1 ft^3
B 401 ft^3
C 270.6 ft^3
D 300.1 ft^3

15. ***SHORT RESPONSE*** The volume of a rectangular prism is 400 cubic centimeters. The prism is 10 centimeters long and 8 centimeters wide. Explain how to determine the height of the prism. Draw the prism, and label all three dimensions.

LESSON 10-9

For problems 16–18, use 3.14 for π.

16. What is the volume in cubic centimeters (cm^3) of the cylinder?

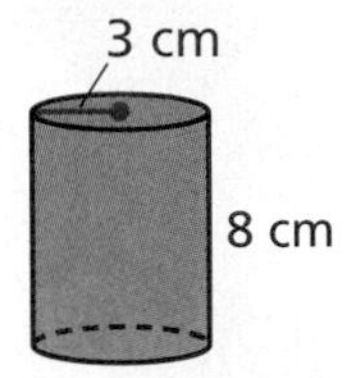

A 315 cm^3
B 226 cm^3
C 245 cm^3
D 298 cm^3

17. What is the volume in cubic feet (ft^3) of the cylinder?

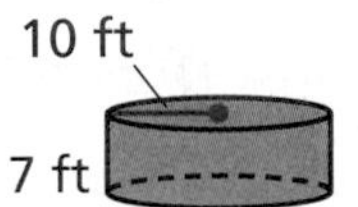

A $2{,}198\text{ ft}^3$
B $1{,}275\text{ ft}^3$
C $3{,}152\text{ ft}^3$
D $3{,}001\text{ ft}^3$

18. What is the volume in cubic inches (in^3) of the cylinder?

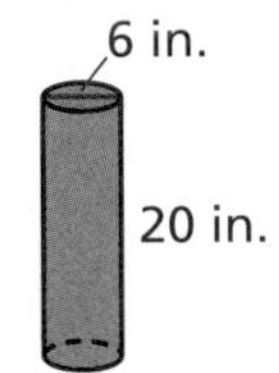

A 395 in^3
B 426 in^3
C 545 in^3
D 565 in^3

19. ***SHORT RESPONSE*** How many cubic feet are in 1 cubic yard? Show your work, or explain in words how you determined your answer.

20. ***SHORT RESPONSE*** Explain the difference between surface area and volume.

EOG Practice

EOG Practice ▪ Chapter 11

11A Understanding Probability

LESSON 11-1

Use the diagram for problems 1–3.

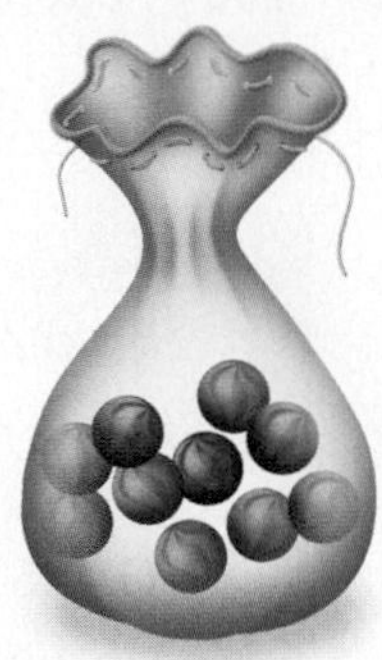

1. What is the likelihood of picking a red marble from this bag of marbles?
   A impossible C likely
   B as likely as not D certain

2. What is the likelihood of picking a red or a blue marble from this bag of marbles?
   A impossible C likely
   B as likely as not D certain

3. What is the likelihood of picking a green marble from this bag of marbles?
   A impossible C as likely as not
   B unlikely D certain

4. ***SHORT RESPONSE*** The chance of winning a sweepstakes is 3%. Explain in words how to write this probability as a decimal and as a fraction. Write the equivalent decimal and fraction.

5. ***SHORT RESPONSE*** A particular brand of cereal is offering a prize in each box. There is a 34% chance the toy will be a rubber ball, a 50% chance it will be a small figurine, and a 16% chance it will be a game. Is it more likely that the prize will be a rubber ball or a game? Explain how you determined your answer.

LESSON 11-2

For one month, Maggie recorded the weather. She organized her results in a frequency table. Use the table for problems 6–8.

Weather	Sunny	Cloudy	Rainy
Frequency	17	6	7

6. What is the experimental probability of cloudy weather?
   A 6% C 20%
   B 12% D 25%

7. What is the experimental probability of sunny weather?
   A 1.3% C 57%
   B 17% D 85%

8. ***SHORT RESPONSE*** Based on Maggie's findings, is it more likely that the weather will be cloudy or rainy? Explain how you determined your answer.

9. ***SHORT RESPONSE*** The spinner below is spun. What is the sample space? What outcome is shown?

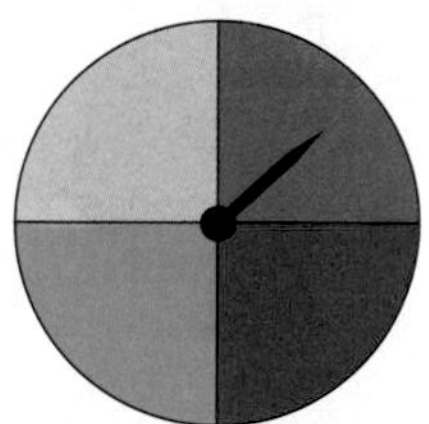

10. ***SHORT RESPONSE*** A marble is drawn from the bag below. What is the sample space? What outcome is shown?

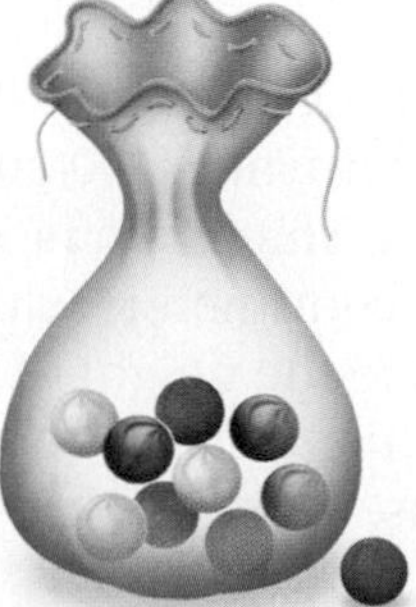

EOG Practice

EOG Practice ■ Chapter 11

11A Understanding Probability

LESSON 11-3

11. Tom flips three coins. What is the probability that all three coins will land heads up?

A $\frac{1}{8}$ C $\frac{1}{3}$

B $\frac{1}{6}$ D $\frac{1}{2}$

12. A spinner is divided into 5 equal sections. Each section is numbered 1 through 5. What is the probability of spinning an odd number?

A $\frac{1}{5}$ C $\frac{3}{5}$

B $\frac{2}{5}$ D $\frac{5}{3}$

13. What is the probability of randomly choosing the letter *T* from the letters *M, A, T, H, E, M, A, T, I, C,* and *S*?

A $\frac{1}{11}$ C $\frac{1}{8}$

B $\frac{2}{11}$ D $\frac{1}{4}$

14. What is the probability of randomly choosing a vowel from the letters *M, A, T, H, E, M, A, T, I, C,* and *S*?

A $\frac{3}{11}$ C $\frac{3}{8}$

B $\frac{4}{11}$ D $\frac{1}{2}$

15. The probability of an event *not* happening is 78%. What is the probability that the event will happen?

A 22% C 78%

B 32% D 100%

16. During its grand opening, a store is giving away prizes. The chance of winning a prize is 0.16. What is the probability of *not* winning a prize?

A 0.16 C 0.94

B 0.84 D 1.00

17. ***SHORT RESPONSE*** A bag contains 3 red beads, 10 blue beads, and 7 green beads. If a single bead is randomly chosen from the bag, what is the sample space of the experiment? What is the probability of choosing a red bead?

18. ***SHORT RESPONSE*** Martin has a 12-sided solid figure that can be rolled like a number cube. What is the probability that Martin does *not* roll a number between 1 and 4? Explain two ways to find this probability.

19. ***SHORT RESPONSE*** Pam and Nancy are going to play a board game. To decide who will go first, they will toss a D-cell battery. If it lands on its side, Nancy will go first. If it lands on its top or bottom, Pam will go first. Is this experiment fair? Explain your answer.

20. ***SHORT RESPONSE*** Tyson writes the letters in the word *probability* on individual pieces of paper and places them in a bag. He then chooses one piece of paper from the bag without looking. What is the probability that Tyson chooses the letter *B*? What is the probability that he chooses the letter *C*? What is the probability that he chooses either the letter *B* or the letter *C*? Explain your answer.

21. ***SHORT RESPONSE*** After winning a contest, Caroline will receive one of three prizes. There is a 45% chance that she will receive a stereo and a 20% chance that she will receive a gift basket. The third prize is a gift certificate. What is the probability that she will win the gift certificate? Explain how you determined your answer.

EOG Practice ■ Chapter 11

11B Using Probability

LESSON 11-4

1. Miguel has 3 choices for the exterior color on a new car: black, silver, or blue. He has 2 choices for the interior color: black or brown. How many color combinations can Miguel choose from?

   A 3   C 6

   B 5   D 36

2. At summer camp, the campers participate in 3 different activities each day: hiking, swimming, and arts and crafts. How many different ways can these 3 activities be arranged?

   A 1   C 5

   B 3   D 6

3. ***SHORT RESPONSE*** For breakfast, Brianna can have oatmeal, cold cereal, or eggs, and then a banana, an apple, or an orange. Make an organized list to help you determine how many different breakfast combinations Brianna can choose from. How many combinations are there?

4. ***SHORT RESPONSE*** Marty, Anna, Josey, and Kurt have been elected to the student council. The four offices in the council are president, vice president, secretary, and treasurer. Make an organized list to help you determine how many different ways these 4 students can hold the 4 different offices. How many ways are there?

5. ***SHORT RESPONSE*** Holly is making ceramic mugs, each of which will have two horizontal bands of color on it. She has 4 colors of paint: red, blue, yellow, and green. To determine how many different mugs she can make, Holly makes the organized list below. Explain why Holly's list is not complete, and then complete it. How many different mugs can she make?

   RB, RY, RG   BR, BY, BG

   YR, YB, YG   GR, GB, GY

LESSON 11-5

Use the diagram for problems 6–8.

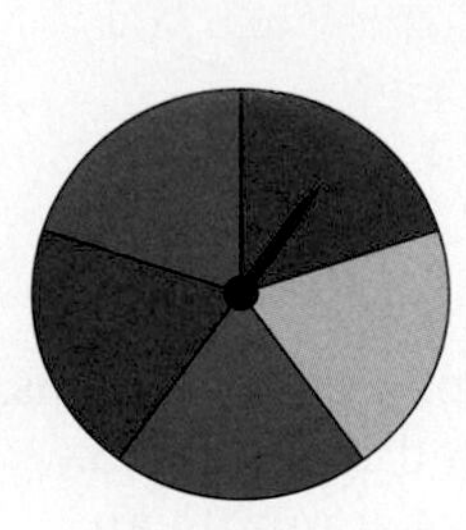

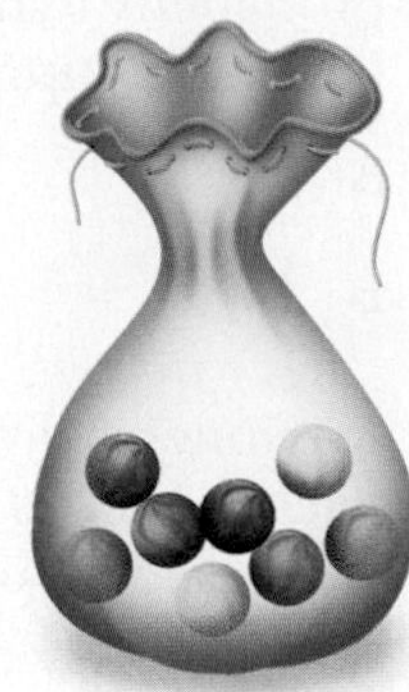

6. What is the probability of spinning red on the spinner and choosing a red marble from the bag?

   A $\frac{1}{40}$ C $\frac{6}{13}$

   B $\frac{1}{5}$ D $\frac{9}{10}$

7. What is the probability of spinning yellow and choosing a marble that is *not* yellow?

   A $\frac{3}{20}$ C $\frac{3}{4}$

   B $\frac{7}{13}$ D $\frac{19}{20}$

8. What is the probability of spinning a color that is *not* blue and choosing a marble that is *not* blue?

   A $\frac{7}{20}$ C $\frac{9}{40}$

   B $\frac{9}{20}$ D $\frac{9}{13}$

9. ***SHORT RESPONSE*** You toss two fair coins and roll a fair number cube. Make an organized list to help you determine the number of possible outcomes. What is the probability that both coins will land heads up and the cube will show a number greater than 4? Explain two ways to determine the probability that this will *not* happen.

EOG Practice ▪ Chapter 11

11B Using Probability

LESSON 11-6

10. A local survey found that 26% of the population have a pet dog. Out of 600 people, how many people do you predict will have a pet dog?

A 26 C 300
B 156 D 444

11. You roll a fair number cube 54 times. How many times do you predict that you will roll a number less than 3?

A 13 C 27
B 18 D 36

12. A commercial claims that 4 out of 5 dentists recommend a certain toothpaste. If the commercial is accurate and 1,500 dentists were interviewed, how many do you predict would recommend the toothpaste?

A 0 C 1,200
B 750 D 1,500

13. You toss two coins 72 times. How many times do you predict that both coins will land heads up?

A 9 C 24
B 18 D 36

14. The newspaper states that only 45% of the local population gets the recommended amount of sleep each night. If there are 45,000 people in town, how many people are sleep deprived?

A 10,000 C 22,500
B 20,250 D 24,750

15. Amy conducts a survey of her classmates and finds that 3 out of 5 students prefer going to the movies to watching television. If there are 650 students in her school, how many total students do you predict would prefer going to the movies?

A 220 C 330
B 260 D 390

16. ***SHORT RESPONSE*** A local theater has seating for 300 guests. Each season, the theater board raffles off 50 free tickets for each performance. The board estimates that the probability of a raffle winner's using his or her free ticket is 90%. The board wants the theater to be full for each performance. Explain in words how to determine the number of tickets the theater should sell for each performance, and then find this number.

Larry surveyed 200 boys at his summer camp to find out their favorite sport. Use the graph for problems 17–19.

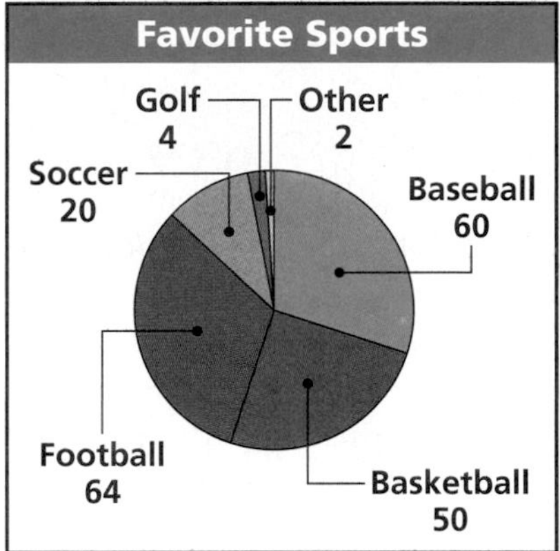

17. ***SHORT RESPONSE*** A random group of boys at camp contains 6 boys whose favorite sport is soccer. Write a proportion that could be used to predict how many boys are in the group. Solve your proportion.

18. ***SHORT RESPONSE*** Larry's horseback-riding class at camp contains 50 boys. Write a proportion that could be used to predict how many boys in horseback-riding class prefer football. Solve your proportion.

19. ***SHORT RESPONSE*** There are 550 students who attend Larry's public school. Based on his survey at summer camp, Larry predicts that baseball is the favorite sport of 165 students at school. Explain why Larry's prediction may not be accurate.

EOG Practice ▪ Chapter 12

12A Introduction to Functions

LESSON 12-1

1. Which equation for a function gives the values in the table?

x	1	2	3	4	5
y	7	9	11	13	15

A $y = 3x - 1$ C $y = 2x + 5$
B $y = 2x + 1$ D $y = 3x + 2$

2. Which equation for a function gives the values in the table?

x	3	5	7	9	11
y	5	11	17	23	29

A $y = 3x - 4$ C $y = 4x + 1$
B $y = x - 3$ D $y = 2x + 3$

3. Which equation for a function gives the values in the table?

x	10	8	6	4	2
y	20	22	24	26	28

A $y = 5x - 3$ C $y = x + 6$
B $y = 14x - 2x$ D $y = 30 - x$

4. What is the missing value of y for the indicated value of x in the function table?

x	0	4	8	12	16
y	2	3	4	5	■

A 5 C 9
B 6 D 10

5. What is the missing value of y for the indicated value of x in the function table?

x	1	2	3	4	5
y	11	12	13	14	■

A 18 C 16
B 17 D 15

6. What are the missing values of y for the indicated values of x in the function table?

x	1	2	3	4	5
y	6	9	12	■	■

A 14, 16 C 16, 17
B 15, 18 D 18, 20

SHORT RESPONSE **Write an equation for the function. Tell what each variable represents.**

7. The speed of the rocket is 8 times faster than the speed of the plane.

8. The length of a rectangle is 4 cm less than 3 times its width.

9. Darren's age is 5 more than 2 times Nicole's age.

10. ***SHORT RESPONSE*** Monica is a hairstylist. She kept track of the number of haircuts she gave for the week. On Thursday, she gave 4 haircuts and earned \$112. On Friday, she gave 7 haircuts and earned \$196. On Saturday, she gave 6 haircuts and earned \$168. Write an equation to represent the function.

11. ***SHORT RESPONSE*** A-Jax T-shirts can produce 15 shirts for \$75, 25 shirts for \$115, 35 shirts for \$155, and 45 shirts for \$195. Make a table of the data. Write an equation to represent the function. How much will it cost A-Jax to produce 150 shirts?

12. ***SHORT RESPONSE*** Party Printers charges \$15 to design an invitation and \$0.50 to print each invitation. Write an equation relating the total cost of printing to the number of invitations. If Joey needs 42 invitations, how much will Party Printers charge him? Show your work.

EOG Practice ■ Chapter 12

12A Introduction to Functions

LESSON 12-2

13. Which ordered pair is a solution to the equation $y = 6x + 2$?

A (4, 24) C (3, 20)
B (2, 15) D (5, 30)

14. Which ordered pair is a solution to the equation $y = 5x - 9$?

A (3, 8) C (2, 2)
B (4, 11) D (5, 18)

15. Which ordered pair is a solution to the equation $y = x + 1$?

A (5, 7) C (7, 4)
B (2, 3) D (9, 8)

16. Which ordered pair is *not* a solution to the equation $y = 3x - 12$?

A (7, 9) C (4, 0)
B (5, 3) D (3, −4)

17. Which ordered pair is *not* a solution to the equation $y = 4x - 10$?

A (2, −2) C (5, 12)
B (3, 2) D (6, 14)

18. The ordered pair (4, 6) is a solution to which equation?

A $y = \frac{1}{2}x + 4$

B $y = 2x + 4$

C $y = x + 3$

D $y = 4x$

19. The ordered pair (3, 3) is a solution to which equation?

A $y = 3x$

B $y = x + 3$

C $y = 5x - 12$

D $y = 2x - 4$

Use the graph of the linear function for problems 20–23.

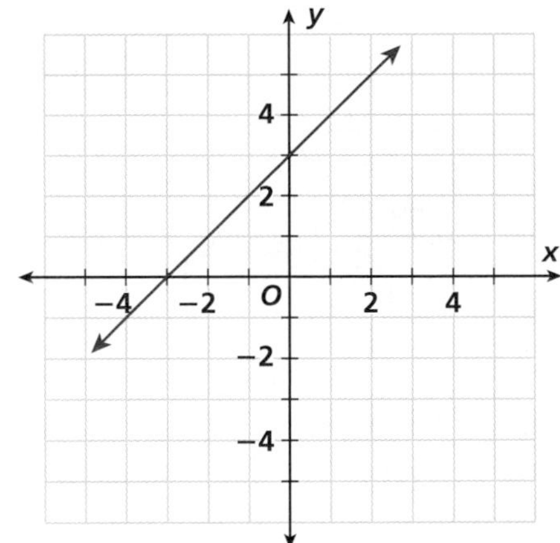

20. Which ordered pair is a solution to the linear function in the graph?

A (−3, 0)
B (−1, −2)
C (0, 4)
D (2, 6)

21. Which ordered pair is *not* a solution to the linear function in the graph.

A (−4, −1)
B (−2, 1)
C (0, −3)
D (1, 4)

22. For which value of y is $(2, y)$ a solution to the linear function in the graph?

A −6 C 5
B −5 D 6

23. For which value of x is $(x, 3)$ a solution to the linear function in the graph?

A −1 C 3
B 0 D 6

24. ***SHORT RESPONSE*** Make a function table for the linear equation $y = 4x - 2$ using x values of −2, −1, 0, 1, and 2. Then graph the function on a coordinate plane.

EOG Practice

EOG Practice ■ Chapter 12

12B Coordinate Geometry

LESSON 12-3

Use the figure for problems 1–5.

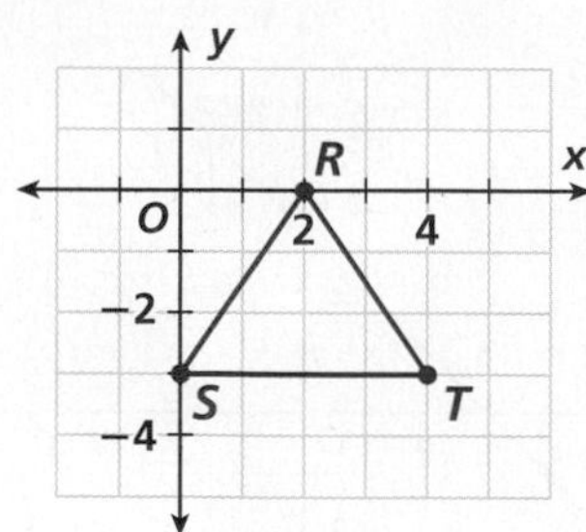

1. Translate triangle RST 3 units right and 4 units up. What are the coordinates of vertex R' after the translation?
   - A (5, 4)
   - B (5, 3)
   - C (4, 4)
   - D (4, −5)

2. Translate triangle RST 3 units right and 4 units up. What are the coordinates of vertex S' after the translation?
   - A (2, 6)
   - B (6, 3)
   - C (3, 1)
   - D (4, 2)

3. Translate triangle RST 3 units right and 4 units up. What are the coordinates of vertex T' after the translation?
   - A (3, 7)
   - B (7, 1)
   - C (1, 4)
   - D (2, −5)

4. ***SHORT RESPONSE*** Describe the translation necessary to move vertex T to the origin.

5. ***SHORT RESPONSE*** Use a sheet of graph paper, and translate triangle RST 2 units left and 3 units down.

LESSON 12-4

Use the figure for problems 6–10.

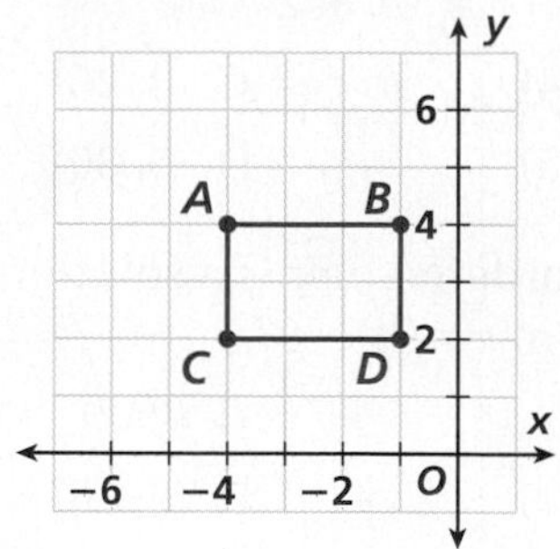

6. Reflect rectangle $ABCD$ across the x-axis. What are the coordinates of vertex A' after the reflection?
   - A (−2, −6)
   - B (−1, 3)
   - C (−3, 1)
   - D (−4, −4)

7. Reflect rectangle $ABCD$ across the x-axis. What are the coordinates of vertex B' after the reflection?
   - A (−1, −4)
   - B (4, 1)
   - C (−1, 4)
   - D (1, −5)

8. Reflect rectangle $ABCD$ across the x-axis. What are the coordinates of vertex C' after the reflection?
   - A (2, −4)
   - B (−4, −2)
   - C (−4, 1)
   - D (−4, 2)

9. What are the coordinates of vertex D' after the reflection?
   - A (−1, −4)
   - B (−1, −2)
   - C (1, −2)
   - D (1, −4)

EOG Practice ■ Chapter 12

12B Coordinate Geometry

LESSON 12-5

Use the figure for problems 10–14.

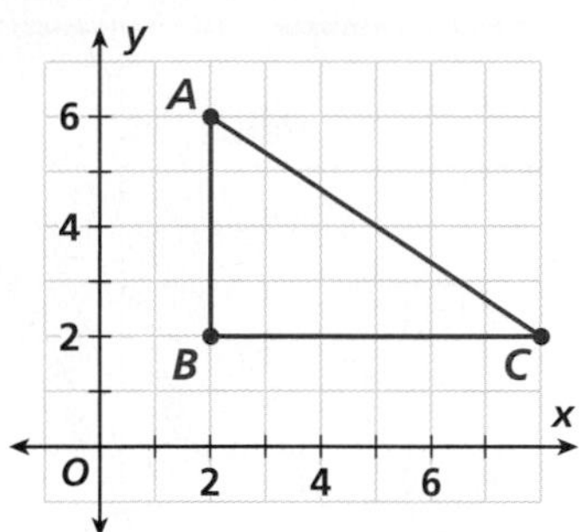

10. Rotate triangle *ABC* clockwise 90° about the origin. What are the coordinates of vertex *A'* after the rotation?

A (2, −6)

B (6, −2)

C (−4, 6)

D (4, −2)

11. Rotate triangle *ABC* clockwise 90° about the origin. What are the coordinates of vertex *B'* after the rotation?

A (−1, 2)

B (1, −2)

C (2, −2)

D (1, 4)

12. Rotate triangle *ABC* clockwise 90° about the origin. What are the coordinates of vertex *C'* after the rotation?

A (2, −6)

B (8, −2)

C (4, −6)

D (2, −8)

13. ***SHORT RESPONSE*** Describe a rotation that will place the original triangle *ABC* in the second quadrant.

14. ***SHORT RESPONSE*** Use a sheet of graph paper, and draw triangle *ABC*. Then rotate it counterclockwise 180°.

LESSON 12-6

Use the figure for problems 15–18.

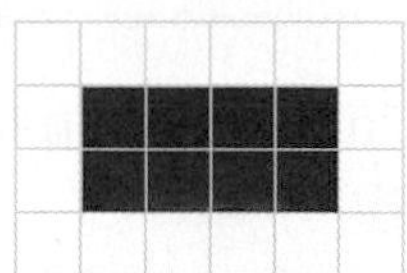

15. When the vertical dimension is increased by a factor of 3, what is the new horizontal dimension?

A 3   C 4

B 12   D 2

16. When the vertical dimension is increased by a factor of 3, what is the new vertical dimension?

A 6   C 4

B 12   D 5

17. When the horizontal dimension is multiplied by $\frac{1}{2}$, what is the new horizontal dimension?

A 8   C 4

B 16   D 2

18. When the horizontal dimension is increased by a factor of 4, what is the new horizontal dimension?

A 8   C 12

B 16   D 24

19. ***SHORT RESPONSE*** The sides of a square are 8 units long. Sketch the square, and calculate its perimeter and area. Now decrease the horizontal dimension of the square by a factor of 4, and increase the vertical dimension by a factor of 2. Sketch the new figure. Calculate the perimeter and area of the new figure, and describe how each has changed.

EOG Practice

Skills Bank Review Skills

Place Value—Hundreds Through Hundred-thousandths

You can use a place-value chart to read and write numbers.

Place Value

Hundreds	Tens	Ones	Tenths	Hundredths	Thousandths	Ten-thousandths	Hundred-thousandths
	1	5 •	2	5	8	3	

EXAMPLE

What is the place value of the digit 3 in 15.2583?

The digit 3 is in the ten-thousandths place.

PRACTICE

Write the place value of the underlined digit.

1. 0.4562<u>9</u>
2. 3<u>4</u>.071
3. 6,<u>1</u>90.05
4. 0.20<u>8</u>19
5. 1<u>0</u>3.526
6. 3.7<u>2</u>11
7. 2.160<u>8</u>
8. <u>9</u>72.8562

Compare and Order Whole Numbers

As you read a number line from left to right, the numbers are ordered from least to greatest.

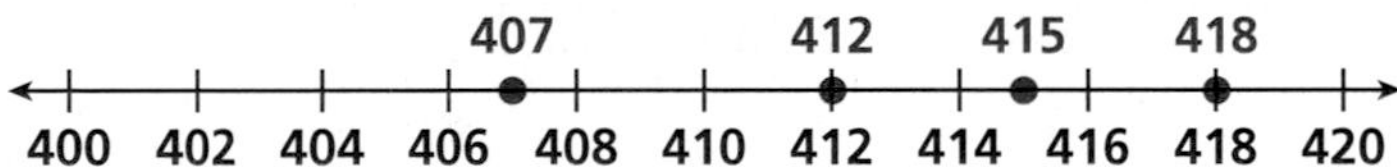

You can use a number line and place value to compare whole numbers. Use the symbols > (is greater than) and < (is less than).

EXAMPLE

Compare. Write <, >, or =.

A 412 ■ 418

418 is to the right of 412 on a number line.

412 < 418

B 415 ■ 407

1 ten is greater than 0 tens.

415 > 407

PRACTICE

Compare. Write <, >, or =.

1. 419 ■ 410
2. 9,161 ■ 8,957
3. 5,036 ■ 5,402
4. 617 ■ 681
5. 700 ■ 698
6. 1,611 ■ 1,489

Round Whole Numbers

You can use a number line or rounding rules to round whole numbers to the nearest 10, 100, 1,000, or 10,000.

EXAMPLE 1

Round 547 to the nearest 10.

Look at the number line.

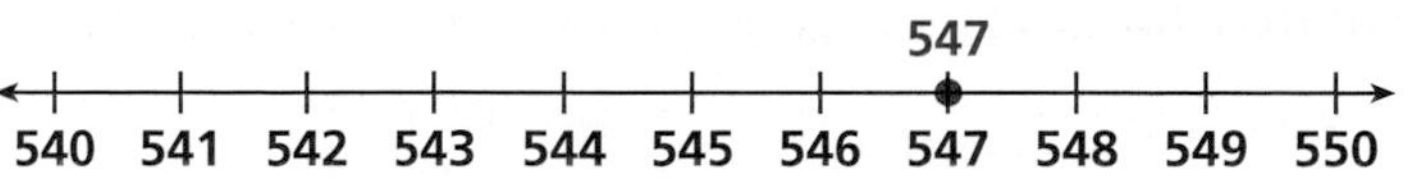

547 is closer to 550 than to 540. So 547 rounded to the nearest 10 is 550.

ROUNDING RULES

If the digit to the right is 5 or greater, increase the digit in the rounding place by 1.

If the digit to the right is less than 5, keep the digit in the rounding place the same.

EXAMPLE 2

Round 12,573 to the nearest 1,000.

12,573 — *Find the digit in the thousands place.*

↑ *Digit is 5 or greater. Add 1.* — *Look at the digit to its right.*

12,573 rounded to the nearest 1,000 is 13,000.

PRACTICE

Round each number to the given place value.

1. 15,638; nearest 100
2. 37,519; nearest 1,000
3. 9,298; nearest 10
4. 69,504; nearest 10,000
5. 852; nearest 1,000
6. 33,449; nearest 100

Round Decimals

You can use rounding rules to round decimals to the nearest whole number, tenth, hundredth, or thousandth.

EXAMPLE

Round each decimal to the given place value.

A **5.16; whole number**

$1 < 5$ So 5.16 rounds to 5.

B **13.4056; hundredth**

$5 \geq 5$ So 13.4056 rounds to 13.41.

PRACTICE

Round each decimal to the given place value.

1. 3.982; tenth
2. 6.3174; hundredth
3. 1.471; whole number
4. 48.1526; hundredth
5. 5.03654; thousandth
6. 0.083; tenth

Place Value Patterns

You can use basic facts and place value to solve math problems mentally.

EXAMPLE

Solve mentally.

A **300 + 200**

Basic fact: 3 + 2 = 5 *Think: 3 hundreds + 2 hundreds*

300 + 200 = 500

B **200 × 600**

Basic fact: 2 × 6 = 12 *Think: There are four zeros in the factors,*

200 × 600 = 120,000 *so place four zeros in the product.*

PRACTICE

Solve mentally.

1. 500 + 400 **2.** 80 − 50 **3.** 700 × 30 **4.** 2,500 ÷ 50

5. 1,200 + 600 **6.** 20 × 9,000 **7.** 650 − 300 **8.** 320 ÷ 8

Roman Numerals

Instead of using place value, as with the decimal system, combinations of letters are used to represent numbers in the Roman numeral system.

I = 1	V = 5	X = 10
L = 50	C = 100	D = 500
M = 1,000		

No letter can be written more than three times in a row. If a letter is written before a letter that represents a larger value, then subtract the first letter's value from the second letter's value.

EXAMPLE

Write each decimal number as a Roman numeral and each Roman numeral as a decimal number.

A 3

3 = I + I + I = III

B 9

9 = X − I = IX

C CLV

CLV = 100 + 50 + 5 = 155

D XC

XC = 100 − 10 = 90

PRACTICE

Write each decimal number as a Roman numeral and each Roman numeral as a decimal number.

1. 12 **2.** 25 **3.** 209 **4.** 54

5. VIII **6.** LXXII **7.** XIX **8.** MMIV

Addition

Addition is used to find the total of two or more quantities. The answer to an addition problem is called the *sum*.

EXAMPLE

4,617 + 5,682

Step 1: Add the ones.	**Step 2:** Add the tens.	**Step 3:** Add the hundreds. Regroup.	**Step 4:** Add the thousands.
4,617 + 5,682 9	4,617 + 5,682 99	1 4,617 + 5,682 299	1 4,617 + 5,682 10,299

The sum is 10,299.

PRACTICE

Find the sum.

1. 711 + 591
2. 2,580 + 2,345
3. 21,470 + 13,329
4. $165 + $304
5. 6,905 + 872
6. 47,231 + 3,254

Subtraction

Subtraction is used to take away one quantity from another quantity or to compare two quantities. The answer to a subtraction problem is called the *difference*. The difference tells how much greater or smaller one number is than the other.

EXAMPLE

780 − 468

Step 1: Subtract the ones. Regroup.	**Step 2:** Subtract the tens.	**Step 3:** Subtract the hundreds.
7 10 7 ~~8~~ ~~0~~ − 4 6 8 2	7 10 7 ~~8~~ ~~0~~ − 4 6 8 1 2	7 10 7 ~~8~~ ~~0~~ − 4 6 8 3 1 2

The difference is 312.

PRACTICE

Find the difference.

1. 6,785 − 2,426
2. 3,000 − 1,930
3. 932 − 868
4. 41,003 − 22,500
5. $1,075 − $918
6. 12,035 − 640

Multiply Whole Numbers

Multiplication is used to combine groups of equal amounts. The answer to a multiplication problem is called the *product.*

EXAMPLE

105 × 214

Step 1: Think of 214 as 2 hundreds, 1 ten, and 4 ones. Multiply by 4 ones.	**Step 2:** Multiply by 1 ten, or 10.	**Step 3:** Multiply by 2 hundreds, or 200.	**Step 4:** Add the partial products.
2 105 × 214 420 ← 4 × 105	105 × 214 420 1050 ← 10 × 105	1 105 × 214 420 1050 21000 ← 200 × 105	105 × 214 420 1050 +21000 22,470

The product is 22,470.

PRACTICE

Find the product.

1. 350 × 112
2. 3,218 × 231
3. 187 × 136
4. 5,028 × 225
5. 642 × 428
6. 2,039 × 570

Skills Bank

Multiply by Powers of Ten

You can use mental math to multiply by powers of ten.

EXAMPLE

4,000 × 100

Step 1: Look for a basic fact using the nonzero part of the factors.	**Step 2:** Add the number of zeros in the factors. Place that number of zeros in the product.
4 × 1 = 4	4,000 × 100 = 400,000

The product is 400,000.

PRACTICE

Multiply.

1. 600 × 100
2. 90 × 1,000
3. 2,000 × 10
4. 400 × 10
5. 10,000 × 1,000
6. 7,100 × 1,000

Divide Whole Numbers

Division is used to separate a quantity into equal groups. The answer to a division problem is known as the *quotient.*

EXAMPLE

672 ÷ 16

Step 1: Write the first number inside the long division symbol and the second number to the left. Place the first digit of the quotient.	**Step 2:** Multiply 4 by 16, and place the product under 67.	**Step 3:** Bring down the next digit of the dividend.
16)672 *16 cannot go into 6, so try 67.*	4 over 16)672; − 64; 3 *Subtract 64 from 67.*	42 over 16)672; − 64; 32; −32; 0 *Divide 32 by 16.*

The quotient is 42.

PRACTICE

Find the quotient.

1. 578 ÷ 34
2. 736 ÷ 8
3. 826 ÷ 118
4. 945 ÷ 45
5. 6,312 ÷ 263
6. 5,989 ÷ 53

Divide with Zeros in the Quotient

Sometimes when dividing, you need to use zeros in the quotient as placeholders.

EXAMPLE

3,648 ÷ 12

Step 1: Divide 36 by 12 because 12 > 3.	**Step 2:** Place a zero in the quotient because 12 > 4.	**Step 3:** Bring down the 8.
3 over 12)3,648	30 over 12)3,648; −36; 04	304 over 12)3,648; −36; 048; −48; 0

The quotient is 304.

PRACTICE

Find the quotient.

1. 424 ÷ 4
2. 5,796 ÷ 28
3. 540 ÷ 18
4. 7,380 ÷ 123
5. 12,045 ÷ 3
6. 10,626 ÷ 21

Compatible Numbers

Compatible numbers are numbers that are easy to compute mentally. They are often based on groups of 10 or on basic facts.

EXAMPLE 1

A $7 + 6 + 3 + 4$

$(7 + 3) + (6 + 4)$ *Make groups of 10.*

$10 + 10$

20

B $2 \times 32 \times 5$

$(2 \times 5) \times 32$ *Make a group of 10.*

10×32

320

EXAMPLE 2

Estimate $358 \div 9$.

Basic fact: $36 \div 9 = 4$ *360 is compatible with 9. $360 \div 9 = 40$*

$358 \div 9 \approx 40$

PRACTICE

Use compatible numbers to solve.

1. $15 + 42 + 38 + 25$
2. $4 \times 3 \times 25$
3. $17 + 51 + 23 + 19$
4. $6 \times 15 \times 4$
5. $11 + 123 + 57 + 9$
6. $2 \times 7 \times 20 \times 5$

Estimate by rounding to find compatible numbers.

7. $473 \div 80$
8. $118 \div 4$
9. $57 \div 11$

Mental Math

You can use the Distributive Property to find products mentally.

EXAMPLE

6×32

Step 1: Write 32 as the sum of a multiple of 10 and a one-digit number. 6×32 $6 \times (30 + 2)$	**Step 2:** Use the Distributive Property. $6 \times (30 + 2)$ $(6 \times 30) + (6 \times 2)$	**Step 3:** Use mental math to multiply and then to add. $(6 \times 30) + (6 \times 2)$ $180 + 12 = 192$

PRACTICE

Use the Distributive Property to find each product.

1. 5×66
2. 3×42
3. 8×21
4. 7×84
5. 5×93
6. 4×75

Properties

Addition and multiplication follow some properties, or laws. Knowing the addition and multiplication properties can help you evaluate expressions.

Addition Properties		
Commutative	You can add numbers in any order.	$5 + 1 = 1 + 5$
Associative	When you are only adding, you can group any of the numbers together.	$(9 + 3) + 2 = 9 + (3 + 2)$
Identity Property of Zero	The sum of any number and zero is equal to the number.	$9 + 0 = 9$

Multiplication Properties		
Commutative	You can multiply numbers in any order.	$5 \times 8 = 8 \times 5$
Associative	When you are only multiplying, you can group any of the numbers together.	$(4 \times 9) \times 7 = 4 \times (9 \times 7)$
Identity Property of One	The product of any number and one is equal to the number.	$6 \times 1 = 6$
Property of Zero	The product of any number and zero is zero.	$5 \times 0 = 0$
Distributive	When you multiply a number times a sum, you can find the sum first and then multiply, or multiply each number in the sum and then add.	$6 \times (4 + 5) = 6 \times 4 + 6 \times 5$

EXAMPLE

Tell which property is shown in the equation $(3 + 4) + 7 = 3 + (4 + 7)$.

The Associative Property of Addition is shown.

PRACTICE

Tell which property is shown.

1. $6 \times (3 \times 2) = (6 \times 3) \times 2$
2. $12 \times 9 = 9 \times 12$
3. $0 + d = d$
4. $k \times 1 = k$
5. $8 + 5 = 5 + 8$
6. $2 \times (3 + 10) = (2 \times 3) + (2 \times 10)$
7. $2 + (3 + 4) = (2 + 3) + 4$
8. $99 \times 0 = 0$
9. $y(3 + 10) = 3y + 10y$

Fractional Part of a Region

You can use fractions to name parts of a whole. The denominator tells how many equal parts are in the whole. The numerator tells how many of those parts are being considered.

EXAMPLE

Tell what fraction of each region is shaded.

A 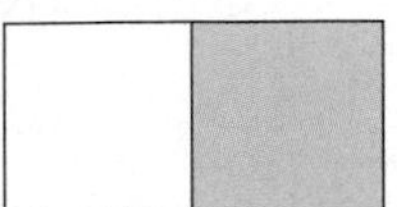

$\frac{1}{2}$

B 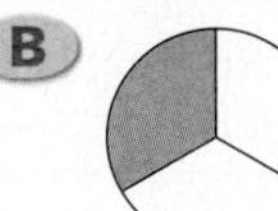

$\frac{1}{3}$

 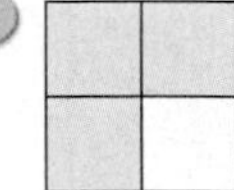

$\frac{3}{4}$

PRACTICE

Tell what fraction of each region is shaded.

1. 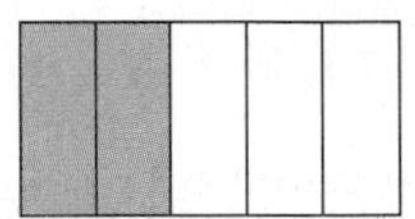

2. 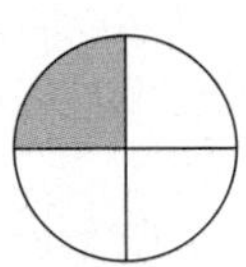

3. 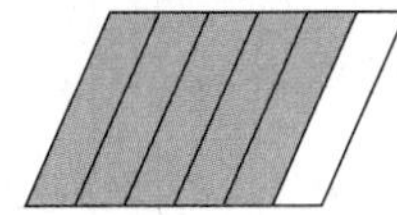

4.

5. 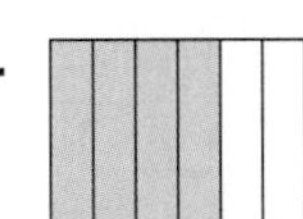

6. 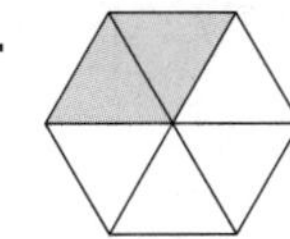

Fractional Part of a Set

You can use fractions to name part of a set. The denominator tells how many items are in the set. The numerator tells how many of those items are being used.

EXAMPLE

Tell what fraction of each set are stars.

A 

3 out of 10 shapes are stars.

$\frac{3}{10}$ of the shapes are stars.

B  ☆●☆☆●☆☆

5 out of 7 shapes are stars.

$\frac{5}{7}$ of the shapes are stars.

PRACTICE

Tell what fraction of each set is shaded.

1. ☆☆☆☆☆☆

2. ■■■■□

3. ●☆○○☆○

4. ■■■□□

5. ●○□□

6. ☆⬢■○▯△♡

Pictographs

Pictographs are graphs that use pictures to display data. Pictographs include a key to tell what each picture represents.

EXAMPLE

How many students chose red as their favorite color?

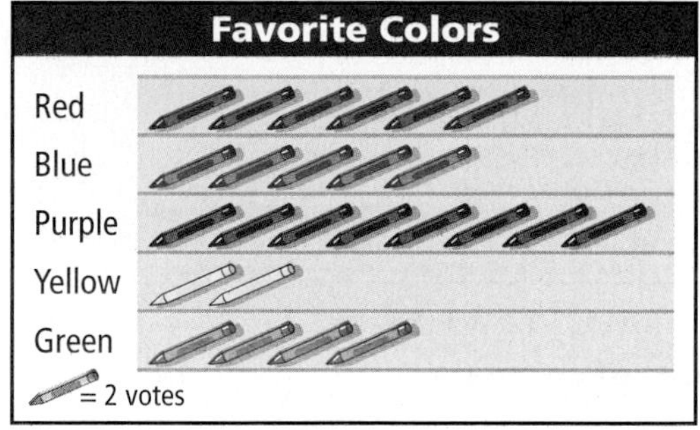

Each stands for 2 students.

There are 6 in the row for red.

$6 \times 2 = 12$

So 12 students chose red as their favorite color.

PRACTICE

Use the pictograph for Exercises 1–4.

1. How many tickets did theater A sell?
2. Which theater sold the most tickets?
3. How many more tickets did theater C sell than theater D?
4. Theater E sold 180 tickets. How would this be shown on the pictograph?

Tickets Sold

Theater A
Theater B
Theater C
Theater D
= 20 tickets

Use the pictograph for Exercises 5–7.

Mr. Carr took a survey of sixth-graders in his school. He asked them which type of pet they have. He recorded the data in a table.

5. How many students have pet birds?
6. How many more students have pet cats than pet fish?
7. How many students were surveyed?
8. Elizabeth took a survey of her neighbors. She recorded the number of children in each family in a table. Use the data to make a pictograph.

Types of Pets

Dog
Cat
Bird
Fish
Other
= 2 students

Children	Families
0	1
1	6
2	4
3 or more	2

Line Plots

You can display data on a line plot. Line plots use a number line and *x*'s or another symbol to show frequency of values. By looking at a line plot, you can quickly see the range of the data, mode, and outliers.

EXAMPLE 1

Use the line plot to answer the following questions.

A What is the range of the data?

The line plot has data plotted from 7 to 15. The range is 7 to 15. $15 - 7 = 8$

B How many campers are age 10?

On the line plot, there are 3 *x*'s above the 10 mark. There are 3 campers who are age 10.

7 8 9 10 11 12 13 14 15

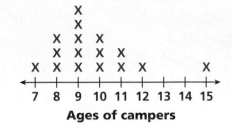

Ages of campers

C What is the mode of the data?

The mode is the number that occurs most often. Age 9 has the most *x*'s. The mode is 9.

EXAMPLE 2

Students in Mr. Gordon's class ran several miles a week. The number of miles run by the students is recorded in the table. Organize the data in a line plot.

Number of Miles Run	3	4	5	6	7	8	9	10
Number of Students	5	0	6	4	3	7	0	2

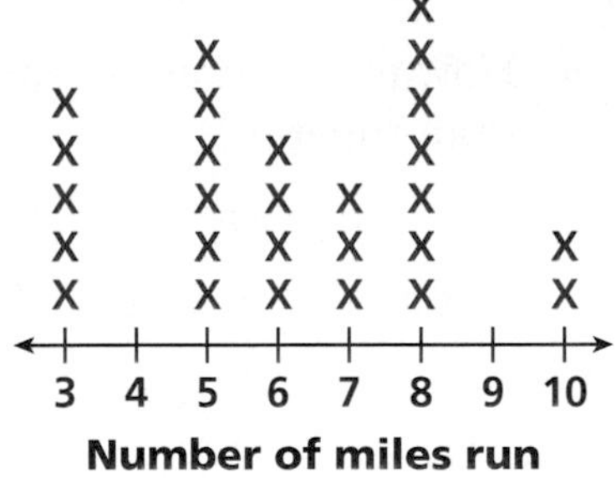

PRACTICE

Mark's baseball coach kept track of the number of hits by each player on the team. He organized the data in a line plot. Use the line plot for Exercises 1–3.

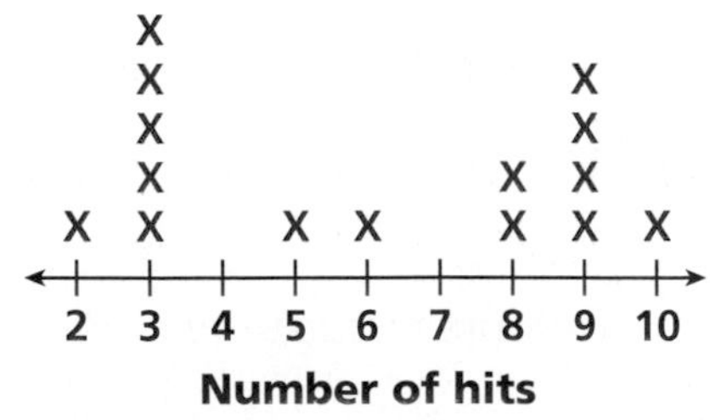

1. How many players had 4 hits?
2. How many players had more than 6 hits?
3. What is the mode of the data?
4. Betty participates in a summer reading program. The frequency table shows the number of books read by the students in the program. Organize the data in a line plot.

Summer Reading Program						
No. of Books Read	5	8	10	11	12	15
No. of Students	4	9	5	10	6	3

Measure Length to the Nearest $\frac{1}{16}$ Inch

Each inch on this ruler is separated into 16 equal parts. Each mark is $\frac{1}{16}$ inch.

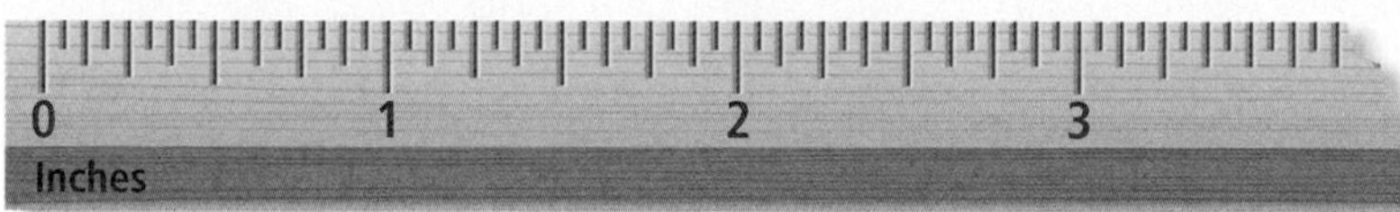

EXAMPLE

What is the length of the pencil?

Count the number of $\frac{1}{16}$ marks after the 5-inch mark. There are 3 marks. The pencil is $5\frac{3}{16}$ inches long.

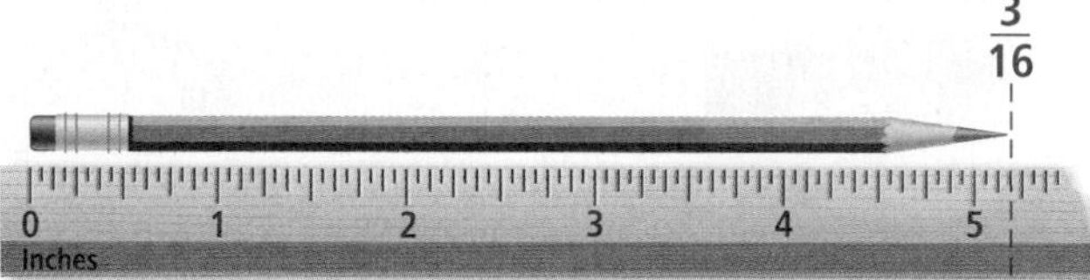

PRACTICE

Use a ruler to find the length of each object to the nearest $\frac{1}{16}$ inch.

1.

2.

3.

Read Scales

A *scale* is similar to a number line with numbers or marks placed at fixed intervals. You can find scales on graphs and on measuring instruments, such as rulers and thermometers.

EXAMPLE

What temperature is shown on the thermometer?

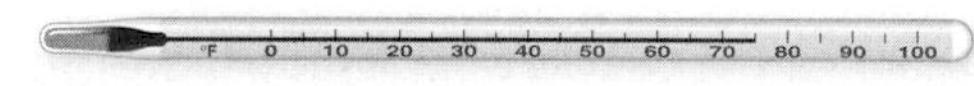

The scale goes from 0°F to 100°F in intervals of 5°F. The temperature shown is 75°F.

PRACTICE

Read each scale.

1.

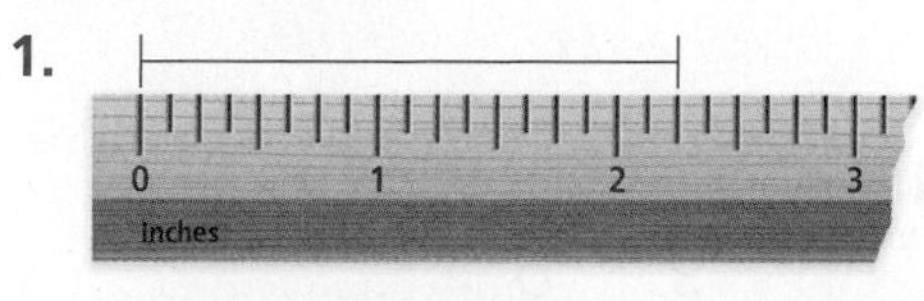

2.

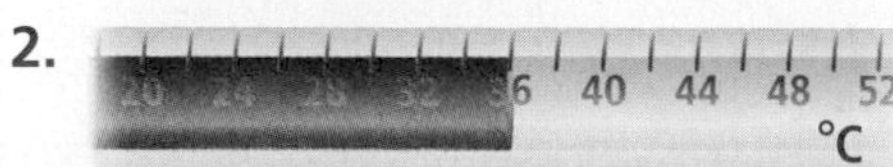

3.

Skills Bank

Time

Seconds, minutes, hours, days, weeks, months, and years are units you can use to measure time.

EXAMPLE

Which instrument would you use to measure how long it takes to read a page in a book?

JANUARY						
S	M	T	W	T	F	S
	1	2	3	4	5	6
7	8	9	10	11	12	13
14	15	16	17	18	19	20
21	22	23	24	25	26	27
28	29	30	31			

A digital clock shows hours and minutes.

An analog clock shows hours, minutes, and seconds.

A calendar shows days, weeks, months, and years.

Since it would take less than a day to read a page in a book, you could use a digital clock or an analog clock.

PRACTICE

Name the appropriate instrument and unit to measure time for each event.

1. completing 6th grade
2. running a mile
3. eating lunch
4. Earth revolving around the Sun

Right Triangle Trigonometry

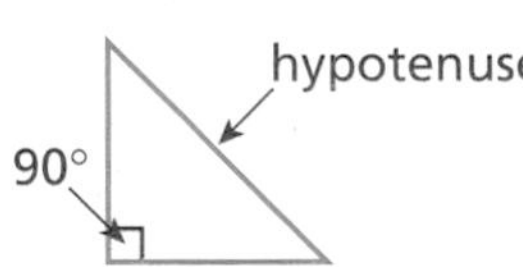

A right triangle has one right angle.
The side opposite the right angle is called the *hypotenuse*.
The hypotenuse is the longest side of a right triangle.
The other sides of a right triangle are called *legs*.

EXAMPLE

Determine if the triangle is a right triangle. If so, identify the hypotenuse.

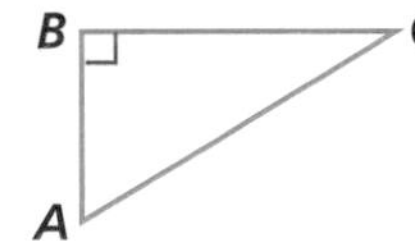

$\triangle ABC$ has a 90° angle.
$\triangle ABC$ is a right triangle.
Line segment CA is the hypotenuse.

PRACTICE

Determine if each triangle is a right triangle. If so, identify the hypotenuse.

1. 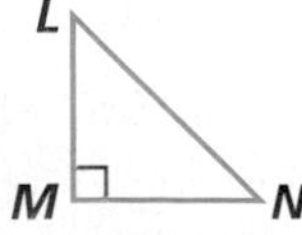

2. 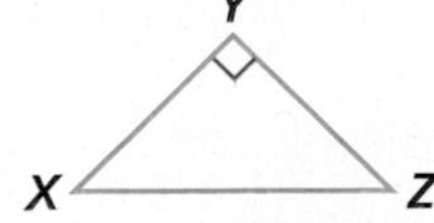

3. 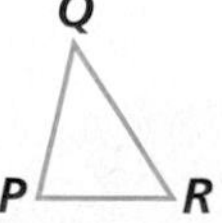

Skills Bank Preview Skills

Graph Cumulative Frequency

You have seen how to make a cumulative frequency table for a data set. You can also graph the cumulative frequencies for a data set.

Midterm Test Scores					
70	86	70	74	77	95
82	62	69	79	7	80
87	68	72	72	91	87
98	73	64	81	77	73
99	76	68	95	85	80

EXAMPLE

The midterm test scores for Mr. Andrews's math class are given in the table at right. Make a cumulative frequency table. Then make a histogram of the cumulative frequencies.

Divide the data into equally sized intervals.

Midterm Score	Frequency	Cumulative Frequency
60–64	2	2
65–69	3	5
70–74	8	13
75–79	4	17
80–84	4	21
85–89	4	25
90–94	1	26
95–99	4	30

The cumulative frequency column shows a running total of all frequencies.

The frequency tells the number of times an event, category, or group occurs.

To make a histogram of the cumulative frequencies, draw a bar for the cumulative frequency for each interval.

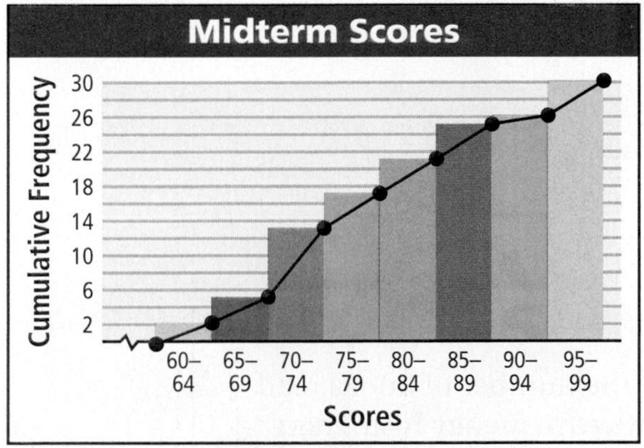

To make a line graph of the cumulative frequencies, place points in the lower left corner of the first bar and upper right corner of every bar. Then connect those points with line segments, as shown.

PRACTICE

1. Make a cumulative frequency histogram and line graph for the data set.

Students' Heights (cm)					
160	130	142	153	164	160
161	162	132	155	140	130
150	145	140	138	166	155
154	155	160	160	155	158

Relative Frequency and Relative Frequency Distributions

In a data set, the relative frequency of a data value is that value's frequency divided by the total number of data values.

$$\text{relative frequency} = \frac{\text{frequency}}{\text{total number of data values}}$$

Relative frequencies can be shown in tables or displayed in histograms.

EXAMPLE

The average class size in 20 schools is given in the table. Make a relative frequency table and a relative frequency histogram of the data.

Average Class Size				
22	25	20	28	31
37	24	19	29	32
38	35	19	32	34
38	25	38	26	33

Divide the data into equally sized intervals.

Class Size	Frequency	Relative Frequency
19–23	4	$\frac{4}{20} = \frac{1}{5}$
24–28	5	$\frac{5}{20} = \frac{1}{4}$
29–33	5	$\frac{5}{20} = \frac{1}{4}$
34–38	6	$\frac{6}{20} = \frac{3}{10}$

There are 20 data points. Divide each frequency by 20 to find the relative frequency.

To make a histogram, draw a bar for each relative frequency.

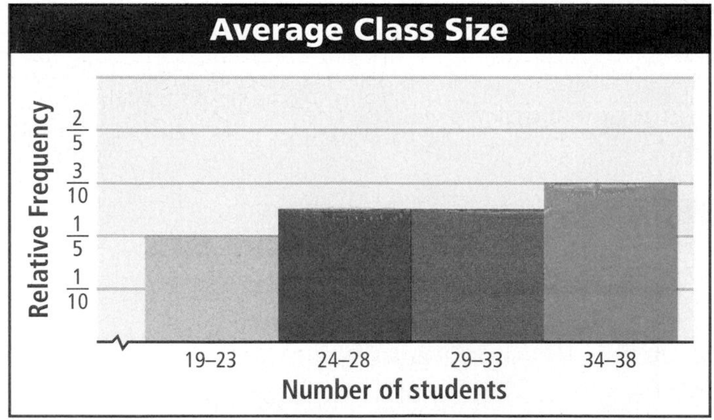

PRACTICE

1. Survey the students in your class and record the number of books read by each student in the past year. Make a relative frequency histogram to display the data.

Independent and Dependent Events 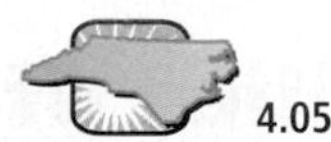 4.05

We say that two events are independent if the occurrence of one event has no effect on the probability that the other event will occur. For example, if Morgan and Melissa each toss a coin and Morgan's coin lands heads up, this has no effect on the probability that Melissa's coin will land heads up.

On the other hand, two events are dependent if the occurrence of one event *does* have an effect on the probability that the other event will occur. For example, suppose a bag of marbles contains 3 red marbles, 2 blue marbles, and 1 white marble. If you reach in and choose one marble without looking, the probability that you will choose a red marble is $\frac{1}{2}$. Suppose that you select a red marble and then pass the bag to your friend without replacing your marble. The probability that your friend will choose a red marble is now $\frac{2}{5}$. (There are now 5 marbles in the bag, and 2 of them are red.)

EXAMPLE

Determine whether the events are dependent or independent.

A **Frank reaches into his sock drawer without looking and removes a white sock. Without replacing this sock, he reaches in and pulls out another sock.**

These two events are dependent. Frank did not replace the first sock, and this changes the probabilities for his second sock.

B **A coin is tossed and lands heads up. The same coin is tossed again and lands tails up.**

These two events are independent. The outcome of one coin toss does not affect the outcome of a second coin toss.

PRACTICE

Determine whether the events are dependent or independent.

1. A tossed coin lands heads up, and a 5 is rolled on a number cube.
2. Ryan draws a black card from a deck of cards and keeps it. Then Laura draws a red card.
3. David chooses a blue mountain bike from a choice of 6 bike colors and he chooses a black helmet from a choice of 4 helmet colors.
4. A bag contains 6 letter tiles labeled *E, V, E, N, T,* and *S*. Patricia selects a tile from the bag and replaces it. She then selects a second tile.
5. A bag contains 6 letter tiles labeled *E, V, E, N, T,* and *S*. Emilio selects a tile and keeps it. He then selects a second tile.
6. A jar contains 1 penny, 1 nickel, 1 dime, and 1 quarter. Chance picks a coin and keeps it. He then picks a second coin.

Probabilities of Independent and Dependent Events

4.05

To find the probability of two independent events, multiply the individual probabilities.

EXAMPLE 1

A nickel and a quarter are tossed. What is the probability that the nickel will land heads up and the quarter will land tails up?

The events are independent.

$P(\text{heads, tails}) = P(\text{heads}) \cdot P(\text{tails})$

$= \frac{1}{2} \cdot \frac{1}{2} = \frac{1}{4}$ *Multiply the individual probabilities.*

To find the probability of two dependent events, you must determine the probability of the second event *after the first event has happened.* Then multiply the probabilities.

EXAMPLE 2

Ten cards numbered from 0 to 9 are placed in a hat. One card is chosen and not replaced. Then a second card is drawn. What is $P(6, 4)$?

The events are dependent.

$P(6) = \frac{1}{10}$ *There is 1 six out of 10 cards.*

$P(4 \text{ after } 6) = \frac{1}{9}$ *There is 1 four out of 9 remaining cards.*

$P(6, 4) = \frac{1}{10} \cdot \frac{1}{9} = \frac{1}{90}$

PRACTICE

Six cards—3 green, 1 red, and 2 black—are placed in a box.

1. One card is chosen and not replaced. Then another card is chosen. What is *P*(green, red)?
2. One card is chosen and replaced. Then another card is chosen. What is *P*(red, green)?
3. One card is chosen and not replaced. Then another card is chosen. What is *P*(black, red)?
4. One card is chosen and replaced. Then another card is chosen. What is *P*(green, black)?
5. One card is chosen and not replaced. Then another card is chosen. What is *P*(black, green)?

Skills Bank

Skills Bank Science Skills

Convert Units

When you know how different units relate, you can use formulas to convert from one unit to another.

Length	Weight	Capacity
1 foot (ft) = 12 inches (in.)	1 pound (lb) = 16 ounces (oz)	1 pint (pt) = 2 cups (c)
1 yard (yd) = 3 feet	1 ton (T) = 2,000 pounds	1 quart (qt) = 2 pints
1 mile (mi) = 1,760 yards		1 gallon (gal) = 4 quarts

EXAMPLE 1

Complete.

A 24 feet = ▢ yards

Think: 3 feet = 1 yard

24 feet ÷ 4 feet per yard = 6 yards

24 feet = 6 yards

B 3 gallons = ▢ quarts

Think: 1 gallon = 4 quarts

3 gallons × 4 quarts per gallon = 12 quarts

3 gallons = 12 quarts

Temperature can be measured in degrees Fahrenheit (°F) or in degrees Celsius (°C). To change from one temperature scale to the other, you can use the following formulas:

$$F = \left(\frac{9}{5} \times C\right) + 32 \qquad C = \frac{5}{9} \times (F - 32)$$

EXAMPLE 2

Convert between temperature scales.

A 50°C = ▢ °F

$F = \left(\frac{9}{5} \times 50\right) + 32$

$F = 122$

B 68°F = ▢ °C

$C = \frac{5}{9} \times (68 - 32)$

$C = 20$

PRACTICE

Complete.

1. 72 in. = ▢ ft
2. 5 lb = ▢ oz
3. 2 mi = ▢ yd
4. 500 lb = ▢ T
5. 8 pt = ▢ c
6. 5 pt = ▢ qt
7. 8 yd = ▢ in.
8. 2 T = ▢ oz
9. 3 gal = ▢ c

Convert between temperature scales.

10. 32°F = ▢ °C
11. 30°C = ▢ °F
12. 50°F = ▢ °C
13. 100°C = ▢ °F
14. 77°F = ▢ °C
15. 41°F = ▢ °C

Skills Bank

Compute Measurements of Combined Units

Sometimes a measurement is given in a combination of units. For example, a piece of wood may measure 3 feet 4 inches. You can add or subtract measurements that are a combination of units.

EXAMPLE 1

4 ft 8 in. + 5 ft 6 in.

Step 1: Line up the units.	Step 2: Add the inches.	Step 3: Add the feet.	Step 4: Rewrite the answer in simplest form.
4 ft 8 in. + 5 ft 6 in.	4 ft 8 in. + 5 ft 6 in. 14 in.	4 ft 8 in. + 5 ft 6 in. 9 ft 14 in.	*Think: 12 in. = 1 ft* 9 ft 14 in. = 10 ft 2 in.

The sum is 10 ft 2 in.

EXAMPLE 2

3 hr 20 min − 1 hr 50 min

Step 1: Line up the units.	Step 2: Regroup if needed.	Step 3: Subtract the minutes.	Step 4: Subtract the hours.
3 hr 20 min − 1 hr 50 min	2 hr 80 min − 1 hr 50 min	2 hr 80 min − 1 hr 50 min 30 min	2 hr 80 min − 1 hr 50 min 1 hr 30 min

The difference is 1 hr 30 min.

PRACTICE

Add.

1. 7 ft 2 in. + 6 ft 8 in.
2. 8 lb 6 oz + 4 lb 12 oz
3. 2 gal 1 qt + 4 gal 1 qt
4. 12 ft 11 in. + 3 ft 4 in.
5. 4 hr 12 min + 3 hr 42 min
6. 152 yd 2 ft + 75 yd 6 in.
7. 5 yd 2 ft 3 in. + 8 yd 1 ft 8 in.
8. 2 hr 36 min 45 s + 5 hr 42 min 20 s

Subtract.

9. 20 ft 8 in. − 7 ft 6 in.
10. 10 yd 1 ft − 5 yd 2 ft
11. 6 lb 5 oz − 2 lb 8 oz
12. 12 h 13 min − 6 h 25 min
13. 5 min 15 s − 4 min 55 s
14. 3 mi 550 yd − 1 mi 760 yd
15. 4 gal 1 c − 3 qt 1 pt
16. 1 day − 8 hr 36 min

Compare Units

When converting area from one unit to another, you must remember that area is measured in square units.

1 ft × 1 ft square

$$\begin{aligned} 1 \text{ square foot} &= 1 \text{ foot} \times 1 \text{ foot} \\ &= 12 \text{ inches} \times 12 \text{ inches} \\ &= 144 \text{ square inches} \end{aligned}$$

Customary Units for Area	
1 square foot (ft^2) = 144 square inches (in^2)	1 acre (a) = 4,850 square yards (yd^2)
1 square yard (yd^2) = 9 square feet (ft^2)	1 acre (a) = 43,560 square feet (ft^2)
1 square yard (yd^2) = 1,296 square inches (in^2)	1 square mile (mi^2) = 640 acres (a)

Multiply to convert from larger units to smaller units.

Divide to convert from smaller units to larger units.

EXAMPLE 1

Find the area of the rectangle in square feet and in square inches.

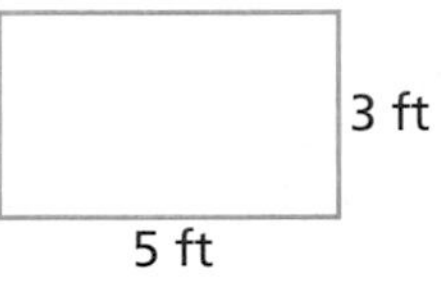

$3 \text{ ft} \times 5 \text{ ft} = 15 \text{ ft}^2$ *Think: $1 \text{ ft}^2 = 144 \text{ in}^2$*

$15 \text{ ft}^2 = 15 \times 144 \text{ in}^2 = 2{,}160 \text{ in}^2$

EXAMPLE 2

Which is the greater area, 3 yd^2 or 25 ft^2?

$3 \text{ yd}^2 = 3 \times 9 \text{ ft}^2 = 27 \text{ ft}^2$ *Think: $1 \text{ yd}^2 = 9 \text{ ft}^2$*

$27 \text{ ft}^2 > 25 \text{ ft}^2$

$3 \text{ yd}^2 > 25 \text{ ft}^2$

PRACTICE

1. Find the area of the rectangle in square yards and square feet.

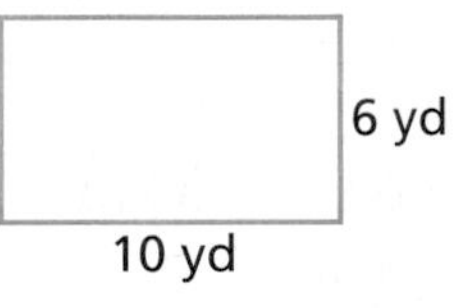

2. A plot of land is 1.5 miles long and 1 mile wide. What is the area of the land in square miles and in acres?

Compare. Write <, >, or =.

3. $12{,}500 \text{ yd}^2$ ▢ 3 acres
4. 6 yd^2 ▢ 42 ft^2
5. 4 ft^2 ▢ 576 in^2
6. 5 yd^2 ▢ $6{,}500 \text{ in}^2$
7. 2.3 mi^2 ▢ 1,430 acres
8. 0.5 acre ▢ $21{,}700 \text{ ft}^2$

Surface Area to Volume Ratio

Surface area is the sum of the areas of all the faces or surfaces of a solid figure. *Volume* is the amount of space within the solid figure. Area is a measurement of two dimensions, length and width. Volume is a measure of three dimensions, length, width, and height. A surface area to volume ratio compares the surface area and volume of a solid.

EXAMPLE 1

Find the surface area and volume of the rectangular prism.

$S = 2wh + 2\ell w + 2\ell h$

$= (2 \times 4 \times 3) + (2 \times 6 \times 4) + (2 \times 6 \times 3)$

$= 24 + 48 + 36$

$= 108 \text{ ft}^2$

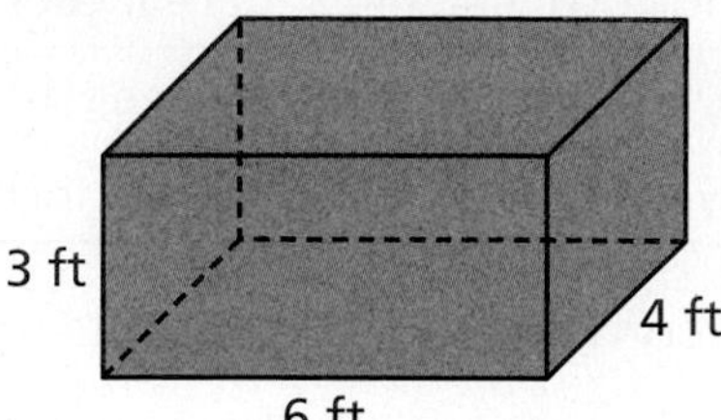

$V = \ell \times w \times h$

$= 6 \times 4 \times 3$

$= 72 \text{ ft}^3$

EXAMPLE 2

What is the surface area to volume ratio for the cube?

$S = 6s^2$

$= 6 \times 5 \times 5$

$= 150 \text{ m}^2$

$V = \ell \times w \times h$

$= 5 \times 5 \times 5$

$= 125 \text{ m}^3$

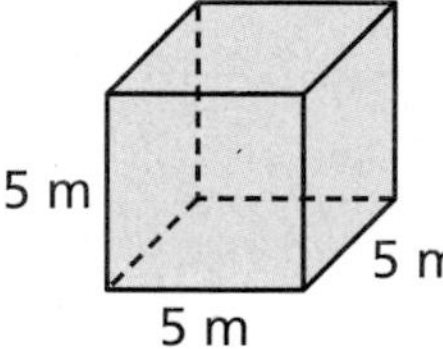

The ratio of surface area to volume for the cube is $150 \text{ m}^2\text{:}125 \text{ m}^3$ or $6 \text{ m}^2\text{:}5 \text{ m}^3$.

PRACTICE

Find the surface area and volume of each rectangular prism.

1.

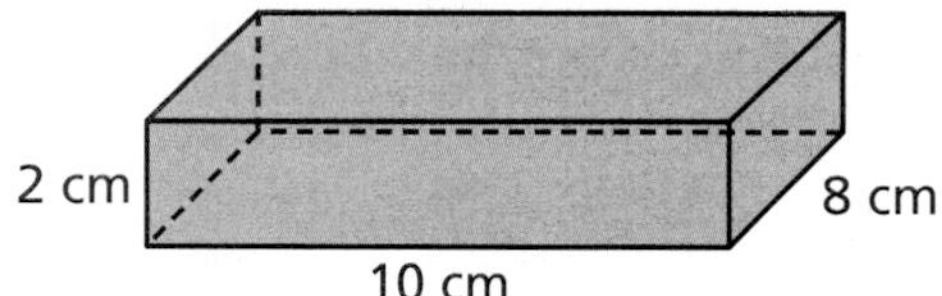

2.

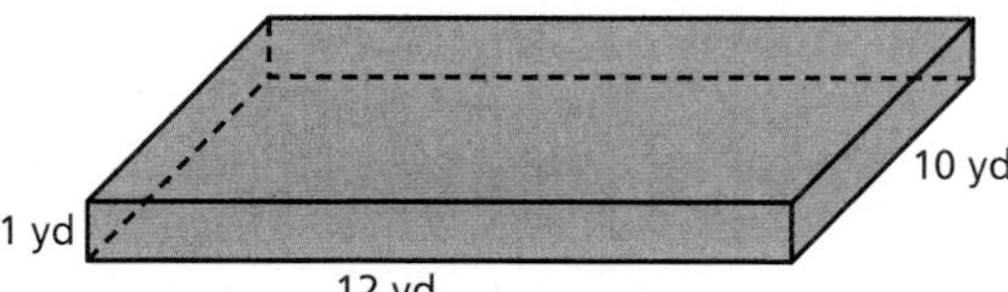

3. a rectangular prism with $\ell = 13$ km, $w = 10$ km, and $h = 3$ km

4. a cube with sides of length 2.5 ft

Write the surface area to volume ratio for each solid.

5.

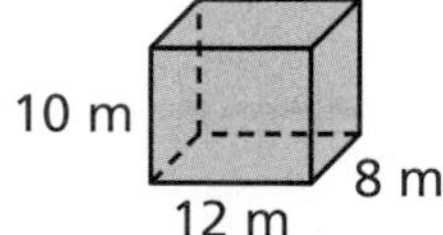

6.

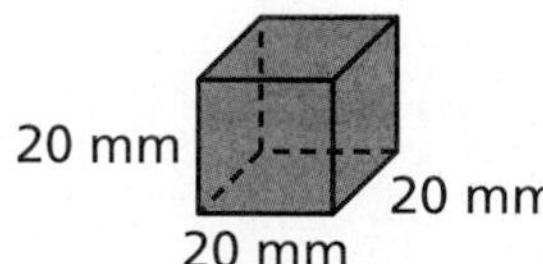
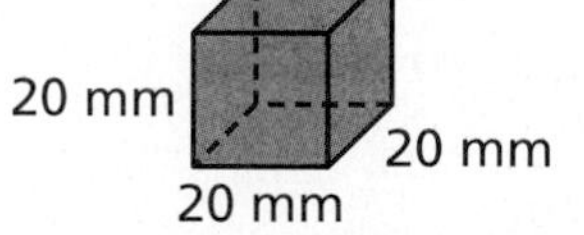

7. a rectangular prism with $\ell = 5$ ft, $w = 4$ ft, and $h = 11$ ft

8. a rectangular prism with $\ell = 8$ dm, $w = 8$ dm, and $h = 4$ dm

Solve Literal Formulas

Formulas are equations that show a relationship between two or more quantities. Formulas can be used to find missing information or to calculate a quantity. For example, the formula $A = \ell w$ is used to find the area of a rectangle. We can solve the formula $A = \ell w$ for w using the same rules used to solve equations.

EXAMPLE

A **Solve $A = \ell w$ for w.**

$A = \ell w$

$\frac{A}{\ell} = \frac{\ell w}{\ell}$ *Divide both sides by ℓ.*

$\frac{A}{\ell} = w$

B **The formula $V = \ell wh$ is used to find the volume of a rectangular prism. Solve $V = \ell wh$ for h.**

$V = \ell wh$

$\frac{V}{\ell} = \frac{\ell wh}{\ell}$ *Divide both sides by ℓ.*

$\frac{V}{\ell} = wh$

$\frac{V}{\ell w} = \frac{wh}{w}$ *Divide both sides by w.*

$\frac{V}{\ell w} = h$

PRACTICE

Solve.

1. The formula $d = rt$ is used to find distance.

Solve $d = rt$ for r.

2. The formula $P = 2\ell + 2w$ is used to find the perimeter of a rectangle.

Solve $P = 2\ell + 2w$ for ℓ.

3. The formula $V = \pi r^2 h$ is used to find the volume of a cylinder.

Solve $V = \pi r^2 h$ for h.

4. The formula $C = \frac{5}{9}(F - 32)$ is used to convert from degrees Fahrenheit to degrees Celsius.

Solve $C = \frac{5}{9}(F - 32)$ for F.

5. The formula $A = \frac{1}{2}bh$ is used to find the area of a triangle.

Solve $A = \frac{1}{2}bh$ for b.

6. The formula $I = Prt$ is used to find simple interest.

Solve $I = Prt$ for P.

Exponential Function Behavior

Data that changes exponentially increases or decreases by a common factor.

The Richter scale is used to express the magnitude of earthquakes. Each counting number represents a magnitude that is 10 times stronger than the one before it.

Magnitude	Relative Strength
0	1
1	10^1
2	10^2
3	10^3
4	10^4
5	10^5
6	10^6
7	10^7
8	10^8

EXAMPLE

How much stronger is an earthquake of magnitude 4 than one of magnitude 2?

An earthquake of magnitude 4 has a relative strength of 10^4. An earthquake of magnitude 2 has a relative strength of 10^2.

An earthquake of magnitude 4 is 10^2, or 100, times stronger than an earthquake of magnitude 2.

PRACTICE

1. In 1976, an earthquake in China registered 8 on the Richter scale. In 1999, an earthquake in Colombia registered 6 on the Richter scale. Which earthquake was weaker and by what factor?
2. Earthquake A registered 3 on the Richter scale. Earthquake B was 10,000 times stronger than earthquake A. What was the magnitude of earthquake B?

You can see exponential population growth by observing bacteria in an environment with unlimited resources. Use the graph of bacteria growth for Exercises 3–5.

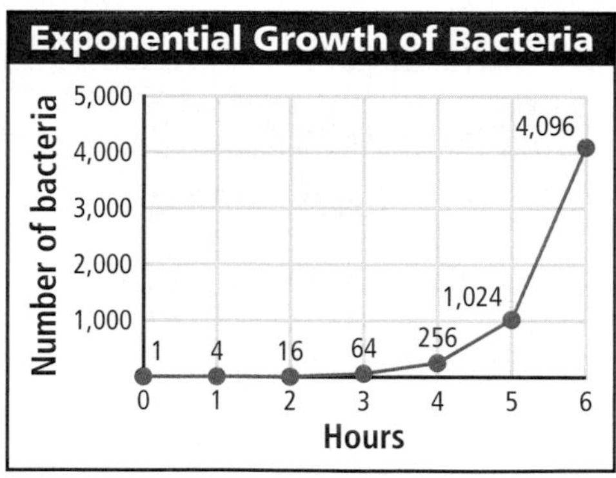

3. By what factor does the graph show the bacteria population increasing each hour?
4. If the bacteria continue to grow at this rate, how many bacteria will there be in 8 hours? Write the answer in both exponential and standard form.
5. How many more bacteria will there be in 10 hours than in 8 hours?

Half-life

Half-life is the time that it takes for half of a certain amount of radioactive material to decay. You can use information about the half-life of an element to determine how much of a sample will remain after a given time or to find the age of a sample.

EXAMPLE 1

The half-life of sodium-24 is 15 hours. If you have a 6 g sample of sodium-24, how much will remain after 45 hours?

Every 15 hours, one-half of the sample decays.

Time	0 hours	15 hours	30 hours	45 hours
Amount of Sample	6 g	3 g	1.5 g	0.75 g

After 45 hours, 0.75 g of sodium-24 will remain.

EXAMPLE 2

The half-life of bismuth-212 is 60.5 minutes. If you have a 4 g sample of bismuth-212 from a sample that was originally 16 g, how old is the sample?

Every 60.5 minutes, one-half of the sample decays.

Time	0 min	60.5 min	121 min
Amount of Sample	16 g	8 g	4 g

The sample is 121 minutes old.

PRACTICE

Solve.

1. Radium-226 has a half-life of 1,600 years. How many years will it take for an 8 g sample to decay to 0.5 g?
2. Cobalt-60 has a half-life of 5.26 years. A 10 g sample of cobalt-60 has decayed to 1.25 g. How old is the sample?
3. Iodine-131 has a half-life of 8.07 days. How much of a 4.4 g sample will there be after 40.35 days?
4. A sample of phosphorus-24 decayed from 12 g to 1.5 g in 42.9 days. What is the half-life of phosphorus-24?
5. You have a 0.6 g sample of sodium-24. The half-life of sodium-24 is 15 hours. The original sample size was 9.6 g. How old is the sample?

Selected Answers

Chapter 1

1-1 Exercises

1. Mount Aconcagua **2.** Mediterranean Sea **3.** 349, 642, 726 **4.** 103, 513, 915 **5.** 497, 809, 1,264 **7.** Mississippi River **9.** 279, 367, 597 **11.** 705, 810, 946 **13.** 111, 1,523, 2,913 **15.** $<$ **17.** $>$ **19.** $<$ **21.** 924, 591, 341 **23.** 911, 747, 439, 291 **25.** 5,480, 5,389, 5,349 **27.** Montana, California, Texas **33.** 300,000 **35.** 6,000 **37.** twenty-four thousand, four hundred ninety-eight **39.** four million, six hundred five thousand, nine hundred twenty-six **41.** 15,903,108

1-2 Exercises

1. 7,000 **2.** 10,000 **3.** 1,500 bottles of water **4.** 300 bottles **5.** 11,000 **7.** 5,000 **9.** 150 softballs **11.** 500 **13.** 3,000 **15.** 40,000 **17.** 900,000 **19.** 10 mi^2 **21.** 40,000 mi^2 **27.** 24,058 **29.** 3,568 **31.** 41 **33.** 10,521 **35.** 70,007

1-3 Exercises

1. 8^3 **2.** 7^2 **3.** 4^4 **4.** 5^5 **5.** 16 **6.** 27 **7.** 625 **8.** 64 **9.** 1,024 people **11.** 9^4 **13.** 6^5 **15.** 3^2 **17.** 243 **19.** 81 **21.** 1 **23.** 100,000,000 **25.** $16 \times 16 \times 16$ **27.** $31 \times 31 \times 31 \times 31 \times 31 \times 31$ **29.** 4 **31.** $17 \times 17 \times 17 \times 17 \times 17 \times 17$ **33.** 1,000,000 **35.** 6,561 **37.** 361 **39.** 57 **41.** $>$ **43.** $<$ **45.** $>$ **47.** 1,024 cells **49.** 8; 2^8 or 256 **51.** 3; 2^3 or 8 **55.** 1,000 + 300 + 50 + 4 **57.** 400,000 + 10,000 + 6,000 + 700 + 90 + 8 **59.** 800 **61.** 20,000 **63.** 2,000,000

1-4 Exercises

1. 33 **2.** 15 **3.** 51 **4.** 50 **5.** 4 **6.** 58 **7.** \$138 **9.** 10 **11.** 25 **13.** 47 **15.** 19 **17.** 24 **19.** 1,250 pages **21.** 18 **23.** 62 **25.** 22 **27.** 40 **29.** $(7 + 2) \times 6 - (4 - 3) = 53$ **31.** $5^2 - 10 + (5 + 4^2) = 36$ **33.** $9^2 - 2 \times (15 + 16) - 8 = 11$ **35.** 60 m^3 **39.** 8,245; 8,452; 8,732 **41.** 11,901; 12,681; 12,751 **43.** 50,000 **45.** 121 **47.** 9 **49.** C

1-5 Exercises

1. 40 **2.** 80 **3.** 280 **4.** 320 **5.** 120 **6.** 416 **7.** 156 **8.** 84 **9.** 99 **10.** 156 **11.** 108 **12.** 174 **13.** 40 **15.** 250 **17.** 108 **19.** 426 **21.** 125 **23.** 147 **25.** 40 **27.** 480 **29.** 111 **31.** 153 **33.** 198 **35.** 56 **37.** 275 **39.** 340 **41.** 192 **43.** \$175 **45.** 70 plants **47.** \$208 **53.** 31,000 **55.** 81 **57.** 64 **59.** B

1-6 Exercises

1. 364 astronauts **2.** 832 medals **3.** 64,890 golf balls **5.** 462 **7.** 111; pencil and paper **9.** 515,844; calculator **11.** 210; mental math **13.** 11,286; calculator **15.** 446,121; calculator **17.** 11,822,400 mi **21.** 4^4 **23.** 10^3 **25.** 14 **27.** C

1-7 Exercises

1. 60, 72, 84 **2.** 30, 15, 0 **3.** 36, 34, 45 **4.** 34, 38, 32 **5.** 18, 486 **6.** 4, 10 **7.** 18, 27 **8.** 64, 4 **9.** 69, 84, 100 **11.** 49, 41, 46 **13.** 24, 720 **15.** 200, 25 **17.** 1, 3, 9, 27, 81 **19.** 100, 93, 86, 79, 72 **21.** 57°F **25.** 3,000 **27.** 14,000 **29.** 125 **31.** 256

Chapter 1 Extension

1. 5 **3.** 7 **5.** 11 **7.** 2 **9.** $(1 \times 16) + (1 \times 8) + (0 \times 4) + (1 \times 2) + (0 \times 1)$ **11.** $(1 \times 16) + (1 \times 8) + (1 \times 4) + (1 \times 2) + (0 \times 1)$ **13.** $(1 \times 16) + (0 \times 8) + (1 \times 4) + (0 \times 2) + (0 \times 1)$ **15.** $(1 \times 16) + (0 \times 8) + (0 \times 4) + (1 \times 2) + (0 \times 1)$ **17.** 1100 **19.** 10010 **21.** 1 **23.** 10110 **25.** $>$ **27.** $<$

Chapter 1 Study Guide and Review

1. sequence, term **2.** base, exponent **3.** order of operations **4.** evaluate **5.** 8,731; 8,735; 8,737; 8,740 **6.** 53,337; 53,341; 53,452; 53,456 **7.** 8,791; 81,790; 87,091; 87,901 **8.** 2,651; 22,561; 25,615; 26,551 **9.** 91,363; 93,613; 96,361; 96,631 **10.** 10,101; 10,110; 11,010; 11,110 **11.** 1,000 **12.** 6,000 **13.** 20,000 **14.** 800 **15.** 5^3 **16.** 3^4 **17.** 7^5 **18.** 8^2 **19.** 4^4 **20.** 1^3 **21.** 256 **22.** 16 **23.** 27 **24.** 1 **25.** 125 **26.** 100 **27.** 59 **28.** 11 **29.** 26 **30.** 17 **31.** 5 **32.** 45 **33.** 9 **34.** 3 **35.** 30 **36.** 520 **37.** 80 **38.** 1,080 **39.** 40 **40.** 320 **41.** 100 **42.** 130 **43.** 168 **44.** 135 **45.** 204 **46.** 152 **47.** 216 **48.** 165 **49.** 52 **50.** 423 **51.** 62° **52.** 186 bars **53.** 17 and 28 **54.** 13 and 16 **55.** 40 and 120 **56.** 32 and 64

Chapter 2

2-1 Exercises

1. 56, 65 **2.** 108, 120 **3.** $x - 5$ **4.** $w + 9$ **5.** 400, 600 **7.** $x \div 8$ **9.** $x - 2$ **11.** 32 **13.** 77 **15.** 56 **17.** 32 zlotys **23.** 876; 972; 1,298 **25.** 40,000 **27.** 3^3 **29.** 10^4

2-2 Exercises

1. 4,028 − m **2.** 279 − 125 **3.** $15x$ **4–7.** Possible answers given. **4.** the sum of r and 87; r plus 87 **5.** the product of 345 and 196; 345 times 196 **6.** the quotient of 476 and 28; 476 divided by 28 **7.** the difference of d and 5; five less than d **9.** $5x$ **11.** 325 ÷ 25 **13.** 137 + 675 **15.** j − 14 **17–23.** Possible answers given. **17.** take away 19 from 243; 243 minus 19 **19.** 75 multiplied by 342; the product of 342 and 75 **21.** the product of 45 and 23; 45 times 23 **23.** the difference of 228 and b; b less than 228 **25.** 15 ÷ d **27.** $67m$ **29.** 678 − 319 **31.** d ÷ 4 **37.** = **39.** 32 **41.** 1,331

2-3 Exercises

1. no **2.** yes **3.** yes **4.** no **5.** yes **6.** no **7.** 53 feet is equal to 636 inches. **9.** no **11.** yes **13.** no **15.** yes **17.** no **19.** yes **21.** no **23.** yes **25.** yes **27.** no **29.** yes **31.** yes **33.** no **35.** 3 **37.** 12 **39.** 13 **41.** no **47.** 81 **49.** 216 **51.** 6 **53.** 75

2-4 Exercises

1. $x = 36$ **2.** $y = 37$ **3.** $n = 19$ **4.** $t = 69$ **5.** $p = 18$ **6.** $c = 16$ **7.** 6 blocks **9.** $r = 7$ **11.** $b = 25$ **13.** $z = 9$ **15.** $g = 16$ **17.** 6 m **19.** $n = 7$ **21.** $y = 19$ **23.** $h = 78$ **25.** $b = 69$ **27.** $t = 26$ **29.** $m = 22$ **31.** 37°C **37.** 20,000 **39.** 100,000 **41.** 36 **43.** C

2-5 Exercises

1. $p = 17$ **2.** $x = 19$ **3.** $a = 31$ **4.** $y = 22$ **5.** $n = 33$ **6.** $d = 41$ **7.** $y = 25$ **9.** $a = 38$ **11.** $a = 97$ **13.** $p = 33$ **15.** $s = 31$ **17.** $x = 36$ **19.** $a = 21$ **21.** $f = 14$ **23.** $r = 154$ **25.** $g = 143$ **27.** $m = 18$ **29.** $k = 64$ **31.** $r − 14{,}162 = 248$; r = 14,410 ft **37.** 81 **39.** 144 **41.** equation **43.** D

2-6 Exercises

1. $x = 3$ **2.** $w = 9$ **3.** $a = 9$ **4.** $b = 8$ **5.** $c = 11$ **6.** $n = 6$ **7.** 45 ft **9.** $a = 4$ **11.** $x = 4$ **13.** $t = 7$ **15.** $m = 11$ **17.** $387w = 104{,}247$; about 250 to 350 miles **19.** $y = 9$ **21.** $y = 8$ **23.** $y = 20$ **25.** $z = 40$ **27.** $y = 23$ **29.** $y = 18$ **31.** $y = 8$ **33.** $a = 14$ **35.** $x = 3$ **37.** 188 segments **39.** 2 times more **43.** < **45.** 17 **47.** 73

2-7 Exercises

1. $y = 12$ **2.** $z = 28$ **3.** $r = 63$ **4.** 90 min **5.** $d = 36$ **7.** $m = 49$ **9.** $c = 96$ **11.** $c = 165$ **13.** $c = 48$ **15.** $c = 180$ **17.** $c = 432$ **19.** 15 seconds **23.** 162, 486 **25.** 14 **27.** 123

Chapter 2 Extension

1. −2 0 2 4 6 8 10

3. −2 0 2 4 6 8 10

5. −2 0 2 4 6 8 10

7. −2 0 2 4 6 8 10

9. −2 0 2 4 6 8 10

11. $y \geq 5$ **13.** $c > 1$ **15.** $r \leq 14$ **17.** $r \leq 3$ **19.** $k < 4$ **21.** $p > 11$;

−2 0 2 4 6 8 10 12

23. $a < 20{,}320$

Chapter 2 Study Guide and Review

1. algebraic expression **2.** equation **3.** variable **4.** constant **5.** 7, 6 **6.** 6, 10 **7.** $p \times 6$ **8.** $s \div 2$ **9.** 15 + b **10.** 6 × 5 **11.** $9t$ **12.** $g \div 9$ **13–19.** Possible answers given. **13.** the product of 4 and z; 4 times z **14.** 54 divided by 6; the quotient of 54 and 6 **15.** 3 minus y; the difference of 3 and y **16.** y minus 3; the difference of y and 3 **17.** 15 plus x; the sum of 15 and x **18.** m divided by 20; the quotient of m and 20 **19.** the sum of 5,100 and 64; 64 added to 5,100 **20.** yes **21.** no **22.** yes **23.** yes **24.** no **25.** no **26.** $x = 6$ **27.** $n = 14$ **28.** $c = 29$ **29.** $y = 6$ **30.** $p = 27$ **31.** $w = 9$ **32.** $b = 11$ **33.** $n = 44$ **34.** $p = 16$ **35.** $d = 57$ **36.** $k = 45$ **37.** $d = 9$ **38.** $p = 63$ **39.** $n = 67$ **40.** $r = 14$ **41.** $w = 144$ **42.** $h = 60$ **43.** $p = 167$ **44.** $v = 8$ **45.** $y = 9$ **46.** $c = 7$ **47.** $n = 2$ **48.** $s = 8$ **49.** $t = 10$ **50.** $a = 8$ **51.** $y = 8$ **52.** $r = 42$ **53.** $t = 15$ **54.** $y = 18$ **55.** $n = 72$ **56.** $z = 52$ **57.** $b = 100$ **58.** $n = 77$ **59.** $p = 90$

Chapter 3

3-1 Exercises

1. 1 + 0.9 + 0.08; one and ninety-eight hundredths **2.** 10.041; 10 + 0.04 + 0.001 **3.** 0.0765; seven hundred sixty-five ten-thousandths **4.** 0.04 + 0.007 + 0.0002; four hundred seventy-two ten-thousandths **5.** osmium **6.** 9.35, 9.5, 9.65 **7.** 4.09; 4.1; 4.18 **8.** 12.09, 12.39, 12.92 **9.** 7 + 0.08 + 0.009 + 0.0003; seven and eight hundred ninety-three ten-thousandths **11.** 7.15; 7 + 0.1 + 0.05 **13.** the Chupaderos meteorite **15.** 1.5, 1.56, 1.62 **17.** nine and seven thousandths **19.** ten and twenty-two thousandths **21.** > **23.** = **25.** three hundredths **27.** one tenth **29.** 4.034, 1.43, 1.424, 1.043, 0.34 **31.** Ross 154 **33.** Alpha Centauri, Proxima Centauri **37.** > **39.** 3^5 **41.** 13^3

3-2 Exercises

1. about 12 miles **2.** 1.8 **3.** 12 **4.** 16.20 **5.** 5.5 **6.** 10 **7.** 120 **8.** 7 **9.** from 44 to 46.5 **10.** from 40 to 42 **11.** about 450 miles **13.** 3.4 **15.** 5.157 **17.** 20 **19.** 6 **21.** from 14 to 17 **23.** 48 **25.** 17 **27.** \$0.22, \$0.10, \$0.08, \$0.04 **29.** $(12 \times 8) - (18 \times 4) = 24$; about 24 cents **37.** 10,000 **39.** 24 **41.** C

3-3 Exercises

1. 20.2 mi **2. a.** 5.95 mi **b.** 12.65 mi **3.** 3 **4.** 5.6 **5.** 8 **6.** 4.9 **7.** 2.98 **8.** 3.55 **9.** 0.5888 **10.** 4.948 **11.** 36.115 **13.** 9 **15.** 4 **17.** 6.4 **19.** 25 **21.** 8.15 **23.** 2.46 **25.** 9.81 **27.** 25.839 **29.** 4.4308 **31.** 52.836 **33.** 29.376 **35.** 84.966 **37.** \$72.42 **39.** 0.196 **43.** 343 **45.** 100,000 **47.** 108 **49.** $4m + 2$

3-4 Exercises

1. 593,700 **2.** 0.71925 **3.** 609,120 **4.** cm **5.** L **6.** g **7.** mL **8.** 7,000 **9.** 5 **10.** 500 **11.** 0.18 **13.** 15,090 **15.** 741,000 **17.** 4.2516 **19.** m **21.** mL **23.** 0.25 **25.** 18 **27.** 10,000 **29.** 0.06087 **31.** 11.18 **33.** 0.06 **35.** 7,540 **37.** table B; 91.6 cm longer **43.** $a = 34$ **45.** $w = 75$ **47.** $p = 9$ **49.** $t = 80{,}000$

3-5 Exercises

1. 6.2×10^4 **2.** 5.0×10^5 **3.** 6.913×10^6 **4.** 1.3×10^5 **5.** 7.015×10^6 **6.** 2.0×10^4 **7.** 6,793,000 **8.** 14,000 **9.** 382,000 **10.** 94,010,000 **11.** 3,300 **12.** 18,850 **13.** 9.0×10^4 **15.** 1.607×10^6 **17.** 6.0×10^6 **19.** 1.8×10^3 **21.** 5.04×10^7 **23.** 1,630,000 **25.** 21,400 **27.** 811,640,000 **29.** 91,060,000 **31.** 7,210 **33.** 720 **35.** 6,954 **37.** 1.1205×10^5 **39.** 4.562×10^3 **41.** 6.5342×10^4 **43.** 4.0×10^3 **45.** 1.8×10^4 **47.** 1.95×10^4 **49.** 3×10^5 km/s; 1.125×10^3 ft/s **51.** $150{,}000 = 1.5 \times 10^5$ **53.** 579,000,000; 5.79×10^8 **59.** Multiply by three, then add 5; 276, 281 **61.** $y = 52$ **63.** D

3-6 Exercises

1. \$1.68 **2.** \$74.32 **3.** 0.24 **4.** 0.0040 **5.** 0.21 **6.** 0.072 **7.** 16.52 **8.** 22.90 **9.** 35.63 **11.** \$1.96 **13.** 2.25 **15.** 0.128 **17.** 0.12 **19.** 0.000015 **21.** 17.227 **23.** 1.148 **25.** 2.5914 **27.** 0.294 **29.** 26.46 **31.** 1.6632 **33.** 12.2122 **35.** 15.662 **37.** 73.5 **39.** Mercury and Mars **41.** 3.42 lb **45.** multiply by 7, subtract 4 **47.** 163 + 24 **49.** $y \div 8$ or $\frac{y}{8}$

3-7 Exercises

1. 0.23 **2.** 0.12 **3.** 0.35 **4.** 0.18 **5.** 0.078 **6.** 0.052 **7.** 0.104 **8.** 0.026 **9.** \$8.82 **11.** 0.22 **13.** 0.27 **15.** 0.171 **17.** 0.076 **19.** 0.9 **21.** 2.1432 **23.** 0.0989 **25.** 0.126 **27.** 14.371 **29.** \$13.25 **35.** 225,971; 2,004,801; 298,500,004 **37.** $<$ **39.** C

3-8 Exercises

1. 5 **2.** 34.5 **3.** 17 **4.** 6 **5.** 6 **6.** 264.125 **7.** 54.6 mi/h **8.** 9 lb **9.** 6 **11.** 8 **13.** $213.\overline{3}$ **15.** $11.3\overline{1}$ **17.** 5 **19.** 11.6 gal **21.** 6.3 **23.** 191.1 **25.** 184.74 **27.** 1,270 **29.** 201,000 **31.** 12.2 **33.** 12.2 **35.** 9.44 **37.** 232 bills; \$4,640 **39.** 63.5 mi/h **41.** 78.38 mi **45.** 360 **47.** 360 **49.** $y = 12$ **51.** 8.304, 8.05, 8.009 **53.** 30.75, 30.709, 30.211

3-9 Exercises

1. 10 belts **2.** 6 packs **3.** 2.25 m **5.** 8 bunches **7.** 3 packs **9.** 4 floors **13.** 45; 54; 63 **15.** 1,366 **17.** 6

3-10 Exercises

1. $a = 7.1$ **2.** $n = 1.4$ **3.** $c = 12.8$ **4.** $x = 6.01$ **5.** $d = 3.488$ **6.** $m = 0.4$ **7.** 60.375 m^2 **8.** 16.2 cm **9.** $b = 9.3$ **11.** $r = 20.8$ **13.** $a = 10.7$ **15.** $f = 6.56$ **17.** $z = 4$ **19.** \$3.00/kg **21.** $t = 51.9$ **23.** $m = 8.1$ **25.** $m = 4.367$ **27.** $w = 78.034$ **29.** $c = 36.14$ **31.** $a = 4.6$ **33. a.** 19.5 units, 21 units **b.** 50.5 units **35. a.** 1,900,000 kg **b.** 1.9×10^6 kg **37.** 9 capsules **43.** 30 **45.** $b = 18$ **47.** A

Chapter 3 Extension

1. 4 **3.** 1 **5.** 5 **7.** 14.3 **9.** 17.72 **11.** 35.61

Chapter 3 Study Guide and Review

1. front-end estimation **2.** Scientific notation **3.** Clustering **4.** 5 + 0.6 + 0.08; five and sixty-eight hundredths **5.** 1 + 0.007 + 0.0006; one and seventy-six ten-thousandths **6.** 1 + 0.2 + 0.003; one and two hundred three thousandths **7.** 20 + 3 + 0.005; twenty-three and five thousandths **8.** 1.12, 1.2, 1.3 **9.** 11.07, 11.17, 11.7 **10.** 0.033, 0.3, 0.303 **11.** 5.009, 5.5, 5.950 **12.** 11.32 **13.** 2.3 **14.** 80 **15.** 9 **16.** 24.85 **17.** 5.3 **18.** 2.58 **19.** 2.8718 **20.** 126,000 **21.** 0.546 **22.** 6,700,000 **23.** 1.806 **24.** 8,900 **25.** 0.18 **26.** 5.5×10^5 **27.** 7.23×10^3 **28.** 1.3×10^6 **29.** 1.48×10^1 **30.** 30,200 **31.** 429,300 **32.** 1,700,000 **33.** 5,390 **34.** 9.44 **35.** 0.865 **36.** 0.0072 **37.** 24.416 **38.** 1.03 **39.** 0.72 **40.** 3.85 **41.** 2.59

42. \$3.64 **43.** 8.1 **44.** $6.1\overline{6}$
45. $3.87\overline{6}$ **46.** 52.275
47. 0.75 m **48.** 14 containers
49. 9 cars **50.** $a = 13.38$
51. $y = 2.62$ **52.** $n = 2.29$
53. $p = 60.2$ **54.** \$5.00

Chapter 4

4-1 Exercises

1. 2, 4 **2.** 2, 3, 4, 6, 9 **3.** none **4.** 3, 9 **5.** composite **6.** prime **7.** composite **8.** composite **9.** 3 **11.** 3, 5, 9 **13.** 2, 4 **15.** 2 **17.** composite **19.** prime **21.** composite **23.** prime **25.** composite **27.** prime **29.** no, no, no, no **31.** yes, no, yes, no, no, no, no **39.** 53, 59, 61, 67, 71, 73, 79, 83, 89, and 97 **41.** 2, 3, 6 **43.** 90 **49.** $<$ **51.** $>$ **53.** $16n$ **55.** $b = 27$ **57.** $y = 29$

4-2 Exercises

1. 1, 2, 3, 4, 6, 12 **2.** 1, 3, 7, 21 **3.** 1, 2, 4, 13, 26, 52 **4.** 1, 3, 5, 15, 25, 75 **5.** $2^4 \cdot 3$ **6.** $2^2 \cdot 5$ **7.** $2 \cdot 3 \cdot 11$ **8.** $2 \cdot 17$ **9.** 1, 2, 3, 4, 6, 8, 12, 24 **11.** 1, 2, 3, 6, 7, 14, 21, 42 **13.** 1, 67 **15.** 1, 5, 17, 85 **17.** 7^2 **19.** $2^2 \cdot 19$ **21.** 3^4 **23.** $2^2 \cdot 5 \cdot 7$ **33.** $3^2 \cdot 11$ **35.** $2^2 \cdot 71$ **37.** $2^3 \cdot 3 \cdot 5 \cdot 7$ **39.** $2^2 \cdot 5 \cdot 37$ **41.** $2^2 \cdot 5^2$ **43.** 7^3 **45.** 340; $2^2 \cdot 5 \cdot 17$ **47.** 142; $2 \cdot 71$ **51.** 7,000 **53.** 40,000 **55.** 625 **57.** 81 **59.** $p = 8$ **61.** $b = 77$

4-3 Exercises

1. 9 **2.** 8 **3.** 7 **4.** 15 **5.** 6 **6.** 9 **7.** 4 arrangements **9.** 14 **11.** 2 **13.** 4 **15.** 12 **17.** 3 teams **19.** 12 **21.** 5 **23.** 2 **25.** 75 **27.** 4 **29.** 5 **31.** 4 **33.** 3 **35.** 6 rows **37.** 12 **41.** 51 **43.** 23 **45.** $n = 3$ **47.** $a = 13$ **49.** C

4-4 Exercises

1. $\frac{3}{20}$ **2.** $1\frac{1}{4}$ **3.** $\frac{43}{100}$ **4.** $2\frac{3}{5}$ **5.** 0.4 **6.** 2.875 **7.** 0.125 **8.** 4.1 **9.** 0.21, $\frac{2}{3}$, 0.78 **10.** $\frac{1}{6}$, $\frac{5}{16}$, 0.67 **11.** $\frac{1}{9}$, 0.3, 0.52 **13.** $5\frac{71}{100}$ **15.** $3\frac{23}{100}$ **17.** $2\frac{7}{10}$ **19.** $6\frac{3}{10}$ **21.** 1.6 **23.** 3.275 **25.** 0.375 **27.** 0.625 **29.** $\frac{1}{9}$, 0.29, $\frac{3}{8}$ **31.** $\frac{1}{10}$, 0.11, 0.13 **33.** 0.31, $\frac{3}{7}$, 0.76 **35.** $92\frac{3}{10}$ **37.** $107\frac{17}{100}$ **39.** $0.1\overline{6}$; repeats **41.** $0.41\overline{6}$; repeats **43.** $>$ **45.** $4\frac{1}{2}$, 4.48, 3.92 **47.** 125.25, 125.205, $125\frac{1}{5}$ **49.** Jill **55.** distributive **57.** 2; 4.32 **59.** 4; 16.7552 **61.** C

4-5 Exercises

1–4. Possible answers given. **1.** $\frac{2}{3}, \frac{8}{12}$ **2.** $\frac{1}{4}, \frac{2}{8}$ **3.** $\frac{1}{2}, \frac{5}{10}$ **4.** $\frac{3}{8}, \frac{9}{24}$ **5.** 25 **6.** 3 **7.** 21 **8.** $\frac{1}{5}$ **9.** $\frac{1}{3}$ **10.** $\frac{1}{4}$ **11.** $\frac{3}{5}$ **13–19.** Possible answers given. **13.** $\frac{1}{5}, \frac{5}{25}$ **15.** $\frac{1}{6}, \frac{2}{12}$ **17.** $\frac{2}{5}, \frac{8}{20}$ **19.** $\frac{3}{5}, \frac{9}{15}$ **21.** 8 **23.** 6 **25.** 140 **27.** $\frac{2}{3}$ **29.** $\frac{3}{7}$ **31.** $\frac{1}{7}$ **33.** $\frac{2}{7}$ **35.** $\frac{1}{4} = \frac{2}{8}$ **37.** $\frac{2}{4} = \frac{1}{2}$ **39.** $\frac{8}{10} = \frac{4}{5}$ **41.** Use two of the $\frac{1}{4}$ tsp measuring spoons; use four of the $\frac{1}{8}$ tsp measuring spoons. **43.** 12 bracelets **45.** 644, 640, 271, 204 **47.** $<$ **49.** $<$ **51.** $<$ **53.** C

4-6 Exercises

1. $>$ **2.** $<$ **3.** $=$ **4.** yes **5.** $\frac{1}{5}, \frac{3}{8}, \frac{2}{3}$ **6.** $\frac{1}{4}, \frac{1}{3}, \frac{2}{5}$ **7.** $\frac{1}{8}, \frac{2}{7}, \frac{5}{9}$ **9.** $<$ **11.** $<$ **13.** $=$ **15.** $\frac{3}{7}, \frac{1}{2}, \frac{3}{5}$ **17.** $\frac{1}{3}, \frac{3}{8}, \frac{4}{9}$ **19.** $\frac{5}{9}, \frac{13}{18}, \frac{5}{6}$ **21.** $<$ **23.** $>$ **25.** $>$ **27.** $\frac{3}{10}, \frac{2}{5}, \frac{1}{2}$ **29.** $\frac{1}{5}, \frac{7}{15}, \frac{2}{3}$ **31.** $\frac{1}{2}, \frac{7}{12}, \frac{5}{8}$ **33.** advertising **35.** United States, China, Brazil **39.** 6.8 **41.** 4.9 **43.** 3.7 **45.** 3.1 **47.** 4.5×10^1 **49.** 3.19×10^2 **51.** 4.05×10^5 **53.** D

4-7 Exercises

1. $2\frac{2}{5}$ **2.** $\frac{5}{4}$ **3.** $\frac{8}{3}$ **4.** $\frac{9}{7}$ **5.** $\frac{12}{5}$ **7.** $8\frac{3}{5}$ **9.** $\frac{20}{9}$ **11.** $\frac{13}{3}$ **13.** $\frac{25}{6}$ **15.** $\frac{19}{5}$ **17.** 4; whole number **19.** $8\frac{3}{5}$; mixed number **21.** $8\frac{7}{10}$; mixed number **23.** 15; whole number **25.** $\frac{53}{11}$ **27.** $\frac{93}{5}$ **29.** ■ = 7; ● = 11 **31.** ■ = 7; ● = 7 **33.** ■ = 18; ● = 3 **35.** ■ = 2; ● = 15 **37.** ■ = 3; ● = 3 **39.** ■ = 1; ● = 17 **41.** $\frac{141}{5}$; $\frac{73}{2}$ **43.** femur, tibia, fibula, humerus, ulna **49.** 81 **51.** 256 **53.** $y = 42$ **55.** B

4-8 Exercises

1. $\frac{1}{2}$ ft **2.** $1\frac{2}{5}$ **3.** $7\frac{1}{7}$ **4.** $3\frac{1}{3}$ **5.** $5\frac{1}{6}$ **6.** $\frac{3}{5}$ **7.** $\frac{2}{5}$ **8.** $\frac{6}{5}$, or $1\frac{1}{5}$ **9.** $\frac{1}{5}$ **11.** $\frac{2}{7}$ **13.** $1\frac{3}{5}$ **15.** $\frac{6}{5}$, or $1\frac{1}{5}$ **17.** $\frac{1}{10}$ **19.** $\frac{5}{8}$ **21.** $\frac{14}{33}$ **23.** $\frac{2}{3}$ **25.** $13\frac{1}{3}$ **27.** $\frac{17}{24}$ **29.** $\frac{2}{13}$ **31.** $\frac{11}{17}$ **33.** $15\frac{1}{4}$ **35.** $1\frac{3}{4}$ hr **37.** 1 ft **43.** 1 **45.** Add 7; 31 **47.** Add 3, then subtract 2; 6

4-9 Exercises

1. $\frac{8}{9}$ **2.** $\frac{2}{5}$ **3.** 3 **4.** $3\frac{1}{9}$ **5.** $\frac{3}{7}$ **6.** $\frac{8}{11}$ **7.** 6 **8.** 6 **9.** 8 **10.** 6 **11.** 9 **12.** 10 **13.** 27 boys **15.** $\frac{3}{4}$ **17.** $\frac{4}{5}$ **19.** $\frac{6}{11}$ **21.** 10 **23.** 6 **25.** 2 **27.** 7 **29.** 15 **31.** 5 **33.** $\frac{48}{5}$, or $9\frac{3}{5}$ **35.** 45 **37.** $>$ **39.** $=$ **41.** $<$ **43.** $>$ **45.** 253, $225\frac{1}{2}$, 231 **51.** base = 4, exponent = 8 **53.** base = 12, exponent = 1 **55.** 38 **57.** 128

Chapter 4 Extension

1. A: 0, 2, 4, 6, … B: 1, 3, 5, 7, …

A
0
2 4
6 8 10
12 …

B
1
3 5
7 9 11
13 …

intersection: empty
union: all whole numbers

3. A: 1, 2, 3, 4, 6, 8, 9, 12, 18, 24, 36, 72 B: 1, 2, 3, 4, 6, 9, 12, 18, 36

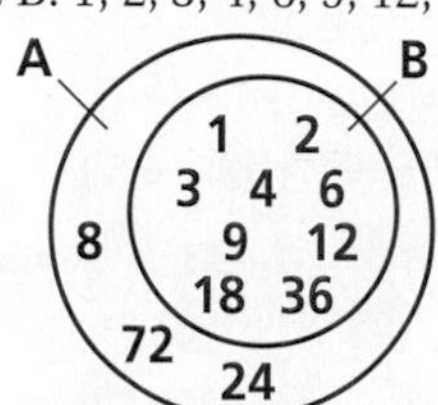

intersection: 1, 2, 3, 4, 6, 9, 12, 18, 36

union: 1, 2, 3, 4, 6, 8, 9, 12, 18, 24, 36, 72

5. yes **7.** no

Chapter 4 Study Guide and Review

1. improper fraction; mixed number **2.** repeating decimal; terminating decimal **3.** prime number; composite number **4.** 2 **5.** 2, 3, 5, 6, 9, 10 **6.** 2, 3, 6, 9 **7.** 2, 4 **8.** 2, 5, 10 **9.** 3 **10.** composite **11.** composite **12.** prime **13.** composite **14.** prime **15.** composite **16.** composite **17.** prime **18.** composite **19.** prime **20.** 1, 2, 3, 4, 5, 6, 10, 12, 15, 20, 30, 60 **21.** 1, 2, 3, 4, 6, 8, 9, 12, 18, 24, 36, 72 **22.** 1, 29 **23.** 1, 2, 4, 7, 8, 14, 28, 56 **24.** 1, 5, 17, 85 **25.** 1, 71 **26.** $5 \cdot 13$ **27.** $2 \cdot 47$ **28.** $2 \cdot 5 \cdot 11$ **29.** 3^4 **30.** $3^2 \cdot 11$ **31.** $2^2 \cdot 19$ **32.** 97 **33.** $5 \cdot 11$ **34.** $2 \cdot 23$ **35.** 12 **36.** 25 **37.** 9 **38.** $\frac{37}{100}$ **39.** $1\frac{4}{5}$ **40.** $\frac{2}{5}$ **41.** 0.875 **42.** 0.4 **43.** $0.\overline{7}$

44–46. Possible answers given.

44. $\frac{2}{3}, \frac{8}{12}$ **45.** $\frac{8}{10}, \frac{16}{20}$ **46.** $\frac{1}{4}, \frac{2}{8}$ **47.** $\frac{7}{8}$ **48.** $\frac{3}{10}$ **49.** $\frac{7}{10}$ **50.** $>$ **51.** $>$ **52.** $\frac{3}{8}, \frac{2}{3}, \frac{7}{8}$ **53.** $\frac{3}{12}, \frac{1}{3}, \frac{4}{6}$ **54.** $\frac{34}{9}$ **55.** $\frac{29}{12}$ **56.** $\frac{37}{7}$ **57.** $3\frac{5}{6}$ **58.** $3\frac{2}{5}$ **59.** $5\frac{1}{8}$ **60.** 1 **61.** $\frac{3}{4}$ **62.** $\frac{3}{5}$ **63.** $6\frac{5}{7}$ **64.** $\frac{5}{7}$ **65.** $\frac{3}{4}$ **66.** $2\frac{4}{7}$ **67.** $\frac{8}{9}$

Chapter 5

5-1 Exercises

1. $\frac{1}{6}$ **2.** $\frac{1}{10}$ **3.** $\frac{3}{7}$ **4.** $\frac{1}{2}$ **5.** $\frac{1}{6}$ **6.** $\frac{4}{33}$ **7.** $\frac{2}{15}$ **9.** $\frac{3}{25}$ **11.** $\frac{1}{15}$ **13.** $\frac{2}{7}$ **15.** $\frac{7}{15}$ **17.** $\frac{9}{20}$ **19.** $\frac{2}{15}$ **21.** $\frac{1}{8}$ **23.** $\frac{3}{20}$ **25.** $\frac{4}{15}$ **27.** $\frac{1}{14}$ **29.** $=$ **31.** $\frac{1}{4}$ cup **33. a.** Multiply by $\frac{1}{4}$. **b.** $\frac{1}{12}$ **35.** $\frac{3}{8}$ **39.** 21 **41.** $>$ **43.** $<$ **45.** D

5-2 Exercises

1. $\frac{5}{6}$ **2.** $\frac{2}{3}$ **3.** $\frac{11}{14}$ **4.** $1\frac{1}{7}$ **5.** $\frac{13}{15}$ **6.** $\frac{8}{11}$ **7.** $2\frac{1}{16}$ **8.** $2\frac{3}{5}$ **9.** $21\frac{5}{7}$ **11.** $\frac{5}{7}$ **13.** $\frac{13}{14}$ **15.** $1\frac{1}{3}$ **17.** $1\frac{1}{3}$ **19.** $2\frac{2}{7}$ **21.** $15\frac{1}{2}$ **23.** $23\frac{1}{2}$ **25.** $3\frac{3}{5}$ **27.** $2\frac{3}{4}$ **29.** $2\frac{1}{3}$ **31.** $\frac{1}{6}$ **33.** $13\frac{3}{4}$ **35.** $16\frac{1}{10}$ **37.** $2\frac{1}{5}$ **39.** 2 **41.** $10\frac{2}{9}$ **43.** $2\frac{2}{15}$ **45.** 240 people **47.** 90 people **53.** $k = 45$ **55.** D

5-3 Exercises

1. $\frac{7}{2}$ **2.** $\frac{9}{5}$ **3.** $\frac{9}{1}$, or 9 **4.** $\frac{11}{3}$ **5.** $\frac{5}{18}$ **6.** $1\frac{5}{7}$ **7.** $\frac{1}{12}$ **8.** 4 **9.** $\frac{9}{50}$ **10.** $\frac{1}{2}$ **11.** $\frac{8}{7}$ **13.** $\frac{8}{3}$ **15.** $\frac{11}{8}$ **17.** $\frac{7}{6}$ **19.** $\frac{7}{32}$ **21.** $\frac{2}{27}$ **23.** $\frac{3}{10}$ **25.** $1\frac{1}{4}$ **27.** $\frac{2}{9}$ **29.** $1\frac{17}{25}$ **31.** $4\frac{2}{3}$ **33.** $3\frac{6}{7}$ **35.** $2\frac{2}{11}$ **37.** $\frac{33}{50}$ **39.** $\frac{4}{5}$ **41.** $1\frac{8}{27}$ **43.** yes **47.** yes **49.** yes **51.** greater than **53.** 25 in. **55.** 16 bags **61.** add 3; 16 **63.** B

5-4 Exercises

1. $z = 16$ **2.** $n = \frac{3}{20}$ **3.** $x = 7\frac{1}{2}$ **4.** 24 **5.** $t = \frac{2}{21}$ **7.** $r = 15$ **9.** $y = 20$ **11.** 4 cans **13.** $m = 1\frac{1}{2}$ **15.** $z = \frac{7}{40}$ **17.** $b = 14$ **19.** $w = 2$ **21.** $d = \frac{1}{32}$ **23.** $n \div 4 = \frac{1}{2}, n = 2$ **25. a.** $\frac{3}{8}$ cup **b.** $1\frac{1}{2}$ cups **27.** 20 pages **29. a.** 200 people **b.** 10 people **33.** 11,000 **35.** 15,000 **37.** 11,000 **39.** 718,000 **41.** 4.2034 **43.** 503 **45.** C

5-5 Exercises

1. 3 packs of pencils and 4 packs of erasers **2.** 15 **3.** 36 **4.** 6 **5.** 20 **6.** 12 **7.** 48 **8.** 24 **9.** 40 **10.** 30 **11.** 63 **12.** 45 **13.** 150 **15.** 8 **17.** 20 **19.** 18 **21.** 12 **23.** 24 **25.** 66 **27.** 60 **29.** 140 **31.** 12 **33. a–b.** Possible answers given. **a.** 16, 20, 28 **b.** 18, 30, 42 **c.** 12 **d.** 120, 144, 168, and 192 **39.** 679; 879; 978 **41.** yes **43.** no **45.** 20.8 **47.** 710,000

5-6 Exercises

1. about 1 **2.** about $\frac{1}{2}$ **3.** about $\frac{1}{2}$ **4.** about 0 **5.** 16 mi **6.** 1 mi **7.** about $\frac{1}{2}$ **9.** about 0 **11.** 4 tons **13.** $3\frac{1}{2}$ tons **15.** $>$ **17.** $<$ **19.** $>$ **21.** $1\frac{1}{2}$ **23.** $6\frac{1}{2}$ **25.** $30\frac{1}{2}$ **27.** $1\frac{1}{4}$ in. **29.** 1 in. **33.** 8^5 **35.** 7^6 **37.** $a \times 2 + 4$ **39.** 16.06 **41.** 3.12

5-7 Exercises

1. $\frac{5}{12}$ ton **2.** $\frac{4}{9}$ **3.** $\frac{3}{10}$ **4.** $\frac{4}{15}$ **5.** $\frac{13}{14}$ **7.** $\frac{1}{6}$ cup **9.** $\frac{7}{12}$ **11.** $\frac{9}{20}$ **13.** $1\frac{2}{15}$ **15.** $1\frac{1}{8}$ **17.** $\frac{7}{15}$ **19.** $\frac{7}{18}$ **21.** $\frac{1}{3}$ **23.** $\frac{28}{33}$ **25.** $\frac{7}{18}$ **27.** $\frac{11}{14}$ **29.** 0 **31.** $\frac{1}{2}$ **33.** $\frac{3}{4}$ **35.** $\frac{13}{18}$ **37.** $\frac{2}{3}$ lb **39.** $\frac{9}{40}$ lb **45.** 16 **47.** 343 **49.** B

5-8 Exercises

1. $10\frac{5}{12}$ **2.** $4\frac{13}{24}$ **3.** $6\frac{1}{12}$ **4.** $5\frac{5}{14}$ **5.** $4\frac{1}{4}$ **7.** $6\frac{7}{12}$ **9.** $6\frac{1}{4}$ **11.** $9\frac{7}{10}$ **13.** $8\frac{1}{3}$ **15.** $34\frac{1}{2}$ **17.** $3\frac{51}{90}$ **19.** $20\frac{13}{36}$ **21.** $12\frac{5}{24}$ **23.** $\frac{7}{10}$ **25.** $2\frac{3}{5}$ **27.** $13\frac{1}{4}$ **29.** 5 **31.** $1\frac{1}{12}$ **33.** $34\frac{2}{3}$ **35. a.** $26\frac{3}{5}$ lb **b.** $2\frac{1}{10}$ lb **c.** $11\frac{1}{10}$ lb **37.** $\frac{1}{2}$ mi **39.** $7\frac{1}{2}$ yd **41.** $5\frac{1}{6}$ yd **45.** 146,500 **47.** 209,467,000 **49.** $\frac{2}{15}$ **51.** $\frac{1}{12}$ **53.** A

5-9 Exercises

1. $\frac{3}{4}$ **2.** $5\frac{4}{9}$ **3.** $1\frac{2}{3}$ **4.** $2\frac{1}{3}$ **5.** $2\frac{3}{5}$ **7.** $3\frac{4}{5}$ **9.** $7\frac{7}{8}$ **11.** $4\frac{13}{18}$ **13.** $2\frac{1}{5}$ **15.** $1\frac{3}{7}$ **17.** $4\frac{1}{4}$ **19.** $2\frac{13}{20}$ **21.** $3\frac{7}{8}$ **23.** $2\frac{7}{8}$ **25.** $3\frac{8}{11}$ **27.** $4\frac{11}{12}$ **29.** $1\frac{5}{6}$ **31.** $\frac{1}{12}$ **33.** $13\frac{5}{12}$ **35.** $\$4\frac{1}{5}$ **37.** $1\frac{1}{12}$ yd^2 **39.** $1\frac{11}{12}$ yd^2 **43.** 16 **45.** $\frac{3}{28}$ **47.** $\frac{3}{14}$ **49.** D

5-10 Exercises

1. $4\frac{1}{2}$ **2.** $8\frac{4}{9}$ **3.** $5\frac{5}{8}$ **4.** $4\frac{1}{10}$ **5.** $57\frac{3}{4}$ in. **7.** $3\frac{5}{8}$ **9.** $4\frac{5}{6}$ **11.** $1\frac{1}{2}$ **13.** $1\frac{1}{6}$ **15.** $9\frac{7}{12}$ **17.** $1\frac{1}{8}$ **19.** $7\frac{7}{18}$ **21.** $4\frac{23}{33}$ **23.** $9\frac{17}{24}$ **25.** $9\frac{9}{10}$ **27.** $12\frac{3}{8}$ **29.** $4\frac{3}{4}$ ft **31.** $53\frac{9}{10}$ **33. a.** $15\frac{3}{4}$ min **b.** yes **39.** 9,198; 10,462; 11,320 **41.** A

Chapter 5 Study Guide and Review

1. reciprocals **2.** least common denominator **3.** $\frac{1}{3}$ **4.** $\frac{15}{28}$ **5.** $\frac{1}{10}$ **6.** $\frac{7}{25}$ **7.** $\frac{5}{81}$ **8.** $\frac{3}{14}$ **9.** $\frac{9}{10}$ **10.** $1\frac{1}{4}$ **11.** 2 **12.** $\frac{4}{21}$ **13.** $\frac{3}{20}$ **14.** $\frac{5}{9}$ **15.** $a = \frac{1}{8}$ **16.** $b = 2$ **17.** $m = 17\frac{1}{2}$ **18.** $g = \frac{2}{15}$ **19.** $r = 10\frac{4}{5}$ **20.** $s = 50$ **21.** 30 **22.** 48 **23.** 27 **24.** 60 **25.** 225 **26.** 660 **27.** about 1 **28.** about $\frac{1}{2}$ **29.** about 11 **30.** about $2\frac{1}{2}$ **31.** $\frac{33}{40}$ **32.** $\frac{3}{4}$ **33.** $\frac{1}{15}$ **34.** $\frac{5}{24}$ **35.** $4\frac{7}{10}$ **36.** $3\frac{1}{18}$ **37.** $11\frac{43}{60}$ **38.** $4\frac{1}{12}$ **39.** $1\frac{13}{20}$ **40.** $2\frac{3}{8}$ **41.** $\frac{11}{30}$ gal **42.** $3\frac{2}{3}$ **43.** $1\frac{1}{2}$ **44.** $5\frac{2}{3}$ **45.** $2\frac{5}{8}$ **46.** $6\frac{13}{14}$ **47.** $1\frac{1}{8}$ **48.** $4\frac{3}{4}$ ft **49.** $30\frac{3}{20}$ **50.** $14\frac{11}{12}$ **51.** $5\frac{5}{12}$ **52.** $3\frac{4}{9}$ **53.** $5\frac{7}{15}$ **54.** 7 oz

Chapter 6

6-1 Exercises

1.

Day	High Temperature (°F)
Mon	72
Tue	75
Wed	68
Thu	62
Fri	55

2. Possible answer: The daily high peaked on Tuesday and then dropped for the remainder of the week. The temperature will continue to drop over the weekend.

3.

Test	Grade
1st	70
2nd	75
3rd	80
4th	85
5th	90

5.

Date	Thickness (in.)
December 3	1
December 18	2
January 3	5
January 18	11
February 3	17

Possible answer: around January 10 **9.** 5.234×10^6 **11.** 1.2078×10 **13.** 0.08 **15.** D

6-2 Exercises

1. range = 19, mean = 54, median = 51, mode = 48 **3.** range = 23, mean = 57.2, median = 55, no mode

5.

State	Mean Score
Connecticut	509
Maine	500
Massachusetts	513
New Hampshire	519
Rhode Island	500
Vermont	508

range = 19, mean = 508.2, median = 508.5, mode = 500 **9.** $\frac{4}{11}, \frac{5}{8}, \frac{2}{3}$ **11.** $\frac{18}{35}$ **13.** $\frac{7}{12}$ **15.** B

6-3 Exercises

1. a. mean = 4.75, median = 5, no mode **b.** mean = 10, median = 7, no mode **2.** with: mean = 45.4, median = 42, no mode; without: mean = 40.2, median = 40, no mode **3.** mean = 225, median = 187.5, mode = 240; median **5.** with: mean = 710.4, median = 788, no mode; without: mean = 877.75, median = 868, no mode **7.** mean ≈ 118.29, median = 128, no mode **13.** 1, 2, 3, 4, 6, 9, 12, 18, 36 **15.** 3 **17.** 4 **19.** B

6-4 Exercises

1. green **2.** black, white, red

3.

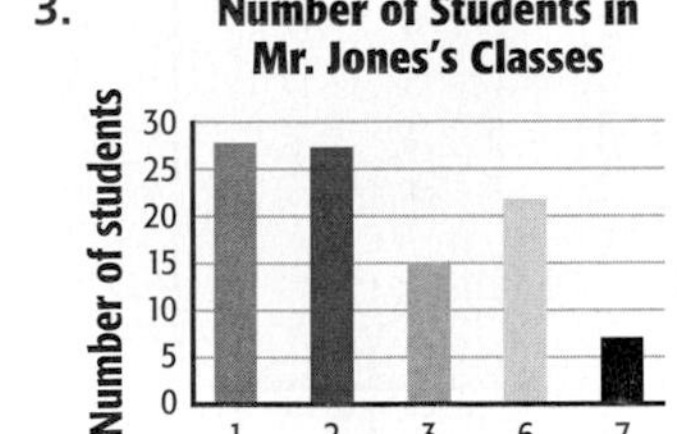

4.

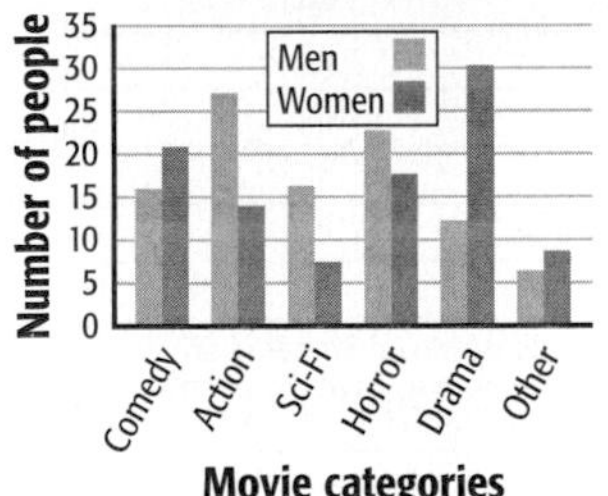

5. orange

7.

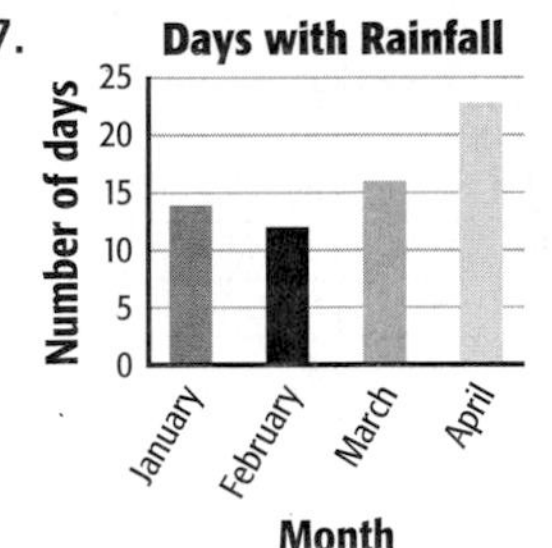

9. 14 million mi^2 **11.** about 8.14 million mi^2 **17.** $7z$ **19.** $\frac{12}{5}$ **21.** $\frac{29}{4}$ **23.** A

6-5 Exercises

1.

Type of Instrument	
Trumpet	𝍸
Drums	II
Tuba	I
Trombone	III
French horn	IIII

2.

Type of Instrument		
Age	Frequency	Cumulative Frequency
Trumpet	5	5
Drums	2	7
Tuba	1	8
Trombone	3	11
French horn	4	15

3.

Number of Years of Each Presidential Term			
Number (Intervals)	0–4	5–8	9–12
Frequency	26	15	1

Selected Answers

4.

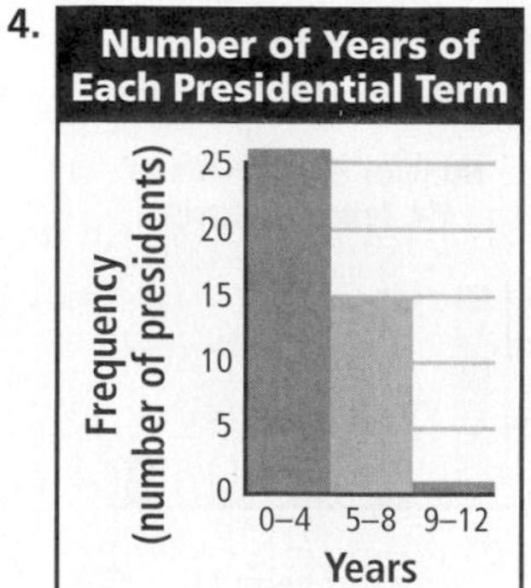

5.

Pets	
Dog	𝍸I
Cat	𝍸
Bird	IIII
Fish	III
Hamster	II

7.

Final Medal Standing at the Summer Olympic Games for the Top 25 Countries					
Number (intervals)	0–20	21–40	41–60	61–80	81–100
Frequency	14	8	3	0	2

9a.

Favorite Sport	
Basketball	II
Football	IIII
Track and field	III
Soccer	II
Hockey	II
Tennis	I
Baseball	I

b.

Favorite Sport		
Age	Frequency	Cumulative Frequency
Basketball	2	2
Football	4	6
Track & field	3	9
Soccer	2	11
Hockey	2	13
Tennis	1	14
Baseball	1	15

11.

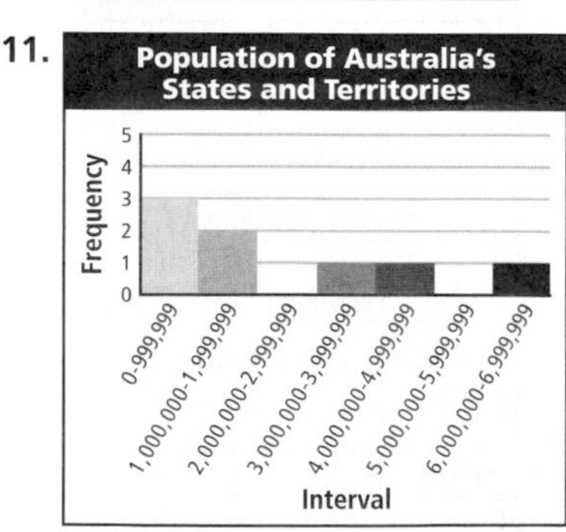

15. 1 + 0.2 + 0.03; one and twenty-three hundredths
17. 20 + 6 + 0.07; twenty-six and seven hundredths **19.** $\frac{1}{6}$
21. $\frac{9}{2}$ **23.** C

6-6 Exercises

1. (2, 3) **2.** (0, 7) **3.** (7, 6) **4.** (9, 1)
5. (4, 5) **6.** (11, 4)
7–10.

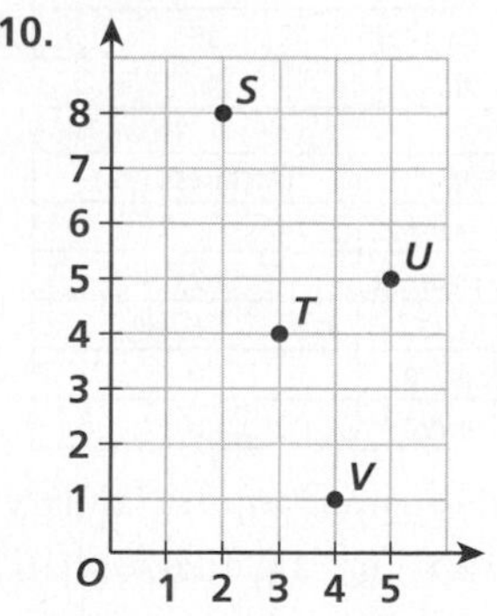

11. (3, 0) **13.** (1, 4) **15.** (11, 7)
17–22.

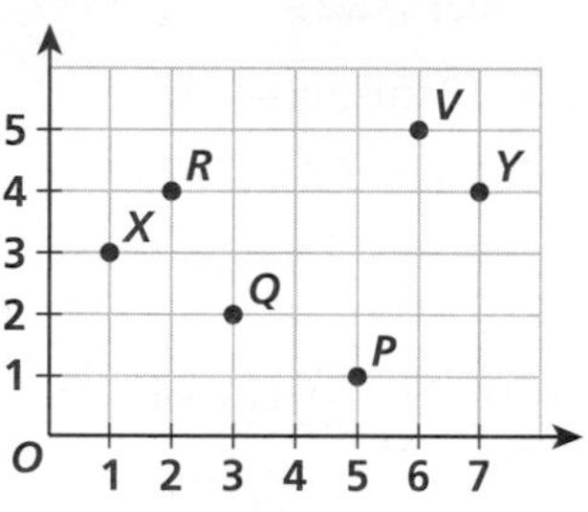

23. A **25.** C **27.** P **29.** (9, 8)
31. (1, 5) **33.** (9, 0) **39.** $2 \cdot 3^2$
41. $3 \cdot 11$ **43.** $1\frac{1}{2}$ min

6-7 Exercises

1.

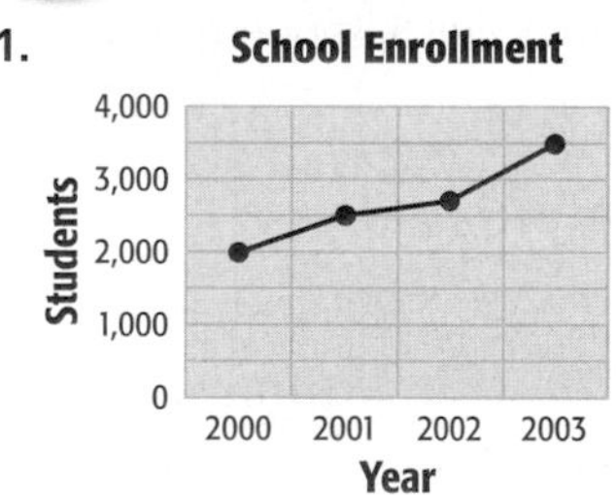

2. 2000 **3.** decrease
4.

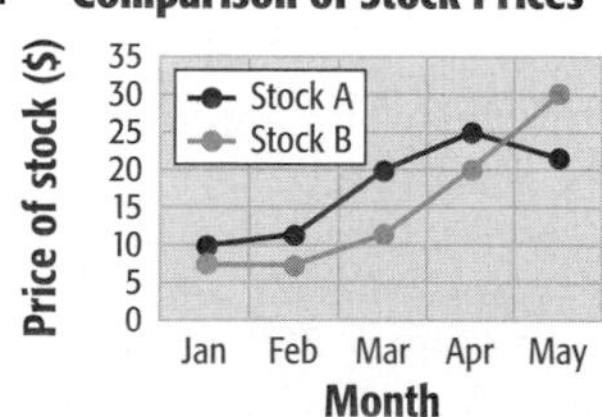

5.

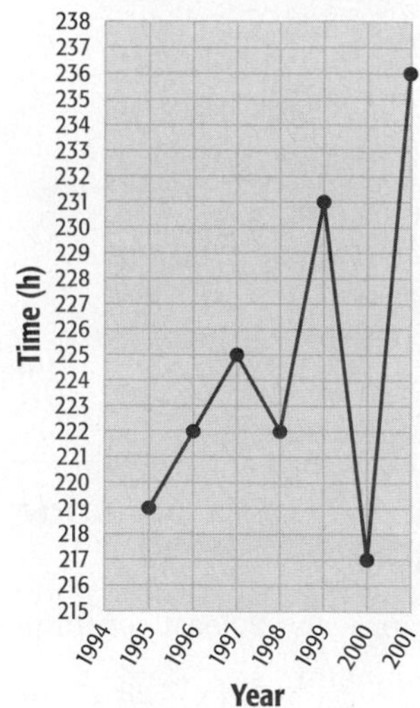

7. 1999 **9.** about 17 lb
11.

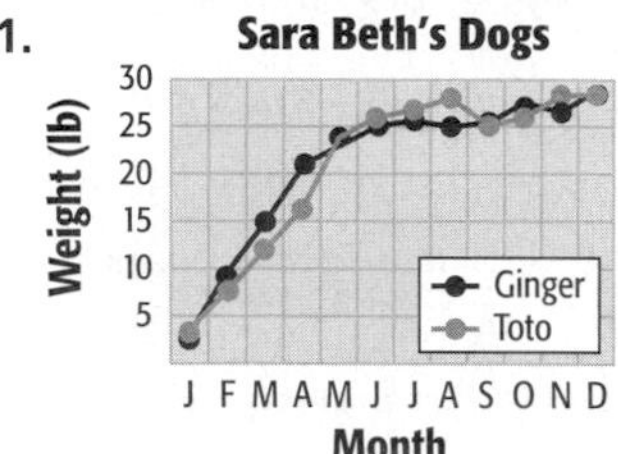

15. Distributive Property
17. $5.00

6-8 Exercises

1. The top bar represents 40 years, but the other bars represent only 10 years. **2.** that the community center had more volunteers in the past than now **3.** Kerry does not begin biking from home.
4. It appears that Kerry biked farther in the 30-minute period when she actually did not. **5.** The vertical axis begins at 430 rather than at zero. **7.** The yearly increments change. **9.** Possible answer: line graph, because you can easily see the changes from month to month **15.** 60 **17.** 120
19. $\frac{3}{28}$ **21.** $\frac{3}{14}$

6-9 Exercises

1.

Stems	Leaves
3	7 9
4	0 5 8
5	1 6

Key: 3|7 means 37

2. 10 **3.** 44 **4.** 27.8 **5.** 32
6. no mode **7.** 34 **9.** 41
11. 52 **13.** 42 **15.** B

17.

Stems	Leaves
8	0 1 2 3 7 8 9
9	2 4 4 5 9
10	0 1 3 9
11	
12	4 5

Key: 8|0 means 80

21. 21.47 **23.** 23.45 **25.** D

Chapter 6 Extension

1. 2 **3.** 13 **5.** 15

7.

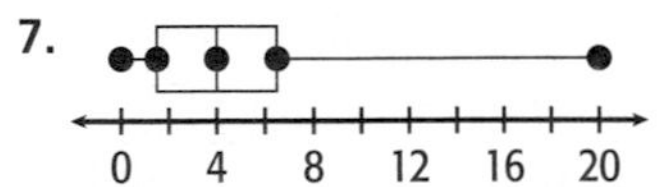

9. Class 1 **11.** 18

Chapter 6 Study Guide and Review

1. histogram, bar graph
2. ordered pair **3.** mode

4.

Snake Lengths (ft)	
Anaconda	35 ft
Diamond python	21 ft
King cobra	19 ft
Boa constrictor	16 ft

5. range = 7; mean = 37; median = 38; mode = 39
6. with outlier: mean ≈ 14.29, median = 11, mode = 12; without outlier: mean ≈ 10.33, median = 10.5, mode = 12
7. with outlier: mean = 31, median = 32, mode = 32; without outlier: mean = 35.75, median = 33, mode = 32
8. with outlier: mean ≈ 19.67; median = 14, mode = none; without outlier: mean = 13.2; median = 13, mode = none
9. 8th grade

10.

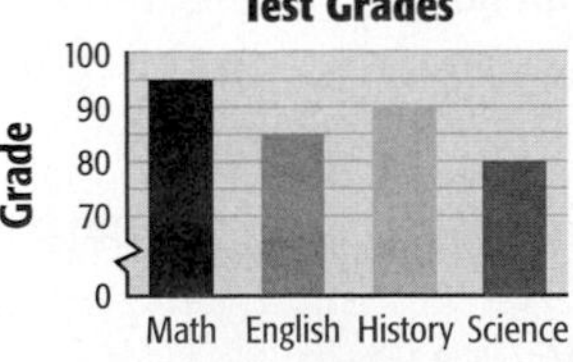

11.

Points Scored			
Points (Intervals)	3–5	6–8	9–11
Frequency	3	2	1

12.

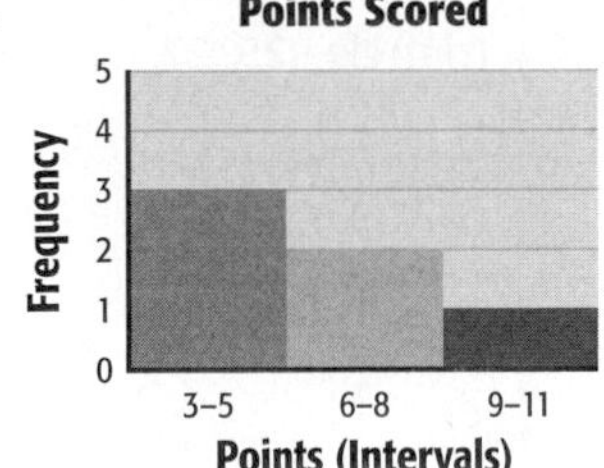

13. (4, 1)
14. (3, 2)

15.

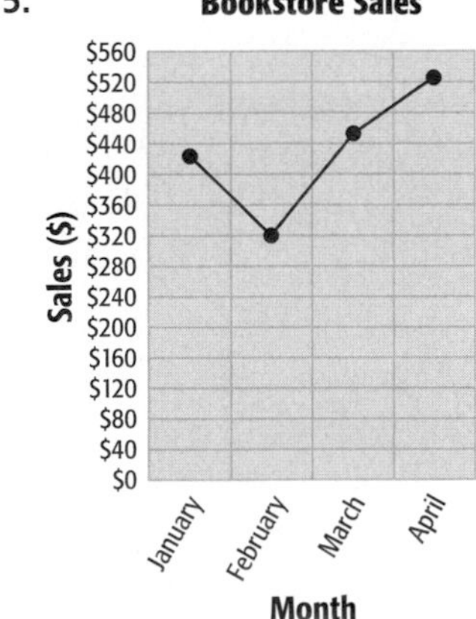

16. April
17. Sales decreased from January to February and then increased from February to April.
18. The scale starts out in increments of 1 mile and then changes to 5 miles.

19. Basketball Scores

Stems	Leaves
2	0 2 6 8
3	4
4	0 4 6

Key: 2|0 means 20

20. least value = 20, greatest value = 46, mean = 32.5, median = 31, no mode, range = 26

Chapter 7

7-1 Exercises

1–2. Possible answers given.
1. M and N **2.** $\overleftrightarrow{KN}$ **3.** K
4–6. Possible answers given.
4. JKL **5.** $\overleftrightarrow{AC}$ and $\overleftrightarrow{AB}$
6. $\overrightarrow{AC}$, $\overrightarrow{BC}$, $\overrightarrow{BA}$, and $\overrightarrow{CA}$ **7.** $\overrightarrow{AB}$
9–13. Possible answers given.
9. $\overleftrightarrow{DF}$; $\overleftrightarrow{ED}$ **11.** FGH **13.** $\overrightarrow{WX}$, $\overrightarrow{XY}$, $\overrightarrow{YZ}$, $\overrightarrow{ZY}$, $\overrightarrow{YW}$, and $\overrightarrow{ZX}$ **15.** C
17–19. Possible answers given.
17. $\overrightarrow{CA}$ and $\overrightarrow{CB}$ **19.** $\overleftrightarrow{BC}$ and $\overleftrightarrow{CB}$
21. 2 mi **25.** $7p$ **27.** 51.85
29. 17

7-2 Exercises

1. 90°, right **2.** 135°, obtuse
3. 60°, acute

4.
5.
6.
7.

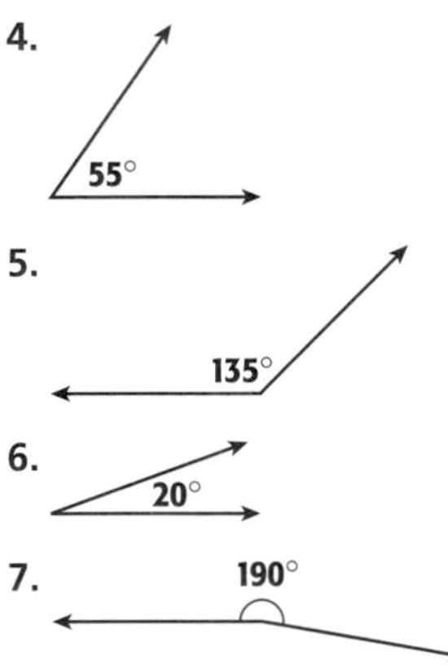

8. actual measure: 140°
9. actual measure: 55°
10. actual measure: 30°
11. 40°, acute **13.** 180°, straight
15. 20°, acute

17. 150°

19.

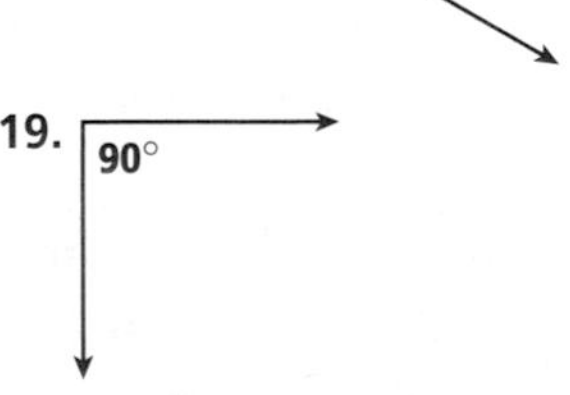

21. 112°

Selected Answers

23. 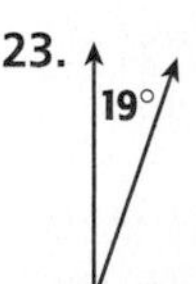

25. actual measure: 53°
27. actual measure: 125°
29–31. Possible answers given.
29.

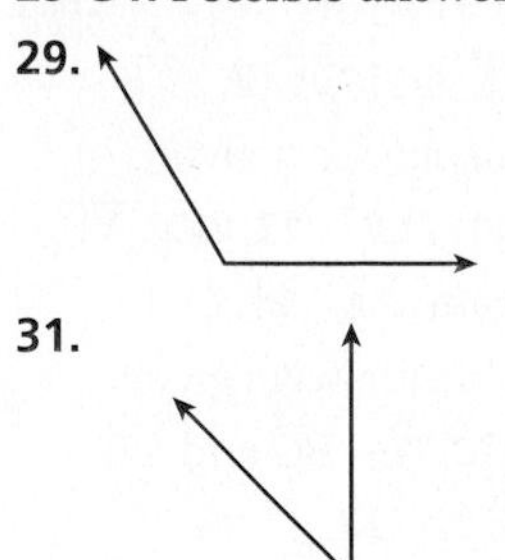

31.

33. obtuse **35.** acute angle
41. 64 **43.** 1 **45.** 2.8 **47.** 2.4

7-3 Exercises

1. adjacent **2.** adjacent
3. $m\angle a = 9°$ **4.** $m\angle b = 30°$
5. adjacent **7.** $m\angle c = 78°$
9. angles 1, 5, 6, 7, and 8
11. 108°, 108°, 72° **13.** 39°
15. 49.91° **17.** 43° **19.** 35°
21. 105° **23.** 44° **25.** 45°
27. a. 47°, 137°
b.

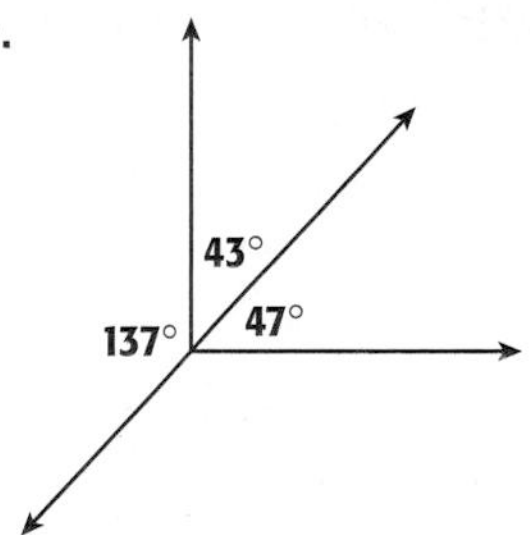

31. $2\frac{1}{4}$ hr

7-4 Exercises

1. intersecting **2.** perpendicular
3. perpendicular **5.** skew
7. intersecting **9.** perpendicular
11. $\overleftrightarrow{BC}$, $\overleftrightarrow{FG}$, and $\overleftrightarrow{EH}$ **13.** Possible answer: $\overleftrightarrow{AD}$ and $\overleftrightarrow{GH}$ **15.** never
17. always **19.** never **25.** 7 **27.** $\frac{5}{6}$
29. $8\frac{5}{7}$ **31.** $6\frac{3}{8}$

7-5 Exercises

1. obtuse triangle **2.** 98° **3.** 82°
4. equilateral **5.** isosceles **7.** 60°
9. scalene **11.** yes; right **13.** no
15. 1.9 cm; isosceles **17.** $1\frac{1}{6}$ ft; equilateral
19–21. Possible answers given.
19.

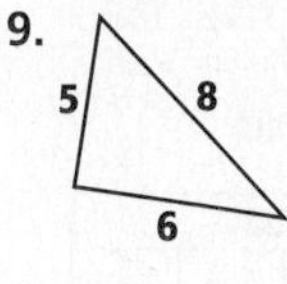

21.

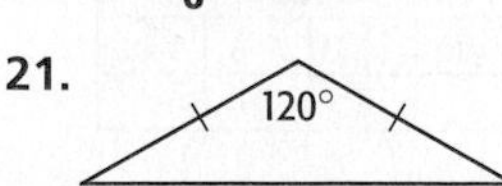

23. right triangle **25.** 27
27. 360° **29.** 3 **31.** 1
32–35.

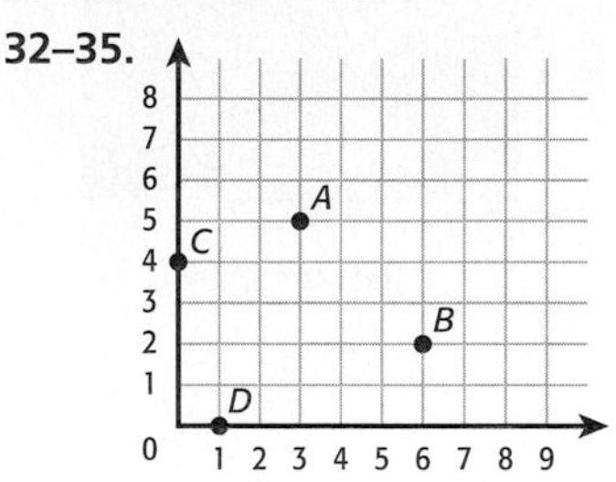

7-6 Exercises

1. rectangle **2.** trapezoid
3. square **4.** quadrilateral
5. squares **6.** right **7.** quadrilateral
9. parallelogram **11.** trapezoid
13. quadrilateral, parallelogram, rectangle, rhombus, square; square **15.** quadrilateral, parallelogram, rhombus; rhombus **17.** never **19.** always
21. sometimes **23.** sometimes
25. not possible **27.** not possible
29. a. If the frame is 10 in. by 13 in., the total length of the sides is 46 in., not 38 in.
b. Possible answer: The dimensions, 10 in. by 13 in., were too long, so try shorter lengths. **c.** 8 in. by 11 in. **35.** 147°

7-7 Exercises

1. polygon, hexagon, regular
2. polygon, quadrilateral, not regular **3.** polygon, triangle, regular **4.** 40 m **5.** not a polygon
7. not a polygon **9.** not formed by line segments **11.** not formed by line segments **13.** hexagon
15. always **17.** never **19. a.** 2
b. 5 **23.** 10, 13, 16, 19, 22 **25.** $\frac{9}{10}$
27. $\frac{1}{5}$ **29.** C

7-8 Exercises

1. Rotate figure 90° clockwise;

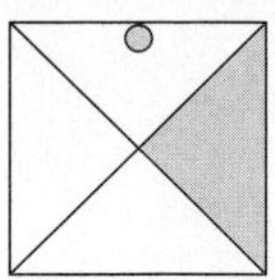

2. Move dot and triangle 1 position counterclockwise;

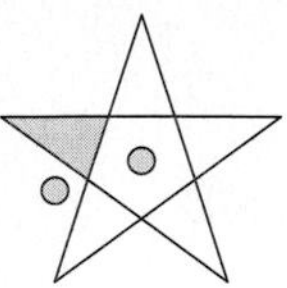

3–9. Possible answers given.
3. purple, purple, red, yellow, green, yellow; next five beads: red, yellow, green, yellow, purple
5. The number of objects doubles each time.

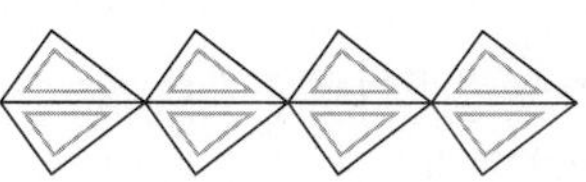

7.

9.

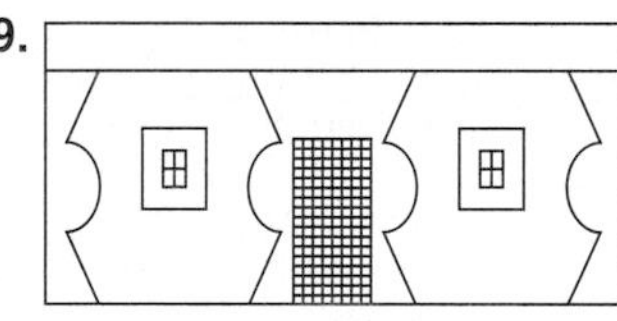

13. $7^2 \times 4^2$ **15.** C

7-9 Exercises

1. not congruent **2.** congruent
3. figure A **5.** The figures are irregular hexagons that are not congruent.
7. Possible answer:

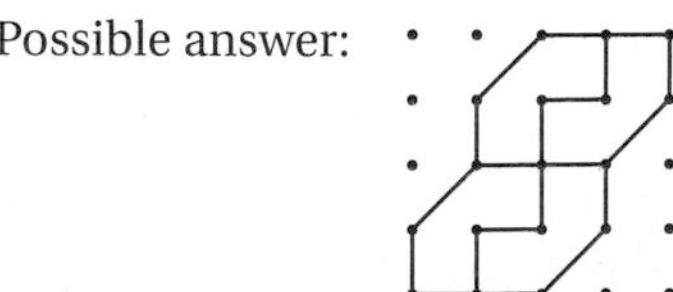

11. $t = 9$ **13.** $s = 48$ **15.** 3 **17.** C

7-10 Exercises

1. reflection **2.** rotation
3. translation

4.

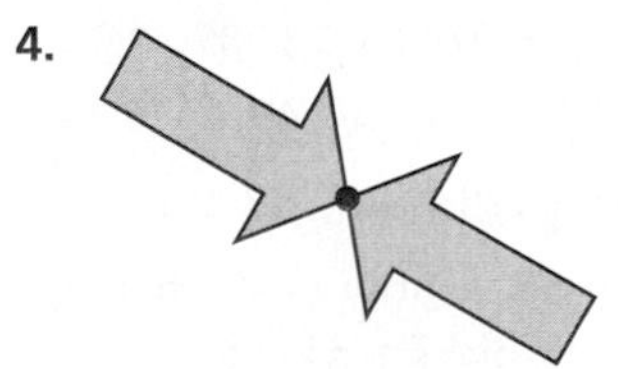

5.

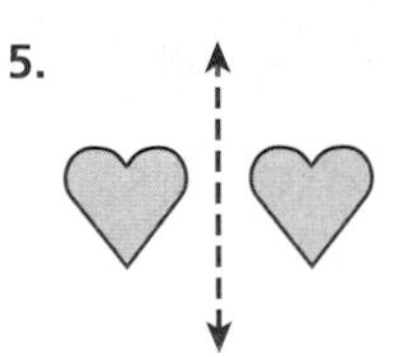

7. translation

9.

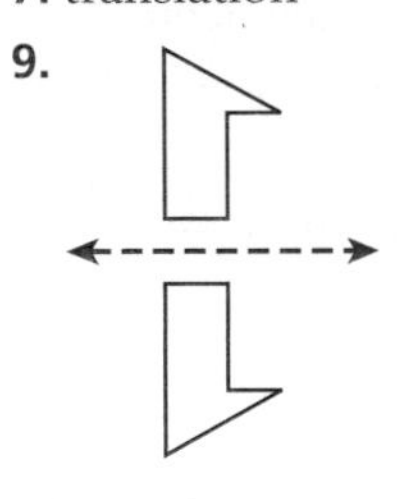

11.

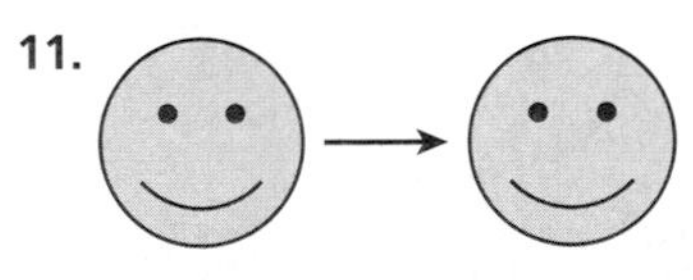

13.

15. A

17.

21. 12 **23.** 72 **25.** 36 **27.** D

7-11 Exercises

1. yes **2.** no **3.** yes **4.** 4 lines of symmetry **5.** 3 lines of symmetry **6.** 5 lines of symmetry **7.** 4 lines of symmetry **8.** 3 lines of symmetry **9.** no **11.** no **13.** 8 lines of symmetry **15.** 1 line of symmetry **17. a.** yes **b.** no **21.** one hundred one and twenty-five hundredths **23.** twelve billion, thirty million, nine hundred twenty-one thousand **25.** C

7-12 Exercises

1. yes

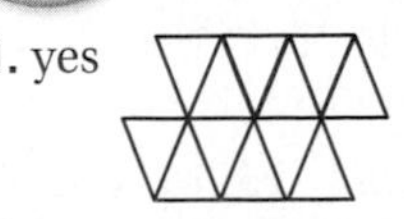

2. yes

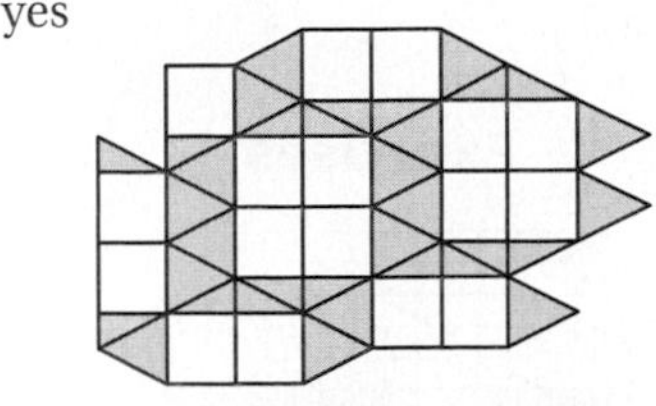

3. yes

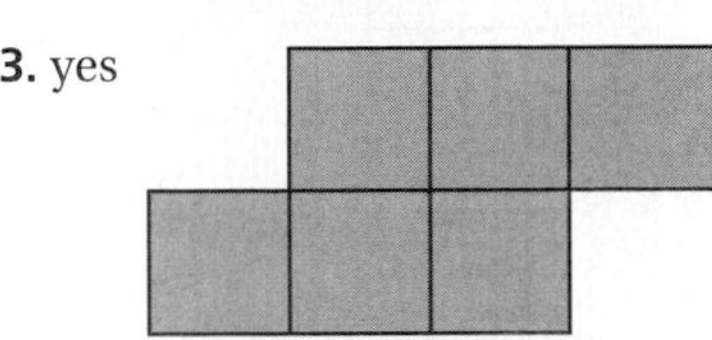

4. no

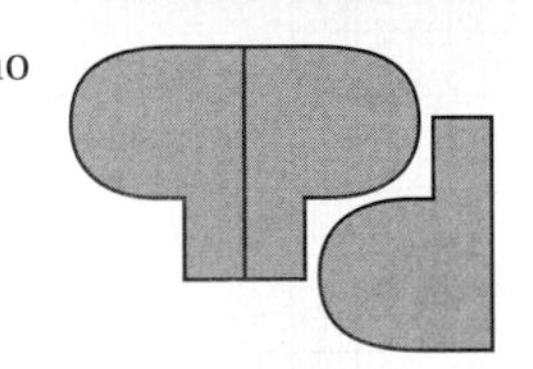

5. no

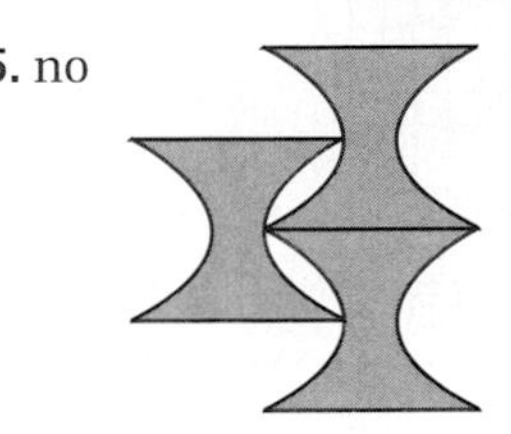

6. yes

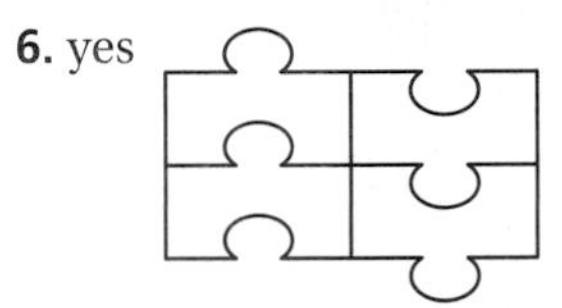

7. yes

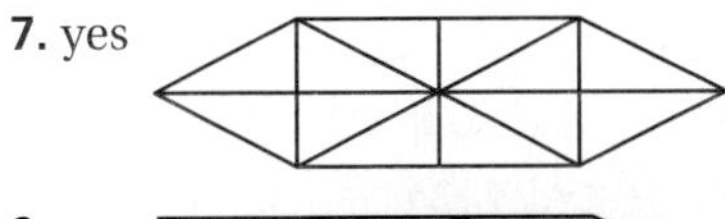

9. yes

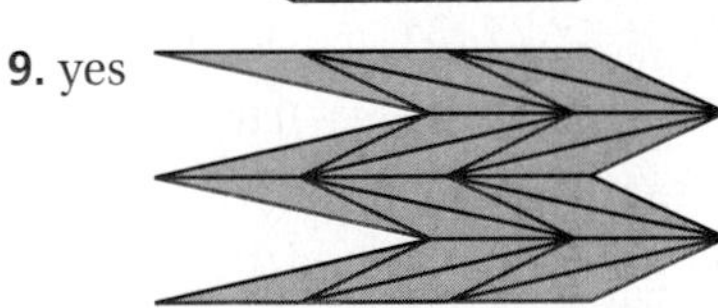

11. no

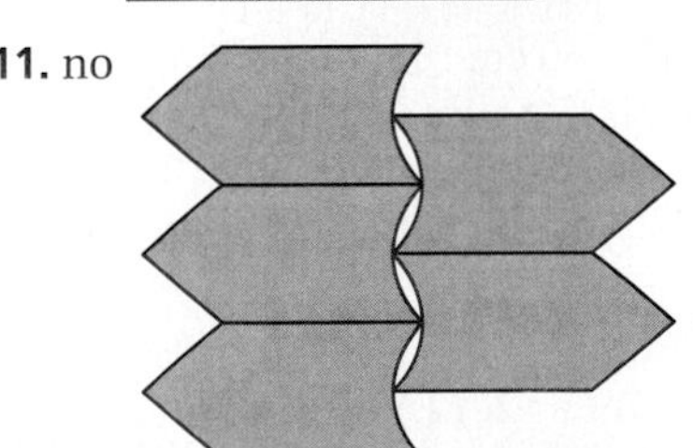

13. B **17–19.** Possible answers given. **17.** b plus 13, 13 more than b **19.** 26 minus c, c less than 26 **21.** D

Chapter 7 Extension

1.

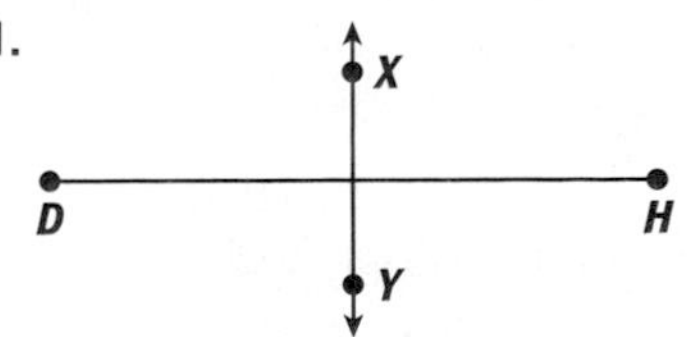

3.

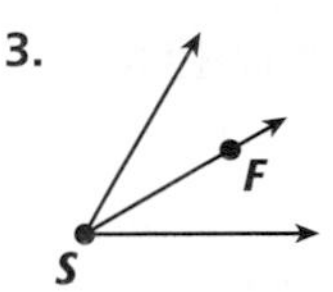

5.

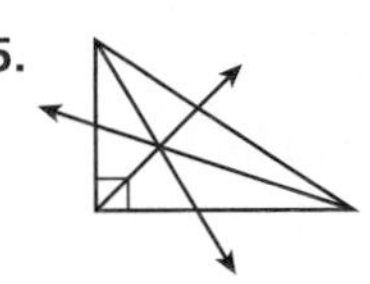

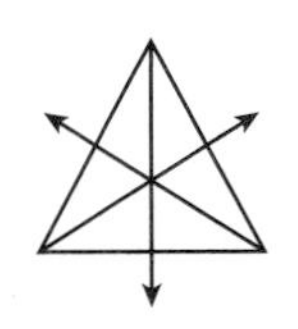

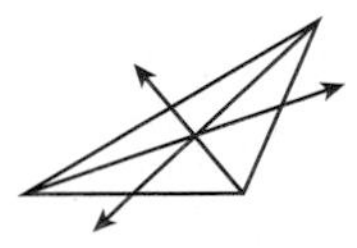

The point of intersection for all three triangles is within the triangles.

Chapter 7 Study Guide and Review

1. trapezoid **2.** polygon **3.** Possible answer: $\overleftrightarrow{ED}$; $\overleftrightarrow{DA}$ **4.** acute **5.** obtuse **6.** acute **7.** $b = 27°$ **8.** $d = 98°$ **9.** perpendicular **10.** skew **11.** obtuse scalene **12.** parallelogram **13.** triangle; regular **14.** rectangle; not regular

15.

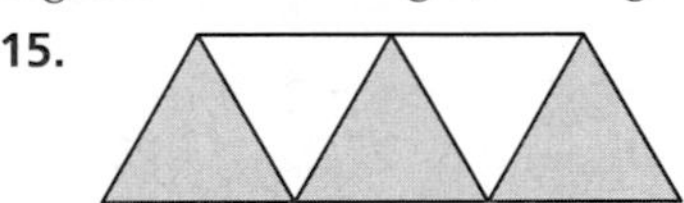

16. not congruent **17.** congruent **18.** translation **19.** yes **20.** yes

Chapter 8

8-1 Exercises

1. 3:10 **2.** 10:41 **3.** 41:16 **4.** Possible answer: $\frac{2}{9}, \frac{4}{18}, \frac{6}{27}$ **5.** the 8-ounce bag **7.** 19:3 **9.** Possible answers: 6:9, 2:3, 12:18 **11.** 10 to 7, 10:7, $\frac{10}{7}$ **13.** four to thirty, 4:30, $\frac{4}{30}$ **15.** 6:5; Possible answer: 12:10, 18:15 **17.** 8:8; Possible answer: 4:4, 1:1 **19.** 4 × 6 and 24 × 36 **21.** 6 million ft^3 **25.** $c = 1.1$ **27.** $t = 51.54$ **29.** $7\frac{7}{9}$ **31.** $12\frac{1}{2}$

8-2 Exercises

1. Possible answer: $\frac{6}{3} = \frac{2}{1}$ **2.** $n = 4$ **3.** $t = 7$ **4.** $c = 2$ **5.** $26.25 **7.** $d = 16$ **9.** $x = 3$ **11.** $p = 2$ **13.** $p = 3$ **15.** $p = 4$ **17.** $p = 75$ **19.** $p = 1.5$ **21.** 113 euros, 160 Canadian dollars, 828 renminbi, 440 shekels, and 910 pesos **27.** mode **29.** C

8-3 Exercises

1. 18 in. **2.** 24 cups **3.** 250,000 lb **5.** 434 yd **7.** 144 **9.** 31 **11.** 14 **13.** $<$ **15.** $=$ **17.** 4 pints **19. a.** 9 yd **b.** 324 in. **c.** about 810 cm **23.** 16 **25.** 35 **27.** B

8-4 Exercises

1. $x = 4$ cm, $m\angle G = 37°$ **2.** 7.5 in. **3.** $n = 3$ in., $m\angle M = 110°$ **5.** sides: $\overline{AC}$ and $\overline{XY}$, $\overline{XW}$ and $\overline{AB}$, $\overline{BC}$ and $\overline{WY}$; angles: X and A, W and B, Y and C **7.** $m\angle E = 78°$, $m\angle L = 78°$, $m\angle M = 51°$; the length of $\overline{ML}$ is 21 in. **9.** yes **11.** 50 ft **15.** 0.4 **17.** 10.007 **19.** D

8-5 Exercises

1. 15 ft **2.** 19.5 ft **3.** 18 ft **5.** 104 in. **7.** 120 m **11.** 30 **13.** 92 **15.** B

8-6 Exercises

1. 300 ft **2.** 3 m **3.** no **5.** 2.5 in. **7.** 3 in. **9.** 1.25 in. **11.** 0.875 in. **13.** 357 km **15.** 297.5 km **17. a.** about 4 cm **b.** which direction Wichita Falls is from San Antonio **21.** $2.71 **23.** $1\frac{1}{5}$ **25.** $\frac{15}{28}$ **27.** C

8-7 Exercises

1.

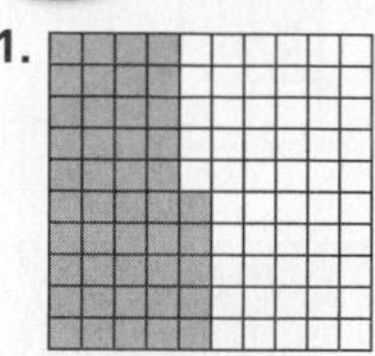

2.

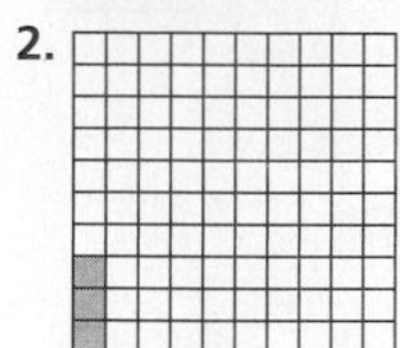

3.

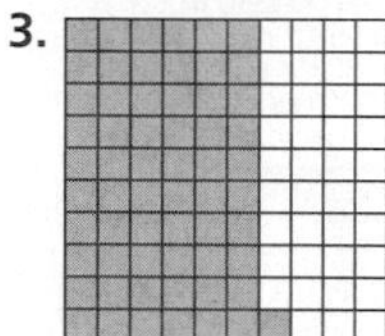

4. $\frac{1}{4}$ **5.** $\frac{4}{5}$ **6.** $\frac{27}{50}$ **7.** $\frac{23}{25}$ **8.** 0.72 **9.** 0.04 **10.** 0.9 **11.** 0.64

13.

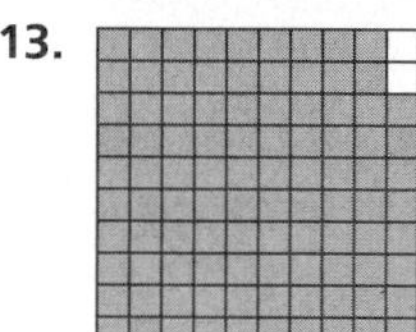

15. $\frac{1}{5}$ **17.** $\frac{11}{100}$ **19.** $\frac{16}{25}$ **21.** $\frac{4}{5}$ **23.** 0.13 **25.** 0.6 **27.** 0.07 **29.** $\frac{23}{100}$, 0.23 **31.** $\frac{49}{100}$, 0.49 **33.** $\frac{37}{100}$, 0.37 **35.** $\frac{2}{25}$, 0.08 **37.** $\frac{47}{50}$, 0.94 **39.** 0, 0 **41.** $\frac{15}{100} = \frac{3}{20}$ **43.** oldies **45.** 1 **49.** 660 cm **51.** range = 71, mean = 33.2, median = 21, no mode **53.** D

8-8 Exercises

1. 39% **2.** 12.5% **3.** 80% **4.** 11.2% **5.** 44% **6.** 87.5% **7.** 70% **8.** 50% **9.** 75% **11.** 55% **13.** 30.8% **15.** 1% **17.** 2% **19.** 30% **21.** 45% **23.** 68.75% **25.** 40% **27.** 4%, $\frac{1}{25}$ **29.** 45%, $\frac{9}{20}$ **31.** 81%, $\frac{81}{100}$ **33.** 39%, $\frac{39}{100}$ **35.** 80%, 0.8 **37.** 83.33%, 0.83 **39.** 6.67%, 0.07 **41.** 72.73%, 0.73 **43.** $<$ **45.** $=$ **47.** $>$ **49.** $\frac{21}{50}$, 0.43, 45% **51.** $\frac{1}{4}$, 26%, 0.7 **53.** 0.125, 14%, $\frac{9}{20}$ **55. a.** about 67% **b.** 0.99 **c.** 38% **61.** 19 **63.** 5

65. G M

67. A

8-9 Exercises

1. 44 T-shirts **2.** 15 hr **3.** 6.72 **4.** 156 **5.** 97.75 **6.** 37.8 **7.** 6 dolls **9.** 30 min **11.** 28.6 **13.** 4.5 **15.** 2.28 **17.** 65.6 **19.** 40.56 **21.** 5 **23. a.** 9 feet **b.** 108 ft^2 **25.** 12 atoms of hydrogen, 6 atoms of carbon, and 6 atoms of oxygen **29.** 7 **31.** 726 **33.** 2 · 19 **35.** $2^3 \cdot 3 \cdot 5$ **37.** B

8-10 Exercises

1. about $7.65 **2.** about $2.40 **3.** about $151.20 **5.** about $18.75 **7.** about $11.00 **9.** about $55.65 **11.** yes **13. a.** Kentucky; about $0.38 **b.** New York; $0.04 less than in North Carolina **17.** 1, 2, 4, 5, 10, 20 **19.** 1, 59 **21.** mean = 24, median = 23, mode = 23

23.

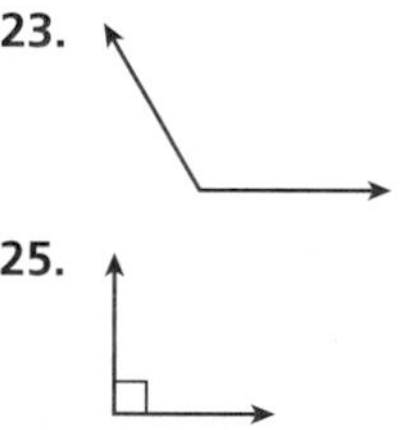

25.

Chapter 8 Extension

1. $148.75 **3.** $32 **5.** $250 **7.** $367.20 **9.** $37,500

Chapter 8 Study Guide and Review

1. discount **2.** equivalent ratios **3.** percent **4.** corresponding angles **5.** Possible answers: 2:4, 3:6, 6:12 **6.** 12 oz for $2.64 **7.** $n = 9$ **8.** $n = 3$ **9.** $n = 14$ **10.** $n = 2$ **11.** 48 cups

12. $2\frac{1}{2}$ or 2.5 **13.** $n = 11$ in.; $m\angle A = 90°$ **14.** 94 ft **15.** 43.75 mi **16.** 3 in. **17.** $\frac{3}{4}$ **18.** $\frac{3}{50}$ **19.** $\frac{3}{10}$ **20.** 0.08 **21.** 0.65 **22.** 0.2 **23.** 89.6% **24.** 70% **25.** 5.7% **26.** 12% **27.** 70% **28.** 25% **29.** 87.5% **30.** 80% **31.** 6.25% **32.** 12 **33.** 5.94 **34.** 117 tickets **35.** about \$19.00 **36.** about \$4.35 **37.** about \$1.08

Chapter 9

9-1 Exercises

1. +5 **2.** −15

3.

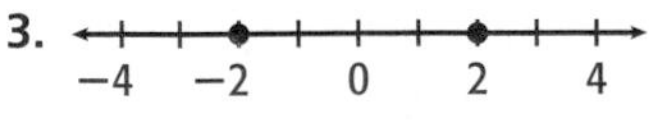

4. −3 −2 −1 0 1 2 3

5. −7 −5 −3 −1 1 3 5 7

6. −9 −6 −3 0 3 6 9

7. 3 **8.** 4 **9.** 1 **10.** 2 **11.** +50 **13.** +7

15. −5 −3 −1 1 3 5

17. −4 −2 0 2 4

19. 4 **21.** 3 **27.** +92 **29.** +25 **31.** 419 **33.** 723 **35.** 35 **37.** 295 **39.** −10,924 **41.** +2; −2 **43.** C **47.** 17,395 < 17,465 < 17,498 < 17,509 **49.** 24

9-2 Exercises

1. > **2.** < **3.** < **4.** −2, 0, 9 **5.** −5, −4, 3, 7 **6.** −6, −1, 8, 10 **7.** 3:30 A.M. **9.** > **11.** −6, −3, 11 **13.** −12, 0, 1, 5 **15.** −10, −6, 4, 7, 24 **17.** Pacific **19.** > **21.** < **23.** < **25.** −31, −18, −9 **27.** −25, −4, 15, 31 **29.** −50, −27, 38, 42, 69 **31.** 68 > −7 or −7 < 68 **37.** 1 **39.** 48 **41.** 0 **43.** 1.1 **45.** 0.125 **47.** D

9-3 Exercises

1. III **2.** no quadrant **3.** (−2, 0) **4.** (1, 2)

5–6.

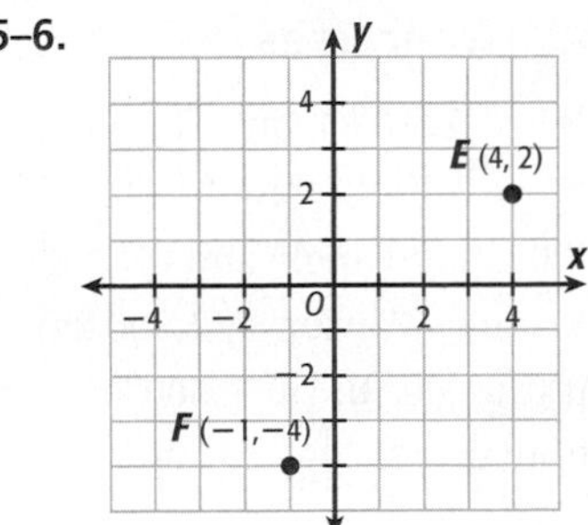

7. no quadrant **9.** I **11.** (3, −4) **13.** (4, 4)

15–18.

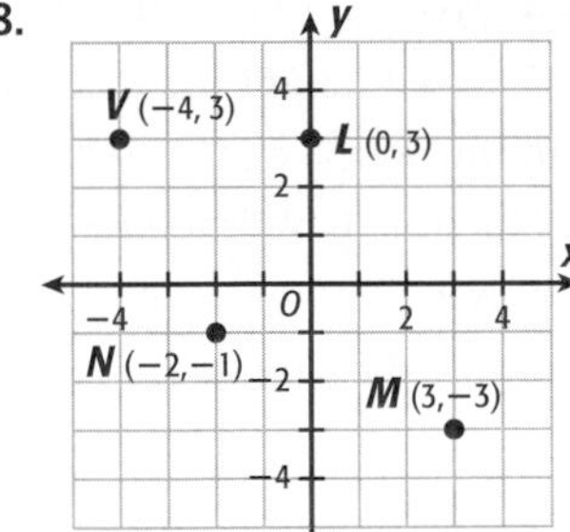

19. IV **21.** II

22–27.

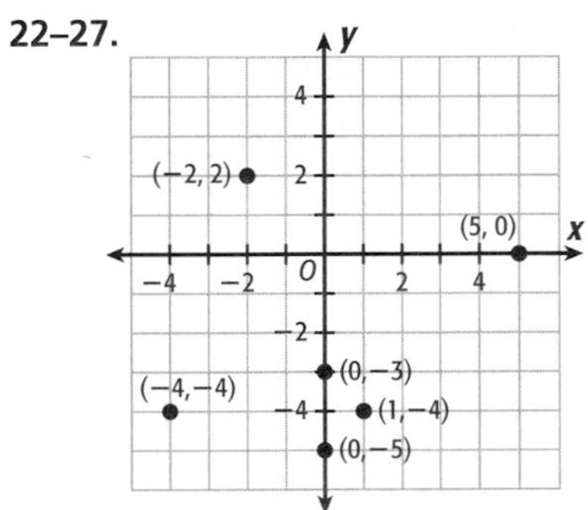

31. Atlantic Ocean **35.** $\frac{3}{40}$ **37.** parallelogram **39.** trapezoid

9-4 Exercises

1. 3 + 2 = 5 **2.** 4 **3.** −5 **4.** 1 **5.** −12 **6.** 0 **7.** −4 **8.** −600 ft **9.** 6 + (−2) = 4 **11.** 11 **13.** 0 **15.** 6 **17.** −16 **19.** −8 **21.** −15 **23.** 29 ft

25. −6 −5 −4 −3 −2 −1 0 1

27. 7 **29.** −11 **31.** 3 **33.** 4 **35.** −1 **37.** −17 **39.** −6°F **45.** 12 **47.** 9 **49.** 60 **51.** 50

9-5 Exercises

1. 6 − 5 = 1 **2.** 3 **3.** −3 **4.** 14 **5.** 2 **6.** 8 **7.** −9 **9.** −4 **11.** 11 **13.** 10 **15.** 7 **17.** −12 **19.** 10 **21.** 21 **23.** 5 **25.** 3 **27.** 19 m **29.** 9°F **33.** 27 **35.** 125 **37.** A

9-6 Exercises

1. 24 **2.** −10 **3.** −21 **4.** 9 **5.** 0 **6.** 16 **7.** 9 **9.** 33 **11.** −36 **13.** 45 **15.** −32 **17.** 48 **19.** −24 **21.** −12 **23.** −28 **25.** −6 **27.** 36 **29.** −24 **31.** −42 **33.** 14 **35.** 54 **37.** −90 **39.** −90 **41.** −42 **45.** $y = 9$ **47.** $n = 48$ **49–51.** Possible answers given: **49.** $\frac{2}{4}, \frac{4}{8}$ **51.** $\frac{4}{6}, \frac{6}{9}$ **53.** C

9-7 Exercises

1. 8 **2.** −5 **3.** 3 **4.** −2 **5.** 10 **6.** −15 **7.** 5 **9.** −8 **11.** −5 **13.** 1 **15.** 3 **17.** −23 **19.** −6 **21.** 1 **23.** −5 **25.** 3 **27.** 8 **29.** 9 **31. a.** −3° F **b.** −2° F **33.** 110; an increase of 110 seals each year from 1971 to 1975 **37.** $n = 2.5$ **39.** $w = 2.4$ **41.** $2\frac{1}{3}$ or $\frac{7}{3}$ **43.** C

9-8 Exercises

1. $m = 12$ **2.** $a = -5$ **3.** $z = 9$ **4.** $b = -8$ **5.** $w = 54$ **6.** $c = -7$ **7.** $g = 4$ **9.** $t = -14$ **11.** $y = -19$ **13.** $j = 8$ **15.** $a = -52$ **17.** $k = -20$ **19.** $x = 17$ **21.** $k = -4$ **23.** $a = 19$ **25.** $g = -3$ **27.** $f = -8$ **29.** $c = -9$ **31.** $b = 32$ **33.** $a = 8$ **35.** $j = -45$ **37.** $n = -4$ **39.** $j = -3$ **41.** −60 ft **43.** $x = -18{,}500$; sponge **45.** −24 ft **49.** $2^2 \times 19$ **51.** 2^4 **53.** 3×7 **55.** $2^2 \times 3 \times 13$ **57.** $1\frac{1}{4}$, or $\frac{5}{4}$ **59.** $\frac{9}{10}$ **61.** $\frac{19}{28}$ **63.** B

Chapter 9 Extension

1. $1, \frac{1}{3}, \frac{1}{9}$ **3.** $1, \frac{1}{6}, \frac{1}{36}$ **5.** 7^{-1} **7.** 8^{-2} **9.** 3^{-1} **11.** 7^{-2} **13.** 4^{-2} **15.** 3^{-4} **17.** $\frac{1}{27}$ **19.** $\frac{1}{729}$ **21.** 256 **23.** $\frac{1}{64}$ **25.** $\frac{1}{125}$ **27.** 1 **29.** 1

Chapter 9 Study Guide and Review

1. opposites **2.** x-axis, y-axis **3.** coordinate plane, quadrants

4. positive numbers, negative numbers **5.** +10 **6.** –50

7.
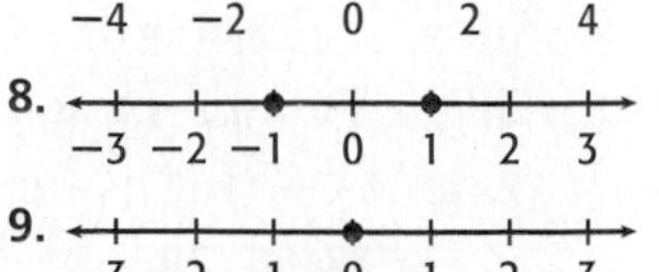

8. **9.**

10. 2 **11.** 1 **12.** 0 **13.** < **14.** < **15.** < **16.** > **17.** > **18.** < **19.** –1, 2, 4 **20.** –3, 0, 4 **21.** –3, –2, 0, 1 **22.** –8, –6, 0 **23.** –7, –4, 7 **24.** –5, –1, 3, 7 **25.** (–2, –3) **26.** (1, 0) **27.** III **28.** II

29–31.

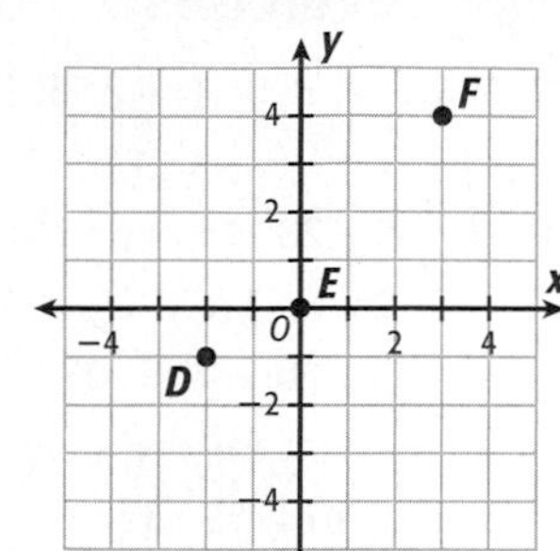

32. –2 **33.** 0 **34.** 1 **35.** –5 **36.** –17 **37.** 8 **38.** –8 **39.** 9 **40.** 10 **41.** 0 **42.** 13 **43.** –6 **44.** –10 **45.** 6 **46.** 6 **47.** –8 **48.** –16 **49.** 45 **50.** –3 **51.** 3 **52.** 2 **53.** –2 **54.** –12 **55.** –9 **56.** $w = 4$ **57.** $a = -12$ **58.** $q = -7$ **59.** $x = -5$

Chapter 10

10-1 Exercises

1. 2 in. **2.** 28 cm **3.** 40 m **4.** 22.6 in. **5.** 7 yd **7.** 96 in. **9.** 7 cm **11.** 42 m **13.** 6 in. **15.** 42 in. **17.** 2 km **19.** 8.5 cm **23.** 24.68 **25.** 74.35 **27.** 3 **29.** 42

10-2 Exercises

1. about 8.5 square units **2.** about 9 square units **3.** 98 mm^2 **4.** 100.1 in^2 **5.** 48 ft^2 **6.** 21 cm^2 **7.** 3 yd^2 **8.** 33 cm^2 **9.** about 6 square units **11.** 125 mi^2 **13.** 260 ft^2 **15.** 37 m^2 **17.** B **21.** 1 min **23.** $\frac{7}{8}$ **25.** $5\frac{13}{15}$

10-3 Exercises

1. 32 in^2 **2.** 18.84 cm^2 **3.** 54 cm^2 **5.** 640 yd^2 **7. a.** Draw a 1,100 km by 600 km rectangle and a 1,100 km by 750 km by 1,300 km right triangle. **b.** 1,072,500 km^2 **c.** about $\frac{1}{7}$ **11.** $1\frac{11}{16}$ **13.** $\frac{1}{21}$

10-4 Exercises

1. The perimeter is divided by 3, and the area is divided by 9, or 3^2. **2.** The perimeter is multiplied by 2, and the area is multiplied by 4, or 2^2. **3.** The perimeter is multiplied by 4, and the area is multiplied by 16, or 4^2. **5. a.** 4,800 ft^2 **b.** 280 ft, 320 ft **7.** 18 × 24 **11.** 19 × 3 **13.** 5 + 9

10-5 Exercises

1. circle G, diameter $\overline{EF}$, and radii $\overline{GF}$, $\overline{GE}$, and $\overline{GD}$ **2.** 31.4 mm **3.** 12.56 in. **4.** 154 ft^2 **5.** 616 cm^2 **7.** 4.71 m **9.** 0.5 in. **11.** 9.63 cm^2 **13.** 14 m, 43.96 m, 153.86 m^2 **15.** 2.25 cm, 1.13 cm, 3.97 cm^2 **17. a.**

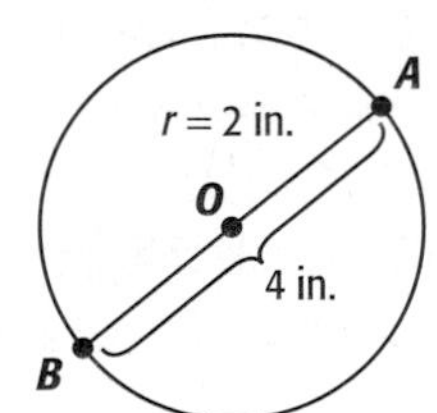

b. 12.56 in. **c.** 12.56 in^2 **19. a.** about 85 in^2 **b.** no **23.** 5.1 **25.** 16.08 **27.** $\frac{5}{8}, \frac{1}{2}, \frac{3}{8}$ **29.** $\frac{7}{10}, \frac{3}{5}, \frac{3}{10}$

10-6 Exercises

1. 5 faces, 8 edges, 5 vertices **2.** 7 faces, 15 edges, 10 vertices **3.** 5 faces, 8 edges, 5 vertices **4.** hexagonal prism **5.** square pyramid **6.** cube **7.** 5 faces, 9 edges, 6 vertices **9.** 6 faces, 12 edges, 8 vertices **11.** rectangular prism **13.** square pyramid, yes **15.** cone, no **17.** B, C, and D **19.** B **21.** true **23.** true **25.** 8; octagonal pyramid **29.** 108, 91, 89, 75, 24, 5 **31.** 19, 18, 17, 15, 13 **33.** 0.03 **35.** 488.41

10-7 Exercises

1. 94 in^2 **2.** 112 in^2 **3.** 132 ft^2 **4.** 2,640 cm^2 **5.** 326.56 ft^2 **6.** 747.32 in^2 **7.** 108 cm^2 **9.** 14,650.75 in^2 **11.** 381.48 ft^2 **13.** 1,274.84 in^2 **15.** 126.39 in^2 **17.** 188.4 m^2 **19.** $13\frac{1}{2}$ km^2 **21.** 345.4 in^2 **23. a.** $\ell = 10$ m; $h = 5$ m; 700 m^2 **b.** $\ell = 9$ in.; $h = 3$ in.; 342 in^2 **27.** 14.2 ft **29.** < **31.** <

10-8 Exercises

1. 162 cm^3 **2.** 64 in^3 **3.** 10 ft^3 **4.** 351 m^3 **5.** 320 ft^3 **6.** 2,500 dm^3 **7.** 1 × 1 × 10 and 2 × 5 × 1 **9.** 79.36 in^3 **11.** 54 m^3 **13.** 71.72 ft^3 **15.** 480 in^3 **17.** 474.375 km^3 **19.** 20 ft **21.** 8.96 g/cm^3, 19.32 g/cm^3, 5.02 g/cm^3, 0.4 g/cm^3, 10.5 g/cm^3 **23.** Check to see if the egg floats in water; if it does, then the egg is spoiled. **27.** 2 **29.** 40 **31.** 3 **33.** D

10-9 Exercises

1. 754 m^3 **2.** 157 cm^3 **3.** 3,140 in^3 **4.** 31 in^3 **5.** cylinder B **7.** 314 ft^3 **9.** 31 cm^3 and 283 cm^3 **11.** 138 in^3 **13.** 4 m^3 **15.** 100.48 in^3 **17.** 659.67 yd^3 **19.** 176 ft^3 **21.** no **23.** 12 mm **27.** no **29–31.** Possible answers given. **29.** $\frac{4}{6}$ **31.** $\frac{3}{4}$

Chapter 10 Study Guide and Review

1. polyhedron **2.** volume **3.** perimeter, circumference **4.** diameter **5.** 33.9 in. **6.** 7 ft **7.** 12 in^2 **8.** 154 in^2 **9.** 135 cm^2 **10.** 175.5 ft^2 **11.** The perimeter is multiplied by 2, and the area is multiplied by 4, or 2^2. **12.** 31.4 ft **13.** 50.24 cm **14.** 9 m **15.** 11 ft **16.** 154 cm^2 **17.** 616 cm^2 **18.** 5 faces, 8 edges, 5 vertices, square pyramid **19.** 6 faces, 12 edges, 8 vertices, rectangular prism **20.** 125 m^2 **21.** 102 cm^2 **22.** 384 cm^3 **23.** 6,300 in^3 **24.** 353 m^3 **25.** 2,308 ft^3

Chapter 11

11-1 Exercises

1. certain **2.** unlikely **3.** 0.4, $\frac{2}{5}$ **4.** boy and girl **5.** likely **7.** unlikely **9.** 0.3, 30% **11.** likely **13.** as likely as not **15.** unlikely **17.** AB negative **21.** $\frac{3}{8}$ **23.** II **25.** III

11-2 Exercises

1. 6; {2, 4, 6, 8} **2.** $\frac{8}{15}$ **3.** no hits; 1 hit **5.** HTH, {HHH, HHT, HTH, THH, HTT, THT, TTH, TTT} **7.** $\frac{13}{25}$ **9.** {yellow T-shirt, red T-shirt, green T-shirt} **11.** {Anna & Joel, Roseann & Anna, Joel & Roseann} **13.** {2, 4, 6} **17.** 9, 11, 13 **19.** 25, 36, 49 **21.** B

11-3 Exercises

1. $\frac{1}{2}$ **2.** $\frac{2}{5}$ **3.** 74% **4.** 0.3 **5.** $\frac{1}{6}$ **7.** $\frac{4}{9}$ **9.** 0.96 **11.** $\frac{5}{6}$ **13.** 1 **15.** $\frac{1}{6}$ **17.** 0 **19.** $<$ **21.** $=$ **23.** $<$ **25.** 0.1 **27.** $\frac{3}{8}$ **29.** 0% **31.** The probability of rolling each number is $\frac{1}{8}$. **33.** $\frac{2}{7}$ **37.** 0.2 m **39.** 9,000,000 m

41.

Month	Number of Days
January	31
February	28 or 29
March	31
April	30
May	31
June	30
July	31
August	31
September	30
October	31
November	30
December	31

11-4 Exercises

1. 6 **2.** 20 **3.** 9 **5.** 27 **7.** 36 **9. a.** 12 **b.** 18 **15.** 169 **17.** 32 **19.** $\frac{4}{9}$ **21.** $\frac{13}{30}$

11-5 Exercises

1. $\frac{1}{4}$ **2.** $\frac{1}{4}$ **3.** $\frac{1}{9}$ **5.** $\frac{1}{3}$ **7.** $\frac{1}{2}$ **9.** $\frac{1}{3}$ **11.** 0 **13.** 1 **15.** $=$ **17.** $<$ **19.** $\frac{1}{5}$ **25.** $m = \frac{1}{14}$ **27.** 142°

11-6 Exercises

1. 600 **2.** 10 **3.** 200 tickets **5.** 32 **7.** 129 donors **9.** about 20 **11.** about 20 **15.** $y = -120$ **17.** $j = 45$ **19.** 6 **21.** 1 **22.** D

Chapter 11 Extension

1. 1 to 29 **3.** 29 to 1 **5.** 1 to 5; 5 to 1 **7.** 2 to 4; 4 to 2 **9.** 1 to 1; 1 to 1 **11.** 7 to 3 **13.** 94 to 6

Chapter 11 Study Guide and Review

1. equally likely **2.** experiment; outcome **3.** probability **4.** theoretical probability **5.** sample space **6.** certain **7.** 0.75; $\frac{3}{4}$ **8.** white **9.** $\frac{4}{15}$ **10.** $\frac{1}{4}$ **11.** $\frac{1}{2}$ **12.** 75% **13.** 18 combinations **14.** $\frac{1}{6}$ **15.** $\frac{1}{8}$ **16.** about 100 items **17.** about 25 times **18.** about 1,575 teenagers **19.** about 100 students

Chapter 12

12-1 Exercises

1. $y = x - 6$; 3 **2.** $y = 5x + 1$; 51 **3.** $j = b - 6$ **4.** $c = \$0.50b$ **5.** $y = 4x$; 28 **7.** $c = 12s - 2$ **9.** $p = 150m$ **11.** $y = x + 3.4$; 2.4, 8.4 **13.** $f = \$2.50 + \$0.90m$ **15.** $58.50 = 6.5h$; 9 hours **21.** –20 **23.** 30

12-2 Exercises

1. (1, 8); (2, 14); (3, 20); (4, 26) **2.** (1, –2); (2, –4); (3, –6); (4, –8) **3.** no **4.** yes **5.** 2 **6.** 1

7.

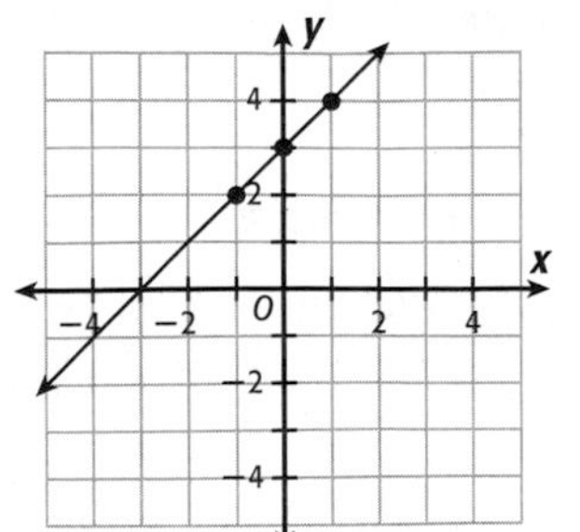

8.

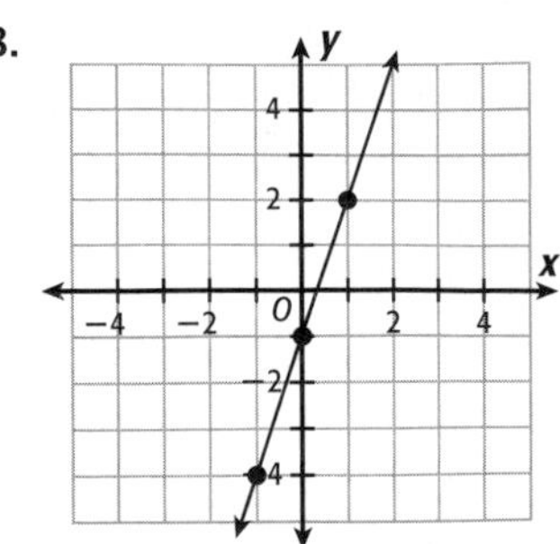

9. (1, –3); (2, –7); (3, –11); (4, –15) **11.** yes **13.** 1 **15.** 3

17.

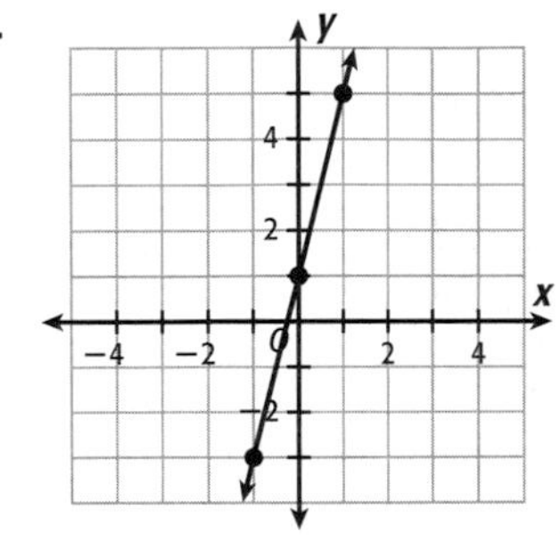

19. –3, –2, –1, 0;

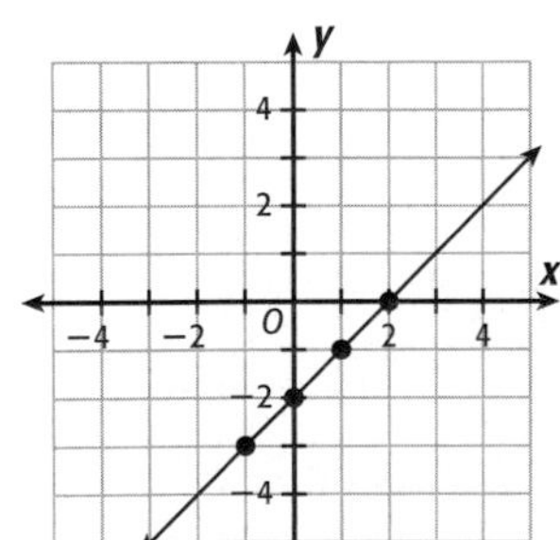

21. (1, 14)

23.

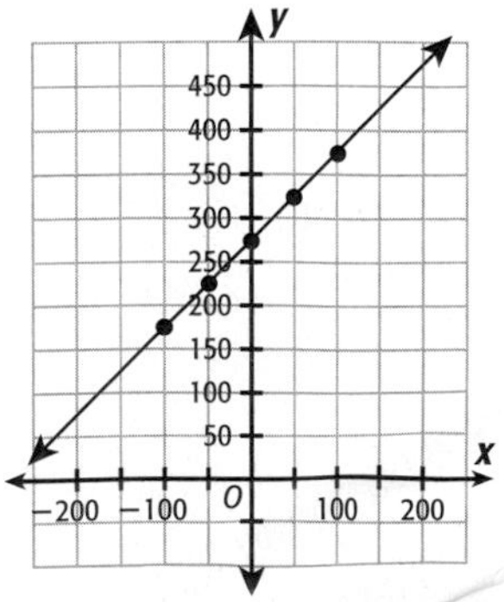

Selected Answers

25. –196°C **29.** $\frac{1}{4}$ **31.** $\frac{2}{5}$ **33.** IV **35.** II **37.** A

12-3 Exercises

1. $R'(2, 5)$, $S'(3, 2)$, $T'(-1, 2)$ **2.** $M'(0, -1)$, $N'(1, 0)$, $O'(2, -1)$, $P'(2, -3)$, $Q'(0, -3)$ **3.** $R'(0, 7)$, $S'(1, 4)$, $T'(-3, 4)$ **5.** $A'(-9, -3)$, $B'(-4, -3)$, $C'(-4, -5)$, $D'(-9, -5)$ **7.** $P'(1, -2)$, $Q'(5, -1)$, $R'(5, -5)$, $S'(1, -4)$ **11.** –12 **13.** –5

12-4 Exercises

1. $P'(3, -1)$, $Q'(1, -1)$, $R'(1, -4)$ **2.** $A'(-2, -2)$, $B'(0, -2)$, $C'(1, 0)$, $D'(-3, 0)$ **3.** $N'(0, -4)$, $Q'(0, -1)$, $P'(2, 0)$ **5.** $A'(-1, 3)$, $B'(-3, 3)$, $C'(-3, 1)$, $D'(-1, 1)$ **7.** WOW **11.** 1×10^2 **13.** A

12-5 Exercises

1. $Q'(-1, 1)$, $R'(-4, 1)$, $S'(-4, 2)$, $T'(-1, 2)$ **2.** $Q'(1, 1)$, $R'(1, 4)$, $S'(2, 4)$, $T'(2, 1)$ **3.** $A'(0, 0)$, $B'(3, -4)$, $C'(0, -4)$ **5.** $P'(1, -1)$, $N'(1, -4)$, $M'(2, -4)$, $L'(2, -1)$ **7.** $A'(3, 0)$, $B'(3, -3)$, $C'(0, -3)$, $D'(0, 0)$ **9.** $B'(3, -3)$, $C'(3, 0)$, $E'(4, 0)$ **11.** math **13.** no **17.** 0.445, 0.9, 1.06, 1.2 **19.** 7.09, 7.9, 7.99, 7.999 **21.** D

12-6 Exercises

1. vertical: 3, horizontal: 5; vertical: 3, horizontal: 15
2. vertical: 3, horizontal: 5; vertical: 30, horizontal: 5
3. vertical: 3, horizontal: 5; vertical: 3, horizontal: 1
4. vertical: 3, horizontal: 5; vertical: 1, horizontal: 15
5. Original dimensions:
top = 1 × 8, center = 3 × 2,
bottom = 1 × 8
New dimensions:
top = 1 × 40, center = 3 × 10,
bottom = 1 × 40
7. Original dimensions:
top = 1 × 8, center = 3 × 2,
bottom = 1 × 8
New dimensions:
top = 1 × 4, center = 3 × 1,
bottom = 1 × 4
9. a.

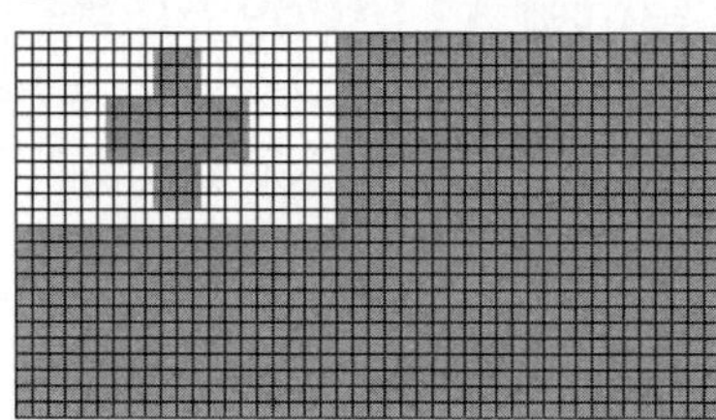

b. 12 units; 48 units **15.** 2 **17.** 5

Chapter 12 Study Guide and Review

1. output, input **2.** linear equation **3.** function
4. $y = 2x + 2$; $y = 18$ when $x = 8$
5. $y = x \div 2 + 1$; $y = 4$ when $x = 6$
6. $y = 5x - 2$; $y = 53$ when $x = 11$
7. ℓ = length; w = width; $\ell = 4w$
8. (1, –3); (2, –1); (3, 1); (4, 3)
9. (1, 8); (2, 9); (3, 10); (4, 11)
10. yes **11.** no **12.** $y = 2$
13.

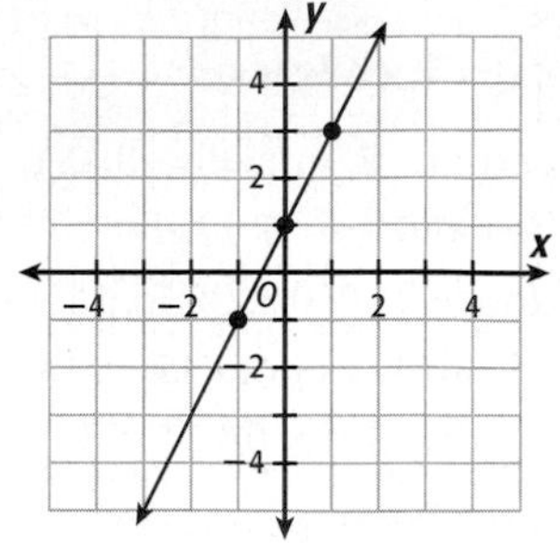

14. $A'(0, 1)$, $B'(2, 1)$, $C'(0, -1)$, $D'(2, -1)$ **15.** $F'(-6, 1)$, $G'(-2, 3)$, $H'(-1, 1)$ **16.** $F'(-3, -4)$, $G'(1, -2)$, $H'(2, -4)$ **17.** $A'(-1, -2)$, $B'(-2, -1)$, $C'(-5, -3)$, $D'(-4, -4)$ **18.** $A'(1, -1)$, $B'(3, -2)$, $C'(3, -5)$, $D'(1, -4)$
19. original dimensions: 6 × 9
new dimensions: 12 × 9

Credits

Photo

Abbreviations used: (t) top, (c) center, (b) bottom, (l) left, (r) right, (bkgd) background

Cover (all), Pronk & Associates.

Title page (all), Pronk & Associates.

Master icons—teens (All): Sam Dudgeon/HRW.

Author photos by Sam Dudgeon/HRW; Jan Scheer photo by Ron Shipper

Front Matter S2 (border), Kim Steele/PhotoDisc/Getty Images; S2 (stamp), United States Postal Service; S2 (squirrel) Susan McElveen/Transparencies, Inc.

Problem Solving Handbook: xxvi (c), Pictor/Alamy Photos; xxxvii (c), Victoria Smith/HRW; xxxviii (tr), Charles Gupton/CORBIS; xxxix (tr), Victoria Smith/HRW; xl (tr), Victoria Smith/HRW; xlii (tr), Cleo Freelance Photography/Painet Inc.; xliii (tr), Sam Dudgeon/HRW; xliii (cl), Victoria Smith/HRW; xliii (bl), Victoria Smith/HRW. **Chapter 1:** 2–3 (bkgd), Getty Images/FPG International; 2 (br), Jeff Greenberg/MR/Photo Researchers, Inc.; 4 (tr), Steve Ewert Photography; 7 (tl), Image Source/elektraVision/PictureQuest; 9 (cr), Iowa State Fair; 11 (tr), National Geographic Image Collection/James Amos; 12 (tr), CORBIS/Bettmann; 15 (tr), CORBIS/Lester V. Bergman; 17 (br), Michael Dunning/Getty Images/FPG International; 21 (b), Sam Dudgeon/HRW; 23 (tr), National Geographic Image Collection/Kenneth Garrett; 23 (tc), National Geographic Image Collection/Kenneth Garrett; 24 (cr), DINODIA/Art Directors & TRIP Photo Library; 24 (tr), PhotoDisc-Digital Image copyright 2004 PhotoDisc; 27 (c), Sam Dudgeon/HRW; 27 (r), PhotoDisc - Digital Image copyright 2004 PhotoDisc; 27 (tl), C.K. Lorenz/Photo Researchers; 28 (tr), NASA; 36 (tr), Jane Faircloth/Transparencies, Inc.; 36 (b), Billy E. Barnes/Transparencies, Inc.; 36 (cr), Ken Taylor/Wildlife Images; 37 (r), © Pat and Chuck Blackley; 37 (c), © George F. Mobley/National Geographic Image Collection/Getty Images; 38 (br), Jenny Thomas/HRW; 44 (cr), Kobal Collection/Lucasfilm/20th Century Fox. **Chapter 2:** 46–47 (bkgd), Christian Michaels/Getty Images/FPG International; 46 (br), Peter Yang/HRW Photo; 51 (tl), Sam Dudgeon/HRW; 55 (tr), AP Photo/NASA; 55 (all patches), NASA; 57 (bl), David A. Northcott/CORBIS; 59 (br), CORBIS/Brandon D. Cole; 62 (tr), Franklin Jay Viola/Viola's Photo Visions; 65 (tl), Peter Yang/HRW; 66 (c - Lincoln), Library of Congress; 66 (tr - Kennedy), AP Photo; 66 (bkgd - flag), Corbis Images; 69 (tr), National Geographic Image Collection/Bianca Lavies; 72 (tr), Darwin Dale/Photo Researchers, Inc.; 73 (tr), Takeshi Takahara/Photo Researchers, Inc.; 73 (c - oyster), Eric Kamp/Index Stock Imagery/PictureQuest; 78 (tr), Chuck Eaton/Transparencies, Inc.; 78 (bc), Jean Higgins/Unicorn Stock Photos; 78 (bl), Digital Image copyright © 2004 PhotoDisc; 79 (t), Chuck Burton/AP/Wide World Photos; 79 (cr), CORBIS Images. **Chapter 3:** 88–89 (bkgd), AP/Wide World Photos; 88 (br), Peter Yang/HRW Photo; 93 (cr), Jerry Schad/Photo Researchers, Inc.; 96 (tr), Getty Images/Stone; 99 (tr), PhotoDisc - Digital Image copyright 2004 PhotoDisc; 102 (tr), Steven E. Sutton/Duomo Photography; 105 (tl), Getty Images/Stone; 110 (c), Peter Van Steen/HRW; 113 (b), Peter Van Steen/HRW; 117 (tl), George Hall/Check Six; 120 (tr), NASA/Photo Researchers, Inc.; 123 (tr), CORBIS; 123 (br), Bettmann/CORBIS; 124 (tr), Peter Van Steen/HRW; 126 (br), Victoria Smith/HRW; 127 (tr), Getty Images/Stone; 130 (tl), Larry Stevens/Nawrocki Stock Photo; 131 (tr), Peter Yang/HRW; 134 (tr), CORBIS/Richard Hamilton Smith; 137 (tr), SuperStock; 138 (tc), Michelle Bridwell/HRW Photo; 140 (bc), Artville - Digital Image copyright 2004 Artville; 140 (br), Layne Kennedy/CORBIS; 141 (t), Ken Taylor/Wildlife Images; 142 (cr), Jenny Thomas/HRW. **Chapter 4:** 150–151 (bkgd), Peter Van Steen/HRW; 150 (br), Digital Image copyright 2004 PhotoDisc; 152 (tr), Darren Carroll/HRW; 155 (tl), Mike Norton/Animals Animals/Earth Scenes; 159 (tr, t), Frans Lanting/Minden Pictures; 161 (tr), PhotoEdit; 163 (tl), Frans Lanting/Minden Pictures; 165 (b), Digital Image copyright 2004 PhotoDisc; 167 (tr), Bettman/Corbis; 167 (tc), ball - Randy Faris/Corbis; signature - Courtesy of the National Baseball Hall of Fame; 167 (frame), Sam Dudgeon/HRW; 170 (cr), Pat Lanza/FIELD/Bruce Coleman, Inc.; 175 (tr), Bob Krist/CORBIS; 175 (br), Wendell Metzen/Bruce Coleman, Inc.; 176 (t), Peter Van Steen/HRW; 177 (egg, shamrock, acorn, shell, rock), PhotoDisc - Digital Image copyright 2004 PhotoDisc; 177 (key, penny), EyeWire - Digital Image copyright 2004 EyeWire; 177(cicada), Artville - Digital Image copyright 2004 Artville; 178 (tr), Steve Cohen/FoodPix; 181 (tl), Peter Van Steen/HRW; 182 (tr), Peter French/Bruce Coleman, Inc.; 185 (tl), Science Photo Library/Photo Researchers, Inc.; 185 (cr), Sam Dudgeon/HRW Photo; 187 (bl), Tom Brakefield/CORBIS; 188 (tr), Mike Norton/Animals Animals/Earth Scenes; 188 (tc), Peter Van Steen/HRW; 191 (tl), Gary Meszaros/Bruce Coleman, Inc.; 195 (tr), David Ryan/Photo 20-20/PictureQuest; 195 (cr), Bob Rowan/Progressive Image/CORBIS; 198 (cr), Michael Meissner, Courtesy Cataloochee Ski Area; 198 (br), Trail Map, Courtesy Cataloochee Ski Area; 199 (r), © Richard T. Nowitz/CORBIS; 200 (br), Randall Hyman/HRW; 206 (br), Peter Van Steen/HRW. **Chapter 5:** 208–209 (bkgd), Peter Van Steen/HRW; 208 (br), Digital Image copyright 2004 PhotoDisc; 215 (tl), Merlin D. Tuttle/Bat Conservation International; 216 (tr), Ken Karp/HRW; 217 (bkgd), Beverly Barrett/HRW; 219 (tr), Corbis Images; 222 (tr), Jenny Thomas Photography/HRW; 225 (tr), Allen Blake Sheldon/Animals Animals/Earth Scenes; 226 (tr), Lori Grinker/Contact Press Images/PictureQuest; 227 (cl), Beverly Barrett/HRW; 227 (tr), Beverly Barrett/HRW; 229 (tl), Frans Lanting/Minden Pictures; 231 (br), Maximilian Stock Ltd./FoodPix; 232 (tr), Pictor International/PictureQuest; 232 (c), Beverly Barrett/HRW; 232 (b), Beverly Barrett/HRW; 236 (tr), James Martin/Getty Images/Stone; 237 (cr), SuperStock; 239 (tl), SuperStock; 239 (tr), Raymond A. Mendez/Animals Animals/Earth Scenes; 239 (tc), Mark Moffett/Minden Pictures; 245 (tr), SuperStock; 245 (br), Gerry Ellis/Minden Pictures; 245 (cr), National Geographic Image Collection/Paul Chesley; 245 (bl), Eric Hosking/CORBIS; 246 (tr), Frans Lanting/Minden Pictures; 247 (c), Frans Lanting/Minden Pictures; 249 (tl), Gerry Ellis/Minden Pictures; 256 (tr), Private Collection/Edmond Von Hoorick/SuperStock; 257 (cr), Alan Pitcairn/Grant Heilman Photography; 259 (t), Victoria Smith/HRW; 260 (b), Steven McBride/Picturesque/PictureQuest; 261 (tr), Ken Taylor/Wildlife Images; 261 (br), © Harrison Shull/shullphoto.com; 262 (br), Ken Karp/HRW; 269 (br), 1998 Image Farm Inc.

Chapter 6: 270–271 (bkgd), Charles W. Campbell/CORBIS; 270 (br), CORBIS; 272 (tc), Reuters NewMedia Inc./CORBIS; 272 (bl), Carl and Ann Purcell/Index Stock Imagery, Inc.; 275 (tr), Jenny Thomas Photography/HRW; 278 (tr), Trent Nelson/The Salt Lake Tribune/CORBIS Sygma; 278 (cl), AFP PHOTO/George FREY/Corbis; 281 (tr), NASA/Science Photo Library/Photo Researchers, Inc.; 283 (piano), Victoria & Albert Museum/Art Resource, NY; 283 (phonograph), U.S. Department of the Interior, National Park Service, Edison National Historic Site; 283 (tape recorder), Index Stock/Alamy Photos; 283 (CD player), Pintail Pictures/Alamy Photos; 284 (cl), Sharon Smith/Bruce Coleman, Inc.; 284 (cr), Tim Davis/Photo Researchers, Inc.; 284 (tl), SuperStock; 284 (tr), Dr. Eckart Pott/Bruce Coleman, Inc.; 290 (t - whorl), Leonard Lessin/Peter Arnold; 290 (t - arch), Federal Bureau of Investigation; 290 (t - loop), Archive Photos; 291 (tl), Rob Crandall/Stock Connection/PictureQuest; 297 (tr), Bettmann/CORBIS; 301 (tc), Corbis Images/PictureQuest; 301 (tr), David Madison/Bruce Coleman, Inc.; 301 (br), Shane Young/AP/Wide World Photos; 304 (tr, c, cr), John Bavosi/Science Photo Library/Photo Researchers, Inc.; 305 (tr), Bryan Berg; 310 (tr), Alan Marler/AP/Wide World Photos; 310 (bl), Jim Reed/Corbis Sygma; 311 (t), © Owaki - Kulla/CORBIS; 311 (br), Pottery from Westmoore Pottery, NC/Photo by Sam Dudgeon/HRW; 312 (br), Randall Hyman/HRW. **Chapter 7:** 320–321 (bkgd), Art by Jane Dixon/HRW; 320 (br), Zhi Xiong China Tourism Press/Getty Images/The Image Bank; 326 (tr), Peter Van Steen/HRW; 332 (c), Michael Kelley/Getty Images/Stone; 332 (bc), (TempSport/CORBIS; 334 (tl, tr), Peter Van Steen/HRW; 334 (cl), Werner Forman Archive/Piers Morris Collection/Art Resource, NY; 334 (cr), P. W. Grace/Photo Researchers, Inc.; 334 (tl), Peter Van Steen/HRW; 336 (t), Walter Bibikow/Index Stock Imagery/PictureQuest; 337 (tl), Emmanuel Faure/SuperStock; 337 (tr), Peter Van Steen/HRW; 337 (cl), Pictor/Alamy Photos; 337 (cr), Peter Van Steen/HRW; 344 (bc), Bob Krist/CORBIS; 344 (t, c), Peter Van Steen/HRW; 347 (tl), Peter Newark's Western Americana; 351 (tl), David Forbert/SuperStock; 352 (c), PhotoDisc - Digital Image copyright 2004 PhotoDisc; 352 (bc), PhotoDisc - Digital Image copyright 2004 PhotoDisc; 353 (tr), PhotoDisc - Digital Image copyright 2004 PhotoDisc; 354 (tl), Beverly Barrett/HRW; 354 (tc, tr), Peter Van Steen/HRW; 354 (cl), Eric Grave/Science Source/Photo Researchers, Inc.; 354 (c), M. Abbey/Photo Researchers, Inc.; 354 (cr), Kim Taylor/Bruce Coleman/PictureQuest; 356 (tr), Lowe Art Museum, The University of Miami/SuperStock; 357 (cr, bl), Peter Van Steen/HRW; 358 (c), Peter Van Steen/HRW; 359 (tr), Steve Vidler/SuperStock; 359 (br), Nicholas DeVore/Getty Images/Stone; 361 (b), Michael Boys/CORBIS; 362 (tr), Ken Karp/HRW Photo; 364 (cl), Original Artwork "Sunflowers" by Mary Backer/Sam Dudgeon/HRW Photo; 364 (c), "Sunflowers" by Mary Backer/Sam Dudgeon/HRW Photo; 364 (cr), "Sunflowers" by Mary Backer/Sam Dudgeon/HRW Photo; 368 (cl), Sam Dudgeon/HRW; 369 (tr), John Greim/Index Stock Imagery/PictureQuest; 372 (cr), corbis images.com; 372 (tr), Anna Clopet/CORBIS; 372 (br), Digital Image copyright 2004 PhotoDisc; 373 (tr), puzzle: http://www.tessellations.com; photo by: Sam Dudgeon/HRW; 373 (c), Ralph A. Clevenger/CORBIS; 375 (tr, cl, c, cr), John Warden/SuperStock; 375 (br), Victoria Smith/HRW; 380 (br), © William A. Bake/CORBIS; 380 (tr), © Raymond Gehman/CORBIS; 381 (t), Greg Loflin/www.loflin.[illegible]om; [illegible] (cr),

Greg Loflin/www.loflin.com; 388 (br), Sam Dudgeon/HRW. **Chapter 8:** 390–391 (bkgd), Ralph A. Clevenger/CORBIS; 390 (br), Beth Davidow; 392 (tr), Reuters NewMedia Inc./CORBIS; 401 (tl), Ronald Zak/AP/Wide World Photos; 401 (cl), Imapress/Jean Claude N'Diaye/The Image Works; 401 (tr), AFP/CORBIS; 401 (cr), Jon Bower/Alamy Photos; 402 (tr), Corbis Images; 404 (l), Pierre Gleizes/Still Pictures; 406 (tr), Mary Cassatt, The Boating Party, Chester Dale Collection, Photograph (c) 2002 Board of Trustees, National Gallery of Art, Washington. 1893/1894, oil on canvas; 407 (tr), Jenny Thomas/HRW; 409 (tr), Sharon McNeill/Bethel Area Chamber of Commerce; 412 (tl), National Geographic Image Collection; 412 (cr), Getty Images/Stone; 412 (tr), W. Perry Conway/CORBIS; 414 (cr), E. R. Degginger/Bruce Coleman, Inc.; 418 (tr), Jenny Thomas/HRW Photo; 421 (tr), Michelle Bridwell/PhotoEdit/PictureQuest; 425 (l), Bettman/CORBIS; 426 (tr), StockFood/Grand; 427 (tr), Corbis Images; 429 (tr), Lucy Schaly/Beaver County Times/AP/Wide World Photos; 432 (t), Peter Van Steen/HRW; 432 (cr), J. & P. Wegner/Animals Animals/Earth Scenes; 435 (cl), Sam Dudgeon/HRW; 438 (tr), Courtesy of the North Carolina Division of Archives and History; 438 (bl), Ernest H. Robl; 439 (tr), Photos Courtesy of NC Division of Tourism, Film, and Sports Development; 439 (b), Robert Bailey/Transparencies, Inc.; 440 (br), Ken Karp/HRW; 440 (cl), Mimmo Jodice/CORBIS; 440 (c), Massimo Listri/CORBIS; 440 (cr), Gianni Dagli Orti/CORBIS; 446 (b), Digital Vision. **Chapter 9:** 448–489 (bkgd), Christopher A. Klein/National Geographic Artist/National Geographic Image Collection; 448 (br), Peter Yang/HRW; 450 (tr - Death Valley), CORBIS/Bill Ross; 450 (tr - snow), Patrick J. Endres/Alaska Photographics.com; 453 (tl), National Geographic Image Collection/Emory Kristof; 454 (tr), CORBIS/AFP; 455 (cl), Chris Trotman/Duomo Photography; 463 (br), AFP/CORBIS; 465 (tr), CORBIS/Douglas Peebles; 468 (tl), Robert Emmett Bright/Photo Researchers; 473 (tr), Dennis Degnan/CORBIS; 476 (tr), AP/Wide World Photos; 479 (cr), Francisco Marquez; 485 (tl), Robert Frerck/Odyssey/Chicago; 488 (tr), Lynne Siler/Focus Group/PictureQuest; 488 (cr), Alan Marler/AP/Wide World Photos; 489 (tr), Dan Burton/www.underwaterimages.co.uk; 489 (cr), Dan Burton/www.underwaterimages.co.uk; 489 (br), Corbis Images; 490 (br), Jenny Thomas/HRW. **Chapter 10:** 498–499 (bkgd), Sam Dudgeon/HRW; 498 (br), Getty Images/FPG International; 500 (tr), Courtesy of Sigma Tau Pharmaceuticals, Inc.; 500 (b), Victoria Smith/HRW; 502 (tr), Barry L. Runk/Grant Heilman Photography; 502 (br), PhotoDisc - Digital Image copyright 2004 PhotoDisc; 504 (tr), Larry Lefever/Grant Heilman Photography; 507 (tr), William Hamilton/SuperStock; 507 (c), NASA/Science Source/Photo Researchers, Inc.; 507 (b), Sam Dudgeon/HRW; 511 (tr), Kwame Zikomo/SuperStock; 512 (tl, tr), Getty Images/The Image Bank; 516 (tr), Sam Dudgeon/HRW; 516 (cr), Corbis Images; 519 (cr), William Sallez/Duomo/CORBIS; 519 (tl), Mark McKeown/The Paris News/AP/Wide World Photos; 521 (bl), Omni Photo Communications Inc./Index Stock Imagery/PictureQuest; 524 (tr), With permission from the Isamu Noguchi Foundation Inc. NY © Michael S. Yamashita/CORBIS; 525 (c), Sam Dudgeon/HRW; 525 (b), Peter Van Steen/HRW; 526 (cl), Barry L. Runk/Grant Heilman Photography; 526 (c, cr), Charles D. Winters/Photo Researchers, Inc.; 526 (bl, bc, br), Peter Van Steen/HRW; 527 (tl), Richard Cummins/CORBIS; 527 (tc), Joseph Sohm; ChromoSohm Inc./CORBIS; 527 (tr), Hisham Ibrahim/CORBIS; 533 (tl), Mark C. Burnett/Photo Researchers, Inc.; 535 (all), Peter Van Steen/HRW; 535 (c, t), Sam Dudgeon/HRW; 537 (tr), TEK Image/Science Photo Library/Photo Researchers, Inc.; 537 (c), Runk/Schoenberger/Grant Heilman Photography; 537 (b), Charles D. Winters/Photo Researchers, Inc.; 538 (cr), U.S. Department of the Interior, National Park Service, Edison National Historic Site; 538 (t), LOC/Science Source/Photo Researchers, Inc.; 542 (cr), NOAA/National Climatic Data Center; 542 (br), NOAA/National Climatic Data Center; 543 (t), © Jim Hargan; 543 (br), © Michael Prince/CORBIS; 544 (c), Tim Davis/HRW Photo; 550 (b), Gianni Dagli Orti/CORBIS. **Chapter 11:** 552–553 (bkgd), SuperStock; 552 (br), Bill Varie/CORBIS; 557 (tr), TEK Image/Science Photo Library/Photo Researchers, Inc.; 558 (penny, nickel), EyeWire - Digital Image copyright 2004 EyeWire; 560 (quarter, dime), EyeWire - Digital Image copyright 2004 EyeWire; 560 (dollar coin), Sam Dudgeon/HRW; 560 (l), Sam Dudgeon/HRW; 565 (l), Peter Van Steen/HRW; 567 (l), CORBIS; 569 (b), Getty Images/Stone; 571 (l), Prof. K. Seddon & Dr. T. Evans, Queen's University, Belfast/Science Photo Library/Photo Researchers, Inc.; 574 (tr), IT Stock International/Index Stock Imagery/PictureQuest; 575 (tl), Digital Image copyright 2004 EyeWire; 577 (r), Pat Doyle/CORBIS; 580 (tr), The Old Farmer's Almanac* and Design Mark* are used with permission of Yankee Publishing Inc., Dublin, New Hampshire; 583 (hockey team), Reuters NewMedia Inc./CORBIS; 583 (tr), Jeff Curtes/CORBIS; 583 (two girls), Yvette Cardozo/Index Stock Imagery/PictureQuest; 583 (bkgd), Fotopic/Index Stock Imagery/PictureQuest; 583 (tc), Richard T. Nowitz/CORBIS; 586 (r), © Pat and Chuck Blackley; 586 (bl), Photos Courtesy of NC Division of Tourism, Film, and Sports Development. 587 (tl), Courtesy of Vought Aircraft Industries Retiree Club; 587 (br), Courtesy of Dare County Airport; 588 (br), Ken Karp/HRW; 592 (penny, nickel), Digital Image copyright (c) 2004 EyeWire; 593 (b), (c)Stockbyte - four food images; frame - Victoria Smith/HRW; 594 (spinner), Sam Dudgeon/HRW; 594 (penny), EyeWire - Digital Image copyright 2004 EyeWire; 594 (cards), Scott Van Osdol/HRW. **Chapter 12:** 596–597 (bkgd), Reuters NewMedia Inc./CORBIS; 596 (br), Maximilian Stock Ltd./Science Photo Library/Photo Researchers, Inc.; 601 (l), Catherine Karnow/CORBIS; 604 (tr), Ken Karp/HRW; 607 (tr), Simon Fraser/MRC Unit, Newcastle General Hospital/Science Photo Library/Photo Researchers, Inc.; 610 (t), Philip Gould/CORBIS; 613 (tc), Owen Franken/CORBIS; 613 (tr), Sam Dudgeon/HRW; 616 (tr), AFP/CORBIS; 619 (tl), A Fish Story by Dutch artist Gustave Verbeek. From his comic strip series Upside-Down, which appeared in the New York Herald from 1903 to 1905./Photo by Sam Dudgeon/HRW; 620 (tr), Peter Van Steen/HRW; 623 (tr), Maxine Cass; 624 (br), Photos Courtesy of NC Division of Tourism, Film, and Sports Development. 624 (bl), © Raymond Gehman/CORBIS; 625 (b), Tim Pflaum; 625 (t), Courtesy of the Lexington Tourist Authority; 632 (br), Sam Dudgeon/HRW.

Back matter: 634(tl), Sam Dugeon/HRW Photo; 635 (br), Sam Dudgeon/HRW Photo.

Glossary

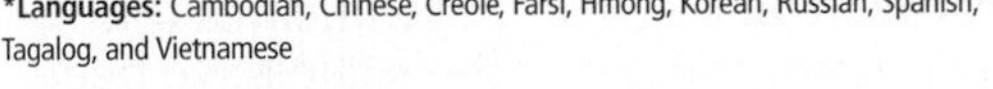

internet connect

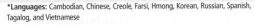

Multilingual Glossary Online:
go.hrw.com
Keyword: MR4 Glossary

*Languages: Cambodian, Chinese, Creole, Farsi, Hmong, Korean, Russian, Spanish, Tagalog, and Vietnamese

A

absolute value The distance of a number from zero on a number line; shown by | |. (p. 451)

Example: $|-5| = 5$

acute angle An angle that measures less than 90°. (p. 326)

acute triangle A triangle with all angles measuring less than 90°. (p. 344)

addend A number added to one or more other numbers to form a sum. For example, in the expression 4 + 6 + 7, 4, 6, and 7 are addends.

Addition Property of Opposites The property that states that the sum of a number and its opposite equals zero.

Example: $12 + (-12) = 0$

adjacent angles Angles in the same plane that have a common vertex and a common side; in the diagram, $\angle a$ and $\angle b$ are adjacent angles. (p. 332)

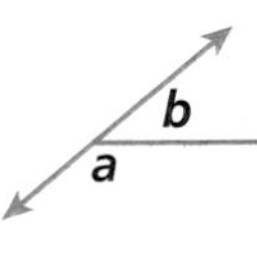

algebraic expression An expression that contains at least one variable. (p. 48)

Example: $x + 8$, $4(m - b)$

algebraic inequality An inequality that contains at least one variable.

Example: $x + 3 > 10$, $5a > b + 3$

alternate exterior angles A pair of angles formed by two lines intersected by a third line; in the diagram, the pairs of alternate exterior angles are $\angle a$ and $\angle d$ and $\angle b$ and $\angle c$. (p. 341)

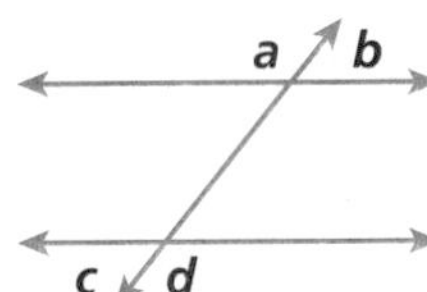

alternate interior angles A pair of angles formed by two lines intersected by a third line; in the diagram, the pairs of alternate interior angles are $\angle r$ and $\angle v$ and $\angle s$ and $\angle t$. (p. 341)

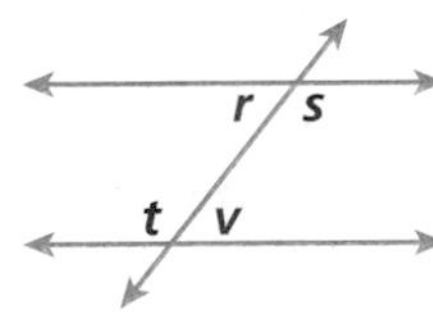

angle A figure formed by two rays with a common endpoint called the vertex. (p. 326)

angle bisector A line, segment, or ray that divides an angle into two congruent angles; in the diagram, $\overrightarrow{MP}$ is an angle bisector of $\angle NMO$.

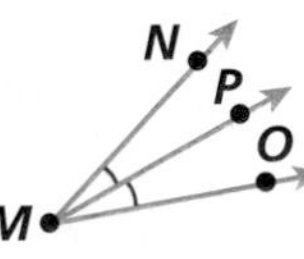

area The number of square units needed to cover a given surface. (p. 504)

Associative Property

Addition: The property that states that for three or more numbers, their sum is always the same, regardless of their grouping. (p. 24)

Example: $2 + 3 + 8 = (2 + 3) + 8 = 2 + (3 + 8)$

Multiplication: The property that states that for three or more numbers, their product is always the same, regardless of their grouping. (p. 24)

Example: $2 \cdot 3 \cdot 8 = (2 \cdot 3) \cdot 8 = 2 \cdot (3 \cdot 8)$

asymmetrical Not identical on either side of a central line; not symmetrical.

average The sum of the items in a set of data divided by the number of items in the set; also called *mean.* (p. 275)

axes The two perpendicular lines of a coordinate plane that intersect at the origin. (p. 458)

B

bar graph A graph that uses vertical or horizontal bars to display data. (p. 284)

base (in numeration) When a number is raised to a power, the number that is used as a factor is the base. (p. 12)

Example: $3^5 = 3 \cdot 3 \cdot 3 \cdot 3 \cdot 3$; 3 is the base.

base (of a polygon or three-dimensional figure) A side of a polygon; a face of a three-dimensional figure by which the figure is measured or classified. (p. 524)

Bases of a cylinder

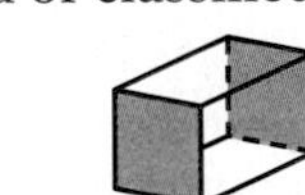
Bases of a prism

Base of a cone

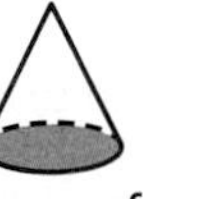

Base of a pyramid

base-10 system A number system in which all numbers are expressed using the digits 0–9. (p. 34)

binary number system A number system in which all numbers are expressed using only two digits, 0 and 1. (p. 34)

bisect To divide into two congruent parts. (p. 378)

box-and-whisker plot A graph that displays the highest and lowest quarters of data as whiskers, the middle two quarters of the data as a box, and the median. (p. 308)

break (graph) A zigzag on a horizontal or vertical scale of a graph that indicates that some of the numbers on the scale have been omitted. (p. 285)

capacity The amount a container can hold when filled.

Celsius A metric scale for measuring temperature in which 0°C is the freezing point of water and 100°C is the boiling point of water; also called *centigrade.*

center (of a circle) The point inside a circle that is the same distance from all the points on the circle. (p. 516)

center (of rotation) The point about which a figure is rotated.

certain (probability) Sure to happen; having a probability of 1. (p. 554)

circle The set of all points in a plane that are the same distance from a given point called the center. (p. 516)

circle graph A graph that uses sections of a circle to compare parts to the whole and parts to other parts. (p. 430)

circumference The distance around a circle. (p. 516)

clockwise A circular movement in the direction shown.

clustering A method used to estimate a sum when all addends are close to the same value. (p. 96)

Example: 27, 29, 24, and 26 all cluster around 25.

combination An arrangement of items or events in which order does not matter. (p. 579)

common denominator A denominator that is the same in two or more fractions. (p. 179)

Example: The common denominator of $\frac{5}{8}$ and $\frac{2}{8}$ is 8.

common factor A number that is a factor of two or more numbers.

Example: 8 is a common factor of 16 and 40.

common multiple A number that is a multiple of each of two or more numbers.

Example: 15 is a common multiple of 3 and 5.

Commutative Property

Addition: The property that states that two or more numbers can be added in any order without changing the sum. (p. 24)

Example: 8 + 20 = 20 + 8

Multiplication: The property that states that two or more numbers can be multiplied in any order without changing the product. (p. 24)

Example: 6 · 12 = 12 · 6

compatible numbers Numbers that are close to the given numbers that make estimation or mental calculation easier. (p. 8)

compensation When a number in a problem is close to another number that is easier to calculate with, the easier number is used to find the answer. Then the answer is adjusted by adding to it or subtracting from it. (p. 28)

complementary angles Two angles whose measures add to 90°. (p. 333)

composite number A number greater than 1 that has more than two whole-number factors. (p. 153)

compound event An event made up of two or more simple events. (p. 574)

cone A three-dimensional figure with one vertex and one circular base. (p. 525)

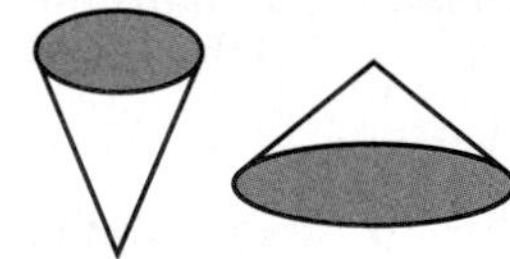

congruent Having the same size and shape. (p. 332)

congruent angles Angles that have the same measure. (p. 331)

congruent segments Segments that have the same length. (p. 330)

constant A value that does not change. (p. 48)

coordinate One of the numbers of an ordered pair that locate a point on a coordinate graph. (p. 458)

coordinate plane (coordinate grid) A plane formed by the intersection of a horizontal number line called the x-axis and a vertical number line called the y-axis. (pp. 294, 458)

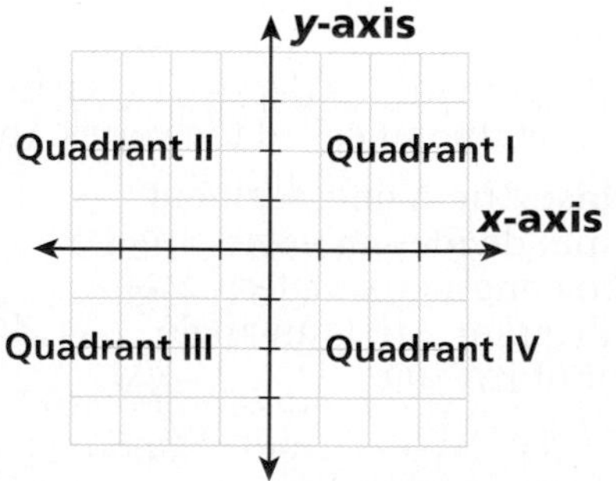

correspondence The relationship between two or more objects that are matched.

corresponding angles (for lines) A pair of angles formed by two lines intersected by a third line; in the diagram, the pairs of corresponding angles are $\angle m$ and $\angle q$, $\angle n$ and $\angle r$, $\angle o$ and $\angle s$, and $\angle p$ and $\angle t$. (p. 341)

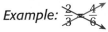

corresponding angles (in polygons) Matching angles of two or more polygons. (p. 405)

corresponding sides Matching sides of two or more polygons. (p. 405)

counterclockwise A circular movement in the direction shown.

cross product The product of numbers on the diagonal when comparing two ratios. (p. 399)

Example: $\frac{2}{3} \times \frac{4}{6}$ $\quad 2 \cdot 6 = 12$

$3 \cdot 4 = 12$

cube (geometric figure) A rectangular prism with six congruent square faces.

cube (in numeration) A number raised to the third power. (p. 12)

cumulative frequency The sum of successive data items. (p. 290)

customary system of measurement The measurement system often used in the United States.

Example: inches, feet, miles, ounces, pounds, tons, cups, quarts, gallons

cylinder A three-dimensional figure with two parallel, congruent circular bases connected by a curved lateral surface. (p. 524)

D

decagon A polygon with ten sides.

degree The unit of measure for angles or temperature.

denominator The bottom number of a fraction that tells how many equal parts are in the whole.

Example: $\frac{3}{4}$ ← denominator

diagonal A line segment that connects two non-adjacent vertices of a polygon.

diameter A line segment that passes through the center of a circle and has endpoints on the circle, or the length of that segment. (p. 516)

difference The result when one number is subtracted from another.

dimension The length, width, or height of a figure.

discount The amount by which the original price is reduced. (p. 432)

Distributive Property The property that states if you multiply a sum by a number, you will get the same result if you multiply each addend by that number and then add the products. (p. 25)

Example: $5(20 + 1) = (5 \cdot 20) + (5 \cdot 1)$

dividend The number to be divided in a division problem.

Example: In $8 \div 4 = 2$, 8 is the dividend.

divisible Can be divided by a number without leaving a remainder. (p. 152)

divisor The number you are dividing by in a division problem.

Example: In $8 \div 4 = 2$, 4 is the divisor.

dodecahedron A polyhedron with 12 faces.

domain The set of all possible input values of a function.

double-bar graph A bar graph that compares two related sets of data. (p. 285)

double-line graph A graph that shows how two related sets of data change over time. (p. 298)

E

edge The line segment along which two faces of a polyhedron intersect. (p. 524)

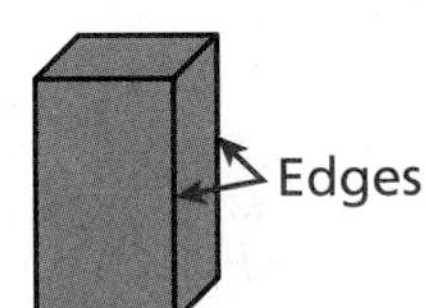

elements The words, numbers, or objects in a set. (p. 196)

empty set A set that has no elements. (p. 196)

endpoint A point at the end of a line segment or ray.

equally likely outcomes Outcomes that have the same probability. (p. 564)

equation A mathematical sentence that shows that two expressions are equivalent. (p. 58)

equilateral triangle A triangle with three congruent sides. (p. 345)

equivalent Having the same value.

equivalent fractions Fractions that name the same amount or part. (p. 172)

equivalent ratios Ratios that name the same comparison. (p. 392)

estimate (n) An answer that is close to the exact answer and is found by rounding or other methods.

estimate (v) To find an answer close to the exact answer by rounding or other methods.

evaluate To find the value of a numerical or algebraic expression. (p. 20)

even number A whole number that is divisible by two.

event An outcome or set of outcomes of an experiment or situation.

expanded form A number written as the sum of the values of its digits.

Example: 236,536 written in expanded form is 200,000 + 30,000 + 6,000 + 500 + 30 + 6.

experiment In probability, any activity based on chance, such as tossing a coin. (p. 558)

experimental probability The ratio of the number of times an event occurs to the total number of trials, or times that the activity is performed. (p. 558)

exponent The number that indicates how many times the base is used as a factor. (p. 12)

Example: $2^3 = 2 \times 2 \times 2 = 8$; 3 is the exponent.

exponential form A number is in exponential form when it is written with a base and an exponent. (p. 12)

Example: 4^2 is the exponential form for $4 \cdot 4$.

expression A mathematical phrase that contains operations, numbers, and/or variables.

face A flat surface of a polyhedron. (p. 524)

factor A number that is multiplied by another number to get a product. (p. 156)

factor tree A diagram showing how a whole number breaks down into its prime factors. (p. 157)

Fahrenheit A temperature scale in which 32°F is the freezing point of water and 212°F is the boiling point of water.

fair When all outcomes of an experiment are equally likely, the experiment is said to be fair. (p. 564)

first quartile The median of the lower half of a set of data; also called *lower quartile.* (p. 308)

formula A rule showing relationships among quantities.

Example: $A = \ell w$ is the formula for the area of a rectangle.

fraction A number in the form $\frac{a}{b}$, where $b \neq 0$.

frequency table A table that lists items together according to the number of times, or frequency, that the items occur. (p. 290)

front-end estimation An estimating technique in which the front digits of the addends are added and then the sum is adjusted for a closer estimate. (p. 97)

function An input-output relationship that has exactly one output for each input. (p. 598)

function table A table of ordered pairs that represent solutions of a function. (p. 598)

graph of an equation A graph of the set of ordered pairs that are solutions of the equation.

greatest common factor (GCF) The largest common factor of two or more given numbers. (p. 160)

height In a triangle or quadrilateral, the perpendicular distance from the base to the opposite vertex or side. (p. 505)

In a prism or cylinder, the perpendicular distance between the bases. (pp. 531, 534)

heptagon A seven-sided polygon.

hexagon A six-sided polygon.

histogram A bar graph that shows the frequency of data within equal intervals. (p. 291)

hypotenuse In a right triangle, the side opposite the right angle.

Identity Property of One The property that states that the product of 1 and any number is that number.

Identity Property of Zero The property that states the sum of zero and any number is that number.

image A figure resulting from a transformation.

impossible (probability) Can never happen; having a probability of 0. (p. 554)

improper fraction A fraction in which the numerator is greater than or equal to the denominator. (p. 182)

Example: $\frac{5}{5}, \frac{5}{3}$

indirect measurement The technique of using similar figures and proportions to find a measure. (p. 409)

inequality A mathematical sentence that shows the relationship between quantities that are not equal. (p. 76)

Example: $5 < 8$, $5x + 2 \geq 12$

input The value substituted into an expression or function. (p. 598)

integers The set of whole numbers and their opposites. (p. 450)

interest The amount of money charged for borrowing or using money, or the amount of money earned by saving money. (p. 436)

interior angles Angles on the inner sides of two lines intersected by a third line. In the diagram, $\angle c$, $\angle d$, $\angle e$, and $\angle f$ are interior angles. (p. 341)

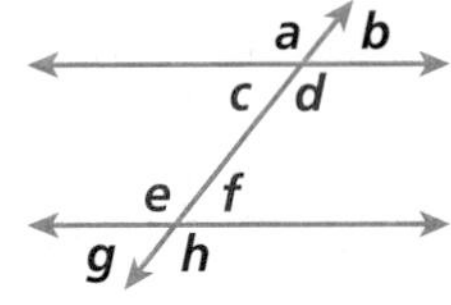

interquartile range The difference between the upper and lower quartiles of a data set. (p. 309)

intersecting lines Lines that cross at exactly one point. (p. 336)

intersection (sets) The set of elements common to two or more sets. (p. 196)

interval The space between marked values on a number line or the scale of a graph.

inverse operations Operations that undo each other: addition and subtraction, or multiplication and division.

isosceles triangle A triangle with at least two congruent sides. (p. 345)

lateral surface In a cylinder, the curved surface connecting the circular bases; in a cone, the curved surface that is not a base.

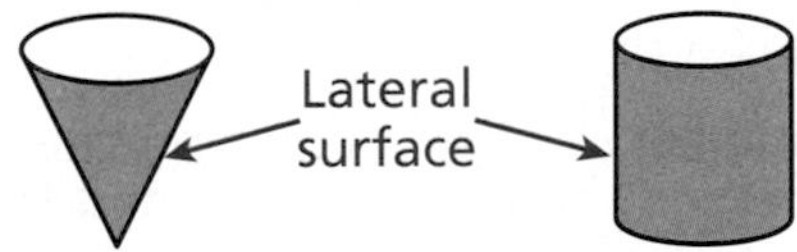

least common denominator (LCD) The least common multiple of two or more denominators. (p. 242)

least common multiple (LCM) The smallest number, other than zero, that is a multiple of two or more given numbers. (p. 232)

like fractions Fractions that have the same denominator. (p. 178)

Example: $\frac{5}{12}$ and $\frac{3}{12}$ are like fractions.

line A straight path that extends without end in opposite directions. (p. 322)

line graph A graph that uses line segments to show how data changes. (p. 297)

line plot A number line with marks or dots that show frequency.

line of reflection A line that a figure is flipped across to create a mirror image of the original figure. (p. 365)

line of symmetry The imaginary "mirror" in line symmetry. (p. 369)

line segment A part of a line between two endpoints. (p. 323)

line symmetry A figure has line symmetry if one half is a mirror-image of the other half. (p. 369)

linear equation An equation whose solutions form a straight line on a coordinate plane. (p. 605)

lower extreme The least number in a set of data. (p. 308)

lower quartile The median of the lower half of a set of data; also called *first quartile.* (p. 308)

mean The sum of the items in a set of data divided by the number of items in the set; also called *average.* (p. 275)

median The middle number or the mean (average) of the two middle numbers in an ordered set of data. (p. 275)

metric system of measurement A decimal system of weights and measures that is used universally in science and commonly throughout the world.

Example: centimeters, meters, kilometers, gram, kilograms, milliliters, liters

midpoint The point that divides a line segment into two congruent line segments.

mixed number A number made up of a whole number that is not zero and a fraction. (p. 167)

mode The number or numbers that occur most frequently in a set of data; when all numbers occur with the same frequency, we say there is no mode. (p. 275)

multiple The product of any number and a whole number is a multiple of that number.

Multiplication Property of Zero The property that states that the product of any number and 0 is 0.

negative number A number less than zero. (p. 450)

net An arrangement of two-dimensional figures that can be folded to form a polyhedron. (p. 530)

numerator The top number of a fraction that tells how many parts of a whole are being considered.

Example: $\frac{4}{5}$ ← numerator

numerical expression An expression that contains only numbers and operations. (p. 20)

obtuse angle An angle whose measure is greater than 90° but less than 180°. (p. 326)

obtuse triangle A triangle containing one obtuse angle. (p. 344)

octagon An eight-sided polygon.

odd number A whole number that is not divisible by two.

odds A comparison of favorable outcomes and unfavorable outcomes. (p. 584)

opposites Two numbers that are an equal distance from zero on a number line. (p. 450)

order of operations A rule for evaluating expressions: first perform the operations in parentheses, then compute powers and roots, then perform all multiplication and division from left to right, and then perform all addition and subtraction from left to right. (p. 20)

ordered pair A pair of numbers that can be used to locate a point on a coordinate plane. (p. 294)

origin The point where the x-axis and y-axis intersect on the coordinate plane; (0, 0). (p. 458)

outcome A possible result of a probability experiment. (p. 558)

outlier A value much greater or much less than the others in a data set. (p. 278)

output The value that results from the substitution of a given input into an expression or function. (p. 598)

overestimate An estimate that is greater than the exact answer. (p. 8)

parallel lines Lines in a plane that do not intersect. (p. 336)

parallelogram A quadrilateral with two pairs of parallel sides. (p. 348)

pentagon A five-sided polygon.

percent A ratio comparing a number to 100. (p. 418)

Example: $45\% = \frac{45}{100}$

perfect square A square of a whole number. (p. 31)

Example: $5 \cdot 5 = 25$, and $7^2 = 49$; 25 and 49 are perfect squares.

perimeter The distance around a polygon. (p. 500)

permutation An arrangement of items or events in which order is important. (p. 578)

perpendicular bisector A line that intersects a segment at its midpoint and is perpendicular to the segment. (p. 378)

perpendicular lines Lines that intersect to form right angles. (p. 336)

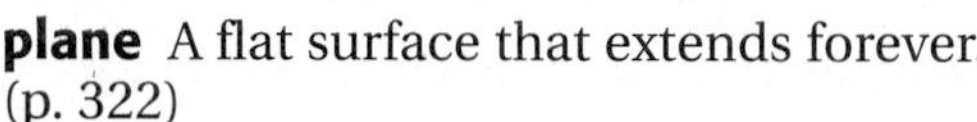

pi (π) The ratio of the circumference of a circle to the length of its diameter; $\pi \approx 3.14$ or $\frac{22}{7}$. (p. 516)

plane A flat surface that extends forever. (p. 322)

point An exact location in space. (p. 322)

polygon A closed plane figure formed by three or more line segments that intersect only at their endpoints. (p. 352)

polyhedron A three-dimensional figure in which all the surfaces or faces are polygons. (p. 524)

positive number A number greater than zero. (p. 450)

power A number produced by raising a base to an exponent.

Example: $2^3 = 8$, so 2 to the 3rd power is 8.

prediction A guess about something that will happen in the future. (p. 580)

prime factorization A number written as the product of its prime factors. (p. 156)

Example: $10 = 2 \cdot 5$, $24 = 2^3 \cdot 3$

prime number A whole number greater than 1 that has exactly two factors, itself and 1. (p. 153)

principal The initial amount of money borrowed or saved. (p. 436)

prism A polyhedron that has two congruent, polygon-shaped bases and other faces that are all rectangles. (p. 524)

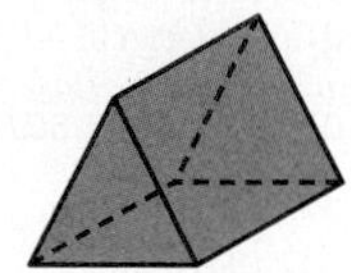

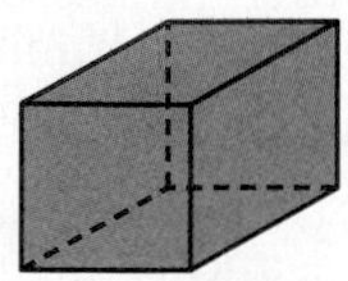

probability A number from 0 to 1 (or 0% to 100%) that describes how likely an event is to occur. (p. 554)

product The result when two or more numbers are multiplied.

proper fraction A fraction in which the numerator is less than the denominator. (p. 182)

Example: $\frac{3}{4}, \frac{1}{12}, \frac{7}{8}$

proportion An equation that states that two ratios are equivalent. (p. 398)

protractor A tool for measuring angles.

pyramid A polyhedron with a polygon base and triangular sides that all meet at a common vertex. (p. 525)

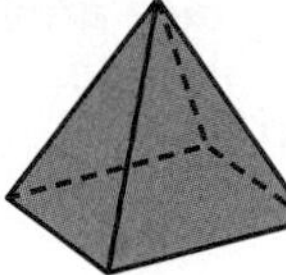

quadrant The x- and y-axes divide the coordinate plane into four regions. Each region is called a quadrant. (p. 458)

quadrilateral A four-sided polygon. (p. 348)

quartile Three values, one of which is the median, that divide a data set into fourths. See also *first quartile, third quartile.* (p. 308)

quotient The result when one number is divided by another.

radius A line segment with one endpoint at the center of a circle and the other endpoint on the circle, or the length of that segment. (p. 516)

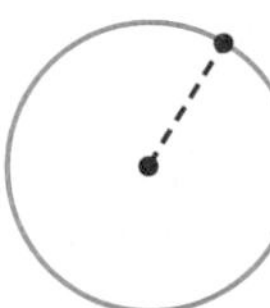

random numbers In a set of random numbers, each number has an equal chance of being selected. (p. 589)

range (in statistics) The difference between the greatest and least values in a data set. (p. 275)

rate A ratio that compares two quantities measured in different units. (p. 393)

Example: The speed limit is 55 miles per hour, or 55 mi/h.

rate of interest The percent charged or earned on an amount of money; see *simple interest.* (p. 436)

ratio A comparison of two quantities by division. (p. 392)

Example: 12 to 25, 12:25, $\frac{12}{25}$

ray A part of a line that starts at one endpoint and extends forever. (p. 323)

reciprocal One of two numbers whose product is 1. (p. 222)

Example: The reciprocal of $\frac{2}{3}$ is $\frac{3}{2}$.

rectangle A parallelogram with four right angles. (p. 348)

rectangular prism A polyhedron whose bases are rectangles and whose other faces are rectangles.

reflection A transformation of a figure that flips the figure across a line. (p. 365)

regular polygon A polygon with congruent sides and angles. (p. 352)

repeating decimal A decimal in which one or more digits repeat infinitely. (p. 168)

Example: $0.757575... = 0.\overline{75}$

rhombus A parallelogram with all sides congruent. (p. 348)

right angle An angle that measures 90°. (p. 326)

right triangle A triangle containing a right angle. (p. 344)

rotation A transformation in which a figure is turned around a point. (p. 365)

rounding Replacing a number with an estimate of that number to a given place value.

Example: 2,354 rounded to the nearest thousand is 2,000; 2,354 rounded to the nearest 100 is 2,400.

sales tax A percent of the cost of an item, which is charged by governments to raise money. (p. 432)

sample space All possible outcomes of an experiment. (p. 558)

scale The ratio between two sets of measurements. (p. 412)

scale drawing A drawing that uses a scale to make an object proportionally smaller than or larger than the real object. (p. 412)

scale model A proportional model of a three-dimensional object.

scalene triangle A triangle with no congruent sides. (p. 345)

scientific notation A method of writing very large or very small numbers by using powers of 10. (p. 114)

second quartile The median of a set of data. (p. 308)

segment A part of a line between two endpoints. (p. 323)

sequence An ordered list of numbers. (p. 31)

set A group of items. (p. 196)

side A line bounding a geometric figure; one of the faces forming the outside of an object.

significant figures The figures used to express the precision of a measurement. (p. 138)

similar Figures with the same shape but not necessarily the same size are similar. (p. 405)

simple interest A fixed percent of the principal. It is found using the formula $I = Prt$, where P represents the principal, r the rate of interest, and t the time. (p. 436)

simplest form (of a fraction) A fraction is in simplest form when the numerator and denominator have no common factors other than 1. (p. 173)

simplify To write a fraction or expression in simplest form.

simulation A model of an experiment, often one that would be too difficult or too time-consuming to actually perform. (p. 562)

skew lines Lines that lie in different planes that are neither parallel nor intersecting. (p. 336)

solid figure A three-dimensional figure.

solution of an equation A value or values that make an equation true. (p. 58)

solution of an inequality A value or values that make an inequality true. (p. 76)

solve To find an answer or a solution.

square (geometry) A rectangle with four congruent sides. (p. 348)

square (numeration) A number raised to the second power. (p. 12)

Example: In 5^2, the number 5 is squared.

square number The product of a number and itself.

Example: 25 is a square number. $5 \cdot 5 = 25$

square root One of the two equal factors of a number. (p. 603)

Example: $16 = 4 \cdot 4$, or $16 = -4 \cdot -4$, so 4 and -4 are square roots of 16.

standard form (in numeration) A way to write numbers by using digits.

Example: Five thousand, two hundred ten in standard form is 5,210.

stem-and-leaf plot A graph used to organize and display data so that the frequencies can be compared. (p. 305)

straight angle An angle that measures 180°. (p. 326)

subset A set contained within another set. (p. 197)

substitute To replace a variable with a number or another expression in an algebraic expression.

sum The result when two or more numbers are added.

supplementary angles Two angles whose measures have a sum of 180°. (p. 333)

surface area The sum of the areas of the faces, or surfaces, of a three-dimensional figure. (p. 530)

term (in a sequence) An element or number in a sequence. (p. 31)

terminating decimal A decimal number that ends or terminates. (p. 168)

Example: 6.75

tessellation A repeating pattern of plane figures that completely cover a plane with no gaps or overlaps. (p. 373)

theoretical probability The ratio of the number of equally likely outcomes in an event to the total number of possible outcomes. (p. 564)

third quartile The median of the upper half of a set of data; also called *upper quartile.* (p. 308)

tip The amount of money added to a bill for service; usually a percent of the bill. (p. 432)

transformation A change in the size or position of a figure. (p. 365)

translation A movement (slide) of a figure along a straight line. (p. 365)

trapezoid A quadrilateral with exactly one pair of parallel sides. (p. 348)

tree diagram A branching diagram that shows all possible combinations or outcomes of an event. (p. 570)

triangle A three-sided polygon.

Triangle Sum Theorem The theorem that states that the measures of the angles in a triangle add to 180°.

triangular prism A polyhedron whose bases are triangles and whose other faces are rectangles.

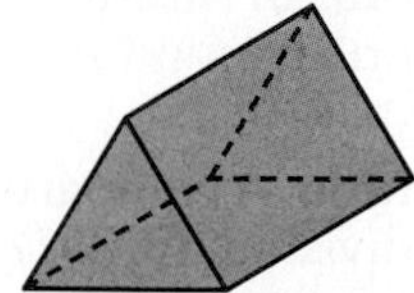

underestimate An estimate that is less than the exact answer. (p. 8)

union The set of all elements that belong to two or more sets. (p. 196)

unit conversion The process of changing one unit of measure to another.

unit rate A rate in which the second quantity in the comparison is one unit. (p. 393)

Example: 10 cm per minute

unlike fractions Fractions with different denominators. (p. 179)

Example: $\frac{3}{4}$ and $\frac{1}{2}$ are unlike fractions.

upper extreme The greatest number in a set of data. (p. 308)

upper quartile The median of the upper half of a set of data; also called *third quartile.* (p. 308)

variable A symbol used to represent a quantity that can change. (p. 48)

Venn diagram A diagram that is used to show relationships between sets. (p. 196)

vertex On an angle or polygon, the point where two sides intersect; on a polyhedron, the intersection of three or more faces; on a cone or pyramid, the top point. (pp. 326, 524)

vertical angles A pair of opposite congruent angles formed by intersecting lines; in the diagram, $\angle a$ and $\angle c$ are vertical angles and $\angle b$ and $\angle d$ are vertical angles. (p. 332)

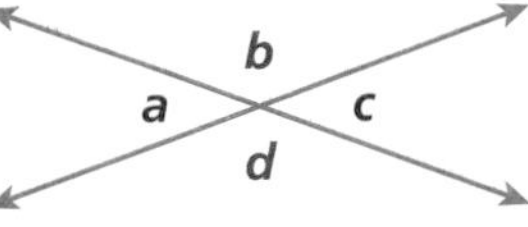

volume The number of cubic units needed to fill a given space. (p. 534)

x-axis The horizontal axis on a coordinate plane. (p. 458)

x-coordinate The first number in an ordered pair; it tells the distance to move right or left from the origin, (0, 0). (p. 458)

Example: 5 is the x-coordinate in (5, 3).

y-axis The vertical axis on a coordinate plane. (p. 458)

y-coordinate The second number in an ordered pair; it tells the distance to move up or down from the origin, (0, 0). (p. 458)

Example: 3 is the y-coordinate in (5, 3).

zero pair A number and its opposite, which add to 0.

Example: 18 and −18

Glossary

Index

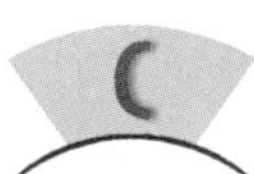

C

E

F

G

Index

H

I

N

O

Index

S

U

V

W

X

Y

Z

Table of Measures

METRIC	CUSTOMARY
Length	
1,000 millimeters (mm) = 1 meter (m)	1 foot (ft) = 12 inches (in.)
100 centimeters (cm) = 1 meter	1 yard (yd) = 36 inches
10 decimeters (dm) = 1 meter	1 yard = 3 feet
1 kilometer (km) = 1,000 meters	1 mile (mi) = 5,280 feet
	1 mile = 1,760 yards
Capacity	
1,000 milliliters (mL) = 1 liter (L)	1 cup (c) = 8 fluid ounces (fl oz)
100 centiliters (cL) = 1 liter	1 pint (pt) = 2 cups
10 deciliters (dL) = 1 liter	1 quart (qt) = 2 pints
1 kiloliter (kL) = 1,000 liters	1 quart = 4 cups
	1 gallon (gal) = 4 quarts
	1 gallon = 16 cups
Mass/Weight	
1,000 milligrams (mg) = 1 gram (g)	1 pound (lb) = 16 ounces (oz)
100 centigrams (cg) = 1 gram	1 ton (T) = 2,000 pounds
10 decigrams (dg) = 1 gram	
1 kilogram (kg) = 1,000 grams	
1 tonne (t) = 1,000 kilograms	

TIME

1 minute (min) = 60 seconds (s)	1 year (yr) = 12 months (mo)
1 hour (hr) = 60 minutes	1 year = 365 days
1 day = 24 hours	1 leap year = 366 days
1 week (wk) = 7 days	